# Delmar's Standard Textbook of Electricity

### Stephen L. Herman

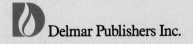 Delmar Publishers Inc.

# NOTICE TO THE READER

Cover Design by: design M design W
Airbrush Art by: Richard D. Skiermont

**Delmar Staff**
New Product Acquisitions Specialist: Mark Huth
Associate Editor: Kimberly Davies
Project Editor: Judith Boyd Nelson
Production Supervisor: Teresa Luterbach
Production Coordinator: Dianne Jensis
Sr. Design Supervisor: Susan C. Mathews
Art/Design Coordinator: Cheri Plasse
Art Coordinator: John Lent

For information, address Delmar Publishers Inc.
3 Columbia Circle, Box 15-015,
Albany, NY 12212-5015

Printed in the United States of America
Published simultaneously in Canada
by Nelson Canada,
a division of The Thomson Corporation

2    3    4    5    6    7    8    9    10    XXX    00    99    98    97    96    95

**Library of Congress Cataloging-in-Publication Data**

Herman, Stephen L.
    Delmar's standard textbook of electricity/Stephen L. Herman.
        p. cm.
    Includes index.
    ISBN 0-8273-6849-6 (textbook)
        1. Electricity. 2. Electric circuits. 3. Electric meters.
    I. Delmar Publishers. II. Title. III. Title: Standard textbook of electricity.
QC522.H47 1993                    92-11410
537—dc20                          CIP

# Contents

## Unit 3   *Static Electricity*

## Unit 4   *Magnetism*

## Unit 5   *Resistors*

*section* **2**

# *basic electrical circuits*

## Unit 6   *Series Circuits*

**Unit 7** *Parallel Circuits*

**Unit 8** *Combination Circuits*

*section* **3**

*meters and wire sizes*

**Unit 9** *Measuring Instruments*

## Unit 10 — Using Wire Tables and Determining Conductor Sizes

*section* **4**

# small sources of electricity

## Unit 11 — Conduction in Liquid and Gases

## Unit 12 — Batteries and Other Sources of Electricity

## Unit 13 — Magnetic Induction

## section 5 — basics of alternating current

## Unit 14 — Basic Trigonometry

## Unit 15 — Alternating Current

# section 6

# alternating current circuits containing inductance

## Unit 21 — Resistive-Capacitive Series Circuits

## Unit 22 — Resistive-Capacitive Parallel Circuits

# section 8
# alternating current circuits containing resistance-inductance-capacitance

## Unit 23 — Resistive-Inductive-Capacitive Series Circuits

## Unit 24 — Resistive-Inductive-Capacitive Parallel Circuits

*section* **11**

# *direct current machines*

*section* **12**

# *alternating current machines*

## Appendix A
### Trigonometric Functions–Natural Sines and Cosines

## Appendix B

## Appendix C

## Appendix D

## Appendix E

## Appendix F

## Appendix G

## Appendix H

## Glossary

## Index

# Preface

DELMAR'S STANDARD TEXTBOOK OF ELECTRICITY is a blend of the practical and theoretical. The text not only explains the different concepts relating to electrical theory, but also provides many practical examples of the common tasks the construction electrician or industrial electrician must perform. It was the author's intention to make this one of those books that an electrician would want to keep at work for reference.

The text was written with the assumption that the student has no knowledge of electricity. Although the text covers many areas of both direct and alternating current theory, **the math level has been kept to basic algebra and basic trigonometry**. Chapter 14 of the text provides an introduction to basic trigonometry for those who are weak in this area. **The appendix contains a section on electrical formulas**. The formulas have been divided into groups that are related to a particular application, such as resistive-inductive series circuits, resistive-capacitive parallel circuits, resistive-inductive-capacitive series circuits, etc.

The subject matter for DELMAR'S STANDARD TEXTBOOK OF ELECTRICITY has been divided into 32 separate chapters. Each chapter is designed to "stand alone." The text has been organized in this manner for several reasons. Although the author has presented the chapters in what he feels is a logic progression, he realizes that other instructors may want to present the information in a different manner. **The "stand alone" concept permits the information to be presented in almost any sequence the instructor desires.**

Another reason for presenting the information in this manner is that the student or electrician does not have to search the entire text to find information concerning a particular subject. **All relevant information concerning a specific subject is located within a given chapter.** It is not assumed that the student has read previous chapters. If the student or electrician desires information concerning the connection of dual-voltage three-phase motors, for example, that is covered in the chapter concerning three-phase motors.

In this revised version of DELMAR'S STANDARD TEXTBOOK OF ELECTRICITY, several new principles and theorems have been added in Units 6, 7, and 8 to make the text more complete. Based upon customer feedback, we have added new material on Kirchhoff's law, Thevenin's theorem, Norton's theorem, general voltage and current dividers, and the superposition theorem. Additional review questions reflect the inclusion of this material.

In addition to this material, the revised version of DELMAR STANDARD TEXTBOOK OF ELECTRICITY has been thoroughly checked for technical accuracy. We thank you, our customers, for helping us make a fine textbook even better.

## Supplements

In developing *Delmar's Standard Textbook of Electricity,* we asked many instructors and students what we could do to improve the package. We listened to their suggestions, and as a result this textbook is part of a complete teaching and learning system. The entire package consists of the *Standard Textbook,* two laboratory manuals, three instructor's guides, and a set of full-color transparencies.

Two laboratory manuals, entitled *Experiments In Electricity for Use with Lab-Bolt EMS Equipment* and *Experiments In Electricity for Use with Standard Electrical Equipment,* provide extensive opportunities for students to apply what they have learned. Both manuals contain multiple hands-on experiments for each chapter for the textbook. Both manuals have been extensively field-tested to ensure that all of the experiments will work as planned. The engineers at Lab-Volt conducted each of the experiments in *Experiments In Electricity for Use with Lab-Volt EMS Equipment* and, following their testing, Lab-Bolt has endorsed this manual. It is the manual they recommend to their customers. *Experiments in Electricity for Use with Standard Electrical Equipment* was field-tested at Shreveport-Bossier Regional Technical School under the direction of Richard Cameron.

A separate instructor's guide is available for the *Standard Textbook* and each of the lab manuals. The instructor's guide to accompany the textbook includes answers to the review questions found at the end of each chapter as well as 15 tests correlated to the textbook. The instructor's guides for the lab manuals include answers to all of the questions asked in the lab manuals as well as suggestions for the instructor in using the experiments.

A package of 100 full-color transparencies rounds out the system. These transparencies are based on key figures within the textbook and will help the instructor explain the important principles in the textbook.

# Acknowledgments

The author would like to express thanks to those people who helped in the production of this text. A special thanks goes to those people of the NJATC (National Joint Apprenticeship Training Committee) for their suggestions concerning the subject matter presented in this text. A list of reviewers and advisors is shown below.

| Reviewer | Affiliation |
| --- | --- |
| Allen R. Beiling | Assabet Valley Regional Vocational High School, Marlboro, MA |
| Jack Bohannon | Long Beach City College, Long Beach, CA |
| Richard Cameron | Shreveport Bossier Regional Tech Institute, Shreveport, LA |
| Robin Carey | Central Florida Electrical JATC |
| Ronald Coffee | Mid-America AVTS, Wayne, OK |
| Joseph H. Cook | Billings Vo-Tech, Billings, MT |
| Thomas J. Cress | Belleville Area College, Belleville, IL |
| Michael Cunningham | South Texas Electrical JATC |
| Dick Cutbirth | NJATC for Electrical Industry, Las Vegas, NV |
| Eric David | Long Beach City College, Long Beach, CA |
| Gaylon Eastridge | Phoenix Electrical JATC |
| Rosser S. Farley III | Dundalk Community College, Baltimore, MD |
| Ben Gaddis | Tulsa County Area Vo-Tech |
| Thomas W. Giblin | SUNY-Delhi, Delhi, NY |
| Wayne L. Grinolds | Rogue Community College, Grant's Pass, OR |
| James M. Guthrie | Milwaukee Vo-Tech College |
| Larry Dan Killebrew | Mid-America Vo-Tech School, Wayne, OK |
| Gary Lane | Carroll Technical College, Carrollton, GA |
| Bob McNeel | Los Angeles Trade Technical College |
| Brent Meyers | Oakland Community College, Auburn Hills, MI |
| Melvin Mize | Pinellas County Technical Institute, FL |
| Ray Mullin | Cooper Industries—Bussman Division, Northbrook, IL |
| A.J. Pearson | Director, National Joint Apprenticeship and Training Committee (NJATC) |
| Larry Phillips | Minneapolis Technological College |
| James Roberts | Tarrant County Community College, Fort Worth, TX |
| James D. Rupe | Eastern Oklahoma Electrical JATC |
| Greg Skudlarek | Minneapolis Technical College |
| Wally Sowers | Spokane Community College, Spokane, WA |
| David Stone | Central Maine Technical College |
| Smokey Stover | Moore-Norman Vo-Tech, Norman, OK |
| James Sullivan | Central Florida Electrical JATC |
| Riddeck Tuten | Colleton Area Vocational Center, Walterboro, SC |
| Ken Weaver | Durham Technical Community College, Durham, NC |
| Ernest M. Shaffer | San Diego Electrical JATC |

*section*  **1**

## basic electricity and Ohm's law

**Electricity is the flow of free electrons. The physical properties of electric current are discussed in this unit.**

# Atomic Structure

## Key Terms

Alternating current (AC)

Atom

Atomic number

Attraction

Bidirectional

Centrifugal force

Conductors

Direct current (DC)

Electron

Electron orbits

Element

Insulators

Matter

Molecules

Negative

Neutron

Nucleus

Positive

Proton

Repulsion

Semiconductors

Unidirectional

Valence electrons

**Objectives**

**Preview**

**direct current**

**alternating current**

**uni-directional**

**bi-directional**

*After studying this unit, you should be able to:*

- List the three major parts of an atom.
- State the law of charges.
- Discuss the law of centrifugal force.
- Discuss the differences between conductors and insulators.

Electricity is the driving force that provides most of the power for the industrialized world. It is used to light homes, cook meals, heat and cool buildings, drive motors, and supply the ignition for most automobiles. The technician who understands electricity can seek employment in almost any part of the world.

Electrical sources are divided into two basic types, **direct current** (DC) and **alternating current** (AC). Direct current is **unidirectional**, which means that it flows in only one direction. The first part of this text will be mainly devoted to the study of direct current. Alternating current is **bidirectional**, which means that it reverses its direction of flow at regular intervals. The latter part of this text is devoted mainly to the study of alternating current.

## 1-1 Early History of Electricity

Although the practical use of electricity has become common only within the last hundred years, it has been known as a force for much longer. The Greeks were the first to discover electricity about 2500 years ago. They noticed that when amber was rubbed with other materials it became charged with an unknown force that had the power to attract objects such as dried leaves, feathers, bits of cloth, or other lightweight materials. The Greeks called amber *elektron*. The word *electric* was derived from it and meant "to be like amber," or to have the ability to attract other objects.

This mysterious force remained little more than a curious phenomenon until about 2000 years later, when other people began to conduct experiments. In the early 1600s, William Gilbert discovered that amber was not the only material that could be charged to attract other objects. He called materials that could be charged *electriks* and materials that could not be charged *nonelektriks*.

About 300 years ago a few men began to study the behavior of various charged objects. In 1733, a Frenchman named Charles DuFay found that a piece of charged glass would repel some charged objects and attract others. These men soon learned that the force of **repulsion** was just as important as the force of **attraction**. From these experiments, two lists were developed *(Figure 1-1)*. It was determined that any material in list A would attract any material in list B, and that all materials in list A would

**repulsion**

**attraction**

| LIST A | LIST B |
|---|---|
| Glass (rubbed on silk) | Hard rubber (rubbed on wool) |
| Glass (rubbed on wool or cotton) | Block of sulfur (rubbed on wool or fur) |
| Mica (rubbed on cloth) | Most kinds of rubber (rubbed on cloth) |
| Asbestos (rubbed on cloth or paper) | Sealing wax (rubbed on silk, wool, or fur) |
| Stick of sealing wax (rubbed on wool) | Glass or mica (rubbed on dry wool) |
| | Amber (rubbed on cloth) |

**Figure 1-1** List of charged materials

repel each other and all materials in list B would repel each other *(Figure 1-2)*. Various names were suggested for the materials in lists A and B. Any opposite-sounding names could have been chosen, such as east and west, north and south, male and female. Benjamin Franklin named the materials in list A **positive** and the materials in list B **negative**. These names are still used today. The first item in each list was used as a standard for determining if a charged object was positive or negative. Any object repelled by a piece of glass rubbed on silk would have a positive charge and any item repelled by a hard rubber rod rubbed on wool would have a negative charge.

**positive**

**negative**

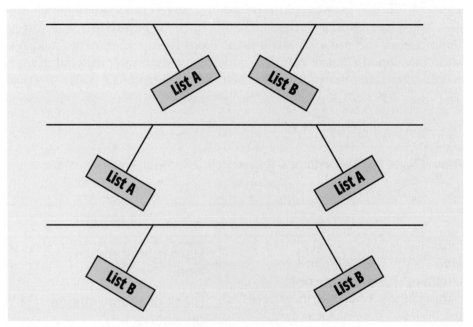

**Figure 1-2** Unlike charges attract and like charges repel.

## 1-2 Atoms

**atom**

**matter**

**element**

To understand electricity, it is necessary to start with the study of atoms. The **atom** is the basic building block of the universe. All **matter** is made from a combination of atoms. Matter is any substance that has mass and occupies space. Matter can exist in any of three states: solid, liquid, or gas. Water, for example, can exist as a solid in the form of ice, as a liquid, or as a gas in the form of steam *(Figure 1-3)*. An **element** is a substance that cannot be chemically divided into a simpler substance. A

**Figure 1-3** Water can exist in three states, depending on temperature and pressure.

table listing both natural and artificial elements is shown in *Figure 1-4*. An atom is the smallest part of an element. The three principal parts of an atom are the **electron**, **neutron**, and **proton**. *Figure 1-5* illustrates those parts of the atom. It is theorized that protons and neutrons are actually made of smaller particles called *quarks*.

Notice that the proton has a positive charge, the electron a negative charge, and the neutron no charge. The neutron and proton combine to form the **nucleus** of the atom. Since the neutron has no charge, the nucleus will have a net positive charge. The number of protons in the nucleus determines what kind of element an atom is. Oxygen, for example, contains eight protons in its nucleus, and gold contains 79. The **atomic number** of an element is the same as the number of protons in the nucleus. The lines of force produced by the positive charge of the proton extend outward in all directions *(Figure 1-6)*. The nucleus may or may not contain as many neutrons as protons. For example, an atom of helium contains two protons and two neutrons in its nucleus, while an atom of copper contains 29 protons and 35 neutrons *(Figure 1-7)*.

electron

neutron

proton

nucleus

atomic number

| ATOMIC NUMBER | NAME | VALENCE ELECTRONS | SYMBOL | ATOMIC NUMBER | NAME | VALENCE ELECTRONS | SYMBOL | ATOMIC NUMBER | NAME | VALENCE ELECTRONS | SYMBOL |
|---|---|---|---|---|---|---|---|---|---|---|---|
| 1 | Hydrogen | 1 | H | 37 | Rubidium | 1 | Rb | 73 | Tantalum | 2 | Ta |
| 2 | Helium | 2 | He | 38 | Strontium | 2 | Sr | 74 | Tungsten | 2 | W |
| 3 | Lithium | 1 | Li | 39 | Yttrium | 2 | Y | 75 | Rhenium | 2 | Re |
| 4 | Beryllum | 2 | Be | 40 | Zirconium | 2 | Zr | 76 | Osmium | 2 | Os |
| 5 | Boron | 3 | B | 41 | Niobium | 1 | Nb | 77 | Iridium | 2 | Ir |
| 6 | Carbon | 4 | C | 42 | Molybdenum | 1 | Mo | 78 | Platinum | 1 | Pt |
| 7 | Nitrogen | 5 | N | 43 | Technetium | 2 | Tc | 79 | Gold | 1 | Au |
| 8 | Oxygen | 6 | O | 44 | Ruthenium | 1 | Ru | 80 | Mercury | 2 | Hg |
| 9 | Fluorine | 7 | F | 45 | Rhodium | 1 | Rh | 81 | Thallium | 3 | Tl |
| 10 | Neon | 8 | Ne | 46 | Palladium | – | Pd | 82 | Lead | 4 | Pb |
| 11 | Sodium | 1 | Na | 47 | Silver | 1 | Ag | 83 | Bismuth | 5 | Bl |
| 12 | Magnesium | 2 | Ma | 48 | Cadmium | 2 | Cd | 84 | Polonium | 6 | Po |
| 13 | Aluminum | 3 | Al | 49 | Indium | 3 | In | 85 | Astatine | 7 | At |
| 14 | Silicon | 4 | Si | 50 | Tin | 4 | Sn | 86 | Radon | 8 | Rd |
| 15 | Phosphorus | 5 | P | 51 | Antimony | 5 | Sb | 87 | Francium | 1 | Fr |
| 16 | Sulfur | 6 | S | 52 | Tellurium | 6 | Te | 88 | Radium | 2 | Ra |
| 17 | Chlorine | 7 | Cl | 53 | Iodine | 7 | I | 89 | Actinium | 2 | Ac |
| 18 | Argon | 8 | A | 54 | Xenon | 8 | Xe | 90 | Thorium | 2 | Th |
| 19 | Potassium | 1 | K | 55 | Cesium | 1 | Cs | 91 | Protactinium | 2 | Pa |
| 20 | Calcium | 2 | Ca | 56 | Barium | 2 | Ba | 92 | Uranium | 2 | U |
| 21 | Scandium | 2 | Sc | 57 | Lanthanum | 2 | La | | | | |
| 22 | Titanium | 2 | Ti | 58 | Cerium | 2 | Ce | | Artifical Elements | | |
| 23 | Vanadium | 2 | V | 59 | Praseodymium | 2 | Pr | | | | |
| 24 | Chromium | 1 | Cr | 60 | Neodymium | 2 | Nd | 93 | Neptunium | 2 | Np |
| 25 | Manganese | 2 | Mn | 61 | Promethium | 2 | Pm | 94 | Plutonium | 2 | Pu |
| 26 | Iron | 2 | Fe | 62 | Samarium | 2 | Sm | 95 | Americium | 2 | Am |
| 27 | Cobalt | 2 | Co | 63 | Europium | 2 | Eu | 96 | Curium | 2 | Cm |
| 28 | Nickel | 2 | Ni | 64 | Gadolinium | 2 | Gd | 97 | Berkelium | 2 | Bk |
| 29 | Copper | 1 | Cu | 65 | Terbium | 2 | Tb | 98 | Californium | 2 | Cf |
| 30 | Zinc | 2 | Zn | 66 | Dysprosium | 2 | Dy | 99 | Einsteinium | 2 | E |
| 31 | Gallium | 3 | Ga | 67 | Holmium | 2 | Ho | 100 | Fermium | 2 | Fm |
| 32 | Germanium | 4 | Ge | 68 | Erbium | 2 | Er | 101 | Mendelevium | 2 | Mv |
| 33 | Arsenic | 5 | As | 69 | Thulium | 2 | Tm | 102 | Nobelium | 2 | No |
| 34 | Selenium | 6 | Se | 70 | Ytterbium | 2 | Yb | 103 | Lawrencium | 2 | Lw |
| 35 | Bromine | 7 | Br | 71 | Lutetium | 2 | Lu | | | | |
| 36 | Krypton | 8 | Kr | 72 | Hafnium | 2 | Hf | | | | |

**Figure 1-4** Table of elements

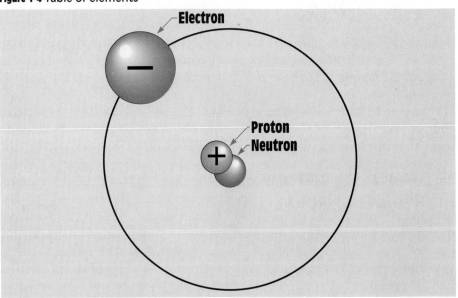

**Figure 1-5** The three principal parts of an atom

The electron orbits the outside of the nucleus. Notice in *Figure 1-5* that the electron is shown to be larger than the proton. Actually, an electron is about three times as large as a proton. The estimated size of a proton is 0.07 trillionth of an inch in diameter, and the estimated size of an electron is 0.22 trillionth of an inch in diameter. Although the electron is larger in size, the proton weighs about 1840 times more. Imagine comparing a soap bubble with a piece of buckshot. Compared with the electron the proton is a very massive particle. Since the electron exhibits a negative charge, the lines of force come in from all directions *(Figure 1-8)*.

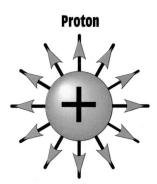

**Proton**

**Figure 1-6** The lines of force extend outward.

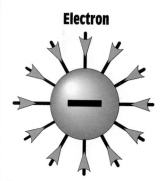

**Electron**

**Figure 1-8** The lines of force come inward.

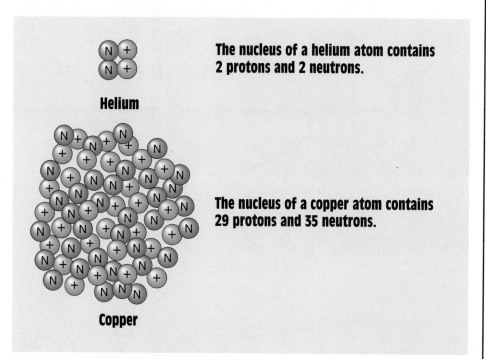

**Helium**

**The nucleus of a helium atom contains 2 protons and 2 neutrons.**

**The nucleus of a copper atom contains 29 protons and 35 neutrons.**

**Copper**

**Figure 1-7** The nucleus may or may not contain the same number of protons and neutrons.

## 1-3 The Law of Charges

To understand atoms, it is necessary first to understand two basic laws of physics. One of these is the law of charges, which states that **opposite charges attract and like charges repel**. In *Figure 1-9*, which illustrates this principle, charged balls are suspended from strings. Notice that the two balls that contain opposite charges are attracted to each other. The two positively charged balls and the two negatively charged balls repel each other. The reason for this is that lines of force can never cross each

**Opposite charges attract and like charges repel.**

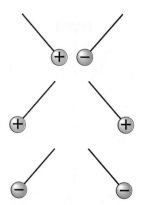

**Figure 1-9** Unlike charges attract and like charges repel.

other. The outward-going lines of force of a positively charged object combine with the inward-going lines of force of a negatively charged object *(Figure 1-10)*. This combining produces an attraction between the two objects. If two objects with like charges come close to each other, the lines of force repel *(Figure 1-11)*. Since the nucleus has a net positive charge and the electron has a negative charge, the electron is attracted to the nucleus.

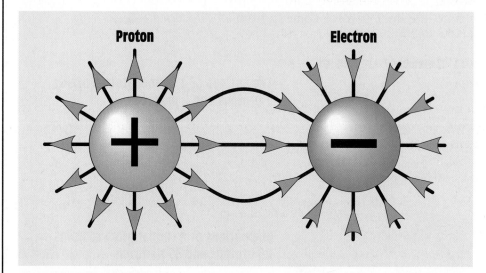

**Figure 1-10** Unlike charges attract each other.

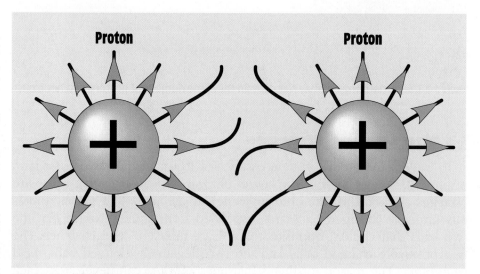

**Figure 1-11** Like charges repel each other.

Because the nucleus of an atom is formed from the combination of protons and neutrons, one might ask why the protons of the nucleus do not repel each other since they all have the same charge. Two theories attempt to explain this. The first asserts that the force of gravity holds the protons and neutrons together. Neutrons, like protons, are extremely massive particles. Their combined mass produces the gravitational force necessary to overcome the repelling force of the positive charges. The second explanation involves a theoretical particle called a *gluon*. A gluon is a subatomic particle that acts as a bonding agent that not only holds quarks together, but also holds the protons and neutrons together.

## 1-4 Centrifugal Force

The second law important to the understanding of atoms is **the law of centrifugal force. It states that a spinning object will pull away from its center point and that the faster it spins, the greater the centrifugal force becomes.** *Figure 1-12* shows an example of this principle. If you tie an object to a string and spin it around, it will try to pull away from you. The faster the object spins, the greater the force that tries

**Figure 1-12** Centrifugal force causes an object to pull away from its axis point.

centrifugal force

The law of centrifugal force states that a spinning object will pull away from its center point and that the faster it spins, the greater the centrifugal force becomes.

to pull the object away. Centrifugal force prevents the electron from falling into the nucleus of the atom. The faster an electron spins, the farther away from the nucleus it will be.

## 1-5 Electron Orbits

Each **electron orbit** of an atom contains a set number of electrons *(Figure 1-13)*. The number of electrons that can be contained in any one orbit, or shell, is found by the formula $(2N^2)$. The letter $N$ represents the

**electron
orbit**

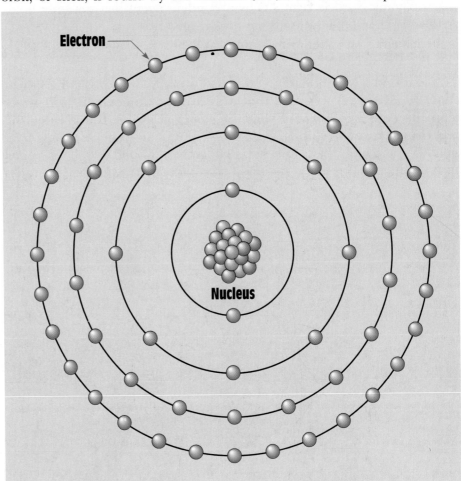

**Figure 1-13** Electron orbits

number of the orbit, or shell. For example, the first orbit can hold no more than two electrons.

$$2 \times (1)^2 \text{ or}$$
$$2 \times 1 = 2$$

The second orbit can hold no more than eight electrons.

$$2 \times (2)^2 \text{ or}$$

$$2 \times 4 = 8$$

The third orbit can contain no more than 18 electrons.

$$2 \times (3)^2 \text{ or}$$

$$2 \times 9 = 18$$

The fourth and fifth orbits cannot hold more than 32 electrons. Thirty-two is the maximum number of electrons that can be contained in any orbit.

$$2 \times (4)^2 \text{ or}$$

$$2 \times 16 = 32$$

Although atoms are often drawn flat, as illustrated in *Figure 1-13*, electrons orbit the nucleus in a spherical fashion, as shown in *Figure 1-14*. Electrons travel at such a high rate of speed that they form a shell around the nucleus. For this reason, electron orbits are often referred to as shells.

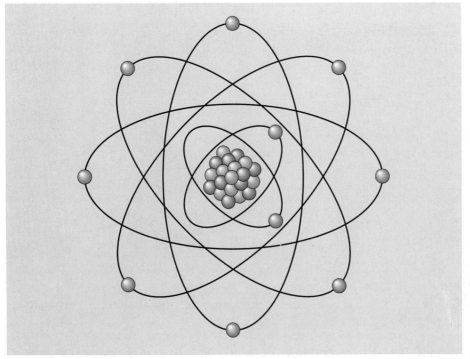

**Figure 1-14** Electrons orbit the nucleus in a circular fashion.

## 1-6 Valence Electrons

The outer shell of an atom is known as the *valence* shell. Any electrons located in the outer shell of an atom are known as **valence electrons** *(Figure 1-15)*. The valence shell of an atom cannot hold more than eight electrons. It is the valence electrons that are of primary concern in the study of electricity, because it is these electrons that explain much of electrical theory. A **conductor**, for instance, is generally made from a material that contains one or two valence electrons. Atoms with one or two valence electrons are unstable and can be made to give up these electrons with little effort. Conductors are materials that permit electrons to flow through them easily. When an atom has only one or two valence electrons, these electrons are loosely held by the atom and are easily given up for current flow. Silver, copper, and gold all contain one valence electron and are excellent conductors of electricity. Silver is the best natural conductor of electricity, followed by copper, gold, and aluminum. An atom of copper is shown in *Figure 1-16*. Although it is known that atoms

**valence electrons**

**conductor**

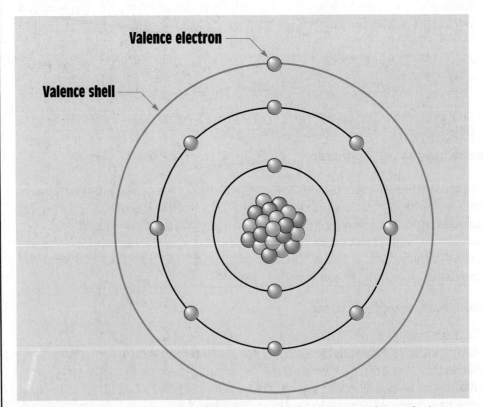

**Figure 1-15** The electrons located in the outer orbit of an atom are valence electrons.

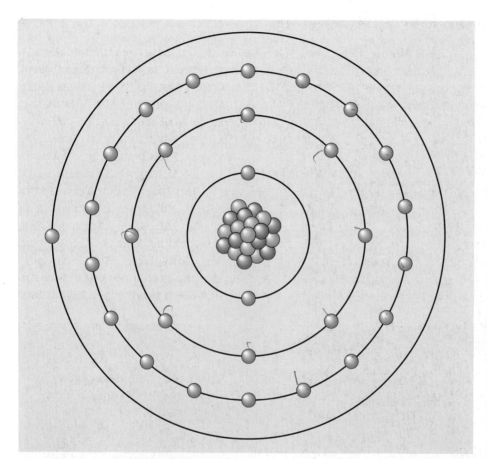

**Figure 1-16**  A copper atom contains 29 electrons and has one valence electron.

containing few valence electrons are the best conductors, it is not known why some of these materials are better conductors than others. Copper, gold, platinum, and silver all contain only one valence electron. Silver, however will conduct electricity more readily than any of the others. Aluminum, which contains three valence electrons, is a better conductor than platinum, which contains only one valence electron.

## 1-7 Electron Flow

Electrical current is the flow of electrons. It is produced when an electron from one atom knocks electrons of another atom out of orbit. *Figure 1-17* illustrates this action. When an atom contains only one valence electron, that electron is easily given up when struck by another electron. The striking electron gives its energy to the electron being struck. The striking electron settles into orbit around the atom, and the electron that was struck moves off to strike another electron. This same effect can be seen in the game of pool. If the moving cue ball strikes a stationary ball exactly right, the energy of the cue ball is given to the stationary ball. The stationary ball then moves off with *most* of the cue ball's energy, and the cue ball stops moving *(Figure 1-18)*. The stationary ball did not move off with all of the energy of the cue ball. It moved off with most of the energy of the cue ball. Some of the cue ball's energy was lost to heat when it struck the stationary ball. Some energy is also lost when one electron strikes another *(Figure 1-17)*. That is why a wire heats when current flows through it. If too much current flows through a wire, overheating will damage the wire and possibly become a fire hazard.

If an atom containing two valence electrons is struck by a moving electron, the energy of the striking electron will be divided between the two valence electrons *(Figure 1-19)*. If the valence electrons are knocked out of orbit, they will contain only half the energy of the striking electron.

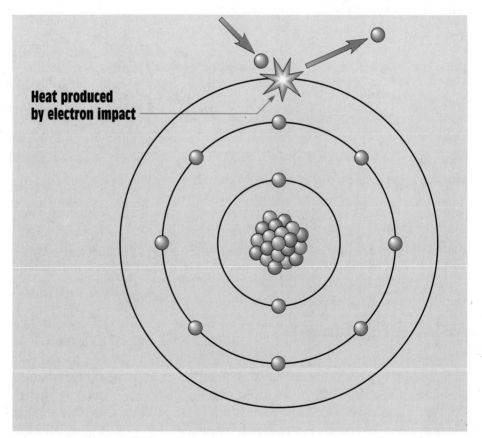

**Heat produced by electron impact**

**Figure 1-17** An electron of one atom knocks an electron of another atom out of orbit.

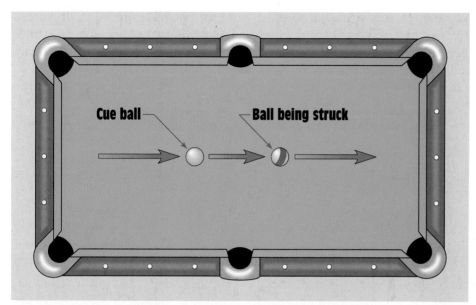

**Figure 1-18**  The energy of the cue ball is given to the ball being struck.

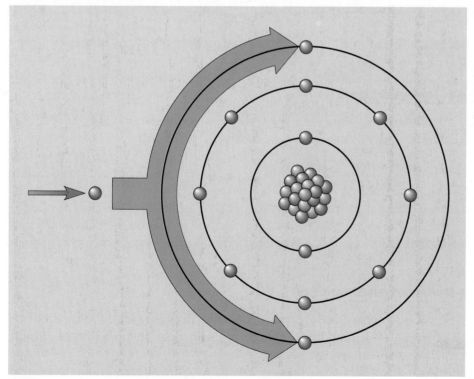

**Figure 1-19**  The energy of the striking electron is divided.

This effect can also be seen in the game of pool *(Figure 1-20)*. If a moving cue ball strikes two stationary balls at the same time, the energy of the cue ball is divided between the two stationary balls. Both stationary balls will move, but with only half the energy of the cue ball.

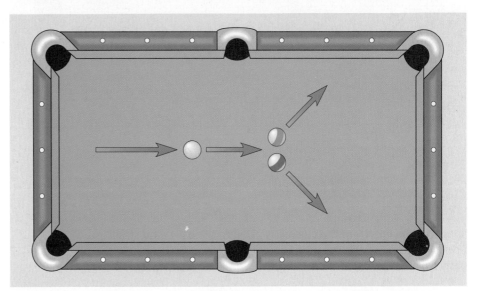

**Figure 1-20** The energy of the cue ball is divided between the two other balls.

### 1-8 Insulators

Materials containing seven or eight valence electrons are known as **insulators**. Insulators are materials that resist the flow of electricity. When the valence shell of an atom is full or almost full, the electrons are held tightly and are not given up easily. Some good examples of insulator materials are rubber, plastic, glass, and wood. *Figure 1-21* illustrates what happens when a moving electron strikes an atom containing eight valence electrons. The energy of the moving electron is divided so many times that it has little effect on the atom. Any atom that has seven or eight valence electrons is extremely stable and does not easily give up an electron.

### 1-9 Semiconductors

**Semiconductors** are materials that are neither good conductors nor good insulators. They contain four valence electrons *(Figure 1-22)* and are characterized by the fact that as they are heated, their resistance decreases. Heat has the opposite effect on conductors, whose resistance *increases* with an increase of temperature. Semiconductors have become

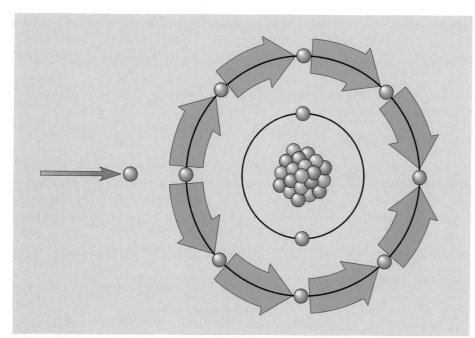

**Figure 1-21**  The energy of the striking electron is divided among the eight electrons.

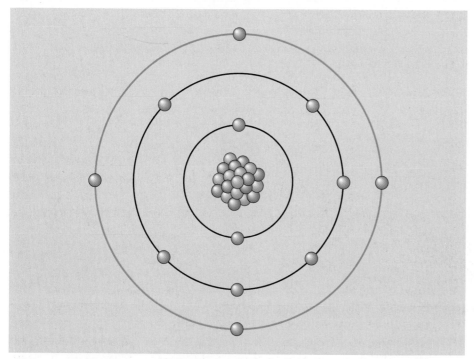

**Figure 1-22**  Semiconductors contain four valence electrons.

extremely important in the electrical industry since the invention of the transistor in 1947. All solid state devices such as diodes, transistors, and integrated circuits are made from combinations of semiconductor materials. The two most common materials used in the production of electronic components are silicon and germanium. Of the two, silicon is used more often because of its ability to withstand heat. Before any pure semiconductor can be used to construct an electronic device, it must be mixed or "doped" with an impurity.

### 1-10 Molecules

**molecules**

Although all matter is made from atoms, atoms should not be confused with **molecules,** which are the smallest part of a compound. Water, for example, is a compound, not an element. The smallest particle of water is a molecule made of two atoms of hydrogen and one atom of oxygen, $H_2O$ *(Figure 1-23)*. If the molecule of water is broken apart, it becomes two hydrogen atoms and one oxygen atom, and is no longer water.

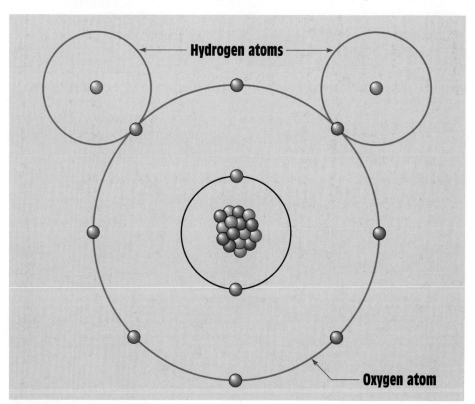

**Figure 1-23** A water molecule

### Summary

1. The atom is the smallest part of an element.

2. The three basic parts of an atom are the proton, electron, and neutron.

3. Protons have a positive charge, electrons a negative charge, and neutrons no charge.

4. Valence electrons are located in the outer orbit of an atom.

5. Conductors are materials that provide an easy path for electron flow.

6. Conductors are made from materials that contain one, two, or three valence electrons.

7. Insulators are materials that do not provide an easy path for the flow of electrons.

8. Insulators are generally made from materials containing seven or eight valence electrons.

9. Semiconductors contain four valence electrons.

10. Semiconductors are used in the construction of all solid state devices such as diodes, transistors, and integrated circuits.

11. A molecule is the smallest part of a compound.

## Review Questions

1. What are the three subatomic parts of an atom, and what charge does each carry?

2. How many times larger is an electron than a proton?

3. How many times more does the proton weigh than the electron?

4. State the law of charges.

5. What force keeps the electron from falling into the nucleus of the atom?

6. How many valence electrons are generally contained in materials used for conductors?

7. How many valence electrons are generally contained in materials used for insulators?

8. What is electricity?

9. What is a gluon?

10. It is theorized that protons and neutrons are actually formed from a combination of smaller particles. What are these particles called?

**The basic measurements and physical laws used to determine the actions and values of an electric circuit are discussed in this unit.**

*Courtesy of New York State Electric and Gas; photo by Kirk van Zandbergen.*

# Electrical Quantities and Ohm's Law

## Key Terms

Amp

British thermal unit (BTU)

Complete path

Conventional current flow theory

Coulomb

Electromotive force

Electron theory

Grounding conductor

Horsepower

Impedance

Joule

Neutral conductor

Ohm ($\Omega$)

Ohm's law

Power

Resistance

Volt

Watt

## Objectives

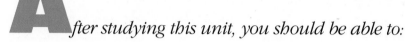

*fter studying this unit, you should be able to:*

- Define a coulomb.
- Define an amp.
- Define a volt.
- Define an ohm.
- Define a watt.
- Compute different electrical values using Ohm's law formulas.
- Discuss different types of electrical circuits.
- Select the proper Ohm's law formula from a chart.

## Preview

Electricity has a standard set of values. Before one can work with electricity, one must know these values and how to use them. Because the values of electrical measurement have been standardized, they are understood by everyone who uses them. For instance, carpenters use a standard system for measuring length, such as the inch, foot, meter, or centimeter. Imagine what a house would look like if it was constructed by two carpenters who used different lengths of measure for an inch or foot. The same holds true for people who work with electricity. The standards of measurement must be the same for everyone. Meters should be calibrated to indicate the same quantity of current flow or voltage or resistance. A volt, an ampere, or an ohm is the same everywhere in the world.

## 2-1 The Coulomb

A **coulomb** is a quantity measurement for electrons. One coulomb contains 6.25 x $10^{18}$, or 6,250,000,000,000,000,000, electrons. To better understand the number of electrons contained in a coulomb, think of comparing one second to 200 billion years. Since the coulomb is a quantity measurement, it is similar to a quart, gallon, or liter. It takes a certain amount of liquid to equal a liter, just as it takes a certain amount of electrons to equal a coulomb.

The coulomb is named for a French scientist who lived in the 1700s named Charles Augustin de Coulomb. Coulomb experimented with electrostatic charges and developed a law dealing with the attraction and repulsion of these forces. The law, known as **Coulomb's law of electrostatic charges, states that the force of electrostatic attraction or repulsion is directly proportional to the product of the two charges and inversely proportional to the square of the distance between them.** The number of electrons contained in the coulomb was determined by the average charge of an electron.

## 2-2 The Amp

The **amp**, or ampere, is named for André Ampère, a scientist who lived from the late 1700s to the early 1800s. Ampère is most famous for his work dealing with electromagnetism, which will be discussed in a later chapter. The amp (A) is defined as one coulomb per second. Notice that the definition of an amp involves a quantity measurement, the coulomb, and a time measurement, the second. One amp of current flows through a wire when one coulomb flows past a point in one second *(Figure 2-1)*. The ampere is a measurement of the amount of electricity that is flowing through a circuit. In a water system, it would be comparable to gallons per minute or gallons per second *(Figure 2-2)*. The letter *I*, which stands for intensity of current, and the letter *A*, which stands for amp, are both used to represent current flow in algebraic formulas. This text will use the letter *I* in formulas to represent current.

---

**coulomb**

**Coulomb's law of electrostatic charges states that the force of electrostatic attraction or repulsion is directly proportional to the product of the two charges and inversely proportional to the square of the distance between them.**

$$F = \frac{Q_1 \times Q_2}{KD^2}$$

F = Force in dynes
Q = Strength of charge in electrostatic units
D = Separation in cm
K = Dielectric constant

---

**amp**

---

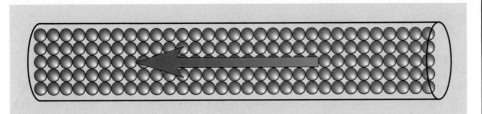

**Figure 2-1** One ampere equals one coulomb per second.

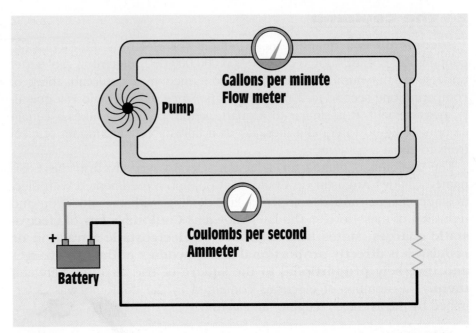

**Figure 2-2** Current in an electrical circuit can be compared to flow rate in a water system.

### 2-3 The Electron Theory

electron theory

There are actually two theories concerning current flow. One theory is known as the **electron theory** and states that since electrons are negative particles, current flows from the most negative point in the circuit to the most positive. The electron theory is the more widely accepted as being correct and is used throughout this text.

### 2-4 The Conventional Current Theory

conventional current flow theory

The second theory, known as the **conventional current flow theory**, is older than the electron theory and states that current flows from the most positive point to the most negative. Although it has been established almost to a certainty that the electron theory is correct, the conventional current theory is still widely used for several reasons. Most electronic circuits use the negative terminal as ground or common. When the negative terminal is used as ground, the positive terminal is considered to be above ground, or hot. It is easier for most people to think of something flowing down rather than up, or from a point above ground to ground. An automobile electrical system is a good example of this type of circuit. Most people consider the positive battery terminal to be the hot terminal.

Many people who work in the electronics field prefer the conventional current flow theory because all the arrows on the semiconductor symbols point in the direction of conventional current flow. If the electron flow theory is used, it must be assumed that current flows against the arrow *(Figure 2-3)*. Another reason that many people prefer using the conventional current flow theory is that most electronic schematics are drawn in such a manner that it assumes current to flow from the more positive to the more negative source. In *Figure 2-4* the positive voltage point is shown at the top of the schematic and the negative (ground) is shown at the bottom. When tracing the flow of current through a circuit, most people find it easier to go from top to bottom than from bottom to top.

**current flow**

**The Electron Theory states that current flows in the direction that electrons move; electrons always move from negative to positive.**

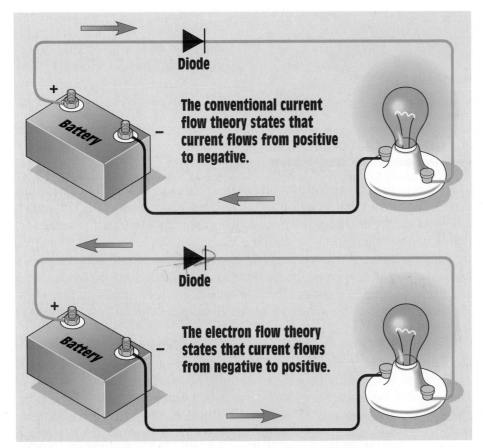

**Figure 2-3** Conventional current flow theory and electron flow theory

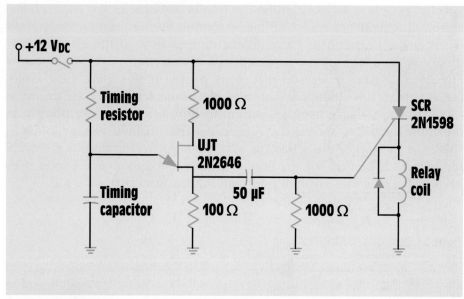

**Figure 2-4** On-delay timer

## 2-5 Speed of Current

To determine the speed of current flow through a wire, one must first establish exactly what is being measured. As stated previously, current is a flow of electrons through a conductive substance. Assume for a moment that it is possible to remove a single electron from a wire and identify it by painting it red. If it were possible to observe the progress of the identified electron as it moved from atom to atom, it would be seen that a single electron moves rather slowly *(Figure 2-5)*. It is estimated that a single electron moves at a rate of about 3 in. per hour at one ampere of current flow.

The impulse of electricity, however, is extremely fast. Assume for a moment that a pipe has been filled with Ping-Pong balls *(Figure 2-6)*. If another ball is forced into one end of the pipe, the ball at the other end of the pipe will be forced out. Each time a ball enters one end of the pipe, another ball is forced out the other end. This principle is also true for electrons in a wire. There are billions of electrons in a wire. If an electron enters one end of a wire, another electron is forced out the other end.

For many years it was assumed that the speed of the electrical impulse had a theoretical limit of 186,000 miles per second, or 300,000,000 meters

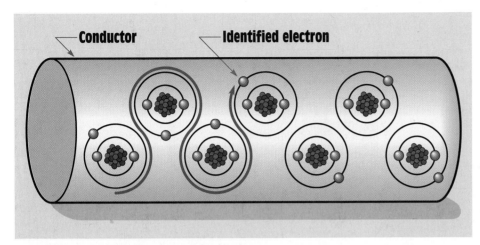

**Figure 2-5** Electrons moving from atom to atom

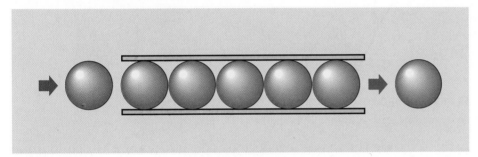

**Figure 2-6** When a ball is pushed into one end, another ball is forced out the other end. This basic principle causes the instantaneous effect of electrical impulses.

per second, which is the speed of light. In recent years, however, it has been shown that the impulse of electricity can actually travel faster than light. Assume that a wire is long enough to be wound around the earth 10 times. If a power source and switch were connected at one end of the wire and a light at the other end *(Figure 2-7)*, the light would turn on at the moment the switch was closed. It would take light approximately 1.3 seconds to travel around the earth 10 times.

## 2-6 Basic Electrical Circuits

A **complete path** must exist before current can flow through a circuit *(Figure 2-8)*. A complete circuit is often referred to as a closed circuit,

**complete path**

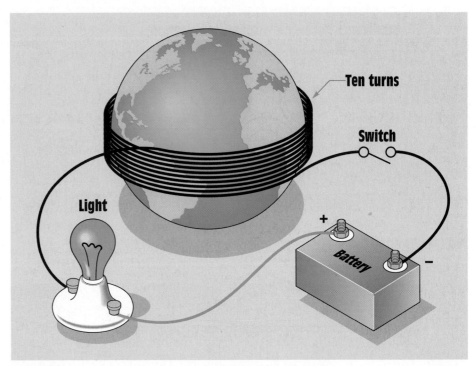

**Figure 2-7** The impulse of electricity can travel faster than light.

because the power source, conductors, and load form a closed loop. In *Figure 2-8* a lamp is used as the load. The load offers resistance to the circuit and limits the amount of current that can flow. If the switch is opened, there is no longer a closed loop and no current can flow. This is often referred to as an incomplete, or open, circuit.

Another type of circuit is the short circuit, which has very little or no resistance. It generally occurs when the conductors leading from and back to the power source become connected together *(Figure 2-9)*. In this example, a separate current path has been established that bypasses the load. Since the load is the device that limits the flow of current, when it is bypassed an excessive amount of current can flow. Short circuits generally cause a fuse to blow or a circuit breaker to open. If the circuit has not been protected by a fuse or circuit breaker, a short circuit can damage equipment, melt wires, and start fires.

Another type of circuit, one that is often confused with a short circuit, is a grounded circuit. Grounded circuits can also cause an excessive amount of current flow. They occur when a path other than the one intended is established to ground. Many circuits contain an extra conductor

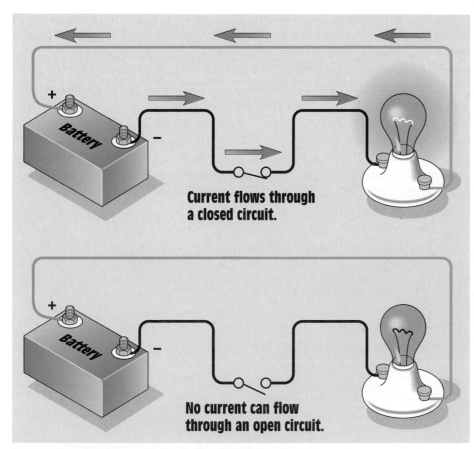

**Current flows through a closed circuit.**

**No current can flow through an open circuit.**

**Figure 2-8** Current flows only through a closed circuit.

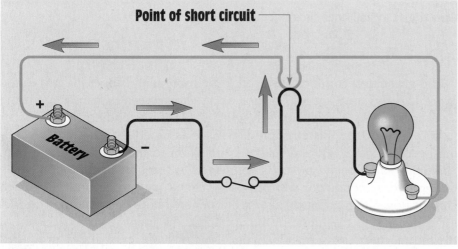

**Point of short circuit**

**Figure 2-9** A short circuit bypasses the load and permits too much current to flow.

**grounding
conductor**

**neutral
conductor**

called the **grounding conductor**. A typical 120-volt appliance circuit is shown in *Figure 2-10*. In this circuit, the ungrounded, or hot, conductor is connected to the fuse or circuit breaker. The hot conductor supplies power to the load. The grounded conductor, or **neutral conductor**, provides the return path and completes the circuit back to the power source. The grounding conductor is generally connected to the case of the appliance to provide a low-resistance path to ground. Although both the neutral and grounding conductors are grounded at the power source, the grounding conductor is not considered to be a circuit conductor, because current will flow through the grounding conductor only when a circuit fault develops. In normal operation, current flows through the hot and neutral conductors only.

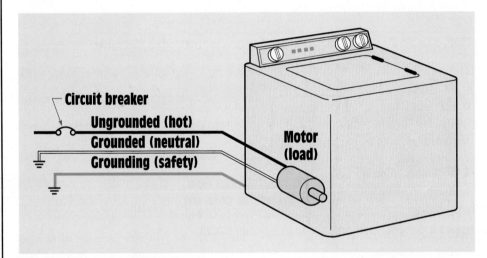

**Figure 2-10** 120-V appliance circuit

The grounding conductor is used to help prevent a shock hazard in the event that the ungrounded, or hot, conductor comes in contact with the case or frame of the appliance *(Figure 2-11)*. This condition could occur in several ways. In this example it will be assumed that the motor winding becomes damaged and makes connection to the frame of the motor. Since the frame of the motor is connected to the frame of the appliance, the grounding conductor will provide a circuit path to ground. If enough current flows, the circuit breaker will open. Without a grounding conductor connected to the frame of the appliance the frame would become hot (in the electrical sense) and anyone touching the case and a grounded

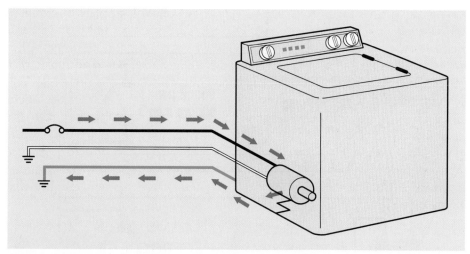

**Figure 2-11** The grounding conductor provides a low-resistance path to ground.

point, such as a water line, would complete the circuit to ground. The resulting shock could be fatal. For this reason the grounding prong of a plug should never be cut off or bypassed.

## 2-7 The Volt

Voltage is defined as **electromotive force,** or EMF. It is the force that pushes the electrons through a wire and is often referred to as electrical pressure. A **volt** is the amount of potential necessary to cause one coulomb to produce one joule of work. One thing to remember is that voltage cannot flow. Voltage in an electrical circuit is like pressure in a water system *(Figure 2-12)*. To say that voltage flows through a circuit is like saying that pressure flows through a pipe. Pressure can push water through a pipe, and it is correct to say that water flows through a pipe, but it is not correct to say that pressure flows through a pipe. The same is true for voltage. Voltage pushes current through a wire, but voltage cannot flow through a wire.

Voltage is often thought of as the potential to do something. For this reason it is frequently referred to as potential, especially in older publications and service manuals. Voltage must be present before current can flow, just as pressure must be present before water can flow. A voltage, or potential, of 120 volts is present at a common wall outlet, but there is no flow until some device is connected and a complete circuit exists. The same is true in a water system. Pressure is present, but water cannot flow

**The grounding prong of a plug should never be cut off or bypassed.**

**electromotive force**

**volt**

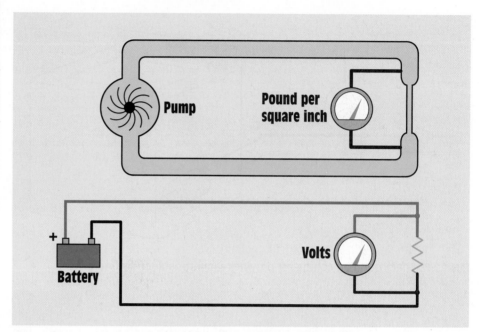

**Figure 2-12** Voltage in an electrical circuit can be compared to pressure in a water system.

until the valve is opened and a path is provided to a region of lower pressure. The letter *E*, which stands for EMF, or the letter *V*, which stands for volt, can be used to represent voltage in an algebraic formula. This text will use the letter *E* to represent voltage in an algebraic formula.

## 2-8 The Ohm

An **ohm** is the unit of **resistance** to current flow. It was named after the German scientist Georg S. Ohm. The symbol used to represent an ohm, or resistance, is the Greek letter omega ($\Omega$). The letter *R*, which stands for resistance, is used to represent ohms in an algebraic formula. An ohm is the amount of resistance that allows 1 amp of current to flow when the applied voltage is 1 volt. Without resistance, every electrical circuit would be a short circuit. All electrical loads, such as heating elements, lamps, motors, transformers, and so on, are measured in ohms. Just as in a water system a reducer can be used to control the flow of water, in an electrical circuit, a resistor can be used to control the flow of electrons. *Figure 2-13* illustrates this concept.

To understand the effect of resistance on an electric circuit, imagine a person running along a beach. As long as the runner stays on the hard,

**ohm**

**resistance**

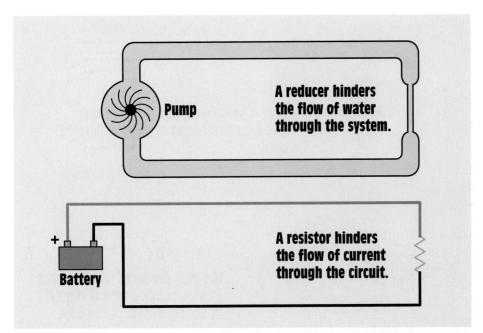

**Figure 2-13** A resistor in an electrical circuit can be compared to a reducer in a water system.

compact sand, he can run easily along the beach. Likewise, current can flow easily through a good conductive material, such as a copper wire. Now imagine that the runner wades out into the water until it is knee deep. He will no longer be able to run along the beach as easily because of the resistance of the water. Now imagine that the runner wades out into the water until it is waist deep. His ability to run along the beach will be hindered to a greater extent because of the increased resistance of the water against his body. The same is true for resistance in an electric circuit. The higher the resistance, the greater the hindrance to current flow.

Another fact an electrician should be aware of is that any time current flows through a resistance, heat is produced *(Figure 2-14)*. That is why a wire becomes warm when current flows through it. The elements of an electric range become hot, and the filament of an incandescent lamp becomes extremely hot because of resistance.

Another term similar in meaning to resistance is **impedance.** Impedance is most often used in calculations of alternating current rather than direct current. Impedance will be discussed to a greater extent later in this text.

**Any time current flows through a resistance, heat is produced**

**impedance**

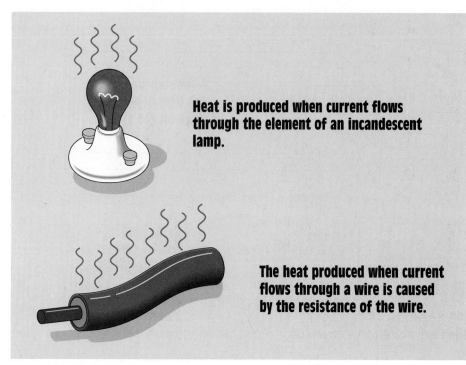

**Heat is produced when current flows through the element of an incandescent lamp.**

**The heat produced when current flows through a wire is caused by the resistance of the wire.**

**Figure 2-14** Heat is produced when current flows through resistance.

### 2-9 The Watt

Wattage is a measure of the amount of power that is being used in a circuit. The **watt** was named in honor of the English scientist, James Watt. In an algebraic formula, wattage is generally represented either by the letter *P*, for power, or *W*, for watts. It is proportional to the amount of voltage and the amount of current flow. To understand watts, return to the example of the water system. Assume that a water pump has a pressure of 120 pounds per square inch (PSI) and causes a flow rate of one gallon per second. Now assume that this water is used to drive a turbine, as shown in *Figure 2-15*. The turbine has a radius of one foot from the center shaft to the rim of the wheel. Since water weighs 8.34 pounds per gallon and it is being forced against the wheel at a pressure of 120 PSI, the turbine would develop a torque of 1000.8 foot-pounds (120 x 8.34 x 1 = 1000.8). If the pressure is increased to 240 PSI, but the flow of water remains constant, the force against the wheel will double (240 x 8.34 x 1 = 2001.6). Notice that the amount of power developed by the turbine is

**watt**

**torque**

**Torque is a force that produces, or tends to produce, rotation or torsion.**

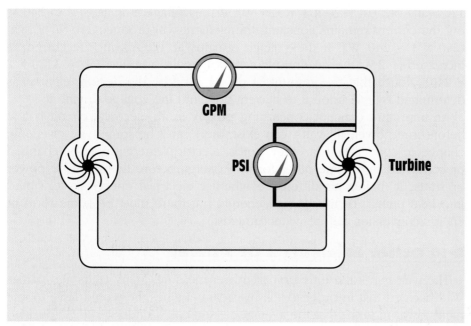

**Figure 2-15** Force equals flow rate times pressure.

determined by both the amount of pressure driving the water and the amount of flow.

The power of an electrical circuit is very similar. *Figure 2-16* shows a resistor connected to a circuit with a voltage of 120 V and a current flow of 1 A. The resistor shown represents an electrical heating element. When 120 V force a current of 1 A through it, the heating element will produce

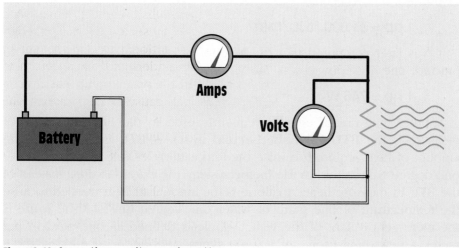

**Figure 2-16** Amps times volts equals watts.

120 watts of heat (120 x 1 = 120 W). If the voltage is increased to 240 V, but the current remains constant, the element will produce 240 W of heat (240 x 1 = 240 W). If the voltage remains at 120 V, but the current is increased to 2 A, the heating element will again produce 240 W (120 x 2 = 240). Notice that the amount of power used by the heating element is determined by the amount of current flow and the voltage driving it.

**power**

An important concept concerning **power** in an electrical circuit is that before true power, or watts, can exist, there must be some type of energy change or conversion. In other words, electrical energy must be changed or converted into some other form of energy before there can be power or watts. It makes no difference whether electrical energy is converted into heat energy or mechanical energy, but there must be some form of energy conversion before watts can exist.

## 2-10 Other Measures Of Power

The watt is not the only unit of power measure. Many years ago, James Watt decided that in order to sell his steam engines, he would have to rate their power in terms that the average person could understand. He decided to compare his steam engines to the horses he hoped his engines would replace. After experimenting, Watt found that the average horse working at a steady rate could do 550 foot-pounds of work per second. A foot-pound (ft-lb) is the amount of force required to raise a one-pound weight one foot. This rate of doing work is the definition of a **horsepower** (hp).

**horsepower**

$$1 \text{ hp} = 550 \text{ ft-lb/s}$$

Horsepower can also be expressed as 33,000 foot-pounds per minute (550 ft-lb x 60 s = 33,000).

$$1 \text{ hp} = 33,000 \text{ ft-lb/min}$$

It was later computed that the amount of electrical energy needed to produce one horsepower was 746 W.

$$1 \text{ hp} = 746 \text{ W}$$

**BTU (British thermal unit)**

Another measure of energy frequently used in the English system of measure is the **BTU (British thermal unit).** A BTU is defined as the amount of heat required to raise the temperature of one pound of water one degree Fahrenheit. In the metric system, the calorie is used instead of the BTU to measure heat. A calorie is the amount of heat needed to raise the temperature of one gram of water one degree Celsius. The **joule** is the metric equivalent of the watt. A joule is defined as one newton per meter. A newton is a force of 100,000 dynes, or about 3-1/2 ounces, and a

**joule**

meter is about 39 inches. The joule can also be expressed as the amount of work done by one coulomb flowing through a potential of one volt, or as the amount of work done by one watt for one second.

### 1 joule = 1 watt/s

The chart in *Figure 2-17* gives some common conversions for different quantities of energy. These quantities can be used to calculate different values.

| | |
|---|---|
| 1 Horsepower = | 746 Watts |
| 1 Horsepower = | 550 Ft-lb./s |
| 1 Watt = | 0.00134 Horsepower |
| 1 Watt = | 3.412 BTU/hr |
| 1 Watt/s = | 1 Joule |
| 1 BTU/s = | 1.055 Watts |
| 1 Cal/s = | 4.19 Watts |
| 1 Ft-lb./s = | 1.36 Watts |
| 1 BTU = | 1050 Joules |
| 1 Joule = | 0.2389 Cal |
| 1 Cal = | 4.186 Joules |

**Figure 2-17** Common power units

An elevator must lift a load of 4000 lb to a height of 50 ft in 20 s. How much horsepower is required to operate the elevator?

**Example 1**

## Solution
Find the amount of work that must be performed, and then convert that to horsepower.

$$4000 \text{ lb} \times 50 \text{ ft} = 200{,}000 \text{ ft-lb}$$

$$\frac{200{,}000 \text{ ft-lb}}{20 \text{ s}} = 10{,}000 \text{ ft-lb/s}$$

$$\frac{10{,}000 \text{ ft-lb./s}}{550 \text{ ft-lb./s}} = 18.18 \text{ Hp}$$

A water heater contains 40 gallons of water. Water weighs 8.34 lb per gallon. The present temperature of the water is 68°F. The water must be raised to a temperature of 160°F in 1 hour. How much power will be required to raise the water to the desired temperature?.

## Solution

First determine the weight of the water in the tank, because a BTU is the amount of heat required to raise the temperature of one pound of water one degree Fahrenheit.

$$40 \text{ gal} \times 8.34 \text{ lb per gal} = 333.6 \text{ lb}$$

The second step is to determine how many degrees of temperature the water must be raised. This amount will be the difference between the present temperature and the desired temperature.

$$160°F - 68°F \ = \ 92°F$$

The amount of heat required in BTUs will be the product of the pounds of water and the desired increase in temperature.

$$333.6 \text{ lb} \times 92° \ = \ 30{,}691.2 \text{ BTU}$$

$$1 \text{ W} \ \ \ \ \ = \ 3.412 \text{ BTU/hr}$$

Therefore

$$\frac{30{,}691 \text{ BTU}}{3.412 \text{ BTU/hr}} = 8995.1 \text{ W/hr}$$

## 2-11 Ohm's Law

In its simplest form, **Ohm's law** states that **it takes one volt to push one amp through one ohm.** Ohm discovered that all electrical quantities are proportional to each other and can therefore be expressed as mathematical formulas. He found if the resistance of a circuit remained constant and the voltage increased, there was a corresponding proportional increase of current. Similarly. if the resistance remained constant and the voltage decreased, there would be a proportional decrease of current. He also found that if the voltage remained constant and the resistance increased, there would be a decrease of current, and if the voltage remained constant and the resistance decreased, there would be an increase of current. This finding lead Ohm to the conclusion that **in a DC circuit, the current is directly proportional to the voltage and inversely proportional to the resistance.**

Since Ohm's law is a statement of proportion, it can be expressed as an algebraic formula when standard values such as the volt, amp, and ohm are used. The three basic Ohm's law formulas are shown.

$$E = I \times R$$

$$I = \frac{E}{R}$$

$$R = \frac{E}{I}$$

where

$$E = \text{EMF, or voltage}$$

$$I = \text{intensity of current, or amperage}$$

$$R = \text{resistance}$$

The first formula states that the voltage can be found if the current and resistance are known. Voltage is equal to amps multiplied by ohms. For example, assume a circuit has a resistance of 50 Ω and a current flow through it of 2 A. The voltage connected to this circuit is 100 V.

$$E = I \times R$$

$$E = 2 \times 50$$

$$E = 100$$

The second formula states that the current can be found if the voltage and resistance are known. In the example shown, 120 V are connected to a resistance of 30 Ω. The amount of current flow will be 4 A.

$$I = \frac{E}{R}$$

$$I = \frac{120}{30}$$

$$I = 4$$

The third formula states that if the voltage and current are known, the resistance can be found. Assume a circuit has a voltage of 240 V and a current flow of 10 A. The resistance in the circuit is 24 Ω.

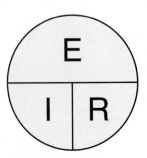

**Figure 2-18** Chart for finding values of voltage, current, and resistance

$$R = \frac{E}{I}$$

$$R = \frac{240}{10}$$

$$R = 24$$

*Figure 2-18* shows a simple chart that can be a great help when trying to remember an Ohm's law formula. To use the chart, cover the quantity that is to be found. For example, if the voltage, E, is to be found, cover the E on the chart. The chart now shows the remaining letters IR *(Figure 2-19)*, thus E = I x R. The same method reveals the formulas for current (I) and resistance (R).

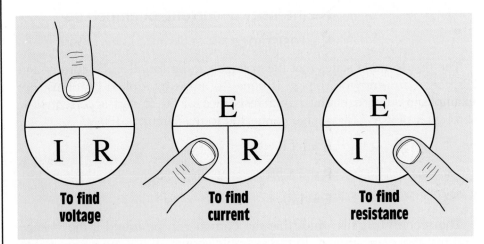

**Figure 2-19** Using the Ohm's law chart

A larger chart, which shows the formulas needed to find watts as well as voltage, amperage, and resistance, is shown in *Figure 2-20*. The letter *P* (power) is used to represent the value of watts. Notice that this chart is divided into four sections and that each section contains three different formulas. To use this chart, select the section containing the quantity to be found and then choose the proper formula from the given quantities.

## Example 3

An electric iron is connected to 120 V and has a current draw of 8 A. How much power is used by the iron?

## Solution

The quantity to be found is watts, or power. The known quantities are

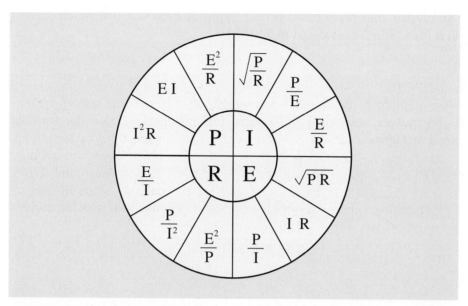

**Figure 2-20** Formula chart for finding values of voltage, current, resistance, and power

voltage and amperage. The proper formula to use is shown in *Figure 2-21.*

$$P = EI$$
$$P = 120 \times 8$$
$$P = 960 \text{ W}$$

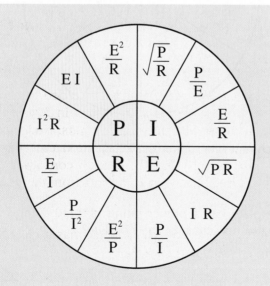

**Figure 2-21** Finding power when voltage and current are known

**Example 4**

An electric hair dryer has a power rating of 1000 W. How much current will it draw when connected to 120 V?

### Solution

The quantity to be found is amperage, or current. The known quantities are power and voltage. To solve this problem, choose the formula shown in *Figure 2-22*.

$$I = \frac{P}{E}$$

$$I = \frac{1000}{120}$$

$$I = 8.33 \text{ A}$$

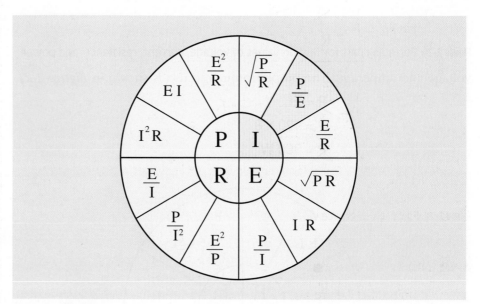

**Figure 2-22** Finding current when power and voltage are known

**Example 5**

An electric hotplate has a power rating of 1440 W and a current draw of 12 A. What is the resistance of the hotplate?

### Solution

The quantity to be found is resistance, and the known quantities are power and current. Use the formula shown in *Figure 2-23*.

$$R = \frac{P}{I^2}$$

$$R = \frac{1440}{12 \times 12}$$

$$R = \frac{1440}{144}$$

$$R = 10\ \Omega$$

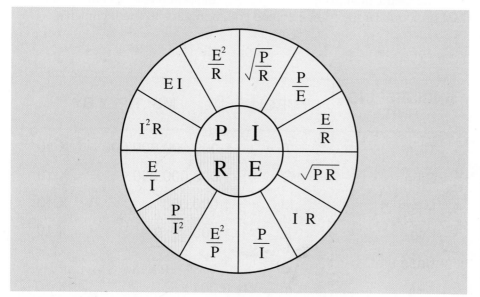

**Figure 2-23** Finding resistance when power and current are known

## 2-12 Metric Units

Metric units of measure are used in the electrical field just as they are in most scientific fields. A special type of metric notation, known as Engineering Notation, is used in electrical measurements. Engineering notation is the same as any other metric measure except that engineering notation is in steps of one thousand instead of ten. The chart in *Figure 2-24* shows standard metric units. The first step above the base unit is deka, which means 10. The second unit is hecto, which means 100, and the third unit is kilo, which means 1000. The first unit below the base unit is deci, which means 1/10; the second unit is centi, which means 1/100; and the third is milli, which means 1/1000.

| | |
|---|---|
| Kilo | 1000 |
| Hecto | 100 |
| Deka | 10 |
| Base unit | 1 |
| Deci | 1/10 or 0.1 |
| Centi | 1/100 or 0.01 |
| Milli | 1/1000 or 0.001 |

**Figure 2-24** Standard units of metric measure

The chart in *Figure 2-25* shows engineering units. The first unit above the base unit is kilo, or 1000; the second unit is mega, or 1,000,000; and the third unit is giga, or 1,000,000,000. Notice that each unit is 1000 times greater than the previous unit. The chart also shows that the first unit below the base unit is milli, or 1/1000; the second is micro, represented by the Greek mu ($\mu$), or 1/1,000,000; and the third is nano, or 1/1,000,000,000.

Metric units are used in almost all scientific measurements for ease of notation. It is much simpler to write a value such as 10 M$\Omega$ than it is to write 10,000,000 ohms, or to write 0.5 ns than to write 0.000,000,000,5 second. Once the metric system has been learned, measurements such as 47 kilohms (k$\Omega$) or 50 milliamps (mA) become commonplace to the technician.

| ENGINEERING UNIT | SYMBOL | MULTIPLY BY | |
|:---:|:---:|:---|:---|
| Tera | T | 1,000,000,000,000 | $\times 10^{12}$ |
| Giga | G | 1,000,000,000 | $\times 10^{9}$ |
| Mega | M | 1,000,000 | $\times 10^{6}$ |
| Kilo | k | 1,000 | $\times 10^{3}$ |
| Base unit | | 1 | |
| Milli | m | 0.001 | $\times 10^{-3}$ |
| Micro | $\mu$ | 0.000,001 | $\times 10^{-6}$ |
| Nano | n | 0.000,000,001 | $\times 10^{-9}$ |
| Pico | p | 0.000,000,000,001 | $\times 10^{-12}$ |

**Figure 2-25** Standard units of engineering notation

## Summary

1. A coulomb is a quantity measurement of electrons.

2. An amp (A) is one coulomb per second.

3. Either the letter *I*, which stands for intensity of current flow, or the letter *A*, which stands for amps, can be used in Ohm's law formulas.

4. Voltage is referred to as electrical pressure, potential difference, or electromotive force. An E or a V can be used to represent voltage in Ohm's law formulas.

5. An ohm (Ω) is a measurement of resistance (R) in an electrical circuit.

6. The watt (W) is a measurement of power in an electrical circuit. It is represented by either a W or a P (power) in Ohm's law formulas.

7. Electrical measurements are generally expressed in engineering notation.

8. Engineering notation differs from the standard metric system in that it uses steps of 1000 instead of steps of 10.

9. Before current can flow, there must be a complete circuit.

10. A short circuit has little or no resistance.

## Review Questions

1. What is a coulomb?

2. What is an amp?

3. Define voltage.

4. Define ohm.

5. Define watt.

6. An electric heating element has a resistance of 16 Ω and is connected to a voltage of 120 V. How much current will flow in this circuit?

7. How many watts of heat are being produced by the heating element in question 6?

8. A 240-V circuit has a current flow of 20 A. How much resistance is connected in the circuit?

9. An electric motor has an apparent resistance of 15 Ω. If 8 A of current are flowing through the motor, what is the connected voltage?

10. A 240-V air conditioning compressor has an apparent resistance of 8 Ω. How much current will flow in the circuit?

11. How much power is being used by the motor in question 8?

12. A 5-kW electric heating unit is connected to a 240-V line. What is the current flow in the circuit?

13. If the voltage in question 12 is reduced to 120 V, how much current would be needed to produce the same amount of power?

14. Is it less expensive to operate the electric heating unit in question 12 on 240 V or 120 V?

## Ohm's Law

Fill in the missing values.

| Volts (E) | Amps (I) | Ohms (R) | Watts (P) |
|---|---|---|---|
| 153 | 0.056 | 2732.142857 | 8.568 |
| 305.5 | 0.65 | 470 Ω | 198.575 |
| 24 | 5.1667 | 4.6451 | 124 |
| | 0.00975 | | 0.035 |
| | | 6.8 kΩ | 0.86 |
| 460 | | 72 Ω | |
| 48 | 1.2 | | |
| | 154 | 0.8 Ω | |
| 277 | | | 760 |
| | 0.0043 | | 0.0625 |
| | | 130 kΩ | 0.0225 |
| 96 | | 2.2 kΩ | |

The word "static" means not moving or at rest. This unit deals with electric charges developed by either an excess or lack of electrons. Static charges can be helpful, harmful, and sometimes dramatic.

Photo by Gary Nelson.

Unit 3

# Static Electricity

## Outline



## Key Terms

- Electroscope
- Electrostatic charge
- Lightning
- Lightning arrestor
- Lightning bolt
- Lightning rod
- Nuisance static charges
- Precipitators
- Selenium
- Static
- Thundercloud
- Useful static charges

## Objectives

**A**fter studying this unit, you should be able to:

- Discuss the nature of static electricity.
- Use an electroscope to determine unknown charges.
- Discuss lightning protection.
- List nuisance charges of static electricity.
- List useful charges of static electricity.

## Preview

Static electrical charges occur often in everyday life. Almost everyone has received a shock after walking across a carpet and then touching a metal object or after sliding across a car seat and touching the door handle. Almost everyone has combed their hair with a hard rubber or plastic comb and then used the comb to attract small pieces of paper or other lightweight objects. Static electrical charges cause clothes to stick together when they are taken out of a clothes dryer. Lightning is without doubt the greatest display of a static electrical discharge.

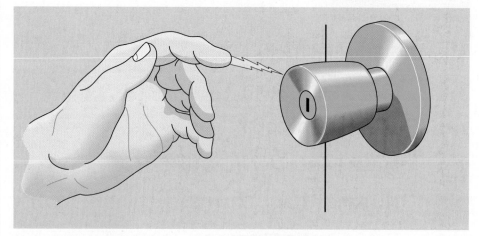

**Figure 3-1** Static electric charges can cause a painful shock.

## 3-1 Static Electricity

Although static charges can be a nuisance *(Figure 3-1),* or even danger-ous, they can also be beneficial. Copy machines, for example, operate on the principle of static electricity. The manufacture of sandpaper also relies on the application of static electricity. Grains of sand receive a static charge to make them stand apart and expose a sharper edge *(Figure 3-2).* Electronic air filters (**precipitators**) use static charges to attract small par-ticles of smoke, dust, and pollen *(Figure 3-3).* The precipitator uses a high-voltage DC power supply to provide a set of wires with a positive charge and a set of plates with a negative charge. As a blower circulates air through the unit, small particles receive a positive charge as they move across the charged wires. The charged particles are then attracted to the negative plates. The negative plates hold the particles until the unit is turned off and the plates are cleaned.

The word **static** means not moving or sitting still. Static electricity refers to electrons that are sitting still and not moving. Static electricity is, there-fore, a charge and not a current. **Electrostatic charges** are built up on insulator materials because insulators are the only materials that can hold the electrons stationary and keep them from flowing to a different

**precipita-tors**

**static**

**electro-static charges**

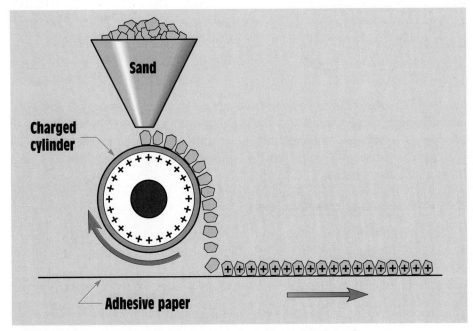

**Figure 3-2** Grains of sand receive a charge to help them stand apart.

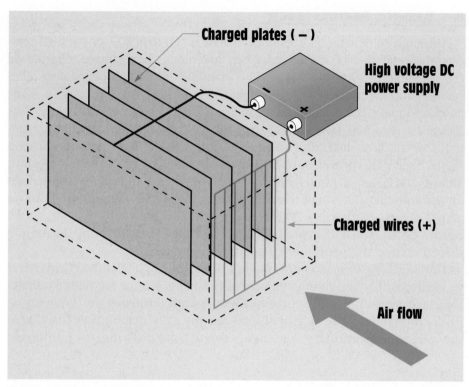

**Figure 3-3** Electronic air cleaner

location. A static charge can be built up on a conductor only if the conductor is electrically insulated from surrounding objects. A static charge can be either positive or negative. If an object has a lack of electrons, it will have a positive charge, and if it has an excess of electrons, it will have a negative charge.

## 3-2 Charging an Object

The charge that accumulates on an object is determined by the materials used to produce the charge. If a hard rubber rod is rubbed on a piece of wool, the wool will deposit excess electrons on the rod and give it a negative charge. If a glass rod is rubbed on a piece of wool, electrons will be removed from the rod, thus producing a positive charge on the rod *(Figure 3-4)*.

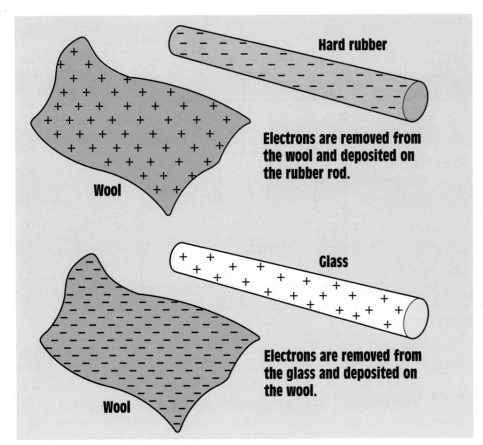

Hard rubber

Electrons are removed from the wool and deposited on the rubber rod.

Wool

Glass

Electrons are removed from the glass and deposited on the wool.

Wool

**Figure 3-4** Producing a static charge

## 3-3 The Electroscope

An early electrical instrument that can be used to determine the polarity of the electrostatic charge of an object is the electroscope *(Figure 3-5)*. An **electroscope** is a metal ball attached to the end of a metal rod. The other end of the rod is attached to two thin metal leaves. The metal leaves are inside a transparent container that permits the action of the leaves to be seen. The metal rod is insulated from the box. The metal leaves are placed inside a container so that air currents cannot affect their movement.

Before the electroscope can be used, it must first be charged. This is done by touching the ball with an object that has a known charge. For this example, assume that a hard rubber rod has been rubbed on a piece

**electro-scope**

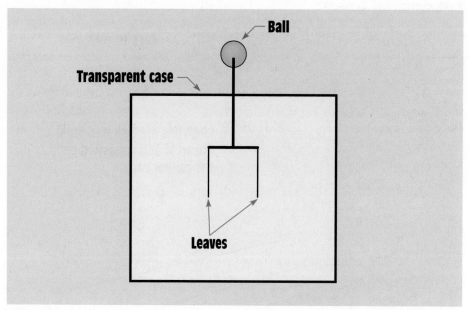

**Figure 3-5** An electroscope

of wool to give it a negative charge. When the rubber rod is wiped against the metal ball, excess electrons are deposited on the metal surface of the electroscope. Since both of the metal leaves now have an excess of electrons, they repel each other as shown in *Figure 3-6*.

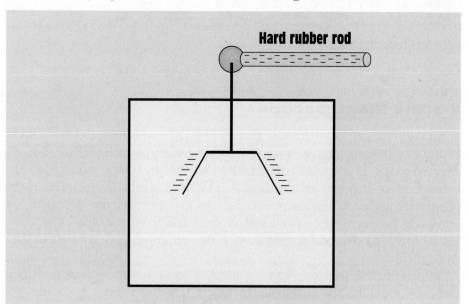

**Figure 3-6** The electroscope is charged with a known static charge.

## Testing an Object

A charged object can now be tested to determine if it has a positive or negative polarity. Assume a ballpoint pen is charged by rubbing the plastic body through a person's hair. Now bring the pen close to but not touching the ball and observe the action of the leaves. If the pen has taken on a negative charge, the leaves will move farther apart as shown in *Figure 3-7.* The field caused by the negative electrons on the pen repels electrons from the ball. These electrons move down the rod to the leaves, causing the leaves to become more negative and to repel each other more, forcing the leaves to move farther apart.

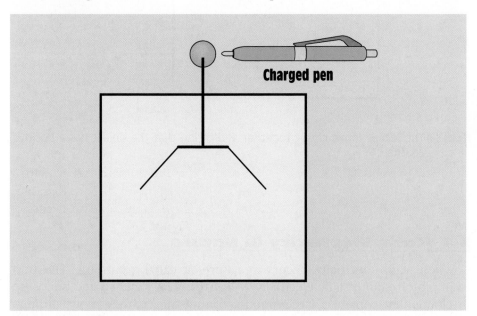

**Charged pen**

**Figure 3-7** The leaves are deflected farther apart, indicating that the object has a negative charge.

If the pen has a positive charge, the leaves will move closer together when the pen is moved near the ball *(Figure 3-8).* This action is caused by the positive field of the pen attracting electrons. When electrons are attracted away from the leaves, they become less negative and move closer together. If the electroscope is charged with a positive charge in the beginning, a negatively charged object will cause the leaves to move closer together, and a positively charged object will cause the leaves to move farther apart.

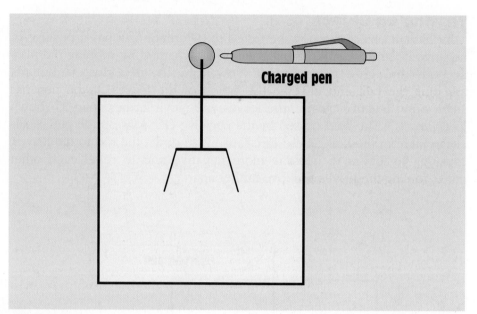

**Figure 3-8** The leaves move closer together, indicating that the object has a positive charge.

## 3-4 Static Electricity in Nature

When static electricity occurs in nature, it can be harmful. The best example of natural static electricity is **lightning**. A static charge builds up in clouds that contain a large amount of moisture as they move through the air. It is theorized that the movement causes a static charge to build up on the surface of drops of water. Large drops become positively charged, and small drops become negatively charged. *Figure 3-9* illustrates a typical **thundercloud**. Notice that both positive and negative charges can be contained in the same cloud. Most lightning discharges, or **lightning bolts**, occur within the cloud. Lightning discharges can also take place between different clouds, between a cloud and the ground, and between the ground and the cloud *(Figure 3-10)*. Whether a lightning bolt travels from the cloud to the ground or from the ground to the cloud is determined by which contains the negative and which the positive charge. Current will always flow from negative to positive. If a cloud is negative and an object on the ground is positive, the lightning discharge will travel from the cloud to the ground. If the cloud has a positive

**lightning**

**thunder-cloud**

**lightning bolts**

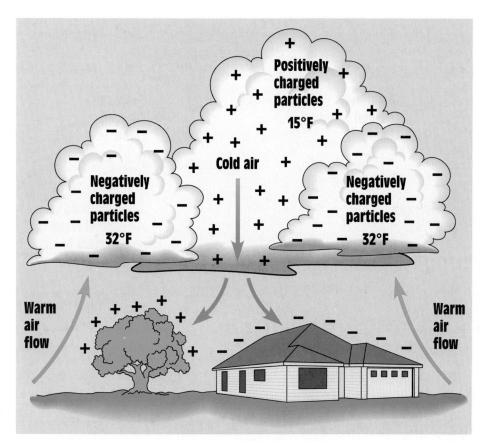

**Figure 3-9** The typical thundercloud contains both negatively and positively charged particles.

charge and the object on the ground has a negative charge, the discharge will be from the ground to the cloud. A lightning bolt has an average voltage of about 15,000,000 V.

## Lightning Protection

**Lightning rods** are sometimes used to help protect objects from lightning. Lightning rods work by providing an easy path to ground for current flow. If the protected object is struck by a lightning bolt, the lightning rod bleeds the lightning discharge to ground before the protected object can be harmed *(Figure 3-11)*. Lightning rods were first invented by Benjamin Franklin.

**lightning rods**

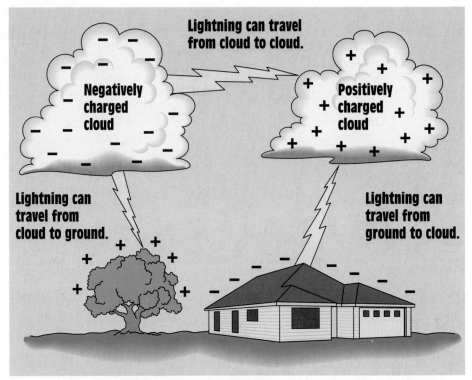

**Figure 3-10** Lightning travels from negative to positive.

Another device used for lightning protection is the lightning arrestor. The **lightning arrestor** works in a manner very similar to the lightning rod except that it is not designed to be struck by lightning itself, and it does not provide a direct path to ground. The lightning arrestor is grounded at one end, and the other end is brought close to but not touching the object to be protected. If the protected object is struck, the high voltage of the lightning arcs across to the lightning arrestor and bleeds to ground.

Power lines are often protected by lightning arrestors that exhibit a very high resistance at the normal voltage of the line. If the power line is struck by lightning, the increase of voltage causes the resistance of the arrestor to decrease and conduct the lightning discharge to ground.

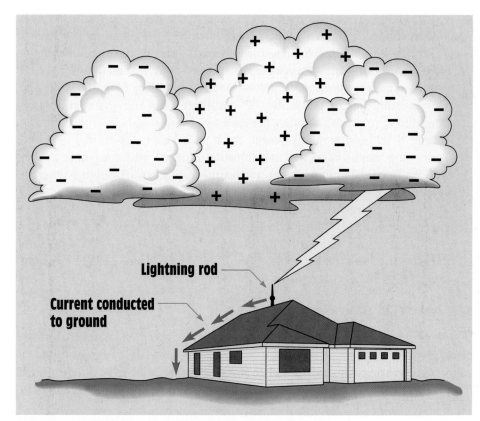

**Figure 3-11**  A lightning rod provides an easy path to ground.

## 3-5 Nuisance Static Charges

Static charges are sometimes a nuisance. Some examples of **nuisance static charges** are:

1. The static charge that accumulates on automobiles as they move through dry air. These static charges can cause dangerous conditions under certain circumstances. For that reason trucks carrying flammable materials such as gasoline or propane use a drag chain. One end of the drag chain is attached to the frame of the vehicle and the other end drags the ground. The chain is used to provide a path to ground while the vehicle is moving and to prevent a static charge from accumulating on the body of the vehicle.

2. The static charge that accumulates on a person's body as he or she walks across a carpet. This charge can cause a painful shock when a metal object is touched and it discharges in the form of an electric

**nuisance static charges**

spark. Most carpets are made from man-made materials that are excellent insulators such as nylon. In the winter, the heating systems of most dwellings remove moisture from the air and cause the air to have a low humidity. The dry air combined with an insulating material provides an excellent setting for the accumulation of a static charge. This condition can generally be eliminated by the installation of a humidifier. A simple way to prevent the painful shock of a static discharge is to hold a metal object, such as a key or coin, in one hand. Touch the metal object to a grounded surface and the static charge will arc from the metal object to ground instead of from your finger to ground.

3. The static charge that accumulates on clothes in a dryer. The static charge is caused by the clothes moving through the dry air. The greatest static charges generally are built up on man-made fabrics because they are the best insulators and retain electrons more readily than natural fabrics such as cotton or wool.

## 3-6 Useful Static Charges

Not all static charges are a nuisance. Some examples of **useful static charges** are:

1. Static electricity is often used in spray painting. A high-voltage grid is placed in front of the spray gun. This grid has a positive charge. The

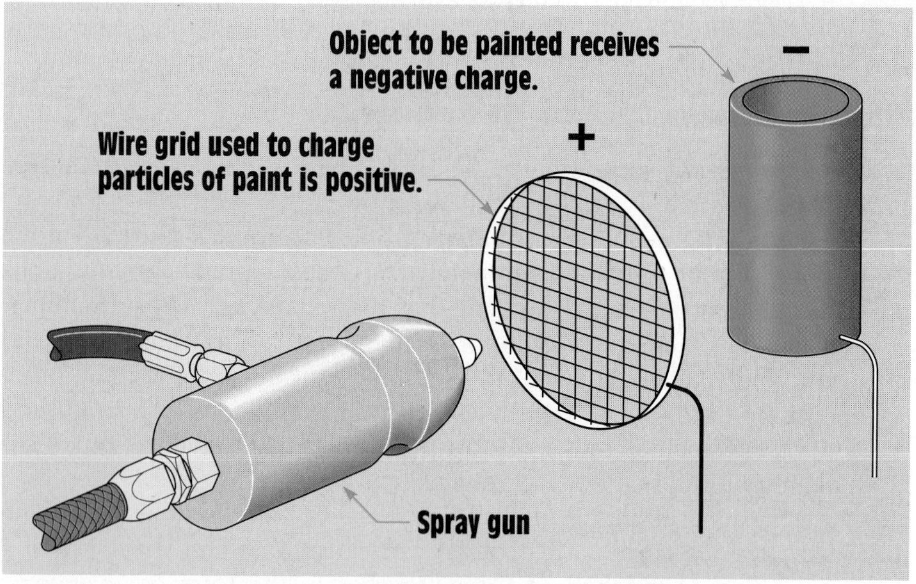

**Figure 3-12** Static electric charges are often used in spray painting.

object to be painted has a negative charge *(Figure 3-12)*. As the droplets of paint pass through the grid, the positive charge causes electrons to be removed from the paint droplets. The positively charged droplets are attracted to the negatively charged object. This static charge helps to prevent waste of the paint and at the same time produces a uniform finish.

2. Another device that depends on static electricity is the dry copy machine. The copy machine uses an aluminum drum coated with selenium *(Figure 3-13)*. **Selenium** is a semiconductor material that changes its conductivity with a change of light intensity. When selenium is in the presence of light, it has a very high conductivity. When it is in darkness, it has a very low conductivity.

A high-voltage wire located near the drum causes the selenium to have a positive charge as it rotates *(Figure 3-14)*. The drum is in dark-

**selenium**

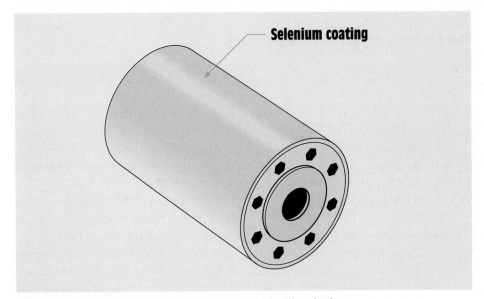

**Selenium coating**

**Figure 3-13** The drum of a copy machine is coated with selenium.

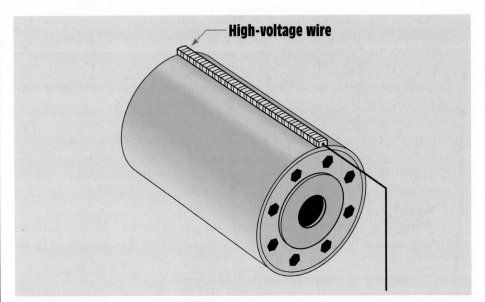

**Figure 3-14** The drum receives a positive charge.

ness when it is charged. An image of the material to be copied is reflected on the drum by a system of lenses and mirrors *(Figure 3-15)*. The light portions of the paper reflect more light than the dark portions. When the reflected light strikes the drum, the conductivity of the selenium increases greatly and negative electrons from the aluminum drum neutralize the selenium charge at that point. The dark area of the paper causes the drum to retain a positive charge.

A dark powder that has a negative charge is applied to the drum *(Figure 3-16)*. The powder is attracted to the positively charged areas on the drum. The powder on the neutral areas of the drum falls away.

A piece of positively charged paper passes under the drum *(Figure 3-17)* and attracts the powder from the drum. The paper then passes under a heating element, which melts the powder into the paper and causes the paper to become a permanent copy of the original.

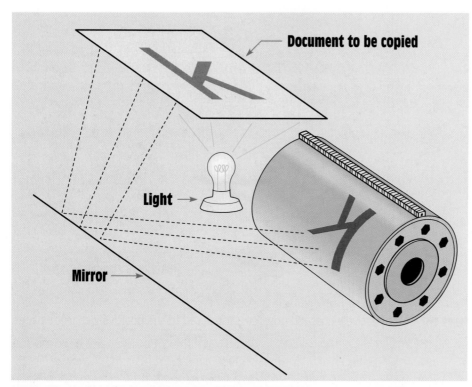

**Figure 3-15** The image is transferred to the selenium drum.

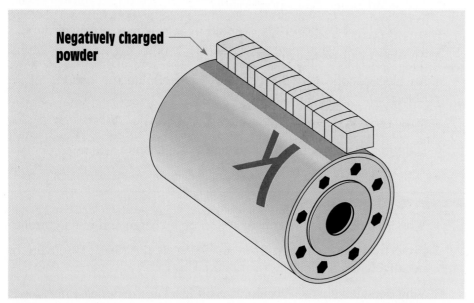

**Figure 3-16** Negatively charged powder is applied to the positively charged drum.

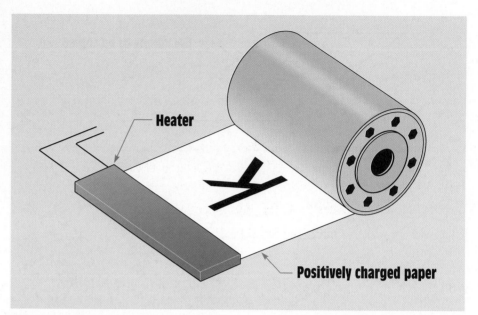

**Figure 3-17** The negatively charged powder is attracted to the positively charged paper.

## Summary

1. The word *static* means not moving.

2. An object can be positively charged by removing electrons from it.

3. An object can be negatively charged by adding electrons to it.

4. An electroscope is a device used to determine the polarity of an object.

5. Static charges accumulate on insulator materials.

6. Lightning is an example of a natural static charge.

## Review Questions

1. Why is static electricity considered to be a charge and not a current?

2. If electrons are removed from an object, is the object positively or negatively charged?

3. Why do static charges accumulate on insulator materials only?

4. What is an electroscope?

5. An electroscope has been charged with a negative charge. An object with an unknown charge is brought close to the electroscope. The leaves of the electroscope come closer together. Does the object have a positive or a negative charge?

6. Can one thundercloud contain both positive and negative charges?

7. A thundercloud has a negative charge, and an object on the ground has a positive charge. Will the lightning discharge be from the cloud to the ground or from the ground to the cloud?

8. Name two devices used for lightning protection.

9. What type of material is used to coat the aluminum drum of a copy machine?

10. What special property does this material have that makes it useful in a copy machine?

**This unit presents the basic laws governing magnetism and how they relate to electricity.**

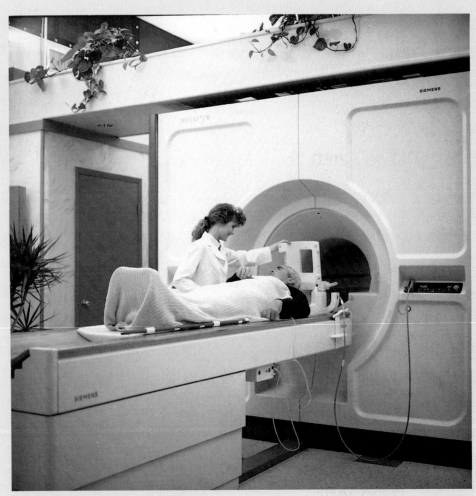

Courtesy of Siemens Medical Systems.

# *Magnetism*

## Key Terms

Ampere-turns

Demagnetized

Electromagnets

Electron spin patterns

Flux

Flux density

Left-hand rule

Lines of flux

Lodestones

Magnetic domains

Magnetic molecules

Magnetomotive force (mmf)

Permeability

Permanent magnets

Reluctance

Residual magnetism

Saturation

## After studying this unit, you should be able to:

**Objectives**

- Discuss the properties of permanent magnets.
- Discuss the difference between the axis poles of the earth and the magnetic poles of the earth.
- Discuss the operation of electromagnets.
- Determine the polarity of an electromagnet when the direction of the current is known.
- Discuss the different systems used to measure magnetism.
- Define terms used to describe magnetism and magnetic quantities.

**Preview**

Magnetism is one of the most important phenomena in the study of electricity. It is the force used to produce most of the electrical power in the world. The force of magnetism has been known for over 2000 years. It was first discovered by the Greeks when they noticed that a certain type of stone was attracted to iron. This stone was first found in Magnesia in Asia Minor and was named magnetite. In the Dark Ages, the strange powers of the magnet were believed to be caused by evil spirits or the devil.

### 4-1 The Earth Is a Magnet

The first compass was invented when it was noticed that a piece of magnetite, a type of stone that is attracted to iron, placed on a piece of wood floating in water always aligned itself north and south *(Figure 4-1)*. Because they are always able to align themselves north and south, natural magnets became known as "leading stones" or **lodestones**. The reason that the lodestone aligned itself north and south is because the earth itself

**lodestones**

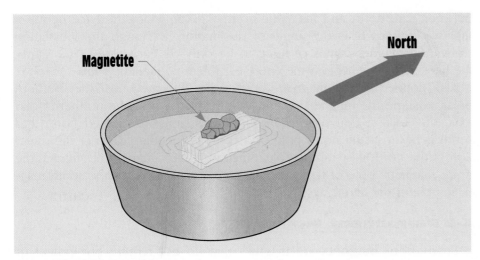

**Figure 4-1** The first compass

contains magnetic poles. *Figure 4-2* illustrates the position of the true North and South poles, or the axis, of the earth and the position of the magnetic poles. Notice that *magnetic* north is not located at the true North Pole of the earth. This is the reason that navigators must distinguish

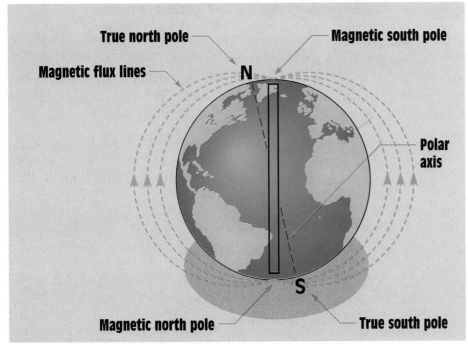

**Figure 4-2** The earth is a magnet.

between true north and magnetic north. The angular difference between the two is known as the angle of declination. Although the illustration shows the magnetic lines of force to be only on each side of the earth, the lines actually surround the entire earth like a magnetic shell.

Also notice that the magnetic north pole is located near the southern polar axis and the magnetic south pole is located near the northern polar axis. The reason that the *geographic* poles (axis) are called north and south is because the north pole of a compass needle points in the direction of the north geographic pole. Since unlike magnetic poles attract, the north magnetic pole of the compass needle is attracted to the south magnetic pole of the earth.

## 4-2 Permanent Magnets

**permanent magnets**

**Energy is required to create a magnetic field, but no energy is required to maintain a magnetic field.**

**Permanent magnets** are magnets that do not require any power or force to maintain their field. They are an excellent example of one of the basic laws of magnetism, which states that **Energy is required to create a magnetic field, but no energy is required to maintain a magnetic field.** Man-made permanent magnets are much stronger and can retain their magnetism longer than natural magnets.

## 4-3 The Electron Theory of Magnetism

There are actually only three substances that form natural magnets: iron, nickel, and cobalt. Why these materials form magnets has been the

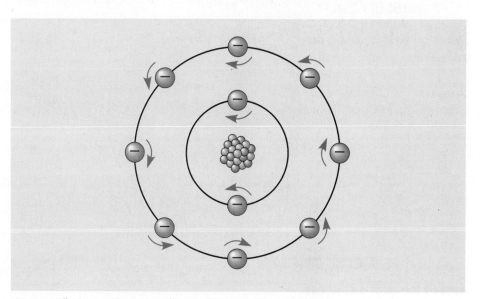

**Figure 4-3** Electron pairs generally spin in opposite directions.

subject of complex scientific investigations, resulting in an explanation of magnetism based on **electron spin patterns**. It is believed that electrons spin on their axis as they orbit around the nucleus of the atom. This spinning motion causes each electron to become a tiny permanent magnet. Although all electrons spin, they do not all spin in the same direction. In most atoms, electrons that spin in opposite directions tend to form pairs *(Figure 4-3)*. Since the electron pairs spin in opposite directions, their magnetic effects cancel each other out as far as having any effect on distant objects. In a similar manner two horseshoe magnets connected together would be strongly attracted to each other, but would have little effect on surrounding objects *(Figure 4-4)*.

An atom of iron contains 26 electrons. Of these 26, 22 are paired and spin in opposite directions, canceling each other's magnetic effect. In the next to the outermost shell, however, four electrons are not paired and spin in the same direction. These four electrons account for the magnetic properties of iron. At a temperature of 1420°F, or 771.1°C, the electron spin patterns rearrange themselves and iron loses its magnetic properties.

When the atoms of most materials combine to form molecules, they arrange themselves in a manner that produces a total of eight valence electrons. The electrons form a spin pattern that cancels the magnetic field of the material. When the atoms of iron, nickel, and cobalt combine, however, the magnetic field is not canceled. Their electrons combine so that they share valence electrons in such a way that their spin patterns are in the same direction, causing their magnetic fields to add instead of cancel. The additive effect forms regions in the molecular structure of the metal called **magnetic domains** or **magnetic molecules**. These magnetic domains act like small permanent magnets.

A piece of nonmagnetized metal has its molecules in a state of disarray as shown in *Figure 4-5*. When the metal is magnetized, its molecules align themselves in an orderly pattern as shown in *Figure 4-6*. In theory,

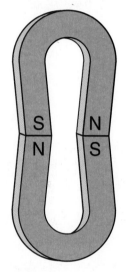

**electron spin patterns**

**Figure 4-4** Two horseshoe magnets attract each other.

**magnetic domains**

**magnetic molecules**

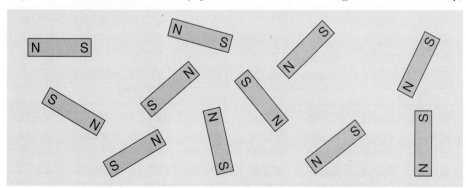

**Figure 4-5** The atoms are disarrayed in a piece of nonmagnetized metal.

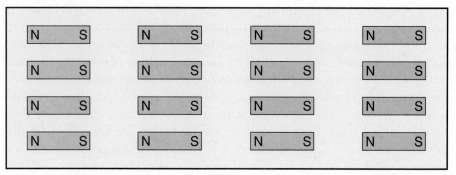

**Figure 4-6** The atoms are aligned in an orderly fashion in a piece of magnetized metal.

each molecule of a magnetic material is itself a small magnet. If a permanent magnet were cut into pieces, each piece would be a separate magnet *(Figure 4-7)*.

## 4-4 Magnetic Materials

Magnetic materials can be divided into three basic classifications. These are:

**Ferromagnetic** materials are metals that are easily magnetized. Examples of these materials are iron, nickel, cobalt, and manganese.

**Paramagnetic** materials are metals that can be magnetized, but not as easily as ferromagnetic materials. Some examples of paramagnetic materials are platinum, titanium, and chromium.

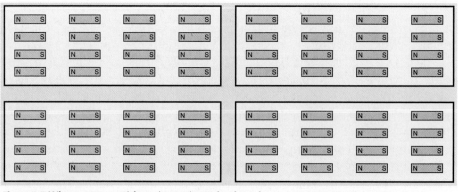

**Figure 4-7** When a magnet is cut apart, each piece becomes a separate magnet.

**Diamagnetic** materials are either metal or nonmetal materials that cannot be magnetized. The magnetic lines of force tend to go around them instead of through them. Some examples of these materials are copper, brass, and antimony.

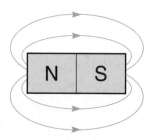

**Figure 4-8** Magnetic lines of force are called lines of flux.

Some of the best materials for the production of permanent magnets are alloys. One of the best permanent magnet materials is Alnico 5, which is made from a combination of aluminum, nickel, cobalt, copper, and iron. Another type of permanent magnet material is made from a combination of barium ferrite and strontium ferrite. Ferrites can have an advantage in some situations because they are insulators and not conductors. They have a resistance of approximately 1,000,000 Ω per centimeter. These two materials can be powdered. The powder is heated to the melting point and then rolled and heat treated. This treatment changes the grain structure and magnetic properties of the material. The new type of material has a property more like stone than metal and is known as a ceramic magnet. Ceramic magnets can be powdered and mixed with rubber, plastic, or liquids. Ceramic magnetic materials mixed with liquids can be used to make magnetic ink, which is used on checks. Another frequently used magnetic material is iron oxide, which is used to make magnetic recording tape and computer diskettes.

## 4-5 Magnetic Lines of Force

Magnetic lines of force are called **flux**. The symbol used to represent flux is the Greek letter phi (Φ). Flux lines can be seen by placing a piece of cardboard on a magnet and sprinkling iron filings on the cardboard. The filings will align themselves in a pattern similar to the one shown in *Figure 4-8*. The pattern produced by the iron filings forms a two-dimensional figure, but the flux lines actually surround the entire magnet (*Figure 4-9*). Magnetic **lines of flux** repel each other and never cross. Although magnetic lines of flux do not flow, it is assumed they are in a direction north to south.

A basic law of magnetism states that **unlike poles attract and like poles repel**. *Figure 4-10* illustrates what happens when a piece of cardboard is placed over two magnets with their north and south poles facing each other and iron filings are sprinkled on the cardboard. The filings form a pattern showing that the magnetic lines of flux are attracted to each other. *Figure 4-11* illustrates the pattern formed by the iron filings when the cardboard is placed over two magnets with like poles facing each other. The filings show that the magnetic lines of flux repel each other.

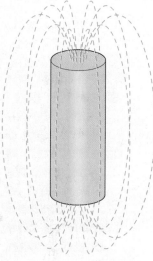

**Figure 4-9** Magnetic lines of flux surround the entire magnet.

**flux**

**lines of flux**

**Unlike poles attract and like poles repel.**

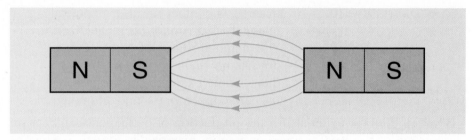

**Figure 4-10** Opposite magnetic poles attract each other.

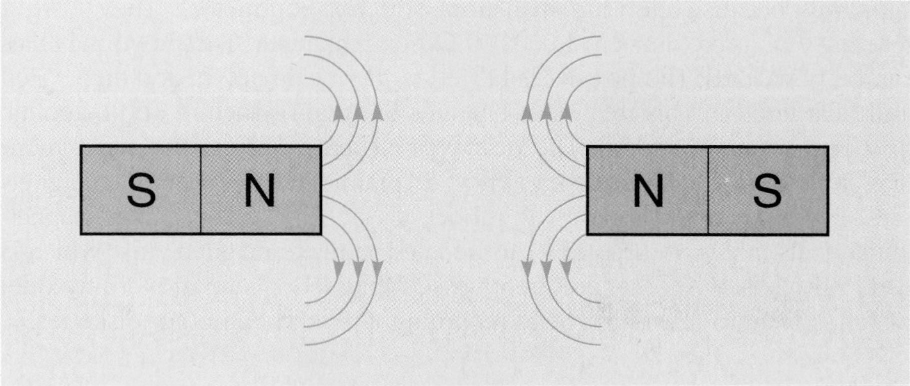

**Figure 4-11** Like magnetic poles repel each other.

Whenever an electric current flows through a conductor, a magnetic field is formed around the conductor.

**electromagnets**

Electromagnets are also known as "temporary magnets" because they are magnets only as long as the current is flowing.

If the opposite poles of two magnets are brought close to each other, they will be attracted to each other as shown in *Figure 4-12.* If like poles of the two magnets are brought together, they will repel each other.

## 4-6 Electromagnetics

A basic law of physics states that **whenever an electric current flows through a conductor, a magnetic field is formed around the conductor**. **Electromagnets** depend on electric current flow to produce a magnetic field. They are generally designed to produce a magnetic field only as long as the current is flowing; they do not retain their magnetism when current flow stops. Electromagnets operate on the principle that current flowing through a conductor produces a magnetic field around the conductor *(Figure 4-13).* If the conductor is wound into a coil as

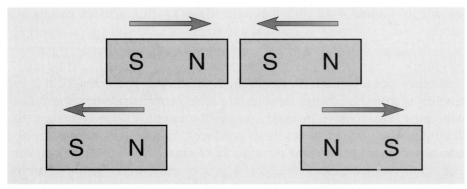

**Figure 4-12**  Opposite poles of a magnet attract and like poles repel.

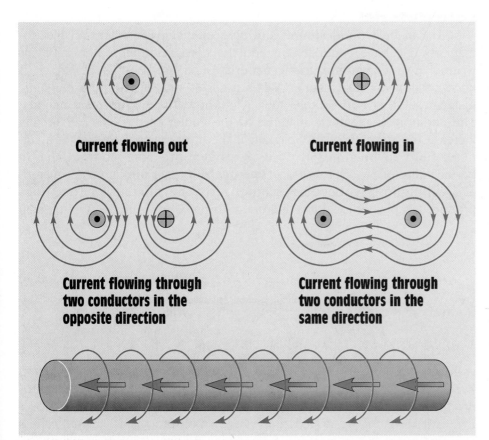

**Figure 4-13**  Current flowing through a conductor produces a magnetic field around the conductor.

shown in *Figure 4-14*, the magnetic lines of flux add to produce a stronger magnetic field. A coil with 10 turns of wire will produce a magnetic field that is 10 times as strong as the magnetic field around a single conductor.

Another factor that affects the strength of an electromagnetic field is the amount of current flowing through the wire. An increase in current flow will cause an increase in magnetic field strength. The two factors that determine the number of flux lines produced by an electromagnet are the number of turns of wire and the amount of current flow through the wire. The strength of an electromagnet is proportional to its **ampere-turns**. Ampere-turns are determined by multiplying the number of turns of wire by the current flow.

**ampere-turns**

## Core Material

Coils can be wound around any type of material to form an electromagnet. The base material is called the core material. When a coil is wound around a nonmagnetic material such as wood or plastic, it is known as an *air core* magnet. When a coil is wound around a magnetic material such as iron or soft steel, it is known as an *iron core* magnet. The addition of magnetic material to the center of the coil can greatly increase the strength of the magnet. If the core material causes the magnetic field to become 50 times stronger, the core material has a permeability of 50 *(Figure 4-15)*. **Permeability** is a measure of a material's willingness to become magnetized. The number of flux lines produced is proportional to the ampere-turns. The magnetic core material

**permeability**

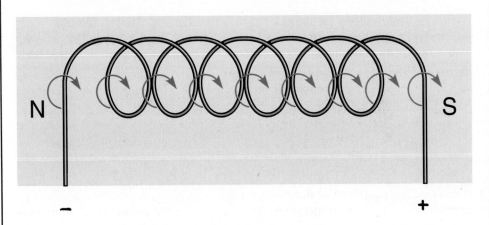

**Figure 4-14** Winding the wire into a coil increases the strength of the magnetic field.

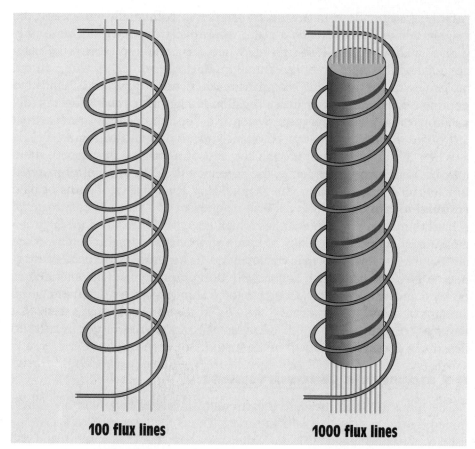

**100 flux lines**          **1000 flux lines**

**Figure 4-15** An iron core increases the number of flux lines per square inch.

provides an easy path for the flow of magnetic lines in much the same way a conductor provides an easy path for the flow of electrons. This increased permeability permits the flux lines to be concentrated in a smaller area, which increases the number of flux lines per square inch or per square centimeter. In a similar manner, a person using a garden hose with an adjustable nozzle attached can adjust the nozzle to spray the water in a fine mist that covers a large area or in a concentrated stream that covers a small area.

Another common magnetic measurement is reluctance. **Reluctance** is resistance to magnetism. A material such as soft iron or steel has a high permeability and low reluctance because it is easily magnetized. A material such as copper has a low permeability and high reluctance.

**reluctance**

If the current flow in an electromagnet is continually increased, the magnet will eventually reach a point where its strength will increase only slightly with an increase in current. When this condition occurs, the magnetic material is at a point of saturation. **Saturation** occurs when all the molecules of the magnetic material are lined up. Saturation is similar to pouring 5 gallons of water into a 5-gallon bucket. Once the bucket is full, it simply cannot hold any more water. If it became necessary to construct a stronger magnet, a larger piece of core material would be required.

When the current flow through the coil of a magnet is stopped, there may be some magnetism left in the core material. The amount of magnetism left in a material after the magnetizing force has stopped is called **residual magnetism**. If the residual magnetism of a piece of core material is hard to remove, the material has a high coercive force. *Coercive force* is a measure of a material's ability to retain magnetism. A high coercive force is desirable in materials that are intended to be used as permanent magnets. A low coercive force is generally desirable for materials intended to be used as electromagnets. Coercive force is measured by determining the amount of current flow through the coil in the direction opposite to that required to remove the residual magnetism. Another term that is used to describe a material's ability to retain magnetism is *retentivity*.

### 4-7 Magnetic Measurement

The terms used to measure the strength of a magnetic field are determined by the system that is being used. There are three different systems used to measure magnetism: the English system, the CGS system, and the MKS system.

### The English System

In the English system of measure, magnetic strength is measured in a term called flux density. Flux density is measured in lines per square inch. The Greek letter phi ($\Phi$) is used to measure flux. The letter $B$ is used to represent flux density. The formula shown below is used to determine flux density.

$$B \text{ (flux density)} = \frac{\Phi \text{ (flux lines)}}{A \text{ (area)}}$$

In the English system, the term used to describe the total force producing a magnetic field, or flux, is **magnetomotive force (mmf)**. Magnetomotive force can be computed using the formula:

$$\text{mmf} = \Phi \times \text{rel (reluctance)}$$

saturation

residual magnetism

magneto-motive force (mmf)

The formula shown below can be used to determine the strength of the magnet.

$$\text{Pull (in pounds)} = \frac{B \times A}{72{,}000{,}000}$$

where B = flux density in lines per square inch
A = area of the magnet.

## The CGS System

In the CGS (centimeter-gram-second) system of measurement, one magnetic line of force is known as a maxwell. A gauss represents a magnetic force of one maxwell per square centimeter. In the English system, magnetomotive force is measured in ampere-turns. In the CGS system, gilberts are used to represent the same measurement. Since the main difference between these two systems of measurement is that one uses English units of measure and the other uses metric units of measure, a conversion factor can be used to help convert one set of units to the other.

1 gilbert = 1.256 ampere-turns

## The MKS System

The MKS (meter-kilogram-second) system uses metric units of measure also. In this system, the main unit of magnetic measurement is the dyne. The dyne is a very weak amount of force. One dyne is equal to 1/27,800 of an ounce, or it requires 27,800 dynes to equal a force of one ounce. In the MKS system, a standard called the unit magnetic pole is used. In *Figure 4-16*, two magnets are separated by a distance of 1 cm. These magnets repel each other with a force of 1 dyne. When two magnets separated by a distance of 1 cm exert a force on each other of 1 dyne, they

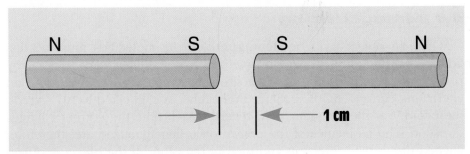

**Figure 4-16** A unit magnetic pole produces a force of one dyne.

are considered to be a unit magnetic pole. Magnetic force can then be determined using the formula

$$\text{Force (in dynes)} = \frac{M_1 \times M_2}{D^2}$$

where $M_1$ = strength of first magnet, in unit magnetic poles

$M_2$ = strength of second magnet, in unit magnetic poles

$D$ = distance between the poles, in centimeters

### 4-8 Magnetic Polarity

The polarity of an electromagnet can be determined using the **left-hand rule**. When the fingers of the left hand are placed around the windings in the direction of electron current flow, the thumb will point to the north magnetic pole *(Figure 4-17)*. If the direction of current flow is reversed, the polarity of the magnetic field will reverse also.

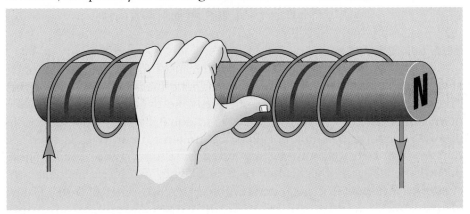

**Figure 4-17**  The left-hand rule can be used to determine the polarity of an electromagnet.

### 4-9 Demagnetizing

When an object is to be **demagnetized**, its molecules must be disarranged as they are in a nonmagnetized material. This can be done by placing the object in the field of a strong electromagnet connected to an alternating current (AC) line. Since the magnet is connected to AC current, the polarity of the magnetic field reverses each time the current changes direction. The molecules of the object to be demagnetized are, therefore, aligned first in one direction and then in the other. If the object is pulled

**left-hand rule**

**demagne-tized**

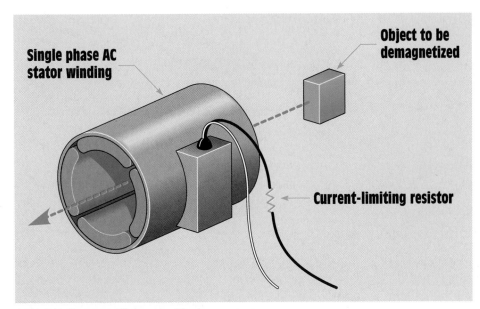

**Figure 4-18** Demagnetizing an object

away from the AC magnetic field, the effect of the field becomes weaker as the object is moved farther away *(Figure 4-18)*. The weakening of the magnetic field causes the molecules of the object to be left in a state of disarray. The ease or difficulty with which an object can be demagnetized depends on the strength of the AC magnetic field and the coercive force of the object.

An object can be demagnetized in two other ways *(Figure 4-19)*. If a magnetized object is struck, the vibration will often cause the molecules to rearrange themselves in a disordered fashion. It may be necessary to strike the object several times. Heating also will demagnetize an object. When the temperature becomes high enough, the molecules will rearrange themselves in a disordered fashion.

## 4-10 Magnetic Devices

A list of devices that operate on magnetism would be very long indeed. Some of the more common devices are electromagnets, measuring instruments, inductors, transformers, and motors.

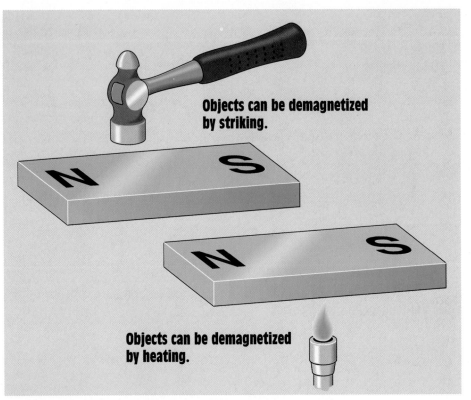

**Figure 4-19** Other methods for demagnetizing objects

## The Speaker

The speaker is a common device that operates on the principle of magnetism *(Figure 4-20)*. The speaker produces sound by moving a cone; the movement causes a displacement of air. The tone is determined by how fast the cone vibrates. Low or bass sounds are produced by vibrations in the range of 20 cycles per second. High sounds are produced when the speaker vibrates in the range of 20,000 cycles per second.

The speaker uses two separate magnets. One is a permanent magnet, and the other is an electromagnet. The permanent magnet is held stationary, and the electromagnet is attached to the speaker cone. When current flows through the coil of the electromagnet, a magnetic field is produced. The polarity of the field is determined by the direction of current flow. When

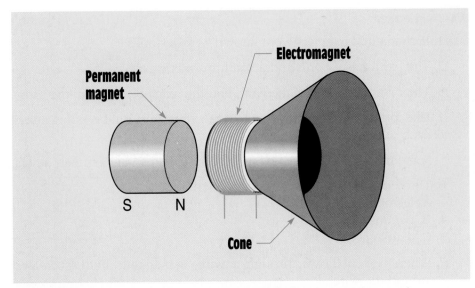

**Figure 4-20** A speaker uses both an electromagnet and a permanent magnet.

the electromagnet has a north polarity, it is repelled away from the permanent magnet, causing the speaker cone to move outward and displace air. When the current flow reverses through the coil, the electromagnet has a south polarity and is attracted to the permanent magnet. The speaker cone then moves inward and again displaces air. The number of times per second that the current through the coil reverses determines the tone of the speaker.

## Summary

1. Early natural magnets were known as lodestones.

2. The earth has a north and a south magnetic pole.

3. The magnetic poles of the earth and the axis poles are not the same.

4. Like poles of a magnet repel each other, and unlike poles attract each other.

5. Some materials have the ability to become better magnets than others.

6. Three basic types of magnetic material are:
   A. ferromagnetic
   B. paramagnetic
   C. diamagnetic

7. When current flows through a wire, a magnetic field is created around the wire.

8. The direction of current flow through the wire determines the polarity of the magnetic field.

9. The strength of an electromagnet is determined by the ampere-turns.

10. The type of core material used in an electromagnet can increase its strength.

11. Three different systems are used to measure magnetic values:
    A. The English system
    B. The CGS system
    C. The MKS system

12. An object can be demagnetized by placing it in an AC magnetic field and pulling it away, by striking, and by heating.

## Review Questions

1. Is the north magnetic pole of the earth a north polarity or a south polarity?

2. What were early natural magnets known as?

3. The south pole of one magnet is brought close to the south pole of another magnet. Will the magnets repel or attract each other?

4. How can the polarity of an electromagnet be determined if the direction of current flow is known?

5. Define the following terms.
   Flux density
   Permeability
   Reluctance
   Saturation
   Coercive force
   Residual magnetism

6. A force of 1 ounce is equal to how many dynes?

Resistors are one of the most common electrical components. This unit presents different types of resistors and how to determine their values.

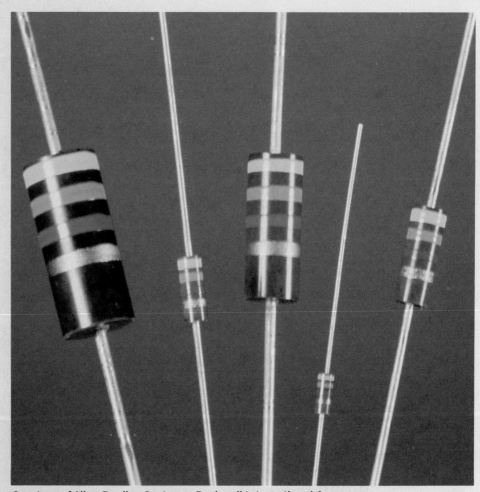

Courtesy of Allen Bradley Co. Inc., a Rockwell International Company.

# Resistors

## Key Terms

Carbon film resistor

Color code

Composition carbon resistor

Fixed resistors

Metal film resistors

Metal glaze resistor

Multiturn variable resistors

Pot

Potentiometer

Rheostat

Short circuit

Tolerance

Variable resistor

Voltage divider

Wire wound resistor

## Objectives

*fter studying this unit, you should be able to:*

■ List the major types of fixed resistors.

■ Determine the resistance of a resistor using the color code.

■ Determine if a resistor is operating within its power rating.

■ Connect a variable resistor for use as a poten-tiometer.

## Preview

Resistors are one of the most common components found in electrical circuits. The unit of measure for resistance (R) is the *ohm*, which was named for a German scientist named Georg S. Ohm. The symbol used to represent resistance is the Greek letter omega (Ω). Resistors come in various sizes, types, and ratings to accommodate the needs of almost any circuit applications.

### 5-1 Uses of Resistors

Resistors are commonly used to perform two functions in a circuit. One is to limit the flow of current through the circuit. In *Figure 5-1* a 30-Ω resistor is connected to a 15-V battery. The current in this circuit is limited to a value of 0.5 A.

$$I = \frac{E}{R}$$

$$I = \frac{15}{30}$$

$$I = 0.5 \text{ A}$$

If this resistor were not present, the circuit current would be limited only by the resistance of the conductor, which would be very low, and a

**Figure 5-1** Resistor used to limit the flow of current

large amount of current would flow. Assume for example that the wire has a resistance of 0.0001 Ω. When the wire is connected across the 15-V power source, a current of 150,000 A would try to flow through the circuit (15/0.0001 = 150,000). This is commonly known as a **short circuit**.

**short circuit**

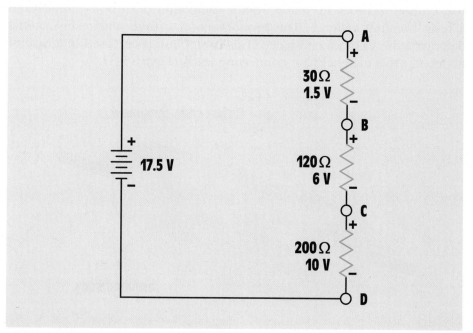

**Figure 5-2** Resistors used as a voltage divider

The second principal function of resistors is to produce a **voltage divider**. The three resistors shown in *Figure 5-2* are connected in series with a 17.5-V battery. If the leads of a voltmeter were connected between different points in the circuit, it would indicate the following voltages:

**A to B**, 1.5 V
**A to C**, 7.5 V
**A to D**, 17.5 V
**B to C**, 6 V
**B to D**, 16 V
**C to D**, 10 V

By connecting resistors of the proper value, almost any voltage desired can be obtained. Voltage dividers were used to a large extent in vacuum tube circuits many years ago. Voltage divider circuits are still used today in applications involving field effect transistors (FETs) and in multirange voltmeter circuits.

## 5-2 Fixed Resistors

**Fixed resistors** have only one ohmic value, which cannot be changed or adjusted. There are several different types of fixed resistors. One of the most common types of fixed resistors is the **composition carbon resistor.** Carbon resistors are made from a compound of carbon graphite and a resin bonding material. The proportions of carbon and resin material determine the value of resistance. This compound is enclosed in a case of nonconductive material with connecting leads *(Figure 5-3)*.

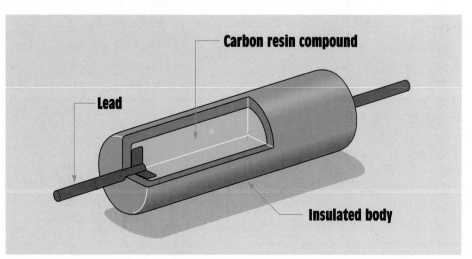

**Figure 5-3** Composition carbon resistor

Carbon resistors are very popular for most applications because they are inexpensive and readily available. They are made in standard values that range from about 1 Ω to about 22 MΩ (M represents meg), and they can be obtained in power ratings of 1/8, 1/4, 1/2, 1, and 2 W. The power rating of the resistor is indicated by its size. A 1/2-W resistor is approximately 3/8 in. in length and 1/8 in. in diameter. A 2-W resistor has a length of approximately 11/16 in. and a diameter of approximately 5/16 in. *(Figure 5-4)*. The 2-W resistor is larger than the 1/2-W or 1-W because it must have a larger surface area to be able to dissipate more heat. Although carbon resistors have a lot of desirable characteristics, they have one characteristic that is not desirable. Carbon resistors will change their value with age or if they are overheated. Carbon resistors generally increase instead of decrease in value.

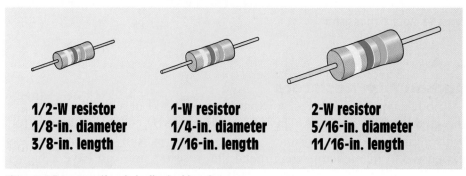

**1/2-W resistor**
**1/8-in. diameter**
**3/8-in. length**

**1-W resistor**
**1/4-in. diameter**
**7/16-in. length**

**2-W resistor**
**5/16-in. diameter**
**11/16-in. length**

**Figure 5-4** Power rating is indicated by size.

## Metal Film Resistors

Another type of fixed resistor is the metal film resistor. **Metal film resistors** are constructed by applying a film of metal to a ceramic rod in a vacuum *(Figure 5-5)*. The resistance is determined by the type of metal used to form the film and the thickness of the film. Typical thicknesses for the film are from 0.00001 to 0.00000001 in. Leads are then attached to the film coating, and the entire assembly is covered with a coating. These resistors are superior to carbon resistors in several respects. Metal film resistors do not change their value with age, and their tolerance is generally better than carbon resistors. **Tolerance** indicates the plus and minus limits of a resistor's ohmic value. Carbon resistors commonly have a tolerance range of 20%, 10%, or 5%. Metal film resistors generally range in tolerance from 2% to 0.1%. The disadvantage of the metal film resistor is that it costs more.

**metal film resistors**

**tolerance**

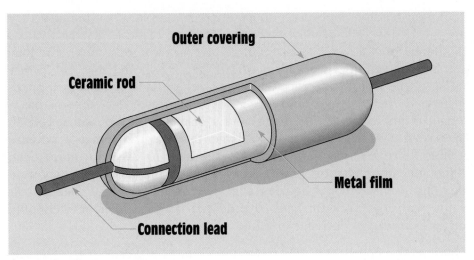

**Figure 5-5** Metal film resistor

### Carbon Film Resistors

**carbon film resistor**

Another type of fixed resistor that is constructed in a similar manner is the **carbon film resistor**. This resistor is made by coating a ceramic rod with a film of carbon instead of metal. Carbon film resistors are less expensive to manufacture than metal film resistors and can have a higher tolerance rating than composition carbon resistors.

### Metal Glaze Resistors

**metal glaze resistor**

The **metal glaze resistor** is also a fixed resistor, similar to the metal film resistor. This resistor is made by combining metal with glass. The compound is then applied to a ceramic base as a thick film. The resistance is determined by the amount of metal used in the compound. Tolerance ratings of 2% and 1% are common.

### Wire Wound Resistors

**wire wound resistors**

**Wire wound resistors** are fixed resistors that are made by winding a piece of resistive wire around a ceramic core *(Figure 5-6)*. The resistance of a wire wound resistor is determined by three factors:

1. the type of material used to make the resistive wire,
2. the diameter of the wire, and
3. the length of the wire.

Wire wound resistors can be found in various case styles and sizes. These resistors are generally used when a high power rating is needed.

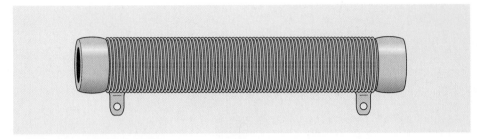

**Figure 5-6** Wire wound resistor

Wire wound resistors can operate at higher temperatures than any other type of resistor. A wire wound resistor that has a hollow center is shown in *Figure 5-7*. This type of resistor should be mounted vertically and not horizontally. The center of the resistor is hollow for a very good reason. When the resistor is mounted vertically, the heat from the resistor produces a chimney effect and causes air to circulate through the center *(Figure 5-8)*. This increase of air flow dissipates heat at a faster rate to help keep the resistor from overheating. The disadvantage of wire wound resistors is they are expensive and generally require a large amount of space for mounting. They can also exhibit an amount of inductance in circuits that operate at high frequencies. This added inductance can cause problems to the rest of the circuit. Inductance will be covered in later units.

**Figure 5-7** Wire wound resistor with hollow core

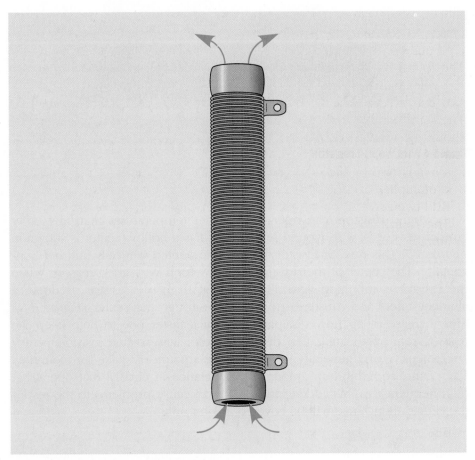

**Figure 5-8** Air flow helps cool the resistor.

### 5-3 Color Code

color code

The values of a resistor can often be determined by the **color code**. Many resistors have bands of color that are used to determine the resistance value, tolerance, and in some cases reliability. The color bands represent numbers. Each color represents a different numerical value. The chart shown in *Figure 5-9* lists the color and the number value assigned to each color. The resistor shown beside the color chart illustrates how to determine the value of a resistor. Resistors can have from three to five bands of color. Resistors that have a tolerance of ±20% have only three color bands. Most resistors contain four bands of color. For resistors with tolerances that range from ±10% to ±2%, the first two color bands represent number values. The third color band is called the multiplier. This means to combine the first two numbers and multiply the resulting two-digit number by the power of 10 indicated by the value of the third band. The fourth band indicates the tolerance. For example, assume a resistor has color bands of brown, green, red, and silver *(Figure 5-10)*. The first two bands represent the numbers 1 and 5 (brown is 1 and green is 5). The third band is red, which has a number value of 2. The number 15 should be multiplied by $10^2$, or 100. The value of the resistor is 1500 $\Omega$. Another method, which is simpler to understand, is to add the number of zeros indicated by the multiplier band to the combined first two numbers. The multiplier band in this example is red, which has a numeric value of 2. Add two zeros to the first two numbers. The number 15 becomes 1500.

The fourth band is the tolerance band. The tolerance band in this example is silver, which means ±10%. This resistor should be 1500 $\Omega$ plus or minus 10%. To determine the value limits of this resistor, find 10% of 1500.

$$1500 \times 0.10 = 150$$

The value can range from 1500 + 10%, or 1500 + 150 = 1650 $\Omega$, to 1500 – 10% or 1500 – 150 = 1350 $\Omega$.

Resistors that have a tolerance of ±1%, as well as some military resistors, contain five bands of color.

**Example 1**

The resistor shown in *Figure 5-11* contains the following bands of color:

**first band** = brown
**second band** = black
**third band** = black
**fourth band** = brown
**fifth band** = brown

## Solution

The brown fifth band indicates that this resistor has a tolerance of ±1%. To determine the value of a 1% resistor, the first three bands are numbers, and the fourth band is the multiplier. In this example, the first band is brown, which has a number value of 1. The next two bands are black, which represents a number value of 0. The fourth band is brown, which means add one 0 to the first three numbers. The value of this resistor is 1000 $\Omega$ ±1%.

**Left chart**

| First Band | Second Band | Multiplier Band | Resistance Ω |
|---|---|---|---|
| Brown | Black | Black | 10 |
| | | Brown | 100 |
| | | Red | 1000 |
| | | Orange | 10000 |
| | | Yellow | 0.1 Meg. |
| | | Green | 1.0 Meg. |
| | | Blue | 10.0 Meg. |
| | Brown | Black | 11 |
| | | Brown | 110 |
| | | Red | 1100 |
| | | Orange | 11000 |
| | | Yellow | 0.11 Meg. |
| | | Green | 1.1 Meg. |
| | | Blue | 11.0 Meg. |
| | Red | Black | 12 |
| | | Brown | 120 |
| | | Red | 1200 |
| | | Orange | 12000 |
| | | Yellow | 0.12 Meg. |
| | | Green | 1.2 Meg. |
| | | Blue | 12.0 Meg. |
| | Orange | Black | 13 |
| | | Brown | 130 |
| | | Red | 1300 |
| | | Orange | 13000 |
| | | Yellow | 0.13 Meg. |
| | | Green | 1.3 Meg. |
| | | Blue | 13.0 Meg. |
| | Green | Black | 15 |
| | | Brown | 150 |
| | | Red | 1500 |
| | | Orange | 15000 |
| | | Yellow | 0.15 Meg. |
| | | Green | 1.5 Meg. |
| | | Blue | 15.0 Meg. |
| | Blue | Black | 16 |
| | | Brown | 160 |
| | | Red | 1600 |
| | | Orange | 16000 |
| | | Yellow | 0.16 Meg. |
| | | Green | 1.6 Meg. |
| | | Blue | 16.0 Meg. |
| | Gray | Black | 18 |
| | | Brown | 180 |
| | | Red | 1800 |
| | | Orange | 18000 |
| | | Yellow | 0.18 Meg. |
| | | Green | 1.8 Meg. |
| | | Blue | 18.0 Meg. |

**Right chart**

| First Band | Second Band | Multiplier Band | Resistance Ω |
|---|---|---|---|
| Yellow | Orange | Gold | 4.3 |
| | | Black | 43 |
| | | Brown | 430 |
| | | Red | 4300 |
| | | Orange | 43000 |
| | | Yellow | 0.43 Meg. |
| | | Green | 4.3 Meg. |
| | Violet | Gold | 4.7 |
| | | Black | 47 |
| | | Brown | 470 |
| | | Red | 4700 |
| | | Orange | 47000 |
| | | Yellow | 0.47 Meg. |
| | | Green | 4.7 Meg. |
| Green | Brown | Gold | 5.1 |
| | | Black | 51 |
| | | Brown | 510 |
| | | Red | 5100 |
| | | Orange | 51000 |
| | | Yellow | 0.51 Meg. |
| | | Green | 5.1 Meg. |
| | Blue | Gold | 5.6 |
| | | Black | 56 |
| | | Brown | 560 |
| | | Red | 5600 |
| | | Orange | 56000 |
| | | Yellow | 0.56 Meg. |
| | | Green | 5.6 Meg. |
| Blue | Red | Gold | 6.2 |
| | | Black | 62 |
| | | Brown | 620 |
| | | Red | 6200 |
| | | Orange | 62000 |
| | | Yellow | 0.62 Meg. |
| | | Green | 6.2 Meg. |
| | Gray | Gold | 6.8 |
| | | Black | 68 |
| | | Brown | 680 |
| | | Red | 6800 |
| | | Orange | 68000 |
| | | Yellow | 0.68 Meg. |
| | | Green | 6.8 Meg. |
| Violet | Green | Gold | 7.5 |
| | | Black | 75 |
| | | Brown | 750 |
| | | Red | 7500 |
| | | Orange | 75000 |
| | | Yellow | 0.75 Meg. |
| | | Green | 7.5 Meg. |
| Gray | Red | Gold | 8.2 |
| | | Black | 82 |
| | | Brown | 820 |
| | | Red | 8200 |
| | | Orange | 82000 |
| | | Yellow | 0.82 Meg. |
| | | Green | 8.2 Meg. |
| White | Brown | Gold | 9.1 |
| | | Black | 91 |
| | | Brown | 910 |
| | | Red | 9100 |
| | | Orange | 91000 |
| | | Yellow | 0.91 Meg. |
| | | Green | 9.1 Meg. |

**Figure 5-9** Resistor color code chart

| First Band | Second Band | Multiplier Band | Resistance Ω | First Band | Second Band | Multiplier Band | Resistance Ω |
|---|---|---|---|---|---|---|---|
| Red | Black | Black | 20 | Orange | Black | Gold | 3.0 |
| | | Brown | 200 | | | Black | 30 |
| | | Red | 2000 | | | Brown | 300 |
| | | Orange | 20000 | | | Red | 3000 |
| | | Yellow | 0.20 Meg. | | | Orange | 30000 |
| | | Green | 2.0 Meg. | | | Yellow | 0.30 Meg. |
| | | Blue | 20.0 Meg. | | | Green | 3.0 Meg. |
| | Red | Black | 22 | | Orange | Gold | 3.3 |
| | | Brown | 220 | | | Black | 33 |
| | | Red | 2200 | | | Brown | 330 |
| | | Orange | 22000 | | | Red | 3300 |
| | | Yellow | 0.22 Meg. | | | Orange | 33000 |
| | | Green | 2.2 Meg. | | | Yellow | 0.33 Meg. |
| | | Blue | 22.0 Meg. | | | Green | 3.3 Meg. |
| | Yellow | Black | 24 | | Blue | Gold | 3.6 |
| | | Brown | 240 | | | Black | 36 |
| | | Red | 2400 | | | Brown | 360 |
| | | Orange | 24000 | | | Red | 3600 |
| | | Yellow | 0.24 Meg. | | | Orange | 36000 |
| | | Green | 2.4 Meg. | | | Yellow | 0.36 Meg. |
| | | | | | | Green | 3.6 Meg. |
| | Violet | Gold | 2.7 | | White | Gold | 3.9 |
| | | Black | 27 | | | Black | 39 |
| | | Brown | 270 | | | Brown | 390 |
| | | Red | 2700 | | | Red | 3900 |
| | | Orange | 27000 | | | Orange | 39000 |
| | | Yellow | 0.27 Meg. | | | Yellow | 0.39 Meg. |
| | | Green | 2.7 Meg. | | | Green | 3.9 Meg. |

# Standard Color Code

* Note Wide Space

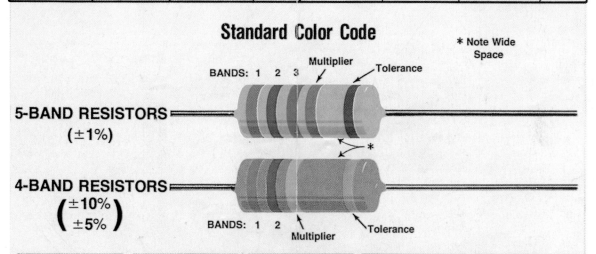

**5-BAND RESISTORS**
(±1%)

BANDS: 1  2  3
Multiplier
Tolerance

**4-BAND RESISTORS**
( ±10%
±5% )

BANDS: 1  2
Multiplier
Tolerance

| Band 1 1st Digit | | Band 2 2nd Digit | | Band 3 (if used) 3rd Digit | | Multiplier | | Resistance Tolerance | |
|---|---|---|---|---|---|---|---|---|---|
| Color | Digit | Color | Digit | Color | Digit | Color | Multiplier | Color | Tolerance |
| Black | 0 | Black | 0 | Black | 0 | Black | 1 | Silver | ±10% |
| Brown | 1 | Brown | 1 | Brown | 1 | Brown | 10 | Gold | ± 5% |
| Red | 2 | Red | 2 | Red | 2 | Red | 100 | Brown | ± 1% |
| Orange | 3 | Orange | 3 | Orange | 3 | Orange | 1,000 | | |
| Yellow | 4 | Yellow | 4 | Yellow | 4 | Yellow | 10,000 | | |
| Green | 5 | Green | 5 | Green | 5 | Green | 100,000 | | |
| Blue | 6 | Blue | 6 | Blue | 6 | Blue | 1,000,000 | | |
| Violet | 7 | Violet | 7 | Violet | 7 | Silver | 0.01 | | |
| Gray | 8 | Gray | 8 | Gray | 8 | Gold | 0.1 | | |
| White | 9 | White | 9 | White | 9 | | | | |

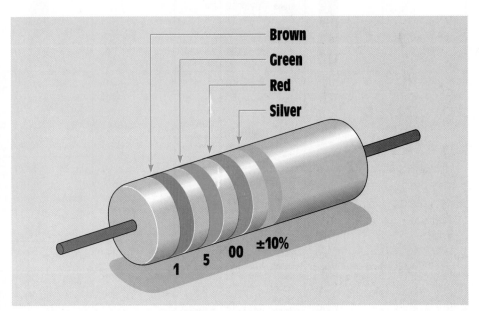

**Figure 5-10** Determining resistor values using the color code

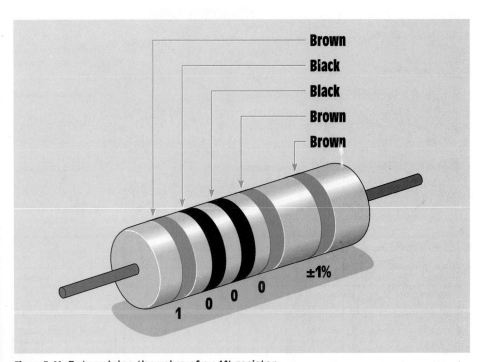

**Figure 5-11** Determining the value of a ±1% resistor

A five-band resistor has the following color bands:

**first band** = red
**second band** = orange
**third band** = violet
**fourth band** = red
**fifth band** = brown

**Example 2**

## Solution

The first three bands represent number values. Red is 2, orange is 3, and violet is 7. The fourth band is the multiplier; in this case, red represents 2. Add two zeros to the number 237. The value of the resistor is 23,700 $\Omega$. The fifth band is brown, which indicates a tolerance of ±1%.

Military resistors often have five bands of color also. These resistors are read in the same manner as a resistor with four bands of color. The fifth band can represent different things. A fifth band of orange or yellow is used to indicate reliability. Resistors with a fifth band of orange have a reliability good enough to be used in missile systems, and a resistor with a fifth band of yellow can be used in space flight equipment. A military resistor with a fifth band of white indicates the resistor has solderable leads.

Resistors with tolerance ratings ranging from 0.5% to 0.1% will generally have their values printed directly on the resistor.

### Gold and Silver as Multipliers

The colors gold and silver are generally found in the fourth band of a resistor, but they can be used in the multiplier band also. When the color gold is used as the multiplier band it means to divide the combined first two numbers by 10. If silver is used as the multiplier band it means to divide the combined first two numbers by 100. For example, assume a resistor has color bands of orange, white, gold, gold. The value of this resistor is 3.9 $\Omega$ with a tolerance of ±5% (orange = 3; white = 9; gold means to divide 39 by 10 = 3.9; and gold in the fourth band means ±5% tolerance).

### 5-4 Standard Resistance Values of Fixed Resistors

Fixed resistors are generally produced in standard values. The higher the tolerance value, the fewer resistance values available. Standard resistor values are listed in the chart shown in *Figure 5-12*. In the column under 10% only 12 values of resistors are listed. These standard values, however, can be multiplied by factors of 10. Notice that one of the standard values listed is 33 $\Omega$. There are also standard values in 10% resistors

### STANDARD RESISTANCE VALUES (Ω)

| 0.1% 0.25% 0.5% | 1% | 0.1% 0.25% 0.5% | 1% | 0.1% 0.25% 0.5% | 1% | 0.1% 0.25% 0.5% | 1% | 0.1% 0.25% 0.5% | 1% |
|---|---|---|---|---|---|---|---|---|---|
| 10.0 | 10.0 | 17.2 | – | 29.4 | 29.4 | 50.5 | – | 86.6 | 86.6 |
| 10.1 | – | 17.4 | 17.4 | 29.8 | – | 51.1 | 51.1 | 87.6 | – |
| 10.2 | 10.2 | 17.6 | – | 30.1 | 30.1 | 51.7 | – | 88.7 | 88.7 |
| 10.4 | – | 17.8 | 17.8 | 30.5 | – | 52.3 | 52.3 | 89.8 | – |
| 10.5 | 10.5 | 18.0 | – | 30.9 | 30.9 | 53.0 | – | 90.9 | 90.9 |
| 10.6 | – | 18.2 | 18.2 | 31.2 | – | 53.6 | 53.6 | 92.0 | – |
| 10.7 | 10.7 | 18.4 | – | 31.6 | 31.6 | 54.2 | – | 93.1 | 93.1 |
| 10.9 | – | 18.7 | 18.7 | 32.0 | – | 54.9 | 54.9 | 94.2 | – |
| 11.0 | 11.0 | 18.9 | – | 32.4 | 32.4 | 55.6 | – | 95.3 | 95.3 |
| 11.1 | – | 19.1 | 19.1 | 32.8 | – | 56.2 | 56.2 | 96.5 | – |
| 11.3 | 11.3 | 19.3 | – | 33.2 | 33.2 | 56.9 | – | 97.6 | 97.6 |
| 11.4 | – | 19.6 | 19.6 | 33.6 | – | 57.6 | 57.6 | 98.8 | – |
| 11.5 | 11.5 | 19.8 | – | 34.0 | 34.0 | 58.3 | – | | |
| 11.7 | – | 20.0 | 20.0 | 34.4 | – | 59.0 | 59.0 | | |
| 11.8 | 11.8 | 20.3 | – | 34.8 | 34.8 | 59.7 | – | | |
| 12.0 | – | 20.5 | 20.5 | 35.2 | – | 60.4 | 60.4 | | |
| 12.1 | 12.1 | 20.8 | – | 35.7 | 35.7 | 61.2 | – | | |
| 12.3 | – | 21.0 | 21.0 | 36.1 | – | 61.9 | 61.9 | | |
| 12.4 | 12.4 | 21.3 | – | 36.5 | 36.5 | 62.6 | – | | |
| 12.6 | – | 21.5 | 21.5 | 37.0 | – | 63.4 | 63.4 | | |
| 12.7 | 12.7 | 21.8 | – | 37.4 | 37.4 | 64.2 | – | **2%,5%** | **10%** |
| 12.9 | – | 22.1 | 22.1 | 37.9 | – | 64.9 | 64.9 | 10 | 10 |
| 13.0 | 13.0 | 22.3 | – | 38.3 | 38.3 | 65.7 | – | 11 | – |
| 13.2 | – | 22.6 | 22.6 | 38.8 | – | 66.5 | 66.5 | 12 | 12 |
| 13.3 | 13.3 | 22.9 | – | 39.2 | 39.2 | 67.3 | – | 13 | – |
| 13.5 | – | 23.2 | 23.2 | 39.7 | – | 68.1 | 68.1 | 15 | 15 |
| 13.7 | 13.7 | 23.4 | – | 40.2 | 40.2 | 69.0 | – | 16 | – |
| 13.8 | – | 23.7 | 23.7 | 40.7 | – | 69.8 | 69.8 | 18 | 18 |
| 14.0 | 14.0 | 24.0 | – | 41.2 | 41.2 | 70.6 | – | 20 | – |
| 14.2 | – | 24.3 | 24.3 | 41.7 | – | 71.5 | 71.5 | 22 | 22 |
| 14.3 | 14.3 | 24.6 | – | 42.2 | 42.2 | 72.3 | – | 24 | – |
| 14.5 | – | 24.9 | 24.9 | 42.7 | – | 73.2 | 73.2 | 27 | 27 |
| 14.7 | 14.7 | 25.2 | – | 43.2 | 43.2 | 74.1 | – | 30 | – |
| 14.9 | – | 25.5 | 25.5 | 43.7 | – | 75.0 | 75.0 | 33 | 33 |
| 15.0 | 15.0 | 25.8 | – | 44.2 | 44.2 | 75.9 | – | 36 | – |
| 15.2 | – | 26.1 | 26.1 | 44.8 | – | 76.8 | 76.8 | 39 | 39 |
| 15.4 | 15.4 | 26.4 | – | 45.3 | 45.3 | 77.7 | – | 43 | – |
| 15.6 | – | 26.7 | 26.7 | 45.9 | – | 78.7 | 78.7 | 47 | 47 |
| 15.8 | 15.8 | 27.1 | – | 46.4 | 46.4 | 79.6 | – | 51 | – |
| 16.0 | – | 27.4 | 27.4 | 47.0 | – | 80.6 | 80.6 | 56 | 56 |
| 16.2 | 16.2 | 27.7 | – | 47.5 | 47.5 | 81.6 | – | 62 | – |
| 16.4 | – | 28.0 | 28.0 | 48.1 | – | 82.5 | 82.5 | 68 | 68 |
| 16.5 | 16.5 | 28.4 | – | 48.7 | 48.7 | 83.5 | – | 75 | – |
| 16.7 | – | 28.7 | 28.7 | 49.3 | – | 84.5 | 84.5 | 82 | 82 |
| 16.9 | 16.9 | 29.1 | – | 49.9 | 49.9 | 85.6 | – | 91 | – |

**Figure 5-12** Standard resistance values

of 0.33, 3.3, 330, 3300, 33,000, 330,000, and 3,300,000 Ω. The 2% and 5% column shows 24 resistor values, and the 1% column list 96 values. All of the values listed in the chart can be multiplied by factors of 10 to obtain other resistance values.

## 5-5 Power Ratings

Resistors also have a power rating in watts that should not be exceeded or the resistor will be damaged. The amount of heat that must be dissipated by (given off to the surrounding air) the resistor can be determined by the use of one of the following formulas.

$$P = I^2R$$

$$P = \frac{E^2}{R}$$

$$P = EI$$

**Example 3**

The resistor shown in *Figure 5-13* has a value of 100 Ω and a power rating of 1/2 W. If the resistor is connected to a 10-V power supply will it be damaged?

**Solution**

Using the formula $P = \frac{E^2}{R}$ determine the amount of heat that will be dissipated by the resistor.

$$P = \frac{E^2}{R}$$

$$P = \frac{10 \times 10}{100}$$

$$P = \frac{100}{100}$$

$$P = 1\ W$$

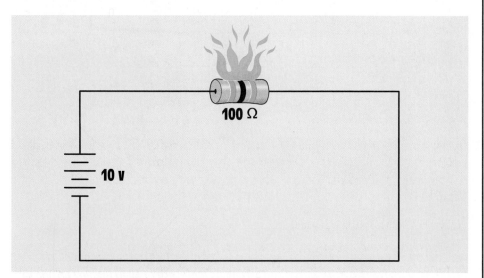

**Figure 5-13** Exceeding the power rating causes damage to the resistor.

Since the resistor has a power rating of 1/2 W, and the amount of heat that will be dissipated is 1 W, the resistor will be damaged.

### 5-6 Variable Resistors

A **variable resistor** is a resistor whose values can be changed or varied over a range. Variable resistors can be obtained in different case styles and power ratings. *Figure 5-14* illustrates how a variable resistor is constructed. In this example, a resistive wire is wound in a circular pattern, and a sliding tap makes contact with the wire. The value of resistance can be adjusted between one end of the resistive wire and the sliding tap. If the resistive wire has a total value of 100 Ω, the resistor can be set between the values of 0 and 100 Ω.

A variable resistor with three terminals is shown in *Figure 5-15*. This type of resistor has a wiper arm inside the case that makes contact with the resistive element. The full resistance value is between the two outside terminals, and the wiper arm is connected to the center terminal. The resistance between the center terminal and either of the two outside terminals can be adjusted by turning the shaft and changing the position of the wiper arm. Wire wound variable resistors of this type can be obtained also (*Figure 5-16*). The advantage of the wire wound type is a higher power rating.

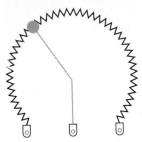

**Figure 5-14** Variable resistor

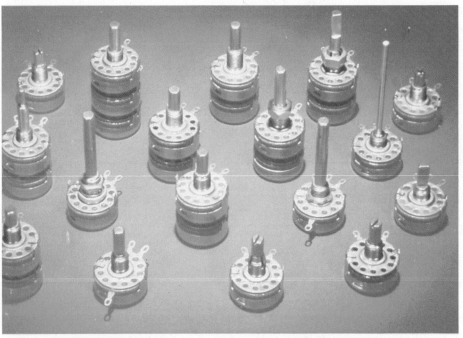

**Figure 5-15** Variable resistors with three terminals *(Courtesy of Allen Bradley Co. Inc., a Rockwell International company)*

**Figure 5-16** Wire wound variable resistor

The resistor shown in *Figure 5-15* can be adjusted from its minimum to maximum value by turning the control approximately three-quarters of a turn. In some types of electrical equipment this range of adjustment may be too coarse to allow for sensitive adjustments. When this becomes a problem, a multiturn resistor *(Figure 5-17)* can be used. **Multiturn variable resistors** operate by moving the wiper arm with a screw of some number of turns. They generally range from 3 turns to 10 turns. If a 10-turn variable resistor is used, it will require 10 turns of the control knob to move the wiper from one end of the resistor to the other end instead of three-quarters of a turn.

## Variable Resistor Terminology

Variable resistors are known by several common names. The most popular name is **pot**, which is shortened from the word *potentiometer*. Another common name is **rheostat**. A rheostat is actually a variable resistor that has two terminals.  They are used to adjust the current in a circuit

**multiturn variable resistors**

**pot**

**rheostat**

**Figure 5-17**  Multiturn variable resistor

**potentio-meter**

to a certain value. A potentiometer is a variable resistor that has three terminals.  Potentiometers can be used as rheostats, by only using two of their three terminals. A **potentiometer** describes how a variable resistor is used rather than some specific type of resistor. The word *potentiometer* comes from the word *potential,* or voltage. A potentiometer is a variable resistor used to provide a variable voltage as shown in *Figure 5-18*. In this example, one end of a variable resistor is connected to +12 V, and the other end is connected to ground. The middle terminal, or wiper, is connected to the positive terminal of a voltmeter and the negative lead is connected to ground. If the wiper is moved to the upper end of the resistor, the voltmeter will indicate a potential of 12 V. If the wiper is moved to the bottom, the voltmeter will indicate a value of 0 V. The wiper can be adjusted to provide any value of voltage between 12 and 0 V.

## 5-7 Schematic Symbols

Electrical schematics use symbols to represent the use of a resistor. Unfortunately, the symbol used to represent a resistor is not standard. *Figure 5-19* illustrates several schematic symbols used to represent both fixed and variable resistors.

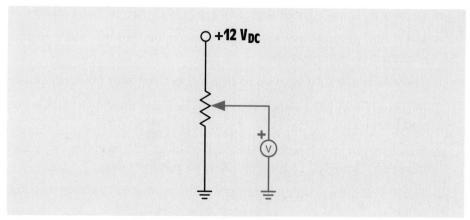

**Figure 5-18** Variable resistor used as a potentiometer

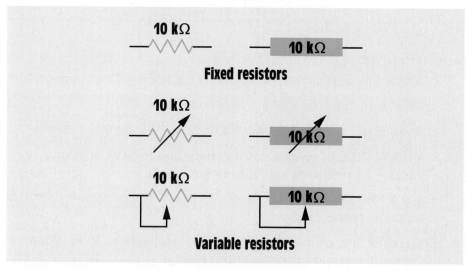

**Figure 5-19** Schematic symbols used to represent resistors

## Summary

1. Resistors are used in two main applications: as voltage dividers and to limit the flow of current in a circuit.

2. The value of fixed resistors cannot be changed.

3. There are several types of fixed resistors such as composition carbon, metal film, and wire wound.

4. Carbon resistors change their resistance with age or if overheated.

5. Metal film resistors never change their value, but are more expensive than carbon resistors.

6. The advantage of wire wound resistors is their high power ratings.

7. Resistors often have bands of color to indicate their resistance value and tolerance.

8. Resistors are produced in standard values. The number of values between 0 and 100 $\Omega$ is determined by the tolerance.

9. Variable resistors can change their value within the limit of their full value.

10. A potentiometer is a variable resistor used as a voltage divider.

## Review Questions

1. Name three types of fixed resistors.

2. What is the advantage of a metal film resistor over a carbon resistor?

3. What is the advantage of a wire wound resistor?

4. How should tubular wire wound resistors be mounted and why?

5. A 0.5-W, 2000-$\Omega$ resistor has a current flow of 0.01 A through it. Is this resistor operating within its power rating?

6. A 1-W, 350-$\Omega$ resistor is connected to 24 V. Is this resistor operating within its power rating?

7. A resistor has color bands of orange, blue, yellow, gold. What are the resistance and tolerance of this resistor?

8. A 10,000-$\Omega$ resistor has a tolerance of 5%. What are the minimum and maximum ratings of this resistor?

9. Is 51,000 $\Omega$ a standard value for a 5% resistor?

10. What is a potentiometer?

## Resistors

Fill in the missing values.

**Practice Problems**

| 1ST Band | 2ND Band | 3RD Band | 4TH Band | Value | %Tol |
|----------|----------|----------|----------|-------|------|
| Red | Yellow | Brown | Silver | | |
| | | | | 6800 Ω | 5 |
| Orange | Orange | Orange | Gold | | |
| | | | | 12 Ω | 2 |
| Brown | Green | Silver | Silver | | |
| | | | | 1.8 MΩ | 10 |
| Brown | Black | Yellow | None | | |
| | | | | 10 kΩ | 5 |
| Violet | Green | Black | Red | | |
| | | | | 4.7 kΩ | 20 |
| Gray | Red | Green | Red | | |
| | | | | 5.6 Ω | 2 |

*section* **2**

*basic electrical circuits*

**Series circuits contain only one path for current flow. This unit discusses the rules governing series circuits.**

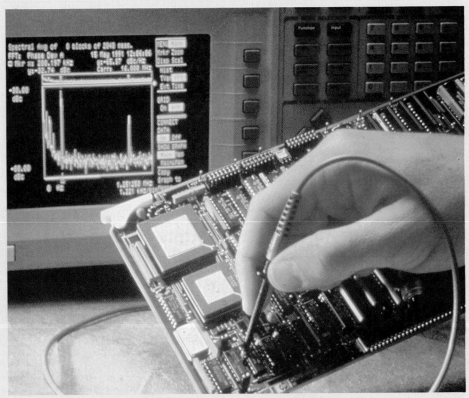

*Courtesy of Hewlett Packard Company.*

# *Series Circuits*

## Key Terms

Chassis ground

Circuit breakers

Earth ground

Fuses

Ground point

General voltage divider formula

Ohm's law

Resistance adds

Series circuit

Series resistance

Voltage divider

Voltage drop

Voltage polarity

## Objectives

**A***fter studying this unit, you should be able to:*

- Discuss the properties of series circuits.
- List three rules for solving electrical values of series circuits.
- Compute values of voltage, current, resistance, and power for series circuits.
- Compute the values of voltage drop in a series circuit using the voltage divider formula.

## Preview

Electrical circuits can be divided into three major types: series, parallel, and combination. Combination circuits are circuits that contain both series and parallel paths. The first type discussed is the series circuit.

### series circuit

### 6-1 Series Circuits

**A series circuit is a circuit that has only one path for current flow** *(Figure 6-1)*. Because there is only one path for current flow, the current is the same at any point in the circuit. Imagine that an electron leaves the negative terminal of the battery. This electron must flow

> A series circuit is a circuit that has only one path for current flow.

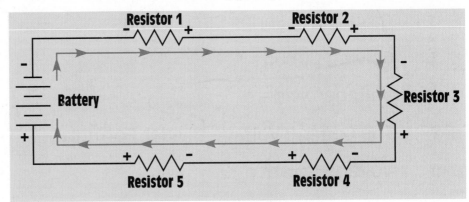

**Figure 6-1** In a series circuit there is only one path for current flow.

through each resistor before it can complete the circuit to the positive battery terminal.

One of the best examples of a series-connected device is a fuse or circuit breaker *(Figure 6-2)*. Since **fuses** and **circuit breakers** are connected in series with the rest of the circuit, all the circuit current must flow through them. If the current becomes excessive, the fuse or circuit breaker will open and disconnect the rest of the circuit from the power source.

**fuses**

**circuit breakers**

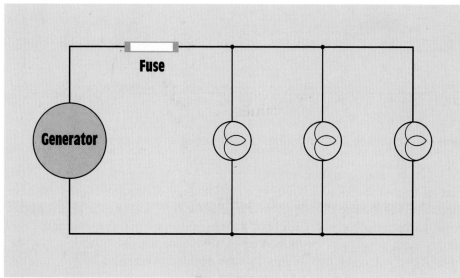

**Figure 6-2** All the current must flow through the fuse.

## 6-2 Voltage Drops in a Series Circuit

Voltage is the force that pushes the electrons through a resistance. The amount of voltage required is determined by the amount of current flow and resistance. If a voltmeter is connected across a resistor *(Figure 6-3)*, the amount of voltage necessary to push the current through that resistor will be indicated by the meter. This amount is known as **voltage drop.** It is similar to pressure drop in a water system. **In a series circuit, the sum of all the voltage drops across all the resistors must equal the voltage applied to the circuit.** The amount of voltage drop across each resistor will be proportional to its resistance and the circuit current.

In the circuit shown in *Figure 6-4*, four resistors are connected in series. It is assumed that all four resistors have the same value. The circuit is connected to a 24-V battery. Since all the resistors have the same value,

**voltage drop**

**In a series circuit, the sum of all the voltage drops across all the resistors must equal the voltage applied to the circuit.**

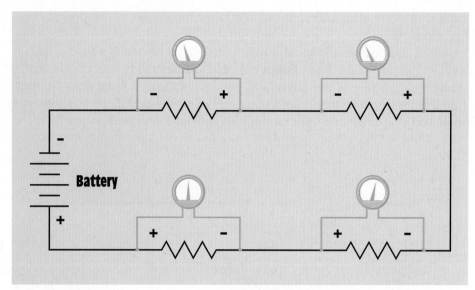

**Figure 6-3** The voltage drops in a series circuit must equal the applied voltage.

the voltage drop across each will be 6 V (24 V/4 resistors = 6 V). Note that all four resistors will have the same voltage drop only if they all have the same value. The circuit shown in *Figure 6-5* illustrates a series circuit comprising resistors of different values. Notice that the voltage drop across each resistor is proportional to its resistance. Also notice that the sum of the voltage drops is equal to the applied voltage of 24 V.

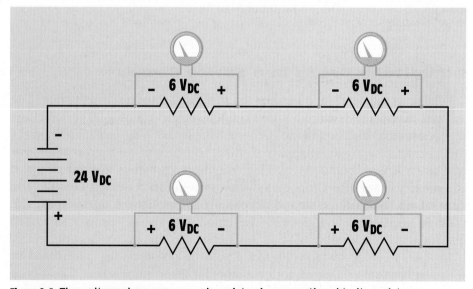

**Figure 6-4** The voltage drop across each resistor is proportional to its resistance.

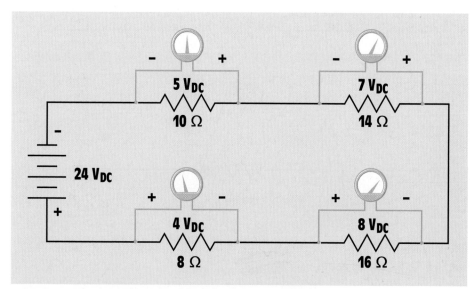

**Figure 6-5** Series circuit with four resistors having different voltage drops

## 6-3 Resistance in a Series Circuit

Because there is only one path for the current to flow through a series circuit, it must flow through each resistor in the circuit *(Figure 6-1)*. Each resistor limits or impedes the flow of current in the circuit. Therefore, the total amount of resistance to current flow in a **series circuit** is equal to the sum of the resistances in that circuit.

**series resistance**

## 6-4 Calculating Series Circuit Values

Three rules can be used with Ohm's law for finding values of voltage, current, resistance, and power in any series circuit.

1. **The current is the same at any point in the circuit.**
2. **The total resistance is the sum of the individual resistors.**
3. **The applied voltage is equal to the sum of the voltage drops across all the resistors.**

The circuit shown in *Figure 6-6* shows the values of current flow, voltage drop, and resistance for each of the resistors. The total resistance ($R_T$) of the circuit can be found by adding the values of the three resistors (**resistance adds**).

**resistance adds**

$$R_T = R_1 + R_2 + R_3$$

$$R_T = 20\ \Omega + 10\ \Omega + 30\ \Omega$$

$$R_T = 60\ \Omega$$

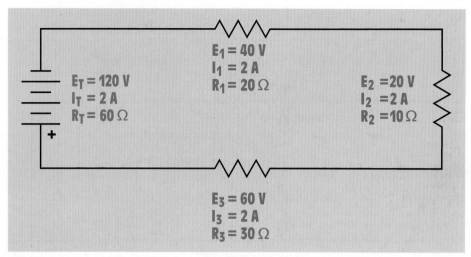

**Figure 6-6** Series circuit values

The amount of current flow in the circuit can be found by using **Ohm's law.**

$$I = \frac{E}{R}$$

$$I = \frac{120}{60}$$

$$I = 2 \text{ A}$$

A current of 2 A flows through each resistor in the circuit.

$$I_T = I_1 = I_2 = I_3$$

Since the amount of current flowing through resistor $R_1$ is known, the voltage drop across the resistor can be found using Ohm's law.

$$E_1 = I_1 \times R_1$$

$$E_1 = 2 \text{ A} \times 20 \text{ } \Omega$$

$$E_1 = 40 \text{ V}$$

In other words, it takes 40 V to push 2 A of current through 20 $\Omega$ of resistance. If a voltmeter were connected across resistor $R_1$, it would indicate a value of 40 V *(Figure 6-7)*. The voltage drop across resistors $R_2$ and $R_3$ can be found in the same way.

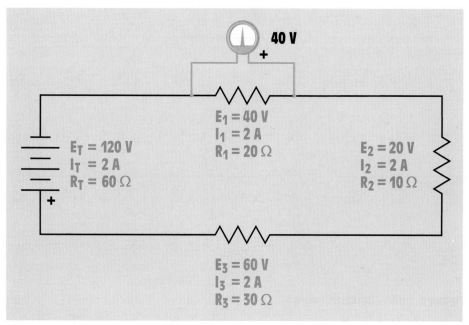

$E_1 = 40\ V$
$I_1 = 2\ A$
$R_1 = 20\ \Omega$

40 V
+

$E_T = 120\ V$
$I_T = 2\ A$
$R_T = 60\ \Omega$

$E_2 = 20\ V$
$I_2 = 2\ A$
$R_2 = 10\ \Omega$

$E_3 = 60\ V$
$I_3 = 2\ A$
$R_3 = 30\ \Omega$

**Figure 6-7**  The voltmeter indicates a voltage drop of 40 V.

$$E_2 = I_2 \times R_2$$
$$E_2 = 2\ A \times 10\ \Omega$$
$$E_2 = 20\ V$$
$$E_3 = I_3 \times R_3$$
$$E_3 = 2\ A \times 30\ \Omega$$
$$E_3 = 60\ V$$

If the voltage drop across all the resistors is added, it equals the total applied voltage ($E_T$).

$$E_T = E_1 + E_2 + E_3$$
$$E_T = 40\ V + 20\ V + 60\ V$$
$$E_T = 120\ V$$

## 6-5 Solving Circuits

In the following problems, circuits that have missing values are shown. The missing values can be found by using the rules for series circuits and Ohm's law.

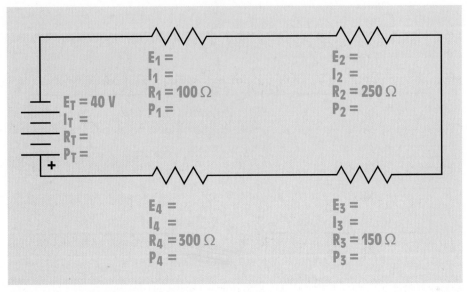

**Figure 6-8** Series circuit, Example 1

Example 1

The first step in finding the missing values in the circuit shown in *Figure 6-8* is to find the total resistance ($R_T$). This can be done using the second rule of series circuits, which states that resistances add to equal the total resistance of the circuit.

$$R_T = R_1 + R_2 + R_3 + R_4$$

$$R_T = 100\ \Omega + 250\ \Omega + 150\ \Omega + 300\ \Omega$$

$$R_T = 800\ \Omega$$

Now that the total voltage and total resistance are known, the current flow through the circuit can be found using Ohm's law.

$$I = \frac{E}{R}$$

$$I = \frac{40}{800}$$

$$I = 0.050\ A$$

The first rule of series circuits states that current remains the same at any point in the circuit. Therefore, 0.050 A flows through each resistor in

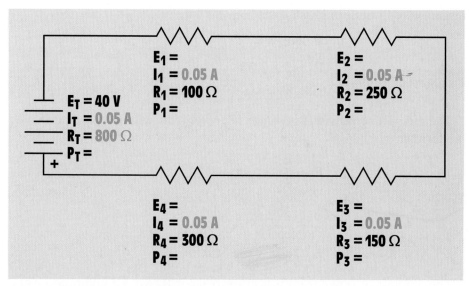

$E_1 =$
$I_1 = 0.05\ A$
$R_1 = 100\ \Omega$
$P_1 =$

$E_2 =$
$I_2 = 0.05\ A$
$R_2 = 250\ \Omega$
$P_2 =$

$E_T = 40\ V$
$I_T = 0.05\ A$
$R_T = 800\ \Omega$
$P_T =$

$E_4 =$
$I_4 = 0.05\ A$
$R_4 = 300\ \Omega$
$P_4 =$

$E_3 =$
$I_3 = 0.05\ A$
$R_3 = 150\ \Omega$
$P_3 =$

**Figure 6-9** The current is the same at any point in a series circuit.

the circuit *(Figure 6-9)*. The voltage drop across each resistor can now be found using Ohm's law *(Figure 6-10)*.

$$E_1 = I_1 \times R_1$$
$$E_1 = 0.050 \times 100$$
$$E_1 = 5\ V$$

$$E_2 = I_2 \times R_2$$
$$E_2 = 0.050 \times 250$$
$$E_2 = 12.5\ V$$

$$E_3 = I_3 \times R_3$$
$$E_3 = 0.050 \times 150$$
$$E_3 = 7.5\ V$$

$$E_4 = I_4 \times R_4$$
$$E_4 = 0.050 \times 300$$
$$E_4 = 15\ V$$

Several formulas can be used to determine the amount of power dissipated (converted into heat) by each resistor. The power dissipation of resistor $R_1$ will be found using the formula

$$P_1 = E_1 \times I_1$$

$$P_1 = 5 \times 0.05$$

$$P_1 = 0.25 \text{ W}$$

The amount of power dissipation for resistor $R_2$ will be computed using the formula

$$P_2 = \frac{E_2^2}{R_2}$$

$$P_2 = \frac{156.25}{250}$$

$$P_2 = 0.375 \text{ W}$$

The amount of power dissipation for resistor $R_3$ will be computed using the formula

$$P_3 = I_3^2 \times R_3$$

$$P_3 = 0.0025 \times 150$$

$$P_3 = 0.375 \text{ W}$$

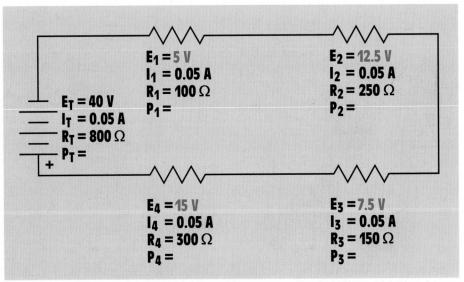

$E_1 = 5 \text{ V}$
$I_1 = 0.05 \text{ A}$
$R_1 = 100 \ \Omega$
$P_1 =$

$E_2 = 12.5 \text{ V}$
$I_2 = 0.05 \text{ A}$
$R_2 = 250 \ \Omega$
$P_2 =$

$E_T = 40 \text{ V}$
$I_T = 0.05 \text{ A}$
$R_T = 800 \ \Omega$
$P_T =$

$E_4 = 15 \text{ V}$
$I_4 = 0.05 \text{ A}$
$R_4 = 300 \ \Omega$
$P_4 =$

$E_3 = 7.5 \text{ V}$
$I_3 = 0.05 \text{ A}$
$R_3 = 150 \ \Omega$
$P_3 =$

**Figure 6-10** The voltage drop across each resistor can be found using Ohm's law.

The amount of power dissipation for resistor $R_4$ will be found using the formula

$$P_4 = E_4 \times I_4$$

$$P_4 = 15 \times 0.05$$

$$P_4 = 0.75 \text{ W}$$

A good rule to remember when calculating values of electrical circuits is that **the total power used in a circuit is equal to the sum of the power used by all parts**. That is, the total power can be found in any kind of a circuit—series, parallel, or combination—by adding the power dissipation of all the parts. The total power for this circuit can be found using the formula

$$P_T = P_1 + P_2 + P_3 + P_4$$

$$P_T = 0.25 + 0.625 + 0.375 + 0.75$$

$$P_T = 2 \text{ W}$$

Now that all the missing values have been found *(Figure 6-11)*, the circuit can be checked by using the third rule of series circuits, which states that voltage drops add to equal the applied voltage.

$$E_T = E_1 + E_2 + E_3 + E_4$$

$$E_T = 5 + 12.5 + 7.5 + 15$$

$$E_T = 40 \text{ V}$$

> The total power used in a circuit is equal to the sum of the power used by all parts.

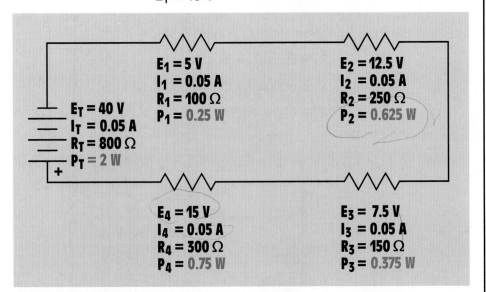

**Figure 6-11** The final values for the circuit in Example 1

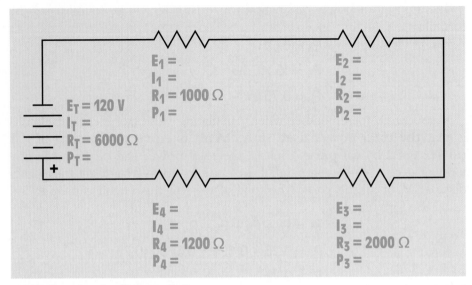

**Figure 6-12** Series circuit, Example 2

The second circuit to be solved is shown in *Figure 6-12*. In this circuit the total resistance is known, but the value of resistor $R_2$ is not. The second rule of series circuits states that resistances add to equal the total resistance of the circuit. Since the total resistance is known, the missing resistance of $R_2$ can be found by adding the values of the other resistors and subtracting their sum from the total resistance of the circuit *(Figure 6-13)*.

$$R_2 = R_T - (R_1 + R_3 + R_4)$$

$$R_2 = 6000 - (1000 + 2000 + 1200)$$

$$R_2 = 6000 - 4200$$

$$R_2 = 1800 \ \Omega$$

The amount of current flow in the circuit can be found using Ohm's law.

$$I = \frac{E}{R}$$

$$I = \frac{120}{6000}$$

$$I = 0.020 \ A$$

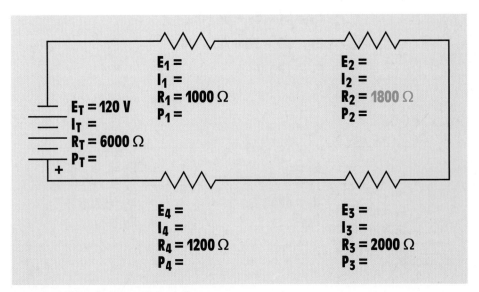

**Figure 6-13** The missing resistor value

Since the amount of current flow is the same through all elements of a series circuit *(Figure 6-14)*, the voltage drop across each resistor can be found using Ohm's law *(Figure 6-15)*.

$$E_1 = I_1 \times R_1$$
$$E_1 = 0.020 \times 1000$$
$$E_1 = 20 \text{ V}$$

$$E_2 = I_2 \times R_2$$
$$E_2 = 0.020 \times 1800$$
$$E_2 = 36 \text{ V}$$

$$E_3 = I_3 \times R_3$$
$$E_3 = 0.020 \times 2000$$
$$E_3 = 40 \text{ V}$$

$$E_4 = I_4 \times R_4$$
$$E_4 = 0.020 \times 1200$$
$$E_4 = 24 \text{ V}$$

The third rule of series circuits can be used to check the answers.

$$E_T = E_1 + E_2 + E_3 + E_4$$

$$E_T = 20 + 36 + 40 + 24$$

$$E_T = 120 \text{ V}$$

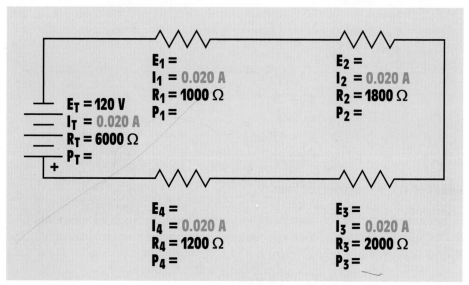

**Figure 6-14** The current is the same through each circuit element.

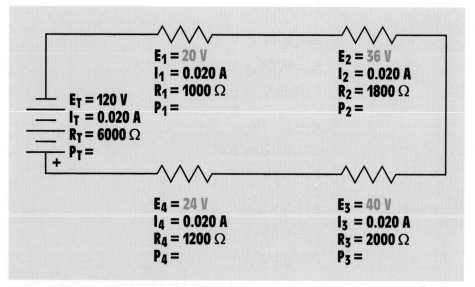

**Figure 6-15** The voltage drops across each resistor

The amount of power dissipation for each resistor in the circuit can be computed using the same method used to solve the circuit in Example 1. The power dissipated by resistor $R_1$ will be computed using the formula

$$P_1 = E_1 \times I_1$$

$$P_1 = 20 \times 0.02$$

$$P_1 = 0.4 \text{ W}$$

The amount of power dissipation for resistor $R_2$ will be found by using the formula

$$P_2 = \frac{E_2{}^2}{R_2}$$

$$P_2 = \frac{1296}{1800}$$

$$P_2 = 0.72 \text{ W}$$

The power dissipation of resistor $R_3$ will be found using the formula

$$P_3 = I_3{}^2 \times R_3$$

$$P_3 = 0.0004 \times 2000$$

$$P_3 = 0.8 \text{ W}$$

The power dissipation of resistor $R_4$ will be computed using the formula

$$P_4 = E_4 \times I_4$$

$$P_4 = 24 \times 0.02$$

$$P_4 = 0.48 \text{ W}$$

The total power will be computed using the formula

$$P_T = E_T \times I_T$$

$$P_T = 120 \times 0.02$$

$$P_T = 2.4 \text{ W}$$

The circuit with all computed values is shown in *Figure 6-16*.

In the circuit shown in *Figure 6-17*, resistor $R_1$ has a voltage drop of 6.4 V, resistor $R_2$ has a power dissipation of 0.102 W, resistor $R_3$ has a power

**Example 3**

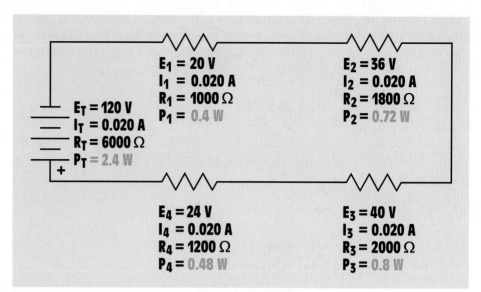

**Figure 6-16** The remaining unknown values for the circuit in Example 2

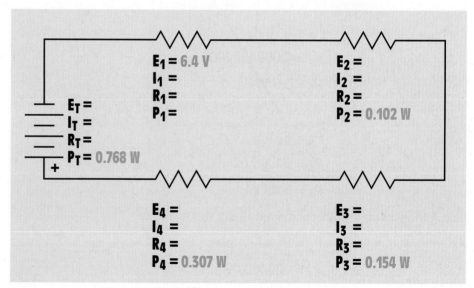

**Figure 6-17** Series circuit, Example 3

dissipation of 0.154 W, resistor $R_4$ has a power dissipation of 0.307 W, and the total power consumed by the circuit is 0.768 W.

The only value that can be found with the given quantities is the amount of power dissipated by resistor $R_1$. Since the total power is known and the power dissipated by the three other resistors is known, the power dissipated by resistor $R_1$ can be found by subtracting the power dissipated by resistors $R_2$, $R_3$, and $R_4$ from the total power used in the circuit.

$$P_1 = P_T - (P_2 + P_3 + P_4)$$

or

$$P_1 = P_T - P_2 - P_3 - P_4$$

$$P_1 = 0.768 - 0.102 - 0.154 - 0.307$$

$$P_1 = 0.205 \text{ W}$$

Now that the amount of power dissipated by resistor $R_1$ and the voltage drop across $R_1$ are known, the current flow through resistor $R_1$ can be found using the formula

$$I = \frac{P}{E}$$

$$I = \frac{0.205}{6.4}$$

$$I = 0.032 \text{ A}$$

Since the current in a series circuit must be the same at any point in the circuit, it must be the same through all circuit components *(Figure 6-18)*.

Now that the power dissipation of each resistor and the amount of cur-

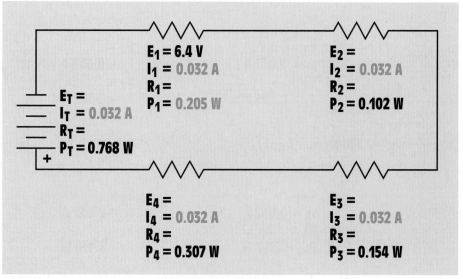

**Figure 6-18** The current flow in the circuit in Example 3

rent flowing through each resistor are known, the voltage drop of each resistor can be computed *(Figure 6-19)*.

$$E_2 = \frac{P_2}{I_2}$$

$$E_2 = \frac{0.102}{0.032}$$

$$E_2 = 3.2 \text{ V}$$

$$E_3 = \frac{P_3}{I_3}$$

$$E_3 = \frac{0.154}{0.032}$$

$$E_3 = 4.8 \text{ V}$$

$$E_4 = \frac{P_4}{I_4}$$

$$E_4 = \frac{0.307}{0.032}$$

$$E_4 = 9.6 \text{ V}$$

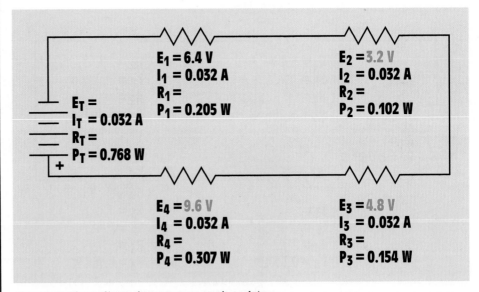

**Figure 6-19** The voltage drops across each resistor

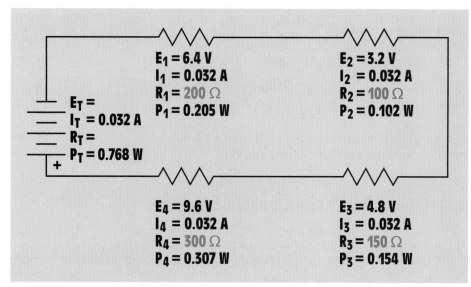

**Figure 6-20** The Ohmic value of each resistor

Ohm's law can now be used to find the ohmic value of each resistor in the circuit *(Figure 6-20)*.

$$R_1 = \frac{E_1}{I_1}$$

$$R_1 = \frac{6.4}{0.032}$$

$$R_1 = 200 \ \Omega$$

$$R_2 = \frac{E_2}{I_2}$$

$$R_2 = \frac{3.2}{0.032}$$

$$R_2 = 100 \ \Omega$$

$$R_3 = \frac{E_3}{I_3}$$

$$R_3 = \frac{4.8}{0.032}$$

$$R_3 = 150 \ \Omega$$

$$R_4 = \frac{E_4}{I_4}$$

$$R_4 = \frac{9.6}{0.032}$$

$$R_4 = 300 \ \Omega$$

The voltage applied to the circuit can be found by adding the voltage drops across the resistor *(Figure 6-21)*.

$$E_T = E_1 + E_2 + E_3 + E_4$$

$$E_T = 6.4 + 3.2 + 4.8 + 9.6$$

$$E_T = 24 \ V$$

The total resistance of the circuit can be found in a similar manner *(Figure 6-21)*. The total resistance is equal to the sum of all the resistive elements in the circuit.

$$R_T = R_1 + R_2 + R_3 + R_4$$

$$R_T = 200 + 100 + 150 + 300$$

$$R_T = 750 \ \Omega$$

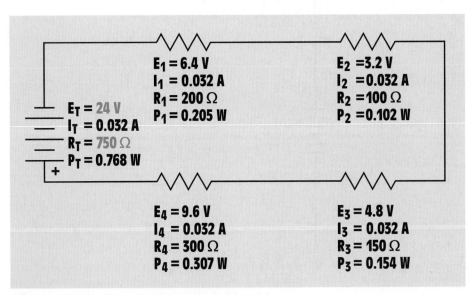

**Figure 6-21** The applied voltage and the total resistance

## 6-6 Voltage Dividers

One common use for series circuits is the construction of voltage dividers. A **voltage divider** works on the principle that the sum of the voltage drops across a series circuit must equal the applied voltage. Voltage dividers are used to provide different voltages between certain points *(Figure 6-22)*. If a voltmeter is connected between points A and B, a voltage of 20 V will be seen. If the voltmeter is connected between points B and D, a voltage of 80 V will be seen.

Voltage dividers can be constructed to produce any voltage desired. For example, assume that a voltage divider is connected to a source of 120 V and is to provide voltage drops of 36 V, 18 V, and 66 V. Notice that the sum of the voltage drops equals the applied voltage. The next step is to decide how much current is to flow through the circuit. Since there is only one path for current flow, the current will be the same through all the resistors. In this circuit, a current flow of 15 mA (0.015 A) will be used. The resistance value of each resistor can now be determined.

**voltage divider**

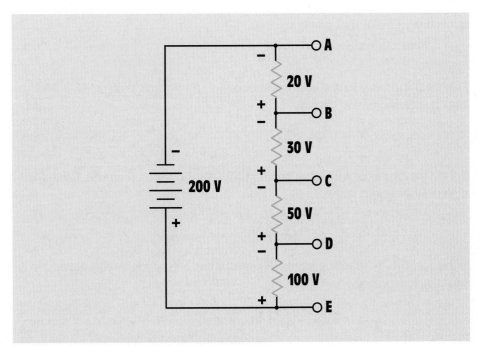

**Figure 6-22** Series circuit used as a voltage divider

$$R = \frac{E}{I}$$

$$R_1 = \frac{36}{0.015}$$

$$R_1 = 2.4 \text{ k}\Omega \text{ (2400 }\Omega)$$

$$R_2 = \frac{18}{0.015}$$

$$R_2 = 1.2 \text{ k}\Omega \text{ (1200 }\Omega)$$

$$R_3 = \frac{66}{0.015}$$

$$R_3 = 4.4 \text{ k}\Omega \text{ (4400 }\Omega)$$

## 6-7 The General Voltage Divider Formula

**general voltage divider formula**

Another method of determining the voltage drop across series elements is to use the **general voltage divider formula**. Since the current flow through a series circuit is the same at all points in the circuit, the voltage drop across any particular resistance is equal to the total circuit current times the value of that resistor.

$$E_X = I_T \times R_X$$

The total circuit current is proportional to the source voltage ($E_T$) and the total resistance of the circuit.

$$I_T = \frac{E_T}{R_T}$$

If the value of $I_T$ is substitute for $E_T/R_T$ in the previous formula, the expression now become:

$$E_X = \left(\frac{E_T}{R_T}\right) R_X$$

If the formula is rearranged, it becomes what is known as the general voltage divider formula.

$$E_X = \left(\frac{R_X}{R_T}\right) E_T$$

The voltage drop across any series component ($E_X$) can be computed by substituting the value of $R_X$ for the resistance value of that component when the source voltage and total resistance are know.

Three resistors are connected in series to a 24 volt source. Resistor $R_1$ has a resistance of 200 Ω, resistor $R_2$ has a value of 300 Ω, and resistor $R_3$ has a value of 160 Ω. What is the voltage drop across each resistor?

**Example 4**

## Solution

Find the total resistance of the circuit.

$$R_T = R_1 + R_2 + R_3$$

$$R_T = 200 + 300 + 160$$

$$R_T = 660 \ \Omega$$

Now use the voltage divider formula to compute the voltage drop across each resistor.

$$E_1 = \left(\frac{R_1}{R_T}\right) E_T$$

$$E_1 = \left(\frac{200}{660}\right) 24$$

$$E_1 = 7.273 \text{ volts}$$

$$E_2 = \left(\frac{R_2}{R_T}\right) E_T$$

$$E_2 = \left(\frac{300}{660}\right) 24$$

$$E_2 = 10.91 \text{ volts}$$

$$E_3 = \left(\frac{R_3}{R_T}\right) E_T$$

$$E_3 = \left(\frac{160}{660}\right) 24$$

$$E_3 = 5.818 \text{ volts}$$

### 6-8 Voltage Polarity

It is often necessary to know the polarity of the voltage developed across a resistor. **Voltage polarity** can be determined by observing the direction of current flow through the circuit. In the circuit shown in *Figure 6-22* it will be assumed that the current flows from the negative terminal of the battery to the positive terminal. Point A is connected to the negative battery terminal, and point E is connected to the positive terminal. If a voltmeter is connected across terminals A and B, terminal B will be positive with respect to A. If a voltmeter is connected across terminals B and C, however, terminal B will be negative with respect to terminal C. Notice that terminal B is closer to the negative terminal of the battery than terminal C is. Consequently, electrons flow through the resistor in a direction that makes terminal B more negative than C. Terminal C would be negative with respect to terminal D for the same reason.

### 6-9 Using Ground as a Reference

Two symbols are used to represent ground *(Figure 6-23)*. The symbol shown in *Figure 6-23A* is an **earth ground** symbol. It symbolizes a **ground point** that is made by physically driving an object such as a rod or a pipe into the ground. The symbol shown in *Figure 6-23B* symbolizes a **chassis ground.** This is a point that is used as a common connection for other parts of a circuit, but it is not actually driven into the ground. Although the symbol shown in *Figure 6-23B* is the accepted symbol for a chassis ground, the symbol shown in *Figure 6-23A* is often used to represent a chassis ground also.

An excellent example of using a chassis ground as a common connection can be found in the electrical system of an automobile. The negative terminal of the battery is grounded to the frame or chassis of the vehicle. The

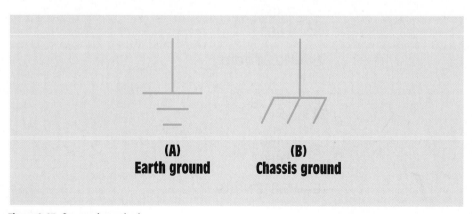

**(A)**
**Earth ground**

**(B)**
**Chassis ground**

**Figure 6-23** Ground symbols

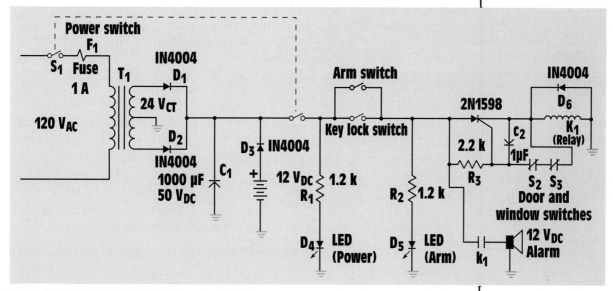

**Figure 6-24** Burglar alarm with battery back-up

frame of the automobile is not connected directly to earth ground; it is insulated from the ground by rubber tires. In the case of an automobile electrical system, the chassis of the vehicle is the negative side of the circuit. An electrical circuit using ground as a common connection point is shown in *Figure 6-24*. This circuit is an electronic burglar alarm. Notice the numerous ground points in the schematic. In practice, when the circuit is connected, all the ground points will be connected together.

In voltage divider circuits, ground is often used to provide a common reference point to produce voltages that are above and below ground *(Figure 6-25)*. An above ground voltage is a voltage that is positive with

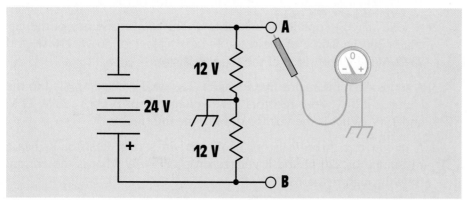

**Figure 6-25** A common ground used to produce above and below ground voltage

respect to ground. A below ground voltage is negative with respect to ground. In *Figure 6-25*, one terminal of a zero-center voltmeter is connected to ground. If the probe is connected to point A, the pointer of the voltmeter will give a negative indication for voltage. If the probe is connected to point B, the pointer will indicate a positive voltage.

## Summary

1. Series circuits have only one path for current flow.

2. The individual voltage drops in a series circuit can be added to equal the applied voltage.

3. The current is the same at any point in a series circuit.

4. The individual resistors can be added to equal the total resistance of the circuit.

5. Fuses and circuit breakers are connected in series with the devices they are intended to protect.

6. The total power in any circuit is equal to the sum of the power dissipated by all parts of the circuit.

7. When the source voltage and total resistance are known, the voltage drop across each element can be computed using the general voltage divider formula.

## Review Questions

1. A series circuit has individual resistor values of 200 $\Omega$, 86 $\Omega$, 91 $\Omega$, 180 $\Omega$, and 150 $\Omega$. What is the total resistance of the circuit?

2. A series circuit contains four resistors. The total resistance of the circuit is 360 $\Omega$. Three of the resistors have values of 56 $\Omega$, 110 $\Omega$, and 75 $\Omega$. What is the value of the fourth resistor?

3. A series circuit contains five resistors. The total voltage applied to the circuit is 120 V. Four resistors have voltage drops of 35 V, 28 V, 22 V, and 15 V. What is the voltage drop of the fifth resistor?

4. A circuit has three resistors connected in series. Resistor $R_2$ has a resistance of 220 $\Omega$ and a voltage drop of 44 V. What is the current flow through resistor $R_3$?

5. A circuit has four resistors connected in series. If each resistor has a voltage drop of 60 V, what is the voltage applied to the circuit?

6. Define a series circuit.

7. State the three rules for series circuits.

8. A series circuit has resistance values of 160 Ω, 100 Ω, 82 Ω, and 120 Ω. What is the total resistance of this circuit?

9. If a voltage of 24 V is applied to the circuit in question 8, what will be the total amount of current flow in the circuit?

10. What will be the voltage drop across each of the resistors?

    160 Ω, _____ V

    100 Ω, _____ V

    82 Ω, _____ V

    120 Ω, _____ V

11. A series circuit contains the following values of resistors:
$R_1$ = 510 Ω; $R_2$ = 680 Ω; $R_3$ = 390 Ω; and $R_4$ = 750 Ω. Assume a source voltage of 48 V. Use the general voltage divider formula to compute the voltage drop across each of the resistors.

    $E_1$ = _____ V    $E_2$ = _____ V    $E_3$ = _____ V    $E_4$ = _____ V

## Series Circuits

Using the three rules for series circuits and Ohm's law, solve for the missing values.

| $E_T$ 120 | $E_1$ | $E_2$ | $E_3$ | $E_4$ | $E_5$ |
|---|---|---|---|---|---|
| $I_T$ | $I_1$ | $I_2$ | $I_3$ | $I_4$ | $I_5$ |
| $R_T$ | $R_1$ 430 $\Omega$ | $R_2$ 360 $\Omega$ | $R_3$ 750 $\Omega$ | $R_4$ 1000 $\Omega$ | $R_5$ 620 $\Omega$ |
| $P_T$ | $P_1$ | $P_2$ | $P_3$ | $P_4$ | $P_5$ |

| $E_T$ | $E_1$ | $E_2$ | $E_3$ 11 | $E_4$ | $E_5$ |
|---|---|---|---|---|---|
| $I_T$ | $I_1$ | $I_2$ | $I_3$ | $I_4$ | $I_5$ |
| $R_T$ | $R_1$ | $R_2$ | $R_3$ | $R_4$ | $R_5$ |
| $P_T$ 0.25 | $P_1$ 0.03 | $P_2$ 0.0825 | $P_3$ | $P_4$ 0.045 | $P_5$ 0.0375 |

| $E_T$ 340 | $E_1$ 44 | $E_2$ 94 | $E_3$ 60 | $E_4$ 40 | $E_5$ |
|---|---|---|---|---|---|
| $I_T$ | $I_1$ | $I_2$ | $I_3$ | $I_4$ | $I_5$ |
| $R_T$ | $R_1$ | $R_2$ | $R_3$ | $R_4$ | $R_5$ |
| $P_T$ | $P_1$ | $P_2$ | $P_3$ | $P_4$ | $P_5$ 0.204 |

2. Use the general voltage divider formula to computed the values of voltage drop for the following series connected resistors. Assume a source voltage of 120 V.

$R_1 = 1K\ \Omega$;    $R_2 = 2.2K\ \Omega$;    $R_3 = 1.8K\ \Omega$;    $R_4 = 1.5K\ \Omega$

$E_1 =$ _____ V    $E_2 =$ _____ V    $E_3 =$ _____ V    $E_4 =$ _____ V

# Parallel Circuits

Courtesy of New York State Electric and Gas; photo by Kirk van Zandbergen.

## Key Terms

Circuit branch

Current adds

Current dividers

Load

Parallel circuits

Product over sum
method

Reciprocal formula

**A**fter studying this unit, you should be able to:

- Discuss the characteristics of parallel circuits.
- State three rules for solving electrical values of parallel circuits.
- Solve the missing values in a parallel circuit using the three rules and Ohm's law.
- Discuss the operation of a current divider circuit.
- Calculate current values using the current divider formula.

Parallel circuits are probably the type of circuit with which most people are familiar. Most devices such as lights and receptacles in homes and office buildings are connected in parallel. Imagine if the lights in your home were wired in series. All the lights in the home would have to be turned on in order for any light to operate, and if one were to burn out, all the lights would go out. The same is true for receptacles. If receptacles were connected in series, some device would have to be connected into each receptacle before power could be supplied to any other device.

## 7-1 Parallel Circuit Values

### Total Current

**Parallel circuits are circuits that have more than one path for current flow** (*Figure 7-1*). If it is assumed that current leaves terminal A and returns to terminal B, it can be seen that the electrons can take three separate paths. In *Figure 7-1*, 3 A of current leave terminal A. One amp flows through resistor $R_1$ and 2 A flow to resistors $R_2$ and $R_3$. At the junction of resistors $R_2$ and $R_3$, 1 A flows through resistor $R_2$, and 1 A flows to resistor $R_3$. Notice that the power supply, terminals A and B, must furnish all the current that flows through each individual resistor, or **circuit branch**. One of the rules for parallel circuits states that **the total current**

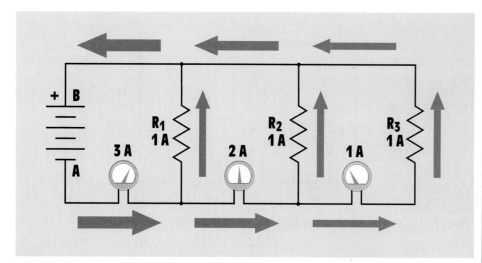

**Figure 7-1** Parallel circuits provide more than one path for current flow.

flow in the circuit is equal to the sum of the currents through all of the branches; this is known as current adds. Notice that the amount of current leaving the source **must** return to the source.

## Voltage Drop

*Figure 7-2* shows another parallel circuit and gives the values of voltage, current, and resistance for each individual resistor or branch. Notice that the voltage drop across all three resistors is the same. If the circuit is traced, it can be seen that each resistor is connected directly to the power source. A second rule for parallel circuits states that **the voltage drop across any branch of a parallel circuit is the same as the applied voltage.**

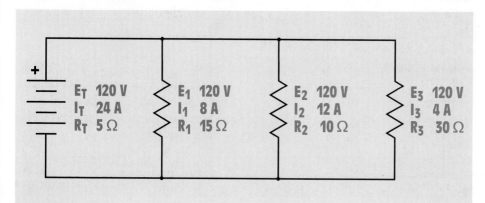

**Figure 7-2** Parallel circuit values

Panel
box

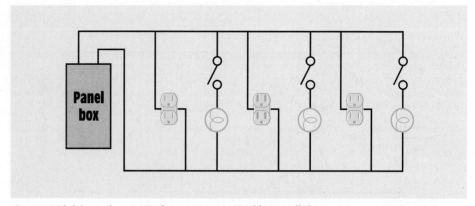

**Figure 7-3** Lights and receptacles are connected in parallel.

For this reason, most electrical circuits in homes are connected in parallel. Each lamp and receptacle is supplied with 120 V *(Figure 7-3).*

The total resistance of a parallel circuit is always less than the resistance of the lowest-value resistor, or branch, in the circuit.

## Total Resistance

In the circuit shown in *Figure 7-4,* three separate resistors have values of 15 Ω, 10 Ω, and 30 Ω. The total resistance of the circuit, however, is 5 Ω. **The total resistance of a parallel circuit is always less than the resistance of the lowest-value resistor, or branch, in the circuit.** Each time another element is connected in parallel, there is less opposition to the flow of current through the entire circuit. Imagine a water system consisting of a holding tank, a pump, and return lines to the tank *(Figure 7-5).* Although large return pipes have less resistance to the flow of water than small pipes, the small pipes do provide a return path to the

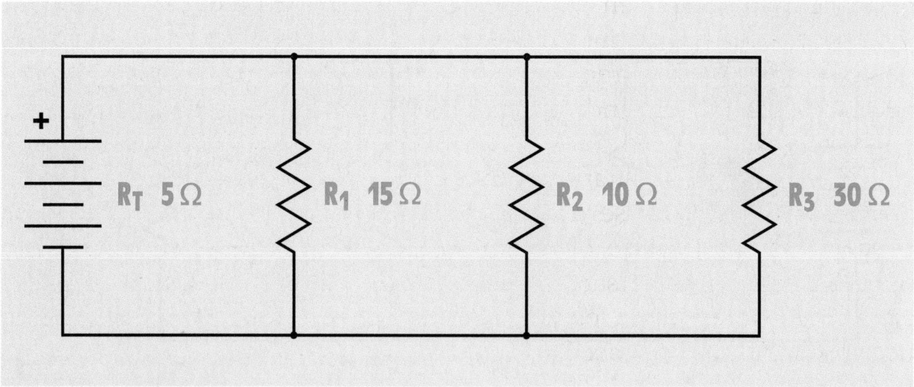

+
$R_T$ 5 Ω
$R_1$ 15 Ω
$R_2$ 10 Ω
$R_3$ 30 Ω

**Figure 7-4** Total resistance is always less than the resistance of any single branch.

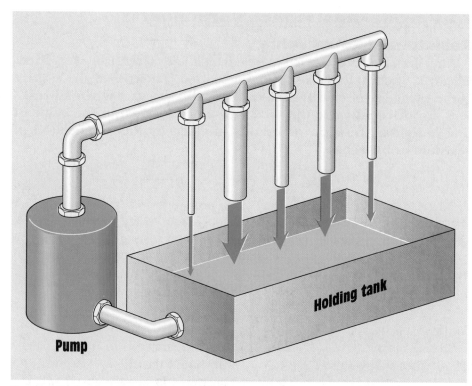

**Figure 7-5** Each new path reduces the total resistance to the flow of water.

holding tank. Each time another return path is added, regardless of size, there is less overall resistance to flow, and the rate of flow increases.

That concept often causes confusion concerning the definition of **load** among students of electricity. Students often think that an increase of resistance constitutes an increase of load. An increase of current, not resistance, results in an increase of load. In laboratory exercises, students often see the circuit current increase each time a resistive element is connected to the circuit, and they conclude that an increase of resistance must, therefore, cause an increase of current. That conclusion is, of course, completely contrary to Ohm's law, which states that an increase of resistance must cause a proportional decrease of current. The false concept that an increase of resistance causes an increase of current can be overcome once the student understands that if the resistive elements are being connected in parallel, the circuit resistance is actually being decreased and not increased.

**load**

When all
resistors are
of equal value
the total
resistance is
equal to the
value of one
individual
resistor, or
branch,
divided by the
number (N) of
resistors or
branches.

## 7-2 Parallel Resistance Formulas

### Resistors of Equal Value

Three formulas can be used to determine the total resistance of a parallel circuit. The first formula shown can be used only when all the resistors in the circuit are of equal value. This formula states that **when all resistors are of equal value the total resistance is equal to the value of one individual resistor, or branch, divided by the number (N) of resistors or branches**.

$$R_T = \frac{R}{N}$$

For example, assume that three resistors, each having a value of 24 Ω, are connected in parallel *(Figure 7-6)*. The total resistance of this circuit can be found by dividing the resistance of one single resistor by the total number of resistors.

$$R_T = \frac{R}{N}$$

$$R_T = \frac{24}{3}$$

$$R_T = 8 \ \Omega$$

### Product over Sum

The second formula used to determine the total resistance in a parallel circuit divides the product of pairs of resistors by their sum sequentially

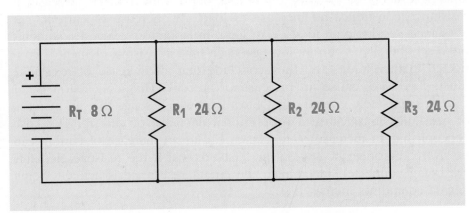

**Figure 7-6** Finding the total resistance when all resistors have the same value.

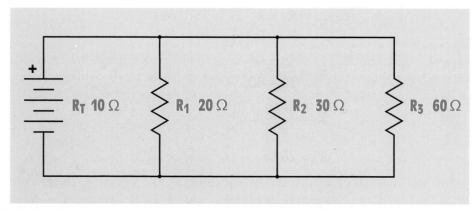

**Figure 7-7** Finding the total resistance of a parallel circuit by dividing the product of two resistors by their sum.

until only one pair is left. This is commonly referred to as the **product over sum method** for finding total resistance.

$$R_T = \frac{R_1 \times R_2}{R_1 + R_2}$$

In the circuit shown in *Figure 7-7,* three branches having resistors with values of 20 Ω, 30 Ω, and 60 Ω are connected in parallel. To find the total resistance of the circuit using the product over sum method, find the total resistance of any two branches in the circuit *(Figure 7-8).*

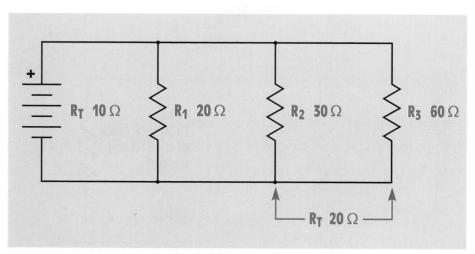

**Figure 7-8** The total resistance of the last two branches

**product over sum method**

$$R_T = \frac{R_2 \times R_3}{R_2 + R_3}$$

$$R_T = \frac{30 \times 60}{30 + 60}$$

$$R_T = \frac{1800}{90}$$

$$R_T = 20 \ \Omega$$

The total resistance of the last two resistors in the circuit is 20 $\Omega$. This 20 $\Omega$, however, is connected in parallel with a 20-$\Omega$ resistor. The total resistance of the last two resistors is now substituted for the value of $R_1$ in the formula, and the value of the first resistor is substituted for the value of $R_2$ *(Figure 7-9)*.

$$R_T = \frac{R_1 \times R_2}{R_1 + R_2}$$

$$R_T = \frac{20 \times 20}{20 + 20}$$

$$R_T = \frac{400}{40}$$

$$R_T = 10 \ \Omega$$

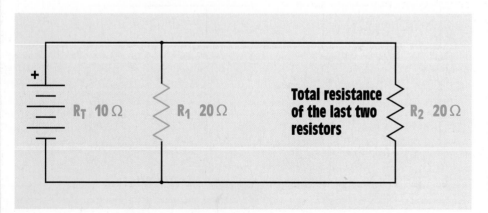

**Figure 7-9** The total value of the first two resistors is used as resistor 2.

## Reciprocal Formula

The third formula used to find the total resistance of a parallel circuit is:

$$\frac{1}{R_T} = \frac{1}{R_1} + \frac{1}{R_2} + \frac{1}{R_3} + \frac{1}{R_N}$$

Notice that this formula actually finds the reciprocal of the total resistance, instead of the total resistance. To make the formula equal to the total resistance it can be rewritten as follows:

$$R_T = \frac{1}{\dfrac{1}{R_1} + \dfrac{1}{R_2} + \dfrac{1}{R_3} + \dfrac{1}{R_N}}$$

The value $R_N$ stands for the number of resistors in the circuit. If the circuit has 25 resistors connected in parallel, for example, the last resistor in the formula would be $R_{25}$.

This formula is known as the **reciprocal formula**. The reciprocal of any number is that number divided into 1. The reciprocal of 4, for example, is 0.25 because 1/4 = 0.25. Another rule of parallel circuits is **the total resistance of a parallel circuit is the reciprocal of the sum of the reciprocals of the individual branches**. A modified version of this formula is used in several different applications to find values other than resistance. Some of those other formulas will be covered later.

Before the invention of hand-held calculators, the slide rule was often employed to help with the mathematical calculations in electrical work. At that time, the product over sum method of finding total resistance was the most popular. Since the invention of calculators, however, the reciprocal formula has become the most popular, because scientific calculators have a reciprocal key (1/X), which makes computing total resistance using the reciprocal method very easy.

**reciprocal formula**

**The total resistance of a parallel circuit is the reciprocal of the sum of the reciprocals of the individual branches.**

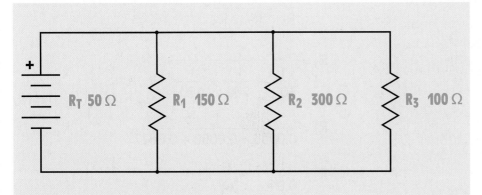

**Figure 7-10**  Finding the total resistance using the reciprocal method

In *Figure 7-10,* three resistors having values of 150 $\Omega$, 300 $\Omega$, and 100 $\Omega$ are connected in parallel. The total resistance can be found using the reciprocal formula.

$$R_T = \frac{1}{\dfrac{1}{R_1} + \dfrac{1}{R_2} + \dfrac{1}{R_3}}$$

$$R_T = \frac{1}{\dfrac{1}{150} + \dfrac{1}{300} + \dfrac{1}{100}}$$

$$R_T = \frac{1}{0.006667 + 0.003333 + 0.01}$$

$$R_T = \frac{1}{0.02}$$

$$R_T = 50 \ \Omega$$

**Example 1**

In the circuit shown in *Figure 7-11,* three resistors having values of 300 $\Omega$, 200 $\Omega$, and 600 $\Omega$ are connected in parallel. The total current flow through the circuit is 0.6 A. Find all the missing values in the circuit.

### Solution

The first step is to find the total resistance of the circuit. The reciprocal formula will be used.

$$R_T = \frac{1}{\dfrac{1}{R_1} + \dfrac{1}{R_2} + \dfrac{1}{R_3}}$$

$$R_T = \frac{1}{\dfrac{1}{300} + \dfrac{1}{200} + \dfrac{1}{100}}$$

$$R_T = \frac{1}{0.0033 + 0.0050 + 0.0017}$$

$$R_T = \frac{1}{0.01}$$

$$R_T = 100 \ \Omega$$

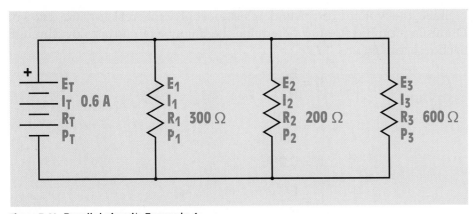

**Figure 7-11** Parallel circuit, Example 1

Now that the total resistance of the circuit is known, the voltage applied to the circuit can be found by using the total current value and Ohm's law.

$$E_T = I_T \times R_T$$

$$E_T = 0.6 \times 100$$

$$E_T = 60 \text{ V}$$

One of the rules for parallel circuits states that the voltage drops across all the parts of a parallel circuit are the same as the total voltage. Therefore, the voltage drop across each resistor is 60 V *(Figure 7-12)*.

$$E_T = E_1 = E_2 = E_3$$

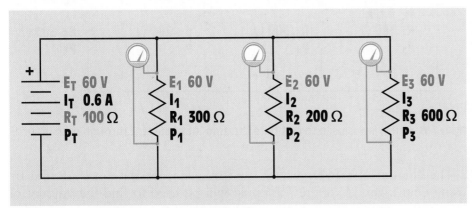

**Figure 7-12** The voltage is the same across all branches of a parallel circuit.

Since the voltage drop and resistance of each resistor are known, Ohm's law can be used to determine the amount of current flow through each resistor *(Figure 7-13)*.

$$I_1 = \frac{E_1}{R_1}$$

$$I_1 = \frac{60}{300}$$

$$I_1 = 0.2 \text{ A}$$

$$I_2 = \frac{E_2}{R_2}$$

$$I_2 = \frac{60}{200}$$

$$I_2 = 0.3 \text{ A}$$

$$I_3 = \frac{E_3}{R_3}$$

$$I_3 = \frac{60}{600}$$

$$I_3 = 0.1 \text{ A}$$

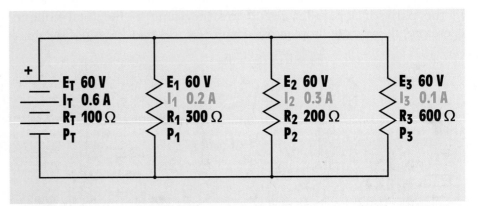

**Figure 7-13** Ohm's law is used to compute the amount of current through each branch.

The amount of power (watts) used by each resistor can be found by using Ohm's law. A different formula will be used to find the amount of electrical energy converted into heat by each of the resistors.

$$P_1 = \frac{E_1^{\,2}}{R_1}$$

$$P_1 = \frac{60 \times 60}{300}$$

$$P_1 = 12 \text{ W}$$

$$P_2 = I_2^{\,2} \times R_2$$

$$P_2 = 0.3 \times 0.3 \times 200$$

$$P_2 = 18 \text{ W}$$

$$P_3 = E_2 \times I_3$$

$$P_3 = 60 \times 0.1$$

$$P_3 = 6 \text{ W}$$

In Unit 6, it was stated that the total amount of power in a circuit is equal to the sum of the power used by all the parts. This is true for any type of circuit. Therefore, the total amount of power used by this circuit can be found by taking the sum of the power used by all the resistors (*Figure 7-14*).

$$P_T = P_1 + P_2 + P_3$$

$$P_T = 12 + 18 + 6$$

$$P_T = 36 \text{ W}$$

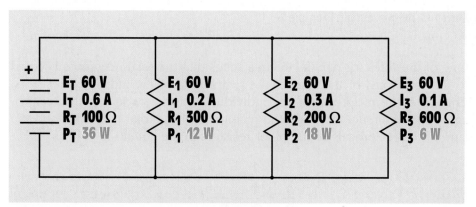

**Figure 7-14** The amount of power used by the circuit

**Example 2**

In the circuit shown in *Figure 7-15*, three resistors are connected in parallel. Two of the resistors have a value of 900 Ω and 1800 Ω. The value of resistor $R_2$ is unknown. The total resistance of the circuit is 300 Ω. Resistor $R_2$ has a current flow of 0.2 A. Find the missing circuit values.

## Solution

The first step in solving this problem is to find the missing resistor value. This can be done by changing the reciprocal formula as shown:

$$\frac{1}{R_2} = \frac{1}{R_T} - \frac{1}{R_1} - \frac{1}{R_3}$$

or

$$R_2 = \frac{1}{\dfrac{1}{R_T} - \dfrac{1}{R_1} - \dfrac{1}{R_3}}$$

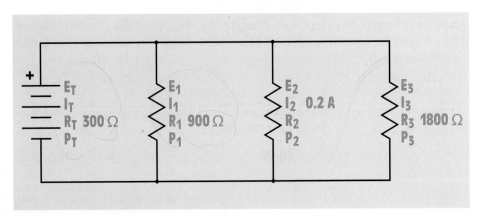

**Figure 7-15** Parallel circuit, Example 2

One of the rules for parallel circuits states that the total resistance is equal to the reciprocal of the sum of the reciprocals of the individual resistors. Therefore, the reciprocal of any individual resistor is equal to the reciprocal of the difference between the reciprocal of the total resistance and the sum of the reciprocals of the other resistors in the circuit.

$$R_2 = \frac{1}{\dfrac{1}{R_T} - \dfrac{1}{R_1} - \dfrac{1}{R_3}}$$

$$R_2 = \cfrac{1}{\cfrac{1}{300} - \cfrac{1}{900} - \cfrac{1}{1800}}$$

$$R_2 = \cfrac{1}{0.003333 - 0.001111 - 0.0005556}$$

$$R_2 = \cfrac{1}{0.001666}$$

$$R_2 = 600 \ \Omega$$

Now that the resistance of resistor $R_2$ has been found, the voltage drop across resistor $R_2$ can be determined using the current flow through the resistor and Ohm's law *(Figure 7-16).*

$$E_2 = I_2 \times R_2$$

$$E_2 = 0.2 \times 600$$

$$E_2 = 120 \ V$$

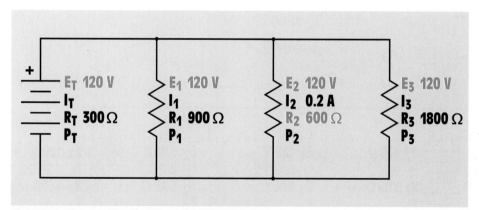

**Figure 7-16** The missing resistor and voltage values

If 120 V is dropped across resistor $R_2$, the same voltage is dropped across each component of the circuit.

$$E_2 = E_T = E_1 = E_3$$

Now that the voltage drop across each part of the circuit is known and the resistance is known, the current flow through each branch can be determined using Ohm's law *(Figure 7-17).*

$$I_T = \frac{E_T}{R_T}$$

$$I_T = \frac{120}{300}$$

$$I_T = 0.4 \text{ A}$$

$$I_1 = \frac{E_1}{R_1}$$

$$I_1 = \frac{120}{900}$$

$$I_1 = 0.1333 \text{ A}$$

$$I_3 = \frac{E_3}{R_3}$$

$$I_3 = \frac{120}{1800}$$

$$I_3 = 0.0666 \text{ A}$$

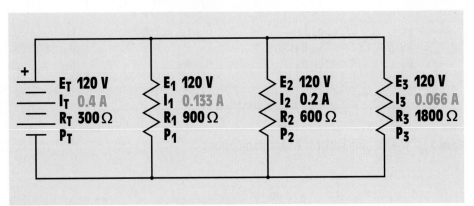

**Figure 7-17** Determining the current using Ohm's law

The amount of power used by each resistor can be found using Ohm's law *(Figure 7-18).*

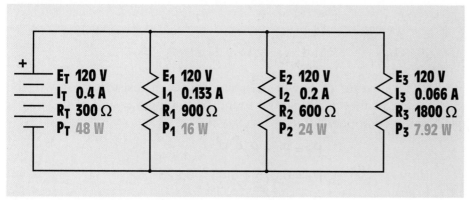

**Figure 7-18**  The values of power for the circuit in Example 2

$$P_1 = \frac{E_1^2}{R_1}$$

$$P_1 = \frac{120 \times 120}{900}$$

$$P_1 = 16 \text{ W}$$

$$P_2 = I_2^2 \times R_2$$

$$P_2 = 0.2 \times 0.2 \times 600$$

$$P_2 = 24 \text{ W}$$

$$P_3 = E_3 \times I_3$$

$$P_3 = 120 \times 0.066$$

$$P_3 = 7.92 \text{ W}$$

$$P_T = E_T \times I_T$$

$$P_T = 120 \times 0.4$$

$$P_T = 48 \text{ W}$$

If the wattage values of the three resistors are added to compute total power for the circuit, it will be seen that their total is 47.92 W instead of the computed 48 W. The small difference in answers is caused by the rounding off of other values. In this instance, the current of resistor $R_3$ was rounded from 0.066666666 to 0.066.

**Example 3**

In the circuit shown in *Figure 17-19,* three resistors are connected in parallel. Resistor $R_1$ is producing 0.075 W of heat, $R_2$ is producing 0.45 W of heat, and $R_3$ is producing 0.225 W of heat. The circuit has a total current of 0.05 A.

## Solution

Since the amount of power dissipated by each resistor is known, the total power for the circuit can be found by finding the sum of the power used by the components.

$$P_T = P_1 + P_2 + P_3$$

$$P_T = 0.075 + 0.45 + 0.225$$

$$P_T = 0.75 \text{ W}$$

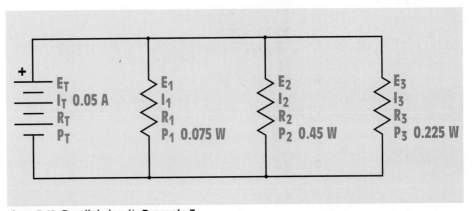

**Figure 7-19** Parallel circuit, Example 3

Now that the amount of total current and total power for the circuit are known, the applied voltage can be found using Ohm's law *(Figure 7-20).*

$$E_T = \frac{P_T}{I_T}$$

$$E_T = \frac{0.75}{0.05}$$

$$E_T = 15 \text{ V}$$

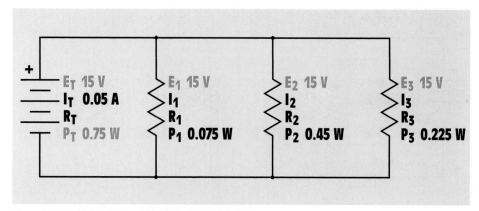

**Figure 7-20** The applied voltage for the circuit

The amount of current flow through each resistor can now be found using Ohm's law *(Figure 7-21)*.

$$I_1 = \frac{P_1}{E_1}$$

$$I_1 = \frac{0.075}{15}$$

$$I_1 = 0.005 \text{ A}$$

$$I_2 = \frac{P_2}{E_2}$$

$$I_2 = \frac{0.45}{15}$$

$$I_2 = 0.03 \text{ A}$$

$$I_3 = \frac{P_3}{E_3}$$

$$I_3 = \frac{0.225}{15}$$

$$I_3 = 0.015 \text{ A}$$

All resistance values for the circuit can now be found using Ohm's law *(Figure 7-22)*.

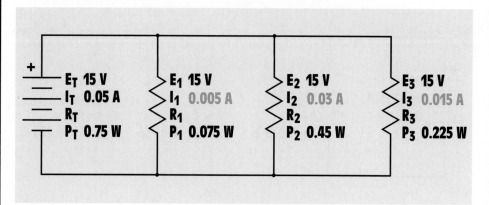

**Figure 7-21** The current through each branch

$$R_1 = \frac{E_1}{I_1}$$

$$R_1 = \frac{15}{0.005}$$

$$R_1 = 3000 \ \Omega$$

$$R_2 = \frac{E_2}{I_2}$$

$$R_2 = \frac{15}{0.03}$$

$$R_2 = 500 \ \Omega$$

$$R_3 = \frac{E_3}{I_3}$$

$$R_3 = \frac{15}{0.015}$$

$$R_3 = 1000 \ \Omega$$

$$R_T = \frac{E_T}{I_T}$$

$$R_T = \frac{15}{0.05}$$

$$R_T = 300 \ \Omega$$

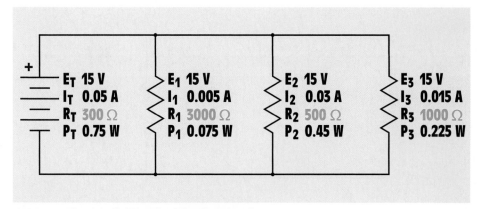

**Figure 7-22** The remaining values for the circuit

## Current Dividers

All parallel circuits are **current dividers** *(Figure 7-23)*. As previously discussed in this unit, the sum of the currents in a parallel circuit must equal the total current. Assume that a current of 1 ampere enters the circuit at point A. This 1 ampere of current will divide between resistors $R_1$ and $R_2$, and then recombine at point B. The amount of current that flows through each resistor is proportional to the resistance value. A greater amount of current will flow through a low value resistor and less current will flow through a high value resistor. In other words, the amount of current flowing through each resistor is inversely proportional to its resistance.

In a parallel circuit, the voltage across each branch must be equal *(Figure 7-24)*. Therefore, the current flow through any branch can be computed by

**current dividers**

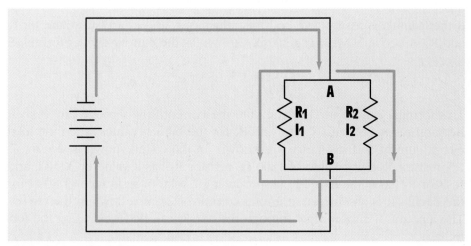

**Figure 7-23** Parallel circuits are current dividers

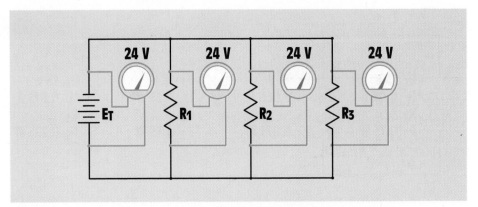

**Figure 7-24** The voltage is the same across all branches of a parallel circuit.

dividing the source voltage ($E_T$) by the resistance of that branch. The current flow through branch #1 can be computed using the formula.

$$I_1 = \frac{E_T}{R_1}$$

It is also true that the total circuit voltage is equal to the product of the total circuit current and the total circuit resistance.

$$E_T = I_T \times R_T$$

If the value of $E_T$ is substituted for ($I_T \times R_T$) in the previous formula, it becomes

$$I_1 = \frac{I_T \times R_T}{R_1}$$

If the formula is rearranged, and the values of $I_1$ and $R_1$ are substitute for $I_X$ and $R_X$, it becomes what is generally known as the current divider formula.

$$I_X = \left(\frac{R_T}{R_X}\right) I_T$$

This formula can be used to compute the current flow through any branch by substituting the values of $I_X$ and $R_X$ for the branch values when the total circuit current and resistance are known. In the circuit shown in *Figure 7-25*, resistor $R_1$ has a value of 1200 $\Omega$, resistor $R_2$ has a value of 300 $\Omega$, and resistor $R_3$ has a value of 120 $\Omega$, producing a total of resistance of 80 $\Omega$ for the circuit. It is assumed that a total current of 2 amps flows in the circuit. The amount of current flow through resistor $R_1$ can be found using the formula.

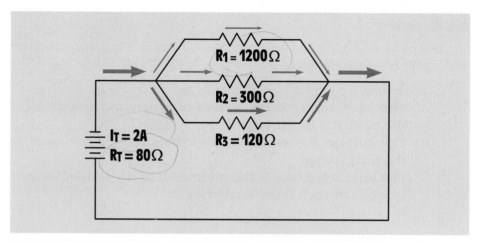

**Figure 7-25** The current divides through each branch of a parallel circuit

$$I_1 = \left(\frac{R_T}{R_1}\right) I_T$$

$$I_1 = \left(\frac{80}{1200}\right) 2$$

$$I_1 = 0.133 \text{ amp}$$

The current flow through each of the other resistors can be found by substituting in the same formula.

$$I_2 = \left(\frac{R_T}{R_2}\right) I_T$$

$$I_2 = \left(\frac{80}{300}\right) 2$$

$$I_2 = 0.533 \text{ amp}$$

$$I_3 = \left(\frac{R_T}{R_3}\right) I_T$$

$$I_3 = \left(\frac{80}{120}\right) 2$$

$$I_3 = 1.333 \text{ amp}$$

---

## Summary

1. A parallel circuit is characterized by the fact that it has more than one path for current flow.

2. Three rules for solving parallel circuits are:
   A. **The total current is the sum of the currents through all of the branches of the circuit.**
   B. **The voltage across any part of the circuit is the same as the total voltage.**
   C. **The total resistance is the reciprocal of the sum of the reciprocals of each individual branch.**

3. Circuits in homes are connected in parallel.

4. The total power in a parallel circuit is equal to the sum of the power dissipation of all the components.

5. Parallel circuits are current dividers.

6. The current flowing through each branch of a parallel circuit can be computed when the total resistance and total current are known.

7. The amount of current flow through each branch of a parallel circuit is inversely proportional to its resistance.

---

## Review Questions

1. What characterizes a parallel circuit?

2. Why are circuits in homes connected in parallel?

3. State three rules concerning parallel circuits.

4. A parallel circuit contains four branches. One branch has a current flow of 0.8 A, another has a current flow of 1.2 A, the third has a current flow of 0.25 A, and the fourth has a current flow of 1.5 A. What is the total current flow in the circuit?

5. Four resistors having a value of 100 $\Omega$ each are connected in parallel. What is the total resistance of the circuit?

6. A parallel circuit has three branches. An ammeter is connected in series with the output of the power supply and indicates a total current flow of 2.8 A. If branch 1 has a current flow of 0.9 A and branch 2 has a current flow of 1.05 A, what is the current flow through branch 3?

7. Four resistors having values of 270 Ω, 330 Ω, 510 Ω, and 430 Ω are connected in parallel. What is the total resistance in the circuit?

8. A parallel circuit contains four resistors. The total resistance of the circuit is 120 Ω. Three of the resistors have values of 820 Ω, 750 Ω, and 470 Ω. What is the value of the fourth resistor?

9. A circuit contains a 1,200 Ω , a 2,200 Ω, and a 3,300 Ω resistor connected in parallel. The circuit has a total current flow of 0.25 amp. How much current flows through each of the resistors?

## Parallel Circuits

Using the rules for parallel circuits and Ohm's law, solve for the missing values.

**Practice Problems**

$P = I \times E$

$\dfrac{I}{} \quad E$

$\dfrac{P}{I} = E$

1.

| $E_T$ 120 | $E_1$ 120 | $E_2$ 120 | $E_3$ 120 | $E_4$ 120 |
|---|---|---|---|---|
| $I_T$ 0.942 | $I_1$ .176 | $I_2$ .146 | $I_3$ .255 | $I_4$ .363 |
| $R_T$ 127.42 | $R_1$ 680 Ω | $R_2$ 820 Ω | $R_3$ 470 Ω | $R_4$ 330 Ω |
| $P_T$ 113.04 | $P_1$ 21.12 | $P_2$ 17.52 | $P_3$ 30.6 | $P_4$ 43.56 |

2.

| $E_T$ 277.05 | $E_1$ 277.05 | $E_2$ 277.05 | $E_3$ 277.05 | $E_4$ 277.05 |
|---|---|---|---|---|
| $I_T$ 0.00639 | $I_1$ .00231 | $I_2$ 0.00139 | $I_3$ 0.00154 | $I_4$ 0.00115 |
| $R_T$ 43.348K | $R_1$ 119.913K | $R_2$ 199.316K | $R_3$ 179.870K | $R_4$ 240.869K |
| $P_T$ 1.77 | $P_1$ 0.640 | $P_2$ .385 | $P_3$ .426 | $P_4$ .318 |

3.

| $E_T$ 48 | $E_1$ 48 | $E_2$ 48 | $E_3$ 48 | $E_4$ 48 |
|---|---|---|---|---|
| $I_T$ 13.4 | $I_1$ 3 | $I_2$ 4.8 | $I_3$ 3.2 | $I_4$ 2.4 |
| $R_T$ 3.582 Ω | $R_1$ 16 Ω | $R_2$ 10 Ω | $R_3$ 15.0 Ω | $R_4$ 20 Ω |
| $P_T$ 643.2 | $P_1$ 144 | $P_2$ 230.4 | $P_3$ 153.6 | $P_4$ 115Ω |

4.

| $E_T$ | $E_1$ | $E_2$ | $E_3$ | $E_4$ |
|---|---|---|---|---|
| $I_T$ | $I_1$ | $I_2$ | $I_3$ | $I_4$ |
| $R_T$ | $R_1$ 82 kΩ | $R_2$ 75 kΩ | $R_3$ 56 kΩ | $R_4$ 62 kΩ |
| $P_T$ 3.436 | $P_1$ | $P_2$ | $P_3$ | $P_4$ |

5. A parallel circuit contains the following resistor values:

$R_1$ = 360 Ω    $R_2$ = 470 Ω    $R_3$ = 300 Ω

$R_4$ = 270 Ω    $I_T$ = 0.05 amp

Find the following missing values:

$R_T$ = _____ Ω   $I_1$ = _____ A    $I_2$ = _____ A

$I_3$ = _____ A   $I_4$ = _____ A

6. A parallel circuit contains the following resistor values:

$R_1$ = 270K Ω    $R_2$ = 360K Ω    $R_3$ = 430K Ω

$R_4$ = 100K Ω    $I_T$ = 0.006 amp

Find the following missing values:

$R_T$ = _____ Ω   $I_1$ = _____ A    $I_2$ = _____ A

$I_3$ = _____ A   $I_4$ = _____ A

**Combination circuits involve both series and parallel elements. This unit discusses the rules for each type of circuit. To determine which components are in parallel and which are in series, trace the flow of current through the circuit.**

*Courtesy of International Business Machines Corporation.*

# Combination Circuits

**Objectives**

**A**fter studying this unit, you should be able to:

- Define a combination circuit.
- List the rules for parallel circuits.
- List the rules for series circuits.
- Solve combination circuits using the rules for parallel circuits, the rules for series circuits, and Ohm's law.
- State Kirchhoff's voltage and current laws.
- Solve problems using Kirchhoff's law.
- Discuss Thevenin's theorem.
- Find the Thevenin equivalent voltage and resistance values for a circuit network.
- Discuss Norton's theorem.
- Find the Norton equivalent current and resistance values for a circuit network.
- Solve circuits using the superposition theorem.

**Preview**

Combination circuits contain a combination of both series and parallel elements. To determine which components are in parallel and which are in series, trace the flow of current through the circuit. Remember that a series circuit is one that has only one path for current flow, and a parallel circuit has more than one path for current flow.

### 8-1 Combination Circuits

**trace the current path**

A simple combination circuit is shown in *Figure 8-1*. It will be assumed that the current in *Figure 8-1* will flow from point A to point B. To identify the series and parallel elements, **trace the current path**. All the current in the circuit must flow through resistor $R_1$. Resistor $R_1$, is therefore, in series

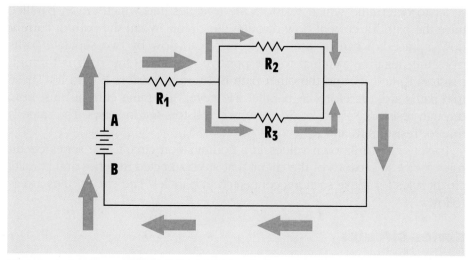

**Figure 8-1** A simple combination circuit

with the rest of the circuit. When the current reaches the junction point of resistors $R_2$ and $R_3$, however, it splits. A junction point such as this is often referred to as a **node**. Part of the current flows through resistor $R_2$ and part flows through resistor $R_3$. These two resistors are in parallel. Because this circuit contains both series and parallel elements, it is a **combination circuit**.

## 8-2 Solving Combination Circuits

The circuit shown in *Figure 8-2* contains four resistors with values of 325 $\Omega$, 275 $\Omega$, 150 $\Omega$, and 250 $\Omega$. The circuit has a total current flow of 1 A. In

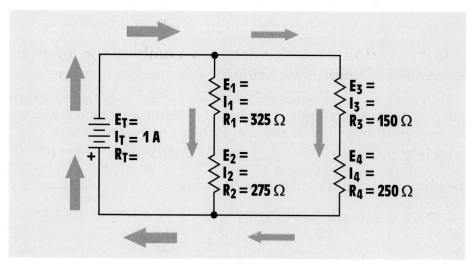

**Figure 8-2** Tracing the current paths through a combination circuit

**node**

**combina- tion circuit**

order to determine which resistors are in series and which are in parallel, trace the path for current flow through the circuit. When the path of current flow is traced, it can be seen that current can flow by two separate paths from the negative terminal to the positive terminal. One path is through resistors $R_1$ and $R_2$, and the other path is through resistors $R_3$ and $R_4$. These two paths are, therefore, in parallel. However, the same current must flow through resistors $R_1$ and $R_2$. So these two resistors are in series. The same is true for resistors $R_3$ and $R_4$.

To solve the unknown values in a combination circuit, use series circuit rules for those sections of the circuit that are connected in series and parallel circuit rules for those sections connected in parallel. The circuit rules are as follows:

## Series Circuits

1. The current is the same at any point in the circuit.

2. The total resistance is the sum of the individual resistances.

3. The sum of the voltage drops of the individual resistors must equal the applied voltage.

## Parallel Circuits

1. The voltage across any circuit branch is the same as the total voltage.

2. The total current is the sum of the current through all of the circuit paths.

3. The total resistance is equal to the reciprocal of the sum of the reciprocals of the branch resistances.

### 8-3 Simplifying the Circuit

The circuit shown in *Figure 8-2* can be reduced or simplified to a **simple parallel circuit** *(Figure 8-3)*. Since resistors $R_1$ and $R_2$ are connected in series, their values can be added to form one equivalent resistor, $R_{c(1\&2)}$, which stands for a combination of resistors 1 and 2. The same is true for resistors $R_3$ and $R_4$. Their values are added to form resistor $R_{c(3\&4)}$. Now that the circuit has been reduced to a simple parallel circuit, the total resistance can be found.

**simple parallel circuit**

$$R_T = \cfrac{1}{\cfrac{1}{R_{c(1\&2)}} + \cfrac{1}{R_{c(3\&4)}}}$$

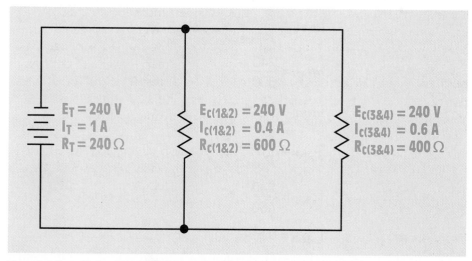

**Figure 8-3** Simplifying the combination circuit

$$R_T = \cfrac{1}{\cfrac{1}{600} + \cfrac{1}{400}}$$

$$R_T = \cfrac{1}{0.0016667 + 0.0025}$$

$$R_T = \cfrac{1}{0.0041667}$$

$$R_T = 240 \ \Omega$$

Now that the total resistance has been found, the other circuit values can be computed. The applied voltage can be found using Ohm's law.

$$E_T = I_T \times R_T$$

$$E_T = 1 \times 240$$

$$E_T = 240 \ V$$

One of the rules for parallel circuits states that the voltage is the same across each branch of the circuit. For this reason, the voltage drops across resistors $R_{c(1\&2)}$ and $R_{c(3\&4)}$ are the same. Since the voltage drop and the resistance are known, Ohm's law can be used to find the current flow through each branch.

$$I_{c(1\&2)} = \frac{E_{c(1\&2)}}{R_{c(1\&2)}}$$

$$I_{c(1\&2)} = \frac{240}{600}$$

$$I_{c(1\&2)} = 0.4 \text{ A}$$

$$I_{c(3\&4)} = \frac{E_{c(3\&4)}}{R_{c(3\&4)}}$$

$$I_{c(3\&4)} = \frac{240}{400}$$

$$I_{c(3\&4)} = 0.6 \text{ A}$$

These values can now be used to solve the missing values in the original circuit. Resistor $R_{c(1\&2)}$ is actually a combination of resistors $R_1$ and $R_2$. The values of voltage and current that apply to $R_{c(1+2)}$ therefore apply to resistors $R_1$ and $R_2$. Resistors $R_1$ and $R_2$ are connected in series. One of the rules for a series circuit states that the current is the same at any point in the circuit. Since 0.4 A of current flows through resistor $R_{c(1+2)}$, the same amount of current flows through resistors $R_1$ and $R_2$. Now that the current flow through these two resistors is known, the voltage drop across each can be computed using Ohm's law.

$$E_1 = I_1 \times R_1$$

$$E_1 = 0.4 \times 325$$

$$E_1 = 130 \text{ V}$$

$$E_2 = I_2 \times R_2$$

$$E_2 = 0.4 \times 275$$

$$E_2 = 110 \text{ V}$$

These values of voltage and current can now be added to the circuit in *Figure 8-2* to produce the circuit shown in *Figure 8-4*.

The values of voltage and current for resistor $R_{c(3\&4)}$ apply to resistors $R_3$ and $R_4$. The same amount of current that flows through resistor $R_{c(3\&4)}$ flows through resistors $R_3$ and $R_4$. The voltage across these two resistors can now be computed using Ohm's law.

$$E_3 = I_3 \times R_3$$

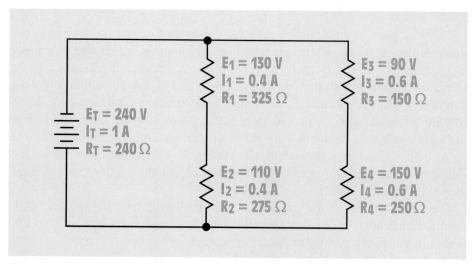

**Figure 8-4** All the missing values for the combination circuit

$$E_3 = 0.6 \times 150$$

$$E_3 = 90\ V$$

$$E_4 = I_4 \times R_4$$

$$E_4 = 0.6 \times 250$$

$$E_4 = 150\ V$$

Solve the combination circuit shown in *Figure 8-5*.

**Example 1**

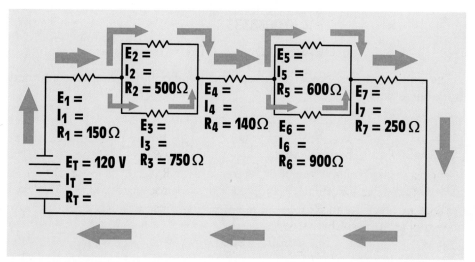

**Figure 8-5** Tracing the flow of current through the combination circuit

## Solution

The first step in finding the missing values is to trace the current path through the circuit to determine which resistors are in series and which are in parallel. All the current must flow through resistor $R_1$. Resistor $R_1$ is, therefore, in series with the rest of the circuit. When the current reaches the junction of resistors $R_2$ and $R_3$, it divides and part flows through each resistor. Resistors $R_2$ and $R_3$ are in parallel. All the current must then flow through resistor $R_4$, which is connected in series, to the junction of resistors $R_5$ and $R_6$. The current path is divided between these two resistors. Resistors $R_5$ and $R_6$ are connected in parallel. All the circuit current must then flow through resistor $R_7$.

The next step in solving this circuit is to **reduce** it to a simpler circuit. If the total resistance of the first **parallel block** formed by resistors $R_2$ and $R_3$ is found, this block can be replaced by a single resistor.

$$R_T = \frac{1}{\dfrac{1}{R_2} + \dfrac{1}{R_3}}$$

$$R_T = \frac{1}{\dfrac{1}{500} + \dfrac{1}{750}}$$

$$R_T = \frac{1}{0.002 + 0.0013333}$$

$$R_T = \frac{1}{0.0033333}$$

$$R_T = 300 \ \Omega$$

The equivalent resistance of the second parallel block can be computed in the same way.

$$R_T = \frac{1}{\dfrac{1}{R_5} + \dfrac{1}{R_6}}$$

$$R_T = \frac{1}{\dfrac{1}{600} + \dfrac{1}{900}}$$

**reduce**

**parallel block**

$$R_T = \frac{1}{0.0016667 + 0.0011111}$$

$$R_T = \frac{1}{0.00277778}$$

$$R_T = 360 \ \Omega$$

Now that the total resistance of the second parallel block is known, you can **redraw** the circuit as a simple series circuit as shown in *Figure 8-6*. The first parallel block has been replaced with a single resistor of 300 $\Omega$ labeled $R_{c(2\&3)}$, and the second parallel block has been replaced with a single 360-$\Omega$ resistor labeled $R_{c(5\&6)}$. Ohm's law can be used to find the missing values in this series circuit.

One of the rules for series circuits states that the total resistance of a series circuit is equal to the sum of the individual resistances. $R_T$ can be computed by adding the resistances of all the resistors.

$$R_T = R_1 + R_{c(2\&3)} + R_4 + R_{c(5\&6)} + R_7$$

$$R_T = 150 + 300 + 140 + 360 + 250$$

$$R_T = 1200 \ \Omega$$

Since the total voltage and total resistance are known, the total current flow through the circuit can be computed.

$$I_T = \frac{E_T}{R_T}$$

**redraw**

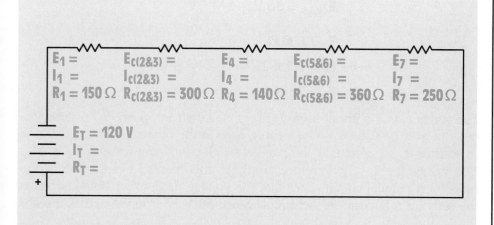

**Figure 8-6** Simplifying the combination circuit

$$I_T = \frac{120}{1200}$$

$$I_T = 0.1 \text{ A}$$

The first rule of series circuits states that the current is the same at any point in the circuit. The current flow through each resistor is, therefore, 0.1 A. The voltage drop across each resistor can now be computed using Ohm's law.

$$E_1 = I_1 \times R_1$$

$$E_1 = 0.1 \times 150$$

$$E_1 = 15 \text{ V}$$

$$E_{c(2\&3)} = I_{c(2\&3)} \times R_{c(2\&3)}$$

$$E_{c(2\&3)} = 0.1 \times 300$$

$$E_{c(2\&3)} = 30 \text{ V}$$

$$E_4 = I_4 \times R_4$$

$$E_4 = 0.1 \times 140$$

$$E_4 = 14 \text{ V}$$

$$E_{c(5\&6)} = I_{c(5\&6)} \times R_{c(5\&6)}$$

$$E_{c(5\&6)} = 0.1 \times 360$$

$$E_{c(5\&6)} = 36 \text{ V}$$

$$E_7 = I_7 \times R_7$$

$$E_7 = 0.1 \times 250$$

$$E_7 = 25 \text{ V}$$

The series circuit with all solved values is shown in *Figure 8-7*. These values can now be used to solve missing parts in the original circuit.

Resistor $R_{c(2\&3)}$ is actually the parallel block containing resistors $R_2$ and $R_3$. The values for $R_{c(2\&3)}$, therefore, apply to this parallel block. One of the rules for a parallel circuit states that the voltage drop of a parallel circuit is the same at any point in the circuit. Since 30 V is dropped across resistor $R_{c(2\&3)}$, the same 30 V is dropped across resistors $R_2$ and $R_3$ *(Figure 8-8)*. The current flow through these resistors can now be computed using Ohm's law.

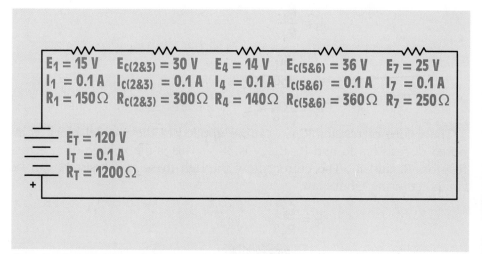

**Figure 8-7** The simplified circuit with all values solved

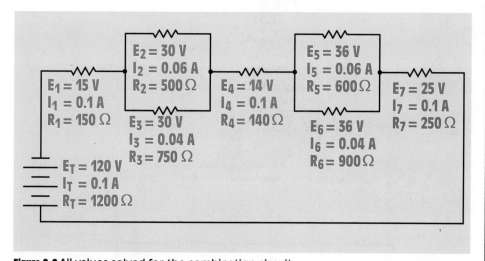

**Figure 8-8** All values solved for the combination circuit

$$I_2 = \frac{E_2}{R_2}$$

$$I_2 = \frac{30}{500}$$

$$I_2 = 0.06 \text{ A}$$

$$I_3 = \frac{E_3}{R_3}$$

$$I_3 = \frac{30}{750}$$

$$I_3 = 0.04 \text{ A}$$

The values of resistor $R_{c(5\&6)}$ can be applied to the parallel block composed of resistors $R_5$ and $R_6$. $E_{c(5\&6)}$ is 36 V. This is the voltage drop across resistors $R_5$ and $R_6$. The current flow through these two resistors can be computed using Ohm's law.

$$I_5 = \frac{E_5}{R_5}$$

$$I_5 = \frac{36}{600}$$

$$I_5 = 0.06 \text{ A}$$

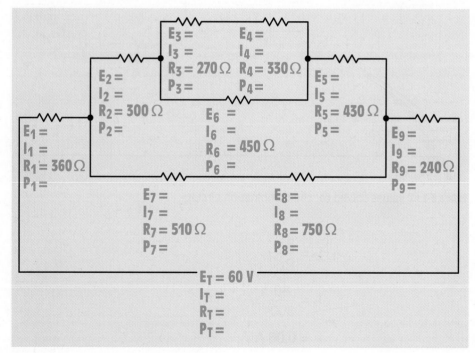

**Figure 8-9** A complex combination circuit

$$I_6 = \frac{E_6}{R_6}$$

$$I_6 = \frac{36}{900}$$

$$I_6 = 0.04 \text{ A}$$

Both of the preceding circuits were solved by first determining which parts of the circuit were in series and which were in parallel. The circuits were then reduced to a simple series or parallel circuit. This same procedure can be used for any combination circuit. The circuit shown in *Figure 8-9* will be reduced to a simpler circuit first. Once the values of the simple circuit are found, they can be placed back in the original circuit to find other values.

**Example 2**

## Solution

The first step will be to reduce the top part of the circuit to a single resistor. This part consists of resistors $R_3$ and $R_4$. Since these two resistors are connected in series with each other, their values can be added to form one single resistor. This combination will form $R_{c1}$ *(Figure 8-10).*

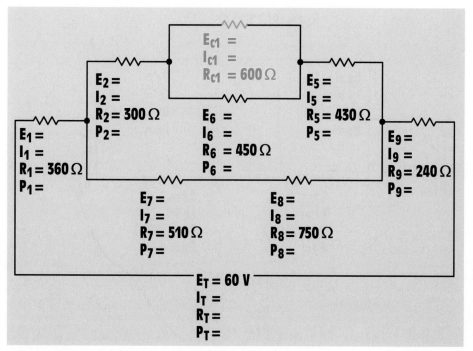

$E_{c1} =$
$I_{c1} =$
$R_{c1} = 600 \, \Omega$

$E_2 =$
$I_2 =$
$R_2 = 300 \, \Omega$
$P_2 =$

$E_6 =$
$I_6 =$
$R_6 = 450 \, \Omega$
$P_6 =$

$E_5 =$
$I_5 =$
$R_5 = 430 \, \Omega$
$P_5 =$

$E_1 =$
$I_1 =$
$R_1 = 360 \, \Omega$
$P_1 =$

$E_9 =$
$I_9 =$
$R_9 = 240 \, \Omega$
$P_9 =$

$E_7 =$
$I_7 =$
$R_7 = 510 \, \Omega$
$P_7 =$

$E_8 =$
$I_8 =$
$R_8 = 750 \, \Omega$
$P_8 =$

$E_T = 60 \text{ V}$
$I_T =$
$R_T =$
$P_T =$

**Figure 8-10** Resistors $R_1$ and $R_2$ are combined to form $R_{c1}$.

$$R_{c1} = R_3 + R_4$$

$$R_{c1} = 270 + 330$$

$$R_{c1} = 600 \ \Omega$$

The top part of the circuit is now formed by resistors $R_{c1}$ and $R_6$. These two resistors are in parallel with each other. If their total resistance is computed, they can be changed into one single resistor with a value of 257.143 $\Omega$. This combination will become resistor $R_{c2}$ *(Figure 8-11)*.

$$R_T = \frac{1}{\dfrac{1}{600} + \dfrac{1}{450}}$$

$$R_T = 257.143 \ \Omega$$

The top of the circuit now consists of resistors $R_2$, $R_{c2}$, and $R_5$. These three resistors are connected in series with each other. They can be combined to form resistor $R_{c3}$ by adding their resistances together *(Figure 8-12)*.

$$R_{c3} = R_2 + R_{c2} + R_5$$

$$R_{c3} = 300 + 257.143 + 430$$

$$R_{c3} = 987.143 \ \Omega$$

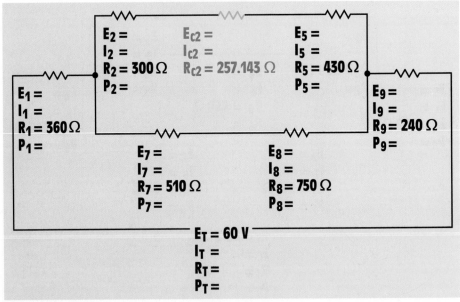

**Figure 8-11** Resistors $R_{c1}$ and $R_6$ are combined to form $R_{c2}$.

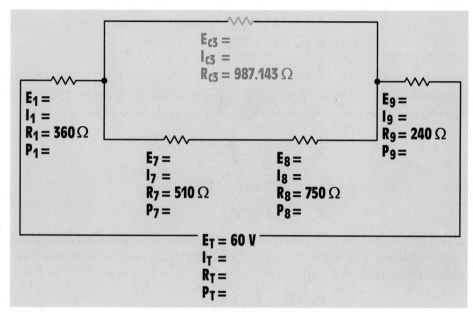

**Figure 8-12** Resistors $R_2$, $R_{c2}$, and $R_5$ are combined to form $R_{c3}$.

Resistors $R_7$ and $R_8$ are connected in series with each other also. These two resistors will be added to form resistor $R_{c4}$ *(Figure 8-13)*.

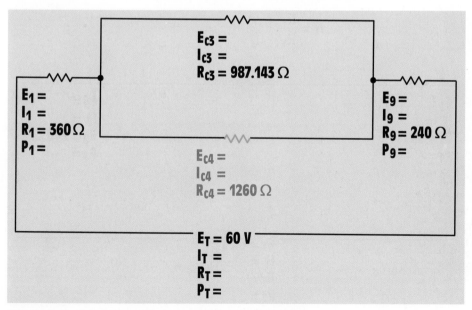

**Figure 8-13** Resistors $R_7$ and $R_8$ are combined to form $R_{c4}$.

$$R_{C4} = R_7 + R_8$$

$$R_{C4} = 510 + 750$$

$$R_{C4} = 1260 \ \Omega$$

Resistors $R_{c3}$ and $R_{c4}$ are connected in parallel with each other. Their total resistance can be computed to form resistor $R_{c5}$ *(Figure 8-14)*.

$$R_{c5} = \cfrac{1}{\cfrac{1}{987.143} + \cfrac{1}{1260}}$$

$$R_{c5} = 553.503 \ \Omega$$

The circuit has now been reduced to a simple series circuit containing three resistors. The total resistance of the circuit can be computed by adding resistors $R_1$, $R_{c5}$, and $R_9$.

$$R_T = R_1 + R_{c5} + R_9$$

$$R_T = 360 + 553.503 + 240$$

$$R_T = 1153.503 \ \Omega$$

Now that the total resistance and total voltage are known, the total circuit current and total circuit power can be computed using Ohm's law.

$$I_T = \frac{E_T}{R_T}$$

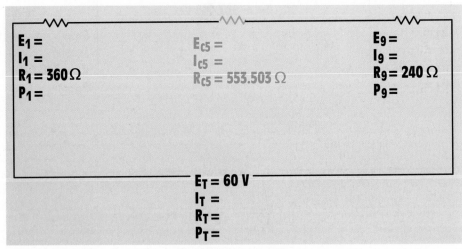

$$E_1 =$$
$$I_1 =$$
$$R_1 = 360 \ \Omega$$
$$P_1 =$$

$$E_{c5} =$$
$$I_{c5} =$$
$$R_{c5} = 553.503 \ \Omega$$

$$E_9 =$$
$$I_9 =$$
$$R_9 = 240 \ \Omega$$
$$P_9 =$$

$$E_T = 60 \ V$$
$$I_T =$$
$$R_T =$$
$$P_T =$$

**Figure 8-14** Resistors $R_{c3}$ and $R_{c4}$ are combined to form $R_{c5}$.

$$I_T = \frac{60}{1153.503}$$

$$I_T = 0.052 \text{ A}$$

$$P_T = E_T \times I_T$$

$$P_T = 60 \times 0.052$$

$$P_T = 3.12 \text{ W}$$

Ohm's law can now be used to find the missing values for resistors $R_1$, $R_{c5}$, and $R_9$ *(Figure 8-15)*.

$$E_1 = I_1 \times R_1$$

$$E_1 = 0.052 \times 360$$

$$E_1 = 18.72 \text{ V}$$

$$P_1 = E_1 \times I_1$$

$$P_1 = 18.72 \times 0.052$$

$$P_1 = 0.973 \text{ W}$$

$$E_{c5} = I_{c5} \times R_{c5}$$

$$E_{c5} = 0.052 \times 553.143$$

$$E_{c5} = 28.763 \text{ V}$$

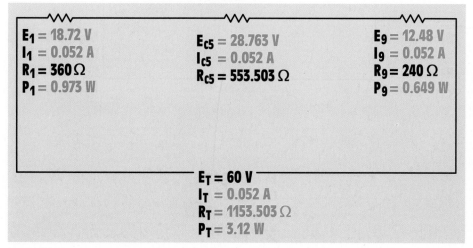

**Figure 8-15** Missing values are found for the first part of the circuit.

$$E_9 = I_9 \times R_9$$

$$E_9 = 0.052 \times 240$$

$$E_9 = 12.48 \text{ V}$$

$$P_9 = E_9 \times I_9$$

$$P_9 = 1.48 \times 0.052$$

$$P_9 = 0.649 \text{ W}$$

Resistor $R_{c5}$ is actually the combination of resistors $R_{c3}$ and $R_{c4}$. The values of $R_{c5}$, therefore, apply to resistors $R_{c3}$ and $R_{c4}$. Since these two resistors are connected in parallel with each other, the voltage drop across them will be the same. Each will have the same voltage drop as resistor $R_{c5}$ *(Figure 8-16)*. Ohm's law can now be used to find the remaining values of $R_{c3}$ and $R_{c4}$.

$$I_{c4} = \frac{E_{c4}}{R_{c4}}$$

$$I_{c4} = \frac{28.763}{1260}$$

$$I_{c4} = 0.0228 \text{ A}$$

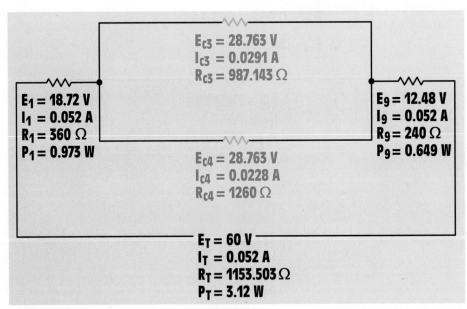

**Figure 8-16** The values for resistors $R_{c3}$ and $R_{c4}$

$$I_{c3} = \frac{E_{c3}}{R_{c3}}$$

$$I_{c3} = \frac{28.763}{987.143}$$

$$I_{c3} = 0.0291 \text{ A}$$

Resistor $R_{c4}$ is the combination of resistors $R_7$ and $R_8$. The values of resistor $R_{c4}$ apply to resistors $R_7$ and $R_8$. Since resistors $R_7$ and $R_8$ are connected in series with each other, the current flow will be the same through both *(Figure 8-17)*. Ohm's law can now be used to compute the remaining values for these two resistors.

$$E_7 = I_7 \times R_7$$

$$E_7 = 0.0228 \times 510$$

$$E_7 = 11.268 \text{ V}$$

$$P_7 = E_7 \times I_7$$

$$P_7 = 11.268 \times 0.0228$$

$$P_7 = 0.265 \text{ W}$$

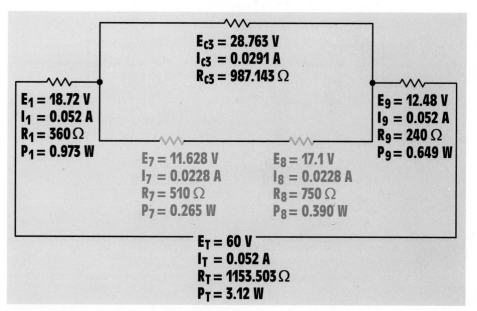

**Figure 8-17** The values for resistors $R_7$ and $R_8$

$$E_8 = I_8 \times R_8$$

$$E_8 = 0.0228 \times 750$$

$$E_8 = 17.1 \text{ V}$$

$$P_8 = E_8 \times I_8$$

$$P_8 = 17.1 \times 0.0228$$

$$P_8 = 0.390 \text{ W}$$

Resistor $R_{c3}$ is the combination of resistors $R_2$, $R_{c2}$, and $R_5$. Since these resistors are connected in series with each other, the current flow through each will be the same as the current flow through $R_{c3}$. The remaining values can now be computed using Ohm's law *(Figure 8-18)*.

$$E_{c2} = I_{c2} \times R_{c2}$$

$$E_{c2} = 0.0292 \times 257.143$$

$$E_{c2} = 7.509 \text{ V}$$

$$E_2 = I_2 \times R_2$$

$$E_2 = 0.0292 \times 300$$

$$E_2 = 8.76 \text{ V}$$

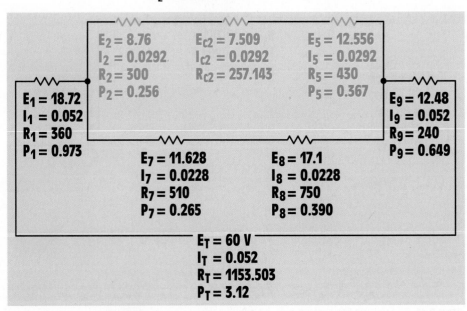

**Figure 8-18** Determining values for $R_2$, $R_{c2}$, and $R_5$

$$P_2 = E_2 \times I_2$$

$$P_2 = 8.76 \times 0.292$$

$$P_2 = 0.256 \text{ W}$$

$$E_5 = I_5 \times R_5$$

$$E_5 = 0.0292 \times 430$$

$$E_5 = 12.556 \text{ V}$$

$$P_5 = E_5 \times I_5$$

$$P_5 = 12.556 \times 0.0292$$

$$P_5 = 0.367 \text{ W}$$

Resistor $R_{c2}$ is the combination of resistors $R_{c1}$ and $R_6$. Resistors $R_{c1}$ and $R_6$ are connected in parallel and will, therefore, have the same voltage drop as resistor $R_{c2}$. Ohm's law can be used to compute the remaining values for $R_{c2}$ and $R_6$ *(Figure 8-19)*.

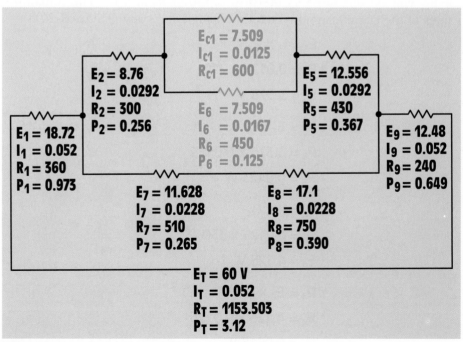

**Figure 8-19** The values of $R_{c1}$ and $R_6$

$$I_{c1} = \frac{E_{c1}}{R_{c1}}$$

$$I_{c1} = \frac{7.509}{600}$$

$$I_{c1} = 0.0125 \text{ A}$$

$$I_6 = \frac{E_6}{R_6}$$

$$I_6 = \frac{7.509}{450}$$

$$I_6 = 0.0167 \text{ A}$$

$$P_6 = E_6 \times I_6$$

$$P_6 = 7.509 \times 0.0167$$

$$P_6 = 0.125 \text{ W}$$

Resistor $R_{c1}$ is the combination of resistors $R_3$ and $R_4$. Since these two resistors are connected in series, the amount of current flow through resistor $R_{c1}$ will be the same as the flow through $R_3$ and $R_4$. The remaining values of the circuit can now be found using Ohm's law *(Figure 8-20)*.

$$E_3 = I_3 \times R_3$$

$$E_3 = 0.0125 \times 270$$

$$E_3 = 3.375 \text{ V}$$

$$P_3 = E_3 \times I_3$$

$$P_3 = 3.375 \times 0.0125$$

$$P_3 = 0.0423 \text{ W}$$

$$E_4 = I_4 \times R_4$$

$$E_4 = 0.0125 \times 330$$

$$E_4 = 4.125 \text{ V}$$

$$P_4 = E_4 \times I_4$$

$$P_4 = 4.125 \times 0.0125$$

$$P_4 = 0.0516 \text{ W}$$

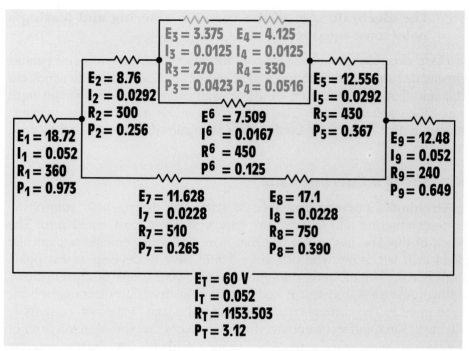

**Figure 8-20** The values for $R_3$ and $R_4$.

## Other Electrical Laws

Thus far in this text electrical values for series, parallel, and combination circuits have been computed using Ohm's law. There are some circuits, however, where Ohm's law either cannot be used or it would be very difficult to use Ohm's law to find unknown values. There are some circuits that do not have clearly defined series or parallel connections. When this is the case, Kirchhoff's law is often employed because it can be used to solve any type of circuit. This is especially true for circuits that contain more than one source of power.

Also discussed in this unit are Thevenin's and Norton's theorems. These theorems are used to simplify circuit networks, making it easier to find electrical quantities for different values of load resistances.

## Kirchhoff's Law

Kirchhoff's law was developed by a German physicist named Gustav R. Kirchhoff. In 1847 Kirchhoff stated two rules for dealing with voltage and current relationships in an electric circuit. These two rules are as follows:

1. **The algebraic sum of the voltage sources and voltage drops in a closed circuit must equal zero.**

**Kirchhoff's law**

2. **The algebraic sum of the currents entering and leaving a point must equal zero.**

These two rules are actually two of the rules used for series and parallel circuits discussed earlier in this text. The first rule is actually the series circuit rule that states the sum of the voltage drops in a series circuit must equal the applied voltage. The second rule is the parallel circuit rule that states the total current will be the sum of the currents through all the circuit branches.

## Kirchhoff's Current Law

**Kirchhoff's current law** basically states that the algebraic sum of the currents entering and leaving any particular point must equal zero. The proof of this law lies in the fact that if more current entered a particular point than left, some type of charge would have to develop at that point. Consider the circuit shown in *Figure 8-21*. Four amps of current flows through resistor $R_1$ to point P, and 6 amps of current flows through resistor $R_2$ to point P. The current leaving point P is the sum of the two currents or 10 amps. Kirchhoff's current law, however, states that the algebraic sum of the currents must equal zero. When using Kirchhoff's law, current entering a point is considered to be positive and current leaving a point is considered to be negative.

$$4A + 6A - 10A = 0$$

Currents $I_1$ and $I_2$ are considered to be positive because they enter point P. Current $I_3$ is negative because it leaves point P.

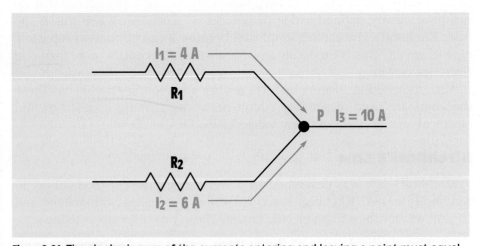

**Figure 8-21** The algebraic sum of the currents entering and leaving a point must equal zero.

A second circuit that illustrates Kirchhoff's current law can be seen in *Figure 8-22*. Consider what happens to the current at point B. Two amps of current flow into point B from resistor $R_1$. The current splits at point B, part flowing to resistor $R_2$ and part to resistors $R_4$ through $R_6$. The current entering point B is positive, and the two currents leaving point B are negative.

$$I_1 - I_2 - I_{4^-6} = 0$$

$$2A - 0.8A - 1.2A = 0$$

Now consider the currents at point E. There is 0.8 amp of current entering point E from resistor $R_2$ and 1.2 amps of current enters point E from resistors $R_4$ through $R_6$. Two amps of current leaves point E and flows through resistor $R_3$.

$$0.8A + 1.2A - 2A = 0$$

## Kirchhoff's Voltage Law

**Kirchhoff's voltage law** is very similar to the current law in that the algebraic sum of the voltages around any closed loop must equal zero. Before determining the algebraic sum of the voltages, it is necessary to first determine which end of the resistive element is positive and which is negative. To make this determination, assume a direction of current flow and mark the end of the resistive element where current enters and where current leaves. It will be assumed that current flows from negative to positive. Therefore, the point at which current enters a resistor will be marked negative, and the point where current leaves the resistor will be marked positive. The voltage drops and the polarity markings have been added to

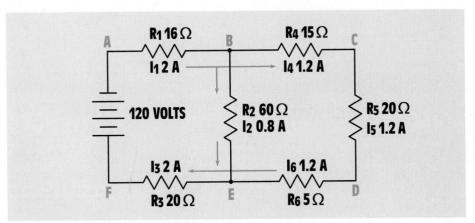

**Figure 8-22** The current splits to separate branches

each of the resistors in the example circuit shown in *Figure 8-22*. The amended circuit is shown in *Figure 8-23*.

To used Kirchhoff's voltage law, start at some point and add the voltage drops around any closed loop. Be certain to return to the starting point. In the circuit shown in *Figure 8-23*, there are actually three separate closed loops to be considered. Loop ACDF contains the voltage drops $E_1$, $E_4$, $E_5$, $E_6$, $E_3$ and $E_T$ (120 V source). Loop ABEF contains voltage drops $E_1$, $E_2$, $E_3$ and $E_T$. Loop BCDE contains voltage drops $E_4$, $E_5$, $E_6$, and $E_2$.

The voltage drops for the first loop are as follows:

$$-E_1 - E_4 - E_5 - E_6 - E_3 + E_T = 0$$

$$-32 - 18 - 24 - 6 - 40 + 120 = 0$$

The positive or negative sign for each number is determined by the assumed direction of current flow. In this example, it is assumed that current leaves point A and returns to point A. Current leaving point A enters resistor $R_1$ at the negative end. Therefore, the voltage is considered to be negative (–32 V). The same is true for resistors $R_4$, $R_5$, $R_6$, and $R_3$. The current enters the voltage source at the positive end, however. Therefore, $E_T$ is assumed to be positive.

For the second loop it is assumed that current will leave point A and return to point A through resistors $R_1$, $R_2$, $R_3$, and the voltage source. The voltage drops are as follows:

$$-E_1 - E_2 - E_3 + E_T = 0$$

$$-32 - 48 - 40 + 120 = 0$$

The current path for the third loop assumes that current leaves point B and returns to point B. The current will flow through resistors $R_4$, $R_5$, $R_6$, and $R_2$.

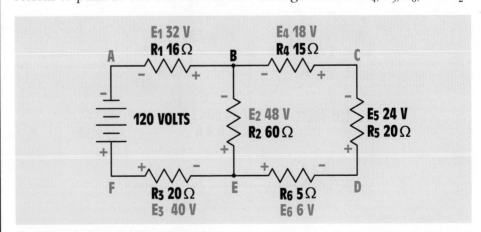

**Figure 8-23** Marking resistors elements

$$-E_4 - E_5 - E_6 + E_2 = 0$$

$$-18 - 24 - 6 + 48 = 0$$

## Solving Problems With Kirchhoff's Law

Up to this point, the values of voltage and current have been given to illustrate how Kirchhoff's law is applied to a circuit. If the values of voltage and current had to be known before Kirchhoff's law could be used, however, there would be little need for it. In the circuit shown in *Figure 8-24* the voltage drops and currents are not shown. This circuit contains three resistors and two voltage sources. Although there are three separate loops in this circuit, only two are needed to find the missing values. The two loops used will be ABEF and CBED as shown in *Figure 8-24*. The resistors have been marked positive and negative to correspond with the assumed direction of current flow. For the first loop it will be assumed that current leaves point A and returns to point A. This equation for this loop is:

$$-E_1 - E_3 + E_{S1} = 0$$

$$-E_1 - E_3 + 60 = 0$$

For the second loop it is assumed that current leaves point C and returns to point C. The equation for the second loop is:

$$-E_2 - E_3 + E_{S2} = 0$$

$$-E_2 - E_3 + 15 = 0$$

To simplify these two equations, the whole numbers will be moved to the other side of the equal sign. This will be done in the first equation by

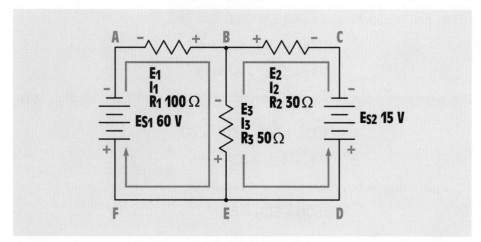

**Figure 8-24** Finding circuit values with Kirchhoff's law

subtracting 60 from both sides. The equation now becomes:

$$-E_1 - E_3 = -60$$

In the second equation, 15 will be subtracted from both sides. The equation becomes

$$-E_2 - E_3 = -15$$

The equations can be further simplified by removing the negative signs. To do this, both equations will be multiplied by negative one (–1). The two equations now become:

$$E_1 + E_3 = 60$$

$$E_2 + E_3 = 15$$

According to Ohm's law, the voltage drop across any resistive element is equal to the amount of current flowing through the element times its resistance (E = I X R). In order to solve the equations presented, it will be necessary to change the values of $E_1$, $E_2$, and $E_3$ to their Ohm's law equivalents.

$$E_1 = I_1 \times R_1 = I_1 \times 100 = 100I_1$$

$$E_2 = I_2 \times R_2 = I_2 \times 30 = 30I_2$$

Although it is true that $E_3 = I_3 \times R_3$, this would produce three unknown currents in the equation. Since Kirchhoff's current law states that the currents entering a point must equal the current leaving a point, $I_3$ is actually the sum of currents $I_1$ and $I_2$. Therefore, the third voltage equation will be written

$$E_3 = (I_1 + I_2) \times R_3 = (I_1 + I_2) \times 50 = 50(I_1 + I_2)$$

The two equations can now be written as

$$100I_1 + 50(I_1 + I_2) = 60$$

$$30I_2 + 50(I_1 + I_2) = 15$$

The parenthesis can be removed by multiplying $I_1$ and $I_2$ by 50. The equations now become

$$100I_1 + 50I_1 + 50I_2 = 60$$

$$30I_2 + 50I_1 + 50I_2 = 15$$

After gathering terms, the equations become

$$150I_1 + 50I_2 = 60$$

$$50I_1 + 80I_2 = 15$$

In order to solve these equations, it is necessary to solve them as simultaneous equations. To solve simultaneous equations, it is necessary to eliminate unknowns until there is only one unknown left. An equation can not be solved if there is more than one unknown. This will be done by multiplying the bottom equation by negative 3 (–3).

$$-3(50I_1 + 80\, I_2 = 15)$$

$$-150I_1 - 240I_2 = -45$$

The two equations can now be added.

$$150I_1 + 50I_2 = 60$$

$$-150I_1 - 240I_2 = -45$$

The positive $150I_1$ and the negative $150I_1$ will cancel each other, leaving $-190I_2$.

$$-190I_2 = 15$$

Dividing both sides of the equation by –190 will produce the answer for $I_2$.

$$I_2 = -0.0789 \text{ amp}$$

The negative answer for $I_2$ indicates that the assumed direction of current flow was incorrect. Current actually flows through the circuit as shown in *Figure 8-25*.

Now the value of $I_2$ is known, that answer can be substituted in either of the equations to find $I_1$.

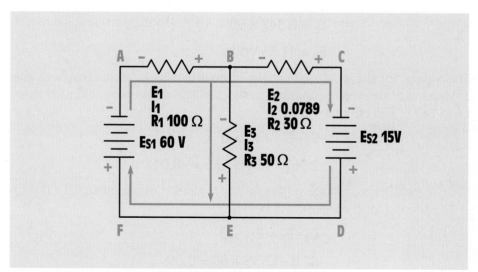

**Figure 8-25** Actual direction of current flow

$$150I_1 + 50(-0.0789) = 60$$

$$150I_1 - 3.945 = 60$$

Now add +3.945 to both sides of the equation.

$$150I_1 = 63.945$$

$$I_1 = 0.426 \text{ amp}$$

There is 0.426 amp leaving point A, flowing through resistor $R_1$, and entering point B. At point B, 0.0789 amp branches to point C through resistor $R_2$ and the remainder of the current branches to point E through resistor $R_3$. The value for $I_3$, therefore, can be found by subtracting 0.0789 amp from 0.426 amp.

$$I_3 = I_1 - I_2$$

$$I_3 = 0.426 - 0.0789$$

$$I_3 = 0.347 \text{ amp}$$

The voltage drops across each resistor can now be determined using Ohm's Law.

$$E_1 = 0.426 \times 100$$

$$E_1 = 42.6 \text{ volts}$$

$$E_2 = 0.0789 \times 30$$

$$E_2 = 2.367 \text{ volts}$$

$$E_3 = 0.347 \times 50$$

$$E_3 = 17.35 \text{ volts}$$

The values for the entire circuit are shown in *Figure 8-26*. To check the answers, add the voltages around the loops. The answers should total zero. Loop BCDE will be added first.

$$-E_2 - E_{S2} + E_3 = 0$$

$$-2.367 - 15 + 17.35 = -0.017$$

(The slight negative voltage in the answer is caused by rounding off values.) The second loop checked will be ABEF.

$$-E_1 - E_3 + E_{S1} = 0$$

$$-42.6 - 17.35 + 60 = 0.05$$

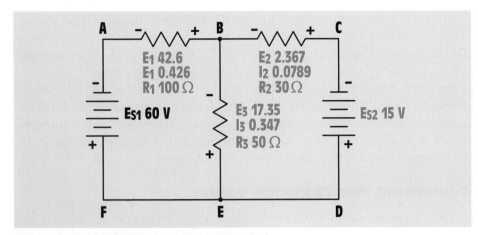

**Figure 8-26** All circuit values have been computed

The third loop checked will be ACDF.

$$-E_1 - E_2 - E_{S2} + E_{S1} = 0$$

$$-42.6 - 2.367 - 15 + 60 = 0.033$$

## Thevenin's Theorem

**Thevenin's theorem** was developed by a French engineer named M. L. Thevenin. It is **used to simplify a circuit network into an equivalent circuit, which contains a single voltage source and series resistor** *(Figure 8-27)*. Imagine a black box that contains an unknown circuit and two output terminals labeled A and B. The output terminals exhibit some amount of voltage and some amount of internal impedance.

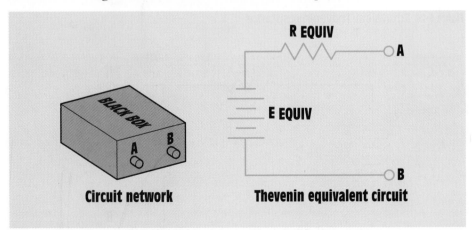

**Figure 8-27** Thevenin's theorem reduces a circuit network to a single power source and a single series resistor.

**Thevenin's theorem**

Thevenin's theorem is used to simplify a circuit network into an equivalent circuit, which contains a single voltage source and series resistor.

Thevenin's theorem reduces the circuit inside the black box to a single source of power and a series resistor equivalent to the internal impedance. The equivalent Thevenin circuit assumes the output voltage to be the open circuit voltage with no load connected. The equivalent Thevenin resistance is the open circuit resistance with no power source connected. Imagine the Thevenin circuit shown in *Figure 8-27* with the power source removed and a single conductor between terminal B and the equivalent resistor *(Figure 8-28)*. If an ohmmeter were to be connected across terminals A and B, the equivalent Thevenin resistance would be measured.

## Computing The Thevenin Values

The circuit shown in *Figure 8-29* has a single power source of 24 V and two resistors connected in series. Resistor $R_1$ has a value of 2 $\Omega$ and resistor $R_2$ has a value of 6 $\Omega$. The Thevenin equivalent circuit will be computed

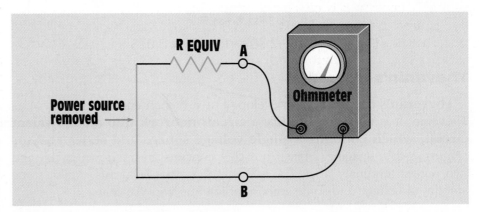

**Figure 8-28** Equivalent Thevenin resistance

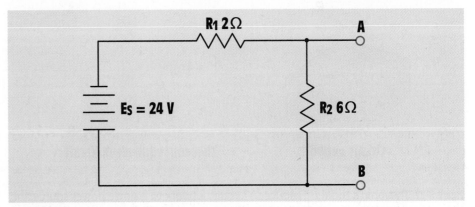

**Figure 8-29** Determining the Thevenin equivalent circuit

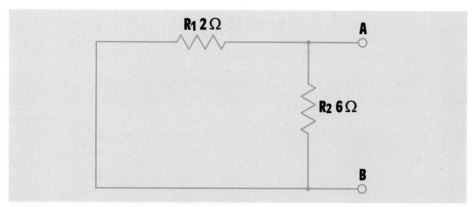

**Figure 8-30** Determining the Thevenin equivalent resistance

across terminals A and B. To do this, determine the voltage drop across resistor $R_2$ because it is connected directly across terminals A and B. The voltage drop across resistor $R_2$ will be the open circuit voltage of the equivalent Thevenin circuit when no load is connected across terminals A and B. Since resistors $R_1$ and $R_2$ form a series circuit, a total of 8 Ω is connected to the 24 V power source. This will produce a current flow 3 amps through resistors $R_1$ and $R_2$ (24/8 = 3). Since 3 amps of current flows through resistor $R_2$, a voltage drop of 18 V will appear across it (3 X 6 = 18). The equivalent Thevenin voltage for this circuit is 18 V.

To determine the equivalent Thevenin resistance, disconnect the power source and replace it with a conductor *(Figure 8-30)*. In this circuit, resistors $R_1$ and $R_2$ are connected in parallel with each other. The total resistance can now be determined using one of the formulas for finding parallel resistance.

$$R_T = \frac{R_1 \times R_2}{R_1 + R_2}$$

$$R_T = \frac{2 \times 6}{2 + 6}$$

$$R_T = 1.5\Omega$$

The Thevenin equivalent circuit is shown in *Figure 8-31*.

Now that the Thevenin equivalent of the circuit is known, the voltage and current values for different load resistances can be quickly computed. Assume, for example, that a load resistance of 10 Ω is connected across terminals A and B *(Figure 8-32)*. The voltage and current value for the circuit can now be easily computed. The total resistance of the circuit is 11.5 Ω

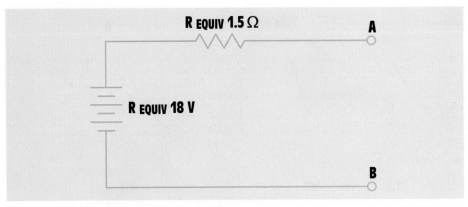

**Figure 8-31** The Thevenin equivalent circuit

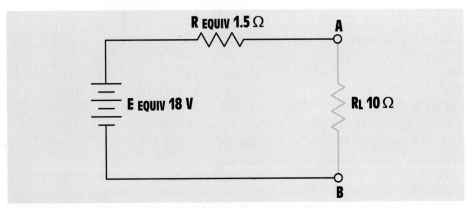

**Figure 8-32** A 10-Ω load resistor is connected across terminals A and B.

(1.5 + 10). This will produce a current flow of 1.565 amps (18/11.5), and a voltage drop of 15.65 V (1.565 X 10) across the 10-Ω load resistor.

## Norton's Theorem

**Norton's theorem** is was developed by an American scientist named E. L. Norton. **Norton's theorem is used to reduce a circuit network into a simple current source and a single parallel resistance.** This is the opposite of Thevenin's theorem, which reduces a circuit network into a simple voltage source and a single series resistor, *Figure 8-33*. Norton's theorem assumes a source of current that is divided among parallel branches. A source of current is often easier to work with, especially when calculating values for parallel circuits, than a voltage source, which drops voltages across series elements.

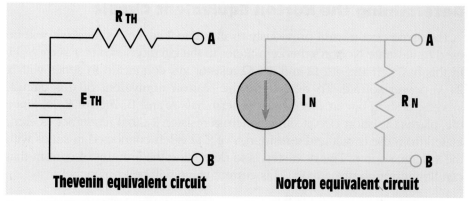

**Figure 8-33** The Thevenin equivalent circuit contains a voltage source and series resistance. The Norton equivalent circuit contains a current source and parallel resistance.

## Current Sources

Power sources can be represented in one of two ways, as a voltage source or as a current source. Voltage sources are generally shown as a battery with a resistance connected in series with the circuit to represent the internal resistance of the source. This is the case when using Thevenin's theorem. Voltage sources are rated with some amount of voltage such as 12 V, 24 V, etc.

Power sources can also be represented by a current source connected to a parallel resistance that delivers a certain amount of current such as 1 amp, 2 amps, 3 amps, etc. Assume that a current source is rated at 1.5 amps *(Figure 8-34)*. This means that 1.5 amps will flow from the power source regardless of the circuit connected. In the circuit shown in *Figure 8-34,* 1.5 amps flow through resistor $R_N$.

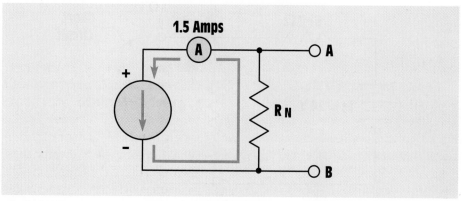

**Figure 8-34** The current source supplies a continuous 1.5 amps.

## Determining the Norton Equivalent Circuit

The same circuit used previously to illustrate Thevenin's theorem will be used to illustrate Norton's theorem. Refer to the circuit shown in *Figure 8-35*. In this basic circuit, a 2-$\Omega$ and a 6-$\Omega$ resistor are connected in series with a 24 V power source. To determine the Norton equivalent of this circuit, imagine a short circuit placed across terminals A and B, *Figure 8-36*. Since this places the short circuit directly across resistor $R_2$, that resistance is eliminated from the circuit and a resistance of 2 $\Omega$ is left connected in series with the voltage source. The next step is to determine the amount of current that can flow through this circuit. This current value will be known as $I_N$.

$$I_N = \frac{E}{R}$$

$$I_N = \frac{24}{2}$$

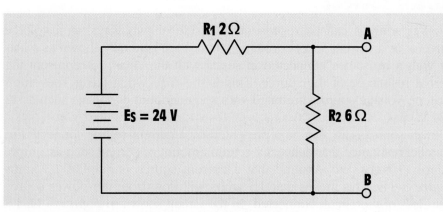

**Figure 8-35** Determining the Norton equivalent circuit

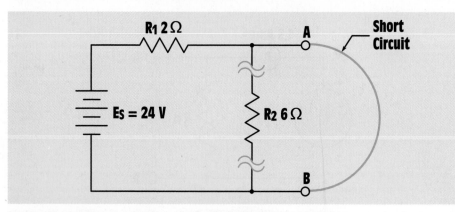

**Figure 8-36** Shorting terminals A and B eliminates the 6-$\Omega$ resistor

$$I_N = 12 \text{ A}$$

$I_N$, or 12 amps is the amount of current available in the Norton equivalent circuit.

The next step is to find the equivalent parallel resistance, $R_N$, connected across the current source. To do this, remove the short circuit across terminals A and B. Now replace the power source with a short circuit just as was done in determining the Thevenin equivalent circuit *(Figure 8-37)*. The circuit now has a 2-$\Omega$ and a 6-$\Omega$ resistor connected in parallel. This produces an equivalent resistance of 1.5-$\Omega$. The Norton equivalent circuit, shown in *Figure 8-38*, is a 1.5-$\Omega$ resistor connected in parallel with a 12 amp current source.

Now that the Norton equivalent for the circuit has been computed, any value of resistance can be connected across terminals A and B and the

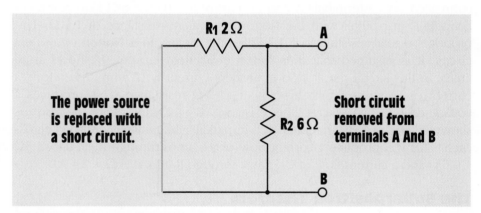

**Figure 8-37** Determining the Norton equivalent resistance

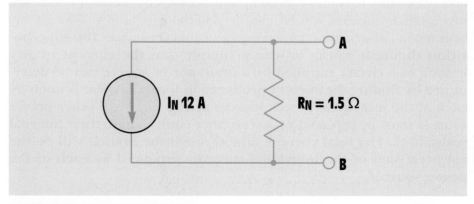

**Figure 8-38** Equivalent Norton circuit

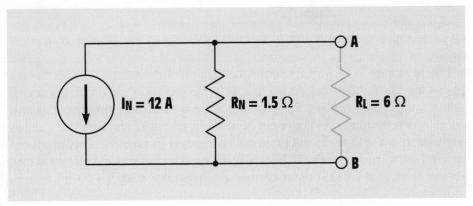

**Figure 8-39** A 6-Ω load resistor is connected to the equivalent Norton circuit

electrical values computed quickly. Assume that a 6-Ω load resistor, $R_L$, is connected across terminals A and B *(Figure 8-39)*. The 6-Ω load resistor is connected in parallel with the Norton equivalent resistance of 1.5 Ω. This produces a total resistance of 1.2-Ω for the circuit. In a Norton equivalent circuit, it is assumed that the Norton equivalent current, $I_N$, flows at all times. In this circuit the Norton equivalent current is 12 amps. Therefore, a current of 12 amps flows through the 1.2-Ω resistance. This produces a voltage drop of 14.4 V across the resistance (E = 12 X 1.2). Since the resistors shown in *Figure 8-39* are connected in parallel, 14.4 volts is dropped across each. This will produce a current flow of 9.6 amps through $R_N$ (14.4V/1.5Ω = 9.6A) and a current flow of 2.4 amps through $R_L$ (14.4V/6Ω = 2.4 A).

## The Superposition Theorem

The **superposition theorem** is somewhat similar to combining Kirchhoff's law, and Thevenin's theorem and Norton's theorem into one. The superposition theorem can be used to find the current flow through any branch of a circuit containing more than one power source. **The superposition theorem works on the principle that the current in any branch of a circuit supplied by a multi-power source can be determined by finding the current produced in that particular branch by each of the individual power sources acting alone. All other power sources must be replaced by a resistance equivalent to their internal resistances. The total current flow through the branch will be the algebraic sum of the individual currents produced by each of the power sources.**

**super-
position
theorem**

An example circuit is shown in *Figure 8-40*. In this example, the circuit contains two voltage sources. The amount of current flowing through resistor $R_2$ will be determined using the superposition theorem. The circuit can be solved by following a procedure through several distinct steps.

## Step #1

Reduce all but one of the voltage sources to zero by replacing it with a short circuit, leaving any internal series resistance. Reduce the current source to zero by replacing it with an open circuit, leaving any internal parallel resistance. Voltage source Es2 will be shorted *(Figure 8-41)*. The circuit now exists as a simple combination circuit with resistor $R_1$ connected in

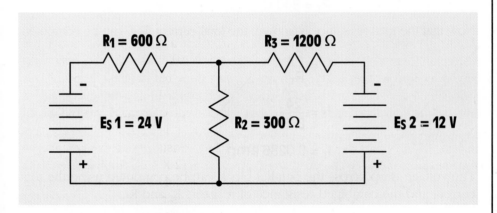

**Figure 8-40** A circuit with two power sources

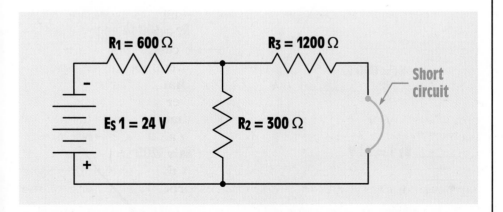

**Figure 8-41** Voltage source ES2 is replaced with a short circuit.

series with resistors $R_2$ and $R_3$, which are in parallel with each other *(Figure 8-42)*. The total resistance of this circuit can now be found by finding the total resistance of the two resistors connected in parallel, $R_2$ and $R_3$, and adding them to $R_1$.

$$R_T = R_1 + \left( \dfrac{1}{\dfrac{1}{R_2} + \dfrac{1}{R_3}} \right)$$

$$R_T = 600 + \left( \dfrac{1}{\dfrac{1}{300} + \dfrac{1}{1200}} \right)$$

$$R_T = 840 \ \Omega$$

Now that the total resistance is known, the total current flow can be computed.

$$I_T = \dfrac{E_T}{R_T}$$

$$I_T = \dfrac{24}{840}$$

$$I_T = 0.0286 \ \text{amp}$$

The voltage drop across the parallel block can be computed using the total current and the combined resistance of resistors $R_2$ and $R_3$.

$$E_{COMBINATION} = 0.0286 \ \text{X} \ 240 \ \Omega$$

$$E_{COMBINATION} = 6.864 \ \text{V}$$

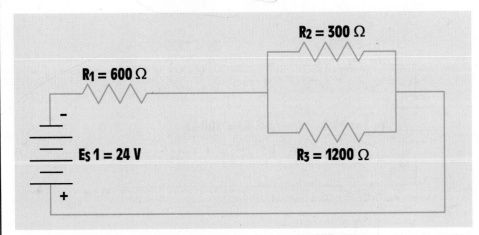

**Figure 8-42** The circuit is reduced to a simple combination circuit.

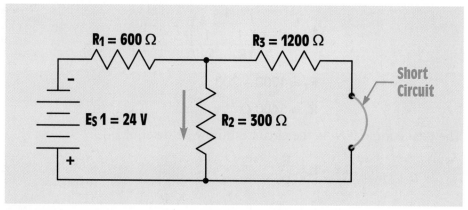

**Figure 8-43** Current flows through the resistor in the direction shown.

The current flowing through resistor $R_2$ can now be computed.

$$I_2 = \frac{6.864}{300}$$

$$I_2 = 0.0229 \text{ amp}$$

Notice that the current is flowing through resistor $R_2$ in the direction of the arrow in *Figure 8-43*.

## Step #2

Find the current flow through resistor $R_2$ by shorting voltage source Es1. Power is now supplied by voltage source $E_S2$ *(Figure 8-44)*. In this circuit, resistors $R_1$ and $R_2$ are in parallel with each other. Resistor $R_3$ is in series with $R_1$ and $R_2$.

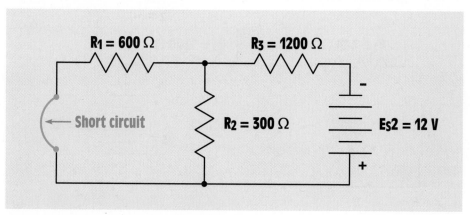

**Figure 8-44** Voltage source $E_s1$ is shorted.

$$R_T = R_3 + \left( \frac{1}{\frac{1}{R_1} + \frac{1}{R_2}} \right)$$

$$R_T = 1200 + 200$$

$$R_T = 1400 \ \Omega$$

The total current flow in the circuit can now be determined.

$$I_T = \frac{E_T}{R_T}$$

$$I_T = \frac{12}{1400}$$

$$I_T = 0.00857 \ A$$

The amount of voltage drop across the parallel combination can now be computed.

$$E_C = 0.00857 \times 200$$

$$E_C = 1.714$$

The amount of current flow through resistor $R_2$ can now be computed using Ohm's law.

$$I_2 = \frac{1.714}{300}$$

$$I_2 = 0.00571 \ amp$$

Notice that the current flowing through resistor $R_2$ is in the same direction as in the previous circuit *(Figure 8-45)*.

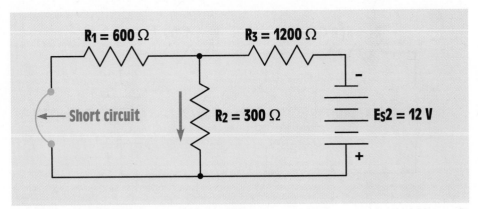

**Figure 8-45** Current flows in the same direction.

## Step #3

The next step is to find the algebraic sum of the two currents. Since both currents flow through resistor $R_2$ in the same direction, the two currents will be added.

$$I_{2(TOTAL)} = 0.0229 + 0.00571$$

$$I_{2(TOTAL)} = 0.0286 \text{ amp.}$$

Example circuit #2 is shown in *Figure 8-46*. This circuit contains a voltage source of 30 V and a current source of 0.2 amp. The amount of current flowing through resistor $R_2$ will be computed.

**Example 2**

## Step #1

The first step will be to find the current flow through resistor $R_2$ using the voltage source only. This is done by replacing the current source with an open circuit *(Figure 8-47)*. Notice the direction of current flow through resistor $R_2$.

When the current source is replaced with an open circuit, resistors $R_1$ and $R_2$ become connected in series with each other. The total resistance is the sum of the two resistances.

$$R_T = R_1 + R_2$$

$$R_T = 250 + 100$$

$$R_T = 350 \ \Omega$$

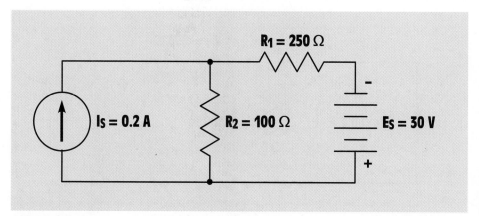

**Figure 8-46** Example circuit #2 contains a current source and a voltage source.

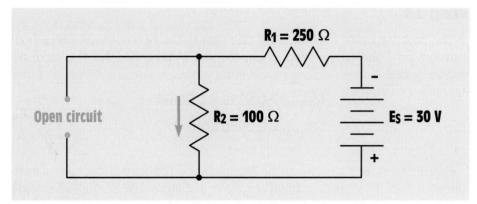

**Figure 8-47** The current source is replaced with an open circuit.

Now that the total resistance is known, the total current flow in the circuit can be found using Ohm's law.

$$I_T = \frac{30}{350}$$

$$I_T = 0.0857 \text{ amp}$$

Since the current flow must be the same in all points of a series circuit, the same amount of current flows through resistor $R_2$.

$$I_T = I_2$$

$$I_2 = 0.0857 \text{ amp}$$

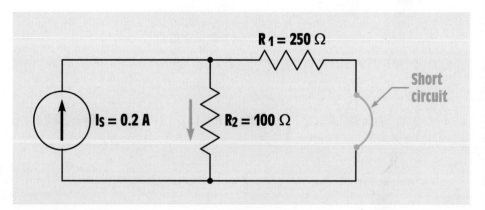

**Figure 8-48** The voltage source is replaced with a short circuit.

## Step #2

The next step will be to find the amount of current flow through resistor $R_2$ that would be supplied by the current source only. This can be done by replacing the voltage source with a short circuit *(Figure 8-48)*. Notice the direction of current flow through resistor $R_2$.

When the voltage source is removed and replaced with a short circuit, resistors $R_1$ and $R_2$ become connected in parallel with each other. The current flow through resistor $R_2$ will be computed using the current divider formula.

$$I_2 = \left(\frac{R_1}{R_1 + R_2}\right) \times I_s$$

$$I_2 = \frac{250}{350} \times 0.2$$

$$I_2 = 0.143 \text{ amp}$$

## Step #3

The total amount of current flow through resistor $R_2$ can now be determined by finding the algebraic sum of both currents.

$$I_{2(TOTAL)} = 0.0857 + 0.143$$

$$I_{2(TOTAL)} = 0.229 \text{ amp}$$

In the third example, a circuit contains two resistors and two current sources *(Figure 8-49)*. The amount of current flowing through resistor $R_2$ will be determined.

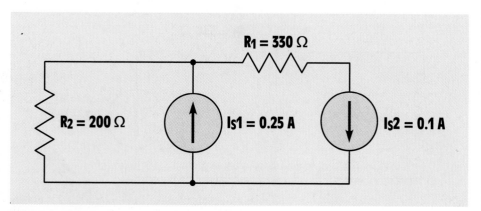

**Figure 8-49** The circuit contains two current sources.

## Step #1

Remove one of the current sources from the circuit and replace it with an open circuit. In this example, current source $I_S2$ will be removed first *(Figure 8-50)*. Resistor $R_1$ is now removed from the circuit and the entire 0.25 amp of current flows through resistor $R_2$. Notice the direction of current flow through the resistor.

## Step #2

The next step is to replace current source $I_S1$ with an open circuit and determine the amount and direction of current flow through resistor $R_2$ produced by current source $I_S2$ *(Figure 8-51)*.

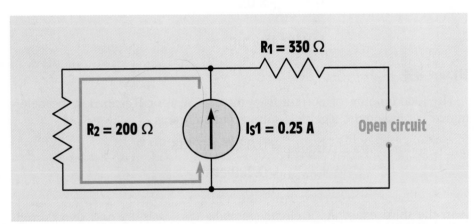

**Figure 8-50** Current source $I_S2$ is replaced with an open circuit.

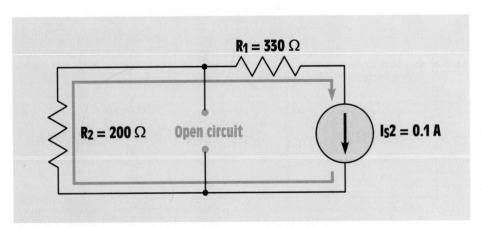

**Figure 8-51** Current sources $I_S1$ is replaced with an open circuit.

When $I_s1$ is replaced with an open circuit, resistors $R_1$ and $R_2$ become connected in series with each other. Since the current is the same in a series circuit, both resistors have a current flow of 0.1 amp through them.

## Step #3

Now that the amount and direction of current flow through resistor $R_2$ for both current sources is known, the total current flow can be determined by adding the two currents together. Since the currents flow in opposite directions, the algebraic sum will be the difference of the two currents and the direction will be determined by the greater *(Figure 8-52)*.

$$I_{2(TOTAL)} = 0.25 - 0.1$$

$$I_{2(TOTAL)} = 0.15 \text{ amp}$$

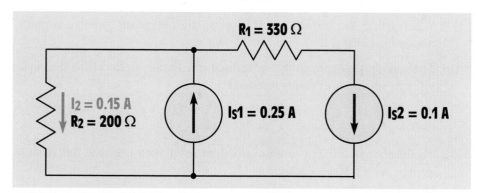

**Figure 8-52** The amount and direction of current flow through resistor $R_2$ has been determined.

## Summary

1. Combination circuits are circuits that contain both series and parallel branches.

2. The three rules for series circuits are:
   A. The current is the same at any point in the circuit.
   B. The total resistance is the sum of the individual resistances.
   C. The applied voltage is equal to the sum of the voltage drops across the individual components.

3. The three rules for parallel circuits are:
   A. The total voltage is the same as the voltage across any branch of a parallel circuit.

B. The total current is the sum of the individual currents through each path in the circuit.

C. The total resistance is the reciprocal of the sum of the reciprocals of the branch resistances.

4. When solving combination circuits, it is generally easier if the circuit is reduced to simpler circuits.

5. Kirchhoff's law can be used to solve any type of circuit.

6. Kirchhoff's voltage law states that the algebraic sum of the voltage drops and voltage sources around any closed path must equal zero.

7. Kirchhoff's current law states that the algebraic sum of the currents entering and leaving any point must equal zero.

8. Kirchhoff's law can be used to solve unknown values for circuits that contain more than one power source.

9. When using Kirchhoff's law it is generally necessary to solve simultaneous equations.

10. Thevenin's theorem involves reducing a circuit network to a simple voltage source and series resistance.

11. The Thevenin equivalent voltage is the open circuit voltage across two points.

12. To determine the Thevenin equivalent resistance, replace the voltage source with a short circuit.

13. Norton's theorem involves reducing a circuit network to a current source and parallel resistance.

14. The Norton equivalent current is determined by shorting the output terminals.

15. The Norton equivalent resistance is determined by replacing the current source with a short circuit.

16. The superposition theorem uses elements of Kirchhoff's law, Thevenin's theorem, and Norton's theorem.

## Review Questions

1. Refer to *Figure 8-2*. Replace the values shown with the following. Solve for all the unknown values.

$$I_T = 0.6 \text{ A}$$

$R_1 = 470 \ \Omega$

$R_2 = 360 \ \Omega$

$R_3 = 510 \ \Omega$

$R_4 = 430 \ \Omega$

2. Refer to *Figure 8-5*. Replace the values shown with the following. Solve for all the unknown values.

$E_T = 63 \ V$

$R_1 = 1000 \ \Omega$

$R_2 = 2200 \ \Omega$

$R_3 = 1800 \ \Omega$

$R_4 = 910 \ \Omega$

$R_5 = 3300 \ \Omega$

$R_6 = 4300 \ \Omega$

$R_7 = 860 \ \Omega$

3. State Kirchhoff's voltage law.

4. State Kirchhoff's current law.

5. What is the purpose of Thevenin's and Norton's theorems?

6. When using Kirchhoff's current law, do the currents entering a point carry a positive or negative sign?

7. When using Kirchhoff's law, if a negative answer is found for current flow, what does this indicate?

To answer questions 8 through 12, refer to the circuit shown in Figure 8-53.

8. How much current flows through resistor $R_1$?
   $I_1 = $ _____

9. How much current flows through resistor $R_2$?
   $I_2 = $ _____

10. How much voltage is dropped across resistor $R_1$?
   $E_1 = $ _____

11. How much voltage is dropped across resistor $R_2$?
   $E_2 = $ _____

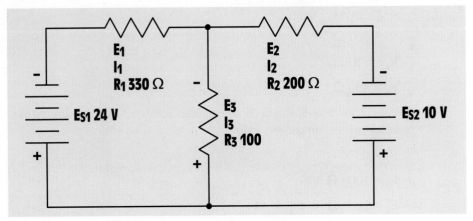

**Figure 8-53** Review Questions 8 through 12.

12. How much voltage is dropped across Resistor $R_3$?

    $E_3 =$ _____

    To answer questions 13 through 17, refer to the circuit shown in Figure 8-54.

13. In the circuit shown in *Figure 8-54*, assume that resistor $R_1$ has a resistance of 4 $\Omega$ and resistor $R_2$ has a resistance of 20 $\Omega$. Battery $E_S$ has a voltage of 48 V. What is the Thevenin equivalent voltage for this circuit across terminals A and B?

    $E_{THEV} =$ _____

14. What is the equivalent Thevenin resistance for the circuit described in question 11?

    $R_{THEV}$ _____$\Omega$

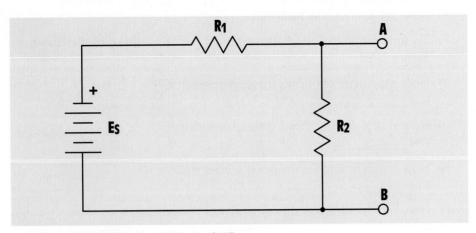

**Figure 8-54** Review Questions 13 through 17.

15. In the circuit shown in *Figure 8-54*, assume that resistor $R_1$ has a resistance of 2.5 Ω and resistor $R_2$ has a resistance of 16 Ω. Power source $E_S$ has a voltage of 20 V. As measured across terminals A and B, what would be the equivalent Norton current for this circuit?

    $I_{NORTON}$ = _____

16. What is the equivalent Norton resistance for the circuit described in question #13?

    $R_{NORTON}$ = _____ Ω

17. Assume that an 8-Ω load resistance is connected across terminals A and B for the circuit described in question 13. How much current will flow through the load resistance?

    $I_{LOAD}$ = _____

## Series-Parallel Circuits

Refer to the circuit shown in *Figure 8-55* to solve the following problems.

1. Find the unknown values in the circuit if the applied voltage is 75 V and the resistors have the following values:

    $R_1$ = 1.5 KΩ; $R_2$ = 910 KΩ; $R_3$ = 2 KΩ; and $R_4$ = 3.6 KΩ.

    $I_T$ _____   $E_1$ _____   $E_2$ _____   $E_3$ _____   $E_4$ _____

    $R_T$ _____   $I_1$ _____   $I_2$ _____   $I_3$ _____   $I_4$ _____

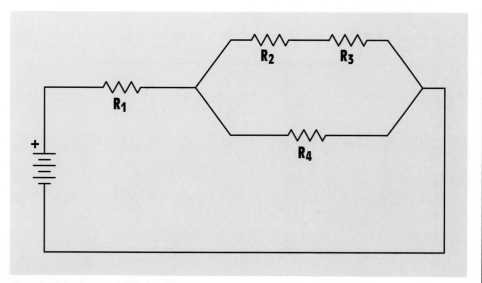

**Figure 8-55** Series-parallel circuit

2. Find the unknown values in the circuit if the applied voltage is 350 V and the resistors have the following values:

$R_1$ = 22 KΩ; $R_2$ = 18 KΩ; $R_3$ = 12 KΩ; and $R_4$ = 30 KΩ.

$I_T$ _____ $E_1$ _____ $E_2$ _____ $E_3$ _____ $E_4$ _____

$R_T$ _____ $I_1$ _____ $I_2$ _____ $I_3$ _____ $I_4$ _____

3. Find the unknown values in the circuit if the applied voltage is 18 V and the resistors have the following values:

$R_1$ = 82 Ω; $R_2$ = 160 Ω; $R_3$ = 220 Ω; and $R_4$ = 470 Ω.

$I_T$ _____ $E_1$ _____ $E_2$ _____ $E_3$ _____ $E_4$ _____

$R_T$ _____ $I_1$ _____ $I_2$ _____ $I_3$ _____ $I_4$ _____

## Parallel-Series Circuits

Refer to the circuit shown in *Figure 8-56* to solve the following problems.

4. Find the unknown values in the circuit if the total current is 0.8 A and the resistors have the following values:

$R_1$ = 1.5 KΩ; $R_2$ = 910 KΩ; $R_3$ = 2 KΩ; and $R_4$ = 3.6 KΩ.

$E_T$ _____ $E_1$ _____ $E_2$ _____ $E_3$ _____ $E_4$ _____

$R_T$ _____ $I_1$ _____ $I_2$ _____ $I_3$ _____ $I_4$ _____

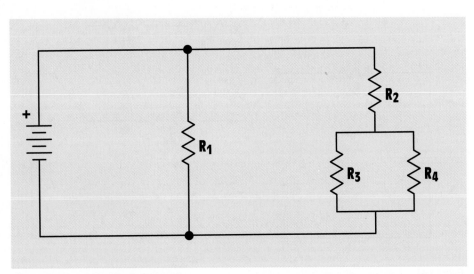

**Figure 8-56** A parallel-series circuit

5. Find the unknown values in the circuit if the total current is 0.65 A and the resistors have the following values:

$R_1$ = 22 KΩ; $R_2$ = 15 KΩ; $R_3$ = 22 KΩ; and $R_4$ = 33 KΩ.

$E_T$ _____ $E_1$ _____ $E_2$ _____ $E_3$ _____ $E_4$ _____

$R_T$ _____ $I_1$ _____ $I_2$ _____ $I_3$ _____ $I_4$ _____

6. Find the unknown values in the circuit if the total current is 1.2 A and the resistors have the following values:

$R_1$ = 75 Ω; $R_2$ = 47 Ω; $R_3$ = 220 Ω; and $R_4$ = 160 Ω.

$E_T$ _____ $E_1$ _____ $E_2$ _____ $E_3$ _____ $E_4$ _____

$R_T$ _____ $I_1$ _____ $I_2$ _____ $I_3$ _____ $I_4$ _____

## Combination Circuits

Refer to the circuit shown in *Figure 8-57* to solve the following problems.

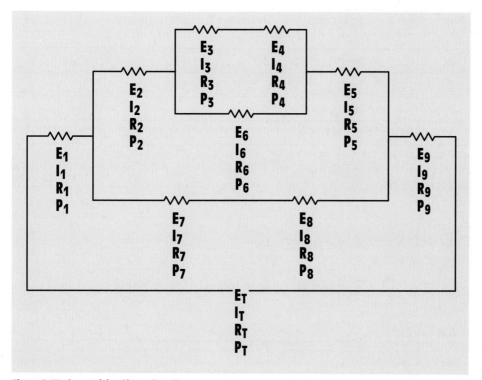

**Figure 8-57** A combination circuit

7.

| $E_T$ 250 | $E_1$ | $E_2$ | $E_3$ | $E_4$ |
|---|---|---|---|---|
| $I_T$ | $I_1$ | $I_2$ | $I_3$ | $I_4$ |
| $R_T$ | $R_1$ 220 Ω | $R_2$ 500 Ω | $R_3$ 470 Ω | $R_4$ 280 Ω |
| $P_T$ | $P_1$ | $P_2$ | $P_3$ | $P_4$ |

| $E_5$ | $E_6$ | $E_7$ | $E_8$ | $E_9$ |
|---|---|---|---|---|
| $I_5$ | $I_6$ | $I_7$ | $I_8$ | $I_9$ |
| $R_5$ 400 Ω | $R_6$ 500 Ω | $R_7$ 350 Ω | $R_8$ 450 Ω | $R_9$ 300 Ω |
| $P_5$ | $P_6$ | $P_7$ | $P_8$ | $P_9$ |

8.

| $E_T$ | $E_1$ | $E_2$ | $E_3$ | $E_4$ 1.248 |
|---|---|---|---|---|
| $I_T$ | $I_1$ | $I_2$ | $I_3$ | $I_4$ |
| $R_T$ | $R_1$ | $R_2$ | $R_3$ | $R_4$ |
| $P_T$ 0.576 | $P_1$ 0.0806 | $P_2$ 0.0461 | $P_3$ 0.00184 | $P_4$ |

| $E_5$ | $E_6$ | $E_7$ | $E_8$ | $E_9$ |
|---|---|---|---|---|
| $I_5$ | $I_6$ | $I_7$ | $I_8$ | $I_9$ |
| $R_5$ | $R_6$ | $R_7$ | $R_8$ | $R_9$ |
| $P_5$ 0.0203 | $P_6$ 0.00995 | $P_7$ 0.0518 | $P_8$ 0.0726 | $P_9$ 0.288 |

**Practice Problems**

To solve the following Kirchhoff's law problems, refer to the circuit shown in *Figure 8-58*.

9.

$E_{S1} = 12$ V  $E_1 =$ _____  $E_2 =$ _____  $E_3 =$ _____

$E_{S2} = 32$ V  $I_1 =$ _____  $I_2 =$ _____  $I_3 =$ _____

$R_1 = 680$ Ω  $R_2 = 1000$ Ω  $R_3 = 500$ Ω

10.

$E_{S1} = 3$ V  $E_1 =$ _____  $E_2 =$ _____  $E_3 =$ _____

$E_{S2} = 1.5$ V  $I_1 =$ _____  $I_2 =$ _____  $I_3 =$ _____

$R_1 = 200$ Ω  $R_2 = 120$ Ω  $R_3 = 100$ Ω

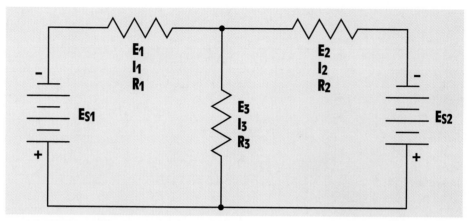

**Figure 8-58** Kirchhoff's law practice problem circuit

11.

$E_{S1}$ = 6 V        $E_1$ = _____     $E_2$ = _____     $E_3$ = _____

$E_{S2}$ = 60 V       $I_1$ = _____     $I_2$ = _____     $I_3$ = _____

                     $R_1$ = 1.6KΩ      $R_2$ = 1.2KΩ      $R_3$ = 2.4KΩ

To answer the following questions, refer to the circuit shown in *Figure 8-58*.

12.

Find the Thevenin equivalent voltage and resistance across terminals A and B.

  $E_S$ = 32 V        $R_1$ = 4 Ω         $R_2$ = 6 Ω         $E_{TH}$ = _____     $R_{TH}$ = _____

13.

  $E_S$ = 18 V        $R_1$ = 2.5 Ω       $R_2$ = 12 Ω        $E_{TH}$ = _____     $R_{TH}$ = _____

14.

Find the Norton equivalent current and resistance across terminals A and B.

  $E_S$ = 10 V        $R_1$ = 3 Ω         $R_2$ = 7 Ω         $I_N$ = _____       $R_N$ = _____

15.

  $E_S$ = 48 V        $R_1$ = 12 Ω        $R_2$ = 64 Ω        $I_N$ = _____       $R_N$ = _____

*section*

## meters and wire sizes

This unit discusses the different types of measuring instruments commonly used in the electrical field. Although it would be almost impossible to present every type of instrument made, a wide variety of instruments will be discussed.

*Courtesy of DeVry Inc.*

Unit 9

# Measuring Instruments

## Outline

9-1 Analog Meters
9-2 The Voltmeter
9-3 Multirange Voltmeters
9-4 Reading a Meter
9-5 The Ammeter
9-6 Ammeter Shunts
9-7 Multirange Ammeters
9-8 The Ayrton Shunt
9-9 AC Ammeters
9-10 Clamp-on Ammeters
9-11 DC-AC Clamp-on Ammeters
9-12 The Ohmmeter
9-13 Shunt Type Ohmmeters
9-14 Digital Meters
9-15 The Low-Impedance Voltage Tester
9-16 The Oscilloscope
9-17 The Wattmeter
9-18 Recording Meters
9-19 Bridge Circuits

## Key Terms

Ammeter
Ammeter shunt
Analog meters
Ayrton shunt
Bridge circuit
Clamp-on ammeter
Current transformer
D'Arsonval movement
Galvanometers
Moving coil meter
Moving iron meter
Multirange ammeters
Multirange voltmeters
Ohmmeter
Oscilloscope
Voltmeter
Wattmeter
Wheatstone bridge

**A**fter studying this unit, you should be able to:

■ Discuss the operation of a d'Arsonval meter movement.

■ Discuss the operation of a moving iron type of movement.

■ Connect a voltmeter to a circuit.

■ Connect and read an analog multimeter.

■ Connect an ammeter.

■ Measure resistance using an ohmmeter.

■ Interpret wave forms shown on the display of an oscilloscope.

■ Connect a wattmeter into a circuit.

**Preview**

**analog meters**

**d'Arsonval movement**

**moving coil meter**

Anyone desiring to work in the electrical and electronics field must become proficient with the common instruments used to measure electrical quantities. These instruments are the voltmeter, ammeter, and ohmmeter. Without meters it would be impossible to make meaningful interpretations of what is happening in a circuit. Meters can be divided into two general types: analog and digital.

### 9-1 Analog Meters

**Analog meters** are characterized by the fact that they use a pointer and scale to indicate their value *(Figure 9-1)*. There are different types of analog meter movements. One of the most common is the **d'Arsonval movement** shown in *Figure 9-2*. This type of movement is often referred to as a **moving coil meter**. A coil of wire is suspended between the poles of a permanent magnet. The coil is suspended either by jeweled

**Figure 9-1** An analog meter *(Courtesy of Simpson Electric.)*

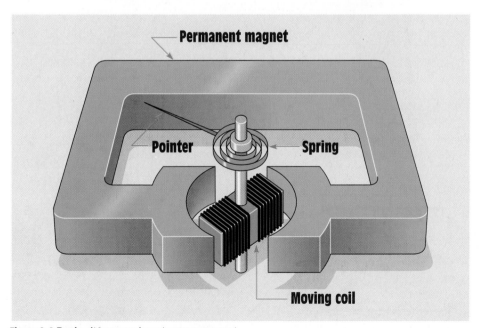

**Figure 9-2** Basic d'Arsonval meter movement

**galvano-
meters**

movements similar to those used in watches or by taut bands. The taut band type offers less turning friction than the jeweled movement. These meters can be made to operate on very small amounts of current and often are referred to as **galvanometers.**

## Principle of Operation

Analog meters operate on the principle that like magnetic poles repel each other. As current passes through the coil, a magnetic field is created around the coil. The direction of current flow through the meter is such that the same polarity of magnetic pole is created around the coil as that of the permanent magnet. This like polarity causes the coil to be deflected away from the pole of the magnet. A spring is used to retard the turning of the coil. The distance the coil turns against the spring is proportional to the strength of the magnetic field developed in the coil. If a pointer is added to the coil and a scale is placed behind the pointer, a meter movement is created.

Since the turning force of this meter depends on the repulsion of magnetic fields, it will operate on DC current only. If an AC current is

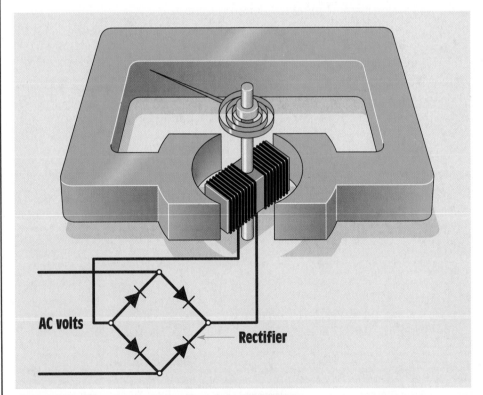

**Figure 9-3** Rectifier changes AC voltage into DC voltage.

connected to the moving coil, the magnetic polarity will change 60 times per second and the net turning force will be zero. For this reason, a DC voltmeter will indicate zero if connected to an AC line. When this type of movement is to be used to measure AC values, the current must be rectified, or changed into DC, before it is applied to the meter *(Figure 9-3)*.

## Moving Iron Meters

Another type of analog meter is the **moving iron meter.** Three common types are the radial vane, the concentric vane, and the plunger. The radial vane meter is constructed by placing two pieces of metal inside a coil of wire *(Figure 9-4)*. One of the pieces of metal is held stationary and the other is connected to a moving pointer. The movement of the pointer is retarded by a spring. When AC current flows through the coil, a magnetic field is produced around the coil. This magnetic field induces a current into the two pieces of iron, producing magnetic fields in the iron. Since these two induced magnetic fields have the same polarity, the movable piece of iron is deflected away from the stationary piece of iron. The amount of deflection is determined by the strength of the two fields. The

**moving iron meter**

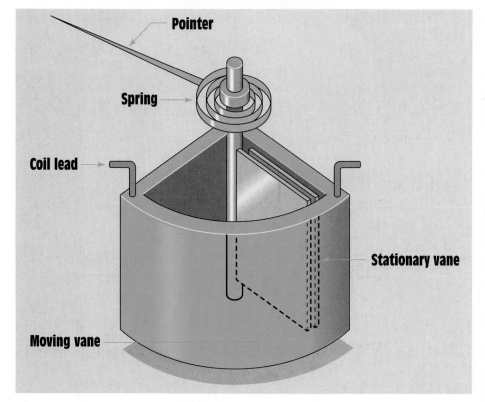

**Figure 9-4** Radial vane meter movement

polarity of the magnetic field in the iron pieces is the same regardless of the polarity of the magnetic field of the coil. For this reason, moving vane type meters are used for measuring values of AC voltage and current.

The concentric vane meter *(Figure 9-5)* operates in a fashion similar to the radial vane meter except that the metal vanes form a semicircle. The movable vane is symmetrical and has the appearance of a cylinder with a slot cut in one side. The stationary vane, however, is tapered to produce a nonuniform magnetic field between the two vanes. The nonuniformity produces the turning force of the pointer. The turning force is proportional to the strength of the magnetic fields between the movable and stationary vanes.

The plunger meter movement *(Figure 9-6)* was the first moving iron movement invented. When an AC current flows through the coil, the

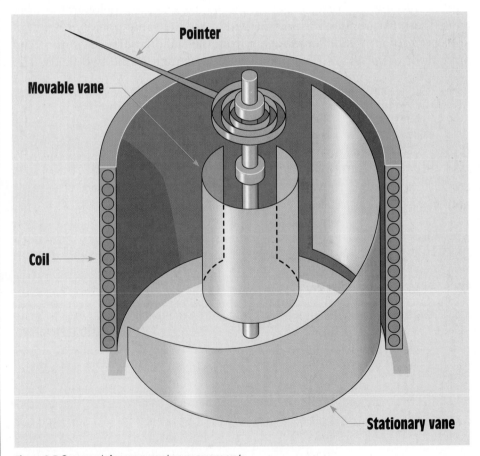

**Figure 9-5** Concentric vane meter movement

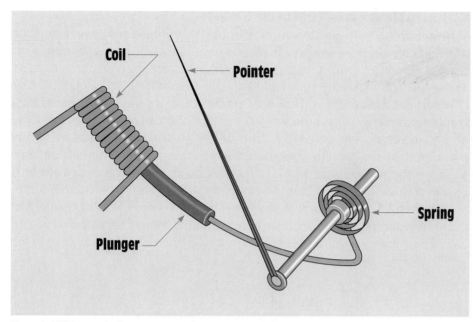

**Figure 9-6** Plunger meter movement

changing magnetic field induces an alternating current into the plunger. This induced current produces a magnetic field around the plunger that is opposite in polarity to the magnetic field of the coil. Because the two magnetic fields have opposite polarity, the plunger is attracted inside the coil. The amount of movement is determined by the strength of the magnetic fields and the tension of the spring.

Although the plunger type movement was the first invented, it is the least used of the moving iron type movements. It requires more power to operate and is less accurate than the vane type movements.

## 9-2 The Voltmeter

The **voltmeter** is designed to be connected directly across the source of power. *Figure 9-7* shows a voltmeter being used to test the voltage of a battery. Notice that the leads of the meter are connected directly across the source of voltage. A voltmeter can be connected directly across the power source because it has a very high resistance connected in series with the meter movement *(Figure 9-8)*. The industrial standard for a voltmeter is 20,000 Ω per volt for DC and 5000 Ω per volt for AC. Assume the voltmeter shown in *Figure 9-8* is an AC meter and has a full scale range of 300 V. The meter circuit (meter plus resistor) would, therefore, have a resistance of 1,500,000 Ω (300 V x 5000 Ω per volt = 1,500,000 Ω).

**voltmeter**

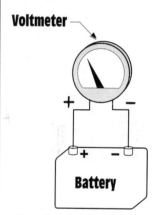

**Figure 9-7** A voltmeter connects directly across the power source.

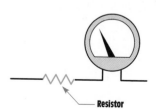

**Figure 9-8** A resistor connects in series with the meter.

## Calculating the Resistor Value

Before the resistor value can be computed, the operating characteristics of the meter must be known. It will be assumed that the meter requires a current of 50 μA and a voltage of 1 V to deflect the pointer full scale. These are known as the *full scale values* of the meter.

When the meter and resistor are connected to a source of voltage, their combined voltage drop must be 300 V. Since the meter has a voltage drop of 1 V, the resistor must have a drop of 299 V. The resistor and meter are connected in series with each other. In a series circuit, the current flow must be the same in all parts of the circuit. If 50 μA of current flow is required to deflect the meter full scale, then the resistor must have a current of 50 μA flowing through it when it has a voltage drop of 299 V. The value of resistance can now be computed using Ohm's law.

$$R = \frac{E}{I}$$

$$R = \frac{299}{0.000050}$$

$$R = 5.98 \text{ M}\Omega \ (5{,}980{,}000 \ \Omega)$$

## 9-3 Multirange Voltmeters

**multirange voltmeters**

Most voltmeters are **multirange voltmeters**, which means that they are designed to use one meter movement to measure several ranges of voltage. For example, one meter may have a selector switch that permits full scale ranges to be selected. These ranges may be 3 V full scale, 12 V full scale, 30 V full scale, 60 V full scale, 120 V full scale, 300 V full scale, and 600 V full scale. Meters are made with that many scales so that they will be as versatile as possible. If it is necessary to check for a voltage of 480 V, the meter can be set on the 600-V range. It would be very difficult, however, to check a 24-V system on the 600-V range. If the meter is set on the 30-V range, it is simple to test for a voltage of 24 V. The meter shown in *Figure 9-9* has multirange selection for voltage.

When the selector switch of this meter is turned, steps of resistance are inserted in the circuit to increase the range or are removed from the circuit to decrease the range. The meter shown in *Figure 9-10* has four range settings for full scale voltage: 30 V, 60 V, 300 V, and 600 V. Notice that when the higher voltage settings are selected, more resistance is inserted in the circuit.

**Figure 9-9** Volt-ohm-milliamp meter with multi-range selection *(Courtesy of Triplett Corp.)*

## Calculating the Resistor Values

The values of the four resistors shown in *Figure 9-10* can be determined using Ohm's law. Assume that the full scale values of the meter are 50 μA and 1 V. The first step is to determine the value for resistor $R_1$, which is used to provide a full scale value of 30 V. Resistor $R_1$, therefore, must have a voltage drop of 29 V when a current of 50 μA is flowing through it.

$$R = \frac{E}{I}$$

$$R = \frac{29}{0.000050}$$

$$R = 580 \text{ k}\Omega \text{ (580,000 } \Omega)$$

When the selector switch is moved to the second position, the meter circuit should have a total voltage drop of 60 V. The meter movement and resistor $R_1$ have a total voltage drop of 30 V, so resistor $R_2$ must have a voltage drop of 30 V when 50 µA of current flow through it. This will provide a total voltage drop of 60 V for the entire circuit.

$$R = \frac{E}{I}$$

$$R = \frac{30}{50 \text{ µA (0.000050 A)}}$$

$$R = 600 \text{ k}\Omega \text{ (600,000 } \Omega)$$

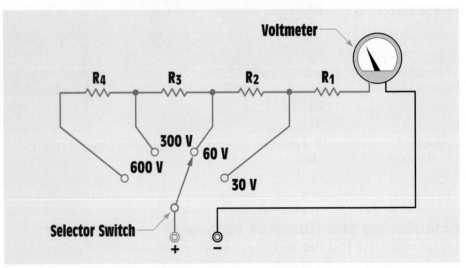

**Figure 9-10** A rotary selector switch is used to change the full range setting.

When the selector switch is moved to the third position, the circuit must have a total voltage drop of 300 V. Resistors $R_1$ and $R_2$ plus the meter movement have a combined voltage drop of 60 V at rated current. Resistor $R_3$, therefore, must have a voltage drop of 240 V at 50 µA.

$$R = \frac{E}{I}$$

$$R = \frac{240}{50\ \mu A}$$

R = 4.8 MΩ (4,800,000 Ω)

When the selector switch is moved to the fourth position, the circuit must have a total voltage drop of 600 V at rated current. Since resistors $R_1$, $R_2$, and $R_3$ plus the meter movement produce a voltage drop of 300 V at rated current, resistor $R_4$ must have a voltage drop of 300 V when 50 μA of current flow through it.

$$R = \frac{E}{I}$$

$$R = \frac{300}{50\ \mu A}$$

R = 6 MΩ (6,000,000 Ω)

## 9-4 Reading a Meter

Learning to read the scale of a multimeter takes time and practice. Most people use meters every day without thinking about it. A common type of meter used daily by most people is shown in *Figure 9-11*. The meter illustrated is a speedometer similar to those seen in automobiles. This meter is designed to measure speed. It is calibrated in miles per hour. The speedometer shown has a full scale value of 80 mph. If the pointer is positioned as shown in *Figure 9-11*, most people would know instantly that the speed of the automobile is 55 mph.

*Figure 9-12* illustrates another common meter used by most people. This meter is used to measure the amount of fuel in the tank of the automobile. Most people can glance at the pointer of the meter and know that the meter is indicating that there is one-quarter of a tank of fuel remaining. Now assume that the tank has a capacity of 20 gallons. The meter is indicating that 5 gallons of fuel remain in the tank.

Learning to read the scale of a multimeter is similar to learning to read a speedometer or fuel gauge. The meter scale shown in *Figure 9-13* has several scales used to measure different quantities and values. The top of the scale is used to measure resistance, or ohms. Notice that the scale begins on the left at infinity and ends at zero on the right. Ohmmeters will be covered later in this unit. The second scale is labeled AC-DC and is used to measure voltage. Notice that this scale has three different full

**Figure 9-11** A speedometer

**Figure 9-12** A fuel gauge

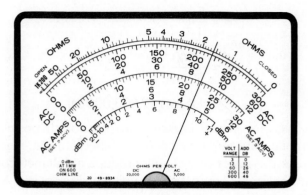

**Figure 9-13** A typical multimeter *(From Herman and Sparkman,* Electricity and Controls for HVAC, *2nd ed., copyright 1991 by Delmar Publishers Inc.)*

scale values. The top scale is 0–300, the second scale is 0–60, and the third scale is 0–12. The scale used is determined by the setting of the range control switch. The third set of scales is labeled AC amps. This scale is used with a clamp-on ammeter attachment that can be used with some meters. The last scale is labeled dBm, which is used to measure decibels.

## Reading a Voltmeter

Notice that the three voltmeter scales use the primary numbers 3, 6, and 12, and are in multiples of 10 of these numbers. Since the numbers are in multiples of 10, it is easy to multiply or divide the readings in your head by moving a decimal point. Remember that any number can be multiplied by 10 by moving the decimal point one place to the right, and any number can be divided by 10 by moving the decimal point one place to the left. For example, if the selector switch were set to permit the meter to indicate a voltage of 3 V full scale, the 300-V scale would be used, and the reading would be divided by 100. The reading can be divided by 100 by moving the decimal point two places to the left. In *Figure 9-14,* the pointer is indicating a value of 250. If the selector switch is set for 3 V full scale, moving the decimal point two places to the left will give a reading of 2.5 V. If the selector switch were set for a full scale value of 30 V, the meter shown in *Figure 9-14* would be indicating a value of 25 V. That reading is obtained by dividing the scale by 10 and moving the decimal point one place to the left.

Now assume that the meter has been set to have a full scale value of 600 V. The pointer in *Figure 9-15* is indicating a value of 44. Since the full scale value of the meter is set for 600 V, use the 60-V range and multiply

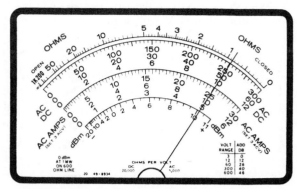

**Figure 9-14** Reading the meter *(From Herman and Sparkman,* Electricity and Controls for HVAC, *2nd ed., copyright 1991 by Delmar Publishers Inc.)*

the reading on the meter by 10, by moving the decimal point one place to the right. The correct reading becomes 440 V.

Three distinct steps should be followed when reading a meter. These steps are especially helpful for someone who has not had a great deal of experience reading a multimeter. The steps are:

1. *Determine what the meter indicates.* Is the meter set to read a value of DC voltage, DC current, AC voltage, AC current, or ohms? It is impossible to read a meter if you don't know what the meter is used to measure.

2. *Determine the full scale value of the meter.* The advantage of a multimeter is that it can measure a wide range of values and quantities. After it has been determined what quantity the meter is set to measure, it must then be determined what the range of the meter is.

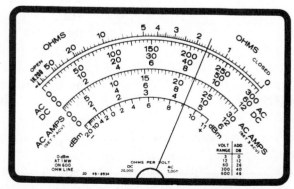

**Figure 9-15** Reading the meter *(From Herman and Sparkman,* Electricity and Controls for HVAC, *2nd ed., copyright 1991 by Delmar Publishers Inc.)*

There is a great deal of difference in reading when the meter is set to indicate a value of 600 V full scale and when it is set for 30 V full scale.

3. *Read the meter.* The last step is to determine what the meter is indicating. It may be necessary to determine the value of the hash marks on the meter face for the range for which the selector switch is set. If the meter in *Figure 9-13* is set for 300 V full scale, each hash mark has a value of 5 V. If the full scale value of the meter is 60 volts, however, each hash mark has a value of 1 V.

### 9-5 The Ammeter

**ammeter**

The **ammeter**, unlike the voltmeter, is a very low-impedance device. The ammeter is used to measure current and must be connected in series with the load to permit the load to limit the current flow *(Figure 9-16)*. An ammeter has a typical impedance of less than 0.1 Ω. If this meter is connected in parallel with the power supply, the impedance of the ammeter is the only thing to limit the amount of current flow in the circuit. Assume that an ammeter with a resistance of 0.1 Ω is connected across a 240-V AC line. The current flow in this circuit would be 2400 A (240/0.1 = 2400). A blinding flash of light would be followed by the destruction of the ammeter. Ammeters connected directly into the circuit as shown in *Figure 9-16* are referred to as in-line ammeters. *Figure 9-17* shows an ammeter of this type.

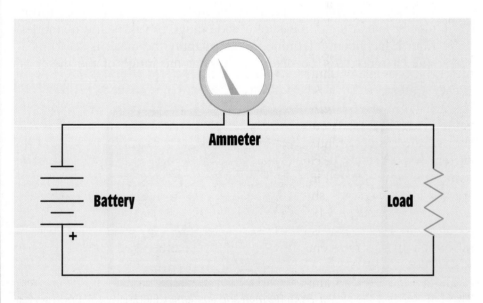

**Figure 9-16** An ammeter connects in series with the load.

**Figure 9-17** In-line ammeter

## 9-6 Ammeter Shunts

DC ammeters are constructed by connecting a common moving coil type of meter across a shunt. An **ammeter shunt** is a low-resistance device used to conduct most of the circuit current away from the meter movement. Since the meter movement is connected in parallel with the shunt, the voltage drop across the shunt is the voltage applied to the meter. Most ammeter shunts are manufactured to have a voltage drop of 50 mV (millivolts). If a 50-mV meter movement is connected across the shunt as shown in *Figure 9-18,* the pointer will move to the full scale value when the rated current of the shunt is flowing. In the example shown, the ammeter shunt is rated to have a 50-mV drop when a 10-A current is flowing in the circuit. Since the meter movement has a full scale voltage of 50 mV, it will indicate the full scale value when 10 A of current are flowing through the shunt. An ammeter shunt is shown in *Figure 9-19.*

Ammeter shunts can be purchased to indicate different values. If the same 50-mV movement is connected across a shunt designed to drop 50 mV when 100 A of current flow through it, the meter will now have a full scale value of 100 A.

The resistance of an ammeter shunt can be computed using Ohm's law. The resistance of a shunt designed to have a voltage drop of 50 mV when

**ammeter shunt**

100 A of current flow through it is:

$$R = \frac{E}{I}$$

$$R = \frac{0.050}{100}$$

$$R = 0.0005\ \Omega,\ \text{or } 0.5\ m\Omega$$

In the above problem, no consideration was given to the electrical values of the meter movement. The reason is that the amount of current needed to operate the meter movement is so small compared with the 100-A circuit current it could have no meaningful effect on the resistance value of the shunt. When computing the value for a low-current shunt, however, the meter values must be taken into consideration. For example, assume the meter has a voltage drop of 50 mV (0.050 V) and requires a current of 1 mA (0.001 A) to deflect the meter full scale. Using Ohm's law it can be found that the meter has an internal resistance of 50 Ω (0.050/0.001 = 50). Now assume that a shunt is to be constructed that will permit the meter to have a full scale value of 10 mA. If a total of 10 mA is to flow through the circuit and 1 mA must flow through the meter, then 9 mA must flow through the shunt *(Figure 9-20)*. Since the shunt must have

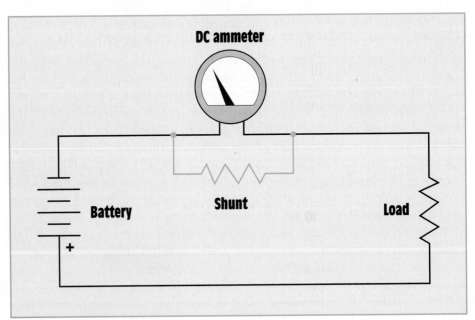

**Figure 9-18** A shunt is used to set the value of the ammeter.

**Figure 9-19** Ammeter shunt

a voltage drop of 50 mV when 9 mA of current are flowing through it, its resistance must be 5.555 Ω (0.050/0.009 = 5.555).

## 9-7 Multirange Ammeters

Many ammeters, called **multirange ammeters**, are designed to operate on more than one range. This is done by connecting the meter movement to different shunts. **When a multirange meter is used, care must be taken that the shunt is never disconnected from the meter.** Disconnection would cause the meter movement to be inserted in series with the circuit, and full circuit current would flow through the meter. Two basic methods are used for connecting shunts to a meter movement. One method is to use a make-before-break switch. This type of switch is

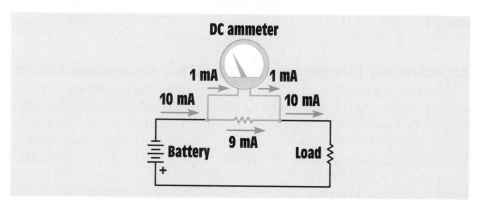

**Figure 9-20** The total current is divided between the meter and the shunt.

**multirange ammeters**

When a multi-range meter is used, care must be taken that the shunt is never disconnected from the meter.

designed so that it will make contact with the next shunt before it breaks connection with the shunt to which it is connected *(Figure 9-21)*. This method does, however, present a problem, contact resistance. Notice in *Figure 9-21* that the rotary switch is in series with the shunt resistors. This arrangement causes the contact resistance to be added to the shunt resistance and can cause inaccuracy in the meter reading.

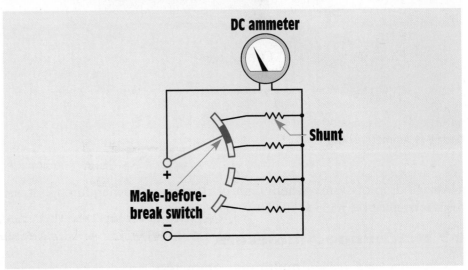

**Figure 9-21** A make-before-break switch is used to change meter shunts.

### 9-8 The Ayrton Shunt

The second method of connecting a shunt to a meter movement is to use an **Ayrton shunt** *(Figure 9-22)*. In this type of circuit, connection is made to different parts of the shunt, and the meter movement is never disconnected from the shunt. Also notice that the switch connections are made external to the shunt and meter. This arrangement prevents contact resistance from affecting the accuracy of the meter.

### Calculating the Resistor Values for an Ayrton Shunt

When an Ayrton shunt is used, the resistors will be connected in parallel with the meter on some ranges and in series with the meter for other ranges. In this example, the meter movement has full scale values of 50 mV, 1 mA, and 50 $\Omega$ of resistance. The shunt will permit the meter to have full scale current values of 100 mA, 500 mA, and 1 A.

To find the resistor values, first compute the resistance of the shunt when the range switch is set to permit a full scale current of 100 mA *(Figure 9-23)*. When the range switch is set in this position, all three shunt

**Ayrton shunt**

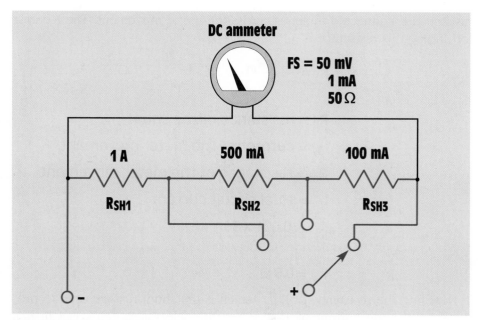

**Figure 9-22** An Ayrton shunt

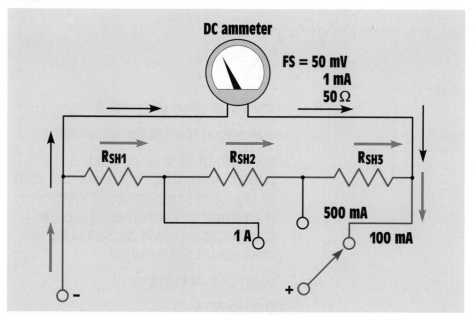

**Figure 9-23** The meter is in parallel with all shunt resistors.

resistors are connected in series across the meter movement. The formula for finding this resistance is

$$R_s = \frac{I_m \times R_m}{I_T}$$

where

$R_s$ = resistance of the shunt

$I_m$ = current of the meter movement

$R_m$ = resistance of the meter movement

$I_T$ = total circuit current

$$R_s = \frac{0.001 \times 50}{0.100}$$

$$R_s = 0.5 \, \Omega$$

Next find the resistance of $R_{SH1}$, which is the shunt resistor used to produce a full scale current of 1 A. When the selector switch is set in this position, resistor $R_{SH1}$ is connected in parallel with the meter and with resistors $R_{SH2}$ and $R_{SH3}$. Resistors $R_{SH2}$ and $R_{SH3}$, however, are connected in series with the meter movement *(Figure 9-24)*. To compute the value of this resistor, a variation of the previous formula will be used. The new formula is

$$R_{SH1} = \frac{I_m \times R_{SUM}}{I_T}$$

where

$R_{SH1}$ = the resistance of shunt 1

$I_m$ = current of the meter movement

$R_{SUM}$ = the sum of all the resistance in the circuit. Note that this is not the sum of the series-parallel combination. It is the sum of all the resistance. In this instance it will be 50.5 $\Omega$ (50 $\Omega$ [meter] + 0.5 $\Omega$ [shunt]).

$I_T$ = total circuit current

$$R_{SH1} = \frac{0.001 \times 50.5}{1}$$

$$R_{SH1} = 0.0505 \, \Omega$$

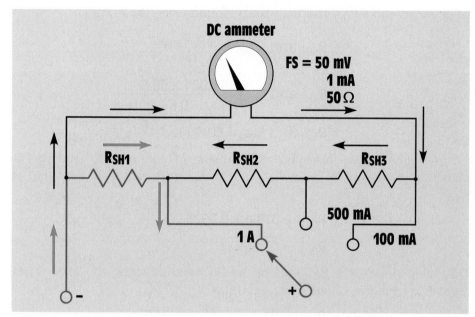

**Figure 9-24** Current path through shunt and meter for a full scale value of 1 A

When the selector switch is changed to the 500-mA position, resistors $R_{SH1}$ and $R_{SH2}$ are connected in series with each other and in parallel with the meter movement and resistor $R_{SH3}$ *(Figure 9-25)*. The combined resistance

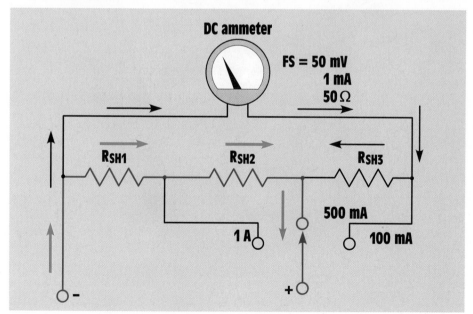

**Figure 9-25** Current path through the meter and shunt for a full scale value of 0.5 A

value for resistors $R_{SH1}$ and $R_{SH2}$ can be found using the formula

$$R_{SH1} \text{ and } R_{SH2} = \frac{I_m \times R_{SUM}}{I_T}$$

$$R_{SH1} \text{ and } R_{SH2} = \frac{0.001 \times 50.5}{0.5}$$

$$R_{SH1} \text{ and } R_{SH2} = 0.101 \ \Omega$$

Now that the total resistance for the sum of resistors $R_{SH1}$ and $R_{SH2}$ is known, the value of $R_{SH2}$ can be found by subtracting it from the value of $R_{SH1}$.

$$R_{SH2} = 0.101 - 0.0505$$

$$R_{SH2} = 0.0505 \ \Omega$$

The value of resistor $R_{SH3}$ can be found by subtracting the total shunt resistance from the values of $R_{SH1}$ and $R_{SH2}$.

$$R_{SH3} = 0.5 - 0.0505 - 0.0505$$

$$R_{SH3} = 0.399 \ \Omega$$

**Figure 9-26** DC ammeter with an Ayrton shunt

The above procedure can be used to find the value of any number of shunt resistors for any value of current desired. It should be noted, however, this type of shunt is not used for large current values because of the problem of switching contacts and contact size. The Ayrton shunt is seldom used for currents above 10 A. An ammeter with an Ayrton shunt is shown in *Figure 9-26*. The Ayrton shunt with all resistor values is shown in *Figure 9-27*.

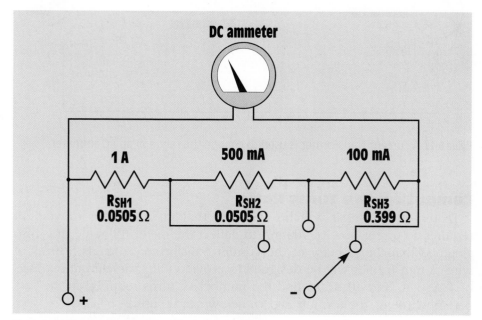

**Figure 9-27** The Ayrton shunt with resistor values

## 9-9 AC Ammeters

Shunts can be used with AC ammeters to increase their range, but cannot be used to decrease their range. Most AC ammeters use a **current transformer** instead of shunts to change scale values. This type of ammeter is shown in *Figure 9-28*. The primary of the transformer is connected in series with the load, and the ammeter is connected to the secondary of the transformer. Notice that the range of the meter is changed by selecting different taps on the secondary of the current transformer. The different taps on the transformer provide different turns ratios between the primary and secondary of the transformer. The turns ratio is the ratio of the number of turns of wire in the primary as compared to the number of turns of wire in the secondary.

**current trans-former**

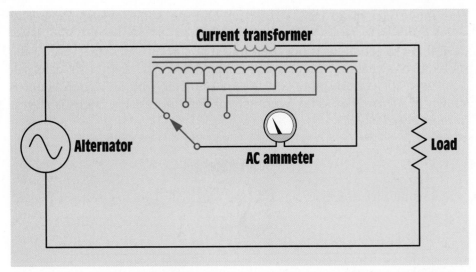

**Figure 9-28** A current transformer is used to change the range of an AC ammeter.

## Computing the Turns Ratio

In this example, it will be assumed that an AC meter movement requires a current flow of 100 mA to deflect the meter full scale. It is also assumed that the primary of the current transformer contains 5 turns of wire. A transformer will be designed to provide full scale current readings of 1 A, 5 A, and 10 A. To find the number of turns required in the secondary winding, the formula shown below can be used.

$$\frac{N_p}{N_s} = \frac{I_s}{I_p}$$

where

$N_p$ = number of turns of wire in the primary

$N_s$ = number of turns of wire in the secondary

$I_p$ = current of the primary

$I_s$ = current of the secondary

The number of turns of wire in the secondary to produce a full scale current reading of 1 A can be computed as follows:

$$\frac{5}{N_s} = \frac{0.1}{1}$$

Cross-multiplication is used to solve the problem. Cross-multiplication is accomplished by multiplying the bottom half of the equation on one side of the equals sign by the top half of the equation on the other side of the equals sign.

$$0.1 \, N_s = 5$$

$$N_s = 50$$

The transformer secondary must contain 50 turns of wire if the ammeter is to indicate a full scale reading when 1 A of current flows through the primary winding.

The number of secondary turns can be found for the other values of primary current in the same way.

$$\frac{5}{N_s} = \frac{0.1}{5}$$

$$0.1 \, N_s = 25$$

$$N_s = 250 \text{ turns}$$

$$\frac{5}{N_s} = \frac{0.1}{10}$$

$$0.1 \, N_s = 50$$

$$N_s = 500 \text{ turns}$$

## Current Transformers (CTs)

When a large amount of AC current must be measured, a different type of current transformer is connected in the power line. These transformers have ratios that start at 200:5 and can have ratios of several thousand to five. These current transformers, generally referred to in industry as CTs, have a standard secondary current rating of 5 A AC. They are designed to be operated with a 5-A AC ammeter connected directly to their secondary winding, which produces a short circuit. CTs are designed to operate with the secondary winding shorted. **The secondary winding of a CT should never be opened when there is power applied to the primary. This will cause the transformer to produce a step-up in voltage that could be high enough to kill anyone who comes in contact with it.**

A current transformer is basically a toroid transformer. A toroid transformer is constructed with a hollow core similar to a doughnut *(Figure 9-29)*. When current transformers are used, the main power line is inserted

**Figure 9-29** A toroid current transformer *(Courtesy of SQUARE D COMPANY.)*

through the opening in the transformer *(Figure 9-30)*. The power line acts as the primary of the transformer and is considered to be 1 turn.

The turns ratio of the transformer can be changed by looping the power wire through the opening in the transformer to produce a primary winding of more than 1 turn. For example, assume a current transformer has a ratio of 600:5. If the primary power wire is inserted through the opening, it will require a current of 600 A to deflect the meter full scale. If the primary power conductor is looped around and inserted through the window a second time, the primary now contains 2 turns of wire instead of 1 *(Figure 9-31)*. It now requires 300 A of current flow in the primary to deflect the meter full scale. If the primary conductor is looped through the opening a third time, it would require only 200 A of current flow to deflect the meter full scale.

## 9-10 Clamp-on Ammeters

**clamp-on ammeter**

Many electricians use the **clamp-on** type of AC **ammeter** *(Figure 9-32)*. The jaw of this type of meter is clamped around one of the conductors

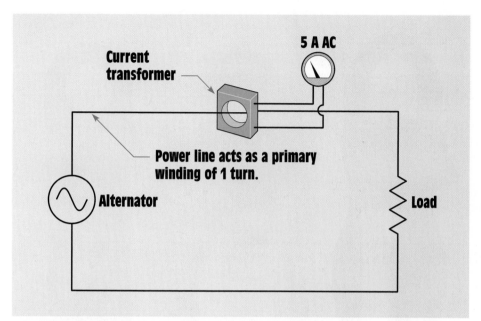

**Figure 9-30** Toroid transformer used to change the scale factor of an AC ammeter

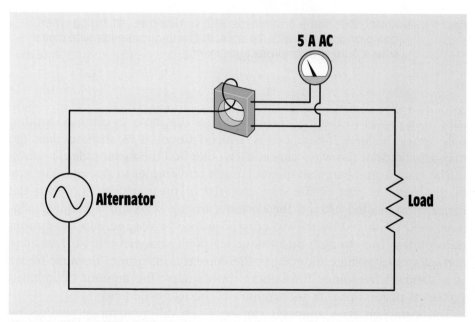

**Figure 9-31** The primary conductor loops through the CT to produce a second turn which changes the ratio.

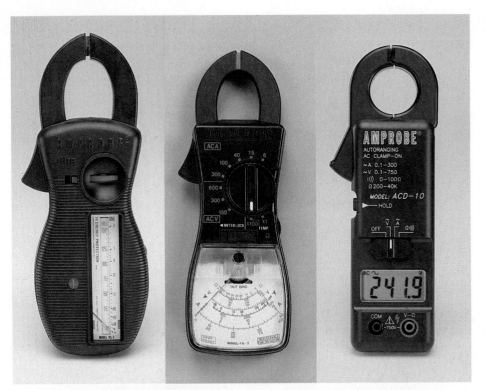

**Figure 9-32 (A)** Analog type clamp-on ammeter with vertical scale. **(B)** Analog type clamp-on ammeter with flat scale. **(C)** Clamp-on ammeter with digital scale *(Courtesy of Amprobe Instrument.)*

supplying power to the load *(Figure 9-33)*. The meter is clamped around only one of the lines. If the meter is clamped around more than one line, the magnetic fields of the wires cancel each other and the meter indicates zero.

The clamp-on meter also uses a current transformer to operate. The jaw of the meter is part of the core material of the transformer. When the meter is connected around the current-carrying wire, the changing magnetic field produced by the AC current induces a voltage into the current transformer. The strength and frequency of the magnetic field determine the amount of voltage induced in the current transformer. Because 60 Hz is a standard frequency throughout the country, the amount of induced voltage is proportional to the strength of the magnetic field.

The clamp-on type ammeter can be given different range settings by changing the turns ratio of the secondary of the transformer just as is done on the in-line ammeter. The primary of the transformer is the con-

**Figure 9-33** The clamp-on ammeter connects around only one conductor.

ductor around which the movable jaw is connected. If the ammeter is connected around one wire, the primary has one turn of wire compared with the turns of the secondary. The turns ratio can be changed in the same manner that the ratio of the CT is changed. If 2 turns of wire are wrapped around the jaw of the ammeter *(Figure 9-34)*, the primary winding now contains 2 turns instead of 1, and the turns ratio of the transformer is changed. The ammeter will now indicate double the amount of current in the circuit. The reading on the scale of the meter would have to be divided by 2 to get the correct reading. The ability to change the turns ratio of a clamp-on ammeter can be useful for measuring low currents. Changing the turns ratio is not limited to wrapping 2 turns of wire around the jaw of the ammeter. Any number of turns can be wrapped around the jaw of the ammeter, and the reading will be divided by that number.

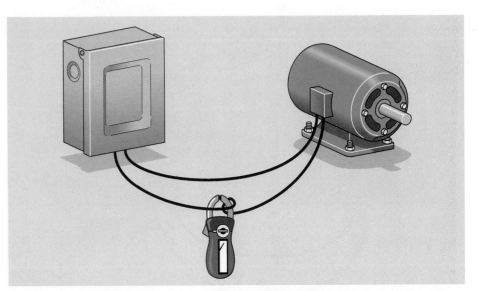

**Figure 9-34** Looping the conductor around the jaw of the ammeter changes the ratio.

## 9-11 DC-AC Clamp-on Ammeters

Most clamp-on ammeters that have the ability to measure both direct and alternating current do not operate on the principle of the current transformer. Current transformers depend on induction, which means that the current in the line must change direction periodically to provide a change of magnetic field polarity. It is the continuous change of field strength and direction that permits the current transformer to operate. The current in a DC circuit is unidirectional and does not change polarity, which would not permit the current transformer to operate.

DC-AC clamp-on ammeters *(Figure 9-35)* use the Hall effect as the basic principle of operation. The Hall effect was discovered by Edward H. Hall at Johns Hopkins University in 1879. Hall originally used a piece of pure gold to produce the Hall effect, but today a semiconductor material is used because it has better operating characteristics and is less expensive. The device is often referred to as a Hall generator. *Figure 9-36* illustrates the operating principle of the Hall generator. A constant-current generator is used to supply a continuous current to the semiconductor chip. The leads of a zero-center voltmeter are connected across the opposite sides of the chip. As long as the current flows through the center of the semiconductor chip, no potential difference or voltage develops across the chip.

If a magnetic field comes near the chip *(Figure 9-37),* the electron path is distorted and the current no longer flows through the center of the

**Figure 9-35** DC-AC clamp-on ammeter *(Courtesy of Amprobe Instrument.)*

chip. A voltage across the sides of the chip is produced. The voltage is proportional to the amount of current flow and the amount of current distortion. Since the current remains constant and the amount of distortion is proportional to the strength of the magnetic field, the voltage produced across the chip is proportional to the strength of the magnetic field.

If the polarity of the magnetic field were reversed *(Figure 9-38)*, the current path would be distorted in the opposite direction, producing a voltage of the opposite polarity. Notice that the Hall generator produces a voltage in the presence of a magnetic field. It makes no difference whether the field is moving or stationary. The Hall effect can, therefore, be used to measure direct or alternating current.

## 9-12 The Ohmmeter

The **ohmmeter** is used to measure resistance. The common VOM (volt-ohm-milliammeter) contains an ohmmeter. The ohmmeter has the

**ohmmeter**

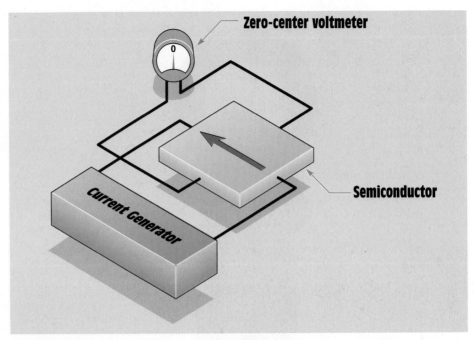

**Figure 9-36** Basic Hall generator

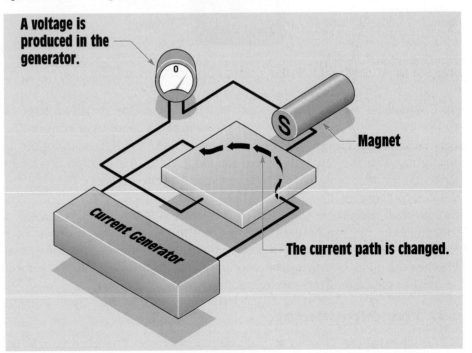

**Figure 9-37** The presence of a magnetic field causes the Hall generator to produce a voltage.

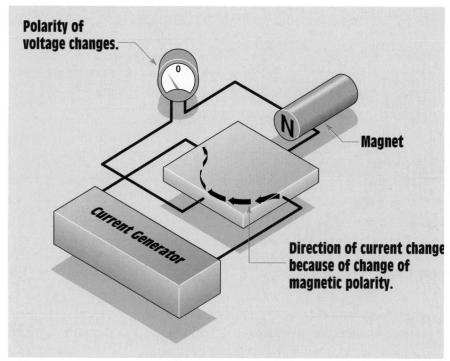

**Polarity of voltage changes.**

**Magnet**

**Current Generator**

**Direction of current change because of change of magnetic polarity.**

**Figure 9-38** If the magnetic field polarity changes, the polarity of the voltage changes.

only scale on a VOM that is nonlinear. The scale numbers increase in value as they progress from right to left. There are two basic types of analog ohmmeters, the series and the shunt. The series ohmmeter is used to measure high values of resistance, and the shunt type is used to measure low values of resistance. Regardless of the type used, the meter must provide its own power source to measure resistance. The power is provided by batteries located inside the instrument.

## The Series Ohmmeter

A schematic for a basic series ohmmeter is shown in *Figure 9-39*. It is assumed that the meter movement has a resistance of 1000 $\Omega$ and requires a current of 50 $\mu$A to deflect the meter full scale. The power source will be a 3-V battery. $R_1$, a fixed resistor with a value of 54 k$\Omega$, is connected in series with the meter movement, and $R_2$, a variable resistor with a value of 10 k$\Omega$, is connected in series with the meter and $R_1$. These resistance values were chosen to ensure there would be enough resistance in the circuit to limit the current flow through the meter movement to 50 $\mu$A. If Ohm's law is used to compute the resistance needed (3 V/0.000050 A = 60,000 $\Omega$), it will be seen that a value of 60 k$\Omega$ is needed.

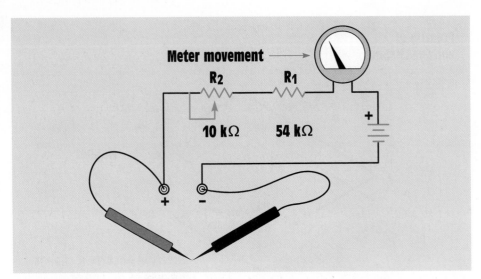

**Figure 9-39** Basic series ohmmeter

This circuit contains a total of 65,000 Ω (1000 [meter] + 54,000 + 10,000). The circuit resistance can be changed by adjusting the variable resistor to a value as low as 55,000 Ω, however, to compensate for the battery as it ages and becomes weaker.

When resistance is to be measured, the meter must first be zeroed. This is done with the ohms-adjust control, the variable resistor located on the front of the meter. To zero the meter, connect the leads *(Figure 9-39)* and turn the ohms-adjust knob until the meter indicates zero at the far right end of the scale *(Figure 9-40)*. When the leads are separated, the meter will again indicate infinity resistance at the left side of the scale. When the leads are connected across a resistance, the meter will again go up the scale. Since resistance has been added to the circuit, less than 50 μA of current will flow, and the meter will indicate some value other than zero. *Figure 9-41* shows a meter indicating a resistance of 2.5 Ω, assuming the range setting is Rx1.

Ohmmeters can have different range settings such as Rx1, Rx100, Rx1000, or Rx10,000. These different scales can be obtained by adding different values of resistance in the meter circuit and resetting the meter to zero. **An ohmmeter should always be readjusted to zero when the scale is changed.** On the Rx1 setting, the resistance is measured straight off the resistance scale located at the top of the meter. If the range is set for Rx1000, however, the reading must be multiplied by 1000. The ohmmeter reading shown in *Figure 9-41* would be indicating a resistance of 2,500 Ω if the range had been set for Rx1000. Notice that the ohmmeter scale is

**An ohmmeter should always be readjusted to zero when the scale is changed.**

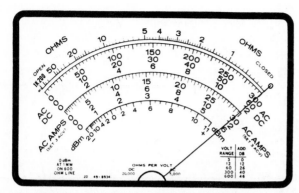

**Figure 9-40** Adjusting the ohmmeter to zero *(From Herman and Sparkman,* Electricity and Controls for HVAC, *2nd ed., copyright 1991 by Delmar Publishers Inc.)*

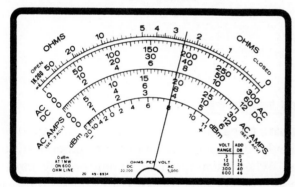

**Figure 9-41** Reading the ohmmeter *(From Herman and Sparkman,* Electricity and Controls for HVAC, *2nd ed., copyright 1991 by Delmar Publishers Inc.)*

read backward from the other scales. Zero ohms is located on the far right side of the scale, and maximum ohms is located at the far left side. It generally takes a little time and practice to read the ohmmeter properly.

## 9-13 Shunt Type Ohmmeters

The shunt type ohmmeter is used for measuring low values of resistance. It operates on the same basic principle as an ammeter shunt. When using a shunt type ohmmeter, place the unknown value of resistance in parallel with the meter movement. This placement causes part of the circuit current to bypass the meter *(Figure 9-42).*

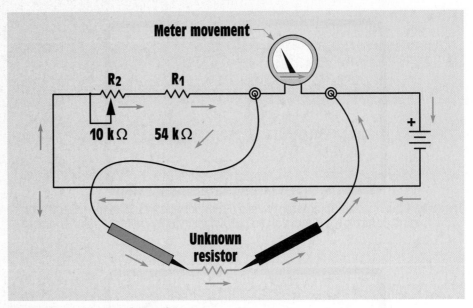

**Figure 9-42** Shunt type ohmmeter

## 9-14 Digital Meters

### Digital Ohmmeters

Digital ohmmeters display the resistance in figures instead of using a meter movement. When using a digital ohmmeter, care must be taken to notice the scale indication on the meter. For example, most digital meters will display a *K* on the scale to indicate kilohms or an *M* to indicate megohms (kilo means 1000 and mega means 1,000,000). If the meter is showing a resistance of 0.200 K, it means 0.200 x 1000, or 200 Ω. If the meter indicates 1.65 M, it means 1.65 x 1,000,000, or 1,650,000 Ω.

Appearance is not the only difference between analog and digital ohmmeters. Their operating principle is different also. Analog meters operate by measuring the amount of current change in the circuit when an unknown value of resistance is added. Digital ohmmeters measure resistance by measuring the amount of voltage drop across an unknown resistance. In the circuit shown in *Figure 9-43,* a constant-current generator is used to supply a known amount of current to a resistor, $R_x$. It will be assumed that the amount of current supplied is 1 mA. The voltage dropped across the resistor is proportional to the resistance of the resistor and the amount of current flow. For example, assume the value of the unknown resistor is 4700 Ω. The voltmeter would indicate a drop of 4.7 V when 1 mA of current flowed through the resistor. The scale factor of the

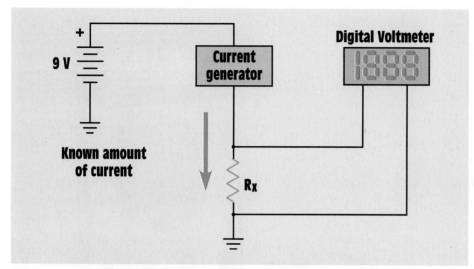

**Figure 9-43** Digital ohmmeters operate by measuring the voltage drop across a resistor when a known amount of current flows through it.

ohmmeter can be changed by changing the amount of current flow through the resistor. Digital ohmmeters generally exhibit an accuracy of about 1%.

**The ohmmeter, whether digital or analog, must never be connected to a circuit when the power is turned on.** Since the ohmmeter uses its own internal power supply, it has a very low operating voltage. Connecting a meter to power when it is set in the ohms position will probably damage or destroy the meter.

## Digital Multimeters

Digital multimeters have become increasingly popular in the past few years. The most apparent difference between digital meters and analog meters is that digital meters display their reading in discrete digits instead of with a pointer and scale. A digital multimeter is shown in *Figure 9-44*. Some digital meters have a range switch similar to the range switch used with analog meters. This switch sets the full range value of the meter. Many digital meters have voltage range settings from 200 mV to 2000 V. The lower ranges are used for accuracy. For example, assume it is necessary to measure a voltage of 16 V. The meter will be able to make a more accurate measurement when set on the 20-V range than when set on the 2000-V range.

Some digital meters do not contain a range setting control. These meters are known as autoranging meters. They contain a function control

The ohmmeter, whether digital or analog, must never be connected to a circuit when the power is turned on.

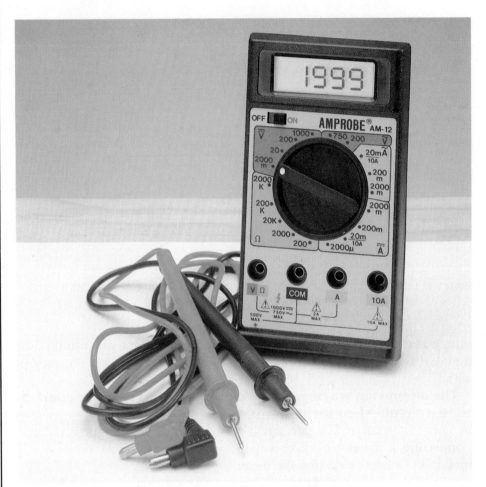

**Figure 9-44** Digital multimeter *(Courtesy of Amprobe Instrument.)*

switch that permits selection of the electrical quantity to be measured, such as AC volts, DC volts, ohms, and so on. When the meter probes are connected to the object to be tested, the meter automatically selects the proper range and displays the value.

Analog meters change scale value by inserting or removing resistance from the meter circuit *(Figure 9-10)*. The typical resistance of an analog meter is 20,000 $\Omega$ per volt for DC and 5000 $\Omega$ per volt for AC. If the meter is set for a full scale value of 60 V, there will be 1.2 M$\Omega$ of resistance connected in series with the meter if it is being used to measure DC (60 x 20,000 = 1,200,000) and 300 k$\Omega$ if it is being used to measure AC (60 x 5000 = 300,000). The impedance of the meter is of little concern if it is used to measure circuits that are connected to a high-current source. For example, assume the voltage of a 480-V panel is to be measured with a multimeter that has a resistance of 5000 Ω per volt. If the meter is set on the 600-V range, the resistance connected in series with the meter is 3 MΩ (600 x 5000 = 3,000,000). This resistance will permit a current of 160 μA to flow in the meter circuit (480/3,000,000 = 0.000160). This 160 μA of current is not enough to affect the circuit being tested.

Now assume that this meter is to be used to test a 24-V circuit that has a current flow of 100 μA. If the 60-V range is used, the meter circuit contains a resistance of 300 kΩ (60 x 5000 = 300,000). Therefore a current of 80 μA will flow when the meter is connected to the circuit (24/300,000 = 0.000080). The connection of the meter to the circuit has changed the entire circuit operation. This phenomenon is known as the *loading effect.*

Digital meters do not have a loading effect. Most digital meters have an input impedance of about 10 MΩ on all ranges. The input impendance is the ohmic value used to limit the flow of current through the meter. This impedance is accomplished by using field effect transistors (FETs) and a voltage divider circuit. A simple schematic for such a circuit is shown in *Figure 9-45.* Notice that the meter input is connected across 10 MΩ of resistance regardless of the range setting of the meter. If this meter is used to measure the voltage of the 24-V circuit, a current of 2.4 μA will flow through the meter. This is not enough current to upset the rest of the circuit, and voltage measurements can be made accurately.

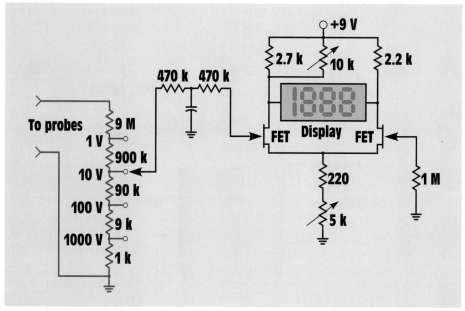

**Figure 9-45** Digital voltmeter

## 9-15 The Low-Impendance Voltage Tester

Another device used to test voltage is often referred to as a voltage tester. This device does measure voltage, but it does not contain a meter movement or digital display. It contains a coil and a plunger. The coil produces a magnetic field that is proportional to the amount of voltage to which the coil of the tester is connected. The higher the voltage to which the tester is connected, the stronger the magnetic field becomes. The plunger must overcome the force of a spring as it is drawn into the coil *(Figure 9-46)*. The plunger acts as a pointer to indicate the amount of voltage to which the tester is connected. The tester has an impedance of approximately 5000 Ω and can generally be used to measure voltages as high as 600 V. **The low-impedance voltage tester has a very large current draw compared with other types of voltmeters and should never be used to test low-power circuits.**

The relatively high current draw of the voltage tester can be an advantage when testing certain types of circuits, however, because it is not susceptible to giving the misleading voltage readings caused by high-impedance ground paths or feedback voltages that affect other types of voltmeters. An example of this advantage is shown in *Figure 9-47*. A transformer is used to supply power to a load. Notice that neither the out-

> **The low-impedance voltage tester has a very large current draw compared with other types of voltmeters and should never be used to test low-power circuits.**

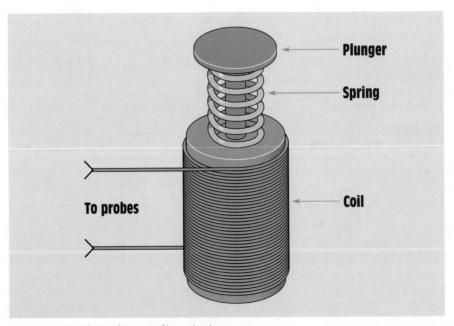

**Figure 9-46** Low-impedance voltage tester

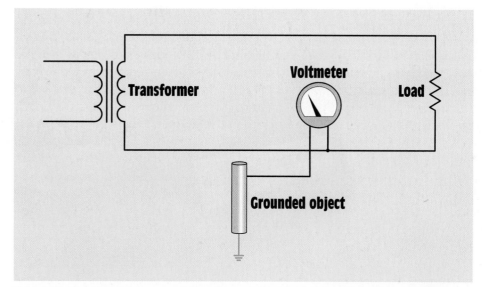

**Figure 9-47** High-impedance ground paths can produce misleading voltage reading.

put side of the transformer nor the load is connected to ground. If a high-impedance voltmeter is used to measure between one side of the transformer and a grounded point, it will most likely indicate some amount of voltage. That is because ground can act as a large capacitor and can permit a small amount of current to flow through the circuit created by the meter. This high-impedance ground path can support only a few microamps of current flow, but it is enough to operate the meter movement. If a voltage tester is used to make the same measurement, it will not show a voltage because there cannot be enough current flow to attract the plunger. A voltage tester is shown in *Figure 9-48*.

## 9-16 The Oscilloscope

Many of the electronic control systems in today's industry produce voltage pulses that are meaningless to a VOM. In many instances, it is necessary to know not only the amount of voltage present at a particular point, but also the length or duration of the pulse and its frequency. Some pulses may be less than one volt and last for only a millisecond. A VOM would be useless for measuring such a pulse. It is therefore necessary to use an oscilloscope to learn what is actually happening in the circuit.

The oscilloscope is a powerful tool in the hands of a trained technician. The first thing to understand is that an **oscilloscope** is a voltmeter. It does not measure current, resistance, or watts. The oscilloscope measures

**oscillo-scope**

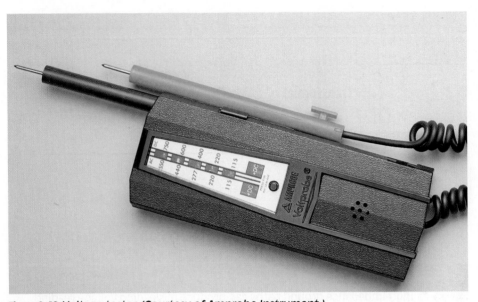

**Figure 9-48** Voltage tester *(Courtesy of Amprobe Instrument.)*

an amount of voltage during a period of time and produces a two-dimensional image.

## Voltage Range Selection

The oscilloscope is divided into two main sections. One section is the voltage section, and the other is the time base. The display of the oscilloscope is divided by vertical and horizontal lines *(Figure 9-49)*. Voltage is measured on the vertical, or *Y,* axis of the display, and time is measured on the horizontal, or *X,* axis. When using a VOM, a range selection switch is used to determine the full scale value of the meter. Ranges of 600 V, 300 V, 60 V, and 12 V are common. The ability to change ranges permits more-accurate measurements to be made. The oscilloscope has a voltage range selection switch also *(Figure 9-50)*. The voltage range selection switch on an oscilloscope selects volts per division instead of volts full scale. Assume that the voltage range switch shown in *Figure 9-50* is set for …at the 1X position. This means that each of the vertical lines on the *Y* axis of the display has a value of 10 mV. Assume the oscilloscope has been adjusted to permit 0 V to be shown on the center line of the display. If the oscilloscope probe were connected to a positive voltage of 30 mV, the trace would rise to the position shown in *Figure 9-51A*. If the probe were connected to a negative 30 mV, the trace will fall to the position shown in *Figure 9-51B*. Notice that the oscilloscope has the ability to display both a positive and a negative voltage. If the range switch were changed to 20 V per division, *Figure 9-51A* would be displaying 60 V positive.

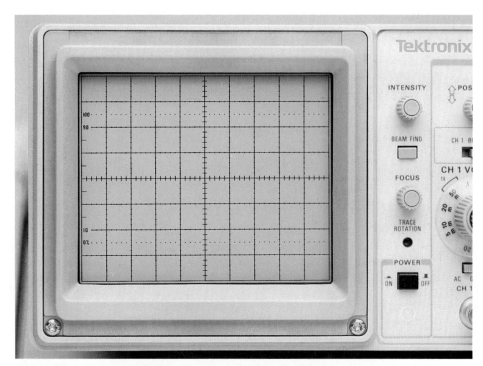

**Figure 9-49** Oscilloscope display *(Courtesy of Tektronix Inc.)*

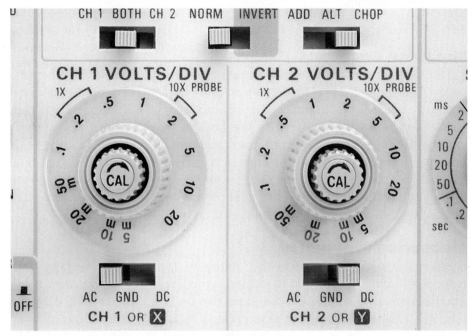

**Figure 9-50** Voltage control *(Courtesy of Tektronix Inc.)*

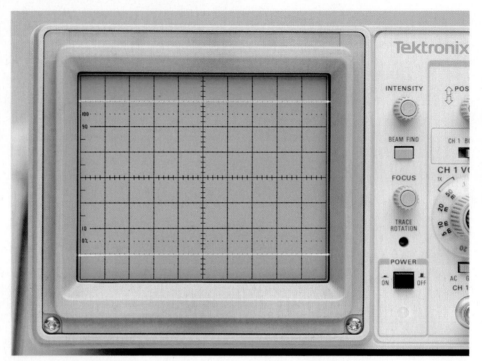

**Figure 9-51** The oscilloscope displays both positive and negative voltages *(Courtesy of Tektronix Inc.)*

### The Time Base

The next section of the oscilloscope to become familiar with is the time base *(Figure 9-52)*. The time base is calibrated in seconds per division and has range values from seconds to microseconds. The time base controls the value of the divisions of the lines in the horizontal direction (*X* axis). If the time base is set for 5 ms (milliseconds) per division, the trace will sweep from one division to the next division in 5 ms. With the time base set in this position, it will take 50 ms to sweep from one side of the display screen to the other. If the time base is set for 2 μs (microseconds) per division, the trace will sweep the screen in 20 μs.

### Measuring Frequency

Since the oscilloscope has the ability to measure the voltage with respect to time, it is possible to compute the frequency of the wave form. The frequency (F) of an AC wave form can be found by dividing 1 by the time (T) it takes to complete one cycle (F = 1/T). For example, assume the time base is set for 0.5 ms per division, and the voltage range is set

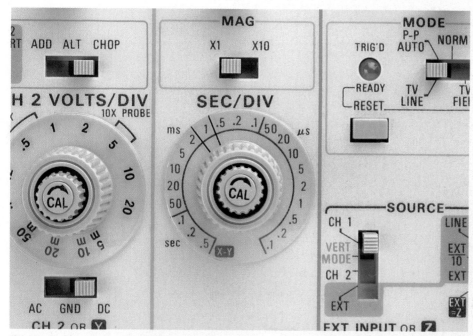

**Figure 9-52** Time base *(Courtesy of Tektronix Inc.)*

for 20 V per division. If the oscilloscope has been set so that the center line of the display is 0 V, the AC wave form shown in *Figure 9-53* has a peak value of 55 V. The oscilloscope displays the peak or peak-to-peak value of voltage and not the root mean square (RMS), or effective, value. These terms will be discussed in greater detail in later units.

To measure the frequency, count the time it takes to complete one full cycle. A cycle is one complete wave form. Begin counting when the wave form starts to rise in the positive direction and stop when it again starts to rise in the positive direction. Since the time base is set for a value of 0.5 ms per division, the wave form shown in *Figure 9-53* takes 4 ms to complete one full cycle. The frequency therefore is 250 Hz (1/0.004 = 250).

## Attenuated Probes

Most oscilloscopes use a probe that acts as an attenuator. An attenuator is a device that divides or makes smaller the input signal *(Figure 9-54)*. An attenuated probe is used to permit higher voltage readings than are normally possible. For example, most attenuated probes are 10 to 1. This means that if the voltage range switch is set for 5 V per division, the display would actually indicate 50 V per division. If the voltage range switch

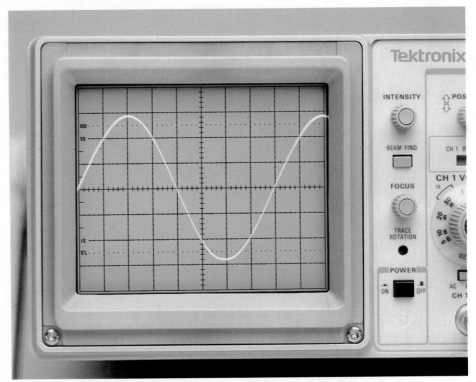

**Figure 9-53** AC sine wave *(Courtesy of Tektronix Inc.)*

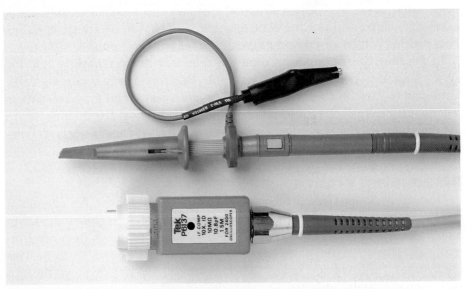

**Figure 9-54** An oscilloscope probe and attenuator *(Courtesy of Tektronix Inc.)*

is set for 2 V per division, each division on the display actually has a value of 20 V per division.

Probe attenuators are made in different styles by different manufacturers. On some probes the attenuator is located in the probe head itself, while on others the attenuator is located at the scope input. Regardless of the type of attenuated probe used, it may have to be compensated or adjusted. In fact, probe compensation should be checked frequently. Different manufacturers use different methods for compensating their probes, so it is generally necessary to follow the procedures given in the operator's manual for the oscilloscope being used.

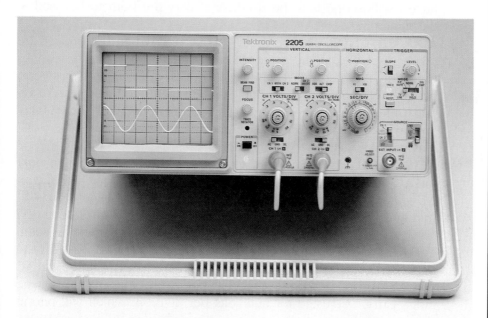

**Figure 9-55** An oscilloscope *(Courtesy of Tektronix Inc.)*

## Oscilloscope Controls

The following is a list of common controls found on the oscilloscope. Refer to the oscilloscope shown in *Figure 9-55*.

1. **Power.** The power switch is used to turn the oscilloscope on or off.
2. **Beam finder.** This control is used to locate the position of the trace if it is off the display. The beam finder button will indicate the approximate location of the trace. The position controls are then used to move the trace back on the display.
3. **Probe adjust** (sometimes called calibrate). This is a reference voltage point used to compensate the probe. Most probe adjust points produce a square wave signal of about 0.5 V.

4. **Intensity and focus.** The intensity control adjusts the brightness of the trace. A bright spot should never be left on the display because it will burn a spot on the face of the CRT (cathode ray tube). This burned spot results in permanent damage to the CRT. The focus control sharpens the image of the trace.

5. **Vertical position.** This is used to adjust the trace up or down on the display. If a dual-trace oscilloscope is being used, there will be two vertical position controls. (A dual-trace oscilloscope contains two separate traces that can be used separately or together.)

6. **CH 1-both-CH 2.** This control determines which channel of a dual-trace oscilloscope is to be used, or if they are both to be used at the same time.

7. **ADD-ALT.-CHOP.** This control is active only when both traces are being displayed at the same time. The ADD adds the two waves together. ALT. stands for alternate. This alternates the sweep between channel 1 and channel 2. The CHOP mode alternates several times during one sweep. This generally makes the display appear more stable. The chop mode is generally used when displaying two traces at the same time.

8. **AC-GRD-DC.** The AC is used to block any DC voltage when only the AC portion of the voltage is to be seen. For instance, assume an AC voltage of a few millivolts is riding on a DC voltage of several hundred volts. If the voltage range is set high enough so that 100 VDC can be seen on the display, the AC voltage cannot be seen. The AC section of this switch inserts a capacitor in series with the probe. The capacitor blocks the DC voltage and permits the AC voltage to pass. Since the 100 VDC has been blocked, the voltage range can be adjusted for millivolts per division, which will permit the AC signal to be seen.

   The GRD section of the switch stands for ground. This section grounds the input so the sweep can be adjusted for 0 V at any position on the display. The ground switch grounds at the scope and does not ground the probe. This permits the ground switch to be used when the probe is connected to a live circuit. The DC section permits the oscilloscope to display all of the voltage, both AC and DC, connected to the probe.

9. **Horizontal position.** This control adjusts the position of the trace from left to right.

10. **Auto-Normal.** This determines whether the time base will be triggered automatically or operated in a free-running mode. If this control is operated in the normal setting, the trigger signal is taken from the line to which the probe is connected. The scope is gener-

ally operated with the trigger set in the automatic position.

11. **Level.** The level control determines the amplitude the signal must be before the scope triggers.

12. **Slope.** The slope permits selection as to whether the trace is triggered by a negative or positive wave form.

13. **Int.-Line-Ext.** The *Int.* stands for internal. The scope is generally operated in this mode. In this setting, the trigger signal is provided by the scope. In the line mode, the trigger signal is provided from a sample of the line. The *Ext.*, or external, mode permits the trigger pulse to be applied from an external source.

These are not all the controls shown on the oscilloscope in *Figure 9-55*, but they are the major controls. Most oscilloscopes contain these controls.

## Interpreting Wave Forms

The ability to interpret the wave forms on the display of the oscilloscope takes time and practice. When using the oscilloscope, one must keep in mind that the display shows the voltage with respect to time.

In *Figure 9-56*, it is assumed that the voltage range has been set for 0.5 V per division, and the time base is set for 2 ms per division. It is also

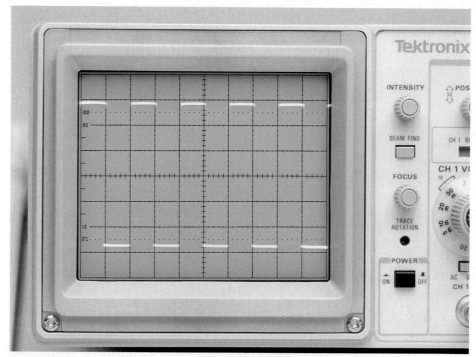

**Figure 9-56** An AC square wave *(Courtesy of Tektronix Inc.)*

assumes that 0 V has been set on the center line of the display. The wave form shown is a square wave. The display shows that the voltage rises in the positive direction to a value of 1.4 V and remains there for 2 ms. The voltage then drops to 1.4 V negative and remains there for 2 ms before going back to positive. Since the voltage changes between positive and negative, it is an AC voltage. The length of one cycle is 4 ms. The frequency is therefore 250 Hz (1/0.004 = 250).

In *Figure 9-57*, the oscilloscope has been set for 50 mV per division and 20 μs (microseconds) per division. The display shows a voltage that is negative to the probe's ground lead and has a peak value of 150 mV. The wave form lasts for 20 μs and produces a frequency of 50 kHz (1/0.000020 = 50,000). The voltage is DC since it never crosses the zero reference and goes in the positive direction. This type of voltage is called *pulsating DC.*

In *Figure 9-58*, assume the scope has been set for a value of 50 V per division and 5 ms per division. The wave form shown rises from 0 to about 45 V in a period of about 1.5 ms. The voltage gradually increases to about 50 V in the space of 1 ms and then rises to a value of about 100 V in the next 2 ms. The voltage then decreases to 0 in the next 4 ms. It then increases to a value of about 10 V in 0.5 ms and remains at that level for

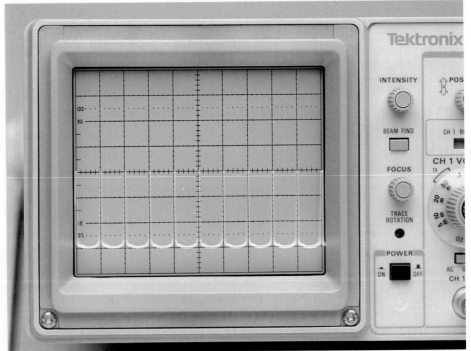

**Figure 9-57** A DC wave form *(Courtesy of Tektronix Inc.)*

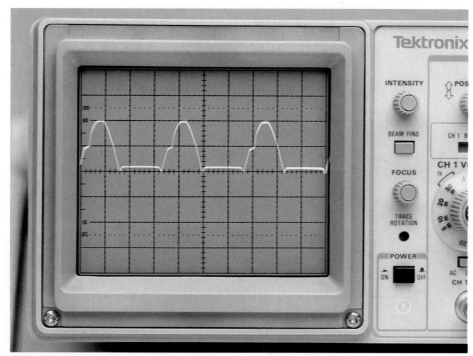

**Figure 9-58** A chopped DC wave form *(Courtesy of Tektronix Inc.)*

about 8 ms. This is one complete cycle for the wave form. The length of one cycle is about 16.6 ms, which is a frequency of 60.2 Hz (1/0.0166 = 60.2). The voltage is DC because it remains positive and never drops below the 0 line.

Learning to interpret the wave forms seen on the display of an oscilloscope will take time and practice, but it is well worth the effort. The oscilloscope is the only means by which many of the wave forms and voltages found in electronic circuits can be understood. Consequently, the oscilloscope is the single most valuable piece of equipment a technician can use.

## 9-17 The Wattmeter

The **wattmeter** is used to measure the power in a circuit. This meter differs from the d'Arsonval type of meter in that it does not contain a permanent magnet. This meter contains a set of electromagnets and a moving coil *(Figure 9-59)*. The electromagnets are connected in series with the load in the same manner that an ammeter is connected. The moving coil has resistance connected in series with it and is connected

**wattmeter**

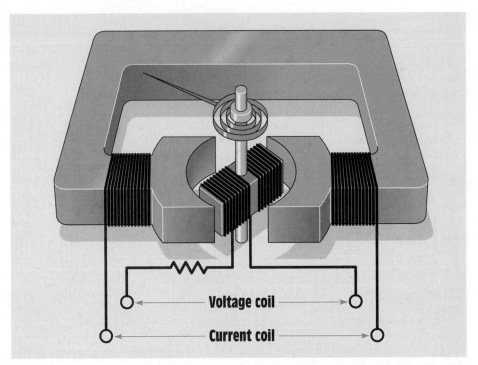

**Figure 9-59** The wattmeter contains two coils—one for voltage and the other for current.

directly across the power source in the same manner as a voltmeter *(Figure 9-60)*.

Since the electromagnet is connected in series with the load, the current flow through the load determines the magnetic field strength of the stationary magnet. The magnetic field strength of the moving coil is determined by the amount of line voltage. The turning force of the coil is proportional to the strength of these two magnetic fields. The deflection of the meter against the spring is proportional to the amount of current flow and voltage.

Since the wattmeter contains an electromagnet instead of a permanent magnet, the polarity of the magnetic field is determined by the direction of current flow. The same is true of the polarity of the moving coil connected across the source of voltage. If the wattmeter is connected into an AC circuit, the polarity of the two coils will reverse at the same time, producing a continuous torque. For this reason, the wattmeter can be used to measure power in either a direct or an alternating current circuit. However, if the connection of the stationary coil or the moving coil is reversed, the meter will attempt to read backward.

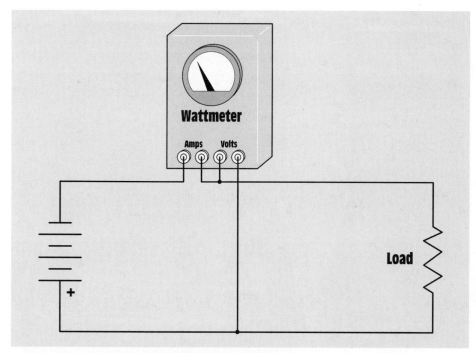

**Figure 9-60** The current section of the wattmeter is connected in series with the load, and the voltage section is connected in parallel with the load.

## 9-18 Recording Meters

On occasion it becomes necessary to make a recording of an electrical value over a long period of time. Recording meters produce a graph of metered values during a certain length of time. They are used to detect spike voltages, or currents of short duration, or sudden drops in voltage, current, or power. Recording meters can show the amount of voltage or current, its duration, and the time of occurrence. Some meters have the ability to store information in memory over a period of several days. This information can be recalled later by the service technician. Several types of recording meters are shown in *Figure 9-61*. The meter shown in *Figure 9-61A* is a single-line-recording volt-ammeter. It will record voltage or current or both for a single phase. A kilowatt-kiloVARs recording meter is shown in *Figure 9-61B*. This meter will record true power (kilowatts) or reactive power (kiloVARs) on a time-share basis. It can be used on single- or three-phase circuits and can also be used to determine circuit power factors. A chartless recorder is shown in *Figure 9-61C*. This instrument can record voltages and currents on single- or three-phase lines. The readings can be stored in memory for as long as 41 days.

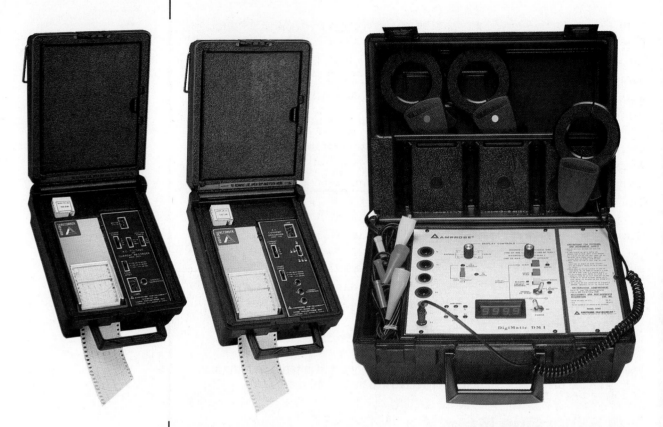

**Figure 9-61 (A)** Single-line-recording volt-ammeter. **(B)** Kilowatt/kiloVARs recorder **(C)** Chartless recorder *(Courtesy of Amprobe Instrument.)*

### 9-19 Bridge Circuits

**bridge circuit**

**Wheatstone bridge**

One of the most common devices used to measure values of resistance, inductance, and capacitance accurately is a **bridge circuit**. A bridge is constructed by connecting four components to form a parallel-series circuit. All four components are of the same type, such as four resistors, four inductors, or four capacitors. The bridge used to measure resistance is called a **Wheatstone bridge**. The basic circuit for a Wheatstone bridge is shown in *Figure 9-62.* The bridge operates on the principle that the sum of the voltage drops in a series circuit must equal the applied voltage. A galvanometer is used to measure the voltage between points B and D. The galvanometer can be connected to different values of resistance or directly between points B and D. Values of resistance are used to determine the sensitivity of the meter circuit. When the meter is connected directly across the two points its sensitivity is maximum.

In *Figure 9-62,* assume the battery has a voltage of 12 V and that resistors $R_1$ and $R_2$ are precision resistors and have the same value of resistance. Since resistors $R_1$ and $R_2$ are connected in series and have the same value, each will have a voltage drop equal to one-half of the applied voltage, or 6 V. This means that point B is 6 V more negative than point A and 6 V more positive than point C.

Resistors $R_V$ (variable) and $R_X$ (unknown) are connected in series with each other. Resistor $R_X$ represents the unknown value of resistance to be measured. Resistor $R_V$ can be adjusted for different resistive values. If the value of $R_V$ is greater than the value of $R_X$, the voltage at point D will be more positive than the voltage at point B. This will cause the pointer of the zero-center galvanometer to move in one direction. If the value of $R_V$ is less than $R_X$, the voltage at point D will be more negative than the voltage at point B, causing the pointer to move in the opposite direction. When the value of $R_V$ becomes equal to that of $R_X$, the voltage at point D will become equal to the voltage at point B. When this occurs, the galvanometer will indicate zero. A Wheatstone bridge is shown in *Figure 9-63.*

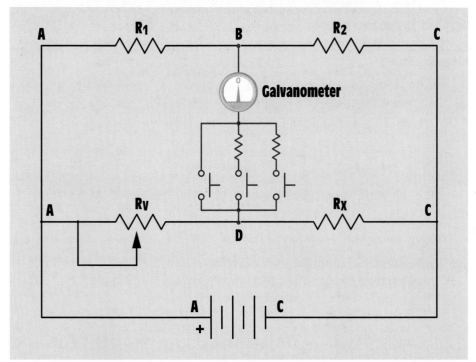

**Figure 9-62** The Wheatstone bridge circuit is used to make accurate measurements of resistance and operates on the principle that the sum of the voltage drops in a series circuit must equal the applied voltage.

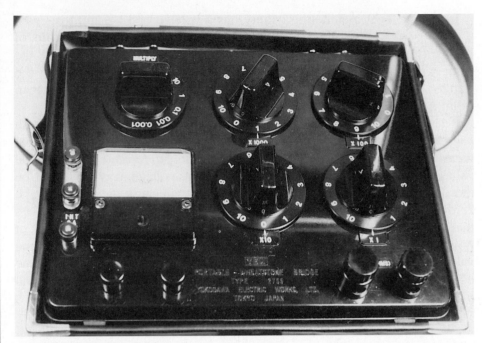

**Figure 9-63** Wheatstone bridge

## Summary

1. The d'Arsonval type of meter movement is based on the principle that like magnetic fields repel.

2. The d'Arsonval movement operates only on DC current.

3. There are other types of meter movements known as moving iron type of meters.

4. The moving vane type of movement will operate on AC current.

5. Voltmeters have a high resistance and are designed to be connected directly across the power line.

6. The steps to reading a meter are:
   A. Determine what quantity the meter is set to measure.
   B. Determine the full range value of the meter.
   C. Read the meter.

7. Ammeters have a low resistance and must be connected in series with a load to limit the flow of current.

8. Shunts are used to change the value of DC ammeters.

9. AC ammeters use a current transformer to change the range setting.

10. Clamp-on ammeters measure the flow of current by measuring the strength of the magnetic field around a conductor.

11. Ohmmeters are used to measure the resistance in a circuit.

12. Ohmmeters contain an internal power source, generally batteries.

13. Ohmmeters must never be connected to a circuit that has power applied to it.

14. Digital multimeters display their value in digits instead of using a meter movement.

15. Digital multimeters generally have an input impedance of 10 MΩ on all ranges.

16. The oscilloscope measures the amplitude of voltage with respect to time.

17. The frequency of a wave form can be determined by dividing 1 by the time of one cycle (F = 1/T).

18. Wattmeters contain a stationary coil and a movable coil.

19. The stationary coil of a wattmeter is connected in series with the load, and the moving coil is connected to the line voltage.

20. The turning force of the wattmeter is proportional to the strength of the magnetic field of the stationary coil and the strength of the magnetic field of the moving coil.

21. Digital ohmmeters measure resistance by measuring the voltage drop across an unknown resistor when a known amount of current flows through it.

22. Low-impedance voltage testers are not susceptible to indicating a voltage caused by a high-impedance ground or a feedback.

23. A bridge circuit can be used to accurately measure values of resistance, inductance, and capacitance.

## Review Questions

1. To what is the turning force of a d'Arsonval meter movement proportional?

2. What type of voltage must be connected to a d'Arsonval meter movement?

3. A DC voltmeter has a resistance of 20,000 Ω per volt. What is the

resistance of the meter if the range selection switch is set on the 250-V range?

4. What is the purpose of an ammeter shunt?

5. Name two methods used to make a DC multirange ammeter.

6. How is an ammeter connected into a circuit?

7. How is a voltmeter connected into a circuit?

8. An ammeter shunt has a voltage drop of 50 mV when 50 A of current flow through it. What is the resistance of the shunt?

9. What type of meter contains its own separate power source?

10. What electrical quantity does the oscilloscope measure?

11. What is measured on the *Y* axis of an oscilloscope?

12. What is measured on the *X* axis of an oscilloscope?

13. A wave form shown on the display of an oscilloscope completes one cycle in 50 μs. What is the frequency of the wave form?

14. What is the major difference between a wattmeter and a d'Arsonval meter?

15. What two factors determine the turning force of a wattmeter?

**Practice Problems**

## Measuring Instruments

1. A d'Arsonval meter movement has a full scale current value of 100 μA. (0.000100 A) and a resistance of 5 kΩ (5000 Ω). What size resistor must be placed in series with this meter to permit it to indicate 10 V full scale?

2. The meter movement described in question 1 is to be used to construct a multirange voltmeter. The meter is to have voltage ranges of 15 V, 60 V, 150 V, and 300 V *(Figure 9-64)*. Find the values of resistors $R_1$, $R_2$, $R_3$, and $R_4$.

3. A meter movement has a full scale value of 500 μA (0.000500 A) and 50 mV (0.050 V). A shunt is to be connected to the meter that permits it to have a full scale current value of 2 A. What is the resistance of the shunt?

4. The meter movement in question 3 is to be used as a multirange ammeter. An Ayrton shunt is to be used to provide full scale current

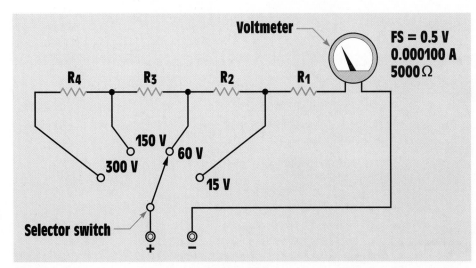

**Figure 9-64** The multirange voltmeter operates by connecting different values of resistance in series with the meter movement.

ranges of 5 A, 1 A, and 0.5 A *(Figure 9-65)*. Find the values of resistors $R_1$, $R_2$, and $R_3$.

5. A digital voltmeter indicates a voltage of 2.5 V when 10 $\mu$A of current flows through a resistor. What is the resistance of the resistor?

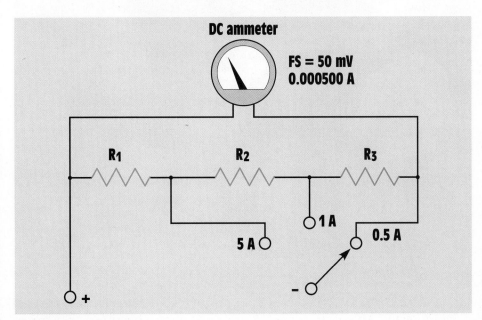

**Figure 9-65** Ayrton shunt

This unit discusses how wire sizes are determined for a particular application. Two methods are presented. One method is to use the National Electrical Code and the other determines the resistance of different types and lengths of wire.

Courtesy of Niagara Mohawk Power Corp.

# Using Wire Tables and Determining Conductor Sizes

## Key Terms

Ambient air temperature

American Wire Gauge (AWG)

Ampacities (current-carrying ability)

Circular mil

Correction factor

Damp locations

Dry locations

Insulation

Maximum operating temperature

MEGGER®

Mil foot

*National Electrical Code®*

Parallel conductors

Voltage drop

Wet locations

**A**fter studying this unit, you should be able to:

- Select a conductor from the proper wire table.
- Discuss the different types of wire insulation.
- Determine insulation characteristics.
- Use correction factors to determine the proper ampacity rating of conductors.
- Determine the resistance of long lengths of conductors.
- Determine the proper wire size for loads located long distances from the power source.
- List the requirements for using parallel conductors.
- Discuss the use of a MEGGER® for testing insulation.

**Preview**

**National Electrical Code®**

The size of the conductor needed for a particular application can be determined by several methods. The ***National Electrical Code*** is used throughout industry to determine the conductor size for most applications. It is imperative that an electrician be familiar with code tables and correction factors. In some circumstances, however, wire tables cannot be used, as in the case of extremely long wire runs or for windings of a transformer or motor. In these instances the electrician should know how to determine the conductor size needed by computing maximum voltage drop and resistance of the conductor.

### 10-1 Using the *National Electrical Code* Charts

Section 310 of the *National Electrical Code* deals with conductors for general wiring. Tables 310-16 through 310-19 are generally used to select a wire size according to the requirements of the circuit. Each of these tables lists different conditions. The table used is determined by the wiring conditions. Table 310-16 *(Figure 10-1)* lists **ampacities (current-carrying ability)** of not more than three single insulated conductors in raceway or cable or buried in the earth based on an **ambient** (surrounding) **air temperature** of 30°C (86°F). Table 310-17 *(Figure 10-2)* lists ampacities of single insulated conductors in free air based on an ambient temperature of 30°C. Table 310-18 *(Figure 10-3)* lists the ampacities of three single insulated conductors in raceway or cable based on an ambient temperature of 40°C (104°F). The conductors listed in Tables 310-18 and 310-19 are generally used for high-temperature locations. The heading at the top of each table lists a different set of conditions.

## 10-2 Factors That Determine Ampacity

### Conductor Material

One of the factors that determines the resistivity of wire is the material from which the wire is made. The wire tables list the current-carrying capacity of both copper and aluminum or copper-clad aluminum conductors. The currents listed in the left-hand half of Table 310-16, for example, are for copper wire. The currents listed in the right-hand half of the table are for aluminum or copper-clad aluminum. The table indicates that a copper conductor is permitted to carry more current than an aluminum conductor of the same size and insulation type. A #8 **AWG (American Wire Gauge)** copper conductor with type TW insulation is rated to carry a maximum of 40 A. A #8 AWG aluminum conductor with type TW insulation is rated to carry only 30 A. One of the columns of Tables 310-18 and 310-19 gives the ampacity rating of nickel or nickel-coated copper conductors.

### Insulation Type

Another factor that determines the amount of current a conductor is permitted to carry is the type of insulation used. This is due to the fact that different types of insulation can withstand more heat than others. The **insulation** is the nonconductive covering around the wire as shown in *Figure 10-4*. The voltage rating of the conductor is also determined by the type of insulation. The amount of voltage a particular type of insulation can withstand without breaking down is determined by the type of material it is made of and its thickness. Table 310-13 lists different types of insulation and certain specifications about each one. The table is divided into seven columns. The first column lists the trade name of the insulation; the second, its identification code letter; the third, its **maximum operating temperature**; the fourth, its applications and where it is permitted to

**ampacities (current-carrying ability)**

**ambient air temperature**

**AWG (American Wire Gauge)**

**insulation**

**maximum operating temperature**

**Table 310-16. Allowable Ampacities of Insulated Conductors,**
**Rated 0-2000 Volts, 60° to 90°C (140° to 194°F)**
**Not More Than Three Conductors in Raceway or Cable or Earth**
**(Directly Buried), Based on Ambient Temperature of 30°C (86°F)**

| Size | Temperature Rating of Conductor. See Table 310-13. | | | | | | Size |
|---|---|---|---|---|---|---|---|
| | 60°C (140°F) | 75°C (167°F) | 90°C (194°F) | 60°C (140°F) | 75°C (167°F) | 90°C (194°F) | |
| AWG kcmil | TYPES TW†  UF† | TYPES FEPW†, RH†, RHW†, THHW†, THW†, THWN†, XHHW†, USE†, ZW† | TYPES TA, TBS, SA, SIS, FEP†, FEPB†, MI, RHH†, RHW-2, THHN†, THHW†, THW-2, THWN-2, USE-2, XHH, XHHW†, XHHW-2, ZW-2 | TYPES TW†  UF† | TYPES RH†, RHW†, THHW†, THW†, THWN†, XHHW†, USE† | TYPES TA, TBS, SA, SIS, THNN†, THHW†, THW-2, THWN-2, RHH†, RHW-2, USE-2, XHH, XHHW, XHHW-2, ZW-2 | AWG kcmil |
| | **COPPER** | | | **ALUMINUM OR COPPER-CLAD ALUMINUM** | | | |
| 18 | . . . . | . . . . | 14 | . . . . | . . . . | . . . . | . . . . |
| 16 | . . . . | . . . . | 18 | . . . . | . . . . | . . . . | . . . . |
| 14 | 20† | 20† | 25† | . . . . | . . . . | . . . . | . . . . |
| 12 | 25† | 25† | 30† | 20† | 20† | 25† | 12 |
| 10 | 30 | 35† | 40† | 25 | 30† | 35† | 10 |
| 8 | 40 | 50 | 55 | 30 | 40 | 45 | 8 |
| 6 | 55 | 65 | 75 | 40 | 50 | 60 | 6 |
| 4 | 70 | 85 | 95 | 55 | 65 | 75 | 4 |
| 3 | 85 | 100 | 110 | 65 | 75 | 85 | 3 |
| 2 | 95 | 115 | 130 | 75 | 90 | 100 | 2 |
| 1 | 110 | 130 | 150 | 85 | 100 | 115 | 1 |
| 1/0 | 125 | 150 | 170 | 100 | 120 | 135 | 1/0 |
| 2/0 | 145 | 175 | 195 | 115 | 135 | 150 | 2/0 |
| 3/0 | 165 | 200 | 225 | 130 | 155 | 175 | 3/0 |
| 4/0 | 195 | 230 | 260 | 150 | 180 | 205 | 4/0 |
| 250 | 215 | 255 | 290 | 170 | 205 | 230 | 250 |
| 300 | 240 | 285 | 320 | 190 | 230 | 255 | 300 |
| 350 | 260 | 310 | 350 | 210 | 250 | 280 | 350 |
| 400 | 280 | 335 | 380 | 225 | 270 | 305 | 400 |
| 500 | 320 | 380 | 430 | 260 | 310 | 350 | 500 |
| 600 | 355 | 420 | 475 | 285 | 340 | 385 | 600 |
| 700 | 385 | 460 | 520 | 310 | 375 | 420 | 700 |
| 750 | 400 | 475 | 535 | 320 | 385 | 435 | 750 |
| 800 | 410 | 490 | 555 | 330 | 395 | 450 | 800 |
| 900 | 435 | 520 | 585 | 355 | 425 | 480 | 900 |
| 1000 | 455 | 545 | 615 | 375 | 445 | 500 | 1000 |
| 1250 | 495 | 590 | 665 | 405 | 485 | 545 | 1250 |
| 1500 | 520 | 625 | 705 | 435 | 520 | 585 | 1500 |
| 1750 | 545 | 650 | 735 | 455 | 545 | 615 | 1750 |
| 2000 | 560 | 665 | 750 | 470 | 560 | 630 | 2000 |
| **AMPACITY CORRECTION FACTORS** | | | | | | | |
| Ambient Temp. °C | For ambient temperatures other than 30°C (86°F), multiply the allowable ampacities shown above by the appropriate factor shown below. | | | | | | Ambient Temp. °F |
| 21-25 | 1.08 | 1.05 | 1.04 | 1.08 | 1.05 | 1.04 | 70-77 |
| 26-30 | 1.00 | 1.00 | 1.00 | 1.00 | 1.00 | 1.00 | 78-86 |
| 31-35 | .91 | .94 | .96 | .91 | .94 | .96 | 87-95 |
| 36-40 | .82 | .88 | .91 | .82 | .88 | .91 | 96-104 |
| 41-45 | .71 | .82 | .87 | .71 | .82 | .87 | 105-113 |
| 46-50 | .58 | .75 | .82 | .58 | .75 | .82 | 114-122 |
| 51-55 | .41 | .67 | .76 | .41 | .67 | .76 | 123-131 |
| 56-60 | . . . . | .58 | .71 | . . . . | .58 | .71 | 132-140 |
| 61-70 | . . . . | .33 | .58 | . . . . | .33 | .58 | 141-158 |
| 71-80 | . . . . | . . . . | .41 | . . . . | . . . . | .41` | 159-176 |

† Unless otherwise specifically permitted elsewhere in this Code, the overcurrent protection for conductor types marked with an obelisk (†) shall not exceed 15 amperes for No. 14, 20 amperes for No. 12, and 30 amperes for No. 10 copper; or 15 amperes for No. 12 and 25 amperes for No. 10 aluminum and copper-clad aluminum after any correction factors for ambient temperature and number of conductors have been applied.

**Figure 10-1** *NEC* Table 310-16. *(Reprinted with permission from NFPA 70-1993, the National Electrical Code®, Copyright © 1993, National Fire Protection Association, Quincy, MA 02269. This reprinted material is not the complete and official position of the National Fire Protection Association, on the referenced subject which is represented only by the standard in its entirety.)*

**Table 310-17. Allowable Ampacities of Insulated Conductors,
Rated 0 through 2000 Volts, In Free Air
Based on Ambient Temperature of 30°C (86°F)**

| Size | Temperature Rating of Conductor. See Table 310-13. | | | | | | Size |
|---|---|---|---|---|---|---|---|
| | 60°C (140°F) | 75°C (167°F) | 90°C (194°F) | 60°C (140°F) | 75°C (167°F) | 90°C (194°F) | |
| AWG kcmil | TYPES TW†, UF† | TYPES FEPW†, RH†, RHW†, THHW†, THW†, THWN†, XHHW†, ZW† | TYPES TA, TBS, SA, SIS, FEP†, FEPB†, MI, RHH†, RHW-2, THHN†, THHW†, THW-2, THWN-2, USE-2, XHH, XHHW†, XHHW-2, ZW-2 | TYPES TW†, UF† | TYPES RH†, RHW†, THHW†, THW†, THWN†, XHHW† | TYPES TA, TBS, SA, SIS, THNN†, THHW†, THW-2, THWN-2, RHH†, RHW-2, USE-2, XHH, XHHW†, XHHW-2, ZW-2 | AWG kcmil |
| | **COPPER** | | | **ALUMINUM OR COPPER-CLAD ALUMINUM** | | | |
| 18 | .... | .... | 18 | .... | .... | .... | .... |
| 16 | .... | .... | 24 | .... | .... | .... | .... |
| 14 | 25† | 30† | 35† | .... | .... | .... | .... |
| 12 | 30† | 35† | 40† | 25† | 30† | 35† | 12 |
| 10 | 40† | 50† | 55† | 35† | 40† | 40† | 10 |
| 8 | 60 | 70 | 80 | 45 | 55 | 60 | 8 |
| 6 | 80 | 95 | 105 | 60 | 75 | 80 | 6 |
| 4 | 105 | 125 | 140 | 80 | 100 | 110 | 4 |
| 3 | 120 | 145 | 165 | 95 | 115 | 130 | 3 |
| 2 | 140 | 170 | 190 | 110 | 135 | 150 | 2 |
| 1 | 165 | 195 | 220 | 130 | 155 | 175 | 1 |
| 1/0 | 195 | 230 | 260 | 150 | 180 | 205 | 1/0 |
| 2/0 | 225 | 265 | 300 | 175 | 210 | 235 | 2/0 |
| 3/0 | 260 | 310 | 350 | 200 | 240 | 275 | 3/0 |
| 4/0 | 300 | 360 | 405 | 235 | 280 | 315 | 4/0 |
| 250 | 340 | 405 | 455 | 265 | 315 | 355 | 250 |
| 300 | 375 | 445 | 505 | 290 | 350 | 395 | 300 |
| 350 | 420 | 505 | 570 | 330 | 395 | 445 | 350 |
| 400 | 455 | 545 | 615 | 355 | 425 | 480 | 400 |
| 500 | 515 | 620 | 700 | 405 | 485 | 545 | 500 |
| 600 | 575 | 690 | 780 | 455 | 540 | 615 | 600 |
| 700 | 630 | 755 | 855 | 500 | 595 | 675 | 700 |
| 750 | 655 | 785 | 885 | 515 | 620 | 700 | 750 |
| 800 | 680 | 815 | 920 | 535 | 645 | 725 | 800 |
| 900 | 730 | 870 | 985 | 580 | 700 | 785 | 900 |
| 1000 | 780 | 935 | 1055 | 625 | 750 | 845 | 1000 |
| 1250 | 890 | 1065 | 1200 | 710 | 855 | 960 | 1250 |
| 1500 | 980 | 1175 | 1325 | 795 | 950 | 1075 | 1500 |
| 1750 | 1070 | 1280 | 1445 | 875 | 1050 | 1185 | 1750 |
| 2000 | 1155 | 1385 | 1560 | 960 | 1150 | 1335 | 2000 |

| **AMPACITY CORRECTION FACTORS** | | | | | | | |
|---|---|---|---|---|---|---|---|
| Ambient Temp. °C | For ambient temperatures other than 30°C (86°F), multiply the allowable ampacities shown above by the appropriate factor shown below. | | | | | | Ambient Temp. °F |
| 21-25 | 1.08 | 1.05 | 1.04 | 1.08 | 1.05 | 1.04 | 70-77 |
| 26-30 | 1.00 | 1.00 | 1.00 | 1.00 | 1.00 | 1.00 | 78-86 |
| 31-35 | .91 | .94 | .96 | .91 | .94 | .96 | 87-95 |
| 36-40 | .82 | .88 | .91 | .82 | .88 | .91 | 96-104 |
| 41-45 | .71 | .82 | .87 | .71 | .82 | .87 | 105-113 |
| 46-50 | .58 | .75 | .82 | .58 | .75 | .82 | 114-122 |
| 51-55 | .41 | .67 | .76 | .41 | .67 | .76 | 123-131 |
| 56-60 | .... | .58 | .71 | .... | .58 | .71 | 132-140 |
| 61-70 | .... | .33 | .58 | .... | .33 | .58 | 141-158 |
| 71-80 | .... | .... | .41 | .... | .... | .41 | 159-176 |

† Unless otherwise specifically permitted elsewhere in this Code, the overcurrent protection for conductor types marked with an obelisk (†) shall not exceed 15 amperes for No. 14, 20 amperes for No. 12, and 30 amperes for No. 10 copper; or 15 amperes for No. 12 and 25 amperes for No. 10 aluminum and copper-clad aluminum.

**Figure 10-2** *NEC* Table 310-17. *(Reprinted with permission from NFPA 70-1993, the National Electrical Code®, Copyright © 1993, National Fire Protection Association, Quincy, MA 02269. This reprinted material is not the complete and official position of the National Fire Protection Association, on the referenced subject which is represented only by the standard in its entirety.)*

**Table 310-18. Allowable Ampacities of Three Single Insulated Conductors Rated 0 through 2000 Volts, 150° to 250°C (302° to 482°F), in Raceway or Cable Based on Ambient Air Temperature of 40°C (104°F)**

| Size | Temperature Rating of Conductor. See Table 310-13. | | | | Size |
|---|---|---|---|---|---|
| | 150°C (302°F) | 200°C (392°F) | 250°C (482°F) | 150°C (302°F) | |
| AWG kcmil | TYPE Z | TYPES FEP, FEPB, PFA | TYPES PFAH, TFE | TYPE Z | AWG kcmil |
| | COPPER | | NICKEL OR NICKEL-COATED COPPER | ALUMINUM OR COPPER-CLAD ALUMINUM | |
| 14 | 34 | 36 | 39 | . . . . | 14 |
| 12 | 43 | 45 | 54 | 30 | 12 |
| 10 | 55 | 60 | 73 | 44 | 10 |
| 8 | 76 | 83 | 93 | 57 | 8 |
| 6 | 96 | 110 | 117 | 75 | 6 |
| 4 | 120 | 125 | 148 | 94 | 4 |
| 3 | 143 | 152 | 166 | 109 | 3 |
| 2 | 160 | 171 | 191 | 124 | 2 |
| 1 | 186 | 197 | 215 | 145 | 1 |
| 1/0 | 215 | 229 | 244 | 169 | 1/0 |
| 2/0 | 251 | 260 | 273 | 198 | 2/0 |
| 3/0 | 288 | 297 | 308 | 227 | 3/0 |
| 4/0 | 332 | 346 | 361 | 260 | 4/0 |
| 250 | . . . . | . . . . | . . . . | . . . . | 250 |
| 300 | . . . . | . . . . | . . . . | . . . . | 300 |
| 350 | . . . . | . . . . | . . . . | . . . . | 350 |
| 400 | . . . . | . . . . | . . . . | . . . . | 400 |
| 500 | . . . . | . . . . | . . . . | . . . . | 500 |
| 600 | . . . . | . . . . | . . . . | . . . . | 600 |
| 700 | . . . . | . . . . | . . . . | . . . . | 700 |
| 750 | . . . . | . . . . | . . . . | . . . . | 750 |
| 800 | . . . . | . . . . | . . . . | . . . . | 800 |
| 1000 | . . . . | . . . . | . . . . | . . . . | 1000 |
| 1500 | . . . . | . . . . | . . . . | . . . . | 1500 |
| 2000 | . . . . | . . . . | . . . . | . . . . | 2000 |

**AMPACITY CORRECTION FACTORS**

| Ambient Temp. °C | For ambient temperatures other than 40°C (104°F), multiply the ampacities shown above by the appropriate factor shown below. | | | | Ambient Temp. °F |
|---|---|---|---|---|---|
| 41-50 | .95 | .97 | .98 | .95 | 106-122 |
| 51-60 | .90 | .94 | .95 | .90 | 123-140 |
| 61-70 | .85 | .90 | .93 | .85 | 141-158 |
| 71-80 | .80 | .87 | .90 | .80 | 159-176 |
| 81-90 | .74 | .83 | .87 | .74 | 177-194 |
| 91-100 | .67 | .79 | .85 | .67 | 195-212 |
| 101-120 | .52 | .71 | .79 | .52 | 213-248 |
| 121-140 | .30 | .61 | .72 | .30 | 249-284 |
| 141-160 | . . . . | .50 | .65 | . . . . | 285-320 |
| 161-180 | . . . . | .35 | .58 | . . . . | 321-356 |
| 181-200 | . . . . | . . . . | .49 | . . . . | 357-392 |
| 201-225 | . . . . | . . . . | .35 | . . . . | 393-437 |

**Figure 10-3** *NEC* Table 310-18. *(Reprinted with permission from NFPA 70-1993, the National Electrical Code®, Copyright © 1993, National Fire Protection Association, Quincy, MA 02293. This reprinted material is not the complete and official position of the National Fire Protection Association, on the referenced subject which is represented only by the standard in its entirety.)*

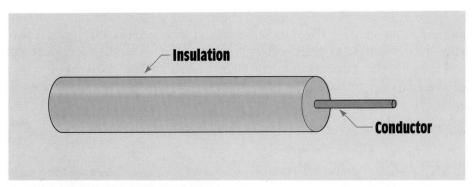

**Figure 10-4** Insulation around conductor

be used; the fifth, the material from which the insulation is made; the sixth, the thickness of the insulation; and the last, the type of outer covering over the insulation.

Example 1

Find the maximum operating temperature of type RHW insulation. (Note: Refer to the *National Electrical Code*.)

## Solution

Find type RHW in the second column of Table 310-13. The third column lists a maximum operating temperature of 75°C or 167°F.

Example 2

Can type THHN insulation be used in wet locations?

## Solution

Locate type THHN insulation in the second column. The fourth column indicates that this insulation can be used in **dry** and **damp locations**. This type of insulation cannot be used in **wet locations**. For an explanation of the difference between damp and wet locations, consult "locations" in article 100 of the *National Electrical Code*.

**dry
locations**

**damp
locations**

**wet
locations**

## 10-3 Correction Factors

One of the main factors that determines the amount of current a conductor is permitted to carry is the ambient, or surrounding, air temperature. Table 310-16, for example, lists the ampacity of not more than three conductors in raceway in free air. These ampacities are based on an ambient air temperature of 30°C or 86°F. If these conductors are to be used in a location that has a higher ambient temperature, the ampacity

**correction factor**

of the conductor must be reduced, because the resistance of copper or aluminum increases with an increase of temperature. The **correction factor** chart located at the bottom of the table is used to make this amendment. The correction factor chart is divided into the same number of columns as the wire table directly above it. The correction factors in each column are used for the conductors listed in the same column of the wire table.

**Example 3**

What is the maximum ampacity of a #4 AWG copper conductor with type THWN insulation used in an area with an ambient temperature of 43°C?

## Solution

Determine the ampacity of a #4 AWG copper conductor with type THWN insulation from the wire table. Type THWN insulation is located in the second column of Table 310-16. The table lists an ampacity of 85 A for this conductor. Follow the second column down to the correction factor chart. Locate 43°C in the far left-hand column of the correction factor chart; 43°C falls between 41° and 45°. The chart lists a correction factor of 0.82. The ampacity of the conductor in the above wire table is to be multiplied by the correction factor.

$$85 \times 0.82 = 69.7 \text{ A}$$

**Example 4**

What is the maximum ampacity of a #1/0 AWG copper-clad aluminum conductor with type RHH insulation if the conductor is to be used in an area with an ambient air temperature of 100°F?

## Solution

Locate the column that contains type RHH insulation in the copper-clad aluminum section of Table 310-16. The table indicates a maximum ampacity of 135 A for this conductor. Follow this column down to the correction factor chart and locate 100°F. Fahrenheit degrees are located in the far right-hand column of the chart. One hundred degrees falls between 97° and 104°F. The correction factor for this temperature is 0.91. Multiply the ampacity of the conductor by this factor.

$$135 \times 0.91 = 122.85 \text{ A}$$

### More Than Three Conductors in a Raceway

Tables 310-16 and 310-18 list three conductors in a raceway. If a raceway is to contain more than three conductors, the ampacity of the conductors must be derated because the heat from each conductor combines with the heat dissipated by the other conductors to produce a higher temperature inside the raceway. Note 8a *(Figure 10-5)* under "Notes to Ampacity Tables of 0 to 2000 Volts" lists these correction factors in a table. If the raceway is used in an area with a greater ambient temperature than that listed in the appropriate wire table, the temperature correction factor must be applied also.

**Example 5**

Twelve #14 AWG copper conductors with type RHW insulation are to be run in a conduit. The conduit is used in an area that has an ambient temperature of 110°F. What is the maximum ampacity of these conductors?

## Solution

Find the ampacity of a #14 AWG copper conductor with type RHW insulation. Type RHW insulation is located in the second column of Table 310-16. A #14 AWG copper conductor has an ampacity of 20 A. Next, use the correction factor for ambient temperature. A correction factor of 0.82 will be used.

$$20 \times 0.82 = 16.4 \text{ A}$$

The correction factor located in the table of note 8a must now be used. The table indicates a correction factor of 50% when 10 through 20 conductors are run in a raceway.

**8. Adjustment Factors.**

**(a) More than Three Current-Carrying Conductors in a Raceway or Cable.** Where the number of current-carrying conductors in a raceway or cable exceeds three, the allowable ampacities shall be reduced as shown in the following table:

| Number of Current-Carrying Conductors | Percent of Values in Tables as Adjusted for Ambient Temperature if necessary |
|---|---|
| 4 through 6 | 80 |
| 7 through 9 | 70 |
| 10 through 20 | 50 |
| 21 through 30 | 45 |
| 31 through 40 | 40 |
| 41 and above | 35 |

**Figure 10-5** Note 8a to wire tables. *(Reprinted with permission from NFPA 70-1993, the National Electrical Code®, Copyright © 1993, National Fire Protection Association, Quincy, MA 02269. This reprinted material is not the complete and official position of the National Fire Protection Association, on the referenced subject which is represented only by the standard in its entirety.)*

$$16.4 \times 0.50 = 8.2 \text{ A}$$

Each #14 AWG conductor has a maximum current rating of 8.2 A.

## Duct Banks

Duct banks are often used when it becomes necessary to bury cables in the ground. An electrical duct can be a single metallic or nonmetallic conduit. An electrical duct bank is a group of electrical ducts buried together as shown in *Figure 10-6*. When a duct bank is used, the center points of individual ducts should be separated by a distance of not less than 7.5 inches.

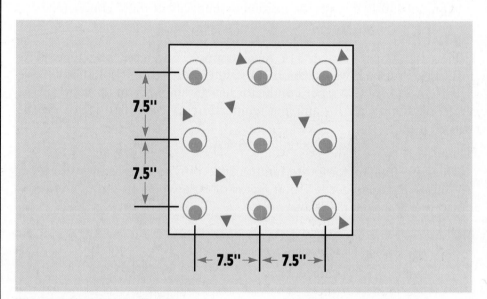

**Figure 10-6** Electrical duct banks

## 10-4 Computing Conductor Sizes and Resistance

Although the wire tables in the *National Electrical Code* are used to determine the proper size wire for most installations, there are instances in which these tables are not used. The formula in 310-15b of the *NEC* is used for ampacities not listed in the wire tables. The formula is shown below:

$$I = \sqrt{\frac{T_C - (T_A + \Delta_{TD})}{R_{DC}(1 + Y_C) R_{CA}}}$$

where

$T_C$ = conductor temperature in °C

$T_A$ = ambient temperature in °C

$\Delta_{TD}$ = dielectric loss temperature rise

$R_{DC}$ = DC resistance of conductor at temperature TC

$Y_C$ = component AC resistance resulting from skin effect and proximity effect

$R_{CA}$ = effective thermal resistance between conductor and surrounding ambient

Although this formula is seldom used by electricians, the *National Electrical Code* does permit its use under the supervision of an electrical engineer.

## Long Wire Lengths

Another situation in which it becomes necessary to compute wire sizes instead of using the tables in the code is when the conductor becomes excessively long. The listed ampacities in the code tables assume that the length of the conductor will not increase the resistance of the circuit by a significant amount. When the wire becomes extremely long, however, it is necessary to compute the size of wire needed.

All wire contains resistance. As wire is added to a circuit, it has the effect of adding resistance in series with the load *(Figure 10-7)*. Four factors determine the resistance of a length of wire.

1. *The type material from which the wire is made.* Different types of material have different wire resistances. A copper conductor will have less resistance than an aluminum conductor of the same size

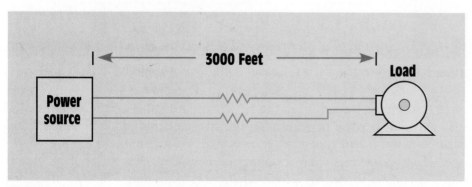

**Figure 10-7** Long wire runs have the effect of adding resistance in series with the load.

and length. An aluminum conductor will have less resistance than a piece of iron wire the same size and length.

2. *The diameter of the conductor.* The larger the diameter, the less resistance it will have. A large-diameter pipe, for example, will have less resistance to the flow of water than will a small-diameter pipe *(Figure 10-8)*. The diameter of round wire is measured in circular mils (CM). One mil equals 0.001 inch. A **circular mil** is the diameter of the wire in mils squared. For example, assume a wire has a diameter of 0.064 inch. Sixty-four thousandths should be written as a whole number, not as a decimal or a fraction ($64^2$ [64 x 64] = 4096 CM).

3. *The length of the conductor.* The longer the conductor, the more resistance it will have. Adding length to a conductor has the same effect as connecting resistors in series.

4. *The temperature of the conductor.* As a general rule, most conductive materials will increase their resistance with an increase of temperature. Some exceptions to this rule are carbon, silicon, and germanium. If the coefficient of temperature for a particular material is known, its resistance at different temperatures can be computed. Materials that increase their resistance with an increase of temperature have a positive coefficient of temperature. Materials that decrease their resistance with an increase of temperature have a negative coefficient of temperature.

**circular mil**

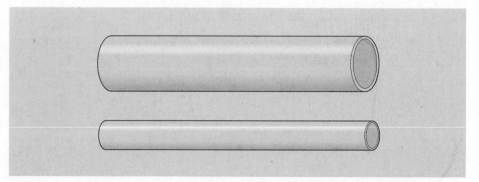

**Figure 10-8** A large pipe has less resistance to the flow of water than a small pipe.

In the English system of measure, a standard value of resistance, called the mil foot, is used to determine the resistance of different lengths and sizes of wire. A **mil foot** is a piece of wire 1 foot long and 1 mil in diameter *(Figure 10-9)*. A chart showing the resistance of a mil foot of wire at 20°C is shown in *Figure 10-10*. Notice the wide range of resistances for

**mil foot**

different materials. The temperature coefficient of the different types of conductors is listed also.

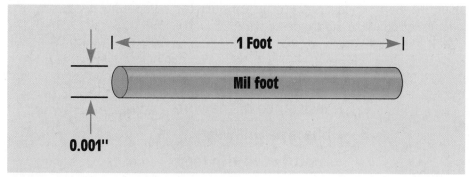

**Figure 10-9** A mil foot is equal to a piece of wire one foot long and one thousandth of an inch in diameter.

## Resistivity (K) of Materials

| Material | Ω per Mil Foot at 20°C | Temp. coeff. (Ω per °C) |
|---|---|---|
| Aluminum | 17 | 0.004 |
| Carbon | 22,000 aprx. | −0.0004 |
| Constantan | 295 | 0.000002 |
| Copper | 10.4 | 0.0039 |
| Gold | 14 | 0.004 |
| Iron | 60 | 0.0055 |
| Lead | 126 | 0.0043 |
| Manganin | 265 | 0.000000 |
| Mercury | 590 | 0.00088 |
| Nichrome | 675 | 0.0002 |
| Nickel | 52 | 0.005 |
| Platinum | 66 | 0.0036 |
| Silver | 9.6 | 0.0038 |
| Tungsten | 33.8 | 0.005 |

**Figure 10-10** Resistivity of materials

## Computing Resistance

Now that a standard measure of resistance for different types of materials is known, the resistance of different lengths and sizes of these materials can be computed. The formula for computing resistance of a certain length, size, and type of wire is

$$R = \frac{K \times L}{CM}$$

where

R = resistance of the wire

K = ohms per mil foot

L = length of wire in feet

CM = circular mil area of the wire

This formula can be converted to compute other values in the formula such as:

to find the SIZE of wire to use:

$$CM = \frac{K \times L}{R}$$

to find the LENGTH of wire to use:

$$L = \frac{R \times CM}{K}$$

to find the TYPE of wire to use:

$$K = \frac{R \times CM}{L}$$

## Example 6

Find the resistance of a piece of #6 AWG copper wire 550 ft long. Assume a temperature of 20°C. The formula to be used is

$$R = \frac{K \times L}{CM}$$

## Solution

The value for K can be found in the table in *Figure 10-10*. The table indicates a value of 10.4 Ω per mil foot for a copper conductor. The length, L, was given as 550 ft, and the circular mil area of #6 AWG wire is listed as 26,250 in the table shown in *Figure 10-11*.

# AMERICAN WIRE GAUGE TABLE

| B & S Gauge No. | Diam. in Mils | Area in Circular Mils | Ohms per 1000 Ft. (ohms per 100 m) | | | Pounds per 1,000 Ft. (kg per 100 m) | |
|---|---|---|---|---|---|---|---|
| | | | Copper* 68°F (20°C) | Copper* 167°F (75°C) | Aluminum 68°F (20°C) | Copper | Aluminum |
| 0000 | 460 | 211 600 | 0.049 ( 0.016 ) | 0.0596 ( 0.0195 ) | 0.0804 ( 0.0263 ) | 640 ( 95.2 ) | 195 ( 29.0 ) |
| 000 | 410 | 167 800 | 0.0618 ( 0.020 ) | 0.0752 ( 0.0246 ) | 0.101 ( 0.033 ) | 508 ( 75.5 ) | 154 ( 22.9 ) |
| 00 | 365 | 133 100 | 0.078 ( 0.026 ) | 0.0948 ( 0.031 ) | 0.128 ( 0.042 ) | 403 ( 59.9 ) | 122 ( 18.1 ) |
| 0 | 325 | 105 500 | 0.0983 ( 0.032 ) | 0.1195 ( 0.0392 ) | 0.161 ( 0.053 ) | 320 ( 47.6 ) | 97 ( 14.4 ) |
| 1 | 289 | 83 690 | 0.1239 ( 0.0406 ) | 0.151 ( 0.049 ) | 0.203 ( 0.066 ) | 253 ( 37.6 ) | 76.9 ( 11.4 ) |
| 2 | 258 | 66 370 | 0.1563 ( 0.0512 ) | 0.190 ( 0.062 ) | 0.526 ( 0.084 ) | 201 ( 29.9 ) | 61.0 ( 9.07 ) |
| 3 | 229 | 52 640 | 0.1970 ( 0.0646 ) | 0.240 ( 0.079 ) | 0.323 ( 0.106 ) | 159 ( 23.6 ) | 48.4 ( 7.20 ) |
| 4 | 204 | 41 740 | 0.2485 ( 0.0815 ) | 0.302 ( 0.099 ) | 0.408 ( 0.134 ) | 126 ( 18.7 ) | 38.4 ( 5.71 ) |
| 5 | 182 | 33 100 | 0.3133 ( 0.1027 ) | 0.381 ( 0.125 ) | 0.514 ( 0.168 ) | 100 ( 14.9 ) | 30.4 ( 4.52 ) |
| 6 | 162 | 26 250 | 0.395 ( 1.29 ) | 0.481 ( 0.158 ) | 0.648 ( 0.212 ) | 79.5 ( 11.8 ) | 24.1 ( 3.58 ) |
| 7 | 144 | 20 820 | 0.498 ( 0.163 ) | 0.606 ( 0.199 ) | 0.817 ( 0.268 ) | 63.0 ( 9.37 ) | 19.1 ( 2.84 ) |
| 8 | 128 | 16 510 | 0.628 ( 0.206 ) | 0.764 ( 0.250 ) | 1.03 ( 0.338 ) | 50.0 ( 7.43 ) | 15.2 ( 2.26 ) |
| 9 | 114 | 13 090 | 0.792 ( 0.260 ) | 0.963 ( 0.316 ) | 1.30 ( 0.426 ) | 39.6 ( 5.89 ) | 12.0 ( 1.78 ) |
| 10 | 102 | 10 380 | 0.999 ( 0.327 ) | 1.215 ( 0.398 ) | 1.64 ( 0.538 ) | 31.4 ( 4.67 ) | 9.55 ( 1.42 ) |
| 11 | 91 | 8 234 | 1.260 ( 0.413 ) | 1.532 ( 0.502 ) | 2.07 ( 0.678 ) | 24.9 ( 3.70 ) | 7.57 ( 1.13 ) |
| 12 | 81 | 6 530 | 1.588 ( 0.520 ) | 1.931 ( 0.633 ) | 2.61 ( 0.856 ) | 19.8 ( 2.94 ) | 6.00 ( 0.89 ) |
| 13 | 72 | 5 178 | 2.003 ( 0.657 ) | 2.44 ( 0.80 ) | 3.29 ( 1.08 ) | 15.7 ( 2.33 ) | 4.8 ( 0.71 ) |
| 14 | 64 | 4 107 | 2.525 ( 0.828 ) | 3.07 ( 1.01 ) | 4.14 ( 1.36 ) | 12.4 ( 1.84 ) | 3.8 ( 0.56 ) |
| 15 | 57 | 3 257 | 3.184 ( 0.043 ) | 3.98 ( 1.27 ) | 5.22 ( 1.71 ) | 9.86 ( 1.47 ) | 3.0 ( 0.45 ) |
| 16 | 51 | 2 583 | 4.016 ( 0.316 ) | 4.88 ( 1.60 ) | 6.59 ( 2.16 ) | 7.82 ( 1.16 ) | 2.4 ( 0.36 ) |
| 17 | 45.3 | 2 048 | 5.06 ( 1.66 ) | 6.16 ( 2.02 ) | 8.31 ( 2.72 ) | 6.20 ( 0.922 ) | 1.9 ( 0.28 ) |
| 18 | 40.3 | 1 624 | 6.39 ( 2.09 ) | 7.77 ( 2.55 ) | 10.5 ( 3.44 ) | 4.92 ( 0.731 ) | 1.5 ( 0.22 ) |
| 19 | 35.9 | 1 288 | 8.05 ( 2.64 ) | 9.79 ( 3.21 ) | 13.2 ( 4.33 ) | 3.90 ( 0.580 ) | 1.2 ( 0.18 ) |
| 20 | 32.0 | 1 022 | 10.15 ( 3.33 ) | 12.35 ( 4.05 ) | 16.7 ( 5.47 ) | 3.09 ( 0.459 ) | 0.94 ( 0.14 ) |
| 21 | 28.5 | 810 | 12.8 ( 4.2 ) | 15.6 ( 5.11 ) | 21.0 ( 6.88 ) | 2.45 ( 0.364 ) | 0.745 ( 0.110 ) |
| 22 | 25.4 | 642 | 16.1 ( 5.3 ) | 19.6 ( 6.42 ) | 26.5 ( 8.69 ) | 1.95 ( 0.290 ) | 0.591 ( 0.09 ) |
| 23 | 22.6 | 510 | 20.4 ( 6.7 ) | 24.8 ( 8.13 ) | 33.4 ( 10.9 ) | 1.54 ( 0.229 ) | 0.468 ( 0.07 ) |
| 24 | 20.1 | 404 | 25.7 ( 8.4 ) | 31.2 ( 10.2 ) | 42.1 ( 13.8 ) | 1.22 ( 0.181 ) | 0.371 ( 0.05 ) |
| 25 | 17.9 | 320 | 32.4 ( 10.6 ) | 39.4 ( 12.9 ) | 53.1 ( 17.4 ) | 0.97 ( 0.14 ) | 0.295 ( 0.04 ) |
| 26 | 15.9 | 254 | 40.8 ( 13.4 ) | 49.6 ( 16.3 ) | 67.0 ( 22.0 ) | 0.77 ( 0.11 ) | 0.234 ( 0.03 ) |
| 27 | 14.2 | 202 | 51.5 ( 16.9 ) | 62.6 ( 20.5 ) | 84.4 ( 27.7 ) | 0.61 ( 0.09 ) | 0.185 ( 0.03 ) |
| 28 | 12.6 | 160 | 64.9 ( 21.3 ) | 78.9 ( 25.9 ) | 106 ( 34.7 ) | 0.48 ( 0.07 ) | 0.147 ( 0.02 ) |
| 29 | 11.3 | 126.7 | 81.8 ( 26.8 ) | 99.5 ( 32.6 ) | 134 ( 43.9 ) | 0.384 ( 0.06 ) | 0.117 ( 0.02 ) |
| 30 | 10.0 | 100.5 | 103.2 ( 33.8 ) | 125.5 ( 41.1 ) | 169 ( 55.4 ) | 0.304 ( 0.04 ) | 0.092 ( 0.01 ) |
| 31 | 8.93 | 79.7 | 130.1 ( 42.6 ) | 158.2 ( 51.9 ) | 213 ( 69.8 ) | 0.241 ( 0.04 ) | 0.073 ( 0.01 ) |
| 32 | 7.95 | 63.2 | 164.1 ( 53.8 ) | 199.5 ( 65.4 ) | 269 ( 88.2 ) | 0.191 ( 0.03 ) | 0.058 ( 0.01 ) |
| 33 | 7.08 | 50.1 | 207 ( 68 ) | 252 ( 82.6 ) | 339 ( 111 ) | 0.152 ( 0.02 ) | 0.046 ( 0.01 ) |
| 34 | 6.31 | 39.8 | 261 ( 86 ) | 317 ( 104 ) | 428 ( 140 ) | 0.120 ( 0.02 ) | 0.037 ( 0.01 ) |
| 35 | 5.62 | 31.5 | 329 ( 108 ) | 400 ( 131 ) | 540 ( 177 ) | 0.095 ( 0.01 ) | 0.029 |
| 36 | 5.00 | 25.0 | 415 ( 136 ) | 505 ( 165 ) | 681 ( 223 ) | 0.076 ( 0.01 ) | 0.023 |
| 37 | 4.45 | 19.8 | 523 ( 171 ) | 636 ( 208 ) | 858 ( 281 ) | 0.0600 ( 0.01 ) | 0.0182 |
| 38 | 3.96 | 15.7 | 660 ( 216 ) | 802 ( 263 ) | 1080 ( 354 ) | 0.0476 ( 0.01 ) | 0.0145 |
| 39 | 3.53 | 12.5 | 832 ( 273 ) | 1012 ( 332 ) | 1360 ( 446 ) | 0.0377 ( 0.01 ) | 0.0115 |
| 40 | 3.15 | 9.9 | 1049 ( 344 ) | 1276 ( 418 ) | 1720 ( 564 ) | 0.0299 ( 0.01 ) | 0.0091 |
| 41 | | | | | | | |
| 42 | 2.50 | 6.3 | | | | | |
| 43 | | | | | | | |
| 44 | 1.97 | 3.9 | | | | | |

*Resistance figures are given for standard annealed copper. For hard-drawn copper add 2%

**Figure 10-11** American wire gauge table

$$R = \frac{10.4 \times 550}{26,250}$$

$$R = \frac{5720}{26,250}$$

$$R = 0.218 \ \Omega$$

**Example 7**

An aluminum wire 2250 ft long cannot have a resistance greater than 0.2 $\Omega$. What size aluminum wire must be used?

### Solution

To find the size of wire use:

$$CM = \frac{K \times L}{R}$$

$$CM = \frac{17 \times 2250}{0.2}$$

$$CM = \frac{38,250}{0.2}$$

$$CM = 191,250$$

The nearest standard size conductor for this installation can be found in the American Wire Gauge table. Since the resistance cannot be greater than 0.2 $\Omega$, the conductor cannot be smaller than 191,250 CM. The nearest standard conductor size is 0000 AWG.

Good examples of when it becomes necessary to compute the wire size for a particular installation can be seen in the following problems.

**Example 8**

A manufacturing plant has decided to install high-pressure sodium vapor lights around the entire perimeter of its buildings and parking area *(Figure 10-12)*. The lighting system is to be single-phase 480 V and has a current draw of 50 A. The voltage drop is to be kept to a maximum of 3% of the applied voltage. It has been determined that the distance around the perimeter is 4000 ft. What size copper conductors should be used in this installation? Assume that the average ambient temperature is 20°C.

### Solution

The first step in the solution of this problem will be to determine the maximum amount of resistance the conductors can have without produc-

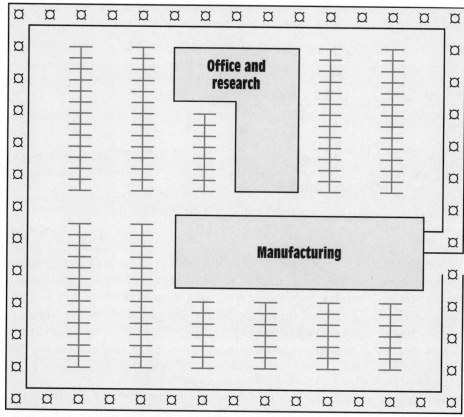

**Figure 10-12** Computing long wire lengths

ing a voltage drop greater than 3% of the applied voltage. The maximum amount of voltage drop can be computed by multiplying the applied voltage by 3%.

$$480 \times 0.03 = 14.4 \text{ V}$$

The maximum amount of resistance can now be computed using Ohm's law.

$$R = \frac{E}{I}$$

$$R = \frac{14.4}{50}$$

$$R = 0.288 \ \Omega$$

The distance around the perimeter is 4000 feet. Since two conductors are used, the resistance of both conductors must be considered. Two conductors

4000 ft long will have the same resistance as one conductor 8000 ft long. For this reason, a length of 8000 ft will be used in the formula.

$$CM = \frac{K \times L}{R}$$

$$CM = \frac{10.4 \times 8000}{0.288}$$

$$CM = \frac{83{,}200}{0.288}$$

$$CM = 288{,}888.9$$

300-kcmil (thousand circular mils) cable will be used for this installation.

## Example 9

The next problem concerns conductors used in a three-phase system. Assume that a motor is located 2500 ft from its power source and operates on 560 V. When the motor starts, it has a current draw of 168 A. The voltage drop at the motor terminals cannot be permitted to be greater than 5% of the source voltage during starting. What size aluminum conductors should be used for this installation?

### Solution

The solution to this problem is very similar to the solution in the previous example. First, find the maximum voltage drop that can be permitted at the load by multiplying the source voltage by 5%.

$$E = 560 \times 0.05$$

$$E = 28 \text{ V}$$

The second step is to determine the maximum amount of resistance of the conductors. To compute this value, the maximum voltage drop will be divided by the starting current of the motor.

$$R = \frac{E}{I}$$

$$R = \frac{28}{168}$$

$$R = 0.166 \ \Omega$$

The next step is to compute the length of the conductors. In the previous example, the lengths of the two conductors were added to find the total amount of wire resistance. In a single-phase system, each conductor must carry the same amount of current. During any period of time, one conductor is supplying current from the source to the load, and the other conductor completes the circuit by permitting the same amount of current to flow from the load to the source.

In a balanced three-phase circuit, three currents are 120° out of phase with each other *(Figure 10-13)*. These three conductors share the flow of current between source and load. In *Figure 10-13,* two lines labeled A and B have been drawn through the three current wave forms. Notice that at position A the current flow in phase 1 is maximum and in a positive direction. The current flow in phases 2 and 3 is less than maximum and in a negative direction. This condition corresponds to the example shown in *Figure 10-14*. Notice that maximum current is flowing in only one conductor. Less than maximum current is flowing in the other two conductors.

Observe the line marking position B in *Figure 10-13*. The current flow in phase 1 is zero, and the currents flowing in phases 2 and 3 are in opposite directions and less than maximum. This condition of current flow is illustrated in *Figure 10-15*. Notice that only two of the three-phase lines are conducting current and that the current in each line is less than maximum.

Since the currents flowing in a three-phase system are never maximum at the same time, and at other times the current is divided between two

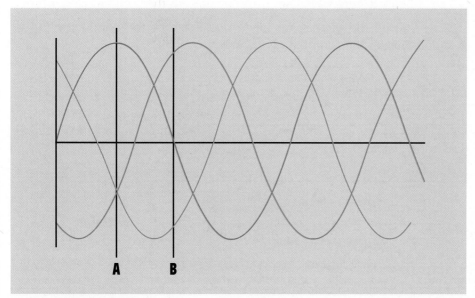

**A**    **B**

**Figure 10-13**   The line currents in a three-phase system are 120° out of phase with each other.

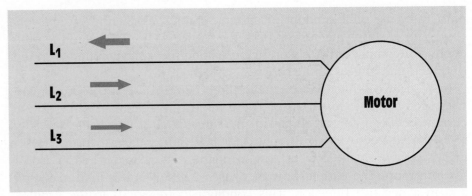

**Figure 10-14** Current flows from lines 2 and 3 to line 1.

phases, the total conductor resistance will not be the sum of two conductors. To compute the resistance of conductors in a three-phase system, a demand factor of 0.866 is used.

In this problem, the motor is located 2500 ft from the source. The conductor length will be computed by doubling the length of one conductor and then multiplying by 0.866.

$$L = 2500 \times 2 \times 0.866$$

$$L = 4330 \text{ feet}$$

Now that all the factors are known, the size of the conductor can be computed using the formula

$$CM = \frac{K \times L}{R}$$

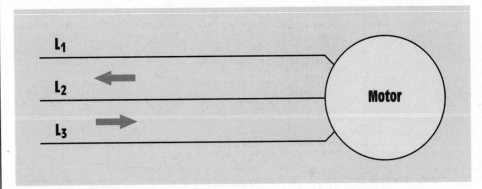

**Figure 10-15** Current flows from line 3 to line 2.

where

$$K = 17 \text{ (ohms per mil foot for aluminum)}$$

$$L = 4330 \text{ ft}$$

$$R = 0.166 \ \Omega$$

$$CM = \frac{17 \times 4330}{0.166}$$

$$CM = \frac{73,601}{0.166}$$

$$CM = 443,433.7$$

Three 500-kcmil conductors will be used.

## 10-5 Computing Voltage Drop

Sometimes it is necessary to compute the **voltage drop** of an installation when the length, size of wire, and current are known. The formula shown below can be used to find the voltage drop of conductors used on a single-phase system.

$$E_D = \frac{2 \ KIL}{CM}$$

where

$$E_D = \text{voltage drop}$$

$$K = \text{ohms per mil foot}$$

$$I = \text{current}$$

$$L = \text{length of conductor in feet}$$

$$CM = \text{circular mil area of the conductor}$$

A single-phase motor is located 250 ft from its power source. The conductors supplying power to the motor are #10 AWG copper. The motor has a full load current draw of 24 A. What is the voltage drop across the conductors when the motor is in operation?

**Solution**

$$E_D = \frac{2 \times 10.4 \times 24 \times 250}{10,380}$$

$$E_D = 12.02 \text{ V}$$

**voltage drop**

**Example 10**

A slightly different formula can be used to compute the voltage drop on a three-phase system. Instead of multiplying KIL by 2, multiply KIL by the square root of three.

$$E_D = \frac{\sqrt{3}\ KIL}{CM}$$

**Example 11**

A three-phase motor is located 175 ft from its source of power. The conductors supplying power to the motor are #1/0 AWG aluminum. The motor has a full load current draw of 88 A. What is the voltage drop across the conductors when the motor is operating at full load?

## Solution

$$E_D = \frac{1.732 \times 17 \times 88 \times 175}{105,500}$$

$$E_D = 4.3\ V$$

## 10-6 Parallel Conductors

**parallel conductors**

Under certain conditions it may become necessary or advantageous to connect conductors in parallel. One such condition for **parallel conductors** is when the conductor is very large as in the example above, where it was computed that the conductors supplying a motor 2500 feet from its source would have to be 500 kcmil. A 500-kcmil conductor is very large and difficult to handle. Therefore, it may be preferable to use parallel conductors for this installation. The *National Electrical Code* lists five conditions that must be met when conductors are connected in parallel (Section 310-4). These conditions are:

1. The conductors must be the same length.
2. The conductors must be made of the same material. For example, all parallel conductors must be either copper or aluminum. It is not permissible to use copper for one of the conductors and aluminum for the other.
3. The conductors must have the same circular mil area.
4. The conductors must use the same type of insulation.
5. The conductors must be terminated or connected in the same manner.

In the example the actual conductor size needed was computed to be 443,433.7 CM. This circular mil area could be obtained by connecting two 250-kcmil conductors in parallel for each phase, or three 000 (3/0) conductors in parallel for each phase. (Note: Each 000 [3/0] conductor has an area of 167,800 CM. This is a total of 503,400 CM.)

Another example of when it may be necessary to connect wires in parallel is when conductors of a large size must be run in a conduit. Conductors of a single phase are not permitted to be run in metallic conduits as shown in *Figure 10-16* (*NEC* Sections 300-5i, and 300-20a, b), because when current flows through a conductor, a magnetic field is produced around the conductor. In an alternating current circuit, the current continuously changes direction and magnitude, which causes the magnetic field to cut through the wall of the metal conduit *(Figure 10-17)*. This

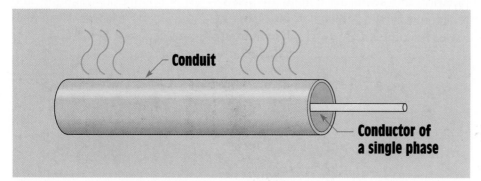

**Figure 10-16**   A single-phase conductor causes heat to be produced in the conduit.

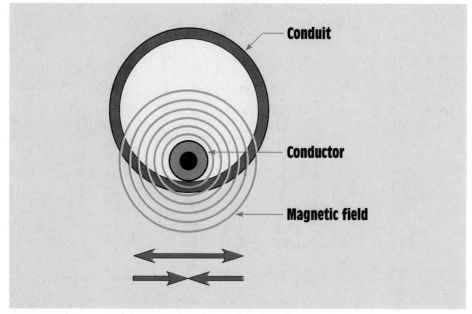

**Figure 10-17**   The magnetic field expands and contracts.

cutting action of the magnetic field induces a current, called an eddy current, into the metal of the conduit. Eddy currents are currents that are induced into metals. They tend to move in a circular fashion similar to the eddies of a river, hence the name eddy currents *(Figure 10-18)*. Eddy currents can produce enough heat in high-current circuits to melt the insulation surrounding the conductors. All metal conduits can have eddy current induction, but conduits made of magnetic materials such as steel have an added problem with hysteresis loss. Hysteresis loss is caused by molecular friction *(Figure 10-19)*. As the direction of the magnetic field reverses, the molecules of the metal are magnetized with the opposite polarity and swing to realign themselves. This continuous aligning and realigning of the molecules produces heat caused by friction. Hysteresis losses become greater with an increase in frequency.

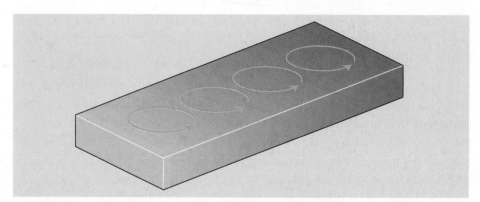

**Figure 10-18** Eddy currents are currents induced in metals.

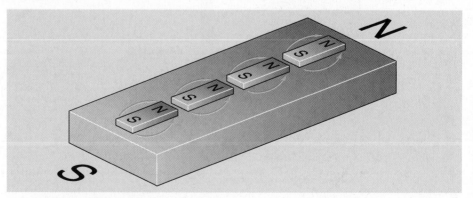

**Figure 10-19** The molecules reverse direction each time the magnetic field changes direction.

To correct this problem, a conductor of each phase must be run in each conduit *(Figure 10-20)*. When all three phases are contained in a single conduit, the magnetic fields of the separate conductors cancel each other resulting in no current being induced in the walls of the conduit.

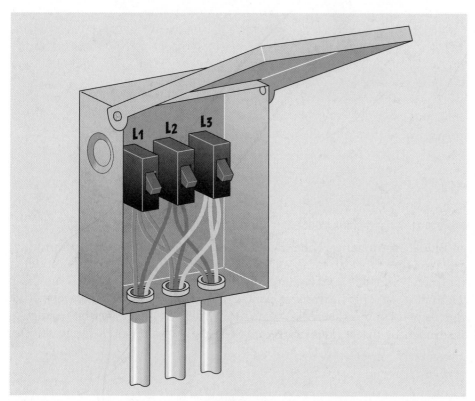

**Figure 10-20**  Each conduit contains a conductor from each phase. This permits the magnetic fields to cancel each other.

## 10-7 Testing Wire Installation

After the conductors have been installed in conduits or raceways, it is accepted practice to test the installation for grounds and shorts. This test requires an ohmmeter, which cannot only measure resistance in millions of ohms, but can also provide a high enough voltage to ensure that the insulation will not break down when rated line voltage is applied to the conductors. Most ohmmeters operate with a maximum voltage that ranges from 1.5 V to about 9 V depending on the type of ohmmeter and the setting of the range scale. To test wire insulation, a special type of ohmmeter, called a MEGGER®, is used. The term **MEGGER®** is a registered trademark

**MEGGER®**

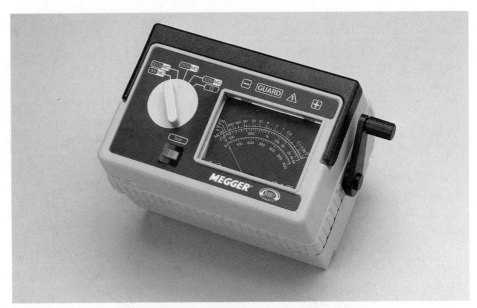

**Figure 10-21**   A hand-crank MEGGER® (*Courtesy of Biddle Instruments.*)

of Biddle Instruments. The MEGGER® is a megohmmeter that can produce voltages that range from about 250 to 5000 V depending on the model of the meter and the range setting. One model of a MEGGER® is shown in *Figure 10-21*. This instrument contains a hand crank that is connected to the rotor of a brushless AC generator. The advantage of this particular instrument is that it does not require the use of batteries. A range selector

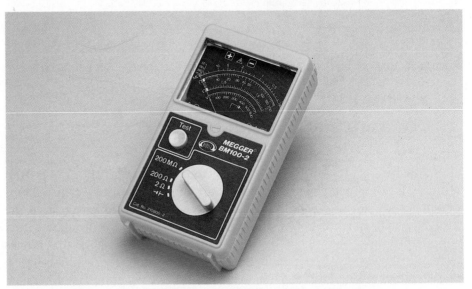

**Figure 10-22**   Battery-operated MEGGER® (*Courtesy of Biddle Instruments.*)

switch permits the meter to be used as a standard ohmmeter or as a megohmmeter. When it is used as a megohmmeter, the selector switch permits the test voltage to be selected. Test voltages of 100 V, 250 V, 500 V, and 1000 V can be obtained.

MEGGER®s can also be obtained in battery-operated models as shown in *Figure 10-22*. These models are small, lightweight, and particularly useful when it becomes necessary to test the dielectric of a capacitor.

Wire installations are generally tested for two conditions, shorts and grounds. Shorts are current paths that exist between conductors. To test an installation for shorts, the MEGGER® is connected across two conductors at a time as shown in *Figure 10-23*. The circuit is tested at rated voltage or slightly higher. The MEGGER® indicates the resistance between the two conductors. Since both conductors are insulated, the resistance between them should be extremely high. Each conductor should be tested against every other conductor in the installation.

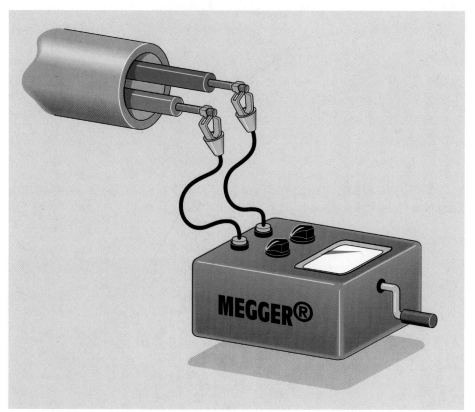

**Figure 10-23**  Testing for shorts with a MEGGER®

To test the installation for grounds, one lead of the MEGGER® is connected to the conduit or raceway as shown in *Figure 10-24*. The other meter lead is connected to one of the conductors. The conductor should be tested at rated voltage or slightly higher. Each conductor should be tested.

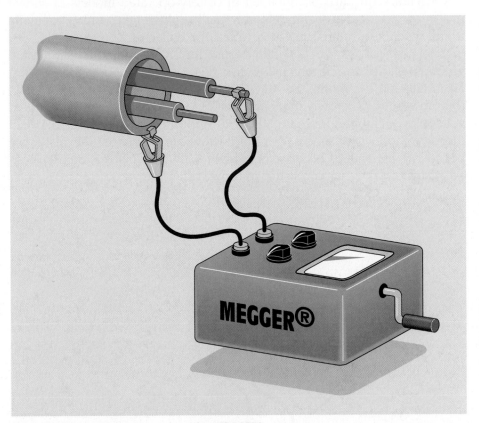

**Figure 10-24** Testing for grounds with a MEGGER®

## Summary

1. The *National Electrical Code* is used to determine the wire size for most installations.

2. The resistance of a wire is determined by four factors:
   A. the type of material from which the conductor is made,
   B. the length of the conductor,
   C. the area of the conductor, and
   D. the temperature of the conductor.

3. When conductors are used in high-temperature locations, their current capacity must be reduced.

4. When there are more than three conductors in a raceway their current-carrying capacity must be reduced.

5. The amount of current a conductor can carry is affected by the type of insulation around the wire.

6. In the English system, the mil foot is used as a standard for determining the resistance of different types of wire.

7. After wires have been installed, they should be checked for shorts or grounds with a MEGGER®.

## Review Questions

Use of the *National Electrical Code* will be required to answer some of the following questions.

1. What is the maximum temperature rating of type XHHW insulation when used in a wet location?

2. Name two types of conductor insulation designed to be used underground.

3. A #10 AWG copper conductor with type THW insulation is to be run in free air. What is the maximum ampacity of this conductor if the ambient air temperature is 40°C?

4. Six #1/0 aluminum conductors are to be run in a conduit. Each conductor has type THWN insulation, and the ambient air temperature is 30°C. What is the ampacity of each conductor?

5. Name five conditions that must be met for running conductors in parallel.

6. What is the largest solid (nonstranded) conductor listed in the wire tables?

7. Can type TW cable be used in an area that has an ambient temperature of 65°C?

8. How is the grounded conductor in a flat multiconductor cable #4 or larger identified?

9. What three colors are ungrounded conductors not permitted to be?

10. Twenty-five #12 AWG copper conductors are run in conduit. Each conductor has type THHN insulation. The conduit is located in an area that has an ambient temperature of 95°F. What is the ampacity of each conductor?

11. A single-phase load is located 2800 ft from its source. The load draws a current of 86 A and operates on 480 V. The maximum voltage drop at the load cannot be greater than 3%. What size aluminum conductors should be installed to operate this load?

12. It is decided to use parallel 0000 conductors to supply the load in question 11. How many 0000 conductors will be needed?

13. A three-phase motor operates on 480 V and is located 1800 ft from the power source. The starting current is 235 A. What size copper conductors will be needed to supply this load if the voltage must not be permitted to drop below 6% of the terminal voltage during starting?

## Using Wire Tables and Determining Conductor Sizes

1. A #2 AWG copper conductor is 450 ft long. What is the resistance of this wire? Assume the ambient temperature to be 20°C.

2. A #8 AWG conductor is 500 ft long and has a resistance of 1.817 Ω. The ambient temperature is 20°C. Of what material is the wire made?

3. Three 500-kCM copper conductors with type RHH insulation are to be used in an area that has an ambient temperature of 58°C. What is the maximum current-carrying capacity of these conductors?

4. Eight #10 AWG aluminum conductors with type THWN insulation are installed in a single conduit. What is the maximum current-carrying capacity of these conductors? Assume an ambient temperature of 30°C.

5. A three-phase motor is connected to 480 V and has a starting current of 522 A. The motor is located 300 ft from the power source. The voltage drop to the motor cannot be greater than 5% during starting. What size copper conductors should be connected to the motor?

**4**

## small sources of electricity

The conduc-
tion of
electric
current
through a gas
or liquid is
somewhat
different than
conduction
through a
wire.
This unit
discusses
some of these
differences.

*Courtesy of PPG Industries Inc.*

# Conduction in Liquids and Gases

## Key Terms

Acids

Alkalies

Anode

Arc

Cathode

Copper sulfate

Cuprous cyanide

Electrolysis

Electrolytes

Electron impact

Electroplating

Ionization potential

Ion

Metallic salt

Sulfuric acid

X-rays

**A**fter studying this unit, you should be able to:

■ Define positive and negative ions.

■ Discuss electrical conduction in a gas.

■ Discuss electrical conduction in a liquid.

■ Discuss several processes that occur as a result of ionization.

The conduction of electrical current is generally thought of as electrons moving through a wire. Many processes, however, depend on electrical current flowing through a gas or liquid. Batteries, for example, would not work if conduction could not take place through a liquid, and fluorescent lighting operates on the principle of conduction through a gas. Conduction through a gas or liquid does not depend on the flow of individual electrons as is the case with metallic conductors. Conduction in gases and liquids depends on the movement of ions.

### 11-1 The Ionization Process: Magnesium and Chlorine

An **ion** is a charged atom. Atoms that have a deficiency of electrons are known as positive ions. Atoms that have gained extra electrons are known as negative ions.

A good example of how ionization occurs can be seen by the combination of two atoms, magnesium and chlorine. The magnesium atom contains two valence electrons and is considered to be a metal. Chlorine contains seven valence electrons and is considered a nonmetal *(Figure 11-1)*.

When magnesium is heated in the presence of chlorine gas, a magnesium atom combines with two chlorine atoms to form a **metallic salt** called magnesium chloride *(Figure 11-2)*. When this process occurs, the

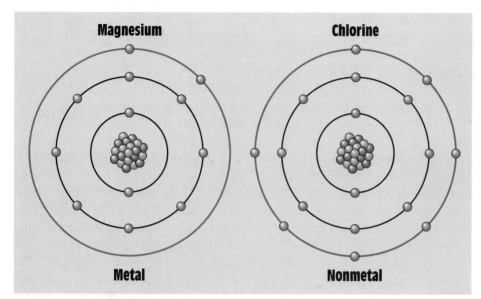

**Figure 11-1**  Magnesium and chlorine atoms

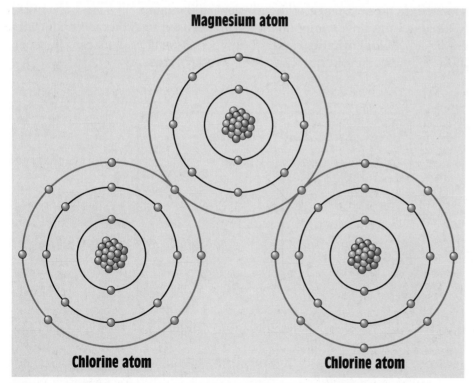

**Figure 11-2**  Magnesium and chlorine atoms combine to form magnesium chloride.

magnesium atom gives up its two valence electrons to the two chlorine atoms. These atoms are no longer called atoms; they are now called ions. Since the magnesium atom gave up two electrons, it has a positive charge and has become a positive ion. The two chlorine atoms have each gained an electron and are now negative ions.

Magnesium chloride is similar to table salt, which is sodium chloride. Both are formed by combining a metal, sodium or magnesium, with chlorine. If either of these salts is mixed with a nonconducting liquid, such as distilled water, it will become a conductor. A simple experiment can be performed to demonstrate conduction through a liquid. In *Figure 11-3*, copper electrodes have been placed in a glass. One electrode is connected to a lamp. The other is connected to one side of a 120-V circuit. The other side of the lamp is connected to the other side of the circuit. The lamp acts as a load to prevent the circuit from drawing excessive current. If pure water is poured into the glass, nothing seems to happen, because pure water is an insulator. If salt is added to the water, the lamp will begin to glow *(Figure 11-4)*. The lamp will glow dimly at first and increase in brightness as salt is added to the water. The glow will continue to brighten until the solution becomes saturated with salt and current is limited by the resistance of the filament in the lamp.

Salt is not the only compound that can be used to promote conduction in a liquid. **Acids**, **alkalies**, and other types of metallic salts can be used. These solutions are often referred to as **electrolytes**.

**acids**

**alkalies**

**electro-
lytes**

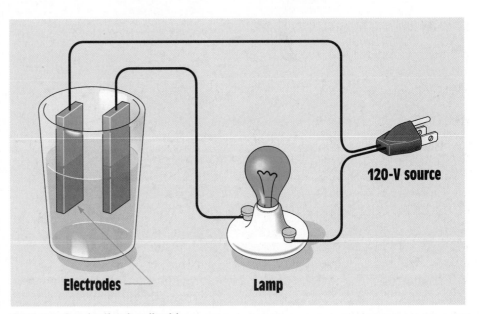

**Figure 11-3** Conduction in a liquid

Electrodes

Lamp

120-V source

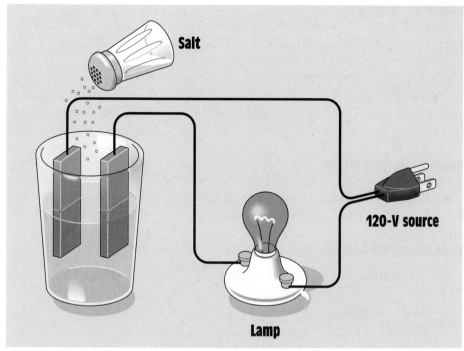

**Figure 11-4** Adding salt to the water causes the lamp to glow.

## 11-2 Other Types of Ions

The ions discussed so far are formed by charging a single atom. There are other types of ions, however, that are formed from groups of atoms. These ions are extremely useful to the electrical industry.

### Sulfuric Acid

One of the most useful ion compounds is **sulfuric acid**. It is the electrolyte for the lead-acid batteries used in automobiles. The chemical formula for sulfuric acid, $H_2SO_4$, indicates that one molecule of sulfuric acid contains two atoms of hydrogen, one atom of sulfur, and four atoms of oxygen. Sulfuric acid in its pure form is not useful as an electrolyte. When it is combined with water, $H_2O$, the molecules separate into ions of $H^+$, $H^+$, and $SO_4^{--}$. The plus sign indicates that each of two hydrogen atoms has lost its electrons and is now a positive ion. The $SO_4^{--}$ is called a sulfate ion and indicates that the two electrons lost by the hydrogen atoms are connected to the sulfur atom and four oxygen atoms. The two extra electrons actually hold the sulfate ion together.

**sulfuric acid**

copper
sulfate

cuprous
cyanide

electro-
plating

cathode

anode

## Copper Sulfate

Another useful compound is **copper sulfate**, $CuSO_4$. In its natural form, copper sulfate is in the form of blue crystals. When it is mixed with water it dissolves and forms two separate ions, $Cu^{++}$ and $SO_4^{--}$. The $Cu^{++}$ is called a cupric ion and indicates that a copper atom has lost two of its electrons. The $SO_4^{--}$ is the same sulfate ion formed when sulfuric acid is mixed with water. This copper sulfate solution is used in copper electroplating processes.

## Cuprous Cyanide

**Cuprous cyanide** is used to electroplate copper to iron. It is a very poisonous solid that becomes $Cu^+$ and $CN^-$ when mixed in solution. The $Cu^+$ is a cuprous ion and $CN^-$ is a cyanide ion.

## 11-3 Electroplating

**Electroplating** is the process of depositing atoms of one type of metal on another. Several factors are always true in an electroplating process.

1. **The electrolyte solution must contain ions of the metal to be plated.**
2. **Metal ions are always positively charged**.
3. **The object to be plated must be connected to the negative power terminal.** The negative terminal is called the **cathode** and refers to the terminal where electrons enter the circuit.
4. **Direct current is used as the power source.**
5. **The positive terminal is made of the same metal that is to form the coating.** The positive terminal is referred to as the **anode** and refers to the terminal where electrons leave the circuit.

An example of the electroplating process is shown in *Figure 11-5*. Note that the object to be plated has been connected to the negative battery terminal, or cathode, and the copper bar has been connected to the positive terminal, or anode. Both objects are submerged in a cuprous cyanide solution. When power is applied to the circuit, positively charged copper ions in the solution move toward the object to be plated, and the negatively charged cyanide ions move toward the copper bar. When the copper ion contacts the object, it receives an electron and becomes a neutral copper atom. These neutral copper atoms form a copper covering over the object to be plated. The thickness of the coating will be determined by the length of time the process is permitted to continue.

When the negatively charged cyanide ions contact the positive copper bar, copper atoms on the surface of the bar lose an electron and become positive ions. The copper ion is attracted into the solution and flows

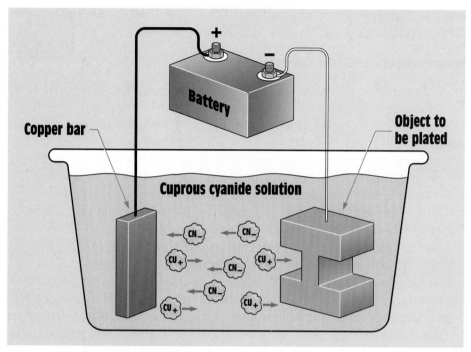

**Figure 11-5** The electroplating process

toward the object to be plated. Just as many copper ions are being formed as are being attracted to the object to be plated. The solution, therefore, remains at a constant strength.

One of the most common uses for electroplating is the production of pure metals. Impure, or raw, copper is connected to the positive terminal, and pure copper is connected to the negative terminal. Since only pure atoms of copper become ions, impurities remain in solution or never leave the positive plate. Impure copper has a high resistance and is not suitable for use as an electrical conductor. Wire manufacturers require electrolytically purified copper because of its low-resistance characteristics.

## 11-4 Electrolysis

The term **electrolysis** refers to the process of separating elements electrically. Elements are often chemically combined with other elements and must be separated. Although aluminum oxide is an abundant compound, aluminum was once very rare because extracting aluminum from aluminum oxide was difficult and expensive. Aluminum became inexpensive when a process was discovered for separating the aluminum and oxygen atoms electrically. In this process, electrons are removed from oxygen

**electrolysis**

ions and returned to aluminum ions. The aluminum ions then become aluminum atoms.

## 11-5 Conduction in Gases

At atmospheric pressure, air is an excellent insulator. In *Figure 11-6,* two electrodes are separated by an air gap. The electrodes are connected to a variable-voltage power supply. Assume that the output voltage of the power supply is zero when the switch is turned on. As the output voltage is increased, the ammeter will remain at zero until the voltage reaches a certain potential. Once this potential has been reached, an **arc** will be established across the electrodes and current will begin to flow *(Figure 11-7).* The amount of current is limited by the impedance of the rest of the circuit because once an arc has been established it has a very low resistance. Once the arc has been established, the voltage can be reduced and the arc will continue, because molecules of air become ionized and conduction is maintained by the ionized gas. One factor that determines the voltage required is the distance between the two conductors. The farther apart they are, the higher the voltage must be. When using an electric arc welder, the arc is first established by touching the electrode to the object to be welded and then withdrawing the electrode once conduction begins. Recall that once an arc has been established, the amount of voltage required to maintain it is reduced.

**arc**

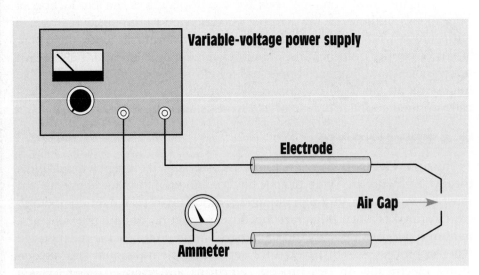

**Figure 11-6** Air acts as an insulator.

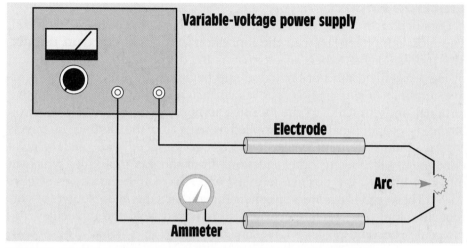

**Figure 11-7** When the voltage becomes high enough an arc is established.

## Ionization Potential in a Gas

The amount of voltage, or potential, an electron must possess to cause ionization is called the **ionization potential**. Several factors determine the amount of voltage required to cause conduction in a gas-filled envelope.

1. **Atmospheric pressure**. The amount of air pressure greatly influences the voltage required to reach ionization potential. Atmospheric pressure is 14.7 PSI (pounds per square inch) at sea level. If the pressure is increased, such as in an automobile engine, the amount of voltage required to reach ionization potential increases greatly. The ignition system of most automobiles produces voltages that range from 20,000 V to 70,000 V.

   If the atmospheric pressure is reduced, the amount of voltage required to reach ionization potential is reduced. This is the principle of operation of the gas-filled tubes that have been used in industry for many years. Gas-filled tubes operate with lower-than-normal atmospheric pressure inside the tube.

2. **The type of gas in the surrounding atmosphere**. The type of gas in the atmosphere can greatly influence the amount of voltage required to cause ionization. Sodium vapor, for example, requires a potential of approximately 5 V. Mercury vapor requires 10.4 V, neon 21.5 V, and helium 24.5 V. Before an electron can ionize an atom of mercury, it must possess a potential difference of 10.4 V.

**ionization potential**

**electron impact**

## Electron Impact

Conduction in a gas is different from conduction in a liquid or metal. The most important factor in the ionization of a gas is **electron impact**. The process begins when an electron is freed from the negative terminal inside a gas-filled tube or envelope and begins to travel toward the positive terminal *(Figure 11-8)*. The electron is repelled from the negative terminal and attracted to the positive terminal. The speed the electron attains is proportional to the applied voltage and the distance it travels before striking a gas molecule. If the electron's speed (potential) is great enough, it will liberate other electrons from the gas molecule, which in turn flow toward the positive terminal and strike other molecules *(Figure 11-9)*. These electrons are very energetic and are the main source of current flow in a gas. When electrons are removed from the molecules, the molecules become positive ions similar to those in a liquid. These positive ions are attracted to the negative terminal, where they receive electrons and become neutral molecules again.

If the electron does not possess enough potential to cause ionization, it bounces off and begins traveling toward the positive terminal again. The pressure inside the tube determines the density of the gas molecules *(Figure 11-10)*. If the pressure is too high, the gas molecules are so dense that the electron cannot gain enough potential to cause ionization. In this case, the electron will bounce from one molecule to another until it finally reaches the positive terminal. If the pressure is low, however, the electron can travel a great enough distance to attain ionization potential *(Figure 11-11)*.

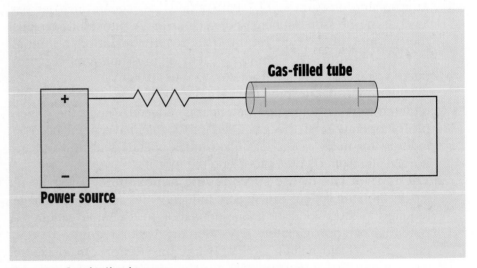

**Figure 11-8** Conduction in a gas

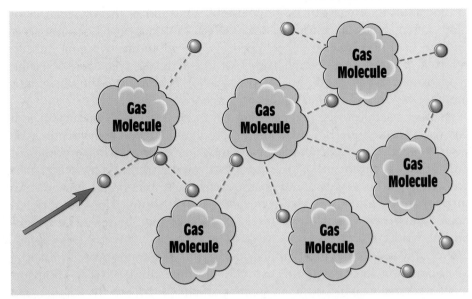

**Figure 11-9**  Electron impact frees other electrons from gas molecules.

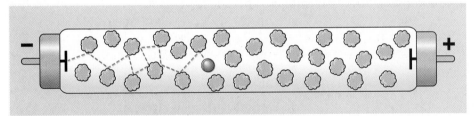

**Figure 11-10**   Under high pressure the electron can travel only a short distance between impacts with gas molecules.

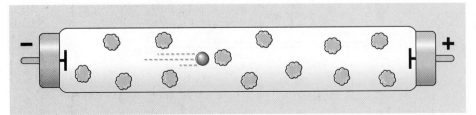

**Figure 11-11**   Under low pressure the electron can travel a greater distance before striking a gas atom.

## Useful Applications

Ionization in a gas can be very useful. One use is the production of light. Often, when an electron strikes a gas molecule that does not have enough energy to release other electrons, the molecule will emit light to rid itself of the excess energy. The color of the light is characteristic of the type of gas. Some examples of this type of lighting are sodium vapor, mercury vapor, and neon.

Another useful device that operates on the principle of conduction in a gas was invented in 1895 by Wilhelm Roentgen. While experimenting, he discovered that when cathode rays struck metal or glass a new kind of radiation was produced. These new rays passed through wood, paper, and air without effect. They passed through thin metal better than through thick and through flesh easier than through bone. Roentgen named these new rays **X-rays**. Today, X-rays are used by doctors throughout the world.

Probably one of the most familiar devices that operates on the principle of conduction in a gas is the cathode ray tube. In 1897 J. J. Thomson measured the weight of the negative particle of the electric charge. He also found that a beam of negative particles could be bent by magnetic fields *(Figure 11-12)*. Today the cathode ray tube is used in television sets, computer display terminals, and oscilloscopes.

**X-rays**

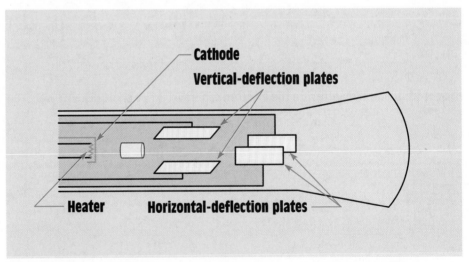

**Cathode**

**Vertical-deflection plates**

**Heater**    **Horizontal-deflection plates**

**Figure 11-12**   Cathode ray tube

## 11-6 Ionization in Nature

One of the most familiar phenomena caused by ionization of gas is the aurora borealis, or northern lights. It is believed that disturbances in the sun cause large amounts of electrons and high-speed protons to be given off and blown into space. When these highly energetic particles reach the earth, they strike the very thin layers of the upper atmosphere and cause ionization to occur. The different colors are caused by different layers of the atmosphere being ionized.

A buildup in the atmosphere of charged particles that are not strong enough to cause lightning will often produce a continuous discharge of light around objects such as the masts of ships or church steeples. This glowing discharge was often referred to as Saint Elmo's fire by sailors.

## Summary

1. Electrical conduction in a liquid is caused by a movement of charged atoms called ions.

2. Pure water is an insulator and becomes a conductor if an acid, an alkali, or a metallic salt is added.

3. Solutions of acids, alkalies, and metallic salts are referred to as electrolytes.

4. Electroplating is the process of depositing the atoms of one type of metal onto another.

5. Several factors that affect the electroplating of metals are:
   A. The electrolyte solution must contain ions of the metal to be plated.
   B. Metal ions are always positively charged.
   C. The object to be plated must be connected to the negative power terminal.
   D. Direct current is used as the power source.
   E. The positive terminal is made of the same metal that is to form the coating.

6. The negative terminal of a device is called the cathode.

7. The positive terminal of a device is called the anode.

8. One of the most common uses for electroplating is the production of pure metals.

9. The term *electrolysis* refers to the process of separating elements electrically.

10. Conduction through a gas depends on the ionization of gas molecules.

11. The amount of voltage required to cause ionization of a gas is called the ionization potential.

12. Gas can be ionized more easily when it is at a low pressure than when it is at a high pressure.

13. The ionization potential is different for different gases.

14. The most important factor in the ionization of a gas is electron impact.

## Review Questions

1. Conduction in a liquid depends on the movement of _____.

2. What is a negative ion?

3. What is a positive ion?

4. Name three basic substances that can be used to produce ionization in a liquid.

5. What is an electrolyte?

6. What is used to hold a sulfate ion together?

7. What is the negative terminal of a power source called?

8. What is the positive terminal of a power source called?

9. What determines the thickness of the coating during an electroplating process?

10. What is electrolysis?

11. What is ionization potential?

12. What is the most important factor in the ionization of a gas?

13. Name two factors that determine the speed an electron attains inside a gas environment.

14. What determines the density of the molecules in a gas environment?

15. What determines the color of light emitted by a gas-filled tube?

**Batteries convert chemical energy into electrical energy. This unit discusses different types of batteries and their characteristics.**

*Courtesy of Power Sonic.*

# Batteries and Other Sources of Electricity

## Key Terms

Battery

Cell

Current capacity

Electromotive series of metals

Hydrometer

Internal resistance

Load test

Nickel-cadmium (ni-cad) cell

Piezoelectricity

Primary cell

Secondary cell

Specific gravity

Thermocouple

Voltaic cell

Voltaic pile

## Outline

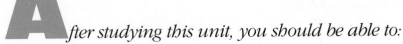

*fter studying this unit, you should be able to:*

- Discuss the differences between primary and secondary cells.
- List voltages for different types of cells.
- Discuss different types of primary cells.
- Construct a cell from simple materials.
- Discuss different types of secondary cells.
- Connect batteries in series and parallel to obtain desired voltage and amp-hour ratings.
- Discuss the operation of solar cells.
- Connect solar cells in series or parallel to produce the desired output voltage and current capacity.
- Discuss the operation of thermocouples.
- Discuss the piezoelectric effect.

**Preview**

Most of the electric power in the world is produced by large rotating machines called alternators. There are other source of electricity, however, that are smaller and are used for emergency situations or for the operation of portable electrical devices. The most common of these small power sources is the battery. Batteries are used to start automobiles, operate toys, flashlights, portable communications equipment, computers, watches, calculators, and hundred of other devices. They range in size from small enough to fit into a wristwatch to large enough to operate electric fork lifts and start diesel trucks.

## 12-1 History of the Battery

In 1791 Luigi Galvani was conducting experiments in anatomy using dissected frog legs preserved in a salt solution. Galvani suspended the frog legs by means of a copper wire. He noticed that when he touched the leg with an iron scalpel the leg would twitch. Galvani realized that the twitch was caused by electricity, but he thought that the electricity was produced by the muscular contraction of the frog's leg. This was the first recorded incident of electricity being produced by chemical action.

In 1800 Alessandro Volta repeated Galvani's experiment. Volta, however, concluded that the electricity was produced by the chemical action of the copper wire, the iron scalpel, and the salt solution. Further experiments led Volta to produce the first practical battery *(Figure 12-1)*. The battery was constructed using zinc and silver discs separated by a piece of cardboard soaked in brine, or saltwater. Volta called his battery a **voltaic pile** because it was a series of individual cells connected together. Each cell produced a certain amount of voltage depending on the materials used to make the cell. A **battery** is actually several cells connected together, although the word *battery* is often used in reference to a single cell. The schematic symbols for an individual **cell** and for a battery are shown in *Figure 12-2.*

**voltaic pile**

**battery**

**cell**

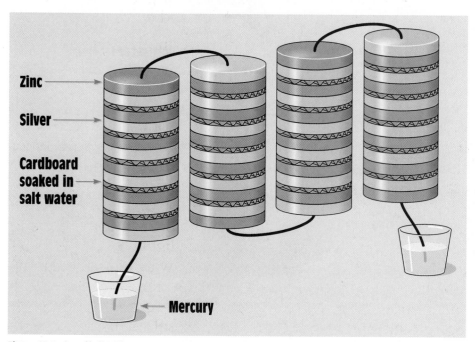

Zinc

Silver

Cardboard soaked in salt water

Mercury

**Figure 12-1** A voltaic pile

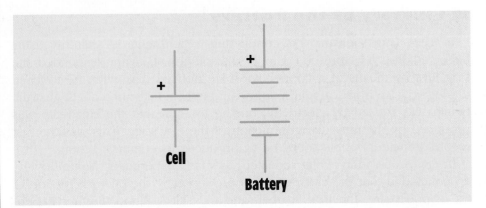

**Figure 12-2** Schematic symbols used to represent an individual cell and a battery

### 12-2 Cells

A **voltaic cell** can be constructed using virtually any two unlike metals and an acid, alkaline, or salt solution. A very simple cell can be constructed as shown in *Figure 12-3*. In this example, a copper wire is inserted in one end of a potato and an aluminum wire is inserted in the other. The

**voltaic cell**

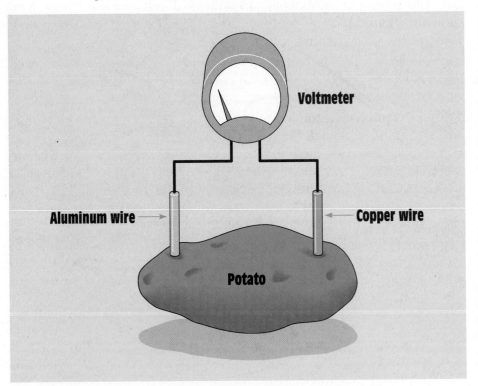

**Figure 12-3** Simple voltaic cell constructed from a potato and two different metals

acid in the potato acts as the electrolyte. If a high-impedance voltmeter is connected to the wires a small voltage can be measured.

Another example of a simple voltaic cell is shown in *Figure 12-4*. In this example two coins of different metal, a nickel and a penny, are separated by a piece of paper. The paper has been wet with saliva from a person's mouth. The saliva contains acid or alkali that acts as the electrolyte. When a high-impedance voltmeter is connected across the two coins it can be seen that a small voltage is produced.

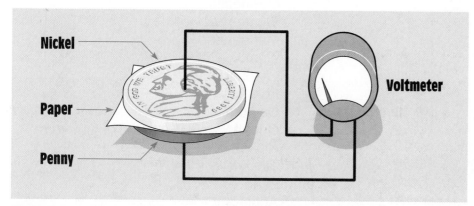

**Figure 12-4** A voltaic cell constructed from two coins and a piece of paper

## 12-3 Cell Voltage

**The amount of voltage produced by an individual cell is determined by the materials from which it is made.** When a voltaic cell is constructed, the plate metals are chosen on the basis of how easily one metal will give up electrons compared with the other. A special list of metals called the **electromotive series of metals** is shown in *Figure 12-5*. This table lists metals in the order of their ability to accept or receive electrons. The metals at the top accept electrons more easily than those at the bottom. The farther apart the metals are on the list, the higher the voltage developed by the cell. One of the first practical cells to be constructed was the zinc-copper cell. This cell uses zinc and copper as the active metals and a solution of water and hydrochloric acid as the electrolyte. Notice that zinc is located closer to the top of the list than copper. Since zinc will accept electrons more readily than copper, zinc will be the negative electrode and copper will be the positive.

Although it is possible to construct a cell from virtually any two unlike metals and an electrolyte solution, not all combinations are practical.

The amount
of voltage
produced by
an individual
cell is determined by the
materials
from which it
is made.

electromotive series
of metals

## ELECTROMOTIVE SERIES OF METALS
### (Partial list)

Lithium
Potassium
Sodium
Calcium
Magnesium
Aluminum
Manganese
Zinc
Chromium
Iron
Cadium
Cobalt
Nickel
Tin
Lead
Antimony
Copper
Mercury
Silver
Platinum
Gold

**Figure 12-5** A partial list of the electromotive series of metals

**primary cell**

**secondary cell**

---

**A primary cell is a cell that cannot be recharged.**

---

**A secondary cell can be recharged.**

Some metals corrode rapidly when placed in an electrolyte solution, and some produce chemical reactions that cause a buildup of resistance. In practice relatively few metals can be used to produce a practical cell.

The table in *Figure 12-6* lists common cells. The table is divided into two sections. One section lists primary cells and the other lists secondary cells. **A primary cell is a cell that *cannot be recharged*.** The chemical reaction of a primary cell causes one of the electrodes to be eaten away as power is produced. When a primary cell becomes discharged, it should be replaced with a new cell. **A secondary cell *can be recharged*.** The recharging process will be covered later in this unit.

| CELL | NEGATIVE PLATE | POSITIVE PLATE | ELECTROLYTE | VOLTS PER CELL |
|------|---------------|---------------|-------------|----------------|
| Primary Cells | | | | |
| Carbon-zinc (Leclanche) | Zinc | Carbon, manganese dioxide | Ammonium chloride | 1.5 |
| Alkaline | Zinc | Manganese dioxide | Potassium hydroxide | 1.5 |
| Mercury | Zinc | Mercuric oxide | Potassium hydroxide | 1.35 |
| Silver-zinc | Zinc | Silver Oxide | Potassium hydroxide | 1.6 |
| Zinc-air | Zinc | Oxygen | Potassium hydroxide | 1.4 |
| Edison-Lalande | Zinc | Copper oxide | Sodium hydroxide | 0.8 |
| Secondary Cells | | | | |
| Lead-acid | Lead | Lead dioxide | Dilute sulfuric acid | 2.2 |
| Nickel-iron (Edison) | Iron | Nickel oxide | Potassium hydroxide | 1.4 |
| Nickel-cadmium | Cadmium | Nickel hydroxide | Potassium hydroxide | 1.2 |
| Silver-zinc | Zinc | Silver oxide | Potassium hydroxide | 1.5 |
| Silver-cadium | Cadmium | Silver oxide | Potassium hydroxide | 1.1 |

**Figure 12-6** Voltaic cells

## 12-4 Primary Cells

An example of a primary cell is the zinc-copper cell shown in *Figure 12-7.* This cell consists of two electrodes (terminals that conduct electricity into or away from a conducting substance), one zinc and one copper, suspended in a solution of water and hydrochloric acid. This cell will produce approximately 1.08 V. The electrolyte contains positive $H^+$ ions, which attract electrons from the zinc atoms, which have two valence electrons. When an electron bonds with the $H^+$ ion it becomes a neutral H atom. The neutral H atoms combine to form $H_2$ molecules of hydrogen gas. This hydrogen can be seen bubbling away in the electrolyte solution. The zinc ions, $Zn^{++}$, are attracted to negative $Cl^-$ ions in the electrolyte solution.

The copper electrode provides another source of electrons that can be attracted by the $H^+$ ions. When a circuit is completed between the zinc and copper electrodes *(Figure 12-8),* electrons are attracted from the zinc electrode to replace the electrons in the copper electrode that are attracted into the solution. After some period of time, the zinc electrode dissolves as a result of the zinc ions being distributed in the solution.

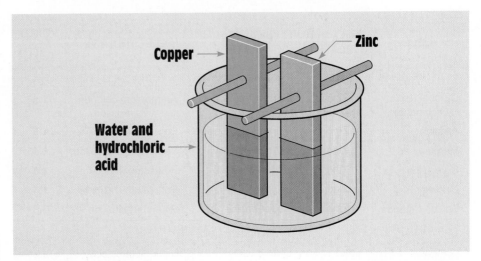

**Figure 12-7** Zinc-copper cell

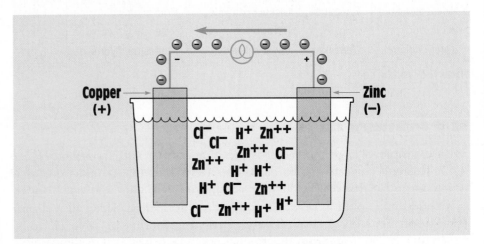

**Figure 12-8** Electrons flow from the zinc to the copper electrode.

## The Carbon-Zinc Cell

One of the first practical primary cells was invented by Leclanché. It is known as the carbon-zinc cell, or the Leclanché cell. Leclanché used a carbon rod as an electrode instead of copper and a mixture of ammonium chloride, manganese dioxide, and granulated carbon as the electrolyte. The mixture was packed inside a zinc container that acted as the negative electrode *(Figure 12-9)*. This cell is often referred to as a dry cell because the electrolyte is actually a paste instead of a liquid. The use of a paste permits the cell to be used in any position without spilling the electrolyte.

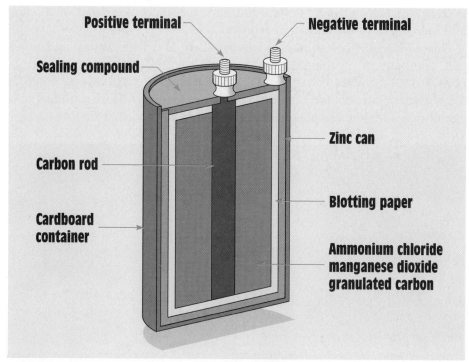

**Figure 12-9** Carbon-zinc cell

As the cell is discharged, the zinc container is eventually dissolved. When the container has dissolved, the cell should be discarded immediately and replaced. Failure to discard the cell can result in damage to equipment.

## Alkaline Cells

Another primary cell very similar to the carbon-zinc cell is the alkaline cell. This cell uses a zinc can as the negative electrode and manganese dioxide, $MnO_2$, as the positive electrode. The major electrolytic ingredient is potassium hydroxide. Like the carbon-zinc cell, the alkaline cell contains a paste electrolyte and is considered a dry cell also. The voltage developed is the same as in the carbon-zinc cell, 1.5 V. The major advantage of the alkaline cell is longer life. The average alkaline cell can supply from three to five times the power of a carbon-zinc cell depending on the discharge rate when the cell is being used. The major disadvantage is cost. Although some special alkaline cells can be recharged, the number of charge and discharge cycles is limited. In general, the alkaline cell is considered to be a nonrechargeable cell. A D-size alkaline-manganese cell is shown in *Figure 12-10*. Several different sizes and case styles of alkaline and mercury cells are shown in *Figure 12-11*.

One of the devices shown in *Figure 12-11* is a true battery and not a cell. The battery is rectangular and has two snap-on type terminal connections at the top. This battery actually contains six individual cells rated at 1.5 V each. The cells are connected in series and provide a terminal voltage of 9 V. The internal make-up of the battery is shown in *Figure 12-12*.

**Figure 12-10** Alkaline-manganese cell *(Courtesy of Duracell Inc.)*

**Figure 12-11** Different sizes and case styles of primary cells *(Courtesy of Duracell Inc.)*

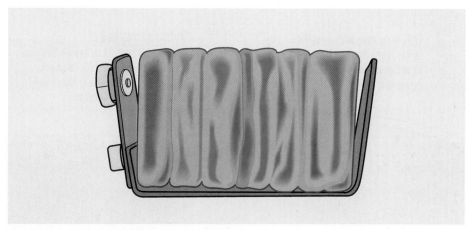

**Figure 12-12**   Internal make-up of a 9-V battery

## Button Cells

Another common type of primary cell is the button cell. The button cell is so named because it resembles a button. Button cells are commonly used in cameras, watches, hearing aids, and hand-held calculators. Most button cells are constructed using mercuric oxide as the cathode, zinc as the anode, and potassium hydroxide as the electrolyte *(Figure 12-13)*. Although these cells are more expensive than other cells, they have a high energy density and a long life. The mercury-zinc cell has a voltage of 1.35 V.

Another type of button cell that is less common because of its higher cost is the silver-zinc cell. This cell is the same as the mercury-zinc cell except that it uses silver oxide as the cathode material instead of mercuric oxide. The silver-zinc cell does have one distinct advantage over the mercury-zinc cell. The silver-zinc cell develops a voltage of 1.6 V as compared with the 1.35 V developed by the mercury-zinc cell. This higher voltage can be of major importance in some electronic circuits.

A variation of the silver-zinc cell uses divalent silver oxide as the cathode material instead of silver oxide. The chemical formula for silver oxide is $Ag_2O$. The chemical formula for divalent silver oxide is $Ag_2O_2$. One of the factors that determines the energy density of a cell is the amount of oxygen in the cathode material. Although divalent silver oxide contains twice as much oxygen as silver oxide, it does not produce twice the energy density. The increased energy density is actually about 10% to 15%. Using divalent silver oxide does have one disadvantage, however. The compound is less stable, which can cause the cell to have a shorter shelf life.

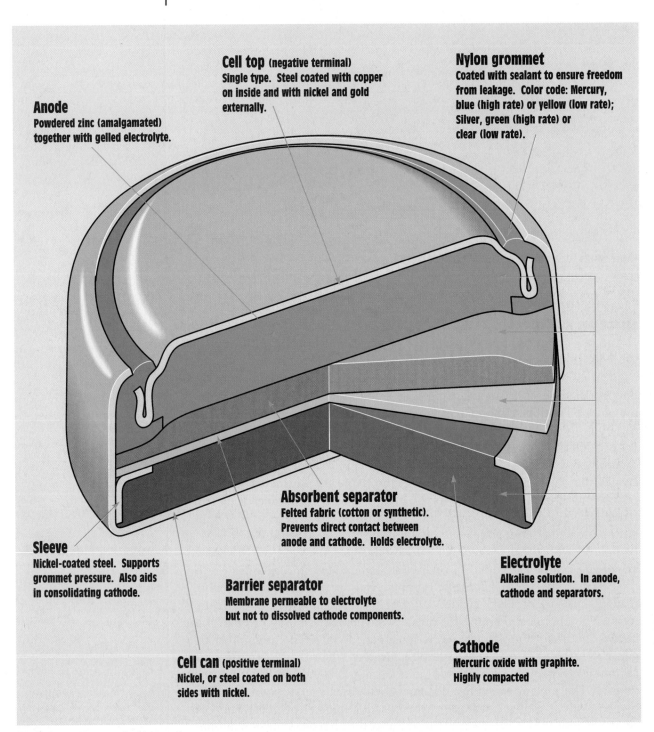

**Anode**
Powdered zinc (amalgamated) together with gelled electrolyte.

**Cell top** (negative terminal)
Single type. Steel coated with copper on inside and with nickel and gold externally.

**Nylon grommet**
Coated with sealant to ensure freedom from leakage. Color code: Mercury, blue (high rate) or yellow (low rate); Silver, green (high rate) or clear (low rate).

**Sleeve**
Nickel-coated steel. Supports grommet pressure. Also aids in consolidating cathode.

**Barrier separator**
Membrane permeable to electrolyte but not to dissolved cathode components.

**Cell can** (positive terminal)
Nickel, or steel coated on both sides with nickel.

**Absorbent separator**
Felted fabric (cotton or synthetic). Prevents direct contact between anode and cathode. Holds electrolyte.

**Electrolyte**
Alkaline solution. In anode, cathode and separators.

**Cathode**
Mercuric oxide with graphite. Highly compacted

**Figure 12-13** Mercury button cell

## Lithium Cells

Lithium cells should probably be referred to as the lithium system because there are several different types of lithium cells. Lithium cells can have voltages that range from 1.9 V to 3.6 V depending on the material used to construct the cell. Lithium is used as the anode material because of its high affinity for oxygen. In fact, for many years lithium had to be handled in an airless and moisture-free environment because of its extreme reactivity with oxygen. Several different cathode materials can be used depending on the desired application. One type of lithium cell uses a solid electrolyte of lithium hydroxide. The use of a solid electrolyte produces a highly stable compound resulting in a shelf life that is measured in decades. This cell produces a voltage of 1.9 V but has an extremely low current capacity. The output current of the lithium cell is measured in microamps (millionths of an amp) and is used to power watches with liquid crystal displays and to maintain memory circuits in computers.

Other lithium cells use liquid electrolytes and can provide current outputs comparable to alkaline-manganese cells. Another lithium system uses a combination electrolyte-cathode material. One of these is sulfur dioxide, $SO_2$. This combination produces a voltage of 2.9 V. In another electrolyte-cathode system sodium chloride, NaCl, is used. This combination provides a terminal voltage of 3.6 V.

Although lithium cells are generally considered to be primary cells, some types are rechargeable. The amount of charging current, however, is critical for these cells. The incorrect amount of charging current can cause the cell to explode.

## Current Capacity and Cell Ratings

**The amount of current a particular type of cell can deliver is determined by the surface area of its plates**. A D cell can deliver more current than a C cell, and a C cell can deliver more current than an AA cell. The amount of power a cell can deliver is called its **current capacity**. To determine a cell's current capacity, several factors must be included, such as the type of cell, the rate of current flow, the voltage, and the length of time involved. Primary cells are generally limited by size and weight, and, therefore, do not contain a large amount of power.

One of the common ratings for primary cells is the milliampere-hour. A milliampere (mA) is 1/1000 of an amp. Therefore, if a cell can provide a current of 1 mA for 1 hr it will have a rating of 1 mA-hr. An average D size alkaline cell has a capacity of approximately 10,000 mA-hr. Some simple calculations would reveal that this cell should be able to supply 100 mA of current for a period of 100 hours, or 200 mA of current for 50 hours.

Another common measure of a primary cell's current capacity is watt-hours.

**The amount of current a particular type of cell can deliver is determined by the surface area of its plates.**

**current capacity**

Watt-hours are determined by multiplying the cell's milliampere-hour rating by its terminal voltage. If the alkaline-manganese cell just discussed has a voltage of 1.5 V, its watt-hour (W-hr) capacity would be 15 W-hr (10,000 mA-hr x 1.5 V = 15,000 mW-hr, or 15 W-hr). The chart in *Figure 12-14* shows the watt-hour capacity for several sizes of alkaline-manganese cells. The chart lists the cell size, the volume of the cell, and the watt-hours per cubic inch.

The amount of power a cell contains depends not only on the volume of the cell, but also on the type of cell being used. The chart in *Figure 12-15* compares the watt-hours per cubic inch for different types of cells.

| ALKALINE-MANGANESE CELL SIZE | VOLUME IN CUBIC INCHES | WATT-HOURS PER CUBIC INCH |
|:---:|:---:|:---:|
| D cell | 3.17 | 4.0 |
| C cell | 1.52 | 4.2 |
| AA cell | 0.44 | 4.9 |
| AAA cell | 0.20 | 5.1 |

**Figure 12-14** Watt-hours per cubic inch for different size cells

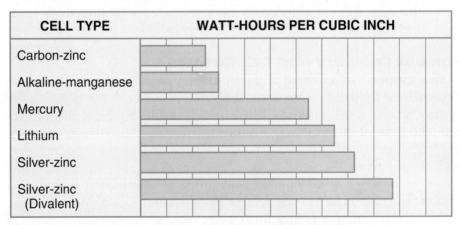

**Figure 12-15** Watt-hours per cubic inch for different types of cells

## Internal Resistance

Batteries actually have two voltage ratings, one at no load and the other at normal load. The cell's rated voltage at normal load is the one used. The no-load voltage of a cell will be greater because of the internal resistance of the cell. All cells have some amount of **internal resistance**. For example, the alkaline cell discussed previously had a rating of 10,000 mA-hr, or 10 A-hr. Theoretically the cell should be able to deliver 10 A of current for 1 hr or 20 A of current for a half-hour. In practice the cell could not deliver 10 A of current even under a short circuit condition. *Figure 12-16* illustrates what happens when a DC ammeter is connected directly across the terminals of a D size alkaline cell. Assume that the ammeter indicates a current flow of 4.5 A and that the terminal voltage of the cell has dropped to 0.5 V. By applying Ohm's law, it can be determined that the cell has an internal resistance of 0.111 $\Omega$ (0.5 V/4.5 A = 0.111 $\Omega$).

As the cell ages and power is used, the electrodes and electrolyte begin to deteriorate. This causes them to become less conductive, which results in an increase of internal resistance. As the internal resistance increases, the terminal voltage decreases.

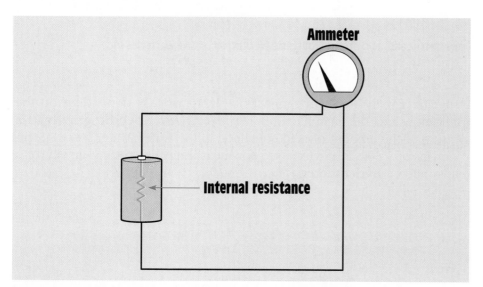

**Figure 12-16**  Short circuit current is limited by internal resistance.

## 12-5 Secondary Cells: Lead-Acid Batteries

The secondary cell is characterized by the fact that once its stored energy has been depleted it can be recharged. One of the most common types of secondary cells is the lead-acid. A string of 30 individual cells connected to form one battery is shown in *Figure 12-17*. A single lead-acid cell consists of one plate made of pure lead, Pb, a second plate of lead dioxide, $PbO_2$, and an electrolyte of dilute sulfuric acid, $H_2SO_4$, with

**Figure 12-17**   Lead-acid storage batteries *(Courtesy of GNB Batteries Inc.)*

**specific gravity**

**hydrometer**

a specific gravity that can range from 1.215 to 1.28 depending on the application and the manufacturer *(Figure 12-18).* **Specific gravity** is a measure of the amount of acid contained in the water. Water has a specific gravity of 1.000. A device used for measuring the specific gravity of a cell is called a **hydrometer** *(Figure 12-19).*

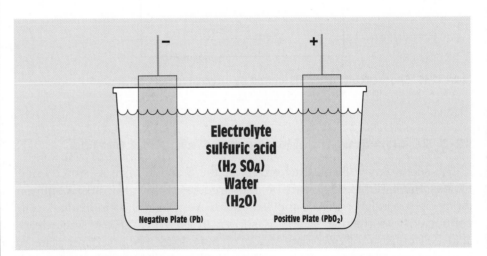

**Figure 12-18**   Basic lead-acid cell

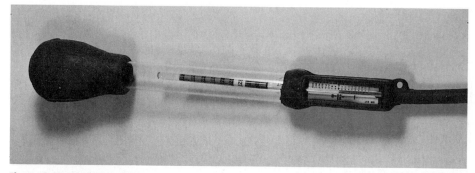

**Figure 12-19**  Hydrometer

## Discharge Cycle

When a load is connected between the positive and negative terminals, the battery begins to discharge its stored energy through the load *(Figure 12-20)*. The process is called the discharge cycle. Each of the lead atoms on the surface of the negative plate loses two electrons to become a $Pb^{++}$ ion. These positive ions attract $SO_4^{--}$ ions from the electrolyte. As a result a layer of lead sulfate, $PbSO_4$, forms on the negative lead plate.

The positive plate is composed of lead dioxide, $PbO_2$. Each of the Pb atoms lacks four electrons that were given to the oxygen atom when the compound was formed and the Pb became $Pb^{++++}$ ion. Each of the $Pb^{++++}$ ions takes two electrons from the load circuit to become a $Pb^{++}$ ion. The $Pb^{++}$ ions cannot hold the oxygen atoms that are released into the electrolyte solution and combine with hydrogen atoms to form molecules of

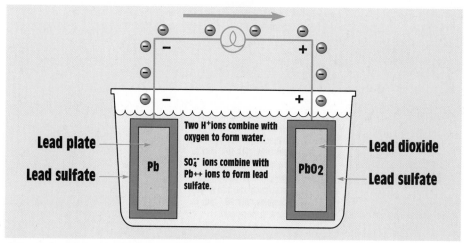

**Figure 12-20**   Discharge cycle of a lead-acid cell

water, $H_2O$. The remaining $Pb^{++}$ ions combine with $SO_4^{--}$ ions and form a layer of lead sulfate around the positive plate also.

As the cell is discharged, two $H^+$ ions contained in the electrolyte combine with an oxygen atom liberated from the lead dioxide to form water. For this reason, the hydrometer can be used to test the specific gravity of the cell to determine the state of charge. The more discharged the cell becomes, the more water is formed in the electrolyte, and the lower the specific gravity reading becomes.

## Charging Cycle

The secondary cell can be recharged by reversing the chemical action that occurred during the discharge cycle. This process, called the charging cycle, is accomplished by connecting a direct current power supply or generator to the cell. The positive output of the power supply connects to the positive terminal of the cell, and the negative output of the power supply connects to the negative terminal of the cell *(Figure 12-21)*.

The terminal voltage of the power supply must be greater than that of the cell or battery. As current flows through the cell, hydrogen is produced at the negative plate and oxygen is produced at the positive plate. If the cell is in a state of discharge, both plates are covered with a layer of lead sulfate. The $H^+$ ions move toward the negative plate and combine with $SO_4^{--}$ ions to form new molecules of sulfuric acid. As a result, $Pb^{++}$ ions are left at the plate. These ions combine with electrons being supplied by the power supply and again become neutral lead atoms.

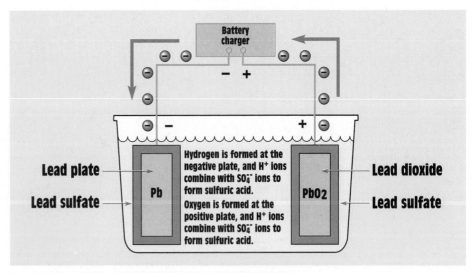

**Figure 12-21** A lead-acid cell during the charging cycle

At the same time, water molecules break down at the positive plate. The hydrogen atoms combine with $SO_4^{--}$ ions in the electrolytic solution to form sulfuric acid. The oxygen atom recombines with the lead dioxide to form a $Pb^{++}$ ion. As electrons are removed from the lead dioxide by the power supply, the $Pb^{++}$ ions become $Pb^{++++}$ ions.

As the cell is charged, sulfuric acid is again formed in the electrolyte. The hydrometer can again be used to determine the state of charge. When the cell is fully charged, the electrolyte should be back to its original strength.

Most lead-acid cells contain multiple plates as shown in *Figure 12-22*. One section of plates is connected together to form a single positive plate, while the other section forms a single negative plate. This arrangement increases the surface area and thus the current capacity of the cell.

**Figure 12-22**  Multiple plates increase the surface area and the amount of current the cell can produce.

## Cautions Concerning Charging

Theoretically a secondary cell should be able to be discharged and recharged indefinitely. In practice, however, it cannot. If the cell is overcharged, that is, the amount of charge current is too great, the lead sulfate does not have a chance to dissolve back into the electrolyte and become acid. High charging current or mechanical shock can cause large flakes of lead sulfate to break away from the plates and fall to the bottom of the cell *(Figure 12-23)*. These flakes can no longer recombine with the $H^+$ ions to become sulfuric acid. Therefore, the electrolyte is permanently weakened. If the flakes build up to a point that they touch the plates, they cause a short circuit and the cell can no longer operate.

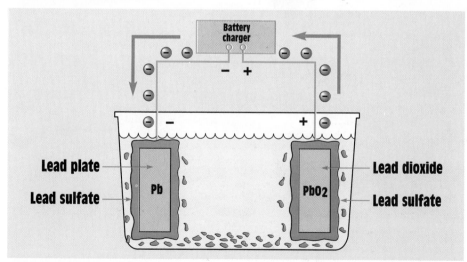

**Figure 12-23**  Improper charging can cause lead sulfate to flake off and fall to the bottom of the cell.

**Overcharging can also cause hydrogen gas to form.**

**Overcharging can also cause hydrogen gas to form**. Because hydrogen is the most explosive element known, sparks or open flames should be kept away from batteries or cells, especially during the charging process. Overcharging also causes excess heat that can permanently damage the cell. The accepted temperature limit for most lead-acid cells is 110°F.

The proper amount of charging current can vary from one type of battery to another, and manufacturers' specifications should be followed when possible. A general rule concerning charging current is that the current should not be greater than 1/10 the amp-hour capacity. An 80-A-hr battery for instance should not be charged with a current greater than 8 A.

## Sealed Lead-Acid Batteries

Sealed lead-acid batteries have become increasingly popular in the past several years. They come in different sizes, voltage ratings, amp-hour ratings, and case styles *(Figure 12-24)*. These batteries are often referred to as gel cells because the sulfuric acid electrolyte is suspended in an immobilized gelatin state. This treatment prevents spillage and permits the battery to be used in any position. Gel cells utilize a cast grid constructed of lead-calcium, which is free of antimony. The calcium is used to add strength to the grid. The negative plate is actually a lead paste material,

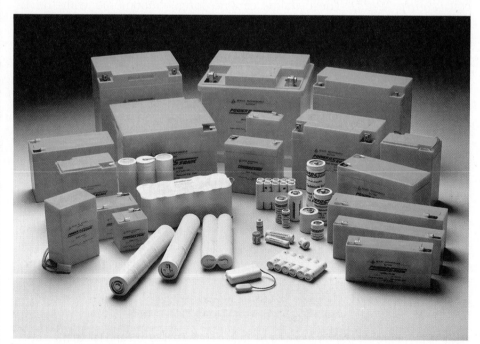

**Figure 12-24**  Sealed lead-acid and NI-CAD cells and batteries *(Courtesy of Power Sonic.)*

and the positive plate is made of lead dioxide paste. A one-way pressure relief valve set to open at 2–6 PSI is used to vent any gas buildup during charging.

## Ratings for Lead-Acid Batteries

One of the most common ratings for lead-acid batteries is the amp-hour rating. The amp-hour rating for lead-acid batteries is determined by measuring the battery's ability to produce current for a 20-hr period at 80°F. A battery with the ability to produce a current of 4 A for 20 hr would have a rating of 80 A-hr.

Another common battery rating, especially for automotive batteries, is cold-cranking amps. This rating has nothing to do with the amp-hour rating of the battery. Cold-cranking amps is the maximum amount of initial current the battery can supply at 20°C (68°F).

## Testing Lead-Acid Batteries

It is sometimes necessary to test the state of charge or condition of a lead-acid battery. The state of charge can often be tested with a hydrometer as previously described. As batteries age, however, the specific gravity remains low even after the battery has been charged. When this happens, it is an indication that the battery has lost part of its materials because of lead sulfate flaking off the plates and falling to the bottom of the battery. When this happens, there is no way to recover the material.

Another standard test for lead-acid batteries is the load test. This test will probably reveal more information concerning the condition of the battery than any other test. To perform a **load test**, the amount of test current should be three times the amp-hour capacity. The voltage should not drop below 80% of the terminal voltage for a period of three minutes. For example, an 80-A-hr, 12-V battery is to be load tested. The test current will be 240 A (80 x 3), and the voltage should not drop below 9.6 V (12 x 0.80) for a period of three minutes.

**load test**

## 12-6 Other Secondary Cells

## Nickel-Iron Batteries (Edison Battery)

The nickel-iron cell is often referred to as the Edison cell, or Edison battery. The nickel-iron battery was developed in 1899 for use in electric cars being built by the Edison Company. The negative plate is a nickeled steel grid that contains powdered iron. The positive plates are nickel tubes that contain nickel oxides and nickel hydroxides. The electrolyte is a solution of 21% potassium hydroxide.

The nickel-iron cell weighs less than lead-acid cells but has a lower energy density. The greatest advantage of the nickel-iron cell is its ability to withstand deep discharges and to recover without harm to the cell. The nickel-iron cell can also be left in a state of discharge for long periods of time without harm. Because these batteries need little maintenance, they are sometimes found in portable and emergency lighting equipment. They are also used to power electric mine locomotives and electric forklifts.

The nickel-iron battery does have two major disadvantages. One is high cost. These batteries cost several times more than comparable lead-acid batteries. The second disadvantage is high internal resistance. Nickel-iron batteries do not have the ability to supply the large initial currents needed to start gasoline or diesel engines.

### Nickel-Cadmium (ni-cad) Batteries

The **nickel-cadmium (ni-cad) cell** was first developed in Sweden by Junger and Berg in 1898. The positive plate is constructed of nickel hydroxide mixed with graphite. The graphite is used to increase conductivity. The negative plate is constructed of cadmium oxide, and the electrolyte is potassium hydroxide with a specific gravity of approximately 1.2. Ni-cad batteries have extremely long life spans. On the average, they can be charged and discharged about 2000 times. Ni-cad batteries can be purchased in a variety of case styles *(Figure 12-25)*.

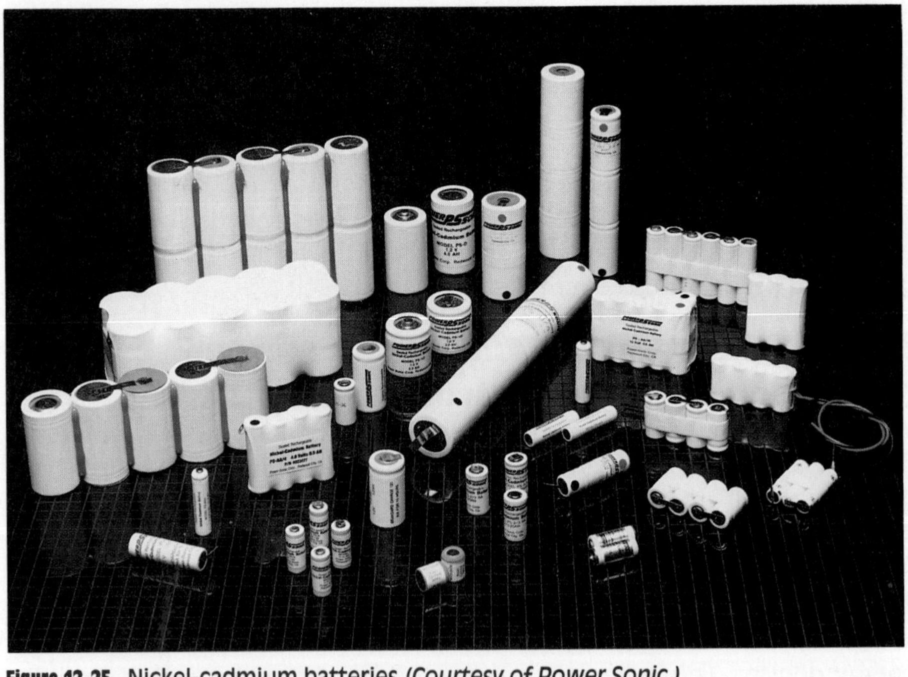

**nickel-cadmium (ni-cad) cell**

**Figure 12-25** Nickel-cadmium batteries *(Courtesy of Power Sonic.)*

Nickel-cadmium batteries have the ability to produce large amounts of current, similar to the lead-acid battery, but do not experience the voltage drop associated with the lead-acid battery *(Figure 12-26)*.

Nickel-cadmium batteries do have some disadvantages:

1. The ni-cad battery develops only 1.2 V per cell as compared with 1.5 V for carbon-zinc and alkaline primary cells, or 2 V for lead-acid cells.
2. Ni-cad batteries cost more initially than lead-acid batteries.
3. Ni-cad batteries remember their charge-discharge cycles. If they are used at only low currents and are permitted to discharge through only part of a cycle and then are recharged, over a long period of time they will develop a characteristic curve to match this cycle.

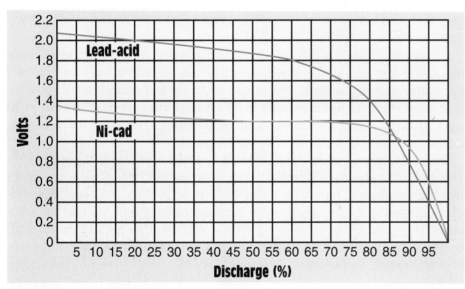

**Figure 12-26**  Typical discharge curves for ni-cad and lead-acid cells

## 12-7 Series and Parallel Battery Connections

When batteries or cells are connected in series, their voltages add and their current capacities remain the same. In *Figure 12-27*, four batteries, each having a voltage of 12 V and 60 A-hr, are connected in series. Connecting them in series has the effect of maintaining the surface area of the plates and increasing the number of cells. The connection shown in *Figure 12-27* will have an output voltage of 48 V and a current capacity of 60 A-hr.

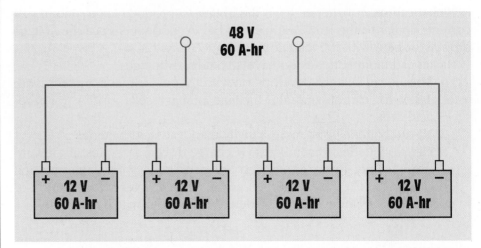

**Figure 12-27**   When batteries are connected in series, their voltages add, and the amp-hour capacities remain the same.

**Batteries of different voltages should never be connected in parallel.**

Connecting batteries or cells in parallel *(Figure 12-28)* has the effect of increasing the area of the plates. In this example, the same four batteries are connected in parallel. The output voltage will remain 12 V, but the amp-hour capacity has increased to 240 A-hr. **Batteries of different voltages should never be connected in parallel.** The batteries could be damaged or one of the batteries could explode.

Batteries can also be connected in a series-parallel combination. In *Figure 12-29*, the four batteries have been connected in such a manner that the output will have a value of 24 V and 120 A-hr. To make this connection, the four batteries were divided into two groups of two batteries each. The batteries of each group were connected in series to produce an output of 24 V at 60 A-hr. These two groups were then connected in parallel to provide an output of 24 V at 120 A-hr.

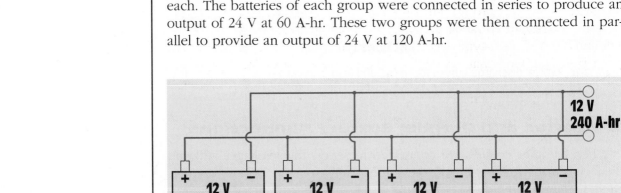

**Figure 12-28**   When batteries are connected in parallel, their voltages remain the same and their amp-hour capacities add.

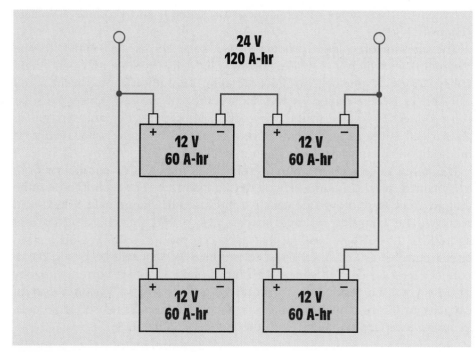

**Figure 12-29** Series-parallel connection

## 12-8 Other Small Sources of Electricity

### Solar Cells

Although batteries are the largest source of electricity after alternators and generators, they are not the only source. One source of electricity is the photovoltaic cell or solar cell. Solar cells are constructed from a combination of P- and N-type semiconductor material. Semiconductors are made from materials that contain four valence electrons. The most common semiconductor materials are silicon and germanium. Impurities must be added to the pure semiconductor material to form P- and N-type materials. If a material containing three valence electrons is added to a pure semiconductor material, P-type material is formed. P-type material has a lack of electrons. N-type material is formed when a material containing five valence electrons is added to a pure semiconductor material. N-type material has an excess of electrons.

Light is composed of particles called photons. A photon is a small package of pure energy that contains no mass. When photons strike the surface of the photocell, the energy contained in the photon is given to a free electron. This additional energy causes the electron to cross the junction between the two types of semiconductor material and produce a voltage *(Figure 12-30)*.

The amount of voltage produced by a solar cell is determined by the material from which it is made. Silicon solar cells produce an open circuit voltage of 0.5 V per cell in direct sunlight. The amount of current a cell can deliver is determined by the surface area of the cell. Since the solar cell produces a voltage in the presence of light, the schematic symbol for a solar cell is the same as that used to represent a single voltaic cell with the addition of an arrow to indicate it is receiving light *(Figure 12-31)*.

It is often necessary to connect solar cells in series or parallel or both to obtain desired amounts of voltage and current. For example, assume that an array of photovoltaic cells is to be used to charge a 12-V lead-acid battery. The charging voltage is to be 14 V and the charging current should be 0.5 A. Now assume that each cell produces 0.5 V with a short circuit current of 0.25 A. In order to produce 14 V, it will be necessary to connect 28 solar cells in series. An output of 14 V with a current capacity of 0.25 A will be produced. To produce an output of 14 V with a current capacity of 0.5 A, it will be necessary to connect a second set of 28 cells in series and parallel this set with the first set *(Figure 12-32)*.

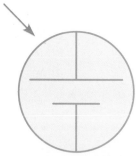

**Figure 12-31** Schematic symbol for a solar cell

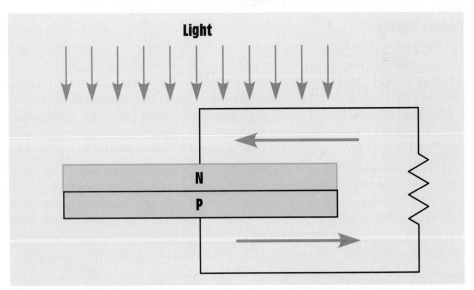

**Figure 12-30** Solar cells are formed by bonding P- and N-type semiconductor materials together.

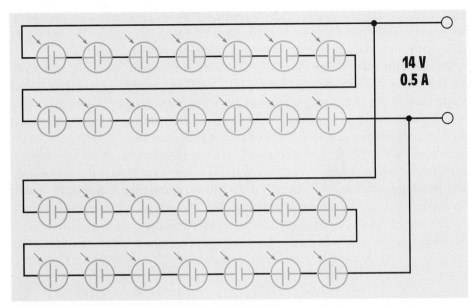

**Figure 12-32**   Series-parallel connection of solar cells

## Thermocouples

In 1822, a German scientist named Seebeck discovered that when two dissimilar metals are joined at one end and that junction is heated, a voltage is produced *(Figure 12-33)*. This is known as the Seebeck effect. The device produced by the joining of two dissimilar metals for the purpose of producing electricity with heat is called a **thermocouple**. The amount of voltage produced by a thermocouple is determined by:

1. *the type of materials used to produce the thermocouple,*
2. *the temperature difference between the two junctions.*

**thermo-couple**

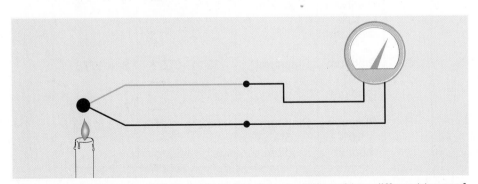

**Figure 12-33**   The thermocouple is made by forming a junction of two different types of metal.

The chart in *Figure 12-34* shows common types of thermocouples. The different metals used in the construction of thermocouples are shown as well as the normal temperature ranges of the thermocouples.

The amount of voltage produced by a thermocouple is small, generally on the order of millivolts (1 mV= 0.001 V). The polarity of the voltage of some thermocouples is determined by the temperature. For example, a type J thermocouple produces 0 V at about 32°F. At temperatures above 32°F, the iron wire is positive and the constantan wire is negative. At temperatures below 32°F, the iron wire becomes negative, and the constantan wire becomes positive. At a temperature of 300°F, a type J thermocouple will produce a voltage of about 7.9 mV. At a temperature of -300°F, it will produce a voltage of about -7.9 mV.

Because thermocouples produce such low voltages, they are often connected in series as shown in *Figure 12-35*. This connection is referred to

| TYPE | MATERIAL | | Degrees F | Degrees C |
|------|----------|--|-----------|-----------|
| J | Iron | Constantan | −328 to +32<br>+32 to +1432 | −200 to 0<br>0 to 778 |
| K | Chromel | Alumel | −328 to +32<br>+32 to +2472 | −200 to 0<br>0 to 1356 |
| T | Copper | Constantan | −328 to +32<br>+32 to 752 | −200 to 0<br>0 to 400 |
| E | Chromal | Constantan | −328 to +32<br>+32 to 1832 | −200 to 0<br>0 to 1000 |
| R | Platinum<br>13%<br>rhodium | Platinum | −32 to +3232 | 0 to 1778 |
| S | Platinum<br>10%<br>rhodium | Platinum | −32 to +3232 | 0 to 1778 |
| B | Platinum<br>10%<br>rhodium | Platinum<br>6%<br>rhodium | +992 to +3352 | 533 to 1800 |

**Figure 12-34**  Thermocouple chart

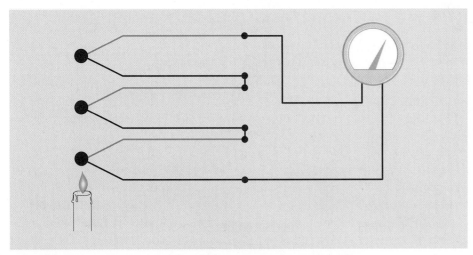

**Figure 12-35**   A thermopile is a series connection of thermocouples.

as a thermopile. Thermocouples and thermopiles are generally used for measuring temperature and are sometimes used to detect the presence of a pilot light in appliances that operate with natural gas. The thermocouple is heated by the pilot light. The current produced by the thermocouple is used to produce a magnetic field, which holds a gas valve open to permit gas to flow to the main burner. If the pilot light goes out, the thermocouple ceases to produce current and the valve closes *(Figure 12-36)*.

## Piezoelectricity

The word *piezo*, pronounced pee-ay´-zo, is derived form the Greek word for pressure. **Piezoelectricity** is produced by some materials when they are placed under pressure. The pressure can be caused by compression, twisting, bending, or stretching. Rochelle salt (sodium potassium tartrate) is often used as the needle, or stylus, of a phonograph. Grooves in the record cause the crystal to be twisted back and forth producing an alternating voltage. This voltage is amplified and is heard as music or speech. Rochelle salt crystals are also used as the pickup for microphones. The vibrations caused by sound waves produce stress on the crystals. This stress causes the crystals to produce an alternating voltage that can be amplified.

Another crystal used to produce the piezoelectric effect is barium titanate. Barium titanate can actually produce enough voltage and current to flash a small neon lamp when struck by a heavy object *(Figure 12-37)*. Industry uses the crystal's ability to produce voltage in transducers for sensing pressure and mechanical vibration of machine parts.

**piezo-electricity**

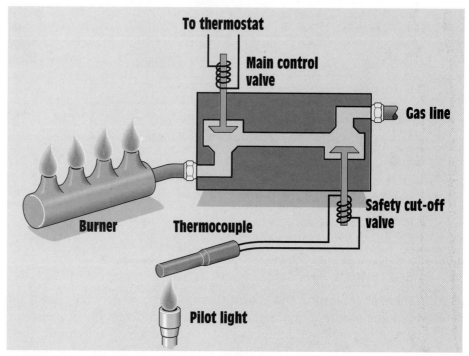

**Figure 12-36** A thermocouple provides power to the safety cut-off valve.

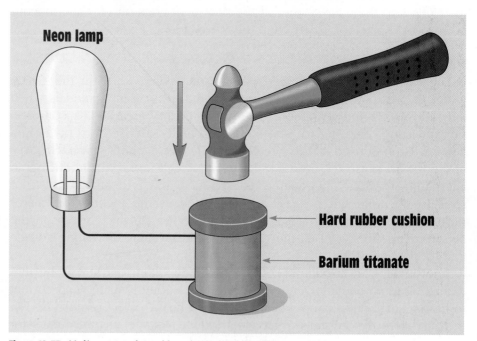

**Figure 12-37** Voltage produced by piezoelectric effect

Quartz crystal has been used for many years as the basis for crystal oscillators. In this application, an AC voltage close to the natural mechanical vibration frequency of a slice of quartz is applied to opposite surfaces of the crystal. This will cause the quartz to vibrate at its natural frequency. The frequency is extremely constant for a particular slice of quartz. Quartz crystals have been used to change the frequency range of two-way radios for many years.

## Summary

1. The first practical battery was invented by Alessandro Volta in 1800.

2. A voltaic cell converts chemical energy into electrical energy.

3. A battery is a group of cells connected together.

4. Primary cells cannot be recharged.

5. Secondary cells can be recharged.

6. The amount of voltage produced by a cell is determined by the materials from which it is made.

7. A voltaic cell can be constructed from almost any two unlike metals and an electrolyte of acid, alkali, or salt.

8. Voltaic cells depend on the movement of ions in solution to produce electricity.

9. The amount of current a cell can provide is determined by the area of the cell plates.

10. Primary cells are often rated in milliampere-hours or watt-hours.

11. The amount of energy a battery can contain is called its capacity or energy density.

12. Secondary cells can be recharged by reversing the current flow through them.

13. A hydrometer is a device for measuring the specific gravity of the electrolyte.

14. When lead-acid batteries are charged, hydrogen gas is produced.

15. Nickel-cadmium batteries have a life span of about 2000 charge-discharge cycles.

16. When cells are connected in series, their voltages add and their amp-hour capacities remain the same.

17. When cells are connected in parallel, their voltages remain the same and their amp-hour capacities add.

18. Cells or batteries of different voltages should never be connected in parallel.

19. Solar cells produce electricity in the presence of light.

20. The amount of voltage produced by a solar cell is determined by the materials from which it is made.

21. The amount of current produced by a solar cell is determined by the surface area.

22. Thermocouples produce a voltage when the junction of two unlike metals is heated.

23. The amount of voltage produced by a thermocouple is determined by the type of materials used and the temperature difference between the ends of the junction.

24. The voltage polarity of some thermocouples is determined by the temperature.

25. Thermocouples connected in series to produce a higher voltage are called a thermopile.

26. Some crystals can produce a voltage when placed under pressure.

27. The production of voltage by application of pressure is called the piezoelectric effect.

## Review Questions

1. What is a voltaic cell?

2. What factors determine the amount of voltage produced by a cell?

3. What determines the amount of current a cell can provide?

4. What is a battery?

5. What is a primary cell?

6. What is a secondary cell?

7. What material is used as the positive electrode in a zinc-mercury cell?

8. What is another name for the Leclanché cell?

9. What is used as the electrolyte in a carbon-zinc cell?

10. What is the advantage of the alkaline cell as compared with the carbon-zinc cell?

11. What material is used as the positive electrode in an alkaline cell?

12. How is the amp-hour capacity of a lead-acid battery determined?

13. What device is used to test the specific gravity of a cell?

14. A 6-V lead-acid battery has an amp-hour rating of 180 A-hr. The battery is to be load tested. What should be the test current, and what are the maximum permissible amount and duration of the voltage drop?

15. Three 12-V, 100 A-hr batteries are connected in series. What are the output voltage and amp-hour capacity of this connection?

16. What is the voltage produced by a silicon solar cell?

17. What determines the current capacity of a solar cell?

18. A solar cell can produce a voltage of 0.5 V and has a current capacity of 0.1 A. How many cells should be connected in series and parallel to produce an output of 6 V at 0.3 A?

19. What determines the amount of voltage produced by a thermocouple?

20. A thermocouple is to be used to measure a temperature of 2800°F. Which type or types of thermocouples can be used to measure this temperature?

21. What materials are used in the construction of a type J thermocouple?

22. What does the word *piezo* mean?

**Magnetic induction is one of the most important concepts in the electrical field. Many of the most common devices, such as generators, motors, and transformers, operate on this principle.**

*Courtesy of Tennessee Valley Authority.*

# Magnetic Induction

## Key Terms

Eddy current

Exponential curve

Henry (H)

Hysteresis loss

Magnetic induction

Lenz's law

Metal oxide varistor (MOV)

R-L time constant

Speed

Strength of magnetic field

Turns of wire

Voltage spike

Weber (Wb)

**Objectives**

**A**fter studying this unit, you should be able to:

■ Discuss magnetic induction.

■ List factors that determine the amount and polarity of an induced voltage.

■ Discuss Lenz's law.

■ Discuss an exponential curve.

■ List devices used to help prevent inductive voltage spikes.

**Preview**

**magnetic induction**

**The principle of magnetic induction states that whenever a conductor cuts through magnetic lines of flux, a voltage is induced into the conductor.**

Magnetic induction is one of the most important concepts in the electrical field. It is the basic operating principle underlying alternators, transformers, and most alternating current motors. It is imperative that anyone desiring to work in the electrical field have an understanding of the principles involved.

### 13-1 Magnetic Induction

In Unit 4, it was stated that one of the basic laws of electricity is that whenever current flows through a conductor, a magnetic field is created around the conductor *(Figure 13-1)*. The direction of the current flow determines the polarity of the magnetic field, and the amount of current determines the strength of the magnetic field.

That basic law in reverse is the principle of **magnetic induction**, which states that **whenever a conductor cuts through magnetic lines of flux, a voltage is induced into the conductor**. The conductor in *Figure 13-2* is connected to a zero-center microammeter, creating a complete circuit. When the conductor is moved downward through the magnetic lines of flux, the induced voltage will cause electrons to flow in the direction indicated by the arrows. This flow of electrons causes the pointer of the meter to be deflected from the center-zero position.

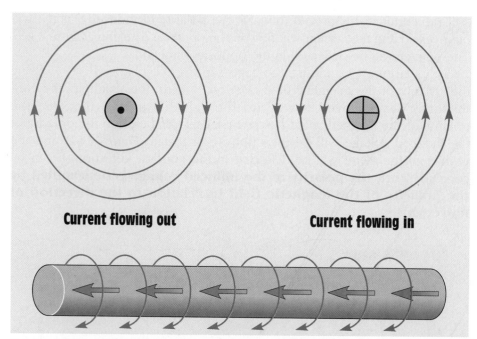

**Current flowing out**　　　　　**Current flowing in**

**Figure 13-1** Current flowing through a conductor produces a magnetic field around the conductor.

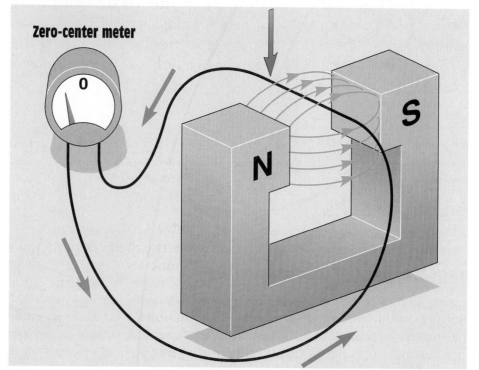

**Figure 13-2** A voltage is induced when a conductor cuts magnetic lines of flux.

**The polarity of the induced voltage is determined by the polarity of the magnetic field in relation to the direction of movement.**

If the conductor is moved upward, the polarity of induced voltage will be reversed and the current will flow in the opposite direction *(Figure 13-3)*. The pointer will be deflected in the opposite direction.

The polarity of the induced voltage can also be changed by reversing the polarity of the magnetic field *(Figure 13-4)*. In this example, the conductor is again moved downward through the lines of flux, but the polarity of the magnetic field has been reversed. Therefore the polarity of the induced voltage will be the opposite of that in *Figure 13-2,* and the pointer of the meter will be deflected in the opposite direction. It can be concluded that **the polarity of the induced voltage is determined by the polarity of the magnetic field in relation to the direction of movement**.

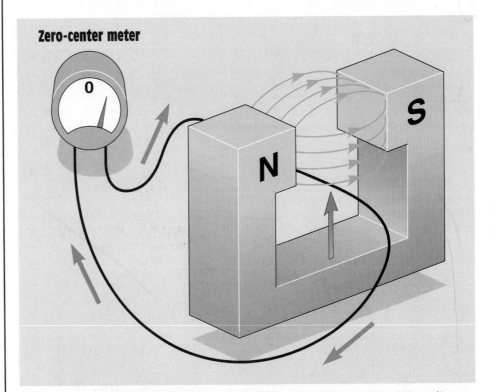

**Zero-center meter**

**Figure 13-3** Reversing the direction of movement reverses the polarity of the voltage.

## 13-2 Moving Magnetic Fields

The important factors concerning magnetic induction are a conductor, a magnetic field, and movement. In practice, it is often desirable to move the magnet instead of the conductor. Most alternating current generators or alternators operate on this principle. In *Figure 13-5,* a coil of wire is

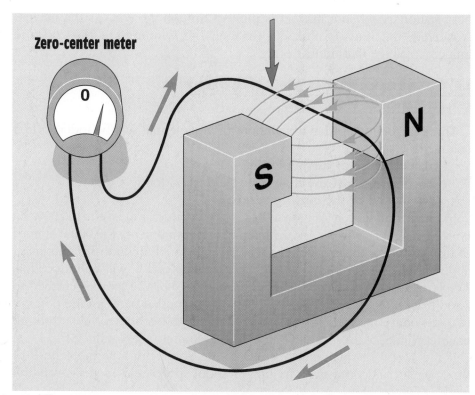

**Figure 13-4** Reversing the polarity of the magnetic field reverses the polarity of the voltage.

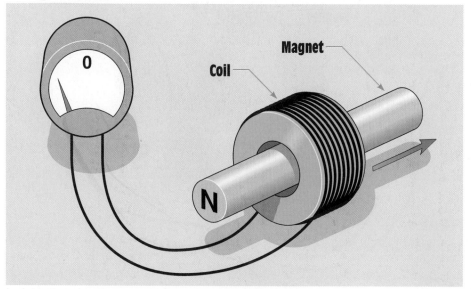

**Figure 13-5** Voltage is induced by a moving magnetic field.

**turns of wire**

**strength of magnetic field**

**speed**

**weber (Wb)**

held stationary while a magnet is moved through the coil. As the magnet is moved, the lines of flux cut through the windings of the coil and induce a voltage into them.

## 13-3 Determining the Amount of Induced Voltage

Three factors determine the amount of voltage that will be induced in a conductor:

1. **the number of turns of wire,**
2. **the strength of the magnetic field** (flux density), and
3. **the speed of the cutting action.**

In order to induce 1 V in a conductor, the conductor must cut 100,000,000 lines of magnetic flux in 1 s. In magnetic measurement, 100,000,000 lines of flux are equal to one **weber (Wb)**. Therefore, if a conductor cuts magnetic lines of flux at a rate of 1 Wb/s, a voltage of 1 V will be induced. A simple one-loop generator is shown in *Figure 13-6*. The loop is attached to a rod that is free to rotate. This assembly is suspended between the poles of two stationary magnets. If the loop is turned, the conductor cuts through magnetic lines of flux and a voltage is induced into the conductor.

If the speed of rotation is increased, the conductor cuts more lines of flux per second, and the amount of induced voltage increases. If the

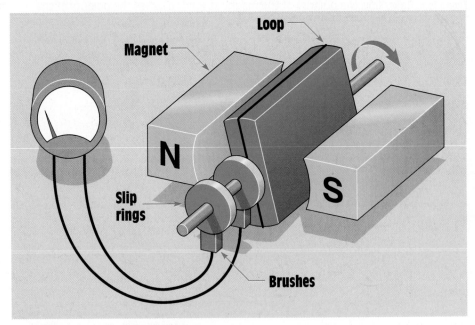

**Figure 13-6** A single-loop generator

speed of rotation remains constant and the strength of the magnetic field is increased, there will be more lines of flux per square inch. When there are more lines of flux, the number of lines cut per second increases and the induced voltage increases. If more turns of wire are added to the loop *(Figure 13-7)*, more flux lines are cut per second and the amount of induced voltage increases again. Adding more turns has the effect of connecting single conductors in series, and the amount of induced voltage in each conductor adds.

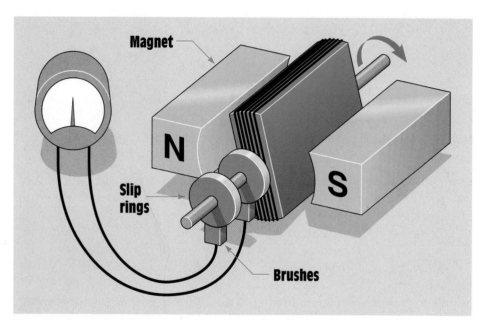

**Figure 13-7** Increasing the number of turns increases the induced voltage.

## 13-4 Lenz's Law

When a voltage is induced in a coil and there is a complete circuit, current will flow through the coil *(Figure 13-8)*. When current flows through the coil, a magnetic field is created around the coil. This magnetic field develops a polarity opposite that of the moving magnet. The magnetic field developed by the induced current acts to attract the moving magnet and pull it back inside the coil.

If the direction of motion is reversed, the polarity of the induced current is reversed, and the magnetic field created by the induced current again opposes the motion of the magnet. This principle was first noticed by Heinrich Lenz many years ago and is summarized in **Lenz's law**, which states that **an induced voltage or current opposes the motion**

**Lenz's law**

**This law stipulates that the polarity of an induced emf will be such that any resulting current will have a magnetic field (flux) which opposes the original action (change in flux) that produced the induced current. In other words, the induced voltage always opposes the original change in current. That is why the induced voltage is known as the counter-emf (cemf), or back-emf.**

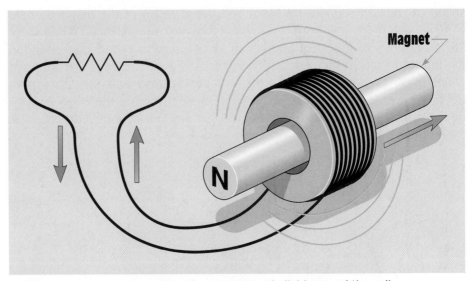

**Figure 13-8** An induced current produces a magnetic field around the coil.

**that causes it**. From this basic principle, other laws concerning inductors have been developed. One is that **inductors always oppose a change of current**. The coil in *Figure 13-9,* for example, has no induced voltage and therefore no induced current. If the magnet is moved toward the coil, however, magnetic lines of flux will begin to cut the conductors of the coil, and a current will be induced in the coil. The induced current causes magnetic lines of flux to expand outward around the coil *(Figure 13-10)*. As this expanding magnetic field cuts through the conductors of the coil,

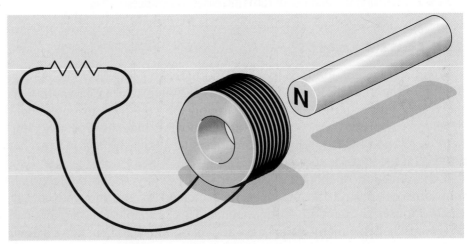

**Figure 13-9** No current flows through the coil.

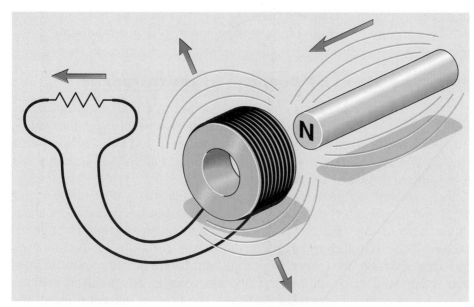

**Figure 13-10** Induced current produces a magnetic field around the coil.

a voltage is induced in the coil. The polarity of the voltage is such that it opposes the induced current caused by the moving magnet.

If the magnet is moved away, the magnetic field around the coil will collapse and induce a voltage in the coil *(Figure 13-11)*. Since the direction

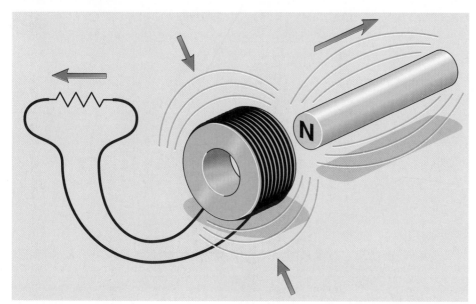

**Figure 13-11** The induced voltage forces current to flow in the same direction.

of movement of the collapsing field has been reversed, the induced voltage will be opposite in polarity, forcing the current to flow in the same direction.

### 13-5 Rise Time of Current in an Inductor

When a resistive load is suddenly connected to a source of direct current *(Figure 13-12)*, the current will instantly rise to its maximum value. The resistor shown in *Figure 13-12* has a value of 10 Ω and is connected to a 20-V source. When the switch is closed the current will instantly rise to a value of 2 A (20 V/10 Ω = 2 A).

If the resistor is replaced with an inductor that has a wire resistance of 10 Ω and the switch is closed, the current cannot instantly rise to its maximum value of 2 A *(Figure 13-13)*. As current begins to flow through an inductor, the expanding magnetic field cuts through the conductors, inducing a voltage into them. In accord with Lenz's law, the induced voltage is opposite in polarity to the applied voltage. The induced voltage, therefore, acts like a resistance to hinder the flow of current through the inductor *(Figure 13-14)*.

**The induced voltage is proportional to the rate of change of current** (speed of the cutting action). When the switch is first closed, current flow through the coil tries to rise instantly. This extremely fast rate of current

> **The induced voltage is proportional to the rate of change of current.**

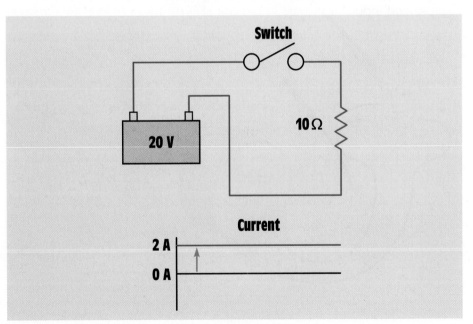

**Figure 13-12** The current rises instantly in a resistive circuit.

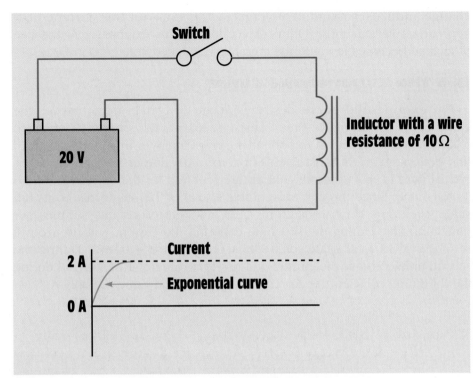

**Figure 13-13**   Current rises through an indicator at an exponential rate.

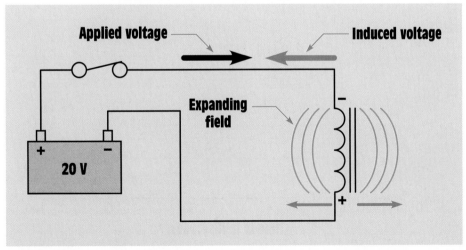

**Figure 13-14**   The applied voltage is opposite in polarity to the induced voltage.

change induces maximum voltage in the coil. As the current flow approaches its maximum Ohm's law value, in this example 2 A, the rate of change becomes less and the amount of induced voltage decreases.

## 13-6 The Exponential Curve

**exponential curve**

The **exponential curve** describes a rate of certain occurrences. The curve is divided into five time constants. Each time constant is equal to 63.2% of some value. An exponential curve is shown in *Figure 13-15*. In this example, current must rise from zero to a value of 1.5 A at an exponential rate. In this example, 100 ms are required for the current to rise to its full value. Since the current requires a total of 100 ms to rise to its full value, each time constant is 20 ms (100 ms/5 time constants = 20 ms per time constant). During the first time constant, the current will rise from 0 to 63.2% of its total value, or 0.948 A (1.5 x 0.632 = 0.948). During the second time constant the current will rise to a value of 1.297 A, and during the third time constant the current will reach a total value of 1.425 A.

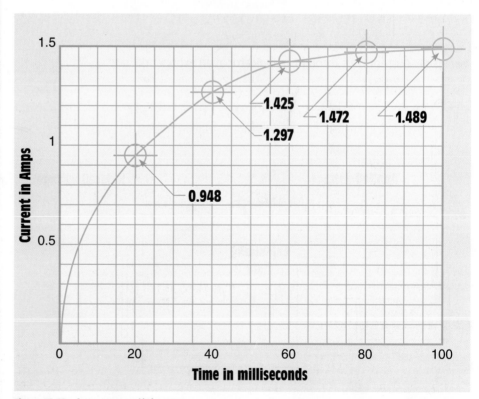

**Figure 13-15** An exponential curve

Because the current increases at a rate of 63.2% during each time constant, it is theoretically impossible to reach the total value of 1.5 A. After five time constants, however, the current has reached approximately 99.3% of the maximum value and for all practical purposes is considered to be complete.

The exponential curve can often be found in nature. If clothes are hung on a line to dry, they will dry at an exponential rate. Another example of the exponential curve can be seen in *Figure 13-16*. In this example, a bucket has been filled to a certain mark with water. A hole has been cut at the bottom of the bucket and a stopper placed in the hole. When the stopper is removed from the bucket, water will flow out at an exponential rate. Assume, for example, it takes 5 min for the water to flow out of the bucket. Exponential curves are always divided into five time constants so in this case each time constant has a value of 1 min. In *Figure 13-17,* if the stopper is removed and water is permitted to drain from the bucket for a period of 1 min before the stopper is replaced, during that first time constant 63.2% of the water in the bucket will drain out. If the stopper is again removed for a period of 1 min, 63.2% of the water remaining in the bucket will drain out. Each time the stopper is removed for a period of one time constant, the bucket will lose 63.2% of its remaining water.

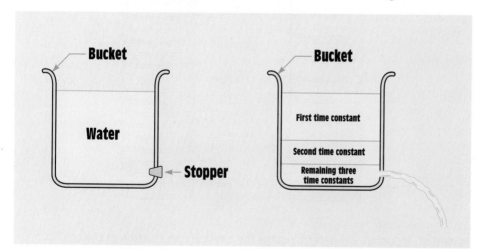

**Figure 13-16**   Exponential curves can be found in nature.

**Figure 13-17**   Water flows from a bucket at an exponential rate.

## 13-7 Inductance

Inductance is measured in units called the **henry (H)** and is represented by the letter *L*. **A coil has an inductance of one henry when a current change of one ampere per second results in an induced voltage of one volt.**

**henry (H)**

A coil has an inductance of one henry when a current change of one ampere per second results in an induced voltage of one volt.

The amount of inductance a coil will have is determined by its physical properties and construction. A coil wound on a nonmagnetic core material such as wood or plastic is referred to as an *air core* inductor. If the coil is wound on a core made of magnetic material such as silicon steel or soft iron it is referred to as an *iron core* inductor. Iron core inductors produce more inductance with fewer turns than air core inductors because of the good magnetic path provided by the core material. Iron core inductors cannot be used for high-frequency applications, however, because of **eddy current** loss and **hysteresis loss** in the core material.

Another factor that determines inductance is how far the windings are separated from each other. If the turns of wire are far apart they will have less inductance than turns wound closer together *(Figure 13-18)*.

**eddy current**

**hysteresis loss**

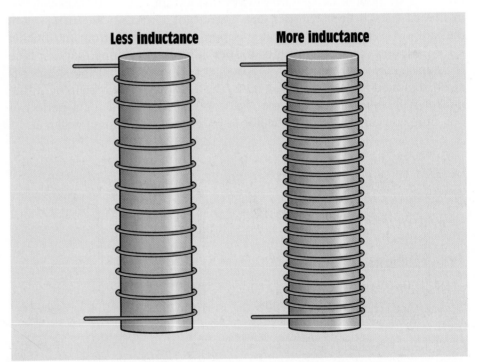

**Figure 13-18** Inductance is determined by the physical construction of the coil.

### 13-8 R-L Time Constants

The time necessary for current in an inductor to reach its full Ohm's law value, called the **R-L time constant**, can be computed using the formula

**R-L time constant**

$$T = \frac{L}{R}$$

where

T = time in seconds

L = inductance in henrys

R = resistance in ohms

This formula computes the time of one time constant.

A coil has an inductance of 1.5 H and a wire resistance of 6 Ω. If the coil is connected to a battery of 3 V, how long will it take the current to reach its full Ohm's law value of 0.5 A (3 V/6 Ω = 0.5 A)?

**Example 1**

## Solution

To find the time of one time constant, use the formula

$$T = \frac{L}{R}$$

$$T = \frac{1.5}{6}$$

$$T = 0.25 \text{ s}$$

The time for one time constant is 0.25 s. Since five time constants are required for the current to reach its full value of 0.5 A, 0.25 s will be multiplied by 5.

$$0.25 \times 5 = 1.25 \text{ s}$$

## 13-9 Induced Voltage Spikes

A **voltage spike** occurs when the current flow through an inductor stops, and the current decreases at an exponential rate also *(Figure 13-19)*. As long as a complete circuit exists when the power is interrupted, there is little or no problem. In the circuit shown in *Figure 13-20*, a resistor and inductor are connected in parallel. When the switch is closed, the battery will supply current to both. When the switch is opened, the magnetic field surrounding the inductor will collapse and induce a voltage into the inductor. The induced voltage will attempt to keep current flowing in the same direction. Recall that inductors oppose a change of current. The amount of current flow and the time necessary for the flow to stop will be determined by the resistor and the properties of the inductor. The amount of voltage produced by the collapsing magnetic field is determined by the maximum current in the circuit and the total resistance

**voltage spike**

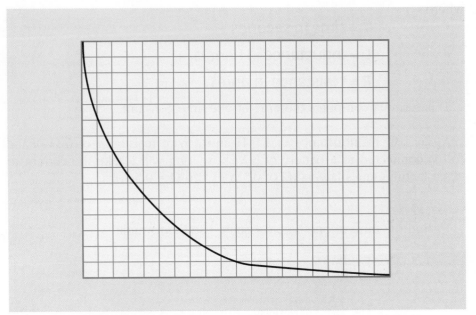

**Figure 13-19** Current flow through an inductor deceases at an exponential rate.

in the circuit. In the circuit shown in *Figure 13-20,* assume that the inductor has a wire resistance of 6 Ω, and the resistor has a resistance of 100 Ω. Also assume that when the switch is closed a current of 2 A will flow through the inductor.

When the switch is opened, a series circuit exists composed of the resistor and inductor *(Figure 13-21)*. The maximum voltage developed in

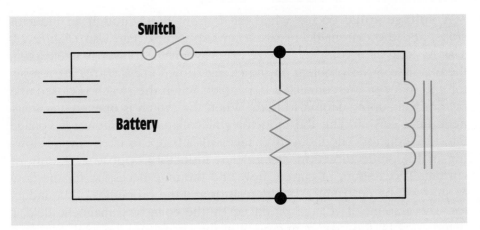

**Figure 13-20** The resistor helps prevent voltage spikes caused by the inductor.

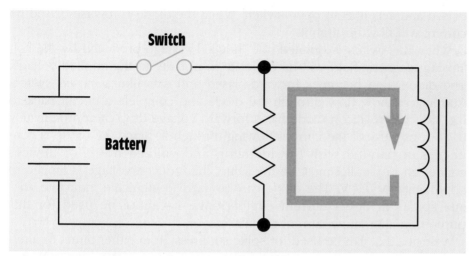

**Figure 13-21**  When the switch is opened, a series path is formed by the resistor and inductor.

this circuit would be 212 V (2 A x 106 $\Omega$ = 212 V). If the circuit resistance were increased, the induced voltage would become greater. If the circuit resistance were decreased, the induced voltage would become less.

Another device often used to prevent induced voltage spikes when the current flow through an inductor is stopped is the diode *(Figure 13-22)*. The diode is an electronic component that operates like an electrical check valve. The diode will permit current to flow through it in only one direction. The diode is connected in parallel with the inductor in such a manner that when voltage is applied to the circuit, the diode is reverse-

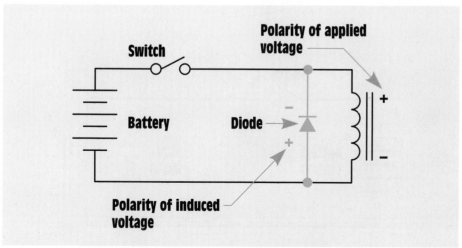

**Figure 13-22**  A diode is used to prevent induced voltage spikes.

biased and acts like an open switch. When the diode is reverse-biased no current will flow through it.

When the switch is opened, the induced voltage produced by the collapsing magnetic field will be opposite in polarity to the applied voltage. The diode then becomes forward-biased and acts like a closed switch. Current can now flow through the diode and complete a circuit back to the inductor. A silicon diode has a forward voltage drop of approximately 0.7 V regardless of the current flowing through it. Since the diode is connected in parallel with the inductor, and voltage drops of devices connected in parallel must be the same, the induced voltage is limited to approximately 0.7 V. The diode can be used to eliminate inductive voltage spikes in direct current circuits only; it cannot be used for this purpose in alternating current circuits.

A device that can be used for spike suppression in either direct or alternating current circuits is the **metal oxide varistor (MOV).** The MOV is a bidirectional device, which means that it will conduct current in either direction, and can, therefore, be used in alternating current circuits. The metal oxide varistor is an extremely fast-acting solid state component that will exhibit a change of resistance when the voltage reaches a certain point. Assume that the MOV shown in *Figure 13-23* has a voltage rating of 140 V, and that the voltage applied to the circuit is 120 V. When the switch is closed and current flows through the circuit, a magnetic field will be established around the inductor *(Figure 13-24)*. As long as the voltage applied to the MOV is less than 140 V, it will exhibit an extremely high resistance, in the range of several hundred thousand ohms.

When the switch is opened, current flow through the coil suddenly stops, and the magnetic field collapses. This sudden collapse of the mag-

**metal oxide varistor (MOV)**

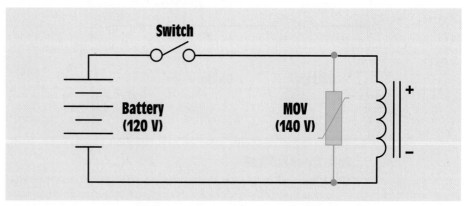

**Figure 13-23** Metal oxide varistor used to suppress a voltage spike

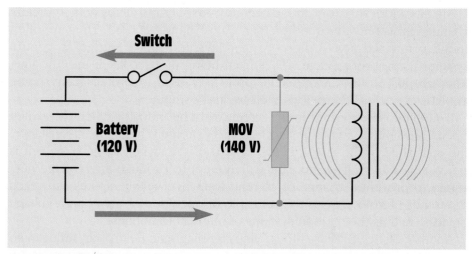

**Figure 13-24**   When the switch is closed, a magnetic field is established around the inductor.

netic field will cause an extremely high voltage to be induced in the coil. When this induced voltage reaches 140 V, however, the MOV will suddenly change from a high resistance to a low resistance, preventing the voltage from becoming greater than 140 V *(Figure 13-25)*.

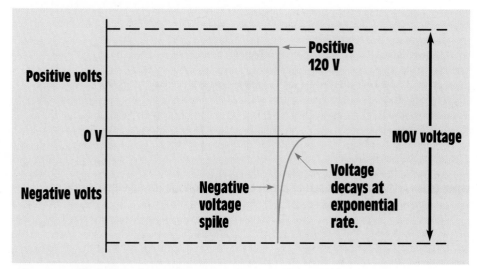

**Figure 13-25**   The MOV prevents the spike from becoming too high.

Metal oxide varistors are extremely fast-acting. They can typically change resistance values in less than 20 ns (nanoseconds). They are often found connected across the coils of relays and motor starters in control systems to prevent voltage spikes from being induced back into the line. They are also found in the surge protectors used to protect many home appliances such as televisions, stereos, and computers.

If nothing is connected in the circuit with the inductor when the switch opens, the induced voltage can become extremely high. In this instance, the resistance of the circuit is the air gap of the switch contacts, which is practically infinite. The inductor will attempt to produce any voltage necessary to prevent a change of current. Inductive voltage spikes can reach thousands of volts. This is the principle of operation of many high-voltage devices such as the ignition systems of many automobiles.

Another device that uses the collapsing magnetic field of an inductor to produce a high voltage is the electric fence charger, shown in *Figure 13-26*. The switch is constructed in such a manner that it will pulse on and

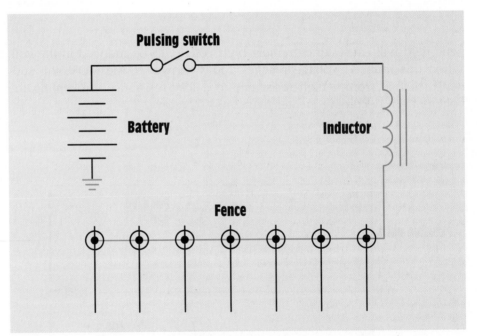

**Figure 13-26**   An inductor is used to produce a high voltage for an electric fence.

off. When the switch closes, current flows through the inductor, and a magnetic field is produced around the inductor. When the switch opens, the magnetic field collapses and induces a high voltage across the inductor. If anything or anyone standing on the ground touches the fence, a circuit is completed through the object or person and the ground. The coil is generally constructed of many turns of very small wire. This construction provides the coil with a high resistance and limits current flow when the field collapses.

## Summary

1. When current flows through a conductor a magnetic field is created around the conductor.

2. When a conductor is cut by a magnetic field a voltage is induced in the conductor.

3. The polarity of the induced voltage is determined by the polarity of the magnetic field in relation to the direction of motion.

4. Three factors that determine the amount of induced voltage are:
   a.  the number of turns of wire,
   b.  the strength of the magnetic field, and
   c.  the speed of the cutting action.

5. One volt is induced in a conductor when magnetic lines of flux are cut at a rate of one weber per second.

6. Induced voltage is always opposite in polarity to the applied voltage.

7. Inductors oppose a change of current.

8. Current rises in an inductor at an exponential rate.

9. An exponential curve is divided into five time constants.

10. Each time constant is equal to 63.2% of some value.

11. Inductance is measured in units called henrys (H).

12. A coil has an inductance of 1 H when a current change of 1 A per second results in an induced voltage of 1 V.

13. Air core inductors are inductors wound on cores of nonmagnetic material.

14. Iron core inductors are wound on cores of magnetic material.

15. The amount of inductance an inductor will have is determined by the number of turns of wire and the physical construction of the coil.

16. Inductors can produce extremely high voltages when the current flowing through them is stopped.

17. Two devices used to help prevent large spike voltages are the resistor and diode.

## Review Questions

1. What determines the polarity of magnetism when current flows through a conductor?

2. What determines the strength of the magnetic field when current flows through a conductor?

3. Name three factors that determine the amount of induced voltage in a coil.

4. How many lines of magnetic flux must be cut in 1 s to induce a voltage of 1 V?

5. What is the effect on induced voltage of adding more turns of wire to a coil?

6. Into how many time constants is an exponential curve divided?

7. Each time constant of an exponential curve is equal to what percentage of the whole?

8. An inductor has an inductance of 0.025 H and a wire resistance of 3 Ω. How long will it take the current to reach its full Ohm's law value?

9. Refer to the circuit shown in *Figure 13-20*. Assume that the inductor has a wire resistance of 0.2 Ω and the resistor has a value of 250 Ω. If a current of 3 A is flowing through the inductor what will be the maximum induced voltage when the switch is opened?

10. What electronic component is often used to prevent large voltage spikes from being produced when the current flow through an inductor is suddenly terminated?

# basics of alternating current

A basic knowledge of trigonometry is required for the study of alternating current. This unit is intended to present trigonometry in a very basic way. The concepts discussed in this unit will be used throughout the remainder of the text.

Courtesy of Cleveland Institute of Electronics.

# *Basic Trigonometry*

## Key Terms

Adjacent

Cosine

Hypotenuse

Opposite

Oscar Had A Heap Of Apples

Pythagoras

Pythagorean theorem

Right angle

Right triangle

Sine

Tangent

## Outline

**Objectives**

**A***fter studying this unit, you should be able to:*

- Define a right triangle.
- Discuss the Pythagorean theorem.
- Solve problems concerning right triangles using the Pythagorean theorem.
- Solve problems using sines, cosines, and tangents.

**Preview**

Before beginning the study of alternating current, a brief discussion of right triangles and the mathematical functions involving them is appropriate. Many alternating current formulas are based on right triangles, because, depending on the type of load, the voltage and current in an AC circuit can be out of phase with each other by approximately 90°. The exact amount of out-of-phase condition is determined by different factors, which will be covered in later units.

### 14-1 Right Triangles

**right triangle**

A **right triangle** is a triangle that contains a **right**, or 90°, **angle** *(Figure 14-1)*. The **hypotenuse** is the longest side of a right triangle and is always opposite the right angle. Several thousand years ago, a Greek mathematician named **Pythagoras** made some interesting discoveries concerning triangles that contain right angles. One of these discoveries was that if the two sides of a right triangle are squared, their sum will equal the square of the hypotenuse. For example, assume a right triangle has one side 3 ft long and the other side is 4 ft long *(Figure 14-2)*. If the side that is 3 ft long is squared, it will produce an area of 9 sq. ft (3 x 3 = 9). If the side that is 4 ft long is squared it will produce an area of 16 sq. ft (4 x 4 = 16). The sum of the areas of these two sides equals the square area formed by the hypotenuse. In this instance, the hypotenuse will have an area of 25 sq. ft (9 + 16 = 25). Now that the area, or square, of the

**right angle**

**hypotenuse**

**Pythagoras**

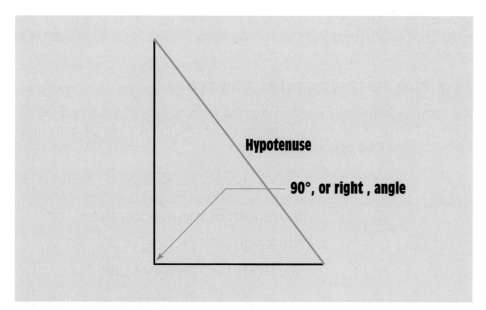

**Figure 14-1** A right triangle

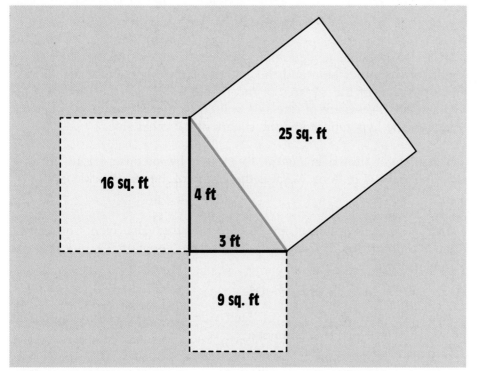

**Figure 14-2** The Pythagorean theorem states that the sum of the squares of the sides of a right triangle is equal to the square of the hypotenuse.

hypotenuse is known, its length can be determined by finding the square root of its area. The length of the hypotenuse will be 5 ft, because the square root of 25 is 5.

## 14-2 The Pythagorean Theorem

**Pytha-gorean theorem**

From this knowledge concerning the relationship of the length of the sides to the length of the hypotenuse, Pythagoras derived a formula known as the **Pythagorean theorem:**

$$c^2 = a^2 + b^2$$

where

$$c = \text{the length of the hypotenuse}$$

$$a = \text{the length of one side}$$

$$b = \text{the length of the other side}$$

If the lengths of the two sides are known, the length of the hypotenuse can be found using the formula

$$c = \sqrt{a^2 + b^2}$$

It is also possible to determine the length of one of the sides if the length of the other side and the length of the hypotenuse are known. Since the sum of the squares of the two sides equals the square of the hypotenuse, the square of one side will equal the difference between the square of the hypotenuse and the square of the other side.

**Example 1**

The triangle shown in *Figure 14-3* has a hypotenuse (c) 18 in. long. One side (b) is 7 in. long. What is the length of the second side (a)?

### Solution

Transpose the formula $c^2 = a^2 + b^2$ to find the value of *a*. This can be done by subtracting $b^2$ from both sides of the equation. The result is the formula shown below.

$$a^2 = c^2 - b^2$$

Now take the square root of each side of the equation so that the answer will be equal to a and not $a^2$.

$$a = \sqrt{c^2 - b^2}$$

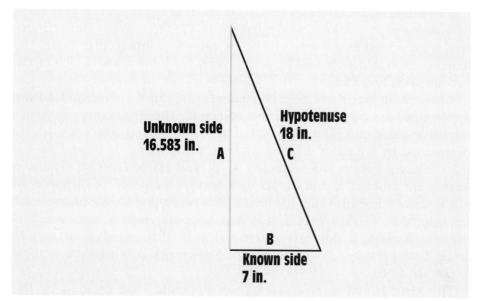

· **Figure 14-3** Using the Pythagorean theorem to find the length of one side

The formula can now be used to find the length of the unknown side.

$$a = \sqrt{18^2 - 7^2}$$

$$a = \sqrt{324 - 49}$$

$$a = \sqrt{275}$$

$$a = 16.583 \text{ in.}$$

## 14-3 Sines, Cosines, and Tangents

A second important concept concerning right triangles is the relationship of the length of the sides to the angles. Since one angle is always 90°, only the two angles that are not right angles are of concern. Another important fact concerning triangles is that the sum of the angles of any triangle equals 180°; since one angle of a right triangle is 90°, the sum of the two other angles must equal 90°.

It was discovered long ago that the number of degrees in each of these two angles is proportional to the lengths of the three sides. This relationship is expressed as the *sine, cosine, or tangent* of a particular angle. The function used is determined by which sides are known and which of the two angles is to be found.

When using sines, cosines, or tangents, the sides are designated the *hypotenuse*, the **opposite**, and the **adjacent**. The hypotenuse is always the longest side of a right triangle, but the opposite and adjacent sides are determined by which of the two angles is to be found. In *Figure 14-4*, a right triangle has its sides labeled A, B, and HYPOTENUSE. The two angles are labeled X and Y. To determine which side is opposite an angle, draw a line bisecting the angle. This bisect line would intersect the opposite side. In *Figure 14-5*, side A is opposite angle X, and side B is opposite angle Y. If side A is opposite angle X, then side B is adjacent to angle X, and if side B is opposite angle Y, then side A is adjacent to angle Y *(Figure 14-6)*.

The **sine** function is equal to the opposite side divided by the hypotenuse.

**opposite**

**adjacent**

**sine**

$$\text{sine} = \frac{\text{opposite}}{\text{hypotenuse}}$$

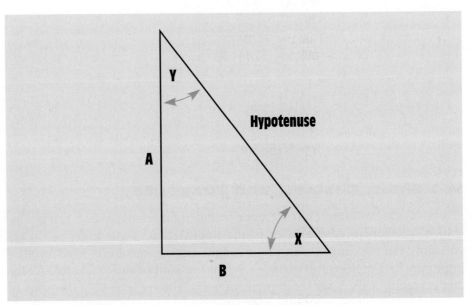

**Figure 14-4** Determining which side is opposite and which is adjacent

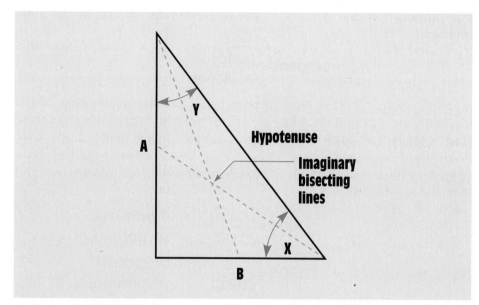

**Figure 14-5** Side A is opposite angle X, and side B is opposite angle Y.

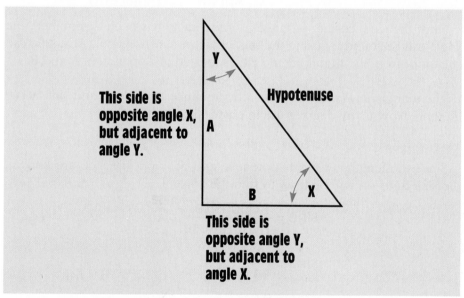

**Figure 14-6** Determining opposite and adjacent sides

The **cosine** function is equal to the adjacent side divided by the hypotenuse.

$$\text{cosine} = \frac{\text{adjacent}}{\text{hypotenuse}}$$

cosine

**tangent**

The **tangent** function is equal to the opposite side divided by the adjacent side.

$$tangent = \frac{opposite}{adjacent}$$

**Oscar Had A Heap Of Apples**

A simple saying is often used to help remember the relationship of the trigonometric function and the side of the triangle. This saying is **Oscar Had A Heap Of Apples**. To use this simple saying, write down *sine, cosine, and tangent.* The first letter of each word becomes the first letter of each of the sides. *O* stands for opposite, *H* stands for hypotenuse, and *A* stands for adjacent.

$$sine = \frac{Oscar}{Had} \qquad \begin{array}{l} \text{(Opposite)} \\ \text{(Hypotenuse)} \end{array}$$

$$cosine = \frac{A}{Heap} \qquad \begin{array}{l} \text{(Adjacent)} \\ \text{(Hypotenuse)} \end{array}$$

$$tangent = \frac{Of}{Apples} \qquad \begin{array}{l} \text{(Opposite)} \\ \text{(Adjacent)} \end{array}$$

Once the sine, cosine, or tangent of the angle has been determined, the angle can be found by using the trigonometric functions on a scientific calculator or from the trigonometric tables located in Appendices A and B.

**Example 2**

The triangle in *Figure 14-7* has a hypotenuse 14 in. long and side A is 9 in. long. How many degrees are in angle X?

**Solution**

Since the lengths of the hypotenuse and the opposite side are known, the sine function will be used to find the angle.

$$sine = \frac{opposite}{hypotenuse}$$

$$sine = \frac{9}{14}$$

$$sine = 0.643$$

0.643 is the sine of the angle, not the angle. To find the angle, use the trigonometric chart in the appendix to determine what angle corresponds to 0.643, or use the SIN function on a scientific calculator. If you are using a scientific calculator it will be necessary to use the ARC SIN function or INV SIN function to find the answer. If the number 0.643 is entered and

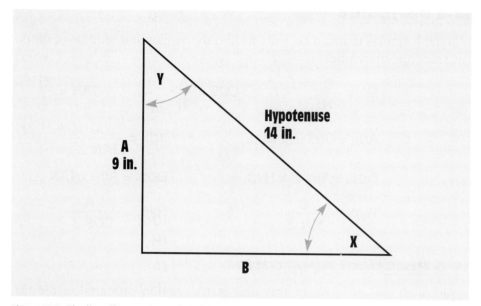

**Figure 14-7** Finding the angles using trigonometric functions

the SIN key is pressed, an answer of 0.0112 will result. This is the sine of a 0.643-degree angle. If the sine, cosine, or tangent of an angle is known, the ARC key or INV (depending on the manufacturer of the calculator) key must be pressed before the SIN, COS, or TAN key is pressed.

$$\text{sine } 0.643 = 40°$$

Using the same triangle *(Figure 14-7)*, determine the number of degrees in angle Y.

**Example 3**

### Solution

In this example, the lengths of the hypotenuse and the adjacent side are known. The cosine function can be used to find the angle.

$$\text{cosine} = \frac{\text{adjacent}}{\text{hypotenuse}}$$

$$\text{cosine} = \frac{9}{14}$$

$$\text{cosine} = 0.643$$

To find what angle corresponds to the cosine of 0.643 use the trigonometric tables in Appendices A and B or the COS function of a scientific calculator.

$$\text{cosine } 0.643 = 50°$$

## 14-4 Formulas

Some formulas that can be used to find the angles and lengths of different sides are shown below.

$$SIN = \frac{O}{H} \qquad COS = \frac{A}{H} \qquad TAN = \frac{O}{A}$$

$$Adj. = COS \angle \times Hyp. \qquad Adj. = \frac{O}{TAN \angle}$$

$$Opp. = SIN \angle \times Hyp. \qquad Opp. = Adj. \times TAN \angle$$

$$Hyp. = \frac{O}{SIN \angle} \qquad Hyp. = \frac{A}{COS \angle}$$

## 14-5 Practical Application

Although the purpose of this unit is to provide preparation for the study of alternating current circuits, basic trigonometry can provide answers to other problems that may be encountered on the job. Assume that it is necessary to know the height of a tall building *(Figure 14-8)*. Now assume that the only tools available to make this measurement are a 1-ft ruler, a tape measure, and a scientific calculator. To make the mea-

**Figure 14-8** Using trigonometry to measure the height of a tall building

surement, find a relatively flat area in the open sunlight. Hold the ruler upright and measure the shadow cast by the sun *(Figure 14-9)*. Assume the length of the shadow to be 7.5 in. Using the length of the shadow as one side of a right triangle and the ruler as the other side, the angle of the sun can now be determined. The two known sides are the opposite and the adjacent. The tangent function corresponds to these two sides. The angle of the sun is

$$\text{TAN} \angle = \frac{\text{O}}{\text{A}}$$

$$\text{TAN} \angle = \frac{12}{7.5}$$

$$\text{TAN} \angle = 1.6$$

$$\angle = 57.99°$$

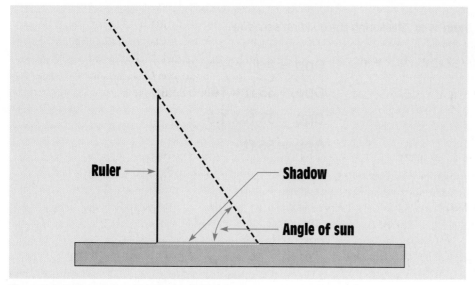

**Figure 14-9** Determining the angle of the sun

The tape measure can now be used to measure the shadow cast by the building on the ground. Assume the length of the shadow to be 35 ft. If the height of the building is used as one side of a right triangle and the shadow is used as the other side, the height can be found since the angle of the sun is known *(Figure 14-10)*.

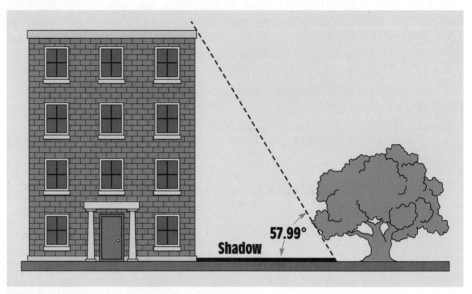

**Figure 14-10**  Measuring the building's shadow

Opp. = Adj. x TAN ∠

Opp. = 35 ft x TAN 57.99°

Opp. = 35 ft x 1.6

Opp. = 56 ft

## Summary

1. Many alternating current formulas are based on right triangles.

2. A right triangle is a triangle that contains a 90° or right angle.

3. The Pythagorean theorem states that the sum of the squares of the sides of a right triangle equals the square of the hypotenuse.

4. The sum of the angles of any triangle is 180°.

5. The relationship of the length of the sides of a right triangle to the number of degrees in its angles can be expressed as the sine, cosine, or tangent of a particular angle.

6. The sine function is the relationship of the opposite side divided by the hypotenuse.

7. The cosine function is the relationship of the adjacent side divided by the hypotenuse.

8. The tangent function is the relationship of the opposite side divided by the adjacent side.

9. The hypotenuse is always the longest side of a right triangle.

10. A simple saying that can be used to help remember the relationship of the trigonometric functions to the sides of a right triangle is "Oscar Had A Heap Of Apples."

## Review Questions

1. Which trigonometric function is used to find the angle if the length of the hypotenuse and of the adjacent side are known?

   Refer to *Figure 14-11* to answer the following questions.

2. If side A has a length of 18.5 ft and side B has a length of 28 ft, what is the length of the hypotenuse?

3. Side A has a length of 12 meters, and angle Y is 12°. What is the length of side B?

4. Side A has a length of 6 in. and angle Y is 45°. What is the length of the hypotenuse?

5. The hypotenuse has a length of 65 in., and side A has a length of 31 in. What is angle X?

6. The hypotenuse has a length of 83 ft and side B has a length of 22 ft. What is the length of side A?

7. Side A has a length of 1.25 in., and side B has a length of 2 in. What is angle Y?

8. Side A has a length of 14 ft, and angle X is 61°. What is the length of the hypotenuse?

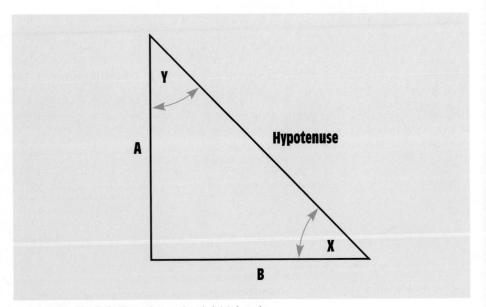

**Figure 14-11** Finding the values of a right triangle

9. Using the dimensions in question 8, what is the length of side B?

10. Angle Y is 36°. What is angle X?

## Basic Trigonometry

Refer to *Figure 14-11* to find the missing values in the chart below.

| ∠ X | ∠ Y | Side A | Side B | Hyp. |
|-----|-----|--------|--------|------|
| 40° |     |        |        | 8    |
|     | 33° | 72     |        |      |
|     |     | 38     |        | 63   |
| 52° |     |        | 14     |      |
|     | 42° |        | 156    |      |

Alternating current is bidirectional, which means that it flows first in one direction and then in the other. Since most of the electric power produced in the world is AC, it is imperative that anyone working in the electrical field have a working knowledge of alternating current.

Courtesy of Niagara Mohawk Power Corp.

# *Alternating Current*

## Key Terms

Average

Cycle

Effective

Frequency

Hertz (Hz)

In phase

Linear wave

Oscillators

Peak

Resistive loads

Ripple

RMS

Sine wave

Skin effect

Triangle wave

True power

Watts

## Outline

*fter studying this unit, you should be able to:*

- Discuss differences between direct and alternating current.

- Be able to compute instantaneous values of voltage and current for a sine wave.

- Be able to compute peak, RMS, and average values of voltage and current.

- Discuss the phase relationship of voltage and current in a pure resistive circuit.

**Preview**

Most of the electrical power produced in the world is alternating current. It is used to operate everything from home appliances such as television sets, computers, microwave ovens, and electric toasters, to the largest motors found in industry. Alternating current has several advantages over direct current that make it a better choice for the large-scale production of electrical power.

## 15-1 Advantages of Alternating Current

Probably the single greatest advantage of alternating current is the fact that AC current can be transformed and DC current cannot. A transformer permits voltage to be stepped up or down. Voltage can be stepped up for the purpose of transmission and then stepped back down when it is to be used by some device. Transmission voltages of 69 kV, 138 kV, and 345 kV are common. The advantage of high-voltage transmission is that less current is required to produce the same amount of power. The reduction of current permits smaller wires to be used, which results in a savings of material.

In the very early days of electric power generation, Thomas Edison, an American inventor, proposed powering the country with low-voltage direct current. He reasoned that low-voltage direct current was safer for people to use than higher-voltage alternating current. A Serbian immigrant named Nikola Tesla, however, argued that direct current was impractical for large-scale applications. The disagreement was finally settled at the 1904 World's Fair held in St. Louis, Missouri. The 1904 World's Fair not only introduced the first ice cream cone and the first iced tea, it was also the first World's Fair to be lighted with "electric candles." At that time, the only two companies capable of providing electric lighting for the World's Fair were the Edison Company, headed by Thomas Edison, and the Westinghouse Company, headed by George Westinghouse, a close friend of Nikola Tesla. The Edison Company submitted a bid of over one dollar per lamp to light the fair with low-voltage direct current. The Westinghouse Company submitted a bid of less than 25 cents per lamp to light the fair using higher-voltage alternating current. This set the precedent for how electric power would be supplied throughout the world.

## 15-2 AC Wave Forms

### Square Waves

Alternating current differs from direct current in that AC current reverses its direction of flow at periodic intervals *(Figure 15-1)*. Alternating current wave forms can vary depending on how the current is produced. One wave form frequently encountered is the square wave *(Figure 15-2)*. It is assumed that the oscilloscope in *Figure 15-2* has been adjusted so that 0 V is represented by the center horizontal line. The wave form shows that the voltage is in the positive direction for some length of time and then changes polarity. The voltage remains negative for some length of time and then changes back to positive again. Each time the voltage reverses polarity, the current flow through the circuit changes direction. A square wave could be produced by a simple single-pole double-throw switch connected to two batteries as shown in *Figure 15-3*. Each time the switch position is changed, current flows through the resistor in a different direction. Although this circuit will produce a square wave alternating

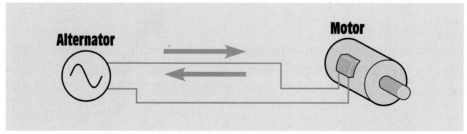

**Figure 15-1** Alternating current flows first in one direction and then in the other.

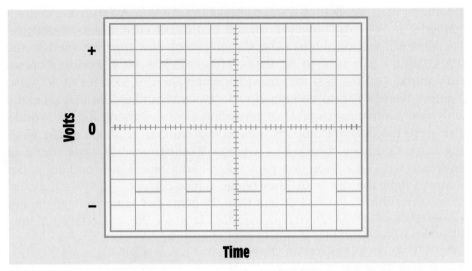

**Figure 15-2** Square wave alternating current

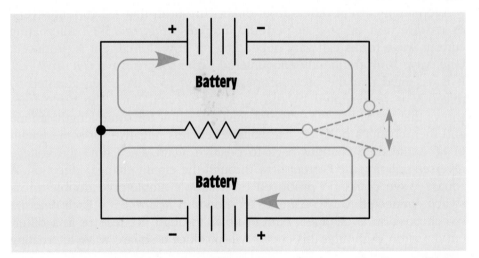

**Figure 15-3** Square wave alternating current produced with a switch and two batteries

current, it is not practical. Square waves are generally produced by electronic devices called **oscillators**. The schematic diagram of a simple square wave oscillator is shown in *Figure 15-4*. In this circuit, two bipolar transistors are used as switches to reverse the direction of current flow through the windings of the transformer.

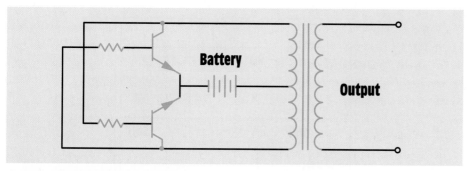

**Figure 15-4** Square wave oscillator

## Triangle Waves

Another common AC wave form is the **triangle wave** shown in *Figure 15-5*. The triangle wave is a linear wave. A **linear wave** is one in which the voltage rises at a constant rate with respect to time. Linear waves form straight lines when plotted on a graph. For example, assume the wave form shown in *Figure 15-5* reaches a maximum positive value of 100 V after 2 ms. The voltage will be 25 V after 0.5 ms, 50 V after 1 ms, and 75 V after 1.5 ms.

## Sine Waves

The most common of all AC wave forms is the **sine wave** *(Figure 15-6)*. They are produced by all rotating machines. The sine wave contains a

**triangle wave**

**linear wave**

**sine wave**

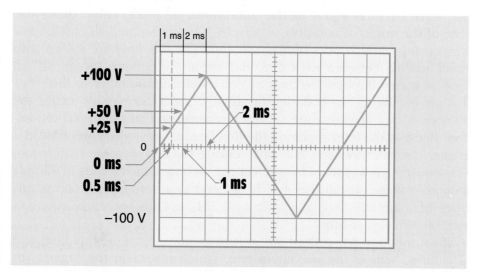

**Figure 15-5** Triangle wave

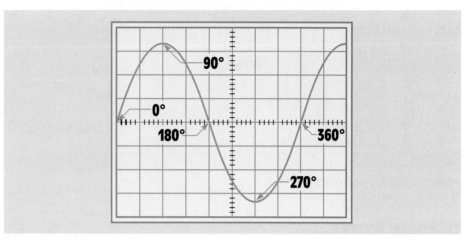

**Figure 15-6** Sine wave

total of 360 electrical degrees. It reaches its peak positive voltage at 90°, returns to a value of 0 V at 180°, increases to its maximum negative voltage at 270°, and returns to 0 V at 360°. Each complete wave form of 360° is called a **cycle**. The number of complete cycles that occur in one second is called the **frequency**. Frequency is measured in **hertz (Hz)**. The most common frequency in the United States and Canada is 60 Hz. This means that the voltage increases from zero to its maximum value in the positive direction, returns to zero, increases to its maximum value in the negative direction, and returns to zero 60 times each second.

**Sine waves are so named because the voltage at any point along the wave form is equal to the maximum, or peak, value times the sine of the angle of rotation.** *Figure 15-7* illustrates one-half of a loop of wire cutting through lines of magnetic flux. The flux lines are shown with equal spacing between each line, and the arrow denotes the arc of the loop as it cuts through the lines of flux. Notice the number of flux lines that are cut by the loop during the first 30° of rotation. Now notice the number of flux lines that are cut during the second and third 30° of rotation. Because the loop is cutting the flux lines at an angle, it must travel a greater distance between flux lines during the first degrees of rotation. Consequently, fewer flux lines are cut per second, which results in a lower induced voltage. Recall that 1 V is induced in a conductor when it cuts lines of magnetic flux at a rate of 1 Wb/s. One weber is equal to 100,000,000 lines of flux.

When the loop has rotated 90°, it is perpendicular to the flux lines and is cutting them at the maximum rate, which results in the highest, or peak, voltage being induced in the loop. The voltage at any point during

---

**cycle**

**frequency**

**hertz (Hz)**

---

**Sine waves are so named because the voltage at any point along the wave form is equal to the maximum, or peak, value times the sine of the angle of rotation.**

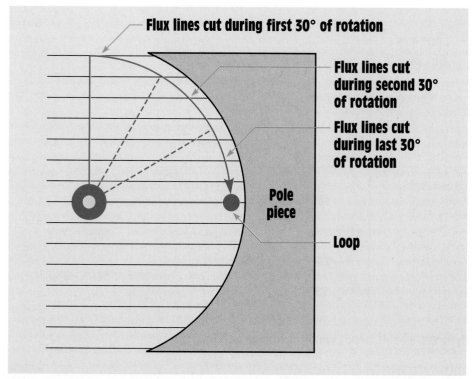

**Flux lines cut during first 30° of rotation**

**Flux lines cut during second 30° of rotation**

**Flux lines cut during last 30° of rotation**

**Pole piece**

**Loop**

**Figure 15-7** As the loop approaches 90° of rotation, the flux lines are cut at a faster rate.

the rotation is equal to the maximum induced voltage times the sine of the angle of rotation. For example, if the induced voltage after 90° of rotation is 100 V, the voltage after 30° of rotation will be 50 V because the sine of a 30° angle is 0.5 (100 x 0.5 = 50 V). The induced voltage after 45° of rotation is 70.7 V because the sine of a 45° angle is 0.707 (100 x 0.707 = 70.7 V). A sine wave showing the instantaneous voltage values after different degrees of rotation is shown in *Figure 15-8*. The instantaneous voltage value is the value of voltage at any instant on the wave form.

The following formula can be used to determine the instantaneous value at any point along the sine wave.

$$E_{(INST)} = E_{(MAX)} \times SIN \angle$$

where

$E_{(INST)}$ = the voltage at any point on the wave form

$E_{(MAX)}$ = the maximum, or peak, voltage

$SIN \angle$ = the sine of the angle of rotation

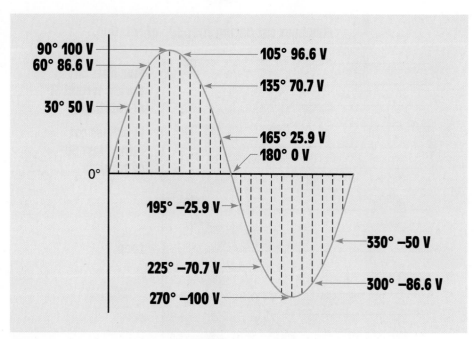

90° 100 V
60° 86.6 V
105° 96.6 V
135° 70.7 V
30° 50 V
165° 25.9 V
180° 0 V
0°
195° −25.9 V
330° −50 V
225° −70.7 V
300° −86.6 V
270° −100 V

**Figure 15-8** Instantaneous values of voltage along a sine wave

**Example 1**

A sine wave has a maximum voltage of 138 V. What is the voltage after 78° of rotation?

**Solution**

$$E_{(INST)} = E_{(MAX)} \times SIN \angle$$

$$E_{(INST)} = 138 \times 0.978 \ (SIN \ of \ 78°)$$

$$E_{(INST)} = 134.96 \ V$$

The formula can be changed to find the maximum value if the instantaneous value and the angle of rotation are known or to find the angle if the maximum and instantaneous values are known.

$$E_{(MAX)} = \frac{E_{(INST)}}{SIN \angle}$$

$$SIN \angle = \frac{E_{(INST)}}{E_{(MAX)}}$$

**Example 2**

A sine wave has an instantaneous voltage of 246 V after 53° of rotation. What is the maximum value the wave form will reach?

## Solution

$$E_{(MAX)} = \frac{E_{(INST)}}{SIN \angle}$$

$$E_{(MAX)} = \frac{246}{0.799}$$

$$E_{(MAX)} = 307.88 \text{ V}$$

A sine wave has a maximum voltage of 350 V. At what angle of rotation will the voltage reach 53 V?

**Example 3**

## Solution

$$SIN \angle = \frac{E_{(INST)}}{E_{(MAX)}}$$

$$SIN \angle = \frac{53}{350}$$

$$SIN \angle = 0.151$$

Note: 0.151 is the *sine* of the angle, not the angle. To find the angle that corresponds to a sine of 0.151 use the trigonometric tables located in Appendices A and B or a scientific calculator.

$$\angle = 8.71°$$

## 15-3 Sine Wave Values

Several measurements of voltage and current are associated with sine waves. These measurements are peak-to-peak, peak, RMS, and average. A sine wave showing peak-to-peak, peak, and RMS measurements is shown in *Figure 15-9*.

## Peak-to-Peak and Peak Values

The peak-to-peak value is measured from the maximum value in the positive direction to the maximum value in the negative direction. The peak-to-peak value is often the simplest measurement to make when using an oscilloscope.

The **peak** value is measured from zero to the highest value obtained in either the positive or negative direction. The peak value is one-half of the peak-to-peak value.

**peak**

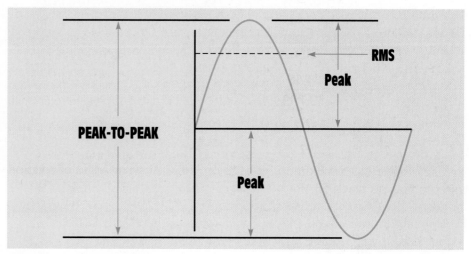

**Figure 15-9** Sine wave values

## RMS Values

In *Figure 15-10*, a 100-V battery is connected to a 100-Ω resistor. This connection will produce 1 A of current flow, and the resistor will dissipate 100 W of power in the form of heat. An AC alternator that produces a peak voltage of 100 V is also shown connected to a 100-Ω resistor. A peak current of 1 A will flow in the circuit, but the resistor will dissipate only 50 W in the form of heat. The reason is that the voltage produced by a pure source of direct current, such as a battery, is one continuous value

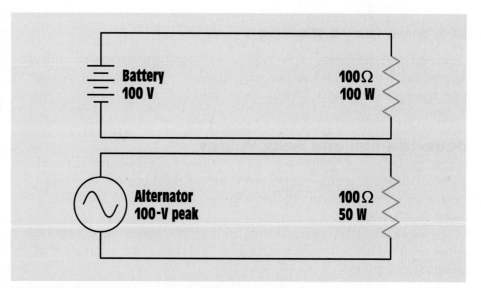

**Figure 15-10** Direct current compared with a sine wave AC current

*(Figure 15-11).* The AC sine wave, however, begins at zero, increases to the maximum value, and decreases back to zero during an equal period of time. Since the sine wave has a value of 100 V for only a short period of time and is less than 100 V during the rest of the half cycle, it cannot produce as much power as 100 V of DC.

The solution to this problem is to use a value of AC voltage that will produce the same amount of power as a like value of DC voltage. This AC value is called the **RMS,** or **effective,** value. It is the value indicated by almost all AC voltmeters and ammeters. **RMS stands for root-mean-square, which is an abbreviation for the square root of the mean of the square of the instantaneous currents.** The RMS value can be found by dividing the peak value by the square root of 2 (1.414) or by multiplying the peak value by 0.707 (the reciprocal of 1.414). The formulas for determining the RMS and peak values are

> RMS = peak x 0.707
> peak = RMS x 1.414

A sine wave has a peak value of 354 V. What is the RMS value?

### Solution

> RMS = peak x 0.707
> RMS = 354 x 0.707
> RMS = 250.3 V

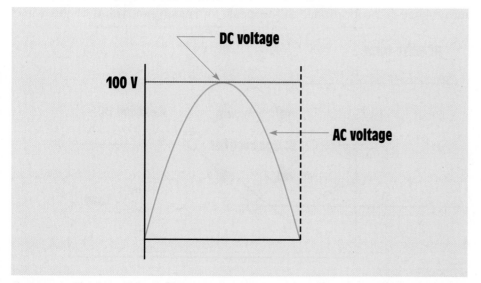

**Figure 15-11**   The DC voltage remains at a constant value during a half cycle of AC voltage.

**RMS**

**effective**

RMS stands for root-mean-square, which is an abbreviation for the square root of the mean of the square of the instantaneous currents.

**Example 4**

An AC voltage has a value of 120 V RMS. What is the peak value of voltage?

**Solution**

$$\text{peak} = \text{RMS} \times 1.414$$

$$\text{peak} = 120 \times 1.414$$

$$\text{peak} = 169.7 \text{ V}$$

When the RMS value of voltage and current is used, it will produce the same amount of power as a like value of DC voltage or current. If 100 V RMS is applied to a 100-$\Omega$ resistor, the resistor will produce 100 W of heat. AC voltmeters and ammeters indicate the RMS value, not the peak value. Oscilloscopes, however, display the peak-to-peak value of voltage. All values of AC voltage and current used from now on in this text will be RMS values unless otherwise stated.

### Average Values

**Average** values of voltage and current are actually direct current values. The average value must be found when a sine wave AC voltage is changed into DC with a rectifier *(Figure 15-12)*. The rectifier shown is a bridge type rectifier that produces full wave rectification. This means that both the positive and negative half of the AC wave form are changed into DC. The average value is the amount of voltage that would be indicated by a DC voltmeter if it were connected across the load resistor. The average voltage is proportional to the peak, or maximum, value of the wave

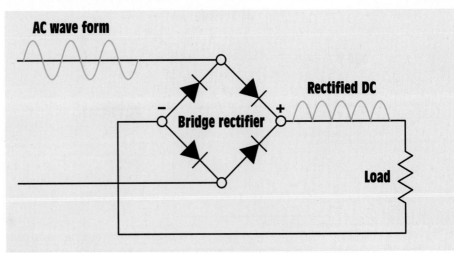

**Figure 15-12** The bridge rectifier changes AC voltage into DC voltage.

form, and to the length of time it is on as compared with the length of time it is off *(Figure 15-13)*. Notice in *Figure 15-13* that the voltage wave form turns on and off, but it never changes polarity. The current, therefore, never reverses direction. This is called pulsating direct current. The pulses are often referred to as **ripple**. The average value of voltage will produce the same amount of power as a nonpulsating source of voltage such as a battery *(Figure 15-14)*. For a sine wave, the average value of

**ripple**

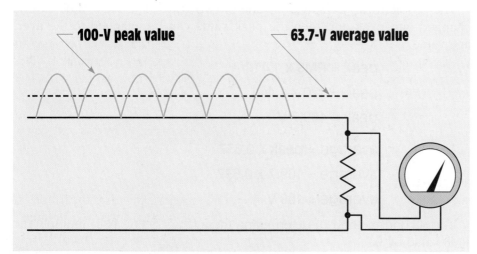

**Figure 15-13**   A DC voltmeter indicates the average value.

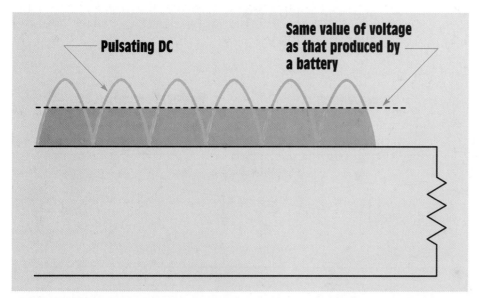

**Figure 15-14**   The average value produces the same amount of power as a nonpulsating source of voltage.

voltage is found by multiplying the peak value by 0.637 or by multiplying the RMS value by 0.9.

**Example 6**

An AC sine wave with an RMS value of 120 V is connected to a full wave rectifier. What is the average DC voltage?

### Solution

The problem can be solved in one of two ways. The RMS value can be changed into peak and then the peak value can be changed to the average value.

$$\text{peak} = \text{RMS} \times 1.414$$
$$\text{peak} = 120 \times 1.414$$
$$\text{peak} = 169.7 \text{ V}$$

$$\text{average} = \text{peak} \times 0.637$$
$$\text{average} = 169.7 \times 0.637$$
$$\text{average} = 108 \text{ V}$$

The second method of determining the average value is to multiply the RMS value by 0.9.

$$\text{average} = \text{RMS} \times 0.9$$
$$\text{average} = 120 \times 0.9$$
$$\text{average} = 108 \text{ V}$$

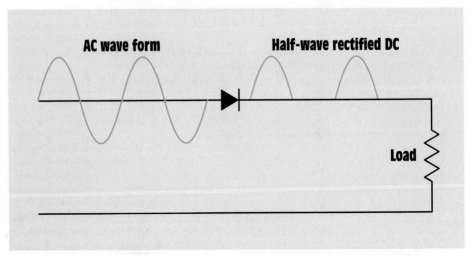

**Figure 15-15** A half-wave rectifier converts only one-half of the AC wave form into DC

The conversion factors given are for full wave rectification. If a half-wave rectifier is used *(Figure 15-15)*, only one-half of the AC wave form is converted into DC. To determine the average voltage for a half-wave rectifier, multiply the peak value by 0.637 or the RMS value by 0.9 and then divide the product by 2. Since only half of the AC wave form has been converted into direct current, the average voltage will be only half that of a full-wave rectifier *(Figure 15-16)*.

A half-wave rectifier is connected to 277 V AC. What is the average DC voltage?

**Example 7**

### Solution

$$\text{average} = \text{RMS} \times \frac{0.9}{2}$$

$$\text{average} = 277 \times \frac{0.9}{2}$$

$$\text{average} = 124.6 \text{ V}$$

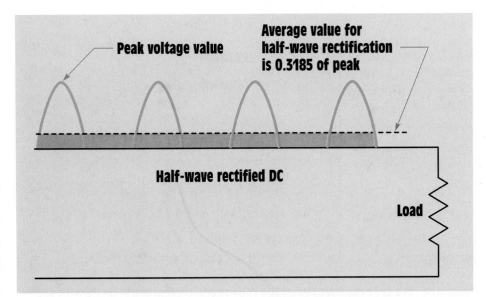

**Peak voltage value**

**Average value for half-wave rectification is 0.3185 of peak**

**Half-wave rectified DC**

**Load**

**Figure 15-16**   The average value for a half-wave rectifier is only half that of a full wave.

## 15-4 Resistive Loads

In direct current circuits, there is only one basic type of load, which is resistive. Even motor loads appear to be resistive because there is a conversion of electrical energy into mechanical energy. In this type of load, the **true power**, or **watts**, is the product of the volts times the amps. In alternating current circuits, the type of load can vary depending on several factors. Alternating current loads are generally described as being resistive, inductive, or capacitive depending on the phase-angle relationship of voltage and current and the amount of true power produced by the circuit. Inductive and capacitive loads will be discussed in later units. **Resistive loads** are loads that contain pure resistance, such as electric heating equipment and incandescent lighting. Resistive loads are characterized by the facts that

    1. they produce heat, and

    2. the current and voltage are in phase with each other.

Any time that a circuit contains resistance, electrical energy will be changed into heat.

When an AC voltage is applied to a resistor, the current flow through the resistor will be a copy of the voltage *(Figure 15-17)*. The current will rise and fall at the same rate as the voltage and will reverse the direction of flow when the voltage reverses polarity. In this condition, the current is said to be **in phase** with the voltage.

**true power**

**watts**

**resistive loads**

**in phase**

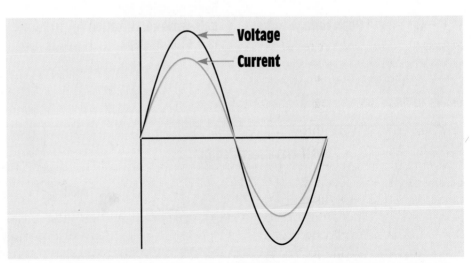

**Figure 15-17** In a pure resistive circuit the voltage and current are in phase with each other.

## 15-5 Power in an AC Circuit

**True power, or watts, can be produced only when both current and voltage are either positive or negative**. When like signs are multiplied, the product is positive (+ × + = +, or − × − = +), and when unlike signs are multiplied the product is negative (+ × − = −). Since the current and voltage are either positive or negative at the same time, the product, watts, will always be positive *(Figure 15-18)*.

**True power, or watts, can be produced only when both current and voltage are either positive or negative.**

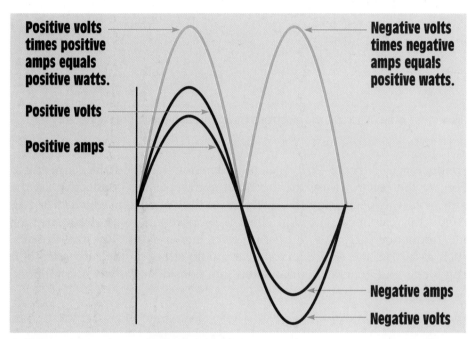

Positive volts times positive amps equals positive watts.

Positive volts

Positive amps

Negative volts times negative amps equals positive watts.

Negative amps

Negative volts

**Figure 15-18**   Power in a pure resistive AC circuit

## 15-6 Skin Effect in AC Circuits

When current flows through a conductor connected to a source of direct current, the electrons flow through the entire conductor *(Figure 15-19)*. The conductor offers some amount of ohmic resistance to the flow of electrons depending on the type of material from which the conductor is made, its length, and its diameter. If that same conductor were connected to a source of alternating current, the resistance of the conductor to the flow of current would be slightly higher because of the **skin effect**. The alternating current induces eddy currents into the conductor. These eddy currents cause the electrons to be repelled toward the outer surface of the

**skin effect**

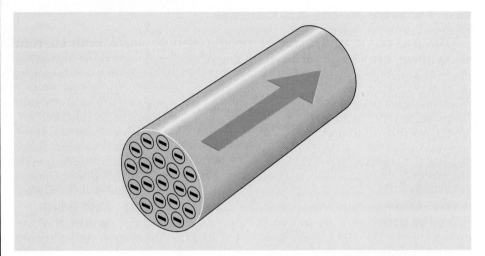

**Figure 15-19** In an DC circuit the electrons travel through the entire conductor.

conductor *(Figure 15-20)*. This phenomenon is called the skin effect. Forcing the electrons toward the outer surface of the conductor has the same effect as decreasing the diameter of the conductor, which increases conductor resistance. The skin effect is proportional to the frequency. As the frequency increases, the skin effect increases. At low frequencies, such as 60 Hz, the skin effect has very little effect on the resistance of a conductor and generally is not taken into consideration when computing

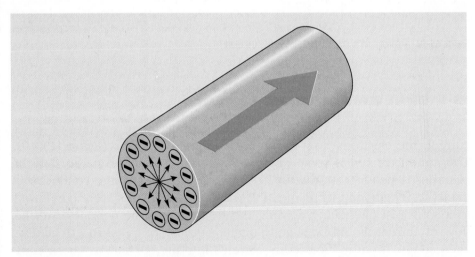

**Figure 15-20** In an AC circuit the electrons are forced to the outside of the conductor. This is called skin effect.

the size wire needed for a particular circuit. At high frequencies, however, the skin effect can have a great effect on the operation of the circuit.

To help overcome the problem of skin effect in high-frequency circuits, conductors with a large amount of surface area must be used. Grounding a piece of equipment operating at 60 Hz, for example, may be as simple as using a grounding rod and a piece of #6 AWG copper conductor. Grounding a piece of equipment operating at 20 MHz, however, may require the use of wide copper tape or a wide, flat, braided cable. Braided cable is less affected by skin effect because it contains many small conductors, which provide a large amount of surface area.

## Summary

1. Most of the electrical power generated in the world is alternating current.

2. Alternating current can be transformed and direct current cannot.

3. Alternating current reverses its direction of flow at periodic intervals.

4. The most common AC wave form is the sine wave.

5. There are 360° in one complete sine wave.

6. One complete wave form is called a cycle.

7. The number of complete cycles that occur in one second is called the frequency.

8. Sine waves are produced by rotating machines.

9. Frequency is measured in hertz (Hz).

10. The instantaneous voltage at any point on a sine wave is equal to the peak, or maximum, voltage times the sine of the angle of rotation.

11. The peak-to-peak voltage is the amount of voltage measured from the positive-most peak to the negative-most peak.

12. The peak value is the maximum amount of voltage attained by the wave form.

13. The RMS value of voltage will produce as much power as a like amount of DC voltage.

14. The average value of voltage is used when an AC sine wave is changed into DC.

15. The current and voltage in a pure resistive circuit are in phase with each other.

16. True power, or watts, can be produced only when current and voltage are both positive or both negative.

17. Resistance in AC circuits is characterized by the fact that the resistive part will produce heat.

18. There are three basic types of AC loads: resistive, inductive, and capacitive.

19. The electrons in an alternating current circuit are forced toward the outside of the conductor by eddy current induction in the conductor itself.

20. Skin effect is proportional to frequency.

21. Skin effect can be reduced by using conductors with a large surface area.

## Review Questions

1. What is the most common type of AC wave form?

2. How many degrees are there in one complete sine wave?

3. At what angle does the voltage reach its maximum negative value on a sine wave?

4. What is frequency?

5. A sine wave has a maximum value of 230 V. What is the voltage after 38° of rotation?

6. A sine wave has a voltage of 63 V after 22° of rotation. What is the maximum voltage reached by this wave form?

7. A sine wave has a maximum value of 560 V. At what angle of rotation will the voltage reach a value of 123 V?

8. A sine wave has a peak value of 433 V. What is the RMS value?

9. A sine wave has a peak-to-peak value of 88 V. What is the average value?

10. A DC voltage has an average value of 68 V. What is the RMS value?

Refer to the alternating current formulas in the appendix to answer the following questions.

## Sine Wave Values

Fill in all the missing values. The table continues on the following page.

| Peak Volts | Inst. Volts | Degrees |
|------------|-------------|---------|
| 347 | 208 | |
| 780 | | 43.5 |
| | 24.3 | 17.6 |
| 224 | 5.65 | |
| 48.7 | | 64.6 |
| | 240 | 45 |
| 87.2 | 23.7 | |
| 156.9 | | 82.3 |

| Peak Volts | Inst. Volts | Degrees |
|------------|-------------|---------|
|            | 62.7        | 34.6    |
| 1256       | 400         |         |
| 15,720     |             | 12      |
|            | 72.4        | 34.8    |

$$E_{(INST)} = E_{(MAX)} \times SIN \angle$$

$$E_{(MAX)} = \frac{E_{(INST)}}{SIN \angle}$$

$$SIN \angle = \frac{E_{(INST)}}{E_{(MAX)}}$$

## Peak, RMS, and Average Values

Fill in all the missing values. The table continues on the next page.

| Peak | RMS  | Average |
|------|------|---------|
| 12.7 |      |         |
|      | 53.8 |         |
|      |      | 164.2   |
| 1235 |      |         |
|      | 240  |         |
|      |      | 16.6    |

| Peak | RMS | Average |
|---|---|---|
| 339.7 | | |
| | 12.6 | |
| | | 9 |
| 123.7 | | |
| | 74.8 | |
| | | 108 |

*section*

# alternating current circuits containing inductance

**The values of voltage, current, and power in an inductive circuit are discussed in this unit.**

Courtesy of Niagara Mohawk Power Corp.

# Inductance in Alternating Current Circuits

## Outline

## Key Terms

Current lags voltage

Impedance (Z)

Induced voltage

Inductance (L)

Inductive reactance ($X_L$)

Quality (Q)

Reactance

Reactive power (VARs)

## Objectives

**A**fter studying this unit, you should be able to:

■ Discuss the properties of inductance in an alternating current circuit.

■ Discuss inductive reactance.

■ Compute values of inductive reactance and inductance.

■ Discuss the relationship of voltage and current in a pure inductive circuit.

■ Be able to compute values for inductors connected in series or parallel.

■ Discuss reactive power (VARs).

■ Determine the Q of a coil.

## Preview

This unit discusses the effects of inductance on alternating current circuits. The unit explains how current is limited in an inductive circuit as well as the effect inductance has on the relationship of voltage and current.

## 16-1 Inductance

**inductance (L)**

**Inductance (L)** is one of the primary types of loads in alternating current circuits. Some amount of inductance is present in all alternating current circuits because of the continually changing magnetic field *(Figure 16-1)*. The amount of inductance of a single conductor is extremely small, and in most instances it is not considered in circuit calculations. Circuits are generally considered to contain inductance when any type of load that contains a coil is used. For circuits that contain a coil, inductance *is* considered in circuit calculations. Loads such as motors, transformers, lighting ballast, and chokes all contain coils of wire.

In Unit 13, it was discussed that whenever current flows through a coil of wire a magnetic field is created around the wire *(Figure 16-2)*. If the

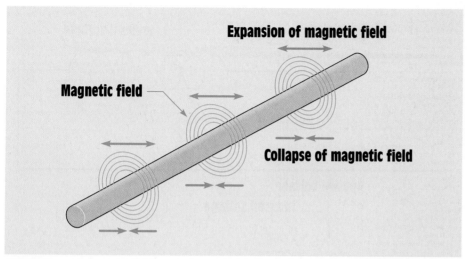

**Figure 16-1** A continually changing magnetic field induces a voltage into any conductor.

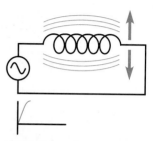

**Figure 16-2** As current flows through a coil, a magnetic field is created around the coil.

amount of current decreases, the magnetic field will collapse *(Figure 16-3)*. Recall from Unit 13 several facts concerning inductance:

1. When magnetic lines of flux cut through a coil, a voltage is induced in the coil.
2. An induced voltage is always opposite in polarity to the applied voltage.
3. The amount of induced voltage is proportional to the rate of change of current.
4. An inductor opposes a change of current.

The inductors in *Figures 16-2* and *16-3* are connected to an alternating voltage. Therefore the magnetic field continually increases, decreases, and reverses polarity. Since the magnetic field continually changes magnitude and direction, a voltage is continually being induced in the coil. This **induced voltage** is 180° out of phase with the applied voltage and is always in opposition to the applied voltage *(Figure 16-4)*. Since the induced voltage is always in opposition to the applied voltage, the applied voltage must overcome the induced voltage before current can flow through the circuit. For example, assume an inductor is connected to a 120-V AC line. Now assume that the inductor has an induced voltage of 116 V. Since an equal amount of applied voltage must be used to overcome the induced voltage, there will be only 4 V to push current through the wire resistance of the coil (120 − 116 = 4).

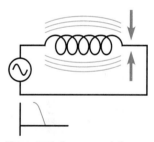

**Figure 16-3** As current flow decreases, the magnetic field collapses.

**induced voltage**

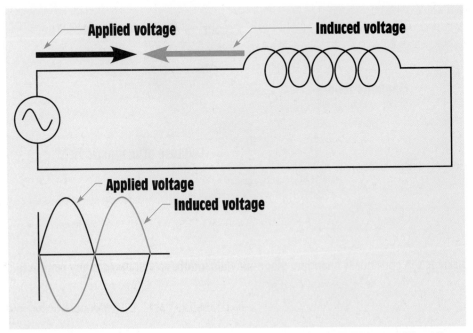

**Figure 16-4** The applied voltage and induced voltage are 180° out of phase with each other.

## Computing the Induced Voltage

The amount of induced voltage in an inductor can be computed if the resistance of the wire in the coil and the amount of circuit current are known. For example, assume that an ohmmeter is used to measure the actual amount of resistance in a coil, and the coil is found to contain 6 $\Omega$ of wire resistance *(Figure 16-5)*. Now assume that the coil is connected to

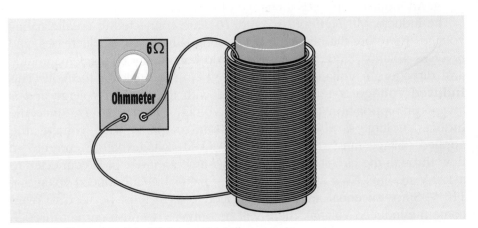

**Figure 16-5** Measuring the resistance of a coil

a 120-V AC circuit and an ammeter measures a current flow of 0.8 A *(Figure 16-6)*. Ohm's law can now be used to determine the amount of voltage necessary to push 0.8 A of current through 6 Ω of resistance.

$$E = I \times R$$

$$E = 0.8 \times 6$$

$$E = 4.8 \text{ V}$$

Since only 4.8 V is needed to push the current through the wire resistance of the inductor, the remainder of the 120 V is used to overcome the coil's induced voltage of 115.2 V (120 − 4.8 = 115.2).

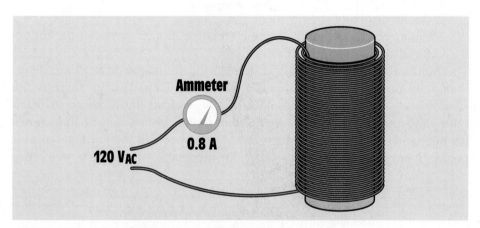

**Figure 16-6** Measuring circuit current with an ammeter

## 16-2 Inductive Reactance

Notice that the induced voltage is able to limit the flow of current through the circuit in a manner similar to resistance. This induced voltage is *not* resistance, but it can limit the flow of current just as resistance does. This current-limiting property of the inductor is called **reactance** and is symbolized by the letter $X$. This reactance is caused by inductance, so it is called **inductive reactance** and is symbolized by $X_L$, pronounced "X sub L." Inductive reactance is measured in ohms just as resistance is and can be computed when the values of inductance and frequency are known. The following formula can be used to find inductive reactance.

$$X_L = 2\pi FL$$

**reactance**

**inductive reactance ($X_L$)**

where

$X_L$ = inductive reactance

2 = a constant

$\pi$ = 3.1416

F = frequency in hertz (Hz)

L = inductance in henrys (H)

Inductive reactance is an induced voltage and is, therefore, proportional to the three factors that determine induced voltage:

1. the **number** of turns of wire,
2. the **strength** of the magnetic field, and
3. the **speed** of the cutting action (relative motion between the inductor and the magnetic lines of flux).

The number of turns of wire and strength of the magnetic field are determined by the physical construction of the inductor. Factors such as the size of wire used, the number of turns, how close the turns are to each other, and the type of core material determine the amount of inductance (in henrys, H) of the coil *(Figure 16-7)*. The speed of the cutting action is proportional to the frequency (Hz). An increase of frequency will cause

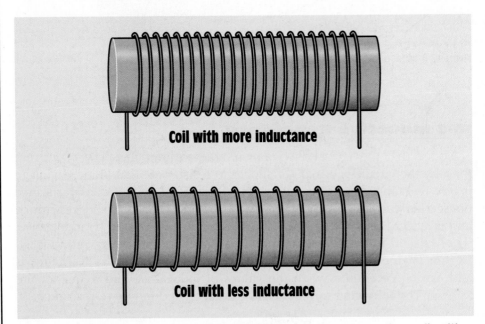

**Coil with more inductance**

**Coil with less inductance**

**Figure 16-7** Coils with turns closer together produce more inductance than coils with turns far apart.

the magnetic lines of flux to cut the conductors at a faster rate, and thus will produce a higher induced voltage or more inductive reactance.

The inductor shown in *Figure 16-8* has an inductance of 0.8 H and is connected to a 120-V, 60-Hz line. How much current will flow in this circuit if the wire resistance of the inductor is negligible?

**Example 1**

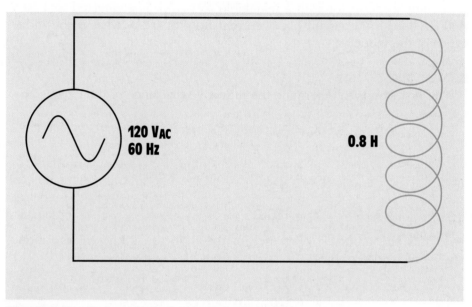

**Figure 16-8** Circuit current is limited by inductive reactance.

## Solution

The first step is to determine the amount of inductive reactance of the inductor.

$$X_L = 2\pi FL$$
$$X_L = 2 \times 3.1416 \times 60 \times 0.8$$
$$X_L = 301.6 \ \Omega$$

Since inductive reactance is the current-limiting property of this circuit, it can be substituted for the value of R in an Ohm's law formula.

$$I = \frac{E}{X_L}$$

$$I = \frac{120}{301.6}$$

$$I = 0.398 \ A$$

If the amount of inductive reactance is known, the inductance of the coil can be determined using the formula

$$L = \frac{X_L}{2\pi F}$$

**Example 2**

Assume an inductor with a negligible resistance is connected to a 36-V, 400-Hz line. If the circuit has a current flow of 0.2 A what is the inductance of the inductor?

### Solution

The first step is to determine the inductive reactance of the circuit.

$$X_L = \frac{E}{I}$$

$$X_L = \frac{36}{0.2}$$

$$X_L = 180 \ \Omega$$

Now that the inductive reactance of the inductor is known, the inductance can be determined.

$$L = \frac{X_L}{2\pi F}$$

$$L = \frac{180}{2 \times 3.1416 \times 400}$$

$$L = 0.0716 \ H$$

**Example 3**

An inductor with negligible resistance is connected to a 480-V, 60-Hz line. An ammeter indicates a current flow of 24 A. How much current will flow in this circuit if the frequency is increased to 400 Hz?

### Solution

The first step in solving this problem is to determine the amount of inductance of the coil. Since the resistance of the wire used to make the inductor is negligible, the current is limited by inductive reactance. The inductive reactance can be found by substituting $X_L$ for R in an Ohm's law formula.

$$X_L = \frac{E}{I}$$

$$X_L = \frac{480}{24}$$

$$X_L = 20 \ \Omega$$

Now that the inductive reactance is known, the inductance of the coil can be found using the formula

$$L = \frac{X_L}{2\pi F}$$

NOTE: When using a frequency of 60 Hz, 2 x π x 60 = 377. Since 60 Hz is the major frequency used throughout the United States and Canada, 377 should be memorized for use when necessary.

$$L = \frac{20}{377}$$

$$L = 0.053 \ H$$

Since the inductance of the coil is determined by its physical construction, it will not change when connected to a different frequency. Now that the inductance of the coil is known, the inductive reactance at 400 Hz can be computed.

$$X_L = 2\pi FL$$

$$X_L = 2 \times 3.1416 \times 400 \times 0.053$$

$$X_L = 133.2 \ \Omega$$

The amount of current flow can now be found by substituting the value of inductive reactance for resistance in an Ohm's law formula.

$$I = \frac{E}{X_L}$$

$$I = \frac{480}{133.2}$$

$$I = 3.6 \ A$$

## 16-3 Schematic Symbols

The schematic symbol used to represent an inductor depicts a coil of wire. Several symbols for inductors are shown in *Figure 16-9*. The symbols

**Air core inductors**

**Iron core inductors**

**Figure 16-9** Schematic symbols for inductors

shown with the two parallel lines represent iron core inductors, and the symbols without the parallel lines represent air core inductors.

## 16-4 Inductors Connected in Series

When inductors are connected in series *(Figure 16-10),* the total inductance of the circuit ($L_T$) equals the sum of the inductances of all the inductors.

$$L_T = L_1 + L_2 + L_3$$

The total inductive reactance ($X_{LT}$) of inductors connected in series equals the sum of the inductive reactances for all the inductors.

$$X_{LT} = X_{L1} + X_{L2} + X_{L3}$$

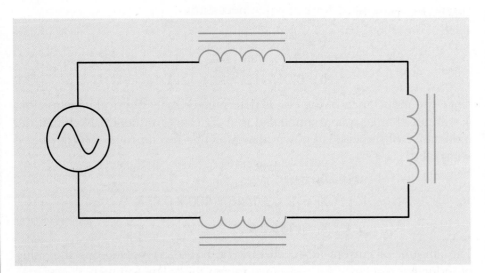

**Figure 16-10** Inductors connected in series

---

**Example 4**

Three inductors are connected in series. Inductor 1 has an inductance of 0.6 H, inductor 2 has an inductance of 0.4 H, and inductor 3 has an inductance of 0.5 H. What is the total inductance of the circuit?

### Solution

$$L_T = 0.6 + 0.4 + 0.5$$

$$L_T = 1.5 \text{ H}$$

Three inductors are connected in series. Inductor 1 has an inductive reactance of 180 Ω, inductor 2 has an inductive reactance of 240 Ω, and inductor 3 has an inductive reactance of 320 Ω. What is the total inductive reactance of the circuit?

**Example 5**

### Solution

$$X_{LT} = 180\ \Omega + 240\ \Omega + 320\ \Omega$$

$$X_{LT} = 740\ \Omega$$

## 16-5 Inductors Connected in Parallel

When inductors are connected in parallel *(Figure 16-11)*, the total inductance can be found in a similar manner to finding the total resistance of a parallel circuit. The reciprocal of the total inductance is equal to the sum of the reciprocals of all the inductors.

$$\frac{1}{L_T} = \frac{1}{L_1} + \frac{1}{L_2} + \frac{1}{L_3}$$

or

$$L_T = \frac{1}{\dfrac{1}{L_1} + \dfrac{1}{L_2} + \dfrac{1}{L_3}}$$

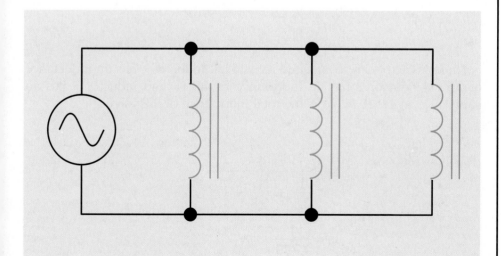

**Figure 16-11** Inductors connected in parallel

Another formula that can be used to find the total inductance of parallel inductors is the product over sum formula.

$$L_T = \frac{L_1 \times L_2}{L_1 + L_2}$$

If the values of all the inductors are the same, total inductance can be found by dividing the inductance of one inductor by the total number of inductors.

$$L_T = \frac{L}{N}$$

Similar formulas can be used to find the total inductive reactance of inductors connected in parallel.

$$\frac{1}{X_{LT}} = \frac{1}{X_{L1}} + \frac{1}{X_{L2}} + \frac{1}{X_{L3}}$$

or

$$X_{LT} = \frac{1}{\dfrac{1}{X_{L1}} + \dfrac{1}{X_{L2}} + \dfrac{1}{X_{L3}}}$$

or

$$X_{LT} = \frac{X_{L1} \times X_{L2}}{X_{L1} + X_{L2}}$$

or

$$X_{LT} = \frac{X_L}{N}$$

**Example 6**

Three inductors are connected in parallel. Inductor 1 has an inductance of 2.5 H, inductor 2 has an inductance of 1.8 H, and inductor 3 has an inductance of 1.2 H. What is the total inductance of this circuit?

**Solution**

$$L_T = \frac{1}{\dfrac{1}{2.5} + \dfrac{1}{1.8} + \dfrac{1}{1.2}}$$

$$L_T = \frac{1}{1.789}$$

$$L_T = 0.559 \text{ H}$$

## 16-6 Voltage and Current Relationships in an Inductive Circuit

In Unit 15, it was discussed that when current flows through a pure resistive circuit, the current and voltage are in phase with each other. **In a pure inductive circuit the current lags the voltage by 90°.** At first this may seem to be an impossible condition until the relationship of applied voltage and induced voltage is considered. How the current and applied voltage can become 90° out of phase with each other can best be explained by comparing the relationship of the current and induced voltage *(Figure 16-12)*. Recall that the induced voltage is proportional to the rate of change of the current (speed of cutting action). At the beginning of the wave form, the current is shown at its maximum value in the negative direction. At this time, the current is not changing, so induced voltage is zero. As the current begins to decrease in value, the magnetic field produced by the flow of current decreases or collapses and begins to induce a voltage into the coil as it cuts through the conductors *(Figure 16-3)*.

The greatest rate of current change occurs when the current passes from negative, through zero, and begins to increase in the positive direction *(Figure 16-13)*. Since the current is changing at the greatest rate, the

**In a pure inductive circuit the current lags the voltage by 90°.**

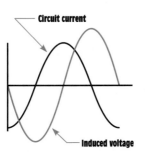

**Figure 16-12**  Induced voltage is proportional to the rate of change of current.

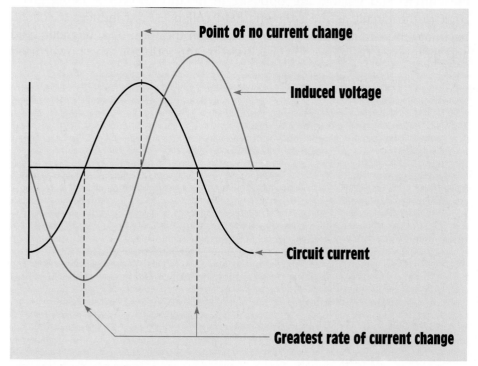

**Figure 16-13**  No voltage is induced when the current does not change.

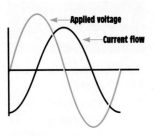

**Figure 16-14** The current lags the applied voltage by 90°

induced voltage is maximum. As current approaches its peak value in the positive direction, the rate of change decreases, causing a decrease in the induced voltage. The induced voltage will again be zero when the current reaches its peak value and the magnetic field stops expanding.

It can be seen that the current flowing through the inductor is leading the induced voltage by 90°. Since the induced voltage is 180° out of phase with the applied voltage, the current will lag the applied voltage by 90° *(Figure 16-14)*.

## 16-7 Power in an Inductive Circuit

In a pure resistive circuit, the true power, or watts, is equal to the product of the voltage and current. In a pure inductive circuit, however, no true power, or watts, is produced. Recall that voltage and current must both be either positive or negative before true power can be produced. Since the voltage and current are 90° out of phase with each other in a pure inductive circuit, the current and voltage will be at different polarities 50% of the time and at the same polarity 50% of the time. During the period of time that the current and voltage have the same polarity, power is being given to the circuit in the form of creating a magnetic field. When the current and voltage are opposite in polarity, power is being given back to the circuit as the magnetic field collapses and induces a voltage back into the circuit. Since power is stored in the form of a magnetic field and then given back, no power is used by the inductor. Any power used

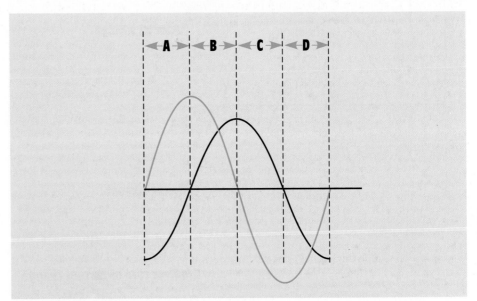

**Figure 16-15** Voltage and current relationships during different parts of a cycle

in an inductor is caused by losses such as the resistance of the wire used to construct the inductor, generally referred to as $I^2R$ losses, eddy current losses, and hysteresis losses.

The current and voltage wave form in *Figure 16-15* has been divided into four sections: A, B, C, and D. During the first time period, indicated by A, the current is negative and the voltage is positive. During this period, energy is being given to the circuit as the magnetic field collapses. During the second time period, section B, both the voltage and current are positive. Power is being used to produce the magnetic field. In the third time period, C, the current is positive and the voltage is negative. Power is again being given back to the circuit as the field collapses. During the fourth time period, D, both the voltage and current are negative. Power is again being used to produce the magnetic field. If the amount of power used to produce the magnetic field is subtracted from the power given back, the result will be zero.

## 16-8 Reactive Power

Although essentially no true power is being used, except by previously mentioned losses, an electrical measurement called **VARs** is used to measure the **reactive power** in a pure inductive circuit. **VARs** is an abbreviation for volt-amps-reactive. VARs can be computed in the same way as watts except that inductive values are substituted for resistive values in the formulas. VARs is equal to the amount of current flowing through an inductive circuit times the voltage applied to the inductive part of the circuit. Several formulas for computing VARs are:

reactive
power
(VARs)

$$VARs = E_L \times I_L$$

$$VARs = \frac{E_L^2}{X_L}$$

$$VARs = I_L^2 \times X_L$$

where

$E_L$ = voltage applied to an inductor

$I_L$ = current flow through an inductor

$X_L$ = inductive reactance

## 16-9 Q of an Inductor

So far in this unit, it has been generally assumed that an inductor has no resistance and that inductive reactance is the only current-limiting factor.

In reality, that is not true. Since inductors are actually coils of wire they all contain some amount of internal resistance. Inductors actually appear to be a coil connected in series with some amount of resistance *(Figure 16-16)*. The amount of resistance compared with the inductive reactance determines the Q of the coil. The letter *Q* stands for **quality**. Inductors that have a higher ratio of inductive reactance to resistance are considered to be inductors of higher quality. An inductor constructed with a large wire will have a low wire resistance and, therefore, a higher Q

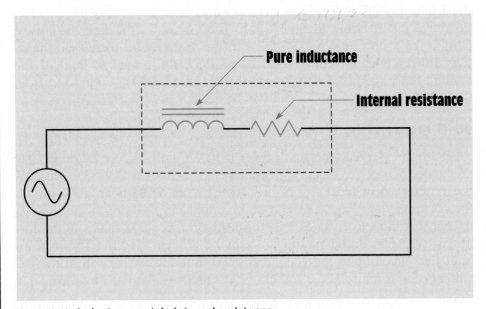

**Figure 16-16**  Inductors contain internal resistance.

*(Figure 16-17)*. Inductors constructed with many turns of small wire have a much higher resistance, and, therefore, a lower Q. To determine the Q of an inductor, divide the inductive reactance by the resistance.

$$Q = \frac{X_L}{R}$$

Although inductors have some amount of resistance, inductors that have a Q of 10 or greater are generally considered to be pure inductors. Once the ratio of inductive reactance becomes 10 times as great as resistance, the amount of resistance is considered negligible. For example, assume an inductor has an inductive reactance of 100 $\Omega$ and a wire resistance of 10 $\Omega$. The inductive reactive component in the circuit is 90° out of phase with

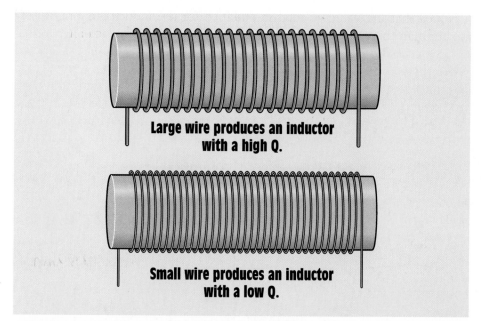

Large wire produces an inductor
with a high Q.

Small wire produces an inductor
with a low Q.

**Figure 16-17**   The Q of an inductor is a ratio of inductive reactance as compared to resistance. The letter Q stands for quality.

the resistive component. This relationship produces a right triangle *(Figure 16-18)*. The total current-limiting effect of the inductor is a combination of the inductive reactance and resistance. This total current-limiting effect is called **impedance** and is symbolized by the letter **Z**. The impedance of the circuit is represented by the hypotenuse of the right triangle formed by the inductive reactance and the resistance. To compute the value of impedance for the coil, the inductive reactance and resistance must be added. Since these two components form the legs of a right triangle and the impedance forms the hypotenuse, the Pythagorean theorem discussed in Unit 14 can be used to compute the value of impedance.

**impedance (Z)**

$$Z = \sqrt{R^2 + X_L^2}$$

$$Z = \sqrt{10^2 + 100^2}$$

$$Z = \sqrt{10,100}$$

$$Z = 100.5 \ \Omega$$

Notice that the value of total impedance for the inductor is only 0.5 Ω greater than the value of inductive reactance.

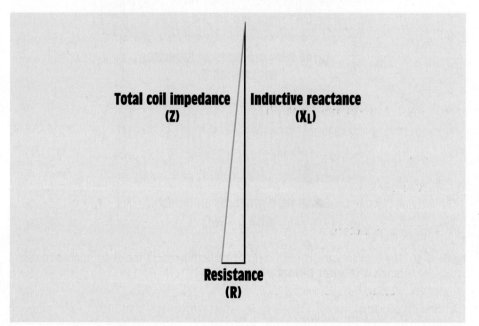

**Total coil impedance (Z)**

**Inductive reactance (X_L)**

**Resistance (R)**

**Figure 16-18**  Coil impedance is a combination of wire resistance and inductive reactance.

## Summary

1. Induced voltage is proportional to the rate of change of current.

2. Induced voltage is always opposite in polarity to the applied voltage.

3. Inductive reactance is a countervoltage that limits the flow of current, as does resistance.

4. Inductive reactance is measured in ohms.

5. Inductive reactance is proportional to the inductance of the coil and the frequency of the line.

6. Inductive reactance is symbolized by $X_L$.

7. Inductance is measured in henrys (H) and is symbolized by the letter $L$.

8. When inductors are connected in series the total inductance is equal to the sum of all the inductors.

9. When inductors are connected in parallel the reciprocal of the total inductance is equal to the sum of the reciprocals of all the inductors.

10. The current lags the applied voltage by 90° in a pure inductive circuit.

11. All inductors contain some amount of resistance.

12. The Q of an inductor is the ratio of the inductive reactance to the resistance.

13. Inductors with a Q of 10 are generally considered to be "pure" inductors.

14. Pure inductive circuits contain no true power, or watts.

15. Reactive power is measured in VARs.

16. VARs is an abbreviation for volt-amps-reactive.

## Review Questions

1. How many degrees are the current and voltage out of phase with each other in a pure resistive circuit?

2. How many degrees are the current and voltage out of phase with each other in a pure inductive circuit?

3. To what is inductive reactance proportional?

4. Four inductors, each having an inductance of 0.6 H, are connected in series. What is the total inductance of the circuit? .24 H

5. Three inductors are connected in parallel. Inductor 1 has an inductance of 0.06 H; inductor 2 has an inductance of 0.05 H; and inductor 3 has an inductance of 0.1 H. What is the total inductance of this circuit? .0214

6. If the three inductors in question 5 were connected in series, what would be the inductive reactance of the circuit? Assume the inductors are connected to a 60-Hz line. √ 8.0635

7. An inductor is connected to a 240-V, 1000-Hz line. The circuit current is 0.6 A. What is the inductance of the inductor? 400  63.6 Ω

8. An inductor with an inductance of 3.6 H is connected to a 480-V, 60-Hz line. How much current will flow in this circuit? 353mA

9. If the frequency in question 8 is reduced to 50 Hz how much current will flow in the circuit? 424mA

10. An inductor has an inductive reactance of 250 Ω when connected to a 60-Hz line. What will be the inductive reactance if the inductor is connected to a 400-Hz line? .6634   60   250

1666.46 Ω

**Practice Problems**

## Inductive Circuits

Fill in all the missing values. Refer to the formulas given below.

$$X_L = 2\pi FL$$

$$L = \frac{X_L}{2\pi F}$$

$$F = \frac{X_L}{2\pi L}$$

| Inductance (H) | Frequency (Hz) | Induct. Rct. (Ω) |
|---|---|---|
| 1.2 | 60 | 452.389 |
| 0.085 | 400.221 | 213.628 |
| .750 | 1000 | 4712.389 |
| 0.65 | 600 | 2449.2 |
| 3.6 | 30.015 | 678.584 |
| 2.620 | 25 | 411.459 |
| 0.5 | 60 | 188.4 |
| 0.85 | 1200.608 | 6408.849 |
| 1.600 | 20 | 201.062 |
| 0.45 | 400 | 1130.4 |
| 4.8 | 80.040 | 2412.743 |
| .00650H | 1000 | 40.841 |

**This unit discusses the behavior of voltage, current, power, and impedance in series circuits that contain elements of both resistance and inductance.**

Courtesy of Simmons Machine Tool Co.

# Resistive-Inductive Series Circuits

## Key Terms

Angle theta ($\angle\theta$)

Apparent power (VA)

Parallelogram method

Power factor (PF)

Quadrature power

Reactive power (VARs)

Total current (I)

Total impedance (Z)

True power (P)

Vector

Vector addition

Voltage drop across the inductor ($E_L$)

Voltage drop across the resistor ($E_R$)

Wattless power

## Objectives

*After studying this unit, you should be able to:*

- Discuss the relationship of resistance and inductance in an alternating current series circuit.

- Define power factor.

- Calculate values of voltage, current, apparent power, true power, reactive power, impedance, resistance, inductive reactance, and power factor in an R-L series circuit.

- Compute the phase angle for current and voltage in an R-L series circuit.

- Connect an R-L series circuit and make measurements using test instruments.

- Discuss vectors and be able to plot electrical quantities using vectors.

## Preview

This unit will cover the relationship of resistance and inductance used in the same circuit. The resistors and inductors will be connected in series. Concepts such as circuit impedance, power factor, and vector addition will be introduced. Although it is true that some circuits are basically purely resistive or purely inductive, many circuits contain a combination of both resistive and inductive elements.

### 17-1 RL Series Circuits

**When a pure resistive load is connected to an alternating current circuit, the voltage and current are in phase with each other.** When a pure inductive load is connected to an alternating current circuit, the voltage and current are 90° out of phase with each other *(Figure 17-1)*. When a circuit containing both resistance, R, and inductance, L, is con-

**When a pure resistive load is connected to an alternating current circuit, the voltage and current are in phase with each other.**

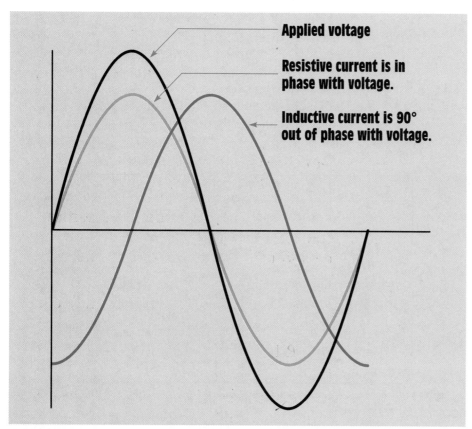

**Applied voltage**

**Resistive current is in phase with voltage.**

**Inductive current is 90° out of phase with voltage.**

**Figure 17-1**  Relationship of resistive and inductive current with voltage

nected to an alternating current circuit, the voltage and current will be out of phase with each other by some amount between 0° and 90°. **The exact amount of phase angle difference is determined by the ratio of resistance as compared to inductance.** In the following example, a series circuit containing 30 Ω of resistance (R) and 40 Ω of inductive reactance ($X_L$) is connected to a 240-V 60-Hz line *(Figure 17-2).* It is assumed the inductor has negligible resistance. The following unknown values will be computed:

Z    —    total circuit impedance

I    —    current flow

$E_R$    —    voltage drop across the resistor

P    —    watts (true power)

L    —    inductance of the inductor

The exact amount of phase angle difference is determined by the ratio of resistance as compared to inductance.

$E_L$ — voltage drop across the inductor

VARs — reactive power

VA — apparent power

PF — power factor

$\angle\theta$ — the angle the voltage and current are out of phase with each other

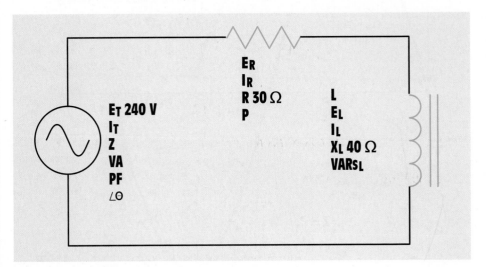

**Figure 17-2**  R-L series circuit

## 17-2 Impedance

In Unit 16 impedance was defined as a measure of the part of the circuit that impedes, or hinders, the flow of current. It is measured in ohms and symbolized by the letter $Z$. In this circuit, impedance will be a combination of resistance and inductive reactance.

In a series circuit the total resistance is equal to the sum of the individual resistors. In this instance, however, the **total impedance (Z)** will be the sum of the resistance and the inductive reactance. It would first appear that the sum of these two quantities should be 70 $\Omega$ (30 $\Omega$ + 40 $\Omega$ = 70 $\Omega$). In practice, however, the resistive part of the circuit and the reactive part of the circuit are out of phase with each other by 90°. To find the sum of these two quantities vector addition must be used. Since these two quantities are 90° out of phase with each other, the resistive and inductive reactance form the two legs of a right triangle, and the impedance is the hypotenuse *(Figure 17-3)*.

**total impedance (Z)**

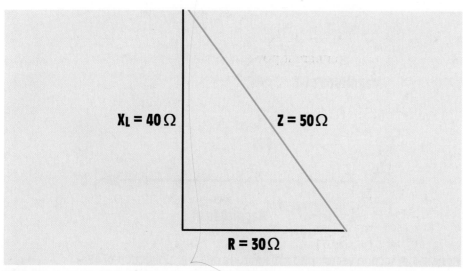

$X_L = 40\,\Omega$    $Z = 50\,\Omega$

$R = 30\,\Omega$

**Figure 17-3**  Impedance is the combination of resistance and inductive reactance.

The total impedance (Z) can be computed using the formula:

$$Z = \sqrt{R^2 + X_L^2}$$

$$Z = \sqrt{30^2 + 40^2}$$

$$Z = \sqrt{900 + 1600}$$

$$Z = \sqrt{2500}$$

$$Z = 50\ \Omega$$

## 17-3 Vectors

Using the right triangle is just one method of graphically showing how out-of-phase quantities can be added. Another method of illustrating this concept is with the use of vectors. A **vector** is a line that indicates both magnitude and direction. The magnitude is indicated by its length, and the direction is indicated by its angle of rotation from 0°. Vectors should not be confused with *scalars*, which are used to represent magnitude

**vector**

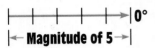

**Figure 17-4** A vector with a magnitude of 5 and an angle of 0°

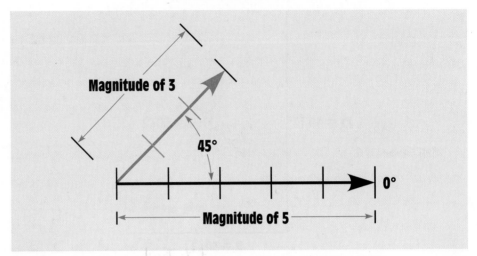

**Figure 17-5** A second vector with a magnitude of 3 and a direction of 45°

only and do not take direction into consideration. Imagine, for example, that you are in a strange city and you ask someone for directions to a certain building. If the person said walk three blocks, that would be a scalar because it contains only the magnitude, three blocks. If the person said walk three blocks south, it would be a vector because it contains both the magnitude, three blocks, and the direction, south.

Zero degrees is indicated by a horizontal line. An arrow is placed at one end of the line to indicate direction. The magnitude can represent any quantity such as inches, meters, miles, volts, amps, ohms, power, and so on. A vector with a magnitude of 5 at an angle of 0° is shown in *Figure 17-4*. Vectors rotate in a counterclockwise direction. Assume that a vector with a magnitude of 3 is to be drawn at a 45° angle from the first vector *(Figure 17-5)*. Now assume that a third vector with a magnitude of 4 is to be drawn in a direction of 120° *(Figure 17-6)*. Notice that the direction of the third vector is referenced from the horizontal 0° line and not from the second vector line, which was drawn at an angle of 45°.

## Adding Vectors

Because vectors are used to represent quantities such as volts, amps, ohms, power, etc., they can be added, subtracted, multiplied, and divided. In electrical work, however, addition is the only function needed, so it will be the only one discussed. Several methods can be used to add vectors. Regardless of the method used, because vectors contain both magnitude and direction, they must be added with a combination of geo-

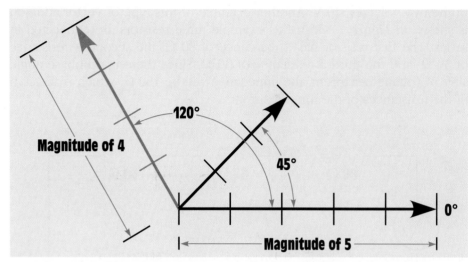

**Figure 17-6** A third vector with a magnitude of 4 and a direction of 120° is added.

metric and algebraic addition. This method is often referred to as **vector addition**.

One method is to connect the starting point of one vector to the end point of another. This method works especially well when all vectors are in the same direction. Consider the circuit shown in *Figure 17-7*. In this circuit two batteries, one rated at 6 V and the other rated at 4 V, are connected in such a manner that their voltages add. If vector addition is used, the starting point of one vector will be placed at the end point of the other. Notice that the sum of the two vector quantities is equal to the sum

**vector addition**

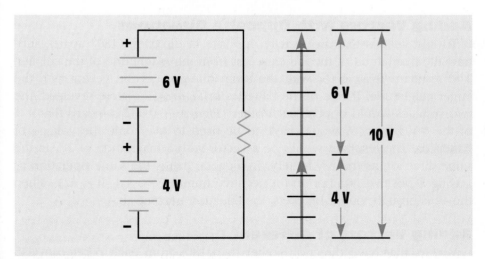

**Figure 17-7** Adding vectors in the same direction

of the two voltages, 10 V. Another example of this type of vector addition is shown in *Figure 17-8*. In this example three resistors are connected in series. The first resistor has a resistance of 80 Ω, the second a resistance of 50 Ω, and the third a resistance of 30 Ω. Since there is no phase angle shift of voltage or current, the impedance will be 160 Ω, which is the sum of the resistances of the three resistors.

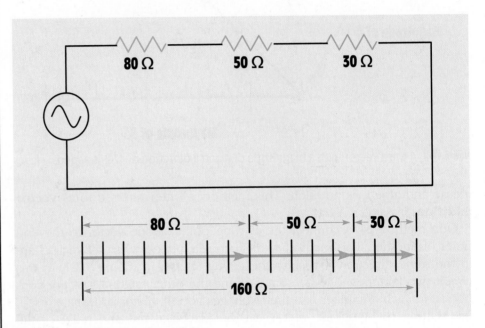

**Figure 17-8** Addition of series resistors

## Adding Vectors with Opposite Directions

To add vectors that are exactly opposite in direction (180° apart), subtract the magnitude of the larger vector from the magnitude of the smaller. The resultant is a vector with the same direction as the vector with the larger magnitude. If one of the batteries in *Figure 17-7* were reversed, the two voltages would oppose each other *(Figure 17-9)*. This means that 4 V of the 6-V battery A would have to be used to overcome the voltage of battery B. The resultant would be a vector with a magnitude of 2 V in the same direction as the 6-V battery. In algebra, this is the same operation as adding a positive number and a negative number (+6 + [–4] = +2). When the –4 is brought out of brackets, the equation becomes 6 – 4 = 2.

## Adding Vectors of Different Directions

Vectors that have directions other than 180° from each other can also be added. *Figure 17-10* illustrates the addition of a vector with a magni-

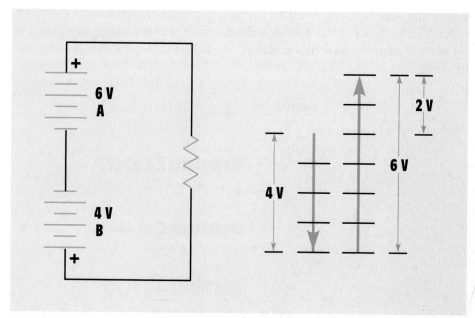

**Figure 17-9** Adding vectors with opposite directions

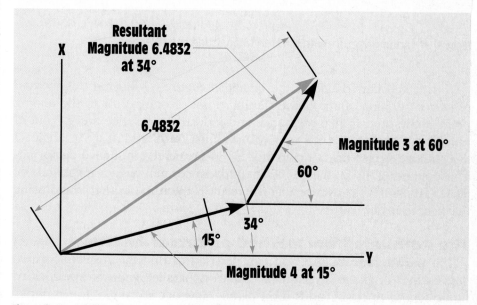

**Figure 17-10** Adding two vectors with different directions

tude of 4 and a direction of 15° to a vector with a magnitude of 3 and a direction of 60°. The addition is made by connecting the starting point of the second vector to the ending point of the first vector. The resultant is drawn from the starting point of the first vector to the ending point of the second. It is possible to add several different vectors using this method. *Figure 17-11* illustrates the addition of several different vectors and the resultant.

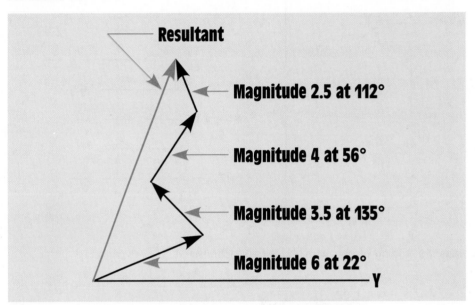

**Figure 17-11**   Adding vectors with different magnitude and direction

To find the impedance of the circuit in *Figure 17-3* using this method of vector addition, connect the starting point of one vector to the ending point of the other. Since resistance and inductive reactance are 90° out of phase with each other, the two vectors must be placed at a 90° angle. If the resistive vector has a magnitude of 30 Ω and the inductive vector has a magnitude of 40 Ω, the resultant (impedance) will have a magnitude of 50 Ω *(Figure 17-12)*. Notice that the result is the same as that found using the right triangle.

## The Parallelogram Method of Vector Addition

The **parallelogram method** can be used to find the resultant of two vectors that originate at the same point. A parallelogram is a four-sided figure whose opposite sides form parallel lines. A rectangle, for example, is a parallelogram with 90° angles. Assume that a vector with a magnitude of 24 and a direction of 26° is to be added to a vector with a magnitude of 18 and a direction of 58°. Also assume that the two vectors originate from the same point *(Figure 17-13)*. To find the resultant of these two vectors, form a parallelogram using the vectors as two of the sides. The

**parallelo-
gram
method**

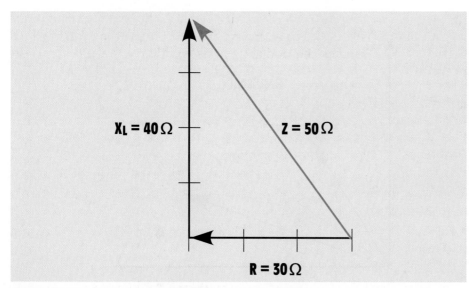

**Figure 17-12** Determining impedance using vector addition

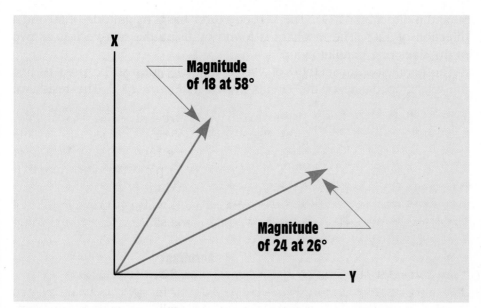

**Figure 17-13** Vectors that originate from the same point

resultant is drawn from the corner of the parallelogram where the two vectors intersect to the opposite corner *(Figure 17-14)*.

Another example of the parallelogram method of vector addition is shown in *Figure 17-15*. In this example, one vector has a magnitude of

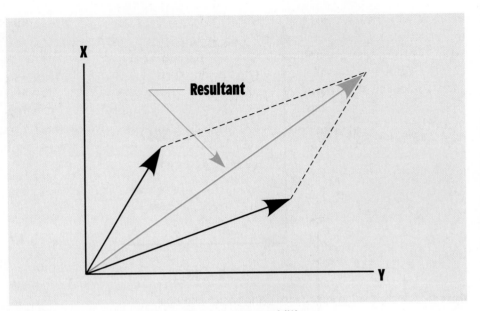

**Figure 17-14**  The parallelogram method of vector addition

50 and a direction of 32°. The second vector has a magnitude of 60 and a direction of 120°. The resultant is found by using the two vectors as two of the sides of a parallelogram.

The parallelogram method of vector addition can also be used to find the total impedance of the circuit shown in *Figure 17-2*. The resistance

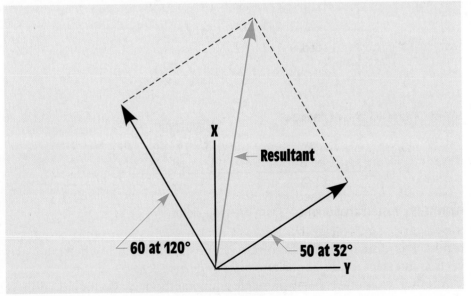

**Figure 17-15**  The parallelogram method of vector addition, Example 2

forms a vector with a magnitude of 30 Ω and a direction of 0°. The inductive reactance forms a vector with a magnitude of 40 Ω and a direction of 90°. When lines are extended to form a parallelogram and a resultant is drawn, the resultant will have a magnitude of 50 Ω, which is the impedance of the circuit *(Figure 17-16)*. Some students of electricity find the right triangle concept easier to understand and others find vectors more helpful. For this reason, this text will use both methods to help explain the relationship of voltage, current, and power in alternating current circuits.

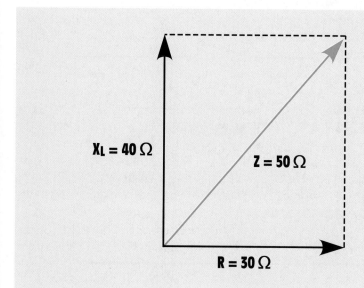

$X_L = 40\ \Omega$

$Z = 50\ \Omega$

$R = 30\ \Omega$

**Figure 17-16** Finding the impedance of the circuit using the parallelogram method of vector addition

## 17-4 Total Current

One of the primary laws for series circuits is that the current must be the same in any part of the circuit. This law holds true of R-L series circuits also. Since the impedance is the total current-limiting component of the circuit, it can be used to replace R in an Ohm's law formula. The **total current (I)** flow through the circuit can be computed by dividing the total applied voltage by the total current-limiting factor. Total current can be found by using the formula

$$I_T = \frac{E_T}{Z}$$

**total current (I)**

The total current for the circuit in *Figure 17-2,* then, is

$$I_T = \frac{240}{50}$$

$$I_T = 4.8 \text{ A}$$

In a series circuit, the current is the same at any point in the circuit. Therefore, 4.8 A of current flow through both the resistor and the inductor. These values can be added to the circuit as shown in *Figure 17-17.*

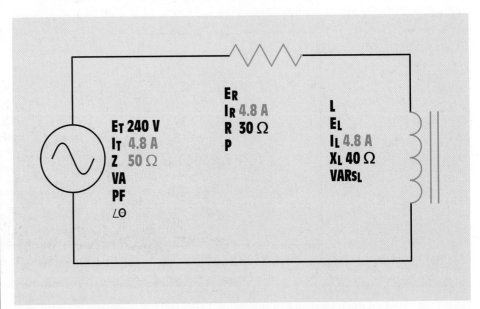

**Figure 17-17**  Total voltage divided by impedance equals total current.

## 17-5 Voltage Drop across the Resistor

Now that the amount of current flow through the resistor is known, the **voltage drop across the resistor ($E_R$)** can be computed using the formula

$$E_R = I_R \times R$$

The voltage drop across the resistor in our circuit is

$$E_R = 4.8 \times 30$$

$$E_R = 144 \text{ V}$$

Notice that the amount of voltage dropped across the resistor was found using quantities that pertained only to the resistive part of the circuit. The amount of voltage dropped across the resistor could not be found using a formula such as

$$E_R = I_R \times X_L$$

or

$$E_R = I_R \times Z$$

Inductive reactance ($X_L$) is an inductive quantity and impedance ($Z$) is a circuit total quantity. These quantities cannot be used with Ohm's law to find resistive quantities. They can, however, be used with vector addition to find like resistive quantities. For example, both inductive reactance and impedance are measured in ohms. The resistive quantity that is measured in ohms is resistance ($R$). If the impedance and inductive reactance of a circuit were known, they could be used with the formula shown below to find the circuit resistance.

$$R = \sqrt{Z^2 - X_L^2}$$

(Note: Refer to the Resistive-Inductive Series Circuits section of the alternating current formulas listed in the appendix.)

## 17-6 Watts

**True power (P)** for the circuit can be computed by using any of the watts formulas with pure resistive parts of the circuit. Watts (W) can be computed, for example, by multiplying the voltage dropped across the resistor ($E_R$) by the current flow through the resistor ($I_R$); or by squaring the voltage dropped across the resistor and dividing by the resistance of the resistor; or by squaring the current flow through the resistor and multiplying by the resistance of the resistor. Watts cannot be computed by multiplying the total voltage ($E_T$) by the current flow through the resistor or by multiplying the square of the current by the inductive reactance. Recall that true power, or watts, can be produced only during periods of time that the voltage and current are both positive or both negative.

In an R-L series circuit, the current is the same through both the resistor and the inductor. The voltage dropped across the resistor, however, is in phase with the current, and the voltage dropped across the inductor is 90° out of phase with the current *(Figure 17-18)*. Since true power, or watts, can be produced only when the current and voltage are both positive or both negative, only resistive parts of the circuit can produce watts.

**true power (P)**

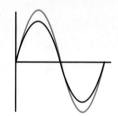

Voltage dropped across the resistor is in phase with the current.

Voltage dropped across the inductor is 90° out of phase with the current.

**Figure 17-18**  Relationship of current and voltage in an R-L series circuit

The formula used in this example will be

$$P = E_R \times I_R$$

$$P = 144 \times 4.8$$

$$P = 691.2 \text{ W}$$

### 17-7 Computing the Inductance

The amount of inductance can be computed using the formula

$$L = \frac{X_L}{2\pi F}$$

$$L = \frac{40}{377}$$

$$L = 0.106 \text{ H}$$

### 17-8 Voltage Drop across the Inductor

The **voltage drop across the inductor (E$_L$)** can be computed using the formula

$$E_L = I_L \times X_L$$

$$E_L = 4.8 \times 40$$

$$E_L = \underline{192 \text{ V}}$$

Notice that only inductive quantities were used to find the voltage drop across the inductor.

### 17-9 Total Voltage

Although the total applied voltage in this circuit is known (240 V) the total voltage is also equal to the sum of the voltage drops, just as it is in any other series circuit. Since the voltage dropped across the resistor is in phase with the current and the voltage dropped across the inductor is 90° out of phase with the current, vector addition must be used. The total voltage will be the hypotenuse of a right triangle, and the resistive and inductive voltage drops will form the legs of the triangle *(Figure 17-19)*. This relationship of voltage drops can also be represented using the parallelogram method of vector addition as shown in *Figure 17-20*. The following formulas can be used to find total voltage or the voltage drops across the resistor or inductor if the other two voltage values are known.

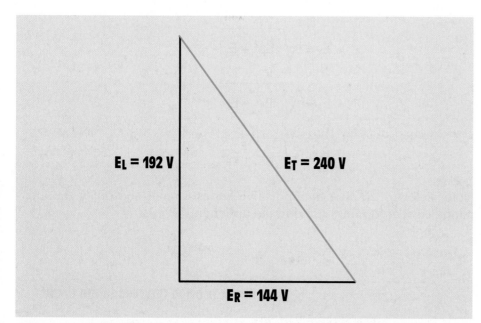

**Figure 17-19**   Relationship of resistive and inductive voltage drops in an R-L series circuit

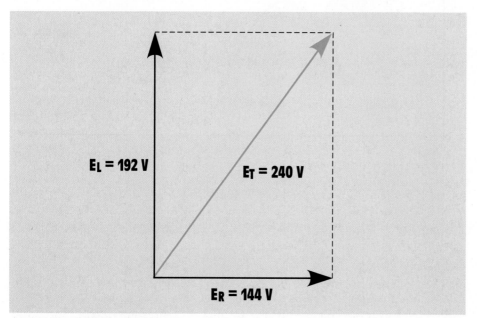

**Figure 17-20**   Graphic representation of voltage drops using the parallelogram method of vector addition

$$E_T = \sqrt{E_R^2 + E_L^2}$$

$$E_R = \sqrt{E_T^2 - E_L^2}$$

$$E_L = \sqrt{E_T^2 - E_R^2}$$

Note: Refer to the Resistive-Inductive Series Circuits section of the alternating current formulas listed in the appendix.

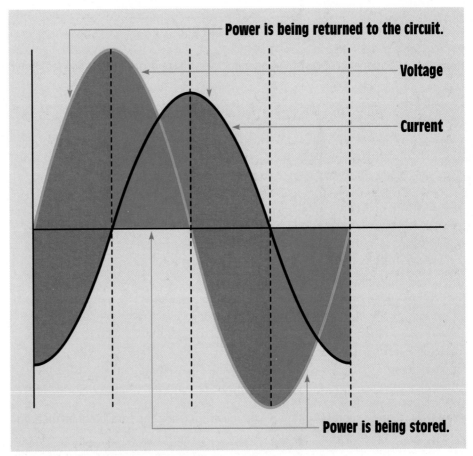

**Figure 17-21**   Power is stored and then returned to the circuit.

## 17-10 Computing the Reactive Power

VARs is an abbreviation for volt-amps-reactive and is the amount of **reactive power (VARs)** in the circuit. VARs should not be confused with watts, which is true power. VARs represents the product of the volts and amps that are 90° out of phase with each other, such as the voltage dropped across the inductor and the current flowing through the inductor. Recall that true power can be produced only during periods of time that the voltage and current are both positive or both negative *(Figure 17-21)*. During these periods, the power is being stored in the form of a magnetic field. During the periods that voltage and current have opposite signs, the power is returned to the circuit. For this reason, VARs is often referred to as **quadrature power**, or **wattless power**. It can be computed in a similar manner as watts except that reactive values of voltage and current are used instead of resistive values. In this example the formula used will be

$$VARs = I_L^2 \times X_L$$
$$VARs = 4.8^2 \times 40$$
$$VARs = 921.6$$

**reactive power (VARs)**

**quadrature power**

**wattless power**

## 17-11 Computing the Apparent Power

Volt-amperes (VA) is the apparent power of the circuit. It can be computed in a similar manner as watts or VARs, except that total values of voltage and current are used. It is called **apparent power (VA)** because it is the value that would be found if a voltmeter and ammeter were used to measure the circuit voltage and current and then these measured values were multiplied together *(Figure 17-22)*. In this example the formula used will be

$$VA = E_T \times I_T$$
$$VA = 240 \times 4.8$$
$$VA = 1152$$

**apparent power (VA)**

The apparent power can also be found using vector addition in a similar manner as impedance or total voltage. Since true power, or watts, is a pure resistive component and VARs is a pure reactive component, they form the legs of a right triangle. The apparent power is the hypotenuse of this triangle *(Figure 17-23)*. This relationship of the three power components can also be plotted using the parallelogram method *(Figure 17-24)*. The formulas shown below can be used to compute the

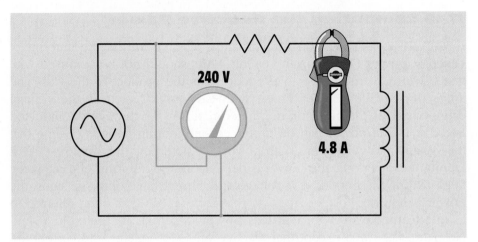

**Figure 17-22** Apparent power is the product of measured values (240 V x 4.8 A = 1152 VA).

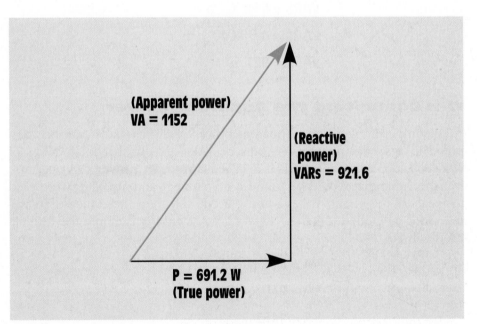

**Figure 17-23** Relationship of true power (watts), reactive power (VARs), and apparent power (volt-amps) in an R-L series circuit

values of apparent power, true power, and reactive power when the other two values are known.

$$VA = \sqrt{P^2 + VARs^2}$$

$$P = \sqrt{VA^2 - VARs^2}$$

$$VARs = \sqrt{VA^2 - P^2}$$

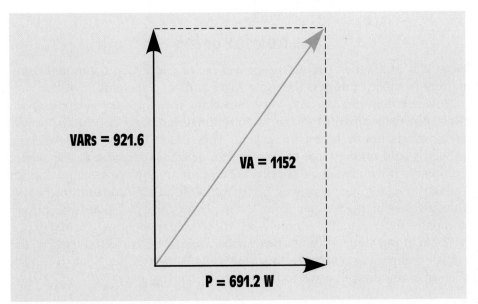

**VARs = 921.6**

**VA = 1152**

**P = 691.2 W**

**Figure 17-24**   Using the parallelogram method to plot the relationship of volt-amps, watts, and VARs

## 17-12 Power Factor

**Power factor (PF)** is a ratio of the true power to the apparent power. It can be computed by dividing any resistive value by its like total value. For example, power factor can be computed by dividing the voltage drop across the resistor by the total circuit voltage; or resistance divided by impedance; or watts divided by volt-amperes.

$$PF = \frac{E_R}{E_T}$$

$$PF = \frac{R}{Z}$$

$$PF = \frac{P}{VA}$$

**power factor (PF)**

Power factor is generally expressed as a percentage. The decimal fraction computed from the division will, therefore, be changed to a percent by multiplying it by 100. In this circuit the formula used will be

$$PF = \frac{W}{VA}$$

$$PF = \frac{691.2}{1152}$$

$$PF = 0.6 \times 100, \text{ or } 60\%$$

Note that in a series circuit, the power factor cannot be computed using current because current is the same in all parts of the circuit.

Power factor can become very important in an industrial application. Most power companies charge a substantial surcharge when the power factor drops below a certain percent. The reason for this is that electric power is sold on the basis of true power, or watts, consumed. The power company, however, must supply the apparent power. Assume that an industrial plant has a power factor of 60% and is consuming 5 MW (megawatts) of power per hour. At a power factor of 60%, the power company must actually supply 8.33 MVA (megavolt-amps) (5 MW/0.6 = 8.33 MVA) per hour. If the power factor were to be corrected to 95%, the power company would have to supply only 5.26 MVA per hour to furnish the same amount of power to the plant.

## 17-13 Angle Theta

**angle theta**
**($\angle\theta$)**

The angular displacement by which the voltage and current are out of phase with each other is called **angle theta ($\angle\theta$)**. Since the power factor is the ratio of true power to apparent power, the phase angle of voltage and current is formed between the resistive leg of the right triangle and the hypotenuse *(Figure 17-25)*. The resistive leg of the triangle is adjacent to the angle, and the total leg is the hypotenuse. The trigonometric function that corresponds to the adjacent side and the hypotenuse is the cosine. Angle theta will be the cosine of watts divided by volt-amps. Watts divided by volt-amps is also the power factor. Therefore, the cosine of angle theta ($\angle\theta$) is the power factor (PF).

$$COS \angle\theta = PF$$

$$COS \angle\theta = 0.6$$

$$\angle\theta = 53.3°$$

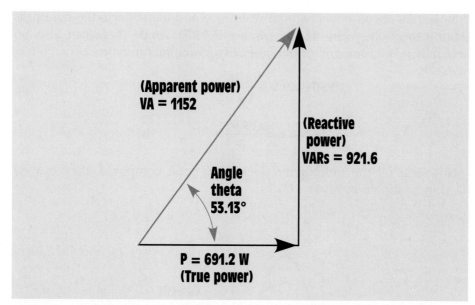

**Figure 17-25**   The angle theta is the relationship of true power to apparent power.

The vectors formed using the parallelogram method of vector addition can also be used to find angle theta as shown in *Figure 17-26*. Notice that the total quantity, volt-amps, and the resistive quantity, watts, are again used to determine angle theta.

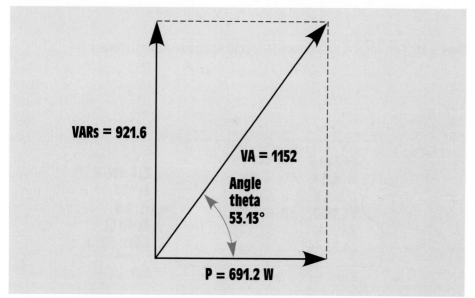

**Figure 17-26**   The angle theta can be found using vectors provided by the parallelogram method.

Since this circuit contains both resistance and inductance, the current is lagging the voltage by 53.13° *(Figure 17-27)*. Angle theta can also be determined by using any of the other trigonometric functions.

$$\text{SIN } \angle\theta = \frac{\text{VARs}}{\text{VA}}$$

$$\text{TAN } \angle\theta = \frac{\text{VARs}}{\text{P}}$$

Now that all the unknown values have been computed, they can be filled in as shown in *Figure 17-28*.

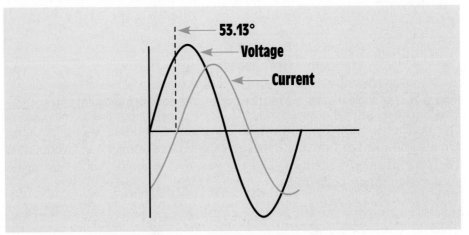

**Figure 17-27** Current and voltage are 53.13° out of phase with each other.

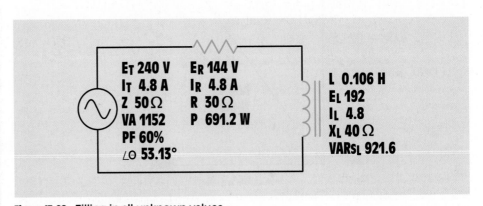

ET 240 V    ER 144 V
IT 4.8 A    IR 4.8 A
Z 50 Ω      R 30 Ω
VA 1152     P 691.2 W
PF 60%
∠θ 53.13°

L 0.106 H
EL 192
IL 4.8
XL 40 Ω
VARSL 921.6

**Figure 17-28** Filling in all unknown values

Two resistors and two inductors are connected in series *(Figure 17-29)*. The circuit is connected to a 130-V, 60-Hz line. The first resistor has a power dissipation of 56 W, and the second resistor has a power dissipation of 44 W. One inductor has a reactive power of 152 VARs and the second a reactive power of 88 VARs. Find the unknown values in this circuit.

**Example 1**

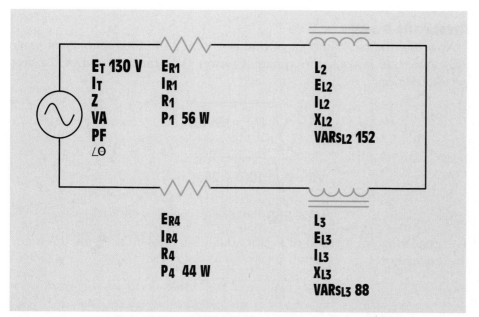

**Figure 17-29** An R-L series circuit containing two resistors and two inductors

## Solution

The first step is to find the total amount of true power and the total amount of reactive power. The total amount of true power can be computed by adding the values of the resistors together. A vector diagram of these two values would reveal that they both have a direction of 0° *(Figure 17-30)*.

$$P_T = P_1 + P_4$$

$$P_T = 56 + 44$$

$$P_T = 100 \text{ W}$$

The total reactive power in the circuit can be found in the same manner. Like watts, the VARs are both inductive and are, therefore, in the same direction *(Figure 17-31)*. The total reactive power will be the sum

**Figure 17-30** The two true power (watts) vectors are in the same direction.

**Figure 17-31** The two reactive power (VARs) vectors are in the same direction.

of the two reactive power ratings.

$$\text{VARS}_T = \text{VARS}_{L2} + \text{VARS}_{L3}$$

$$\text{VARS}_T = 152 + 88$$

$$\text{VARS}_T = 240$$

## Apparent Power

Now that the total amount of true power and the total amount of reactive power are known, the apparent power (VA) can be computed using vector addition.

$$\text{VA} = \sqrt{P_T{}^2 + \text{VARS}_T{}^2}$$

$$\text{VA} = \sqrt{100^2 + 240^2}$$

$$\text{VA} = 260$$

The parallelogram method of vector addition is shown in *Figure 17-32* for this calculation.

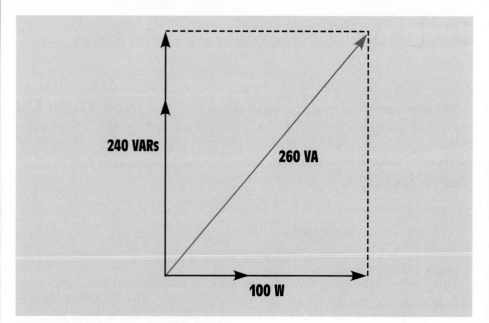

**Figure 17-32**   Power vector for the circuit

## Total Circuit Current

Now that the apparent power is known, the total circuit current can be found using the applied voltage and Ohm's law.

$$I_T = \frac{VA}{E_T}$$

$$I_T = \frac{260}{130}$$

$$I_T = 2\ A$$

Since this is a series circuit, the current must be the same at all points in the circuit. The known values added to the circuit are shown in *Figure 17-33*.

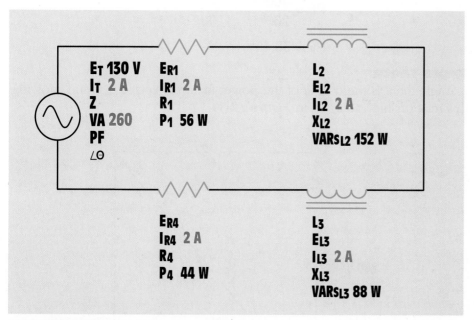

**Figure 17-33**   Adding circuit values

## Other Circuit Values

Now that the total circuit current has been found, other values can be computed using Ohm's law.

## Impedance

$$Z = \frac{E_T}{I_T}$$

$$Z = \frac{130}{2}$$

$$Z = 65 \ \Omega$$

## Power Factor

$$PF = \frac{W_T}{VA}$$

$$PF = \frac{100}{260}$$

$$PF = 38.46\%$$

## Angle Theta

Angle theta is the cosine of the power factor. A vector diagram showing this relationship is shown in *Figure 17-34*.

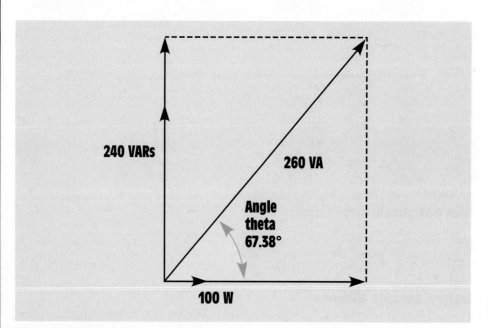

**240 VARs**

**260 VA**

**Angle theta 67.38°**

**100 W**

**Figure 17-34** Angle theta for the circuit

$$COS \angle\theta = PF$$

$$COS \angle\theta = 0.3846$$

$$\angle\theta = 67.38°$$

The relationship of voltage and current for this circuit is shown in *Figure 17-35*.

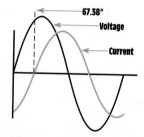

**Figure 17-35** Voltage and current are 67.38° out of phase with each other.

**E$_{R1}$**

$$E_{R1} = \frac{P_1}{I_{R1}}$$

$$E_{R1} = \frac{56}{2}$$

$$E_{R1} = 28 \text{ V}$$

**R$_1$**

$$R_1 = \frac{E_{R1}}{I_{R1}}$$

$$R_1 = \frac{28}{2}$$

$$R_1 = 14 \text{ } \Omega$$

**E$_{L2}$**

$$E_{L2} = \frac{VARS_{L2}}{I_{L2}}$$

$$E_{L2} = \frac{152}{2}$$

$$E_{L2} = 76 \text{ V}$$

**X$_{L2}$**

$$X_{L2} = \frac{E_{L2}}{I_{L2}}$$

$$X_{L2} = \frac{76}{2}$$

$$X_{L2} = 38 \text{ } \Omega$$

**L₂**

$$L_2 = \frac{X_{L2}}{2\pi F}$$

$$L_2 = \frac{38}{377}$$

$$L_2 = 0.101 \text{ H}$$

**E_{L3}**

$$E_{L3} = \frac{VARS_{L3}}{I_{L3}}$$

$$E_{L3} = \frac{88}{2}$$

$$E_{L3} = 44 \text{ V}$$

**X_{L3}**

$$X_{L3} = \frac{E_{L3}}{I_{L3}}$$

$$X_{L3} = \frac{44}{2}$$

$$X_{L3} = 22 \ \Omega$$

**L₃**

$$L_3 = \frac{X_{L3}}{2\pi F}$$

$$L_3 = \frac{22}{377}$$

$$L_3 = 0.058 \text{ H}$$

**E_{R4}**

$$E_{R4} = \frac{P_4}{I_{R4}}$$

$$E_{R4} = \frac{44}{2}$$

$$E_{R4} = 22 \text{ V}$$

**R₄**

$$R_4 = \frac{E_{R4}}{I_{R4}}$$

$$R_4 = \frac{22}{2}$$

$$R_4 = 11\ \Omega$$

The complete circuit with all values is shown in *Figure 17-36.*

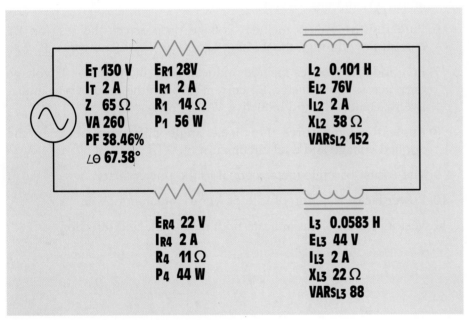

ET 130 V    ER1 28V
IT 2 A    IR1 2 A
Z 65 Ω    R1 14 Ω
VA 260    P1 56 W
PF 38.46%
∠θ 67.38°

L2 0.101 H
EL2 76V
IL2 2 A
XL2 38 Ω
VARSL2 152

ER4 22 V
IR4 2 A
R4 11 Ω
P4 44 W

L3 0.0583 H
EL3 44 V
IL3 2 A
XL3 22 Ω
VARSL3 88

**Figure 17-36** All values for the circuit

## Summary

1. In a pure resistive circuit the voltage and current are in phase with each other.

2. In a pure inductive circuit the voltage and current are 90° out of phase with each other.

3. In an R-L series circuit the voltage and current will be out of phase with each other by some value between 0° and 90°.

4. The amount the voltage and current are out of phase with each other is determined by the ratio of resistance to inductance.

5. Total circuit values include: total voltage, $E_T$; total current, $I_T$; volt-amps, VA; and impedance, Z.

6. Pure resistive values include: voltage drop across the resistor, $E_R$; current flow through the resistor, $I_R$; resistance, R; and watts, P.

7. Pure inductive values include: inductance of the inductor, L; voltage drop across the inductor, $E_L$; current through the inductor, $I_L$; inductive reactance, $X_L$; and inductive VARs, $VARs_L$.

8. Angle theta measures the phase angle difference between the applied voltage and total circuit current.

9. The cosine of angle theta is equal to the power factor.

10. Power factor is a ratio of true power to apparent power.

11. Vectors are lines that indicate both magnitude and direction.

## Review Questions

1. What is the relationship of voltage and current (concerning phase angle) in a pure resistive circuit?

2. What is the relationship of voltage and current (concerning phase angle) in a pure inductive circuit?

3. What is power factor?

4. A circuit contains a 20-$\Omega$ resistor and an inductor with an inductance of 0.093 H. If the circuit has a frequency of 60 Hz what is the total impedance of the circuit?

5. An R-L series circuit has a power factor of 86%. How many degrees are the voltage and current out of phase with each other?

6. An R-L series circuit has an apparent power of 230 VA and a true power of 180 W. What is the reactive power?

7. The resistor in an R-L series circuit has a voltage drop of 53 V, and the inductor has a voltage drop of 28 V. What is the applied voltage of the circuit?

8. An R-L series circuit has a reactive power of 1234 VARs and an apparent power of 4329 VA. How many degrees are voltage and current out of phase with each other?

9. An R-L series circuit contains a resistor and an inductor. The resistor has a value of 6.5 $\Omega$. The circuit is connected to 120 V and has a current flow of 12 A. What is the inductive reactance of this circuit?

10. What is the voltage drop across the resistor in the circuit in question 9?

**Practice Problems**

Refer to the circuit shown in *Figure 17-2* and the Resistive-Inductive Series Circuits section of the alternating current formulas listed in the appendix.

1. Assume that the circuit shown in *Figure 17-2* is connected to a 480-V, 60-Hz line. The inductor has an inductance of 0.053 H, and the resistor has a resistance of 12 Ω.

| | | |
|---|---|---|
| $E_T$ 480 | $E_R$ 247 | $E_L$ 40994 |
| $I_T$ 20.6 | $I_R$ 20.6 | $I_L$ 20.6 |
| Z 23.2 | R 12 | $X_L$ 19.9 |
| VA 9888 | P 5088.2 | VARs$_L$ 4/2594 |
| PF 51% | $\angle\theta$ 59° | L 0.053 H |

2. Assume that the voltage drop across the resistor, $E_R$, is 78 V, that the voltage drop across the inductor, $E_L$, is 104 V, and the circuit has a total impedance, Z, of 20 Ω. The frequency of the AC voltage is 60 Hz.

| | | |
|---|---|---|
| $E_T$ 130 | $E_R$ 78 | $E_L$ 104 V |
| $I_T$ 6.5 | $I_R$ 6.5 | $I_L$ 6.5 |
| Z 20 | R 12Ω | $X_L$ 16Ω |
| VA 845 | P 507 | VARs$_L$ 676 |
| PF 60% | $\angle\theta$ 53.1° | L .042 H |

3. Assume the above circuit has an apparent power of 144 VA and a true power of 115.2 W. The inductor has an inductance of 0.15915 H, and the frequency is 60 Hz.

| | | |
|---|---|---|
| $E_T$ 120 | $E_R$ 96 | $E_L$ 71.98 |
| $I_T$ 1.2 | $I_R$ 1.2 | $I_L$ 1.2 |
| Z 99.99 | R 80 | $X_L$ 59.99 |
| VA 144 | P 115.2 W | VARs$_L$ 86.66 |
| PF 80% | $\angle\theta$ 36.86 | L 0.15915 H |

4. Assume the above circuit has a power factor of 78%, an apparent power of 374.817 VA, and a frequency of 400 Hz. The inductor has an inductance of 0.0382 H.

$E_T$ __240__          $E_R$ __187.40__          $E_L$ __2149.76__

$I_T$ __1.56__          $I_R$ __1.56__          $I_L$ __1.56__

Z. __153.67__          R __1200__          $X_L$ __96.00__

VA 374.817          P __292.35__          $VARS_L$ __234.56__

PF 78%          $\angle\theta$ __38.7%__          L 0.0382 H

$$\frac{P}{VA} = PF \times VA$$

**This unit discusses the behavior of voltage, current, power, and impedance in parallel circuits that contain elements of both resistance and inductance.**

Courtesy of Toyota Motor Manufacturing, USA, Inc.

# Unit 18

# Resistive-Inductive Parallel Circuits

## Outline

18-1 Resistive-Inductive Parallel Circuits

18-2 Computing Circuit Values

## Key Terms

Angle theta ($\angle\theta$)

Apparent power (VA)

Current flow through the inductor ($I_L$)

Current flow through the resistor ($I_R$)

Power factor (PF)

Reactive power (VARs)

Total current ($I_T$)

Total impedance (Z)

True power (P)

Watts (P)

*fter studying this unit, you should be able to:*

■ Discuss the operation of a parallel circuit containing resistance and inductance.

■ Compute circuit values of an R-L parallel circuit.

■ Connect an R-L parallel circuit and measure circuit values with test instruments.

This unit discusses circuits that contain resistance and inductance connected in parallel with each other. Mathematical calculations will be used to show the relationship of current and voltage on the entire circuit, and the relationship of current through different branches of the circuit.

### 18-1 Resistive-Inductive Parallel Circuits

A circuit containing a resistor and an inductor connected in parallel is shown in *Figure 18-1*. Since the voltage applied to any device in parallel must be the same, the voltage applied to the resistor and inductor must be in phase and have the same value. The current flow through the inductor will be 90° out of phase with the voltage, and the current flow through the resistor will be in phase with the voltage *(Figure 18-2)*. This configuration produces a phase angle difference of 90° between the current flow through a pure inductive load and a pure resistive load *(Figure 18-3)*.

The amount of phase angle shift between the total circuit current and voltage is determined by the ratio of the amount of resistance to the amount of inductance. The circuit power factor is still determined by the ratio of apparent power to true power.

### 18-2 Computing Circuit Values

In the circuit shown in *Figure 18-4*, a resistance of 15 Ω is connected in parallel with an inductive reactance of 20 Ω. The circuit is connected to a voltage of 240 V AC and a frequency of 60 Hz. In this example problem,

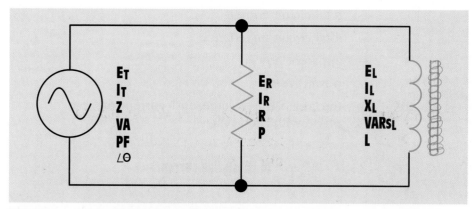

**Figure 18-1** A resistive-inductive parallel circuit

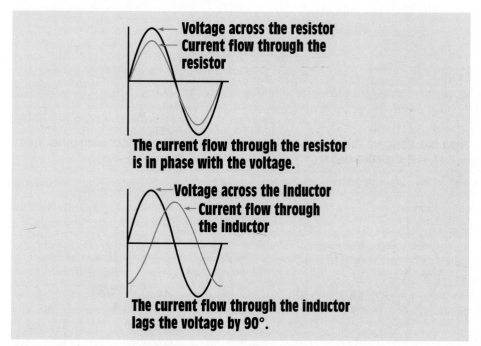

**Figure 18-2** Relationship of voltage and current in an R-L parallel circuit

the following circuit values will be computed:

$I_R$ — current flow through the resistor

P — watts (true power)

$I_L$ — current flow through the inductor

VARs — reactive power

$I_T$     —     total circuit current

Z     —     total circuit impedance

VA     —     apparent power

PF     —     power factor

$\angle\theta$     —     the angle the voltage and current are out of phase with each other.

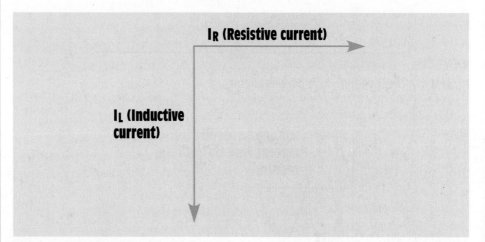

**Figure 18-3** Resistive and inductive currents are 90° out of phase with each other in an R-L parallel circuit.

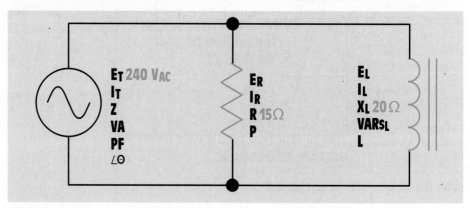

**Figure 18-4** Typical R-L parallel circuit

## Resistive Current

In any parallel circuit, the voltage is the same across each component in the circuit. Therefore, 240 V are applied across both the resistor and the inductor. Since the amount of voltage applied to the resistor is known, the amount of **current flow through the resistor ($I_R$)** can be computed by using the formula

$$I_R = \frac{E}{R}$$

$$I_R = \frac{240}{15}$$

$$I_R = 16 \text{ A}$$

## Watts

**True power (P)**, or **watts**, can be computed using any of the watts formulas and pure resistive values. The amount of true power in this circuit will be computed using the formula

$$P = E_R \times I_R$$

$$P = 240 \times 16$$

$$P = 3840 \text{ W}$$

## Inductive Current

Since the voltage applied to the inductor is known, the current flow can be found by dividing the voltage by the inductive reactance. The amount of **current flow through the inductor ($I_L$)** will be computed using the formula

$$I_L = \frac{E}{X_L}$$

$$I_L = \frac{240}{20}$$

$$I_L = 12 \text{ A}$$

## VARs

The amount of **reactive power, VARs**, will be computed using the formula

$$VARs = E_L \times I_L$$

$$VARs = 240 \times 12$$

$$VARs = 2880$$

current flow through the resistor ($I_R$)

true power (P)

watts

current flow through the inductor ($I_L$)

reactive power (VARs)

## Inductance

Since the frequency and the inductive reactance are known, the inductance of the coil can be found using the formula

$$L = \frac{X_L}{2\pi F}$$

$$L = \frac{20}{377}$$

$$L = 0.053 \text{ H}$$

## Total Current

The **total current ($I_T$)** flow through the circuit can be computed by adding the current flow through the resistor and the inductor. Since these two currents are 90° out of phase with each other, vector addition will be used. If these current values were plotted, they would form a right triangle similar to the one shown in *Figure 18-5*. Notice that the current flow

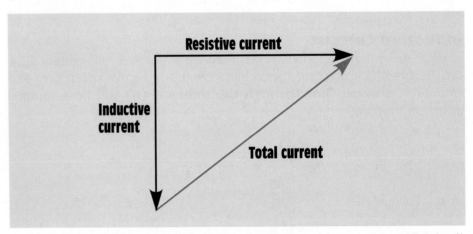

**Figure 18-5** Relationship of resistive, inductive, and total current in an R-L parallel circuit

through the resistor and inductor form the legs of a right triangle, and the total current is the hypotenuse. Since the resistive and inductive currents form the legs of a right triangle and the total current forms the hypotenuse, the Pythagorean theorem can be used to add these currents together.

$$I_T = \sqrt{I_R^2 + I_L^2}$$

total current ($I_T$)

$$I_T = \sqrt{16^2 + 12^2}$$

$$I_T = \sqrt{256 + 144}$$

$$I_T = \sqrt{400}$$

$$I_T = 20 \text{ A}$$

The parallelogram method for plotting the total current is shown in *Figure 18-6*.

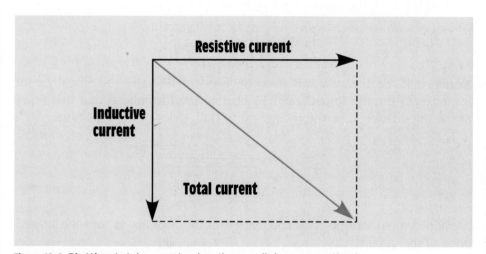

**Figure 18-6** Plotting total current using the parallelogram method

## Impedance

Now that the total current and total voltage are known, the **total impedance (Z)** can be computed by substituting Z for R in an Ohm's law formula. The total impedance of the circuit can be computed using the formula

$$Z = \frac{E}{I_T}$$

$$Z = \frac{240}{20}$$

$$Z = 12 \ \Omega$$

**total impedance (Z)**

The value of impedance can also be found if total current and voltage are not known. In a parallel circuit, the reciprocal of the total resistance is equal to the sum of the reciprocals of each resistor. This same rule can be amended to permit a similar formula to be used in an R-L parallel circuit. Since resistance and inductive reactance are 90° out of phase with each other, vector addition must be used when the reciprocals are added. The initial formula is

$$\left(\frac{1}{Z}\right)^2 = \left(\frac{1}{R}\right)^2 + \left(\frac{1}{X_L}\right)^2$$

This formula states that the square of the reciprocal of the impedance is equal to the sum of the squares of the reciprocals of resistance and inductive reactance. To remove the square from the reciprocal of the impedance, take the square root of both sides of the equation.

$$\frac{1}{Z} = \sqrt{\left(\frac{1}{R}\right)^2 + \left(\frac{1}{X_L}\right)^2}$$

Notice that the formula can now be used to find the reciprocal of the impedance, not the impedance. To change the formula so that it is equal to the impedance, take the reciprocal of both sides of the equation.

$$Z = \frac{1}{\sqrt{\left(\frac{1}{R}\right)^2 + \left(\frac{1}{X_L}\right)^2}}$$

Numeric values can now be substituted in the formula to find the impedance of the circuit.

$$Z = \frac{1}{\sqrt{\left(\frac{1}{15}\right)^2 + \left(\frac{1}{20}\right)^2}}$$

$$Z = \frac{1}{\sqrt{0.004444 + 0.0025}}$$

$$Z = \frac{1}{\sqrt{0.006944}}$$

$$Z = \frac{1}{0.08333}$$

$$Z = 12\ \Omega$$

## Apparent Power

The **apparent power (VA)** can computed by multiplying the circuit voltage by the total current flow. The relationship of volt-amps, watts, and VARs is the same for an R-L parallel circuit as it is for an R-L series circuit. The reason is that power adds in any type of circuit. Since the true power and reactive power are 90° out of phase with each other they form a right triangle with apparent power as the hypotenuse *(Figure 18-7)*.

$$VA = E_T \times I_T$$

$$VA = 240 \times 20$$

$$VA = 4800$$

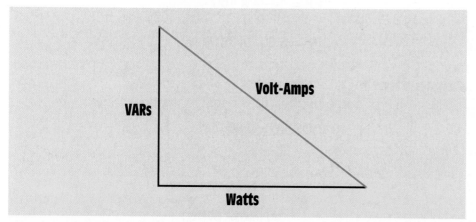

**Figure 18-7** Relationship of apparent power (volt-amps), true power (watts), and reactive power (VARs) in an R-L parallel circuit

## Power Factor

**Power factor (PF)** in an R-L parallel circuit is the relationship of apparent power to the true power just as it was in the R-L series circuit. There are some differences in the formulas used to compute power factor in a parallel circuit, however. In an R-L series circuit, power factor could be computed by dividing the voltage dropped across the resistor by the total, or applied, voltage. In a parallel circuit, the voltage is the same, but the currents are different. Therefore, power factor can be computed by dividing

**apparent power (VA)**

**power factor (PF)**

the current flow through the resistive parts of the circuit by the total circuit current.

$$PF = \frac{I_R}{I_T}$$

Another formula that changes involves resistance and impedance. In a parallel circuit, the total circuit impedance will be less than the resistance. Therefore, if power factor is to be computed using impedance and resistance, the impedance must be divided by the resistance.

$$PF = \frac{Z}{R}$$

The circuit power factor in this example will be computed using the formula

$$PF = \frac{P}{VA} \times 100$$

$$PF = \frac{3840}{4800} \times 100$$

$$PF = 0.80, \text{ or } 80\%$$

## Angle Theta

The cosine of **angle theta ($\angle\theta$)** is equal to the power factor.

$$COS \angle\theta = 0.80$$

$$\angle\theta = 36.87°$$

**angle theta**
**($\angle\theta$)**

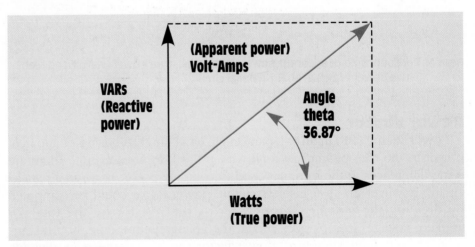

**Figure 18-8** Angle theta

A vector diagram using apparent power, true power, and reactive power is shown in *Figure 18-8*. Notice that angle theta is the angle produced by the apparent power and the true power. The relationship of current and voltage for this circuit is shown in *Figure 18-9*. The circuit with all values is shown in *Figure 18-10*.

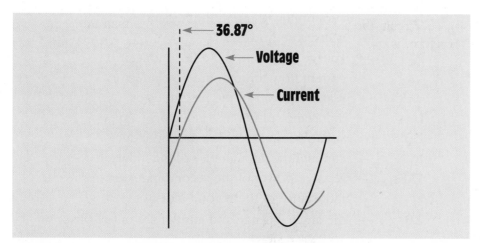

**Figure 18-9** The current is 36.87° out of phase with the voltage.

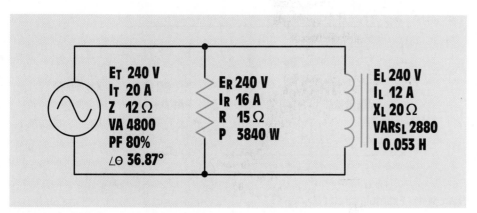

**Figure 18-10**   All values have been found.

In this circuit, one resistor is connected in parallel with two inductors *(Figure 18-11)*. The frequency is 60 Hz. The circuit has an apparent power of 6120 VA, the resistor has a resistance of 45 Ω, the first inductor has an inductive reactance of 40 Ω, and the second inductor has an inductive reactance of 60 Ω. It is assumed that both inductors have a Q greater than

**Example 1**

10 and their resistance is negligible. Find the following missing values.

| | | |
|---|---|---|
| Z | — | total circuit impedance |
| $I_T$ | — | total circuit current |
| $E_T$ | — | applied voltage |
| $E_R$ | — | voltage drop across the resistor |
| $I_R$ | — | current flow through the resistor |
| P | — | watts (true power) |
| $E_{L1}$ | — | voltage drop across the first inductor |
| $I_{L1}$ | — | current flow through the first inductor |
| $VARS_{L1}$ | — | reactive power of the first inductor |
| $L_1$ | — | inductance of the first inductor |
| $E_{L2}$ | — | voltage drop across the second inductor |
| $I_{L2}$ | — | current flow through the second inductor |
| $VARS_{L2}$ | — | reactive power of the second inductor |
| $L_2$ | — | inductance of the second inductor |
| $\angle\theta$ | — | the angle, voltage and current are out of phase with each other |

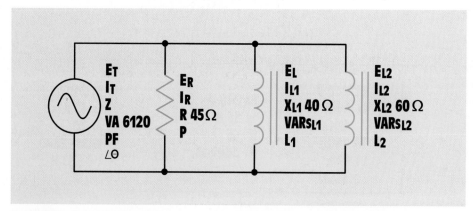

**Figure 18-11** Example circuit 2

## Solution

### Impedance

Before it is possible to compute the impedance of the circuit, the total amount of inductive reactance for the circuit must be found. Since these two inductors are connected in parallel, the reciprocal of their inductive reactances must be added. This will give the reciprocal of the total inductive reactance:

$$\frac{1}{X_{LT}} = \frac{1}{X_{L1}} + \frac{1}{X_{L2}}$$

To find the total inductive reactance, take the reciprocal of both sides of the equation:

$$X_{LT} = \frac{1}{\dfrac{1}{X_{L1}} + \dfrac{1}{X_{L2}}}$$

Refer to the formulas for pure inductive circuits shown in the alternating current formulas section of the appendix. Numeric values can now be substituted in the formula to find the total inductive reactance.

$$X_{LT} = \frac{1}{\dfrac{1}{40} + \dfrac{1}{60}}$$

$$X_{LT} = \frac{1}{0.025 + 0.01667}$$

$$X_{LT} = 24 \ \Omega$$

Now that the total amount of inductive reactance for the circuit is known, the impedance can be computed using the formula

$$Z = \frac{1}{\sqrt{\left(\dfrac{1}{R}\right)^2 + \left(\dfrac{1}{X_{LT}}\right)^2}}$$

$$Z = \frac{1}{\sqrt{\left(\dfrac{1}{45}\right)^2 + \left(\dfrac{1}{24}\right)^2}}$$

$$Z = 21.176 \ \Omega$$

A diagram showing the relationship of resistance, inductive reactance, and impedance is shown in *Figure 18-12*.

**E$_T$**

Now that the circuit impedance and the apparent power are known, the applied voltage can be computed using the formula

$$E_T = \sqrt{VA \times Z}$$

$$E_T = \sqrt{6120 \times 21.176}$$

$$E_T = 360 \text{ V}$$

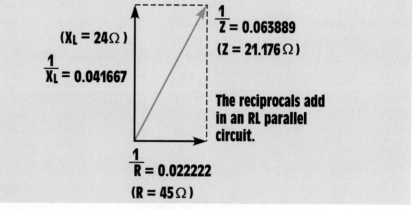

$\dfrac{1}{Z} = 0.063889$

$(Z = 21.176\,\Omega)$

$(X_L = 24\,\Omega)$

$\dfrac{1}{X_L} = 0.041667$

The reciprocals add in an RL parallel circuit.

$\dfrac{1}{R} = 0.022222$

$(R = 45\,\Omega)$

**Figure 18-12**   Relationship of resistance, inductive reactance, and impedance for circuit 2

### $E_R$, $E_{L1}$, and $E_{L2}$

In a parallel circuit, the voltage must be the same across any leg or branch. Therefore, 360 V is dropped across the resistor, the first inductor, and the second inductor.

$$E_R = 360 \text{ V}$$

$$E_{L1} = 360 \text{ V}$$

$$E_{L2} = 360 \text{ V}$$

### $I_T$

The total current of the circuit can now be computed using the formula

$$I_T = \frac{E_T}{Z}$$

$$I_T = \frac{360}{21.176}$$

$$I_T = 17 \text{ A}$$

The remaining values of the circuit can be found using Ohm's law. Refer to the Resistive-Inductive Parallel Circuits listed in the alternating current formula section in the appendix.

**$I_R$**

$$I_R = \frac{E_R}{R}$$

$$I_R = \frac{360}{45}$$

$$I_R = 8 \text{ A}$$

**P**

$$P = E_R \times I_R$$

$$P = 360 \times 8$$

$$P = 2880 \text{ W}$$

**$I_{L1}$**

$$I_{L1} = \frac{E_{L1}}{X_{L1}}$$

$$I_{L1} = \frac{360}{40}$$

$$I_{L1} = 9 \text{ A}$$

**$VARS_{L1}$**

$$VARS_{L1} = E_{L1} \times I_{L1}$$

$$VARS_{L1} = 360 \times 9$$

$$VARS_{L1} = 3240$$

**$L_1$**

$$L_1 = \frac{X_{L1}}{2\pi F}$$

$$L_1 = \frac{360}{377}$$

$$L_1 = 0.106 \text{ H}$$

**I$_{L2}$**

$$I_{L2} = \frac{E_{L2}}{X_{L2}}$$

$$I_{L2} = \frac{360}{60}$$

$$I_{L2} = 6 \text{ A}$$

**VARS$_{L2}$**

$$VARS_{L2} = E_{L2} \times I_{L2}$$

$$VARS_{L2} = 360 \times 6$$

$$VARS_{L2} = 2160$$

**L$_2$**

$$L_2 = \frac{X_{L2}}{2\pi F}$$

$$L_2 = \frac{360}{377}$$

$$L_2 = 0.159 \text{ H}$$

**PF**

$$PF = \frac{W}{VA}$$

$$PF = \frac{2880}{6120}$$

$$PF = 47.06\%$$

**∠θ**

$$COS \angle\theta = PF$$

$$COS \angle\theta = 0.4706$$

$$\angle\theta = 61.93°$$

A vector diagram showing angle theta is shown in *Figure 18-13*. The vectors used are those for apparent power, true power, and reactive power. The phase relationship of voltage and current for this circuit are shown in *Figure 18-14*, and the circuit with all completed values is shown in *Figure 18-15*.

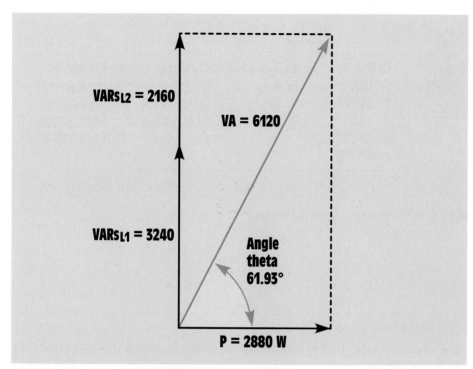

**Figure 18-13**  Angle theta determined by power vectors

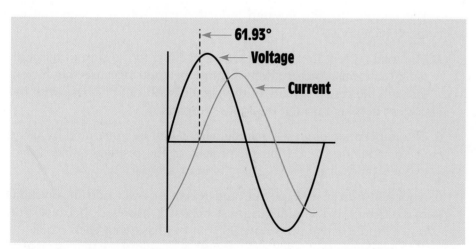

**Figure 18-14**  Voltage and current are 61.93° out of phase with each other.

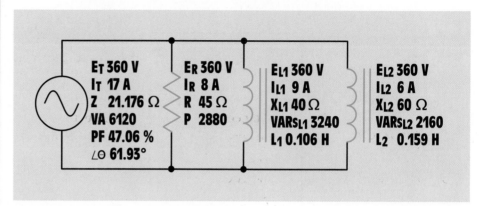

**Figure 18-15** Completed values for circuit 2

## Summary

1. The voltage applied across components in a parallel circuit must be the same.

2. The current flowing through resistive parts of the circuit will be in phase with the voltage.

3. The current flowing through inductive parts of the circuit will lag the voltage by 90°.

4. The total current in a parallel circuit is equal to the sum of the individual currents. Vector addition must be used because the current through the resistive parts of the circuit is 90° out of phase with the current flowing through the inductive parts.

5. The impedance of an R-L parallel circuit can be computed by using vector addition to add the reciprocals of the resistance and inductive reactance.

6. Apparent power, true power, and reactive power add in any kind of a circuit. Vector addition must be used, however, because true power and reactive power are 90° out of phase with each other.

## Review Questions

1. When an inductor and resistor are connected in parallel, how many degrees out of phase are the current flow through the resistor and the current flow through the inductor?

2. An inductor and resistor are connected in parallel to a 120-V, 60-Hz line. The resistor has a resistance of 50 Ω, and the inductor has an inductance of 0.2 H. What is the total current flow through the circuit?

3. What is the impedance of the circuit in question 2?

4. What is the power factor of the circuit in question 2?

5. How many degrees out of phase are the current and voltage in question 2?

6. In the circuit shown in *Figure 18-1,* the resistor has a current flow of 6.5 A, and the inductor has a current flow of 8 A. What is the total current in this circuit?

7. A resistor and inductor are connected in parallel. The resistor has a resistance of 24 Ω, and the inductor has an inductive reactance of 20 Ω. What is the impedance of this circuit?

8. The R-L parallel circuit shown in *Figure 18-1* has an apparent power of 325 VA. The circuit power factor is 66%. What is the true power in this circuit?

9. The R-L parallel circuit shown in *Figure 18-1* has an apparent power of 465 VA and a true power of 320 W. What is the reactive power?

10. How many degrees out of phase are the total current and voltage in question 9?

**Practice Problems**

Refer to the circuit shown in *Figure 18-1*. Use the alternating current formulas in the Resistive-Inductive Parallel Circuits section of the appendix.

1. Assume that the circuit shown in *Figure 18-1* is connected to a 60-Hz line and has a total current flow of 34.553 A. The inductor has an inductance of 0.02122 H, and the resistor has a resistance of 14 Ω.

$E_T$ 240.73      $E_R$ 240      $E_L$ 240

$I_T$ 34.553      $I_R$ 17.14      $I_L$ 30.37

Z 6.8            R 14            $X_L$ 7.9

VA 8292.72       P 4113.6        $VARS_L$ 7288.8

PF 49°           ∠θ 60 %        L 0.02122

2. Assume that the current flow through the resistor, $I_R$, is 15 A, the current flow through the inductor, $I_L$, is 36 A, and the circuit has an apparent power of 10,803 VA. The frequency of the AC voltage is 60 Hz.

$E_T$ 277         $E_R$ 277        $E_L$ 277

$I_T$ 39          $I_R$ 15         $I_L$ 36

Z 7.1            R 18.46          $X_L$ 7.69

VA 10,803        P 4155           $VARS_L$ 9972

PF 38%           ∠θ 67.3°         L 20.3 mh

3. Assume that the circuit in *Figure 18-1* has an apparent power of 144 VA and a true power of 115.2 W. The inductor has an inductance of 0.15915 H, and the frequency is 60 Hz. The resistor has a voltage drop of 78 V.

$E_T$ 78          $E_R$ 78         $E_L$ 78

$I_T$ 1.84        $I_R$ 1.47       $I_L$ 1.13

Z 42.39          R 53.06          $X_L$ 59.99

VA 144           P 115.2          $VARS_L$ 86.4

PF 80%           ∠θ 36.86°        L 0.15915

4. Assume that the circuit in *Figure 18-1* has a power factor of 78%, an apparent power of 374.817 VA, and a frequency of 400 Hz. The inductor has an inductance of 0.0382 H.

$E_T$ __150__        $E_R$ __150__        $E_L$ __150__

$I_T$ __2.49__       $I_R$ __1.94__       $I_L$ __1.56__

Z __60.24__        R __77.3__          $X_L$ __96__

VA 374.817         P __292.35__        $VARS_L$ __234.56__

PF 78%             $\angle\theta$ __38.7°__        L 0.0382

$$X_L = 2\pi F H$$

$39^2 \, R =$

Series        Pall

Power         Power
Voltage       Current
Resistance

$\dfrac{O}{H}$ $\dfrac{A}{H}$ $\dfrac{O}{A}$

section **7**

# alternating current circuits containing capacitors

Capacitors are devices that oppose a change of voltage and have the ability to store an electric charge. This unit discusses their construction and basic operation.

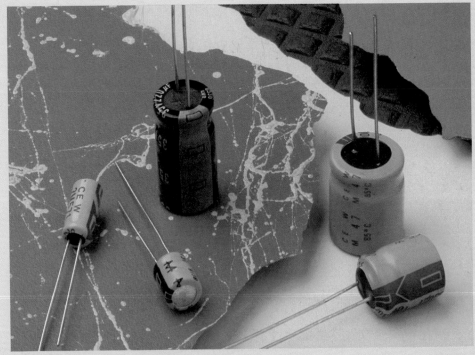

Courtesy of Mallory Capacitor Co.

# *Capacitors*

## Outline

## Key Terms

Dielectric

Dielectric constant

Dielectric stress

Electrolytic

Electrostatic charge

Exponential

Farad

HIPOT

JAN standard

Leakage current

Nonpolarized capacitors

Plates

Polarized capacitors

RC time constant

Surface area

Variable capacitors

### Objectives

*fter studying this unit, you should be able to:*

- List the three factors that determine the capacitance of a capacitor.

- Discuss the electrostatic charge.

- Discuss the differences between nonpolarized and polarized capacitors.

- Compute values for series and parallel connections of capacitors.

- Compute an RC time constant

### Preview

Capacitors perform a variety of jobs such as power factor correction, storing an electrical charge to produce a large current pulse, timing circuits, and electronic filters. Capacitors can be nonpolarized or polarized depending on the application. Nonpolarized capacitors can be used in both AC and DC circuits, while polarized capacitors can be used in DC circuits only. Both types will be discussed in this unit.

## 19-1 Capacitors

**Caution: It is the habit of some people to charge a capacitor to high voltage and then hand it to another person. While some people think this is comical, it is an extremely dangerous practice. Capacitors have the ability to supply an almost infinite amount of current. Under some conditions, a capacitor can have enough energy to cause a person's heart to go into fibrillation.** This statement is not intended to strike fear into the heart of anyone working in the electrical field. It is intended to make you realize the danger that capacitors can pose under certain conditions.

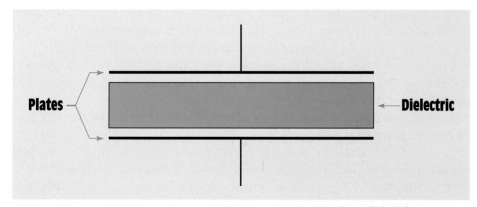

**Figure 19-1** A capacitor is made by separating two metal plates by a dielectric.

**Capacitors are devices that oppose a change of voltage.** The simplest type of capacitor is constructed by separating two metal **plates** by some type of insulating material called the **dielectric** *(Figure 19-1)*. Three factors determine the capacitance of a capacitor:

1. the area of the plates,
2. the distance between the plates, and
3. the type of dielectric used.

The greater the **surface area** of the plates, the more capacitance a capacitor will have. If a capacitor is charged by connecting it to a source of direct current *(Figure 19-2)*, electrons are removed from the plate connected to the positive battery terminal and are deposited on the plate connected to the negative terminal. This flow of current will continue

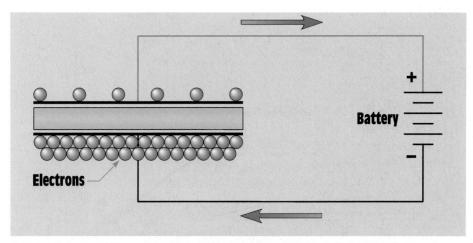

**Figure 19-2** A capacitor can be charged by removing electrons from one plate and depositing electrons on the other plate.

**Capacitors are devices that oppose a change of voltage.**

**plates**

**dielectric**

**surface area**

**Current can flow only during the period of time that a capacitor is either charging or discharging.**

**leakage current**

until a voltage equal to the battery voltage is established across the plates of the capacitor *(Figure 19-3)*. When these two voltages become equal, the flow of electrons will stop. The capacitor is now charged. If the battery is disconnected from the capacitor, the capacitor will remain charged as long as there is no path by which the electrons can move from one plate to the other *(Figure 19-4)*. A good rule to remember concerning a capacitor and current flow is that **current can flow only during the period of time that a capacitor is either charging or discharging**.

In theory, it should be possible for a capacitor to remain in a charged condition forever. In actual practice, however, it cannot. No dielectric is a perfect insulator, and electrons eventually move through the dielectric from the negative plate to the positive, causing the capacitor to discharge *(Figure 19-5)*. This current flow through the dielectric is called **leakage**

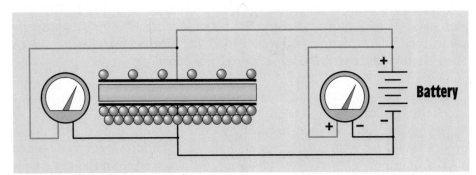

**Figure 19-3** Current flows until the voltage across the capacitor is equal to the voltage of the battery.

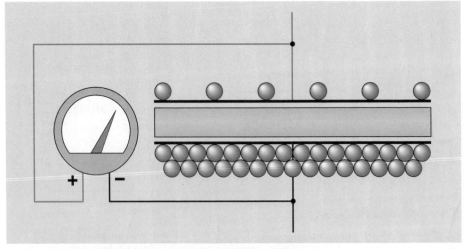

**Figure 19-4** The capacitor remains charged after the battery is removed from the circuit.

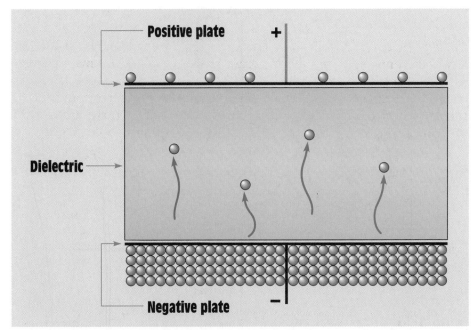

**Figure 19-5** Electrons eventually leak through the dielectric. This flow of electrons is known as leakage current.

**current** and is proportional to the resistance of the dielectric and the charge across the plates. If the dielectric of a capacitor becomes weak, it will permit an excessive amount of leakage current to flow. A capacitor in this condition is often referred to as a leaky capacitor.

## 19-2 Electrostatic Charge

Two other factors that determine capacitance are the type of dielectric used and the distance between the plates. To understand these concepts, it is necessary to understand how a capacitor stores energy. In previous units it was discussed that an inductor stores energy in the form of an electromagnetic field. A capacitor stores energy in an electrostatic field.

The term *electrostatic* refers to electrical charges that are stationary, or not moving. They are very similar to the static electric charges that form on objects that are good insulators, as discussed in Unit 3. The electrostatic field is formed when electrons are removed from one plate and deposited on the other.

### Dielectric Stress

When a capacitor is not charged, the atoms of the dielectric are uniform as shown in *Figure 19-6*. The valence electrons orbit the nucleus in

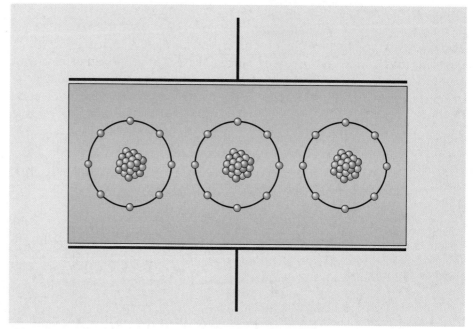

**Figure 19-6** Atoms of the dielectric in an uncharged capacitor

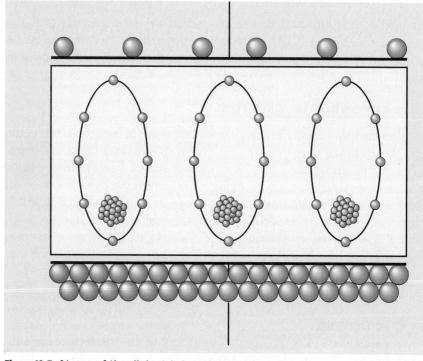

**Figure 19-7** Atoms of the dielectric in a charged capacitor

a circular pattern. When the capacitor becomes charged, however, a potential exists between the plates of the capacitor. The plate with the lack of electrons has a positive charge, and the plate with the excess of electrons has a negative charge. Since electrons are negative particles, they are repelled away from the negative plate and attracted to the positive plate. This attraction causes the electron orbit to become stretched as shown in *Figure 19-7.* This stretching of the atoms of the dielectric is

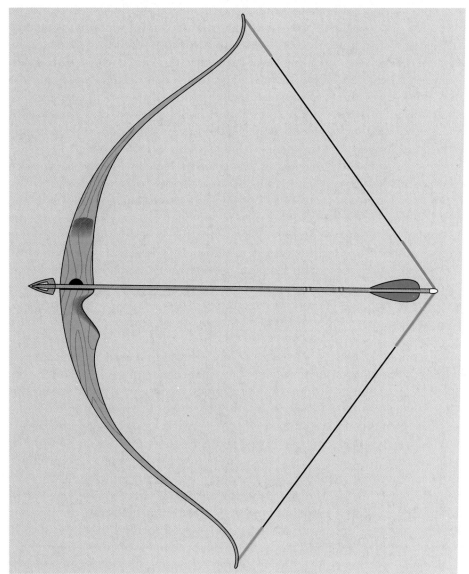

**Figure 19-8** Dielectric stress is similar to drawing back a bowstring with an arrow and holding it.

**dielectric stress**

**electro- static charge**

called **dielectric stress**. Placing the atoms of the dielectric under stress has the same effect as drawing back a bowstring with an arrow and holding it *(Figure 19-8)*; that is, it stores energy.

The amount of dielectric stress is proportional to the voltage difference between the plates. The greater the voltage, the greater the dielectric stress. If the voltage becomes too great, the dielectric will break down and permit current to flow between the plates. At this point the capacitor becomes shorted. Capacitors have a voltage rating that should not be exceeded. The voltage rating indicates the maximum amount of voltage the dielectric is intended to withstand without breaking down. The amount of voltage applied to a capacitor is critical to its life span. Capacitors operated above their voltage rating will fail relatively quickly. Many years ago, the U.S. military made a study of the voltage rating of a capacitor relative to its life span. The results showed that a capacitor operated at one-half its rated voltage will have a life span approximately eight times longer than a capacitor operated at the rated voltage.

The energy of the capacitor is stored in the dielectric in the form of an **electrostatic charge**. It is this electrostatic charge that permits the capacitor to produce extremely high currents under certain conditions. If the leads of a capacitor are shorted together, it has the effect of releasing the drawn-back bowstring *(Figure 19-8)*. When the bowstring is released, the arrow will be propelled forward at a high rate of speed. The same is true for the capacitor. When the leads are shorted, the atoms of the dielectric snap back to their normal position. Shorting causes the electrons on the negative plate to be literally blown off and attracted to the positive plate. Capacitors can produce currents of thousands of amperes for short periods of time.

This principle is used to operate the electronic flash of many cameras. Electronic flash attachments contain a small glass tube filled with a gas called xenon. Xenon produces a very bright white light similar to sunlight when the gas is ionized. A large amount of power is required, however, to produce a bright flash. A battery capable of directly ionizing the xenon would be very large, expensive, and have a potential of about 500 V. The simple circuit shown in *Figure 19-9* can be used to overcome the problem. In this circuit, two small 1.5-V batteries are connected to an oscillator. The oscillator changes the direct current of the batteries into square wave alternating current. The alternating current is then connected to a transformer, and the voltage is increased to about 500 V peak. A diode changes the AC voltage back into DC and charges a capacitor. The capacitor charges to the peak value of the voltage wave form. When the switch is closed, the capacitor suddenly discharges through the xenon tube and supplies the power needed to ionize the gas. It may take several seconds to store

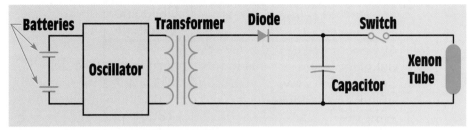

**Figure 19-9** Energy is stored in a capacitor.

enough energy in the capacitor to ionize the gas in the tube, but the capacitor can release the stored energy in a fraction of a second.

To understand how the capacitor can supply the energy needed, consider the amount of gunpowder contained in a 0.357 cartridge. If the powder were to be removed from the cartridge and burned in the open air, it would be found that the actual amount of energy contained in the powder is very small. This amount of energy would not even be able to raise the temperature by a noticeable amount in a small enclosed room. If this same amount of energy is converted into heat in a fraction of a second, however, enough force is developed to propel a heavy projectile with great force. This same principle is at work when a capacitor is charged over some period of time and then discharged in a fraction of a second.

## 19-3 Dielectric Constants

Since much of the capacitor's energy is stored in the dielectric, the type of dielectric used is extremely important in determining the amount of capacitance a capacitor will have. Different materials are assigned a number called the **dielectric constant**. Air is assigned the number 1 and is used as a reference for comparison. For example, assume a capacitor uses air as the dielectric and its capacitance value is found to be 1 microfarad (μF). Now assume that some dielectric material is placed between the plates without changing the spacing and the capacitance value becomes 5 μF. This material has a dielectric constant of 5. A chart showing the dielectric constant of different materials is shown in *Figure 19-10*.

## 19-4 Capacitor Ratings

The basic unit of capacitance is the **farad** and is symbolized by the letter *F*. It receives its name from a famous scientist named Michael Faraday. **A capacitor has a capacitance of one farad when a change of one volt across its plates results in a movement of one coulomb.**

**dielectric constant**

**farad**

A capacitor has a capacitance of one farad when a change of one volt across its plates results in a movement of one coulomb.

| Material | Dielectric constant |
|---|---|
| Air | 1 |
| Bakelite | 4.0 –10.0 |
| Castor oil | 4.3 – 4.7 |
| Cellulose acetate | 7.0 |
| Ceramic | 1200 |
| Dry paper | 3.5 |
| Hard rubber | 2.8 |
| Insulating oils | 2.2 – 4.6 |
| Lucite | 2.4 – 3.0 |
| Mica | 6.4 – 7.0 |
| Mycalex | 8.0 |
| Paraffin | 1.9 – 2.2 |
| Porcelain | 5.5 |
| Pure water | 81 |
| Pyrex glass | 4.1 – 4.9 |
| Rubber compounds | 3.0 – 7.0 |
| Teflon | 2 |
| Titanium dioxide compounds | 90 – 170 |

**Figure 19-10** Dielectric constant of different materials

$$Q = C \times V$$

where

$Q$ = charge in coulombs

$C$ = capacitance in farads

$V$ = charging voltage

Although the farad is the basic unit of capacitance, it is seldom used because it is an extremely large amount of capacitance. The formula shown below can be used to determine the capacitance of a capacitor when the area of the plates, the dielectric constant, and the distance between the plates are known.

$$C = \frac{K \times A}{4.45\,D}$$

where

$C$ = capacitance in pF (picofarads)

$K$ = dielectric constant

$A$ = area of one plate

$D$ = distance between the plates

What would be the plate area of a 1-F (farad) capacitor if air is used as the dielectric and the plates are separated by a distance of 1 in.?

**Example 1**

## Solution

The first step is to convert the above formula to solve for area.

$$A = \frac{C \times 4.45 \times D}{K}$$

$$A = \frac{1,000,000,000,000 \times 4.45 \times 1}{1}$$

$$A = 4,450,000,000,000 \text{ square inches}$$

$$A = 1108.5 \text{ square miles}$$

Since the basic unit of capacitance is so large, other units such as the microfarad ($\mu$F), nanofarad (nF), and picofarad (pF) are generally used.

$$\mu F = \frac{1}{1,000,000} \ (1 \times 10^{-6}) \text{ of a farad}$$

$$nF = \frac{1}{1,000,000,000} \ (1 \times 10^{-9}) \text{ of a farad}$$

$$pF = \frac{1}{1,000,000,000,000} \ (1 \times 10^{-12}) \text{ of a farad}$$

The picofarad is sometimes referred to as a micro-microfarad and is symbolized by $\mu\mu$F.

## 19-5 Capacitors Connected in Parallel

Connecting capacitors in parallel *(Figure 19-11)* has the same effect as increasing the plate area of one capacitor. In the example shown, three capacitors having a capacitance of 20 $\mu$F, 30 $\mu$F, and 60 $\mu$F are connected in parallel. The total capacitance of this connection is:

$$C_T = C_1 + C_2 + C_3$$

$$C_T = 20 + 30 + 60$$

$$C_T = 110 \ \mu F$$

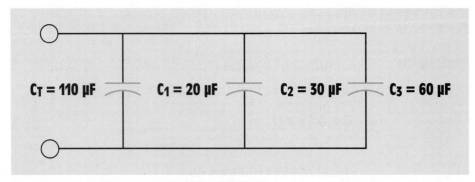

**Figure 19-11** Capacitors connected in parallel

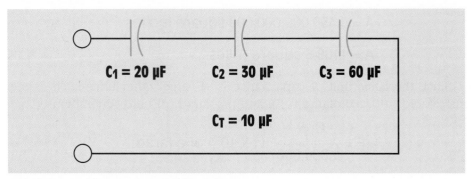

**Figure 19-12** Capacitors connected in series

## 19-6 Capacitors Connected in Series

Connecting capacitors in series *(Figure 19-12)* has the effect of increasing the distance between the plates, thus reducing the total capacitance of the circuit. The total capacitance can be computed in a manner similar to computing parallel resistance. The formulas shown below can be used to find the total capacitance when capacitors are connected in series.

$$C_T = \frac{1}{\dfrac{1}{C_1} + \dfrac{1}{C_2} + \dfrac{1}{C_3}}$$

or

$$C_T = \frac{C_1 \times C_2}{C_1 + C_2}$$

or

$$C_T = \frac{C}{N}$$

where

C = capacitance of one capacitor

N = number of capacitors connected in series

(Note: The last formula can be used only when all the capacitors connected in series are of the same value.)

What is the total capacitance of three capacitors connected in series if $C_1$ has a capacitance of 20 μF, $C_2$ has a capacitance of 30 μF, and $C_3$ has a capacitance of 60 μF?

**Example 2**

**Solution**

$$C_T = \frac{1}{\dfrac{1}{C_1} + \dfrac{1}{C_2} + \dfrac{1}{C_3}}$$

$$C_T = \frac{1}{\dfrac{1}{20} + \dfrac{1}{30} + \dfrac{1}{60}}$$

$$C_T = 10 \ \mu F$$

## 19-7 Capacitive Charge and Discharge Rates

Capacitors charge and discharge at an **exponential** rate. A charge curve for a capacitor is shown in *Figure 19-13*. The curve is divided into five time constants, and each time constant is equal to 63.2% of the whole value. In *Figure 19-13* it is assumed that a capacitor is to be charged to a total of 100 V. At the end of the first time constant the voltage has reached 63.2% of 100, or 63.2 V. At the end of the second time constant, the voltage reaches 63.2% of the remaining voltage, or 86.4 V. This pattern continues until the capacitor has been charged to 100 V.

The capacitor discharges in the same manner *(Figure 19-14)*. At the end of the first time constant the voltage will decrease to 63.2% of its charged value. In this example the voltage will decrease from 100 V to 36.8 V in the first time constant. At the end of the second time constant the voltage will drop to 13.6 V, and by the end of the third time constant the voltage will drop to 5 V. The voltage will continue to drop at this rate until it reaches approximately 0 after five time constants.

**exponential**

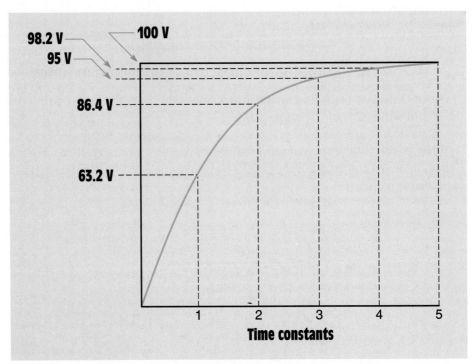

**Figure 19-13**  Capacitors charge at an exponential rate.

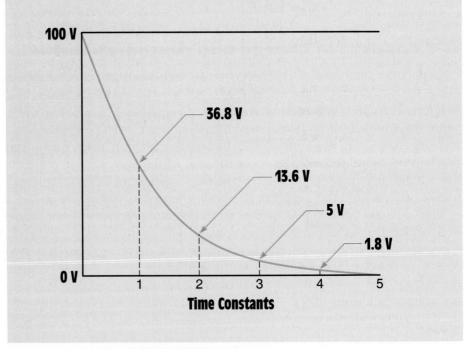

**Figure 19-14**  Capacitor discharge curve

## 19-8 RC Time Constants

When a capacitor is connected in a circuit with a resistor, the amount of time needed to charge the capacitor (that is, the **RC time constant**) can be determined very accurately *(Figure 19-15)*. The formula for determining charge time is:

$$\tau = R \times C$$

where

$\tau$ = the time for one time constant in seconds

R = resistance in ohms

C = capacitance in farads

The Greek letter $\tau$ (tau) is used to represent the time for one time constant. It is not unusual, however, for the letter *T* to be used to represent time.

How long will it take the capacitor shown in *Figure 19-15* to charge if it has a value of 50 μF and the resistor has a value of 100 kΩ?

### Solution

$$\tau = R \times C$$

$$\tau = 0.000050 \text{ F} \times 100,000 \ \Omega$$

$$\tau = 5 \text{ s}$$

The formula is used to find the time for one time constant. Five time constants are required to charge the capacitor.

Total $\tau$ = 5 s x 5 time constants

Total $\tau$ = 25 s

How much resistance should be connected in series with a 100-pF capacitor to give it a total charge time of 0.2 s?

### Solution

Change the above formula to solve for resistance.

$$R = \frac{\tau}{C}$$

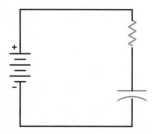

**Figure 19-15**  The charge time of the capacitor can be determined very accurately.

**RC time constant**

**Example 3**

**Example 4**

The total charge time is to be 0.2 s. The value of $\tau$ is therefore $0.2/5 =$ 0.04 s. Substitute these values in the formula.

$$R = \frac{0.04}{100 \times 10^{-12}}$$

$$R = 400 \text{ M}\Omega$$

**Example 5**

A 500-k$\Omega$ resistor is connected in series with a capacitor. The total charge time of the capacitor is 15 s. What is the capacitance of the capacitor?

## Solution

Change the base formula to solve for the value of capacitance.

$$C = \frac{\tau}{R}$$

Since the total charge time is 15 s, the time of one time constant will be 3 s (15/5 = 3).

$$C = \frac{3}{500,000}$$

$$C = 0.000006 \text{ F}$$

or

$$C = 6 \text{ }\mu\text{F}$$

### 19-9 Applications for Capacitors

Capacitors are among the most used of electrical components. They are used for power factor correction in industrial applications, in the start windings of many single-phase AC motors, to produce phase shifts for SCR and Triac circuits, to filter pulsating DC, and in RC timing circuits. (SCRs and Triacs are solid state electronic devices used throughout industry to control high-current circuits.) Capacitors are used extensively in electronic circuits for control of frequency and pulse generation. The type of capacitor used is dictated by the circuit application.

### 19-10 Nonpolarized Capacitors

Capacitors can be divided into two basic groups, nonpolarized and polarized. **Nonpolarized capacitors** are often referred to as AC capacitors, because they are not sensitive to polarity connection. Nonpolarized capacitors can be connected to either DC or AC circuits without harm to

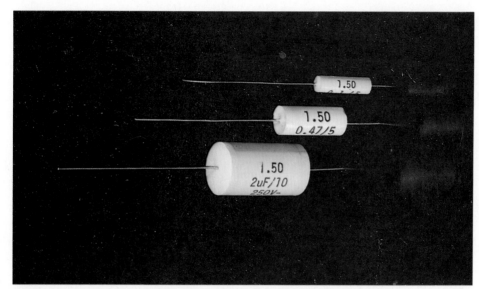

**Figure 19-16**   Nonpolarized capacitors *(Courtesy of Mallory Capacitor Co.)*

the capacitor. Nonpolarized capacitors are constructed by separating metal plates by some type of dielectric *(Figure 19-1)*. These capacitors can be obtained in many different styles and case types *(Figure 19-16)*.

A common type of AC capacitor called the paper capacitor or oil-filled capacitor is often used in motor circuits and for power factor correction *(Figure 19-17)*. It derives its name from the type of dielectric used. This

**Figure 19-17**   Oil-filled paper capacitors

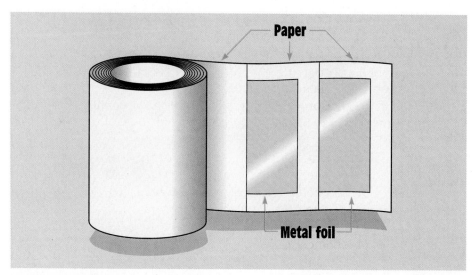

**Figure 19-18** Oil-filled paper capacitor

capacitor is constructed by separating plates made of metal foil with thin sheets of paper soaked in a dielectric oil *(Figure 19-18)*. These capacitors are often used as the run or starting capacitor for single-phase motors. Many manufacturers of oil-filled capacitors will identify one terminal with

**Figure 19-19** Marks indicate plate nearest capacitor case.

an arrow, a painted dot, or by stamping a dash in the capacitor can *(Figure 19-19)*. This identified terminal marks the connection to the plate that is located nearer to the metal container or can. It has long been known that when a capacitor's dielectric breaks down and permits a short circuit to ground, the plate nearer to the outside case most often becomes grounded. For this reason, it is generally desirable to connect the identified capacitor terminal to the line side instead of to the motor start winding.

In *Figure 19-20,* the run capacitor has been connected in such a manner that the identified terminal is connected to the start winding of a single-phase motor. If the capacitor should become shorted to ground, a current path exists through the motor start winding. The start winding is an inductive type load, and inductive reactance will limit the value of current flow to ground. Since the flow of current is limited, it will take the circuit breaker or fuse some time to open the circuit and disconnect the motor from the power line. This time delay can permit the start winding to overheat and become damaged.

In *Figure 19-21,* the run capacitor has been connected in the circuit in such a manner that the identified terminal is connected to the line side. If the capacitor should become shorted to ground, a current path exists directly to ground, bypassing the motor start winding. When the capacitor is connected in this manner, the start winding does not limit current flow and permits the fuse or circuit breaker to open the circuit almost immediately.

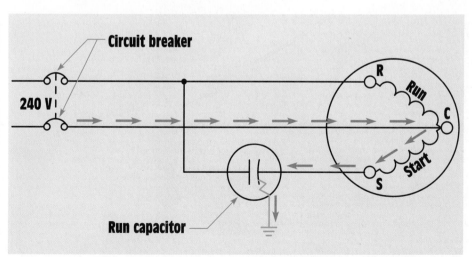

**Figure 19-20** Identified capacitor terminal connected to motor start winding (incorrect connection)

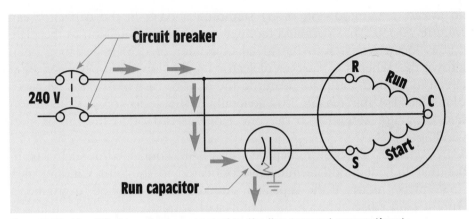

**Figure 19-21** Identified terminal connected to the line (correct connections)

## 19-11 Polarized Capacitors

**polarized
capacitors**

**electrolytic**

**Polarized capacitors** are generally referred to as **electrolytic** capacitors. These capacitors are sensitive to the polarity they are connected to and will have one terminal identified as positive or negative *(Figure 19-22)*. Polarized capacitors can be used in DC circuits only. If their polarity connection is reversed, the capacitor can be damaged and will sometimes explode. The advantage of electrolytic capacitors is that they can have very high capacitance in a small case.

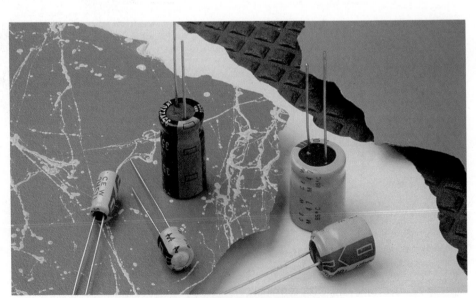

**Figure 19-22** Polarized capacitors *(Courtesy of Mallory Capacitor Co.)*

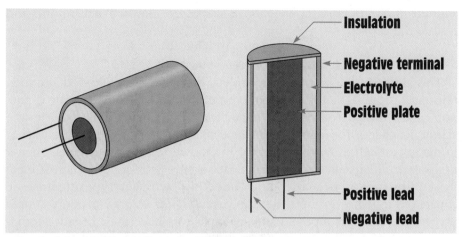

**Figure 19-23**   Wet type electrolytic capacitor

There are two basic types of electrolytic capacitors, the wet type and the dry type. The wet type electrolytic *(Figure 19-23)* has a positive plate made of aluminum foil. The negative plate is actually an electrolyte made from a borax solution. A second piece of aluminum foil is placed in contact with the electrolyte and becomes the negative terminal. When a source of direct current is connected to the capacitor, the borax solution forms an insulating oxide film on the positive plate. This film is only a few molecules thick and acts as the insulator to separate the plates. The capacitance is very high because the distance between the plates is so small.

If the polarity of the wet type electrolytic capacitor becomes reversed, the oxide insulating film will dissolve and the capacitor will become

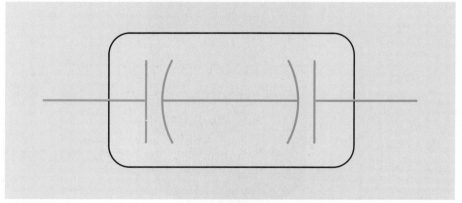

**Figure 19-24**   Two wet type electrolytic capacitors connect to form an AC electrolytic capacitor

shorted. If the polarity connection is corrected, the film will reform and restore the capacitor.

## AC Electrolytic Capacitors

This ability of the wet type electrolytic capacitor to be shorted and then reformed is the basis for a special type of nonpolarized electrolytic capacitor called the AC electrolytic capacitor. This capacitor is used as the starting capacitor for many small single-phase motors, as the run capacitor in many ceiling fan motors, and for low-power electronic circuits when a nonpolarized capacitor with a high capacitance is required. The AC electrolytic capacitor is made by connecting two wet type electrolytic capacitors together inside the same case *(Figure 19-24)*. In the example shown, the two wet type electrolytic capacitors have their negative terminals connected. When alternating current is applied to the leads, one capacitor will be connected to reverse polarity and become shorted. The other capacitor will be connected to the correct polarity and will form. During the next half cycle, the polarity changes and forms the capacitor that was shorted and shorts the other capacitor. An AC electrolytic capacitor is shown in *Figure 19-25*.

**Figure 19-25** An AC electrolytic capacitor

## Dry Type Electrolytic Capacitors

The dry type electrolytic capacitor is very similar to the wet type except that gauze is used to hold the borax solution. This prevents the capacitor from leaking. Although the dry type electrolytic has the advantage of being relatively leak proof, it does have one disadvantage. If the polarity connection should become reversed and the oxide film is broken down, it will not reform when connected to the proper polarity. Reversing the polarity of a dry type electrolytic capacitor will permanently damage the capacitor.

## 19-12 Variable Capacitors

**Variable capacitors** are constructed in such a manner that their capacitance value can be changed over a certain range. They generally contain a set of movable plates, which are connected to a shaft, and a set of stationary plates *(Figure 19-26)*. The movable plates can be interleaved with the stationary plates to increase or decrease the capacitance value. Since

**variable
capacitors**

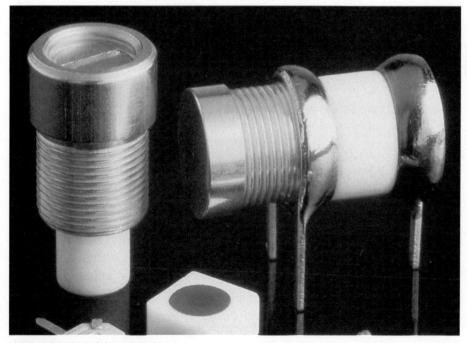

**Figure 19-26**  A variable capacitor

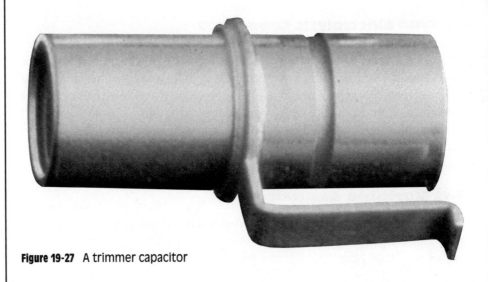

**Figure 19-27** A trimmer capacitor

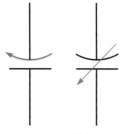

**Figure 19-28** Variable capacitor symbols

air is used as the dielectric and the plate area is relatively small, variable capacitors are generally rated in picofarads. Another type of small variable capacitor is called the trimmer capacitor *(Figure 19-27)*. This capacitor has one movable plate and one stationary plate. The capacitance value is changed by turning an adjustment screw that moves the movable plate closer to or farther away from the stationary plate. *Figure 19-28* shows schematic symbols used to represent variable capacitors.

## 19-13 Capacitor Markings

Different types of capacitors are marked in different ways. Large AC oil-filled paper capacitors generally have their capacitance and voltage values written on the capacitor. The same is true for most electrolytic and small nonpolarized capacitors. Other types of capacitors, however, depend on color codes or code numbers and letters to indicate the capacitance value, tolerance, and voltage rating. Although color coding for capacitors has been abandoned in favor of direct marking by most manufacturers, it is still used by some. Also, many older capacitors with color codes are still in use. For this reason, we will discuss color coding for several types of capacitors. Unfortunately, there is no actual set standard used by all manufacturers. The color codes presented are probably the most common. An identification chart for postage stamp (so called because of their size and shape) mica capacitors and tubular paper or

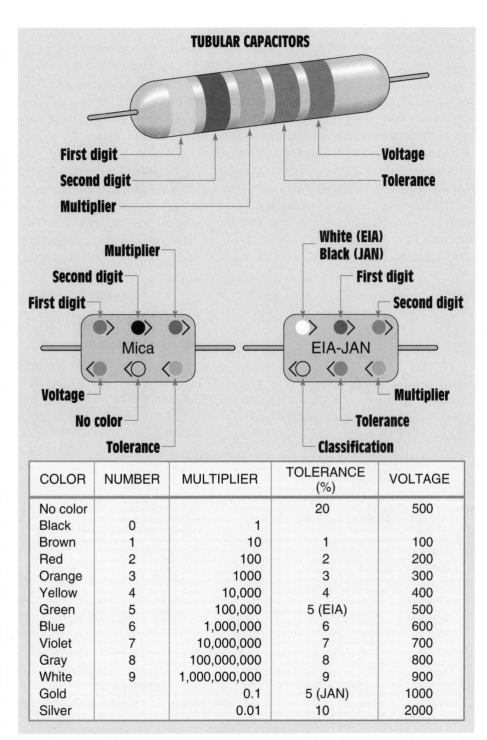

**TUBULAR CAPACITORS**

First digit
Second digit
Multiplier
Voltage
Tolerance

Multiplier
Second digit
First digit
White (EIA)
Black (JAN)
First digit
Second digit

Mica
EIA-JAN

Voltage
No color
Tolerance
Multiplier
Tolerance
Classification

| COLOR | NUMBER | MULTIPLIER | TOLERANCE (%) | VOLTAGE |
|---|---|---|---|---|
| No color | | | 20 | 500 |
| Black | 0 | 1 | | |
| Brown | 1 | 10 | 1 | 100 |
| Red | 2 | 100 | 2 | 200 |
| Orange | 3 | 1000 | 3 | 300 |
| Yellow | 4 | 10,000 | 4 | 400 |
| Green | 5 | 100,000 | 5 (EIA) | 500 |
| Blue | 6 | 1,000,000 | 6 | 600 |
| Violet | 7 | 10,000,000 | 7 | 700 |
| Gray | 8 | 100,000,000 | 8 | 800 |
| White | 9 | 1,000,000,000 | 9 | 900 |
| Gold | | 0.1 | 5 (JAN) | 1000 |
| Silver | | 0.01 | 10 | 2000 |

**Figure 19-29**  Identification of mica and tubular capacitors

tubular mica capacitors is shown in *Figure 19-29*. It should be noted that most postage stamp mica capacitors use a five-dot color code. There are six-dot color codes, however. When a six-dot color code is used the third dot represents a third digit, and the rest of the code is the same as a five-dot code. The capacitance values given are in picofarads. Although these markings are typical, there is no actual standard and it may be necessary to use the manufacturer's literature to determine the true values.

A second method for color coding mica capacitors is called the EIA (Electronic Industries Association) standard, or the **JAN** (Joint Army-Navy) **standard**. The JAN standard is used for electronic components intended for military use. When the EIA standard is employed, the first dot will be white. In some instances, the first dot may be silver instead of white. This indicates that the capacitor's dielectric is paper instead of mica. When the JAN standard is used, the first dot will be black. The second and third

**JAN standard**

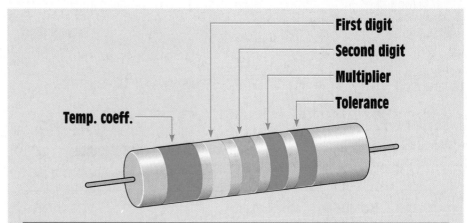

| COLOR | NUMBER | MULTIPLIER | TOLERANCE OVER 10 pF | 10 pF OR LESS | TEMP. COEFF. |
|-------|--------|-----------|------------|-----------|--------------|
| Black | 0 | 1 | 20% | 2.0 pF | 0 |
| Brown | 1 | 10 | 1% | | N30 |
| Red | 2 | 100 | 2% | | N80 |
| Orange | 3 | 1000 | | | N150 |
| Yellow | 4 | | | | N220 |
| Green | 5 | | | | N330 |
| Blue | 6 | | 5% | 0.5 pF | N470 |
| Violet | 7 | | | | N750 |
| Gray | 8 | 0.01 | | 0.25 pF | P30 |
| White | 9 | 0.1 | 10% | 1.0 pF | P500 |

**Figure 19-30**  Color codes for ceramic capacitors

dots represent digits, the fourth dot is the multiplier, the fifth dot is the tolerance, and the sixth dot indicates classes A to E of temperature and leakage coefficients.

## 19-14 Temperature Coefficients

The temperature coefficient indicates the amount of capacitance change with temperature. Temperature coefficients are listed in parts per million (ppm) per degree Celsius. A positive temperature coefficient indicates that the capacitor will increase its capacitance with an increase in temperature. A negative temperature coefficient indicates that the capacitance will decrease with an increase in temperature.

## 19-15 Ceramic Capacitors

Another capacitor that often uses color codes is the ceramic capacitor *(Figure 19-30)*. This capacitor will generally have one band that is wider than the others. The wide band indicates the temperature coefficient, and the other bands are first and second digits, multiplier, and tolerance.

## 19-16 Dipped Tantalum Capacitors

A dipped tantalum capacitor is shown in *Figure 19-31*. This capacitor has the general shape of a match head but is somewhat larger. Color bands and dots determine the value, tolerance, and voltage. The capacitance value is given in picofarads.

## 19-17 Film Capacitors

Not all capacitors use color codes to indicate values. Some capacitors use numbers and letters. A film type capacitor is shown in *Figure 19-32*. This capacitor is marked 105K. The value can be read as follows:
1. The first two numbers indicate the first two digits of the value.
2. The third number is the multiplier. Add the number of zeros to the first two numbers indicated by the multiplier. In this example, add 5 zeros to 10. The value is given in picofarads. This capacitor has a value of 1,000,000 pF or 1 μF.
3. The K is the tolerance. In this example, K indicates a tolerance of ±10%.

## 19-18 Testing Capacitors

Testing capacitors is difficult at best. Small electrolytic capacitors are generally tested for shorts with an ohmmeter. If the capacitor is not shorted, it should be tested for leakage using a variable DC power supply and

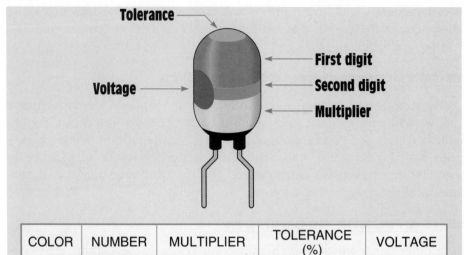

| COLOR | NUMBER | MULTIPLIER | TOLERANCE (%) | VOLTAGE |
|---|---|---|---|---|
| | | | No dot 20 | |
| Black | 0 | | | 4 |
| Brown | 1 | | | 6 |
| Red | 2 | | | 10 |
| Orange | 3 | | | 15 |
| Yellow | 4 | 10,000 | | 20 |
| Green | 5 | 100,000 | | 25 |
| Blue | 6 | 1,000,000 | | 35 |
| Violet | 7 | 10,000,000 | | 50 |
| Gray | 8 | | | |
| White | 9 | | | 3 |
| Gold | | | 5 | |
| Silver | | | 10 | |

**Figure 19-31**  Dipped tantalum capacitors

a microammeter *(Figure 19-33)*. When rated voltage is applied to the capacitor, the microammeter should indicate zero current flow.

Large AC oil-filled capacitors can be tested in a similar manner. To test the capacitor accurately, two measurements must be made. One is to measure the capacitance value of the capacitor to determine if it is the same or approximately the same as the rate value. The other is to test the strength of the dielectric.

The first test should be made with an ohmmeter. With the power disconnected, connect the terminals of an ohmmeter directly across the capacitor terminals *(Figure 19-34)*. This test determines if the dielectric is shorted. When the ohmmeter is connected, the needle should swing up

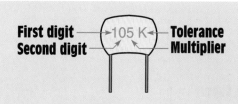

| NUMBER | MULTIPLIER | TOLERANCE | | |
| --- | --- | --- | --- | --- |
| | | | 10 pf or less | Over 10 pf |
| 0 | 1 | B | 0.1 pf | |
| 1 | 10 | C | 0.25 pf | |
| 2 | 100 | D | 0.5 pf | |
| 3 | 1000 | F | 1.0 pf | 1% |
| 4 | 10,000 | G | 2.0 pf | 2% |
| 5 | 100,000 | H | | 3% |
| 6 | | J | | 5% |
| 7 | | K | | 10% |
| 8 | 0.01 | M | | 20% |
| 9 | 0.1 | | | |

**Figure 19-32** Film type capacitors

scale and return to infinity. The amount of needle swing is determined by the capacitance of the capacitor. Then reverse the ohmmeter connection, and the needle should move twice as far up scale and return to the infinity setting.

If the ohmmeter test is successful, the dielectric must be tested at its rated voltage. This is called a dielectric strength test. To make this test, a dielectric test set must be used *(Figure 19-35)*. This device is often referred to as a **HIPOT** because of its ability to produce a high voltage or high potential. The dielectric test set contains a variable voltage control, a voltmeter, and a microammeter. To use the HIPOT, connect its terminal leads to the capacitor terminals. Increase the output voltage until rated voltage is applied to the capacitor. The microammeter indicates any current flow between the plates of the dielectric. If the capacitor is good, the microammeter should indicate zero current flow.

The capacitance value must be measured to determine if there are any open plates in the capacitor. To measure the capacitance value of the capacitor, connect some value of AC voltage across the plates of the

**HIPOT**

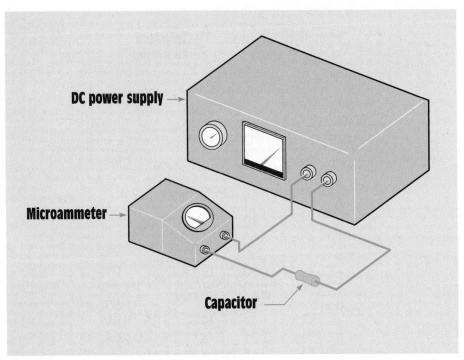

**Figure 19-33** Testing a capacitor for leakage

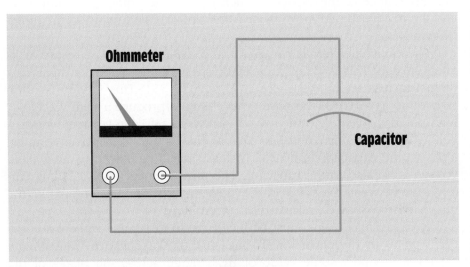

**Figure 19-34** Testing the capacitor with an ohmmeter

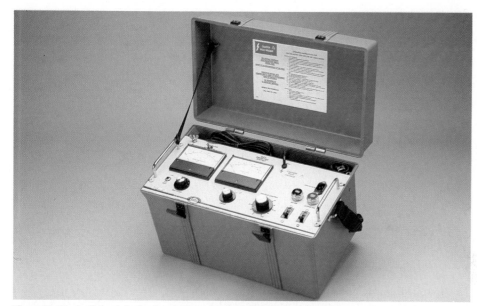

**Figure 19-35**   A dielectric test set *(Courtesy of Biddle Instruments.)*

capacitor *(Figure 19-36)*. This voltage must not be greater than the rated capacitor voltage. Then measure the amount of current flow in the circuit. Now that the voltage and current flow are known, the capacitive reactance of the capacitor can be computed using the formula

$$X_C = \frac{E}{I}$$

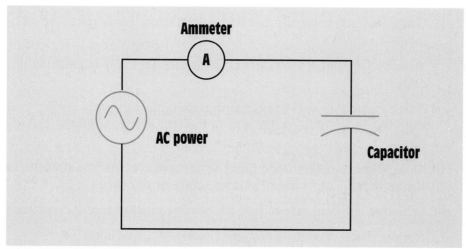

**Figure 19-36**   Determining the capacitance value

After the capacitive reactance has been determined, the capacitance can be computed using the formula

$$C = \frac{1}{2\pi F X_C}$$

(Note: Capacitive reactance is measured in ohms and limits current flow in a manner similar to inductive reactance. Capacitive reactance is covered fully in Unit 20.)

## Summary

1. Capacitors are devices that oppose a change of voltage.

2. Three factors that determine the capacitance of a capacitor are:
   A. The surface area of the plates.
   B. The distance between the plates.
   C. The type of dielectric.

3. A capacitor stores energy in an electrostatic field.

4. Current can flow only during the time a capacitor is charging or discharging.

5. Capacitors charge and discharge at an exponential rate.

6. The basic unit of capacitance is the farad.

7. Capacitors are generally rated in microfarads, nanofarads, or picofarads.

8. When capacitors are connected in parallel their capacitance values add.

9. When capacitors are connected in series, the reciprocal of the total capacitance is equal to the sum of the reciprocals of all the capacitors.

10. The charge and discharge times of a capacitor are proportional to the amount of capacitance and resistance in the circuit.

11. Five time constants are required to charge or discharge a capacitor.

12. Nonpolarized capacitors are often called AC capacitors.

13. Nonpolarized capacitors can be connected to direct or alternating current circuits.

14. Polarized capacitors are often referred to as electrolytic capacitors.

15. Polarized capacitors can be connected to direct current circuits only.

16. There are two basic types of electrolytic capacitors, the wet type and the dry type.

17. Wet type electrolytic capacitors can be reformed if reconnected to the correct polarity.

18. Dry type electrolytic capacitors will be permanently damaged if connected to the incorrect polarity.

19. Capacitors are often marked with color codes or with numbers and letters.

20. To test a capacitor for leakage, a microammeter should be connected in series with the capacitor and rated voltage applied to the circuit.

## Review Questions

1. What is the dielectric?

2. List three factors that determine the capacitance of a capacitor.

3. A capacitor uses air as a dielectric and has a capacitance of 3 μF. A dielectric material is inserted between the plates without changing the spacing, and the capacitance becomes 15 μF. What is the dielectric constant of this material?

4. In what form is the energy of a capacitor stored?

5. Four capacitors having values of 20 μF, 50 μF, 40 μF, and 60 μF are connected in parallel. What is the total capacitance of this circuit?

6. If the four capacitors in question 5 were to be connected in series, what would be the total capacitance of the circuit?

7. A 22-μF capacitor is connected in series with a 90-kΩ resistor. How long will it take this capacitor to charge?

8. A 450-pF capacitor has a total charge time of 0.5 s. How much resistance is connected in series with the capacitor?

9. Can a nonpolarized capacitor be connected to a direct current cir-

cuit?

10. Explain how an AC electrolytic capacitor is constructed.

11. What type of electrolytic capacitor will be permanently damaged if connected to the incorrect polarity?

12. A 500-nF capacitor is connected to a 300-kΩ resistor. What is the total charge time of this capacitor?

13. A film type capacitor is marked 253H. What are the capacitance value and tolerance of this capacitor?

14. A postage stamp mica capacitor has the following color marks starting at the upper left dot: yellow, violet, brown, green, no color, and blue. What are the capacitance value, tolerance, and voltage rating of this capacitor?

15. A postage stamp capacitor has the following color marks starting at the upper left dot: black, orange, orange, black, silver, and white. What are the capacitance value and tolerance of this capacitor?

## RC Time Constants

Fill in all the missing values. Refer to the formulas given below.

**Practice Problems**

| Resistance | Capacitance | Time Constant | Total Time |
|---|---|---|---|
| 150 kΩ | 100 µF | | |
| 350 kΩ | | | 35 s |
| | 350 pF | 0.05 s | |
| | 0.05 µF | | 10 s |
| 1.2 MΩ | 0.47 µF | | |
| | 12 µF | 0.05 s | |
| 86 kΩ | | | 1.5 s |
| 120 kΩ | 470 pF | | |
| | 250 nF | | 100 ms |
| | 8 µF | | 150 µs |
| 100 kΩ | | 150 ms | |
| 33 kΩ | 4 µF | | |

$$\tau = RC$$

$$R = \frac{\tau}{C}$$

$$C = \frac{\tau}{R}$$

Total time = $\tau \times 5$

Although capacitors are an open circuit, they appear to permit current to flow through them when connected to alternating current. This unit discusses voltage, current, and power in pure capacitive circuits.

*Courtesy of Niagara Mohawk Power Corp.*

# Capacitance in Alternating Current Circuits

## Outline

## Key Terms

Appear to flow

Capacitive reactance ($X_c$)

Frequency

Out of phase

Quality (Q)

Reactive power (VARs)

Voltage rating

**A**fter studying this unit, you should be able to:

Objectives

- Explain why current appears to flow through a capacitor when it is connected to an alternating current circuit.

- Discuss capacitive reactance.

- Compute the value of capacitive reactance in an AC circuit.

- Compute the value of capacitance in an AC circuit.

- Discuss the relationship of voltage and current in a pure capacitive circuit.

Preview

In Unit 19, it was discussed that a capacitor is composed of two metal plates separated by an insulating material called the dielectric. Since there is no complete circuit between the plates, current cannot flow through the capacitor. The only time that current can flow is during the period of time that the capacitor is being charged or discharged.

### 20-1 Connecting the Capacitor into an AC Circuit

appear to flow

When a capacitor *(Figure 20-1)* is connected to an alternating current circuit, current will **appear to flow** through the capacitor. The reason is that in an AC circuit, the current continually changes direction and polarity. To understand this concept, consider the hydraulic circuit shown in *Figure 20-2.* Two tanks are connected to a common pump. Assume tank A to be full and tank B to be empty. Now assume that the pump pumps water from tank A to tank B. When tank B becomes full, the pump

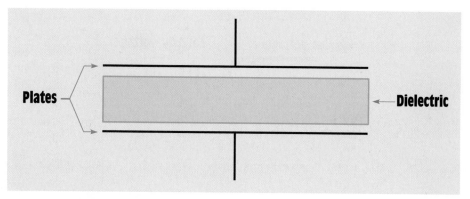

**Figure 20-1** A basic capacitor

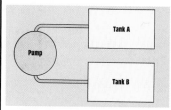

**Figure 20-2** Water can flow continuously, but not between the two tanks.

reverses and pumps the water from tank B back into tank A. Each time a tank becomes filled, the pump reverses and pumps water back into the other tank. Notice that water is continually flowing in this circuit, but there is no direct connection between the two tanks.

A similar action takes place when a capacitor is connected to an alternating current circuit *(Figure 20-3)*. In this circuit, the AC generator or alternator charges one plate of the capacitor positive and the other plate negative. During the next half cycle, the voltage will change polarity and the capacitor will discharge and recharge to the opposite polarity also. As long as the voltage continues to increase, decrease, and change polarity, current will flow from one plate of the capacitor to the other. If an ammeter were placed in the circuit, it would indicate a continuous flow of current, giving the appearance that current is flowing through the capacitor.

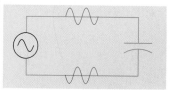

**Figure 20-3** A capacitor connected to an AC circuit

## 20-2 Capacitive Reactance

As the capacitor is charged, an impressed voltage is developed across its plates as an electrostatic charge is built up *(Figure 20-4)*. The impressed voltage is the voltage provided by the electrostatic charge. This impressed voltage opposes the applied voltage and limits the flow of current in the circuit. This counter-voltage is similar to the counter-voltage produced by an inductor. The counter-voltage developed by the capacitor is called reactance also. Since this counter-voltage is caused by capacitance, it is called **capacitive reactance ($X_C$)** and is measured in ohms. The formula for finding capacitive reactance is

$$X_C = \frac{1}{2\pi FC}$$

**capacitive reactance ($X_C$)**

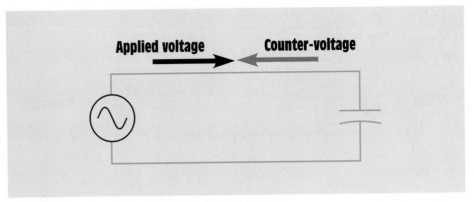

**Figure 20-4** Counter-voltage limits the flow of current.

where

$$X_C \ = \text{capacitive reactance}$$
$$\pi \ = 3.1416$$
$$F \ = \text{frequency in hertz}$$
$$C \ = \text{capacitance in farads}$$

A 35-µF capacitor is connected to a 120-V 60-Hz line. How much current will flow in this circuit?

## Solution

The first step is to compute the capacitive reactance. Recall that the value of C in the formula is given in farads. This must be changed to the capacitive units being used—in this case, microfarads.

$$X_C = \frac{1}{2 \times 3.1416 \times 60 \times (35 \times 10^{-6})}$$

$$X_C = 75.79 \ \Omega$$

Now that the value of capacitive reactance is known, it can be used like resistance in an Ohm's law formula. Since capacitive reactance is the current-limiting factor, it will replace the value of R.

$$I = \frac{E}{X_C}$$

$$I = \frac{120}{75.79}$$

$$I = 1.58 \ A$$

## 20-3 Computing Capacitance

If the value of capacitive reactance is known, the capacitance of the capacitor can be found using the formula

$$C = \frac{1}{2\pi F X_C}$$

A capacitor is connected into a 480-V 60-Hz circuit. An ammeter indicates a current flow of 2.6 A. What is the capacitance value of the capacitor?

**Example 2**

## Solution

The first step is to compute the value of capacitive reactance. Since capacitive reactance, like resistance, limits current flow, it can be substituted for R in an Ohm's law formula.

$$X_C = \frac{E}{I}$$

$$X_C = \frac{480}{2.6}$$

$$X_C = 184.61 \ \Omega$$

Now that the capacitive reactance of the circuit is known, the value of capacitance can be found.

$$C = \frac{1}{2\pi F X_C}$$

$$C = \frac{1}{2 \times 3.1416 \times 60 \times 184.61}$$

$$C = \frac{1}{69,596.49}$$

$$C = 0.00001437 \ F = 14.37 \ \mu F$$

## 20-4 Voltage and Current Relationships in a Pure Capacitive Circuit

Earlier in this text it was shown that the current in a pure resistive circuit is in phase with the applied voltage and that current in a pure inductive circuit lags the applied voltage by 90°. In this unit it will be shown that in a pure capacitive circuit the current will *lead* the applied voltage by 90°.

When a capacitor is connected to an alternating current, the capacitor will charge and discharge at the same rate and time as the applied voltage. The charge in coulombs is equal to the capacitance of the capacitor times the applied voltage (Q = C x V). When the applied voltage is zero, the charge in coulombs and impressed voltage will be zero also. When the applied voltage reaches its maximum value, positive or negative, the charge in coulombs and impressed voltage will reach maximum also *(Figure 20-5)*. The impressed voltage will follow the same curve as the applied voltage.

In the wave form shown, voltage and charge are both zero at 0°. Since there is no charge on the capacitor, there is no opposition to current flow, which is shown to be maximum. As the applied voltage increases from zero toward its positive peak at 90°, the capacitor begins to charge at the same time. The charge produces an impressed voltage across the plates of the capacitor that opposes the flow of current. The impressed voltage is 180° **out of phase** with the applied voltage *(Figure 20-6)*. When the applied voltage reaches 90° in the positive direction, the charge reaches maximum, the impressed voltage reaches peak in the negative direction, and the current flow is zero.

As the applied voltage begins to decrease, the capacitor begins to discharge, causing the current to flow in the opposite or negative direction. When the applied voltage and charge reach zero at 180°, the impressed voltage is zero also and the current flow is maximum in the negative direction. As the applied voltage and charge increase in the negative direction, the increase of the impressed voltage across the capacitor again causes the current to decrease. The applied voltage and charge reach

**out of phase**

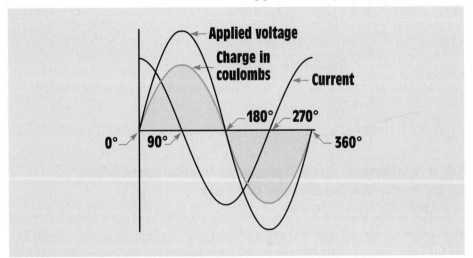

**Figure 20-5** Capacitive current leads the applied voltage by 90°.

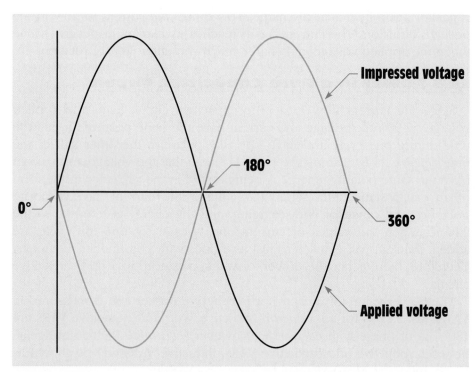

**Figure 20-6** The impressed voltage is 180° out of phase with the applied voltage.

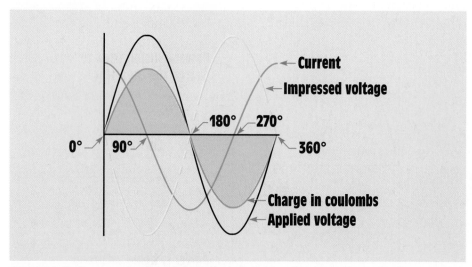

**Figure 20-7** Voltage, current, and charge relationships for a capacitive circuit

maximum negative after 270° of rotation. The impressed voltage reaches maximum positive and the current has decreased to zero *(Figure 20-7)*. As the applied voltage decreases from its maximum negative value, the capacitor again begins to discharge. This causes the current to flow in the positive direction. The current again reaches its maximum positive value when the applied voltage and charge reach zero after 360° of rotation.

### 20-5 Power in a Pure Capacitive Circuit

Since the current flow in a pure capacitive circuit leads the applied voltage by 90°, the voltage and current have the same polarity for half the time during one cycle and have opposite polarities the other half of the time *(Figure 20-8)*. During the period of time that the voltage and current have the same polarity, energy is being stored in the capacitor in the form of an electrostatic field. When the voltage and current have opposite polarities, the capacitor is discharging, and the energy is returned to the circuit. When the values of current and voltage for one full cycle are added, the sum will equal zero just as it does with pure inductive circuits. Therefore, there is no true power, or watts, produced in a pure capacitive circuit.

The power value for a capacitor is **reactive power** and is measured in **VARs**, just as it is for an inductor. Inductive VARs and capacitive VARs are 180° out of phase with each other, however *(Figure 20-9)*. To distinguish between inductive and capacitive VARs, inductive VARs will be shown as VARs$_L$ and capacitive VARs will be shown as VARs$_C$.

> **reactive power (VARs)**

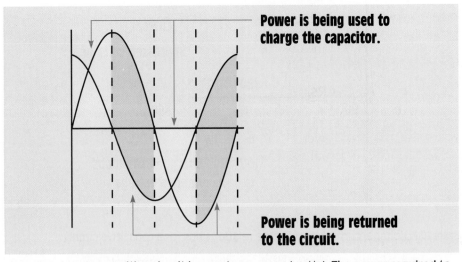

**Power is being used to charge the capacitor.**

**Power is being returned to the circuit.**

**Figure 20-8** A pure capacitive circuit has no true power (watts). The power required to charge the capacitor is returned to the circuit when the capacitor discharges.

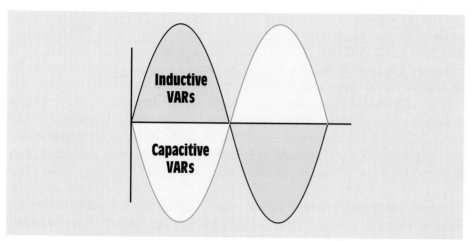

**Figure 20-9** Inductive VARs and capacitive VARs are 180° out of phase with each other.

## 20-6 Quality of a Capacitor

The **quality (Q)** of a capacitor is generally very high. As with inductors, it is a ratio of the resistance to capacitive reactance.

$$Q = \frac{R}{X_C}$$

The R value for a capacitor is generally very high because it is the equivalent resistance of the dielectric between the plates of the capacitor. If a capacitor is leaky, however, the dielectric will appear to be a much lower resistance and the Q rating will decrease.

Q for a capacitor can also be found by using other formulas. One of these formulas is shown below.

$$Q = \frac{VARs_C}{W}$$

where the reactive power is represented by VARs. Another formula that can be used sets Q equal to the reciprocal of the power factor:

$$Q = \frac{1}{PF}$$

## 20-7 Capacitor Voltage Rating

The **voltage rating** of a capacitor is actually the voltage rating of the dielectric. **Voltage rating is extremely important concerning the life of the capacitor and should never be exceeded.** Unfortunately, there are no set standards concerning how voltage ratings are marked. It is not unusual to see capacitors marked VOLTS AC, VOLTS DC, PEAK VOLTS, and WVDC (WORKING VOLTS DC). The voltage rating of electrolytic or polarized capacitors will always be given in DC volts. The voltage rating of nonpolarized capacitors, however, can be given as AC or DC volts.

If a nonpolarized capacitor has a voltage rating given in AC volts, the voltage indicated is the RMS value. If the voltage rating is given as PEAK or as DC volts, it indicates the peak value of AC volts. If a capacitor is to be connected to an AC circuit, it is necessary to compute the peak value if the voltage rating is given as DC volts.

**Example 3**

An AC oil-filled capacitor has a voltage rating of 300 WVDC. Will the voltage rating of the capacitor be exceeded if the capacitor is connected to a 240-V 60-Hz line?

### Solution

The DC voltage rating of the capacitor indicates the peak value of voltage. To determine if the voltage rating will be exceeded, find the peak value of 240 V by multiplying by 1.414.

$$\text{Peak} = 240 \times 1.414$$
$$\text{Peak} = 339.36 \text{ V}$$

The answer is that the capacitor voltage rating will be exceeded.

## 20-8 Effects of Frequency in a Capacitive Circuit

One of the factors that determine the capacitive reactance of a capacitor is the **frequency.** Capacitive reactance is inversely proportional to frequency. As the frequency increases, the capacitive reactance decreases. The chart in *Figure 20-10* shows the capacitive reactance for different values of capacitance at different frequencies. Frequency has an effect on capacitive reactance because the capacitor charges and discharges faster at a higher frequency. Recall that current is a rate of electron flow. A current of 1 A is one coulomb per second.

**frequency**

$$I = \frac{C}{t}$$

where

I = current

C = charge in coulombs

t = time in seconds

| Capacitance | Capacitive reactance | | | |
|---|---|---|---|---|
| | 30 Hz | 60 Hz | 400 Hz | 1000 Hz |
| 10 pF | 530.5 M Ω | 265.26 M Ω | 39.79 M Ω | 15.91 M Ω |
| 350 pF | 15.16 M Ω | 7.58 M Ω | 1.14 M Ω | 454.73 k Ω |
| 470 nF | 112.88 k Ω | 56.44 k Ω | 8.47 k Ω | 3.39 k Ω |
| 750 nF | 22.22 k Ω | 11.11 k Ω | 1.67 k Ω | 666.67 Ω |
| 1 μF | 5.31 k Ω | 2.65 k Ω | 397.89 Ω | 15.915 Ω |
| 25 μF | 212.21 Ω | 106.1 Ω | 159.15 Ω | 6.37 Ω |

**Figure 20-10**  Capacitive reactance is inversely proportional to frequency.

Assume that a capacitor is connected to a 30-Hz line, and 1 coulomb of charge flows each second. If the frequency is doubled to 60 Hz, 1 coulomb of charge will flow in 0.5 s because the capacitor is being charged and discharged twice as fast *(Figure 20-11)*. This means that in a period of 1 s, 2 coulombs of charge will flow. Since the capacitor is being

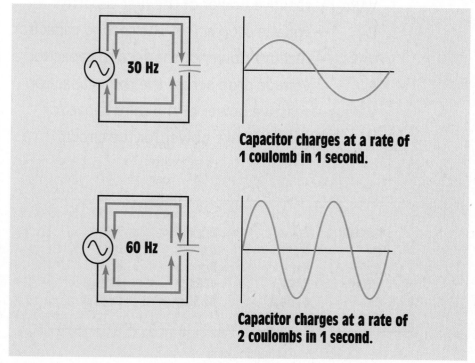

Capacitor charges at a rate of
1 coulomb in 1 second.

Capacitor charges at a rate of
2 coulombs in 1 second.

**Figure 20-11**  The rate of charge increases with frequency.

charged and discharged at a faster rate, the opposition to current flow is decreased.

## 20-9 Series Capacitors

**Example 4**

Three capacitors with values of 10 μF, 30 μF, and 15 μF are connected in series to a 480-V 60-Hz line *(Figure 20-12)*. Find the following circuit values.

| | |
|---|---|
| $X_{C1}$ | — capacitive reactance of the first capacitor |
| $X_{C2}$ | — capacitive reactance of the second capacitor |
| $X_{C3}$ | — capacitive reactance of the third capacitor |
| $X_{CT}$ | — total capacitive reactance for the circuit |
| $C_T$ | — total capacitance for the circuit |
| $I_T$ | — total circuit current |
| $E_{C1}$ | — voltage drop across the first capacitor |
| $VARS_{C1}$ | — reactive power of the first capacitor |
| $E_{C2}$ | — voltage drop across the second capacitor |
| $VARS_{C2}$ | — reactive power of the second capacitor |
| $E_{C3}$ | — voltage drop across the third capacitor |
| $VARS_{C3}$ | — reactive power of the third capacitor |
| $VARS_{CT}$ | — total reactive power for the circuit |

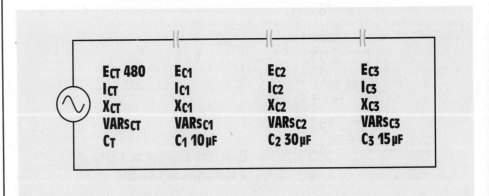

**Figure 20-12** Capacitors connected in series

## Solution

Since the frequency and the capacitance of each capacitor are known, the capacitive reactance for each capacitor can be found using the formula

$$X_C = \frac{1}{2\pi FC}$$

Recall that the value for C in the formula is in farads, and the capacitors in this problem are rated in microfarads.

$$X_{C1} = \frac{1}{2\pi FC}$$

$$X_{C1} = \frac{1}{377 \times 0.000010}$$

$$X_{C1} = 265.25 \ \Omega$$

$$X_{C2} = \frac{1}{2\pi FC}$$

$$X_{C2} = \frac{1}{377 \times 0.000030}$$

$$X_{C2} = 88.417 \ \Omega$$

$$X_{C3} = \frac{1}{2\pi FC}$$

$$X_{C3} = \frac{1}{377 \times 0.000015}$$

$$X_{C3} = 176.83 \ \Omega$$

Since there is no phase angle shift between any of the three capacitive reactances, the total capacitive reactance will be the sum of the three reactances *(Figure 20-13)*.

$$X_{CT} = X_{C1} + X_{C2} + X_{C3}$$

$$X_{CT} = 265.25 + 88.417 + 176.83$$

$$X_{CT} = 530.497 \ \Omega$$

The total capacitance of a series circuit can be computed in a manner similar to that used for computing parallel resistance. Refer to the Pure Capacitive Circuits Formula section of the appendix. Total capacitance in

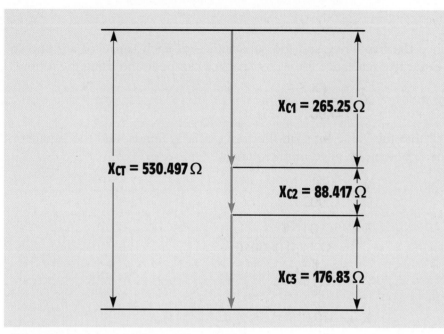

$X_{CT} = 530.497\,\Omega$

$X_{C1} = 265.25\,\Omega$

$X_{C2} = 88.417\,\Omega$

$X_{C3} = 176.83\,\Omega$

**Figure 20-13**  Vector sum for capacitive reactance

this circuit will be computed using the formula

$$C_T = \frac{1}{\dfrac{1}{C_1} + \dfrac{1}{C_2} + \dfrac{1}{C_3}}$$

$$C_T = \frac{1}{\dfrac{1}{10} + \dfrac{1}{30} + \dfrac{1}{15}}$$

$$C_T = \frac{1}{0.2}$$

$$C_T = 5\,\mu F$$

The total current can be found by using the total capacitive reactance to substitute for R in an Ohm's law formula.

$$I_T = \frac{E_{CT}}{X_{CT}}$$

$$I_T = \frac{480}{530.497}$$

$$I_T = 0.905\ A$$

Since the current is the same at any point in a series circuit, the voltage drop across each capacitor can now be computed using the capacitive reactance of each capacitor and the current flowing through it.

$$E_{C1} = I_{C1} \times X_{C1}$$
$$E_{C1} = 0.905 \times 265.25$$
$$E_{C1} = 240.051 \text{ V}$$

$$E_{C2} = I_{C2} \times X_{C2}$$
$$E_{C2} = 0.905 \times 88.417$$
$$E_{C2} = 80.017 \text{ V}$$

$$E_{C3} = I_{C3} \times X_{C3}$$
$$E_{C3} = 0.905 \times 176.83$$
$$E_{C3} = 160.031 \text{ V}$$

Now that the voltage drops of the capacitors are known, the reactive power of each capacitor can be found.

$$VARS_{C1} = E_{C1} \times I_{C1}$$
$$VARS_{C1} = 240.051 \times 0.905$$
$$VARS_{C1} = 217.246$$

$$VARS_{C2} = E_{C2} \times I_{C2}$$
$$VARS_{C2} = 80.017 \times 0.905$$
$$VARS_{C2} = 72.415$$

$$VARS_{C3} = E_{C3} \times I_{C3}$$
$$VARS_{C3} = 160.031 \times 0.905$$
$$VARS_{C3} = 144.828$$

Power, whether true power, apparent power, or reactive will add in any type of circuit. The total reactive power in this circuit can be found by taking the sum of all the VARs for the capacitors or by using total values of voltage and current and Ohm's law.

$$VARS_{CT} = VARS_{C1} + VARS_{C2} + VARS_{C3}$$

$$VARS_{CT} = 217.246 + 72.415 + 144.828$$

$$VARS_{CT} = 434.489$$

The circuit with all computed values is shown in *Figure 20-14*.

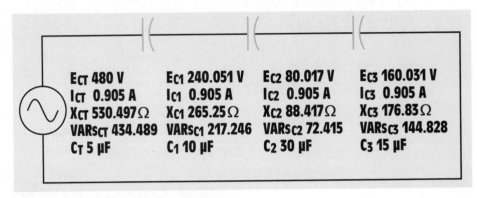

**Figure 20-14** Series circuit 1 with all values

## 20-10 Parallel Capacitors

**. Example 5**

Three capacitors having values of 50 μF, 75 μF, and 20 μF are connected in parallel to a 60-Hz line. The circuit has a total reactive power of 787.08 VARs *(Figure 20-15)*. Find the following unknown values.

$X_{C1}$ — capacitive reactance of the first capacitor

$X_{C2}$ — capacitive reactance of the second capacitor

$X_{C3}$ — capacitive reactance of the third capacitor

$X_{CT}$ — total capacitive reactance for the circuit

$E_T$ — total applied voltage

$I_{C1}$ — current flow through the first capacitor

$VARS_{C1}$ — reactive power of the first capacitor

$I_{C2}$ — current flow through the second capacitor

$VARS_{C2}$ — reactive power of the second capacitor

$I_{C3}$ — current flow through the third capacitor

$VARS_{C3}$ — reactive power of the third capacitor

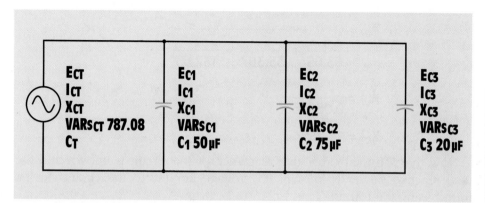

**Figure 20-15**  Capacitors connected in parallel

Since the frequency of the circuit and the capacitance of each capacitor are known, the capacitive reactance of each capacitor can be computed using the formula.

$$X_C = \frac{1}{2\pi FC}$$

(NOTE: Refer to the Pure Capacitive Circuits Formula section of the appendix.

$$X_{C1} = \frac{1}{377 \times 0.000050}$$

$$X_{C1} = 53.05 \ \Omega$$

$$X_{C2} = \frac{1}{377 \times 0.000075}$$

$$X_{C2} = 35.367 \ \Omega$$

$$X_{C3} = \frac{1}{377 \times 0.000020}$$

$$X_{C3} = 132.63 \ \Omega$$

The total capacitive reactance can be found in a manner similar to finding the resistance of parallel resistors.

$$X_{CT} = \frac{1}{\dfrac{1}{X_{C1}} + \dfrac{1}{X_{C2}} + \dfrac{1}{X_{C3}}}$$

$$X_{CT} = \cfrac{1}{\cfrac{1}{53.05} + \cfrac{1}{35.367} + \cfrac{1}{132.63}}$$

$$X_{CT} = \frac{1}{0.05466}$$

$$X_{CT} = 18.295 \ \Omega$$

Now that the total capacitive reactance of the circuit is known and the total reactive power is known, the voltage applied to the circuit can be found using the formula

$$E_T = \sqrt{VARS_{CT} \times X_{CT}}$$

$$E_T = \sqrt{787.08 \times 18.295}$$

$$E_T = 120 \ V$$

In a parallel circuit the voltage must be the same across each branch of the circuit. Therefore, 120 V is applied across each capacitor.

Now that the circuit voltage is known, the amount of total current for the circuit and the amount of current in each branch can be found using Ohm's law.

$$I_{CT} = \frac{E_{CT}}{X_{CT}}$$

$$I_{CT} = \frac{120}{18.295}$$

$$I_{CT} = 6.559 \ A$$

$$I_{C1} = \frac{E_{C1}}{X_{C1}}$$

$$I_{C1} = \frac{120}{53.05}$$

$$I_{C1} = 2.262 \ A$$

$$I_{C2} = \frac{E_{C2}}{X_{C2}}$$

$$I_{C2} = \frac{1}{35.367}$$

$$I_{C2} = 3.393 \text{ A}$$

$$I_{C3} = \frac{E_{C3}}{X_{C3}}$$

$$I_{C3} = \frac{120}{132.63}$$

$$I_{C3} = 0.905 \text{ A}$$

The amount of reactive power for each capacitor can now be computed using Ohm's law.

$$VARS_{C1} = E_{C1} \times I_{C1}$$

$$VARS_{C1} = 120 \times 2.262$$

$$VARS_{C1} = 271.442$$

$$VARS_{C2} = E_{C2} \times I_{C2}$$

$$VARS_{C2} = 120 \times 3.393$$

$$VARS_{C2} = 407.159$$

$$VARS_{C3} = E_{C3} \times I_{C3}$$

$$VARS_{C3} = 120 \times 0.905$$

$$VARS_{C3} = 108.573$$

To make a quick check of the circuit values, add the VARs for all the capacitors and see if they equal the total circuit VARs.

$$VARS_{CT} = VARS_{C1} + VARS_{C2} + VARS_{C3}$$

$$VARS_{CT} = 271.442 + 407.159 + 108.573$$

$$VARS_{CT} = 787.174$$

The slight difference in answers is caused by rounding off of values. The circuit with all values is shown in *Figure 20-16.*

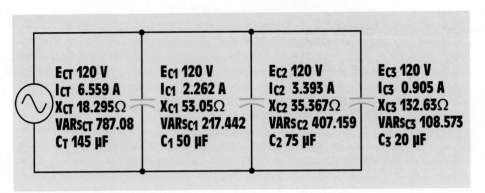

**Figure 20-16** Parallel circuit with completed values

## Summary

1. When a capacitor is connected to an alternating current circuit, current will appear to flow through the capacitor.

2. Current appears to flow through a capacitor because of the continuous increase and decrease of voltage and because of the continuous change of polarity in an AC circuit.

3. The current flow in a pure capacitive circuit is limited by capacitive reactance.

4. Capacitive reactance is proportional to the capacitance of the capacitor and the frequency of the AC line.

5. Capacitive reactance is measured in ohms.

6. In a pure capacitive circuit, the current leads the applied voltage by 90°.

7. There is no true power, or watts, in a pure capacitive circuit.

8. Capacitive power is reactive and is measured in VARs, as is inductance.

9. Capacitive and inductive VARs are 180° out of phase with each other.

10. The Q of a capacitor is the ratio of the resistance to the capacitive reactance.

11. Capacitor voltage ratings are given as volts AC, peak volts, and volts DC.

12. A DC voltage rating for an AC capacitor indicates the peak value of voltage.

## Review Questions

1. Can current flow through a capacitor?

2. What two factors determine the capacitive reactance of a capacitor?

3. How many degrees are the current and voltage out of phase in a pure capacitive circuit?

4. Does the current in a pure capacitive circuit lead or lag the applied voltage?

5. A 30-μF capacitor is connected into a 240-V, 60-Hz circuit. What is the current flow in this circuit?

6. A capacitor is connected into a 1250-V, 1000-Hz circuit. The current flow is 80 A. What is the capacitance of the capacitor?

7. A capacitor is to be connected into a 480-V, 60-Hz line. If the capacitor has a voltage rating of 600 VDC, will the voltage rating of the capacitor be exceeded?

8. On the average, by what factor is the life expectancy of a capacitor increased if the capacitor is operated at half its voltage rating?

9. A capacitor is connected into a 277-V, 400-Hz circuit. The circuit current is 12 A. What is the capacitance of the capacitor?

10. A capacitor has a voltage rating of 350 VAC. Can this capacitor be connected into a 450-VDC circuit without exceeding the voltage rating of the capacitor?

## Practice Problems

## Capacitive Circuits

Fill in all the missing values. Refer to the formulas given below.

$$X_c = \frac{1}{2\pi FC}$$

$$C = \frac{1}{2\pi FX_c}$$

$$F = \frac{1}{2\pi CX_c}$$

| Capacitance | $X_c$ | Frequency |
|---|---|---|
| 38 μF | 69.80 | 60 Hz |
| 5.04 uf | 78.8 Ω | 400 Hz |
| 250 pF | 4.5 kΩ | 141.47Hz |
| 234 μF | 68 mΩ | 10 kHz |
| 1326 uf | 240 Ω | 50 Hz |
| 10 μF | 36.8 Ω | 432.4 8Hz |
| 560 nF | 142 mΩ | 2 MHz |
| 177 nf | 15 kΩ | 60 Hz |
| 75 μF | 560 Ω | 3.78 Hz |

.ovv ovv ovv

| Capacitance | $X_C$ | Frequency |
|---|---|---|
| 470 pF | 1.693 kΩ | 200 kHz |
| 59 pF | 6.8 kΩ | 400 Hz |
| 34 µF | 450 Ω | 10.4 kHz |

**This unit discusses the values of voltage, current, power, and impedance in series circuits that contain elements of both resistance and capacitance.**

Courtesy of Allen-Bradley Co. Inc., a Rockwell International company.

# Resistive-Capacitive Series Circuits

## Outline

## Key Terms

Angle theta $\angle\theta$

Apparent power (VA)

Capacitance (C)

Power factor (PF)

Reactive power (VARs$_C$)

Total current flow (I)

Total impedance (Z)

Total voltage (E$_T$)

True power (P)

Voltage drop across the capacitor (E$_C$)

Voltage drop across the resistor (E$_R$)

## Objectives

*fter studying this unit, you should be able to:*

■ Discuss the relationship of resistance and capacitance in an alternating current series circuit.

■ Calculate values of voltage, current, apparent power, true power, reactive power, impedance, resistance, inductive reactance, and power factor in an RC series circuit.

■ Compute the phase angle for current and voltage in an RC series circuit.

■ Connect an RC series circuit and make measurements using test instruments.

## Preview

In this unit, the relationships of voltage, current, impedance, and power in a resistive-capacitive series circuit will be discussed. As with any type of series circuit, the current flow must be the same through all parts of the circuit. This unit will explore the effect of voltage drop across each component, the relationship of resistance, reactance, and impedance, and the differences between true power, reactive power, and apparent power.

### 21-1 Resistive-Capacitive Series Circuits

When a pure capacitive load is connected to an alternating current circuit, the voltage and current are 90° out of phase with each other. In a capacitive circuit, the current leads the voltage by 90 electrical degrees. When a circuit containing both resistance and capacitance is connected to an alternating current circuit, the voltage and current will be out of phase with each other by some amount between 0° and 90°. The exact amount of phase angle difference is determined by the ratio of resistance to capacitance. Resistive-capacitive series circuits are similar to resistive-inductive series circuits, covered in Unit 17. Other than changing a few formulas, the procedure for solving circuit values is the same.

In the following example, a series circuit containing 12 Ω of resistance and 16 Ω of capacitive reactance is connected to a 240-V, 60-Hz line *(Figure 21-1)*. The following unknown values will be computed.

| | |
|---|---|
| Z | — total circuit impedance |
| I | — total current |
| $E_R$ | — voltage drop across the resistor |
| P | — watts (true power) |
| C | — capacitance |
| $E_C$ | — voltage drop across the capacitor |
| $VARS_C$ | — volt-amperes-reactive (reactive power) |
| VA | — volt-amperes (apparent power) |
| PF | — power factor |
| $\angle\theta$ | — angle theta (the angle the voltage and current are out of phase with each other) |

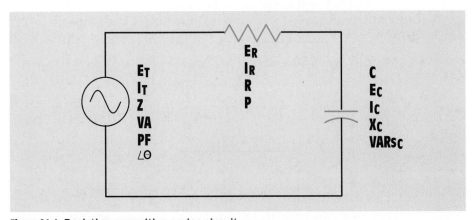

**Figure 21-1**  Resistive-capacitive series circuit

## 21-2 Impedance

The **total impedance (Z)** is the total current-limiting element in the circuit. It is a combination of both resistance and capacitive reactance. Since this is a series circuit, the current-limiting elements must be added. Resistance and capacitive reactance are 90° out of phase with each other,

total
impedance
(Z)

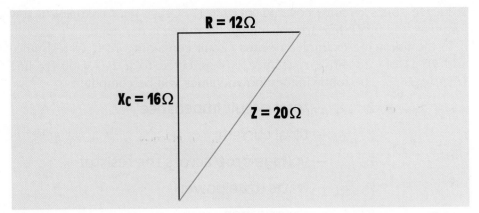

**Figure 21-2** Resistance and capacitive reactance are 90° out of phase with each other.

forming a right triangle with impedance being the hypotenuse *(Figure 21-2)*. A vector diagram illustrating this relationship is shown in *Figure 21-3*. Impedance can be computed using the formula

$$Z = \sqrt{R^2 + X_C^2}$$

$$Z = \sqrt{12^2 + 16^2}$$

$$Z = \sqrt{144 + 256}$$

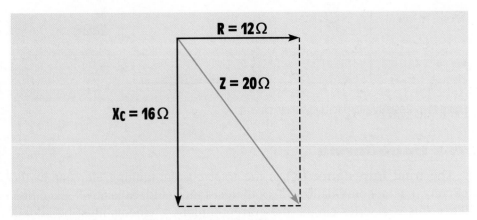

**Figure 21-3** Impedance vector for example circuit 1

$$Z = \sqrt{400}$$

$$Z = 20 \ \Omega$$

## 21-3 Total Current

Now that the total impedance of the circuit is known, the **total current flow (I)** can be computed using the formula

$$I = \frac{E}{Z}$$

$$I = \frac{240}{20}$$

$$I = 12 \ A$$

## 21-4 Voltage Drop across the Resistor

In a series circuit, the current is the same at any point in the circuit. Therefore, 12 A of current flow through both the resistor and the capacitor. The **voltage drop across the resistor ($E_R$)** can be computed by using the formula

$$E_R = I \times R$$

$$E_R = 12 \times 12$$

$$E_R = 144 \ V$$

## 21-5 True Power

**True power (P)** for the circuit can be computed by using any of the watts formulas as long as values that apply only to the resistive part of the circuit are used. Recall that current and voltage must be in phase with each other for true power to be produced. The formula used in this example will be:

$$P = E_R \times I$$

$$P = 144 \times 12$$

$$P = 1728 \ W$$

### 21-6 Capacitance

capaci-
tance (C)

The amount of **capacitance (C)** can be computed using the formula

$$C = \frac{1}{2\pi F X_C}$$

$$C = \frac{1}{377 \times 16}$$

$$C = \frac{1}{6032}$$

$$C = 0.0001658 \text{ F} = 165.8 \ \mu F$$

### 21-7 Voltage Drop across the Capacitor

voltage
drop across
the capac-
itor (E$_C$)

The **voltage drop across the capacitor (E$_C$)** can be computed using the formula

$$E_C = I \times X_C$$

$$E_C = 12 \times 16$$

$$E_C = 192 \text{ V}$$

### 21-8 Total Voltage

total
voltage (E$_T$)

Although the amount of **total voltage (E$_T$)** applied to the circuit is given as 240 V in this circuit, it is possible to compute the total voltage if it is not known by adding the voltage drop across the resistor and the voltage drop across the capacitor together. In a series circuit, the voltage drops across the resistor and capacitor are 90° out of phase with each other, and vector addition must be used. These two voltage drops form the legs of a right triangle, and the total voltage forms the hypotenuse *(Figure 21-4)*. The total voltage can be computed using the formula shown below.

$$E_T = \sqrt{E_R^2 + E_C^2}$$

$$E_T = \sqrt{144^2 + 192^2}$$

$$E_T = 240 \text{ V}$$

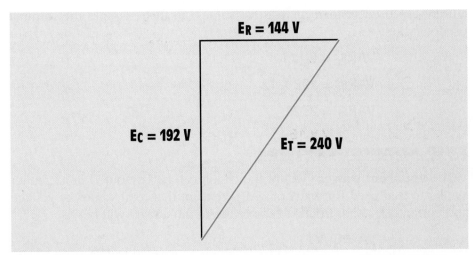

**Figure 21-4** The voltage drops across the resistor and capacitor are 90° out of phase with each other.

A vector diagram illustrating the voltage relationships for this circuit is shown in *Figure 21-5.*

## 21-9 Reactive Power

The **reactive power (VARs$_C$)** in the circuit can be computed in a manner similar to that used for watts except that reactive values of voltage and

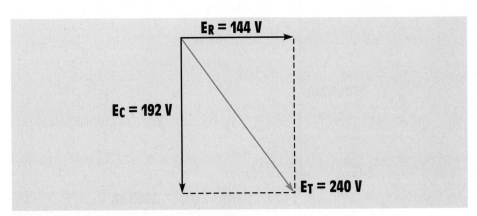

**Figure 21-5** Voltage vector for example circuit

current are used instead of resistive values. In this example the formula used will be

$$VARS_C = E_C \times I$$

$$VARS_C = 192 \times 12$$

$$VARS_C = 2304$$

### 21-10 Apparent Power

**apparent power (VA)**

The **apparent power (VA)** of the circuit can be computed in a manner similar to that used for watts or $VARS_C$, except that total values of voltage and current are used. In this example the formula used will be

$$VA = E_T \times I$$

$$VA = 240 \times 12$$

$$VA = 2880$$

The apparent power can also be determined by vector addition of the true power and reactive power *(Figure 21-6)*.

$$VA = \sqrt{P^2 + VARS_C^2}$$

### 21-11 Power Factor

**power factor (PF)**

**Power factor (PF)** is a ratio of the true power to the apparent power.

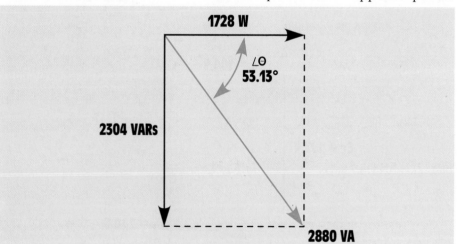

**Figure 21-6** Apparent power vector for example circuit

It can be computed by dividing any resistive value by its like total value. In this circuit the formula used will be

$$PF = \frac{P}{VA}$$

$$PF = \frac{1728}{2880}$$

PF = 0.6 x 100, or 60%

## 21-12 Angle Theta

The power factor of a circuit is the cosine of the phase angle. Since the power factor of this circuit is 0.6, **angle theta ($\angle\theta$)** will be:

**angle theta ($\angle\theta$)**

$$\cos \angle\theta = PF$$

$$\cos \angle\theta = 0.6$$

$$\angle\theta = 51.13°$$

In this circuit, the current leads the applied voltage by 53.13° *(Figure 21-7)*.

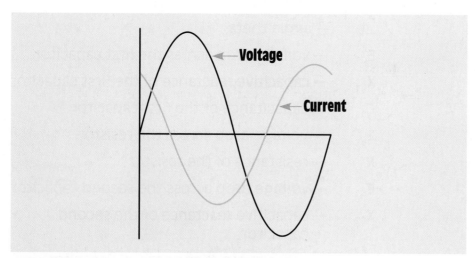

**Figure 21-7** The current leads the voltage by 53.13°.

One resistor and two capacitors are connected in series to a 42.5-V, 60-Hz line *(Figure 21-8)*. Capacitor 1 has a reactive power of 3.75 VARs, the

**Example 1**

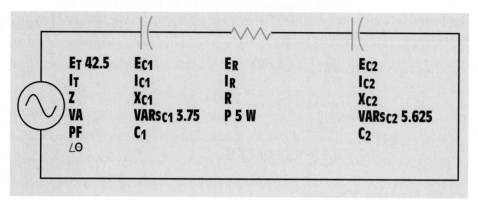

**Figure 21-8** Example circuit

resistor has a true power of 5 W, and the second capacitor has a reactive power of 5.625 VARs. Find the following values.

| | |
|---|---|
| VA | — apparent power |
| $I_T$ | — total circuit current |
| Z | — impedance of the circuit |
| PF | — power factor |
| $\angle\theta$ | — angle theta |
| $E_{C1}$ | — voltage drop across the first capacitor |
| $X_{C1}$ | — capacitive reactance of the first capacitor |
| $C_1$ | — capacitance of the first capacitor |
| $E_R$ | — voltage drop across the resistor |
| R | — resistance of the resistor |
| $E_{C2}$ | — voltage drop across the second capacitor |
| $X_{C2}$ | — capacitive reactance of the second capacitor |
| $C_2$ | — capacitance of the second capacitor |

## Solution

Since the reactive power of the two capacitors is known and the true power of the resistor is known, the apparent power can be found using the formula

$$VA = \sqrt{P^2 + VARs_C{}^2}$$

In this circuit, $VARs_C$ will be sum of the VARs of the two capacitors. A power triangle for this circuit is shown in *Figure 21-9*.

$$VA = \sqrt{5^2 + (3.75 + 5.625)^2}$$

$$VA = \sqrt{25 + 87.891}$$

$$VA = 10.625$$

Now that the apparent power and applied voltage are known, the total circuit current can be computed using the formula

$$I_T = \frac{VA}{E_T}$$

$$I_T = \frac{10.625}{42.5}$$

$$I_T = 0.25 \text{ A}$$

In a series circuit, the current must be the same at any point in the circuit. Therefore, $I_{C1}$, $I_R$, and $I_{C2}$ will all have a value of 0.25 A.

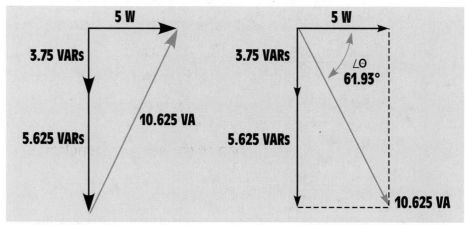

**Figure 21-9**  Power triangle for example circuit

**Figure 21-10**  Vector relationship of reactive, true, and apparent power

The impedance of the circuit can now be computed using the formula

$$Z = \frac{E_T}{I_T}$$

$$Z = \frac{42.5}{0.25}$$

$$Z = 170 \ \Omega$$

The power factor will be computed using the formula

$$PF = \frac{W}{VA}$$

$$PF = \frac{W}{10.625}$$

$$PF = 0.4706, \text{ or } 47.06\%$$

The cosine of angle theta is the power factor.

$$COS \ \angle\theta = 0.4706$$

$$\angle\theta = 61.93°$$

A vector diagram is shown in *Figure 21-10* illustrating the relationship of angle theta to the reactive power, true power, and apparent power.

Now that the current through each circuit element is known and the power of each element is known, the voltage drop across each element can be computed.

$$E_{C1} = \frac{VA}{I_{C1}}$$

$$E_{C1} = \frac{10.625}{0.25}$$

$$E_{C1} = 15 \ V$$

$$E_R = \frac{VA}{I_R}$$

$$E_R = \frac{10.625}{0.25}$$

$$E_R = 20 \ V$$

$$E_{C2} = \frac{VA}{I_{C2}}$$

$$E_{C2} = \frac{10.625}{0.25}$$

$$E_{C2} = 22.5 \text{ V}$$

The capacitive reactance of the first capacitor will be

$$X_{C1} = \frac{E_{C1}}{I_{C1}}$$

$$X_{C1} = \frac{15}{0.25}$$

$$X_{C1} = 60 \ \Omega$$

The capacitance of the first capacitor will be

$$C_1 = \frac{1}{2\pi F X_{C1}}$$

$$C_1 = \frac{1}{377 \times 60}$$

$$C_1 = 0.0000442 \text{ F, or } 44.2 \ \mu\text{F}$$

The resistance of the resistor will be

$$R = \frac{E_R}{I_R}$$

$$R = \frac{20}{0.25}$$

$$R = 80 \ \Omega$$

The capacitive reactance of the second capacitor will be

$$X_{C2} = \frac{E_{C2}}{I_{C2}}$$

$$X_{C2} = \frac{22.5}{0.25}$$

$$X_{C2} = 90 \ \Omega$$

The capacitance of the second capacitor will be

$$C_2 = \frac{1}{2\pi F X_{C2}}$$

$$C_2 = \frac{1}{377 \times 90}$$

$$C_2 = 0.0000295 \text{ F, or } 29.5 \ \mu\text{F}$$

The completed circuit with all values is shown in *Figure 21-11*.

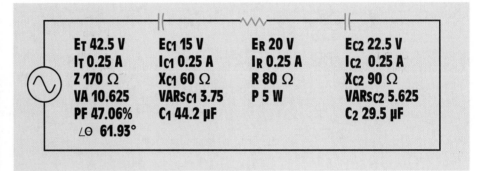

**Figure 21-11** Example circuit with completed values

**Example 2**

A small indicating lamp has a rating of 2 W when connected to 120 V. The lamp must be connected to a voltage of 480 V at 60 Hz. A capacitor will be connected in series with the lamp to reduce the circuit current to the proper value. What value of capacitor will be needed to perform this job?

## Solution

The first step is to determine the amount of current the lamp will normally draw when connected to a 120-V line.

$$I_{LAMP} = \frac{W}{E}$$

$$I_{LAMP} = \frac{2}{120}$$

$$I_{LAMP} = 0.1667 \text{ A}$$

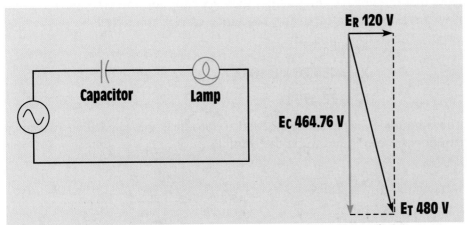

**Figure 21-12**   Determining voltage drop across the capacitor

**Figure 21-13**   The capacitor reduces the current to the lamp.

The next step is to determine the amount of voltage that must be dropped across the capacitor when a current of 0.1667 A flows through it. Since the voltage dropped across the resistor and the voltage dropped across the capacitor are 90° out of phase with each other, vectors must be used to determine the voltage drop across the capacitor *(Figure 21-12)*. The voltage drop across the capacitor can be computed using the formula

$$E_C = \sqrt{E_T^2 - E_R^2}$$

$$E_C = \sqrt{480^2 - 120^2}$$

$$E_C = \sqrt{216,000}$$

$$E_C = 464.76 \text{ V}$$

Now that the voltage drop across the capacitor and the amount of current flow are known, the capacitive reactance can be computed.

$$X_C = E_C \times I$$

$$X_C = 464.76 \times 0.1667$$

$$X_C = 77.475 \ \Omega$$

The amount of capacitance needed to produce this capacitive reactance can now be computed using the formula

$$C = \frac{1}{2\pi F X_C}$$

$$C = \frac{1}{377 \times 77.475}$$

$$C = 0.0000342 \ F, \ \text{or} \ 34.2 \ \mu F$$

The circuit containing the lamp and capacitor is shown in *Figure 21-13*.

## Summary

1. In a pure capacitive circuit the voltage and current are 90° out of phase with each other.

2. In a pure resistive circuit the voltage and current are in phase with each other.

3. In a circuit containing resistance and capacitance the voltage and current will be out of phase with each other by some amount between 0° and 90°.

4. The amount of phase angle difference between voltage and current in an RC series circuit is the ratio of resistance to capacitance.

5. In a series circuit, the current flow through all components is the same. Therefore, the voltage drops across the resistive and capacitive parts become out of phase with each other.

6. True power can be produced by resistive parts of the circuit only.

7. Power factor is the ratio of true power to apparent power.

## Review Questions

Refer to the formulas in the Resistive-Capacitive Series Circuits Formula section of the appendix.

1. In a pure capacitive circuit, does the current lead or lag the voltage?

2. A series circuit contains a 20-$\Omega$ resistor and a capacitor with a capacitance of 110.5 $\mu$F. If the circuit has a frequency of 60 Hz what is the total impedance of the circuit?

3. An RC series circuit has a power factor of 76%. How many degrees are the voltage and current out of phase with each other?

4. An RC series circuit has a total impedance of 84 $\Omega$. The resistor has a value of 32 $\Omega$. What is the capacitive reactance of the capacitor

$$X_C = \sqrt{Z^2 - R^2} \quad ?$$

5. A capacitor has a capacitive reactance of 50 $\Omega$ when connected to a 60-Hz line. What will be the capacitive reactance if the capacitor is connected to a 1000-Hz line?

**Practice Problems**

Refer to the formulas in the Resistive-Capacitive Series Circuits Formula section of the appendix and to *Figure 21-1*.

1. Assume that the circuit shown in *Figure 21-1* is connected to a 480-V, 60-Hz line. The capacitor has a capacitance of 165.782 µF, and the resistor has a resistance of 12 Ω. Find the missing values.

$E_T$ 480 _____     $E_R$ _288_     $E_C$ _384_

$I_T$ _24_     $I_R$ _24_     $I_C$ _24_

Z _20Ω_     R 12 Ω     $X_C$ _16.0_

VA _11520_     P _6912_     $VARS_C$ _9216_

PF _60%_     ∠θ _53.1_     C 165.782 µF _

2. Assume that the voltage drop across the resistor, $E_R$, is 78 V, the voltage drop across the capacitor, $E_C$, is 104 V, and the circuit has a total impedance, Z, of 20 Ω. The frequency of the AC voltage is 60 Hz. Find the missing values.

$E_T$ _130_     $E_R$ 78     $E_C$ 104

$I_T$ _6.5_     $I_R$ _6.5_     $I_C$ _6.5_

Z 20 Ω     R _12_     $X_C$ _16_

VA _845_     P _507_     $VARS_C$ _676_

PF _60_     ∠θ _53.1_     C _165uf_

3. Assume the circuit shown in *Figure 21-1* has an apparent power of 144 VA and a true power of 115.2 W. The capacitor has a capacitance of 6.2833 µF, and the frequency is 60 Hz. Find the missing values.

$E_T$ _318.3Ω_     $E_R$ _254.86_     $E_C$ _190.8_

$I_T$ _.452_     $I_R$ _.452_     $I_C$ _.452_

Z _704.8_     R _563.86_     $X_C$ _422.18_

VA 144     P 115.2     $VARS_C$ _86.4_

PF _80_     ∠θ _36.8°_     C 6.2833 µF

4. Assume the circuit in *Figure 21-1* has a power factor of 68%, an apparent power of 300 VA, and a frequency of 400 Hz. The capacitor has a capacitance of 4.7125 µF. Find the missing values.

$E_T$ _186.3_     $E_R$ _126.7_     $E_C$ _135.8_

$I_T$ _1.61_     $I_R$ _1.61_     $I_C$ _1.619_

$Z$ _115.9_     $R$ _78.7_     $X_C$ _84.38_

VA 300     $P$ _204_     VARs$_C$ _219.96_

PF 68%     $\angle\theta$ _47.1_     C 4.7125 µF

$\acute{E}_L$ series

.00000 4.7125

$I^2 R = P$

$I - 8.43E\ 219.96$
$8.43k$

$I_C$

$E_C$

$I_L$

$Ralld\ I_L$

$E_L$

$I_C$

$E_C$

$I_C$

$I^2 R^{86.4} = 86.4$

This unit discusses the values of voltage, current, power, and impedance in parallel circuits that contain elements of both resistance and capacitance.

*Courtesy of FAA.*

# *Resistive-Capacitive Parallel Circuits*

## Key Terms

Angle theta ($\angle\theta$)

Apparent power (VA)

Circuit impedance (Z)

Current flow through the capacitor ($I_C$)

Current flow through the resistor ($I_R$)

Phase angle shift

Power factor (PF)

Reactive power (VARs)

Total circuit current ($I_T$)

True power (P)

**Objectives**

**A**fter studying this unit, you should be able to:

■ Discuss the operation of a parallel circuit containing resistance and capacitance.

■ Compute circuit values of an RC parallel circuit.

■ Connect an RC parallel circuit and measure circuit values with test instruments.

**Preview**

This unit discusses the relationship of different electrical quantities such as voltage, current, impedance, and power in a circuit that contains both resistance and capacitance connected in parallel. Since all components connected in parallel must share the same voltage, the current flow through different components will be out of phase with each other. The effect this condition has on other circuit quantities will be explored.

### 22-1 Operation of RC Parallel Circuits

When resistance and capacitance are connected in parallel, the voltage across all the devices will be in phase and will have the same value. The current flow through the capacitor, however, will be 90° out of phase with the current flow through the resistor *(Figure 22-1)*. The amount of

**phase
angle shaft**

**phase angle shift** between the total circuit current and voltage is determined by the ratio of the amount of resistance to the amount of capacitance. The circuit power factor is still determined by the ratio of resistance and capacitance.

### 22-2 Computing Circuit Values

**Example 1**

In the RC parallel circuit shown in *Figure 22-2*, assume that a resistance of 30 Ω is connected in parallel with a capacitive reactance of 20 Ω. The circuit is connected to a voltage of 240 VAC and a frequency of 60 Hz. Compute the following circuit values.

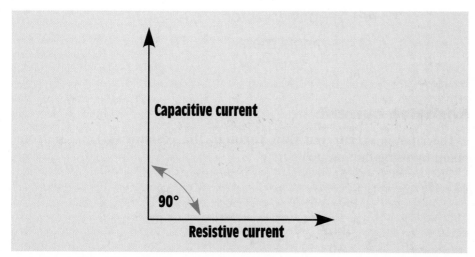

**Figure 22-1** Current flow through the capacitor is 90° out of phase with current flow through the resistor.

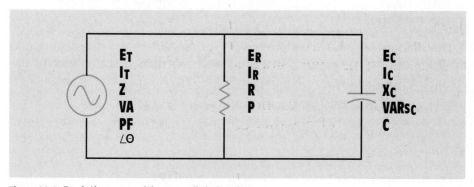

**Figure 22-2** Resistive-capacitive parallel circuit

$I_R$ — current flow through the resistor

P — watts (true power)

$I_C$ — current flow through the capacitor

VARs — volt-amperes-reactive (reactive power)

C — capacitance of the capacitor

$I_T$ — total circuit current

Z — total impedance of the circuit

VA — volt-amperes (apparent power)

$$PF \quad \text{— power factor}$$

$$\angle\theta \quad \text{— angle theta}$$

## Solution

### Resistive Current

The amount of **current flow through the resistor ($I_R$)** can be computed by using the formula

$$I_R = \frac{E}{R}$$

$$I_R = \frac{240}{30}$$

$$I_R = 8 \text{ A}$$

### True Power

The amount of total **true power (P)** in the circuit can be determined by using any of the values associated with the pure resistive part of the circuit. In this example, true power will be found using the formula

$$P = E \times I_R$$

$$P = 240 \times 8$$

$$P = 1920 \text{ W}$$

### Capacitive Current

The amount of **current flow through the capacitor ($I_C$)** will be computed using the formula

$$I_C = \frac{E}{X_C}$$

$$I_C = \frac{240}{20}$$

$$I_C = 12 \text{ A}$$

### Reactive Power

The amount of **reactive power (VARs)** can be found using any of the

---

**current flow through the resistor ($I_R$)**

**true power (P)**

**current flow through the capacitor ($I_C$)**

**reactive power (VARs)**

total capacitive values. In this example VARs will be computed using the formula

$$\text{VARs} = E \times I_C$$

$$\text{VARs} = 240 \times 12$$

$$\text{VARs} = 2880$$

## Capacitance

The capacitance of the capacitor can be computed using the formula

$$C = \frac{1}{2\pi F X_C}$$

$$C = \frac{1}{377 \times 20}$$

$$C = \frac{1}{7540}$$

$$C = 0.0001326 \text{ F} = 132.6 \, \mu\text{F}$$

## Total Current

The voltage is the same across all legs of a parallel circuit. The current flow through the resistor is in phase with the voltage, and the current flow through the capacitor is leading the voltage by 90° *(Figure 22-3)*. The 90°

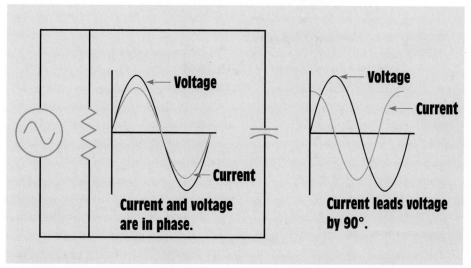

**Figure 22-3** Phase relationship of current and voltage in an RC parallel circuit

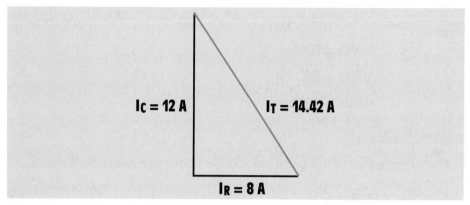

**Figure 22-4** Resistance current and capacitive current are 90° out of phase with each other.

difference in capacitive and resistive current forms a right triangle as shown in *Figure 22-4*. Since these two currents are connected in parallel, vector addition can be used to find the total current flow in the circuit (*Figure 22-5*). The **total circuit current ($I_T$)** flow can be computed by using the formula

**total circuit current ($I_T$)**

$$I_T = \sqrt{I_R^2 + I_C^2}$$

$$I_T = \sqrt{8^2 + 12^2}$$

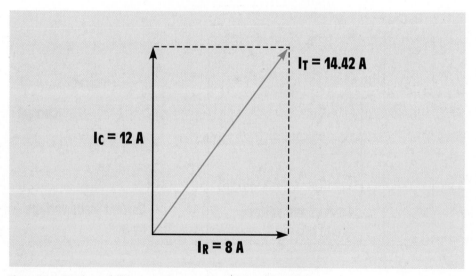

**Figure 22-5** Vector addition can be used to find total current.

$$I_T = \sqrt{64 + 144}$$

$$I_T = \sqrt{208}$$

$$I_T = 14.42 \text{ A}$$

## Impedance

The total **circuit impedance (Z)** can be found by using any of the total values and substituting Z for R in an Ohm's law formula. The total impedance of this circuit will be computed using the formula

$$Z = \frac{E}{I_T}$$

$$Z = \frac{240}{14.42}$$

$$Z = 16.64 \ \Omega$$

circuit
impedance
(Z)

The impedance can also be found by adding the reciprocals of the resistance and capacitive reactance. Since the resistance and capacitive reactance are 90° out of phase with each other, vector addition must be used.

$$Z = \frac{1}{\sqrt{\left(\dfrac{1}{R}\right)^2 + \left(\dfrac{1}{X_C}\right)^2}}$$

## Apparent Power

The **apparent power (VA)** can computed by multiplying the circuit voltage by the total current flow.

$$VA = E \times I_T$$

$$VA = 240 \times 14.42$$

$$VA = 3460.8$$

apparent
power (VA)

## Power Factor

The **power factor (PF)** is the ratio of true power to apparent power.

power
factor (PF)

The circuit power factor can be computed using the formula

$$PF = \frac{W}{VA} \times 100$$

$$PF = \frac{1920}{3460.8}$$

$$PF = 0.5548, \text{ or } 55.48\%$$

### Angle Theta

The cosine of **angle theta (∠θ)** is equal to the power factor.

$$COS \angle\theta = 0.5548$$

$$\angle\theta = 56.3°$$

A vector diagram of apparent, true, and reactive power is shown in *Figure 22-6*. Angle theta is the angle developed between the apparent and true power. The complete circuit with all values is shown in *Figure 22-7*.

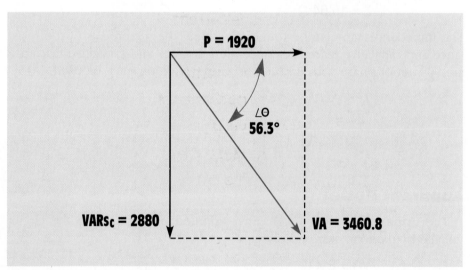

**Figure 22-6** Vector relationship of apparent, true, and reactive power

In this circuit a resistor and a capacitor are connected in parallel to a 400-Hz line. The power factor is 47.05%, the apparent power is 4086.13 VA, and the capacitance of the capacitor is 33.15 μF *(Figure 22-8)*. Find the following unknown values.

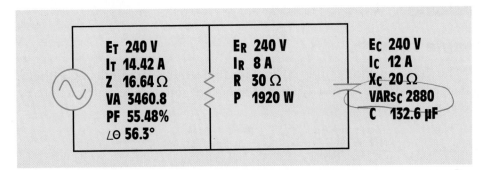

**Figure 22-7** Example circuit with all computed values

$\angle\theta$ — angle theta

P — true power

$VARS_C$ — capacitive VARs

$X_C$ — capacitive reactance

$E_C$ — voltage drop across the capacitor

$I_C$ — capacitive current

$E_R$ — voltage drop across the resistor

$I_R$ — resistive current

R — resistance of the resistor

$E_T$ — applied voltage

$I_T$ — total circuit current

Z — impedance of the circuit

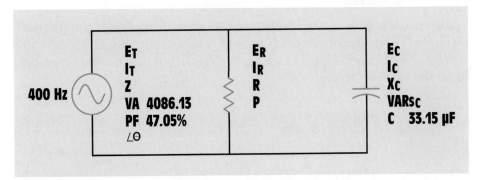

**Figure 22-8** Example circuit

## Solution

### Angle Theta

The power factor is the cosine of angle theta. To find angle theta, change the power factor from a percentage into a decimal fraction by dividing by 100.

$$PF = \frac{47.05}{100}$$

$$PF = 0.4705$$

$$COS \angle\theta = 0.4705$$

$$\angle\theta = 61.93°$$

### True Power

The power factor is determined by the ratio of true power to apparent power.

$$PF = \frac{P}{VA}$$

This formula can be changed to compute the true power when the power factor and apparent power are known (refer to the Resistive-Capacitive Parallel Circuits Formula Section of the appendix).

$$P = VA \times PF$$

$$P = 4086.13 \times 0.4705$$

$$P = 1922.53 \text{ W}$$

### Reactive Power

The apparent power, true power, and reactive power form a right triangle as shown in *Figure 22-9*. Since these powers form a right triangle, the Pythagorean theorem can be used to find the leg of the triangle represented by the reactive power.

$$VARs_C = \sqrt{VA^2 - P^2}$$

$$VARs_C = \sqrt{4086.13^2 - 1922.53^2}$$

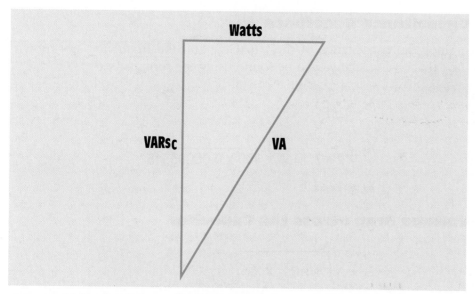

**Figure 22-9** Right triangle formed by the apparent, true, and reactive powers

$$\text{VARs}_C = \sqrt{13,000,337}$$

$$\text{VARs}_C = 3605.60$$

A vector diagram showing the relationship of apparent power, true power, reactive power, and angle theta is shown in *Figure 22-10*.

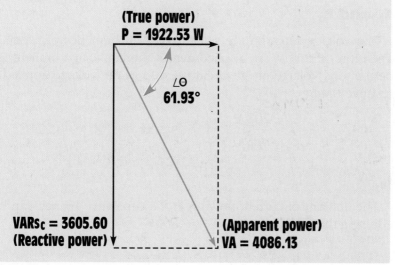

**Figure 22-10** Vector diagram of apparent power, true power, reactive power, and angle theta

## Capacitance Reactance

Since the capacitance of the capacitor and the frequency are known, the capacitive reactance can be found using the formula

$$X_C = \frac{1}{2\pi FC}$$

$$X_C = \frac{1}{2 \times 3.1416 \times 400 \times 0.00003315}$$

$$X_C = 12 \, \Omega$$

## Voltage Drop across the Capacitor

$$E_C = \sqrt{VARS_C \times X_C}$$

$$E_C = \sqrt{3605.60 \times 12}$$

$$E_C = \sqrt{43{,}267.2}$$

$$E_C = 208 \, V$$

## $E_R$ and $E_T$

The voltage must be the same across all branches of a parallel circuit. Therefore, if 208 V are applied across the capacitive branch, 208 V must be the total voltage of the circuit as well as the voltage applied across the resistive branch.

$$E_T = 208 \, V$$

$$E_R = 208 \, V$$

## $I_C$

The amount of current flowing in the capacitive branch can be computed using the formula

$$I_C = \frac{E_C}{X_C}$$

$$I_C = \frac{208}{12}$$

$$I_C = 17.33 \text{ A}$$

## $I_R$

The amount of current flowing through the resistor can be computed using the formula

$$I_R = \frac{P}{E_R}$$

$$I_R = \frac{1922.53}{208}$$

$$I_R = 9.243 \text{ A}$$

## Resistance

The amount of resistance can be computed using the formula

$$R = \frac{E_R}{I_R}$$

$$R = \frac{208}{9.243}$$

$$R = 22.5 \; \Omega$$

## Total Current

The total current can be computed using Ohm's law or by vector addition since both the resistive and capacitive currents are known. Vector addition will be used in this example.

$$I_T = \sqrt{I_R^2 + I_C^2}$$

$$I_T = \sqrt{9.243^2 + 17.33^2}$$

$$I_T = \sqrt{385.762}$$

$$I_T = 19.641 \text{ A}$$

## Impedance

The impedance of the circuit will be computed using the formula

$$Z = \frac{1}{\sqrt{\left(\dfrac{1}{R}\right)^2 + \left(\dfrac{1}{X_C}\right)^2}}$$

$$Z = \frac{1}{\sqrt{\left(\dfrac{1}{22.5}\right)^2 + \left(\dfrac{1}{12}\right)^2}}$$

$$Z = \frac{1}{\sqrt{0.00892}}$$

$$Z = 10.588 \ \Omega$$

The complete circuit with all values is shown in *Figure 22-11*.

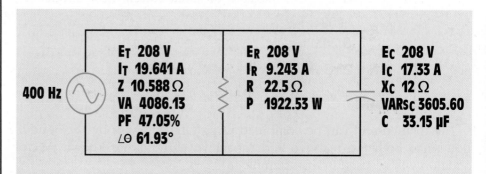

**Figure 22-11** Example circuit with completed values

## Summary

1. The current flow in the resistive part of the circuit is in phase with the voltage.

2. The current flow in the capacitive part of the circuit leads the voltage by 90°.

3. The amount the current and voltage are out of phase with each other is determined by the ratio of resistance to capacitance.

4. The voltage is the same across any leg of a parallel circuit.

5. The circuit power factor is the ratio of true power to apparent power.

## Review Questions

1. When a capacitor and resistor are connected in parallel, how many degrees out of phase are the current flow through the resistor and the current flow through the capacitor?

2. A capacitor and resistor are connected in parallel to a 120-V, 60-Hz line. The resistor has a resistance of 40 Ω, and the capacitor has a capacitance of 132.6 µF. What is the total current flow through the circuit?

3. What is the impedance of the circuit in question 2?

4. What is the power factor of the circuit in question 2?

5. How many degrees out of phase are the current and voltage in question 2?

**Practice Problems**

Refer to the formulas in the Resistive-Capacitive Parallel Circuits section of the appendix and to *Figure 22-2*.

1. Assume that the circuit shown in *Figure 22-2* is connected to a 60-Hz line and has a total current flow of 10.463 A. The capacitor has a capacitance of 132.626 µF, and the resistor has a resistance of 14 Ω. Find the missing values.

$E_T$ _120v_          $E_R$ _120_          $E_C$ _120_

$I_T$ 10.463          $I_R$ _8.57_          $I_C$ _6_

Z _11.4_             R 14 Ω             $X_C$ _20.0_

VA _1255.56_         P _1028.4_         $VARS_C$ _720_

PF _81%_             ∠θ _35°_           C 132.626 µF

2. Assume that the circuit is connected to a 400-Hz line and has a total impedance of 21.6 Ω. The resistor has a resistance of 36 Ω, and the capacitor has a current flow of 2 A through it. Find the missing values.

$E_T$ _54_           $E_R$ _54_           $E_C$ _54_

$I_T$ _2.5_          $I_R$ _1.5_          $I_C$ 2

Z 21.6 Ω            R 36 Ω              $X_C$ _27_

VA _135_            P _81_              $VARS_C$ _108_

PF _60%_            ∠θ _53.1_           C _44.7µF_

3. Assume that the circuit shown in *Figure 22-2* is connected to a 600-Hz line and has a current flow through the resistor of 65.6 A and a current flow through the capacitor of 124.8 A. The total impedance of the circuit is 2.17888 Ω. Find the missing values.

$E_T$ _307_          $E_R$ _307_         $E_C$ _307_

$I_T$ _140.9_        $I_R$ 65.6 A        $I_C$ 124.8

Z 2.17888 Ω         R _4.67_            $X_C$ _2.45_

VA 1325603     P 20139.2     VARs$_C$ 38313.6

PF 46     ∠θ 62%     C 1.08uF

C = $\frac{XL}{}$

4. Assume that the circuit shown in *Figure 22-2* is connected to a 1000-Hz line and has a true power of 486.75 W and a reactive power of 187.5 VARs. The total current flow in the circuit is 7.5 A. Find the missing values.

$E_T$ 69.54     $E_R$ 69.54     $E_C$ 69.54

$I_T$ 7.5 A     $I_R$ 6.99     $I_C$ 2.696

Z 9.27     R 9.94     $X_C$ 25.85

VA 521.6     P 486.75     VARs$_C$ 187.5

PF 93%     ∠θ 21.0°     C 6.15uF

$I^2 \times R = P$

$I^2$     $I^2$

$C = 2.\pi \cdot 1000 \cdot 25.85$

$21.6 = \frac{36 \times X_C}{\sqrt{36^2 + X_C^2}}$  21.59

C

$21.6 = \frac{1}{\sqrt{\left(\frac{1}{36}\right)^2 + \left(\frac{1}{X_C}\right)^2} - \left(\frac{1}{36}\right)^2}$

$21.6\left(\frac{1}{36}\right)^2 =$

*section* **8**

# alternating current circuits containing resistance-inductance-capacitance

**This unit discusses the values of voltage, current, power, and impedance in series circuits that contain elements of resistance, inductance, and capacitance.**

*Courtesy of Motoman, Inc.*

# Resistive-Inductive-Capacitive Series Circuits

## Key Terms

Bandwidth

Lagging power factor

Leading power factor

Resonance

**A**fter studying this unit, you should be able to:

■ Discuss alternating current circuits that contain resistance, inductance, and capacitance connected in series.

■ Connect an RLC series circuit.

■ Compute values of impedance, inductance, capacitance, power, VARs, reactive power, voltage drop across individual components, power factor, and phase angle of voltage and current.

■ Discuss series resonant circuits.

Circuits containing resistance, inductance, and capacitance connected in series will be presented in this unit. Electrical quantities for voltage drop, impedance, and power will be computed for the total circuit values and for individual components. Circuits that become resonant at a certain frequency will be presented as well as the effect a resonant circuit has on electrical quantities such as voltage, current, and impedance.

### 23-1 RLC Series Circuits

When an alternating current circuit contains elements of resistance, inductance, and capacitance connected in series, the **current is the same** through all components, but the **voltages dropped across the elements are out of phase** with each other. The voltage dropped across the resistance will be in phase with the current; the voltage dropped across the inductor will lead the current by 90°; and the voltage dropped across the capacitor will lag the current by 90° *(Figure 23-1)*. An RLC series circuit is shown in *Figure 23-2*. The ratio of resistance, inductance, and capacitance will determine how much the applied voltage will lead or lag the circuit current. If the circuit contains more inductive VARs than capacitive

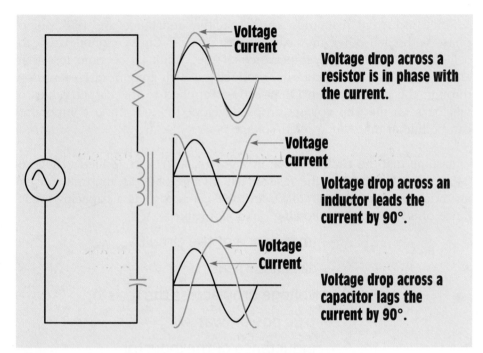

**Figure 23-1** Voltage and current relationship in an RLC series circuit

VARs, the current will lag the applied voltage and the power factor will be a **lagging power factor**. If there are more capacitive VARs than inductive VARs, the current will lead the voltage, and the power factor will be a **leading power factor**.

lagging
power
factor

leading
power
factor

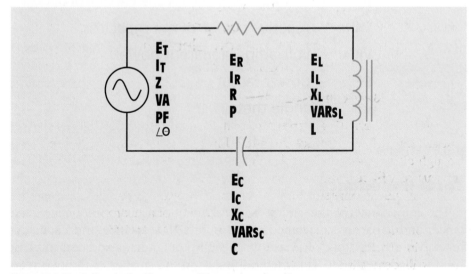

**Figure 23-2** Resistive-inductive-capacitive series circuit

Since inductive reactance and capacitive reactance are 180° out of phase with each other, they cancel each other in an AC circuit. This cancellation can permit the impedance of the circuit to become less than either or both of the reactances, producing a high amount of current flow through the circuit. When Ohm's law is applied to the circuit values, it will be seen that the voltage drops developed across these components can be higher than the applied voltage.

**Example 1**

Assume that the circuit shown in *Figure 23-2* has an applied voltage of 240 V at 60 Hz and that the resistor has a value of 12 Ω, the inductor has an inductive reactance of 24 Ω, and the capacitor has a capacitive reactance of 8 Ω. Find the following unknown values.

| | |
|---|---|
| Z | — impedance of the circuit |
| I | — circuit current |
| $E_R$ | — voltage drop across the resistor |
| P | — true power (watts) |
| L | — inductance of the inductor |
| $E_L$ | — voltage drop across the inductor |
| $VARS_L$ | — reactive power of the inductor |
| C | — capacitance |
| $E_C$ | — voltage drop across the capacitor |
| $VARS_C$ | — reactive power of the capacitor |
| VA | — volt-amps (apparent power) |
| PF | — power factor |
| $\angle\theta$ | — angle theta |

## Solution

### Total Impedance

The impedance of the circuit is the sum of resistance, inductive reactance, and capacitive reactance. Since inductive reactance and capacitive reactance are 180° out of phase with each other, vector addition must be used to find their sum. This method results in the smaller of the two reactive values being subtracted from the larger *(Figure 23-3)*. The smaller

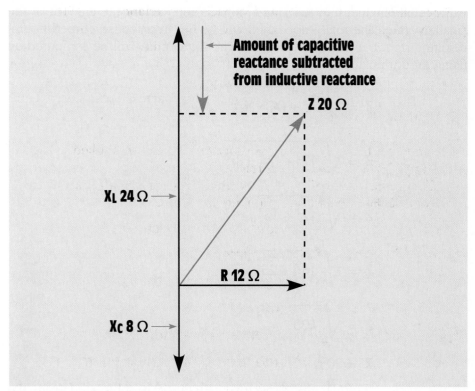

**Figure 23-3**  Vector addition is used to determine impedance.

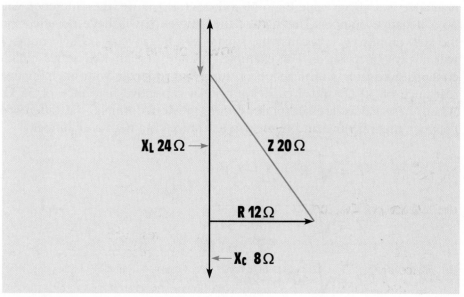

**Figure 23-4**  Right triangle formed by circuit impedance

value is eliminated, and the larger value is reduced by the amount of the smaller value. The total impedance will be the hypotenuse formed by the resulting right triangle *(Figure 23-4)*. The impedance will be computed by using the formula

$$Z = \sqrt{R^2 + (X_L - X_C)^2}$$

$$Z = \sqrt{12^2 + (24 - 8)^2}$$

$$Z = \sqrt{12^2 + 16^2}$$

$$Z = \sqrt{144 + 256}$$

$$Z = \sqrt{400}$$

$$Z = 20 \ \Omega$$

In the above formula, the capacitive reactance is subtracted from the inductive reactance and then the difference is squared. If the capacitive reactance is a larger value than the inductive reactance, the difference will be a negative number. The sign of the difference will have no effect on the answer, however, because the square of a negative or positive number will always be positive. For example, assume that an RLC series circuit contains a resistor with a value of 10 $\Omega$, an inductor with an inductive reactance of 30 $\Omega$, and a capacitor with a capacitive reactance of 54 $\Omega$. When these values are substituted in the previous formula, the difference between the inductive and capacitive reactances is a negative number.

$$Z = \sqrt{R^2 + (X_L - X_C)^2}$$

$$Z = \sqrt{10^2 + (30 - 54)^2}$$

$$Z = \sqrt{10^2 + (-24)^2}$$

$$Z = \sqrt{100 + 576}$$

$$Z = \sqrt{676}$$

$$Z = 26\ \Omega$$

## Current

The total current flow through the circuit can now be computed using the formula

$$I = \frac{E}{Z}$$

$$I = \frac{240}{20}$$

$$I = 12\ A$$

In a series circuit, the current flow is the same at any point in the circuit. Therefore, 12 A flow through each of the circuit components.

## Resistive Voltage Drop

The voltage drop across the resistor can be computed using the formula

$$E_R = I \times R$$

$$E_R = 12 \times 12$$

$$E_R = 144\ V$$

## Watts

The true power of the circuit can be computed using any of the pure resistive values. In this example, true power will be found using the formula

$$P = E_R \times I$$

$$P = 144 \times 12$$

$$P = 1728\ W$$

## Inductance

The amount of inductance in the circuit can be computed using the formula

$$L = \frac{X_L}{2\pi F}$$

$$L = \frac{24}{377}$$

$$L = 0.0637 \text{ H}$$

## Voltage Drop across the Inductor

The amount of voltage drop across the inductor can be computed using the formula

$$E_L = I \times X_L$$

$$E_L = 12 \times 24$$

$$E_L = 288 \text{ V}$$

Notice that the voltage drop across the inductor is greater than the applied voltage.

## Inductance VARs

The amount of reactive power of the inductor can be computed by using inductive values.

$$VARS_L = E_L \times I$$

$$VARS_L = 288 \times 12$$

$$VARS_L = 3456$$

## Capacitance

The amount of capacitance in the circuit can be computed by using the formula

$$C = \frac{1}{2\pi F X_C}$$

$$C = \frac{1}{377 \times 8}$$

$$C = \frac{1}{3016}$$

$$C = 0.0003316 \text{ F, or } 331.6 \,\mu\text{F}$$

## Voltage Drop across the Capacitor

The voltage dropped across the capacitor can be computed using the formula

$$E_C = I \times X_C$$

$$E_C = 12 \times 8$$

$$E_C = 96 \text{ V}$$

## Capacitive VARs

The amount of capacitive VARs can be computed using the formula

$$VARS_C = E_C \times I$$

$$VARS_C = 96 \times 12$$

$$VARS_C = 1152$$

## Apparent Power

The volt-amps (apparent power) can be computed by multiplying the applied voltage and the circuit current.

$$VA = E_T \times I$$

$$VA = 240 \times 12$$

$$VA = 2880$$

The apparent power can also be found by vector addition of true power, inductive VARs, and capacitive VARs *(Figure 23-5)*. As with the addition of resistance, inductive reactance, and capacitive reactance, inductive VARs, $VARS_L$ and capacitive VARs, $VARS_C$ are 180° out of phase with each other. The result is the elimination of the smaller and a reduction of the larger. The formula shown below can be used to determine apparent power.

$$VA = \sqrt{P^2 + (VARS_L - VARS_C)^2}$$

$$VA = \sqrt{1728^2 + (3456 - 1152)^2}$$

$$VA = \sqrt{1728^2 + 2304^2}$$

$$VA = \sqrt{8{,}294{,}400}$$

$$VA = 2880$$

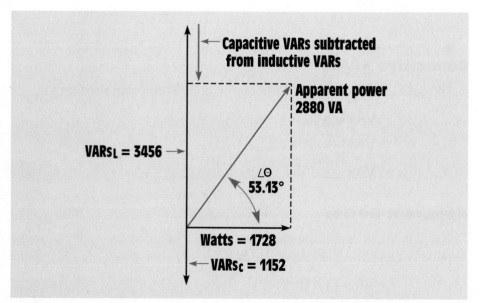

**Figure 23-5** Vector addition of apparent power, true power, and reactive power

## Power Factor

The power factor can be computed by dividing the true power of the circuit by the apparent power. The answer is multiplied by 100 to change the decimal into a percent.

$$PF = \frac{W}{VA} \times 100$$

$$PF = \frac{1728}{2880} \times 100$$

PF = 0.60 x 100

PF = 60%

## Angle Theta

The power factor is the cosine of angle theta.

COS $\angle\theta$ = 0.60

$\angle\theta$ = 53.13°

The circuit with all computed values is shown in *Figure 23-6.*

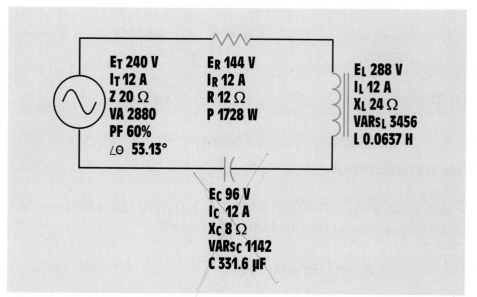

**Figure 23-6** Example circuit with computed values

**Example 2**

An RLC series circuit contains a capacitor with a capacitance of 66.3 µF, an inductor with an inductance of 0.0663 H, and a resistor with a value of 8 Ω connected to a 120-V, 60-Hz line *(Figure 23-7).* How much current will flow in this circuit?

## Solution

The first step in solving this problem is to find the values of capacitive and inductive reactance.

$$X_C = \frac{1}{2\pi FC}$$

$$X_C = \frac{1}{377 \times 0.0000663}$$

$$X_C = 40 \ \Omega$$

$$X_L = 2\pi FL$$

$$X_L = 377 \times 0.0663$$

$$X_L = 24.995 \ \Omega$$

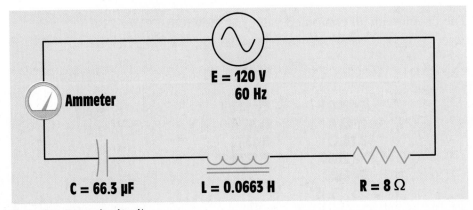

**Figure 23-7** Example circuit

Now that the capacitive and inductive reactance values are known, the circuit impedance can be found using the formula

$$Z = \sqrt{R^2 + (X_L - X_C)^2}$$

$$Z = \sqrt{8^2 + (24.995 - 40)^2}$$

$$Z = \sqrt{64 + 225.15}$$

$$Z = 17 \ \Omega$$

Now that the circuit impedance is known, the current flow can be found using Ohm's law.

$$I_T = \frac{E_T}{Z}$$

$$I_T = \frac{120}{17}$$

$$I_T = 7.059 \text{ A}$$

The RLC series circuit shown in *Figure 23-8* contains an inductor with an inductive reactance of 62 Ω and a capacitor with an inductive reactance of 38 Ω. The circuit is connected to a 208-V, 60-Hz line. How much resistance should be connected in the circuit to limit the circuit current to a value of 8 A?

**Example 3**

## Solution

The first step is to determine the total impedance necessary to limit the circuit current to a value of 8 A.

$$Z = \frac{E_T}{I_T}$$

$$Z = \frac{208}{8}$$

$$Z = 26 \text{ } \Omega$$

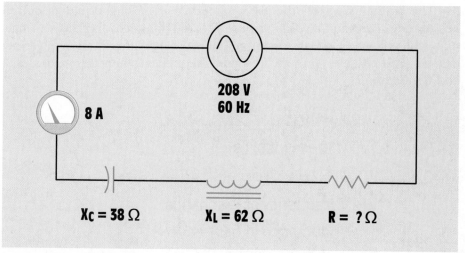

208 V
60 Hz

8 A

$X_C = 38 \text{ } \Omega$     $X_L = 62 \text{ } \Omega$     $R = ? \Omega$

**Figure 23-8** Example circuit

The formula for finding impedance in an RLC series circuit can now be adjusted to find the missing resistance value (refer to the Resistive-

Inductive Capacitive Series Circuits section in appendix D).

$$R = \sqrt{Z^2 - (X_L - X_C)^2}$$

$$R = \sqrt{26^2 - (62 - 38)^2}$$

$$R = \sqrt{26^2 - 24^2}$$

$$R = \sqrt{100}$$

$$R = 10 \ \Omega$$

## 23-2 Series Resonant Circuits

When an inductor and capacitor are connected in series *(Figure 23-9)*, there will be one frequency at which the inductive reactance and capacitive reactance will become equal. The reason for this is that as frequency increases, inductive reactance increases and capacitive reactance decreases. The point at which the two reactances become equal is called **resonance**. Resonant circuits are used to provide great increases of current and voltage at the resonant frequency. The formula shown below can be used to determine the resonant frequency when the values of L and C are known.

resonance

$$F_R = \frac{1}{2\pi \sqrt{LC}}$$

where

$F_R$ = frequency at resonance

L = inductance in henrys

C = capacitance in farads

In the circuit shown in *Figure 23-9*, an inductor has an inductance of 0.0159 H and a wire resistance in the coil of 5 $\Omega$. The capacitor connected in series with the inductor has a capacitance of 1.59 $\mu$F. This circuit will reach resonance at 1000 Hz, when both the inductor and capacitor

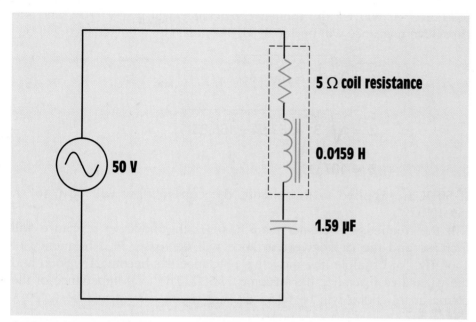

**Figure 23-9** LC series circuit

produce reactances of 100 Ω. At this point, the two reactances are equal and opposite in direction, and the only current-limiting factor in the circuit is the 5 Ω of wire resistance in the coil *(Figure 23-10)*.

During the period of time that the circuit is not at resonance, current flow is limited by the combination of inductive reactance and capacitive

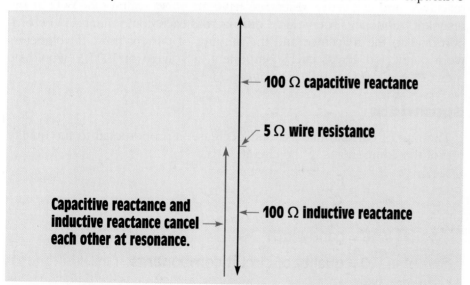

**Figure 23-10**   Inductive reactance and capacitive reactance become equal at resonance.

reactance. At 600 Hz the inductive reactance will be 59.94 $\Omega$, and the capacitive reactance will be 166.83 $\Omega$. The total circuit impedance will be

$$Z = \sqrt{R^2 + (X_L - X_C)^2}$$

$$Z = \sqrt{5^2 + (59.94 - 166.83)^2}$$

$$Z = 107\ \Omega$$

If 50 V are applied to the circuit, the current flow will be 0.467 A (50/107).

If the frequency is greater than 1000 Hz the inductive reactance will increase and the capacitive reactance will decrease. At a frequency of 1400 Hz, for example, the inductive reactance has become 139.86 $\Omega$ and the capacitive reactance has become 71.5 $\Omega$. The total impedance of the circuit at this point will be 68.54 $\Omega$. The circuit current will be 0.729 A (50/68.54).

When the circuit reaches resonance, the current will suddenly increase to 10 A because the only current-limiting factor is the 5 $\Omega$ of wire resistance (50 V/5 $\Omega$ = 10 A). A graph illustrating the effect of current in a resonant circuit is shown in *Figure 23-11*.

Although inductive and capacitive reactance cancel each other at resonance, each is still a real value. In this example, both the inductive reactance and capacitive reactance have an ohmic value of 100 $\Omega$ at the resonant frequency. The voltage drop across each component will be proportional to the reactance and the amount of current flow. If voltmeters were connected across each component, a voltage of 1000 V would be seen (10 A x 100 $\Omega$ = 1000 V) *(Figure 23-12)*.

## Bandwidth

The rate of current increase and decrease is proportional to the quality (Q) of the components in the circuit.

$$B = \frac{F_R}{Q}$$

where

B = bandwidth

Q = quality of circuit components

$F_R$ = frequency at resonance

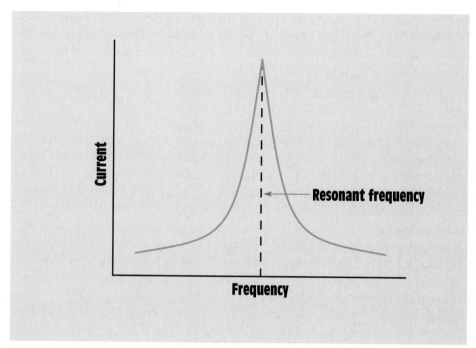

**Figure 23-11**   The current increases sharply at the resonant frequency.

High-Q components result in a sharp increase of current as illustrated by the curve of *Figure 23-11*. Not all series resonant circuits produce as sharp an increase or decrease of current as illustrated in *Figure 23-11*.

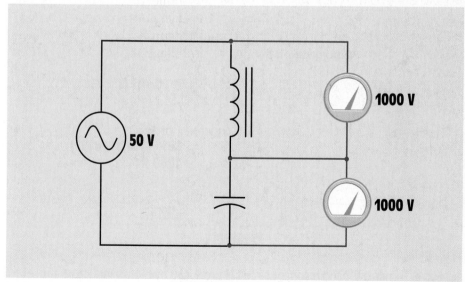

**Figure 23-12**   The voltage drops across the inductor and capacitor increase at resonance.

**bandwidth**

The term used to describe this rate of increase or decrease is **bandwidth**. Bandwidth is a frequency measurement. It is the difference between the two frequencies at which the current is at a value of 0.707 of the maximum current value *(Figure 23-13)*.

$$B = F_2 - F_1$$

Assume the circuit producing the curve in *Figure 23-13* reaches resonance at a frequency of 1000 Hz. Also assume that the circuit reaches a maximum value of 1 A at resonance. The bandwidth of this circuit can be determined by finding the lower and upper frequencies on either side of 1000 Hz at which the current reaches a value of 0.707 of the maximum value. In this illustration that is 0.707 A (1 x 0.707 = 0.707). Assume the lower frequency value to be 995 Hz and the upper value to be 1005 Hz. This circuit has a bandwidth of 10 Hz (1005 – 995 = 10). When resonant circuits are constructed with components that have a relatively high Q, the difference between the two frequencies is small. These circuits are said to have a *narrow* bandwidth.

In a resonant circuit using components with a lower Q rating, the current will not increase as sharply, as shown in *Figure 23-14*. In this circuit, it is assumed that resonance is reached at a frequency of 1200 Hz, and the maximum current flow at resonance is 0.8 A. The bandwidth is deter-

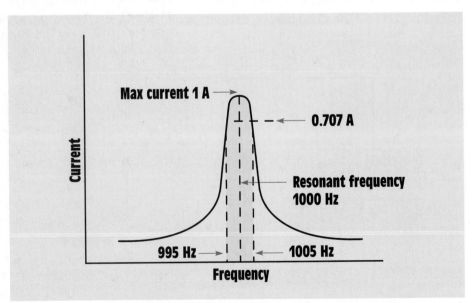

**Figure 23-13** A narrow-band resonant circuit is produced by high-Q inductors and capacitors.

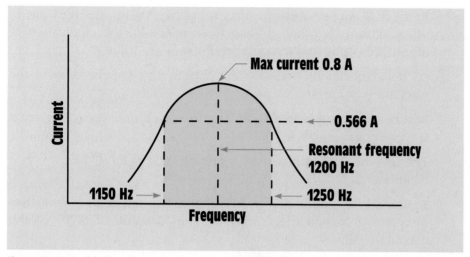

**Figure 23-14** A wide-band resonant circuit is produced by low-Q inductors and capacitors.

mined by the difference between the two frequencies at which the current is at a value of 0.566 A (0.8 x 0.707 = 0.566). Assume the lower frequency to be 1150 Hz, and the upper frequency to be 1250 Hz. This circuit has a bandwidth of 100 Hz (1250 – 1150 = 100). This circuit is said to have a *wide* bandwidth.

## Summary

1. The voltage dropped across the resistor in an RLC series circuit will be in phase with the current.

2. The voltage dropped across the inductor in an RLC series circuit will lead the current by 90°.

3. The voltage dropped across the capacitor in an RLC series circuit will lag the current by 90°.

4. Vector addition can be used in an RLC series circuit to find values of total voltage, impedance, and apparent power.

5. In an RLC circuit, inductive and capacitive values are 180° out of phase with each other. Adding them results in the elimination of the smaller value and a reduction of the larger value.

6. LC resonant circuits increase the current and voltage drop at the resonant frequency.

7. Resonance occurs when inductive reactance and capacitive reactance become equal.

8. The rate the current increases is proportional to the Q of the circuit components.

9. Bandwidth is determined by computing the upper and lower frequencies at which the current reaches a value of 0.707 of the maximum value.

10. Bandwidth is proportional to the Q of the components in the circuit.

## Review Questions

1. What is the phase angle relationship of current and the voltage dropped across a pure resistance?

2. What is the phase angle relationship of current and the voltage dropped across an inductor?

3. What is the phase angle relationship of current and the voltage dropped across a capacitor?

4. An alternating current circuit has a frequency of 400 Hz. A 16-$\Omega$ resistor, a 0.0119-H inductor, and a 16.6-$\mu$F capacitor are connected in series. What is the total impedance of the circuit?

5. If 440 V are connected to the circuit, how much current will flow?

6. How much voltage would be dropped across the resistor, inductor, and capacitor in this circuit?

$E_R$ = _____ V

$E_L$ = _____ V

$E_C$ = _____ V

7. What is the true power of the circuit in question 6?

8. What is the apparent power of the circuit in question 6?

9. What is the power factor of the circuit in question 6?

10. How many degrees are the voltage and current out of phase with each other in the circuit in question 6?

**Practice Problems**

Refer to the Resistive-Inductive-Capacitive Series Circuits Formula section of the appendix and to *Figure 23-2*.

Find all the missing values in the problems below.

1. The circuit shown in *Figure 23-2* is connected to a 120-V, 60-Hz line. The resistor has a resistance of 36 $\Omega$; the inductor has an inductive reactance of 100 $\Omega$; and the capacitor has a capacitive reactance of 52 $\Omega$.

$E_T$ 120        $E_R$ _____        $E_L$ _____        $E_C$ _____

$I_T$ _____        $I_R$ _____        $I_L$ _____        $I_C$ _____

$Z$ _____        $R$ 36 $\Omega$        $X_L$ 100 $\Omega$        $X_C$ 52 $\Omega$

$VA$ _____        $P$ _____        $VARs_L$ _____        $VARs_C$ _____

$PF$ _____        $\angle\theta$ _____        $L$ _____        $C$ _____

2. The circuit is connected to a 400-Hz line with an applied voltage of 35.678 V. The resistor has a true power of 14.4 W, and there are 12.96 inductive VARs and 28.8 capacitive VARs.

$E_T$ 35.678    $E_R$ _____    $E_L$ _____    $E_C$ _____

$I_T$ _____    $I_R$ _____    $I_L$ _____    $I_C$ _____

$Z$ _____    $R$ _____    $X_L$ _____    $X_C$ _____

VA _____    P 14.4    $VARs_L$ 12.96    $VARs_C$ 28.8

PF _____    $\angle\theta$ _____    L _____    C _____

3. The circuit is connected to a 60-Hz line. The apparent power in the circuit is 29.985 VA, and the power factor is 62.5%. The resistor has a voltage drop of 14.993 V; the inductor has an inductive reactance of 60 Ω; and the capacitor has a capacitive reactance of 45 Ω.

$E_T$ _____    $E_R$ 14.993    $E_L$ _____    $E_C$ _____

$I_T$ _____    $I_R$ _____    $I_L$ _____    $I_C$ _____

$Z$ _____    $R$ _____    $X_L$ 60 Ω    $X_C$ 45 Ω

VA 29.985    P _____    $VARs_L$ _____    $VARs_C$ _____

PF 62.5%    $\angle\theta$ _____    L _____    C _____

4. This circuit is connected to a 1000-Hz line. The resistor has a voltage drop of 185 V; the inductor has a voltage drop of 740 V; and the capacitor has a voltage drop of 444 V. The circuit has an apparent power of 51.8 VA.

$E_T$ _____   $E_R$ 185   $E_L$ 740   $E_C$ 444

$I_T$ _____   $I_R$ _____   $I_L$ _____   $I_C$ _____

$Z$ _____   $R$ _____   $X_L$ _____   $X_C$ _____

$VA$   $P$ _____   $VARS_L$ _____   $VARS_C$ _____

$PF$ _____   $\angle\theta$ _____   $L$ _____   $C$ _____

**This unit discusses the values of voltage, current, power, and impedance in parallel circuits that contain elements of resistance, inductance, and capacitance.**

*Courtesy of Tennessee Valley Authority.*

# Resistive-Inductive-Capacitive Parallel Circuits

## Key Terms

Bandwidth

Power factor correction

Tank circuits

Unity

**A**fter studying this unit, you should be able to:

■ Discuss parallel circuits that contain resistance, inductance, and capacitance.

■ Compute the values of an RLC parallel circuit.

■ Compute values of impedance, inductance, capacitance, power, reactive power, current flow through individual components, power factor, and phase angle from measurements taken.

■ Discuss the operation of a parallel resonant circuit.

■ Compute the power factor correction for an AC motor.

Circuits containing elements of resistance, inductance, and capacitance connected in parallel will be discussed in this unit. Electrical quantities of current, impedance, and power will be computed for the entire circuit as well as for individual components. Series resonant circuits, or tank circuits, and their effect on voltage, current, and impedance will also be presented.

### 24-1 RLC Parallel Circuits

When an alternating current circuit contains elements of resistance, inductance, and capacitance connected in parallel, the **voltage dropped across each element is the same**. The **currents flowing through each branch, however, will be out of phase with each other** *(Figure 24-1)*. The current flowing through a pure resistive element will be in phase with the applied voltage. The current flowing through a pure inductive element lags the applied voltage by 90 electrical degrees, and the current flowing through a pure capacitive element will lead the voltage by 90 electrical degrees. The phase angle difference between the

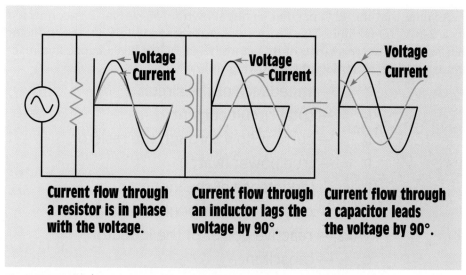

**Current flow through a resistor is in phase with the voltage.**

**Current flow through an inductor lags the voltage by 90°.**

**Current flow through a capacitor leads the voltage by 90°.**

**Figure 24-1** Voltage and current relationship in an RLC parallel circuit. The voltage is the same across each branch, but the currents are out of phase.

applied voltage and the total current is determined by the ratio of resistance, inductance, and capacitance connected in parallel. As with an RLC series circuit, if the inductive VARs is greater than the capacitive VARs, the current will lag the voltage and the power factor will be lagging. If the capacitive VARs is greater the current will lead the voltage and the power factor will be leading.

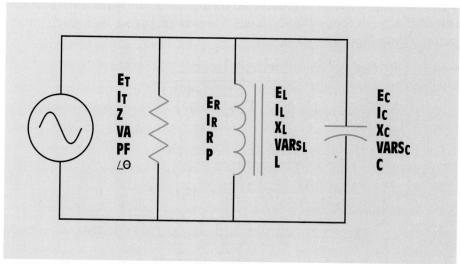

**Figure 24-2** RLC parallel circuit

**Example 1**

Assume that the RLC parallel circuit shown in *Figure 24-2* is connected to a 240-V, 60-Hz line. The resistor has a resistance of 12 Ω, the inductor has an inductive reactance of 8 Ω, and the capacitor has a capacitive reactance of 16 Ω. Complete the following unknown values.

| | |
|---|---|
| Z | — impedance of the circuit |
| $I_T$ | — total circuit current |
| $I_R$ | — current flow through the resistor |
| P | — true power (watts) |
| L | — inductance of the inductor |
| $I_L$ | — current flow through the inductor |
| VARS$_L$ | — reactive power of the inductor |
| C | — capacitance |
| $I_C$ | — current flow through the capacitor |
| VARS$_C$ | — reactive power of the capacitor |
| VA | — volt-amps (apparent power) |
| PF | — power factor |
| $\angle\theta$ | — angle theta |

## Solution

### Impedance

The impedance of the circuit is the reciprocal of the sum of the reciprocals of the legs. Since these values are out of phase with each other, vector addition must be used.

$$Z = \frac{1}{\sqrt{\left(\frac{1}{R}\right)^2 + \left(\frac{1}{X_L} - \frac{1}{X_C}\right)^2}}$$

$$Z = \frac{1}{\sqrt{\left(\frac{1}{12}\right)^2 + \left(\frac{1}{8} - \frac{1}{16}\right)^2}}$$

$$Z = \frac{1}{\sqrt{0.006944 + 0.003906}}$$

$$Z = \frac{1}{\sqrt{0.08085}}$$

$$Z = \frac{1}{0.10416}$$

$$Z = 9.6 \ \Omega$$

## Resistive Current

The next unknown value to be found will be the current flow through the resistor. This can be computed by using the formula

$$I_R = \frac{E}{R}$$

$$I_R = \frac{240}{12}$$

$$I_R = 20 \ A$$

## True Power

The true power, or watts, can be computed using the formula

$$P = E \times I_R$$

$$P = 240 \times 20$$

$$P = 4800 \ W$$

## Inductive Current

The amount of current flow through the inductor can be computed using the formula

$$I_L = \frac{E}{X_L}$$

$$I_L = \frac{240}{8}$$

$$I_L = 30 \ A$$

## Inductive VARs

The amount of reactive power, or VARs, produced by the inductor can be computed using the formula

$$VARS_L = E \times I_L$$

$$VARS_L = 240 \times 30$$

$$VARS_L = 7200$$

## Inductance

The amount of inductance in the circuit can be computed using the formula

$$L = \frac{X_L}{2\pi F}$$

$$L = \frac{8}{377}$$

$$L = 0.0212 \text{ H}$$

## Capacitive Current

The current flow through the capacitor can be computed using the formula

$$I_C = \frac{E}{X_C}$$

$$I_C = \frac{240}{16}$$

$$I_C = 15 \text{ A}$$

## Capacitance

The amount of circuit capacitance can be computed using the formula

$$C = \frac{1}{2\pi F X_C}$$

$$C = \frac{1}{377 \times 16}$$

$$C = 0.0001658 \text{ F} = 165.8 \text{ } \mu F$$

## Capacitive VARs

The capacitive VARs can be computed using the formula

$$VARS_C = E \times I_C$$

$$\text{VARS}_C = 240 \times 15$$

$$\text{VARS}_C = 2400$$

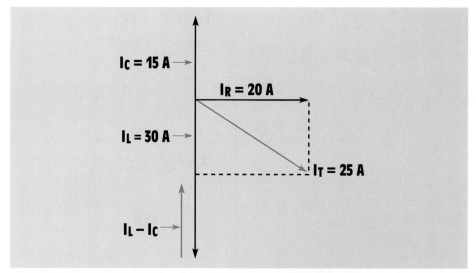

**Figure 24-3** Vector diagram of resistive, inductive, and capacitive currents in example circuit

## Total Circuit Current

The amount of total current flow in the circuit can be computed by vector addition of the current flowing through each leg of the circuit *(Figure 24-3)*. The inductive current is 180° out of phase with the capacitive current. These two currents will tend to cancel each other, resulting in the elimination of the smaller and reduction of the larger. The total circuit current is the hypotenuse of the resulting right triangle *(Figure 24-4)*. The formula shown below can be used to find total circuit current.

$$I_T = \sqrt{I_R^2 + (I_L - I_C)^2}$$

$$I_T = \sqrt{20^2 + (30 - 15)^2}$$

$$I_T = \sqrt{20^2 + 15^2}$$

$$I_T = \sqrt{400 + 225}$$

$$I_T = \sqrt{625}$$

$$I_T = 25 \text{ A}$$

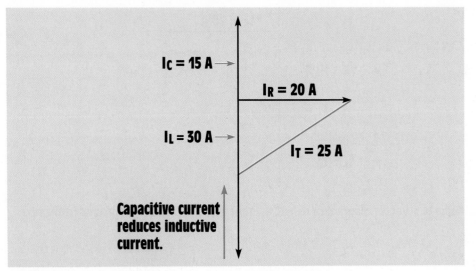

**Figure 24-4** Inductive and capacitive currents cancel each other.

The total current could also be computed by using the value of impedance found earlier in the problem.

$$I_T = \frac{E}{Z}$$

$$I_T = \frac{240}{9.6}$$

$$I_T = 25 \text{ A}$$

## Apparent Power

Now that the total circuit current has been computed, the apparent power, or volt-amps, can be found using the formula

$$VA = E \times I_T$$

$$VA = 240 \times 25$$

$$VA = 6000$$

The apparent power can also be found by vector addition of the true power and reactive power.

$$VA = \sqrt{P^2 + (VARS_L - VARS_C)^2}$$

## Power Factor

The power factor can now be computed using the formula

$$PF = \frac{W}{VA} \times 100$$

$$PF = \frac{4800}{6000} \times 100$$

$$PF = 0.80 \times 100$$

$$PF = 80\%$$

## Angle Theta

The power factor is the cosine of angle theta. Angle theta is therefore

$$COS \angle\theta = 0.80$$

$$\angle\theta = 36.87°$$

The circuit with all completed values is shown in *Figure 24-5.*

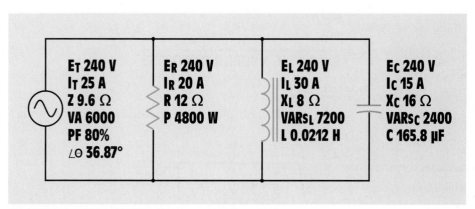

**Figure 24-5** Example circuit with all computed values

**Example 2**

In the circuit shown in *Figure 24-6*, a resistor, an inductor, and a capacitor are connected to a 1200-Hz power source. The resistor has a resistance of 18 Ω, the inductor has an inductance of 9.76 mH (0.00976 H), and the capacitor has a capacitance of 5.5 μF. What is the impedance of this circuit?

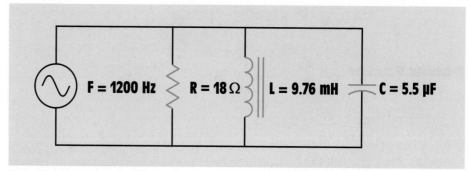

**Figure 24-6** Example circuit

## Solution

The first step in finding the impedance of this circuit is to find the values of inductive and capacitive reactance.

$$X_L = 2\pi FL$$

$$X_L = 2 \times 3.1416 \times 1200 \times 0.00976$$

$$X_L = 73.589 \ \Omega$$

$$X_C = \frac{1}{2\pi FC}$$

$$X_C = \frac{1}{2 \times 3.1416 \times 1200 \times 0.0000055}$$

$$X_C = 24.114 \ \Omega$$

The impedance of the circuit can now be computed using the formula

$$Z = \frac{1}{\sqrt{\left(\dfrac{1}{R}\right)^2 + \left(\dfrac{1}{X_L} - \dfrac{1}{X_C}\right)^2}}$$

$$Z = \cfrac{1}{\sqrt{\left(\cfrac{1}{18}\right)^2 + \left(\cfrac{1}{74.589} - \cfrac{1}{24.114}\right)^2}}$$

$$Z = \cfrac{1}{\sqrt{0.0555^2 + (0.0134 - 0.0415)^2}}$$

$$Z = \cfrac{1}{\sqrt{0.00387}}$$

$$Z = 16.075 \ \Omega$$

## 24-2 Parallel Resonant Circuits

When values of inductive reactance and capacitive reactance become equal, they are said to be resonant. In a parallel circuit, inductive current and capacitive current cancel each other because they are 180° out of phase with each other. This produces minimum line current at the point of resonance. An LC parallel circuit is shown in *Figure 24-7*. LC parallel

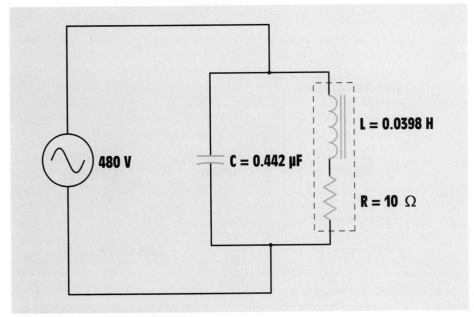

**Figure 24-7** Parallel resonant circuit

**tank circuits**

circuits are often referred to as **tank circuits**. In the example circuit the inductor has an inductance of 0.0398 H and a wire resistance of 10 Ω. The capacitor has a capacitance of 0.442 μF. This circuit will reach resonance at 1200 Hz, when both the capacitor and inductor exhibit reactances of 300 Ω each.

Computing the values for a parallel resonant circuit is a bit more involved than computing the values for a series resonant circuit. In theory, when a parallel circuit reaches resonance, the total circuit current should reach zero and total circuit impedance should become infinite because the capacitive current and inductive currents cancel each other. In practice, the quality (Q) of the circuit components determines total circuit current and, therefore, total circuit impedance. Since capacitors generally have an extremely high Q by their very nature, the Q of the inductor is the determining factor *(Figure 24-8)*.

$$Q = \frac{I_{TANK}}{I_{LINE}}$$

or

$$Q = \frac{X_L}{R}$$

In the example shown in *Figure 24-7*, the inductor has an inductive reactance of 300 Ω at the resonant frequency and a wire resistance of 10 Ω.

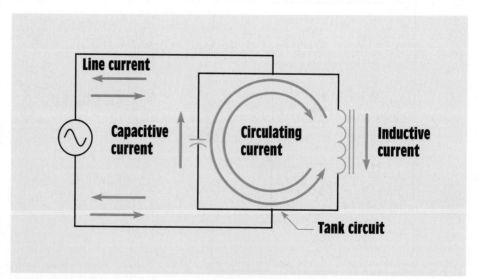

**Line current**

**Capacitive current**

**Circulating current**

**Inductive current**

**Tank circuit**

**Figure 24-8** The amount of current circulating inside the tank is equal to the product of the line current and the Q of the circuit.

The Q of this inductor is 30 at the resonant frequency ($Q = X_L/R$). To determine the total circuit current at resonance it is first necessary to determine the amount of current flow through each of the components at the resonant frequency. Since this is a parallel circuit, the inductor and capacitor will have the alternator voltage of 480 V applied to them. At the resonant frequency of 1200 Hz both the inductor and capacitor will have a current flow of 1.6 A (480 V/300 Ω = 1.6 A). The total current flow in the circuit will be the in-phase current caused by the wire resistance of the coil *(Figure 24-9)*. This value can be computed by dividing the circulating current inside the LC parallel loop by the Q of the circuit. The total current in this circuit will be 0.0533 A (1.6/30 = 0.0533). Now that the total circuit current is known, the total impedance at resonance can be found using Ohm's law.

$$Z = \frac{E}{I_T}$$

$$Z = \frac{480}{0.0533}$$

$$Z = 9006.63 \ \Omega$$

Another method of computing total current for a tank circuit is to determine the true power in the circuit caused by the resistance of the coil. The resistive part of a coil is considered to be in series with the reactive part *(Figure 24-7)*. Since this is a series connection, the current flow

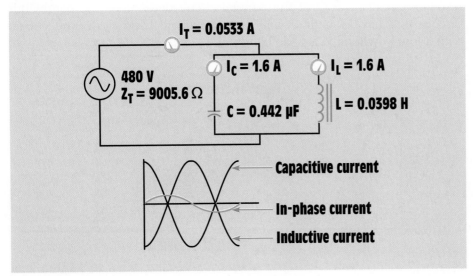

**Figure 24-9** Total current is equal to the in-phase current.

through the inductor is the same for both the reactive and resistive elements. The coil has a resistance of 10 Ω. The true power produced by the coil can be computed by using the formula

$$P = I^2 R$$

$$P = 1.6^2 \times 10$$

$$P = 25.6 \text{ W}$$

Now that the true power is known, the total circuit current can be found using Ohm's law.

$$I_T = \frac{P}{E}$$

$$I_T = \frac{25.6}{480}$$

$$I_T = 0.0533 \text{ A}$$

Graphs illustrating the decrease of current and increase of impedance in a parallel resonant circuit are shown in *Figure 24-10*.

## Bandwidth

The **bandwidth** for a parallel resonant circuit is determined in a manner similar to that used for a series resonant circuit. The bandwidth of a parallel circuit is determined by computing the frequency on either side of resonance at which the impedance is 0.707 of maximum. As in series resonant circuits, the Q of the parallel circuit determines the bandwidth. Circuits that have a high Q will have a narrow bandwidth, and circuits with a low Q will have a wide bandwidth *(Figure 24-11)*.

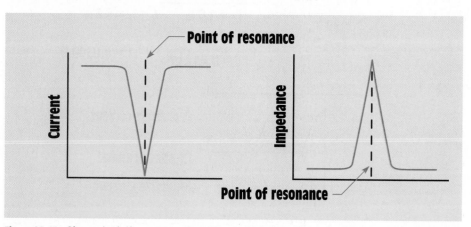

**Point of resonance**

Current

Impedance

**Point of resonance**

**Figure 24-10** Characteristic curves of an LC parallel circuit at resonance

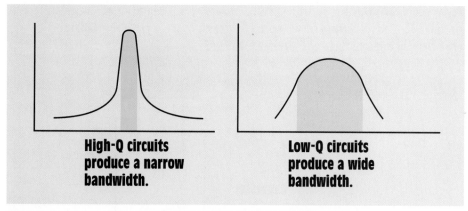

**High-Q circuits produce a narrow bandwidth.**

**Low-Q circuits produce a wide bandwidth.**

**Figure 24-11**  The Q of the circuit determines the bandwidth.

## Induction Heating

The tank circuit is often used when a large amount of current flow is needed. Recall that the formula for Q of a parallel resonant circuit is

$$Q = \frac{I_{TANK}}{I_{LINE}}$$

If this formula is changed, it will be seen that the current circulating inside the tank is equal to the line current times the Q of the circuit:

$$I_{TANK} = I_{LINE} \times Q$$

A high-Q circuit can produce an extremely high current inside the tank with very little line current. A good example of this is an induction heater used to heat pipe for tempering *(Figure 24-12)*. In this example, the coil is the inductor, and the pipe acts as the core of the inductor. The capacitor is connected in parallel with the coil to produce resonance at a desired frequency. The pipe is heated by eddy current induction. Assume that the coil has a Q of 10. If this circuit has a total current of 100 A, then 1000 A of current flow in the tank. This 1000 A is used to heat the pipe. Since the pipe acts as a core for the inductor, the inductance of the coil will change when the pipe is not in the coil. Therefore, the circuit is resonant only during the times that the pipe is in the coil.

Induction heaters of this type have another advantage over other methods that heat pipe with flames produced by oil- or gas-fired furnaces. When induction heating is used, the resonant frequency can be changed by adding or subtracting capacitance in the tank circuit. This ability to control the frequency greatly affects the tempering of metal. If the

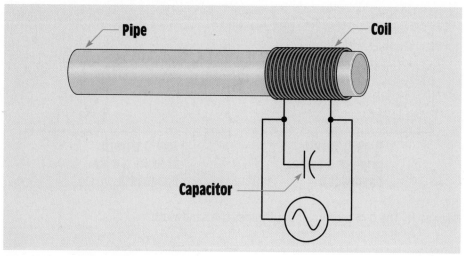

**Figure 24-12** Induction heating system

frequency is relatively low, 400 Hz or less, the metal is heated evenly. If the frequency is increased to 1000 Hz or greater, skin effect causes most of the heating effect to localize at the surface of the metal. This localization at the surface permits a hard coating to develop at the outer surface of the metal without greatly changing the temper of the inside of the metal *(Figure 24-13)*.

## Power Factor Correction

Another common application for LC parallel circuits is **power factor correction**. Assume that a motor is connected to a 240-V single-phase line with a frequency of 60 Hz *(Figure 24-14)*. An ammeter indicates a current flow of 10 A, and a wattmeter indicates a true power of 1630 W

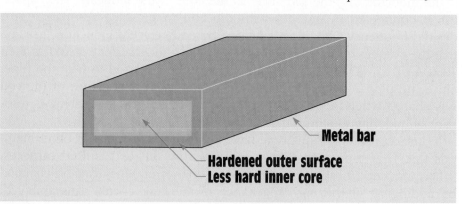

**Figure 24-13** Frequency controls depth of heat penetration.

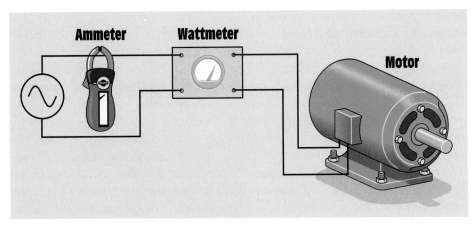

**Figure 24-14**  Determining motor power factor

when the motor is at full load. In this problem, the existing power factor will be determined and then the amount of capacitance needed to correct the power factor will be computed.

Although an AC motor is an inductive device, when it is loaded it must produce true power to overcome the load connected to it. For this reason, the motor appears to be a resistance connected in series with an inductance *(Figure 24-15)*. Also, the inductance of the motor remains constant regardless of the load connected to it. Recall that true power, or watts, can exist only when electrical energy is converted into some other form. A

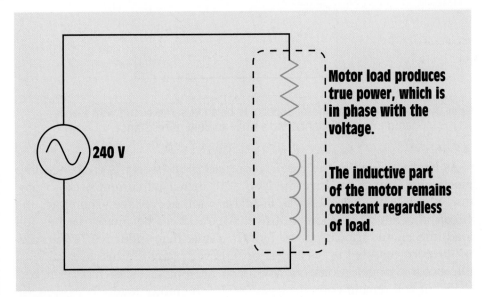

**Figure 24-15**  Equivalent motor circuit

resistor produces true power because it converts electrical energy into heat energy. In the case of a motor, electrical energy is being converted into both heat energy and mechanical energy. When a motor is operated at a no-load condition, the current is relatively small compared to the full-load current. At no load, most of the current is used to magnetize the iron core of the stator and rotor. This current is inductive and is 90° out of phase with the voltage. The only true power produced at no load is caused by motor losses, such as eddy currents being induced into the iron core, the heating effect caused by the resistance of the wire in the windings, hysteresis losses, and the small amount of mechanical energy being produced to overcome the losses of bearing friction and windage. At no load, the motor would appear to be a circuit containing a large amount of inductance and a small amount of resistance *(Figure 24-16)*.

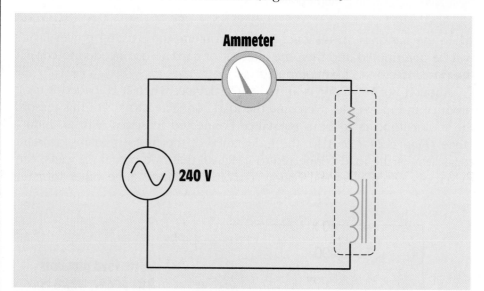

**Figure 24-16** At no load, the motor appears to be an RL series circuit with a large amount of inductance and a small amount of resistance.

As load is added to the motor, more electrical energy is converted into mechanical energy to drive the load. The increased current used to produce the mechanical energy is in phase with the voltage. Therefore, the circuit appears to be more resistive. By the time the motor reaches full load, the circuit appears to be more resistive than inductive *(Figure 24-17)*. Notice that as load is added or removed, only the resistive value of the motor changes. Consequently, once the power factor has been corrected, it will remain constant regardless of the motor load.

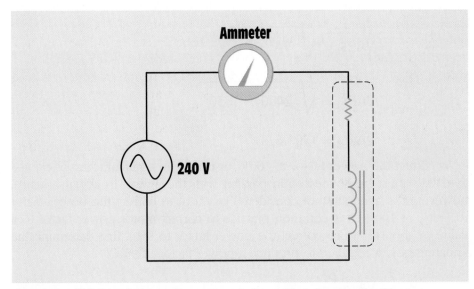

**Figure 24-17**  At full load, the motor appears to be an RL series circuit with a large amount of resistance and a smaller amount of inductance.

To determine the existing motor power factor, compute the apparent power by multiplying the applied voltage and total current.

$$VA = E \times I$$

$$VA = 240 \times 10$$

$$VA = 2400$$

Now that the apparent power is known, the power factor can be computed using the formula

$$PF = \frac{P}{VA}$$

$$PF = \frac{1630}{2400}$$

$$PF = 67.9\%$$

Before the power factor can be corrected, the part of the circuit that is reactive must be determined. The reactive part of the circuit can be determined by finding the reactive power produced by the inductance. Inductive VARs can be computed using the formula

$$VARS_L = \sqrt{VA^2 - P^2}$$

$$VARS_L = \sqrt{2400^2 - 1630^2}$$

$$VARS_L = 1761.6$$

**unity**

To correct the power factor to 100%, or **unity**, an equal amount of capacitive VARs must be connected in parallel with the motor. In actual practice, however, it is generally not considered practical to correct the power factor to unity, or 100%. It is common practice to correct motor power factor to a value of about 95%. To correct the power factor to 95%, first determine the apparent power required to produce a power factor of 95%.

$$VA = \frac{P}{PF}$$

$$VA = \frac{1630}{0.95}$$

$$VA = 1715.789$$

The amount of inductive VARs needed to produce this amount of apparent power can now be determined using the formula

$$VARS_L = \sqrt{VA^2 - P^2}$$

$$VARS_L = \sqrt{1715.789^2 - 1630^2}$$

$$VARS_L = \sqrt{287,031.89}$$

$$VARS_L = 535.754$$

At the present time, the inductive VARs are 1761.6. To find the capacitive VARs needed to produce a total reactive power of 535.754 in the circuit, subtract the amount of reactive power needed from the present amount.

$$VARS_C = VARS_L - 535.754$$

$$VARS_C = 1761.6 - 535.754$$

$$VARS_C = 1225.846$$

To determine the capacitive reactance needed to produce the required reactive power at 240 V, the formula shown below can be used.

$$X_C = \frac{E^2}{VARS_C}$$

$$X_C = \frac{240^2}{1225.846}$$

$$X_C = 46.988\ \Omega$$

The amount of capacitance needed to produce the required capacitive reactance at 60 Hz can be computed using the formula

$$C = \frac{1}{2\pi F X_C}$$

$$C = \frac{1}{2 \times 3.1416 \times 60 \times 46.988}$$

$$C = 56.5\ \mu F$$

The power factor will be corrected to 95% when a capacitor with a capacitance of 56.5 μF is connected in parallel with the motor *(Figure 24-18)*.

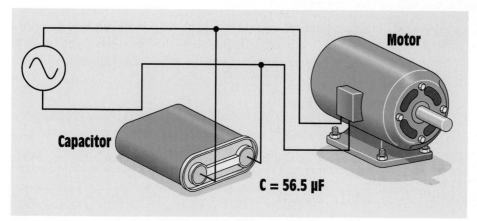

**Motor**

**Capacitor**

$C = 56.5\ \mu F$

**Figure 24-18**   The capacitor corrects the power factor to 95%.

## Summary

1. The voltage applied to all legs of an RLC parallel circuit is the same.

2. The current flow in the resistive leg will be in phase with the voltage.

3. The current flow in the inductive leg will lag the voltage by 90°.

4. The current flow in the capacitive leg will lead the voltage by 90°.

5. Angle theta for the circuit is determined by the amounts of inductance and capacitance.

6. An LC resonant circuit is often referred to as a tank circuit.

7. When an LC parallel circuit reaches resonance, the line current drops and the total impedance increases.

8. When an LC parallel circuit becomes resonant, the total circuit current is determined by the amount of pure resistance in the circuit.

9. Total circuit current and total impedance in a resonant tank circuit are proportional to the Q of the circuit.

10. Motor power factor can be corrected by connecting capacitance in parallel with the motor. The same amount of capacitive VARs must be connected as inductive VARs.

## Review Questions

1. An AC circuit contains a 24-$\Omega$ resistor, a 15.9-mH inductor, and a 13.3-$\mu$F capacitor connected in parallel. The circuit is connected to a 240-V, 400-Hz power supply. Find the following values.

   $X_L =$ _____ $\Omega$

   $X_C =$ _____ $\Omega$

   $I_R =$ _____ A

   $I_L =$ _____ A

$I_C$ = _____ A

P = _____ W

$VARS_L$ = _____

$VARS_C$ = _____

$I_T$ = _____ A

VA = _____

PF = _____ %

$\angle\theta$ = _____ °

2. An RLC parallel circuit contains a resistor with a resistance of 16 Ω, an inductor with an inductive reactance of 8 Ω, and a capacitor with a capacitive reactance of 20 Ω. What is the total impedance of this circuit?

3. The circuit shown in *Figure 24-2* has a current of 38 A flowing through the resistor, 22 A flowing through the inductor, and 7 A flowing through the capacitor. What is the total circuit current?

4. A tank circuit contains a capacitor and an inductor that produce 30 Ω of reactance at the resonant frequency. The inductor has a Q of 15. The voltage of 277 V is connected to the circuit. What is the total circuit current at the resonant frequency?

5. A 0.796-mH inductor produces an inductive reactance of 50 Ω at 10 kHz. What value of capacitance will be needed to produce a resonant circuit at this frequency?

6. An AC motor is connected to a 560-V, 60-Hz line. The motor has a current draw at full load of 53 A. A wattmeter indicates a true power of 18,700 W. Find the power factor of the motor and the amount of capacitance that should be connected in parallel with the motor to correct the power factor to 100%, or unity.

**Practice Problems**

Refer to the Resistive-Inductive-Capacitive Parallel Circuits Formula section of the appendix and to *Figure 24-2*.

Find all the missing values in the problems below.

1. The circuit in *Figure 24-2* is connected to a 120-V, 60-Hz line. The resistor has a resistance of 36 Ω, the inductor has an inductive reactance of 40 Ω, and the capacitor has a capacitive reactance of 50 Ω.

| | | | |
|---|---|---|---|
| $E_T$ 120 | $E_R$ _____ | $E_L$ _____ | $E_C$ _____ |
| $I_T$ _____ | $I_R$ _____ | $I_L$ _____ | $I_C$ _____ |
| Z _____ | R 36 Ω | $X_L$ 40 Ω | $X_C$ 50 Ω |
| VA _____ | P _____ | $VARS_L$ _____ | $VARS_C$ _____ |
| PF _____ | $\angle\theta$ _____ | L _____ | C _____ |

2. The circuit in *Figure 24-2* is connected to a 400-Hz line with a total current flow of 22.627 A. There is a true power of 3840 W, and the inductor has a reactive power of 1920 VARs. The capacitor has a reactive power of 5760 VARs.

| | | | |
|---|---|---|---|
| $E_T$ _____ | $E_R$ _____ | $E_L$ _____ | $E_C$ _____ |
| $I_T$ 22.267 | $I_R$ _____ | $I_L$ _____ | $I_C$ _____ |
| Z _____ | R _____ | $X_L$ _____ | $X_C$ _____ |
| VA _____ | P 3840 | $VARS_L$ 1920 | $VARS_C$ 5760 |
| PF _____ | $\angle\theta$ _____ | L _____ | C _____ |

3. The circuit in *Figure 24-2* is connected to a 60-Hz line. The apparent power in the circuit is 48.106 VA. The resistor has a resistance of 12 Ω. The inductor has an inductive reactance of 60 Ω, and the capacitor has a capacitive reactance of 45 Ω.

| | | | |
|---|---|---|---|
| $E_T$ _____ | $E_R$ _____ | $E_L$ _____ | $E_C$ _____ |
| $I_T$ _____ | $I_R$ _____ | $I_L$ _____ | $I_C$ _____ |
| Z _____ | R 12 Ω | $X_L$ 60 Ω | $X_C$ 45 Ω |
| VA 48.106 | P _____ | $VARS_L$ _____ | $VARS_C$ _____ |
| PF _____ | $\angle\theta$ _____ | L _____ | C _____ |

4. This circuit in *Figure 24-2* is connected to a 1000-Hz line. The resistor has a current flow of 60 A, the inductor has a current flow of 150 A, and the capacitor has a current flow of 70 A. The circuit has a total impedance of 4.8 $\Omega$.

| $E_T$ _____ | $E_R$ _____ | $E_L$ _____ | $E_C$ _____ |
|---|---|---|---|
| $I_T$ _____ | $I_R$ 60 | $I_L$ 150 | $I_C$ 70 |
| Z 4.8 $\Omega$ | R _____ | $X_L$ _____ | $X_C$ _____ |
| VA _____ | P _____ | VARS$_L$ _____ | VARS$_C$ _____ |
| PF _____ | $\angle\theta$ _____ | L _____ | C _____ |

*section*

*three-phase power*

Most of the
electric power
produced in
the world is
three phase.
This unit
discusses the
basics of
three-phase
power gener-
ation and the
most common
three-phase
connections.

*Courtesy of American Petroleum Institute.*

# Unit | 25

# Three-Phase Circuits

## Outline

## Key Terms

Correcting the power factor

Delta connection

Line current

Line voltage

1.732

Phase current

Phase voltage

Star connection

Three-phase watts

Three-phase VARs

Wye connection

*fter studying this unit, you should be able to:*

■ Discuss the differences between three-phase and single-phase voltages.

■ Discuss the characteristics of delta and wye connections.

■ Compute voltage and current values for delta and wye circuits.

■ Connect delta and wye circuits and make measurements with measuring instruments.

■ Compute the amount of capacitance needed to correct the power factor of a three-phase motor.

**Preview**

Most of the electrical power generated in the world today is three-phase. Three-phase power was first conceived by Nikola Tesla. In the early days of electrical power generation, Tesla not only led the battle concerning whether the nation should be powered with low-voltage direct current or high-voltage alternating current, but he also proved that three-phase power was the most efficient way that electricity could be produced, transmitted, and consumed.

### 25-1 Three-Phase Circuits

There are several reasons why three-phase power is superior to single-phase power.

1. The horsepower rating of three-phase motors and the kilovolt-amp rating of three-phase transformers are about 150% greater than for single-phase motors or transformers with a similar frame size.

2. The power delivered by a single-phase system pulsates *(Figure 25-1)*. The power falls to zero three times during each cycle. The power delivered by a three-phase circuit pulsates also, but it never

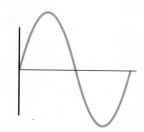

**Figure 25-1** Single-phase power falls to zero three times each cycle.

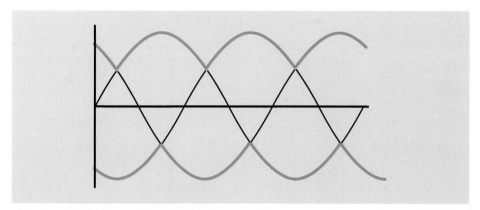

**Figure 25-2** Three-phase power never falls to zero.

falls to zero *(Figure 25-2)*. In a three-phase system, the power delivered to the load is the same at any instant. This produces superior operating characteristics for three-phase motors.

3. In a balanced three-phase system, the conductors need be only about 75% the size of conductors for a single-phase two-wire system of the same KVA (kilovolt-amp) rating. This savings helps offset the cost of supplying the third conductor required by three-phase systems.

A single-phase alternating voltage can be produced by rotating a magnetic field through the conductors of a stationary coil as shown in *Figure 25-3*.

Since alternate polarities of the magnetic field cut through the conductors of the stationary coil, the induced voltage will change polarity at the same speed as the rotation of the magnetic field. The alternator shown in *Figure 25-3* is single-phase because it produces only one AC voltage.

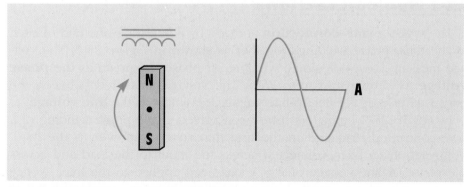

**Figure 25-3**  Producing a single-phase voltage

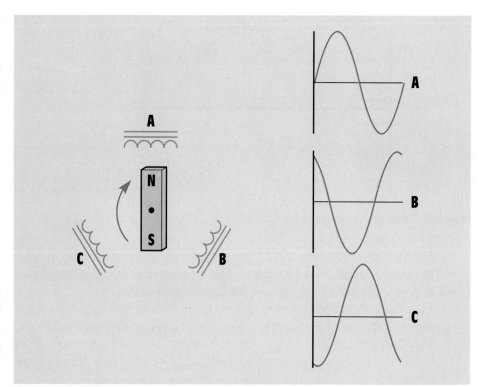

**Figure 25-4** The voltages of a three-phase system are 120° out of phase with each other.

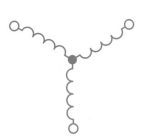

**Figure 25-5** A wye connection is formed by joining one end of each of the windings together.

**wye connection**

**star connection**

**phase voltage**

**line voltage**

If three separate coils are spaced 120° apart as shown in *Figure 25-4*, three voltages 120° out of phase with each other will be produced when the magnetic field cuts through the coils. This is the manner in which a three-phase voltage is produced. There are two basic three-phase connections: the wye, or star, and the delta.

## 25-2 Wye Connections

The **wye**, or **star**, **connection** is made by connecting one end of each of the three-phase windings together as shown in *Figure 25-5*. The voltage measured across a single winding, or phase, is known as the **phase voltage** as shown in *Figure 25-6*. The voltage measured between the lines is known as the line-to-line voltage, or simply as the **line voltage**.

In *Figure 25-7*, ammeters have been placed in the phase winding of a wye-connected load and in the line that supplies power to the load. Voltmeters have been connected across the input to the load and across the phase. A line voltage of 208 V has been applied to the load. Notice that the voltmeter connected across the lines indicates a value of 208 V,

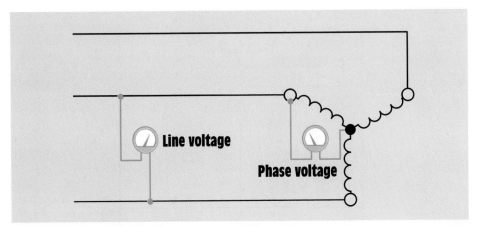

**Figure 25-6** Line and phase voltages are different in a wye connection.

but the voltmeter connected across the phase indicates a value of 120 V.

**In a wye-connected system, the line voltage is higher than the phase voltage by a factor of the square root of 3 (1.732).** Two formulas used to compute the voltage in a wye-connected system are

$$E_{Line} = E_{Phase} \times 1.732$$

and

$$E_{Phase} = \frac{E_{Line}}{1.732}$$

Notice in *Figure 25-7* that 10 A of current flow in both the phase and the line. **In a wye-connected system, phase current and line current are the same**.

$$I_{Line} = I_{Phase}$$

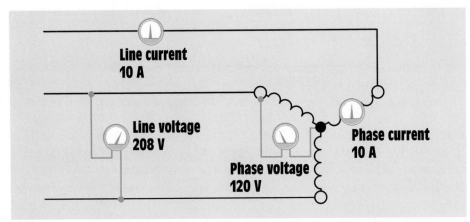

**Figure 25-7** Line current and phase current are the same in a wye connection.

In a wye-connected system, the line voltage is higher than the phase voltage by a factor of the square root of 3 (1.732).

1.732

In a wye-connected system, phase current and line current are the same.

phase current

line current

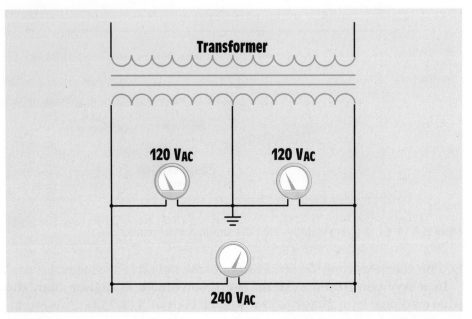

**Figure 25-8** Single-phase transformer with grounded center tap

## Voltage Relationships in a Wye Connection

Many students of electricity have difficulty at first understanding why the line voltage of the wye connection used in this illustration is 208 V instead of 240 V. Since line voltage is measured across two phases that have a voltage of 120 V each, it would appear that the sum of the two voltages should be 240 V. One cause of this misconception is that many students are familiar with the 240/120-V connection supplied to most

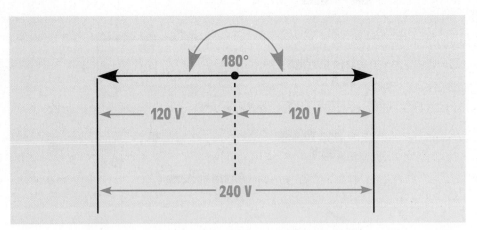

**Figure 25-9** The voltages of a single phase system are 180° out of phase with each other.

homes. If voltage is measured across the two incoming lines, a voltage of 240 V will be seen. If voltage is measured from either of the two lines to the neutral, a voltage of 120 V will be seen. The reason for this is that this is a single-phase connection derived from the center tap of a transformer *(Figure 25-8)*. If the center tap is used as a common point, the two line voltages on either side of it will be 180° apart and opposite in polarity *(Figure 25-9)*. The vector sum of these two voltages would be 240 V.

Three-phase voltages are 120° apart, not 180°. If the three voltages are drawn 120° apart, it will be seen that the vector sum of these voltages is 208 V *(Figure 25-10)*. Another illustration of vector addition is shown in *Figure 25-11*. In this illustration two phase-voltage vectors are added, and the resultant is drawn from the starting point of one vector to the end point of the other. The parallelogram method of vector addition for the voltages in a wye-connected three-phase system is shown in *Figure 25-12*.

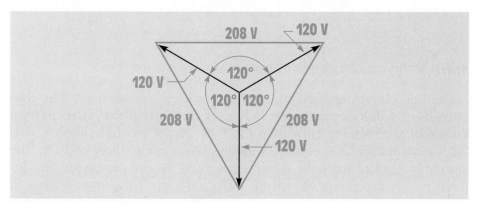

**Figure 25-10**   Vector sum of the voltages in a three-phase wye connection

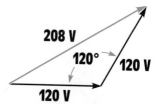

**Figure 25-11**   Adding voltage vectors of two-phase voltage values

## 25-3 Delta Connections

In *Figure 25-13*, three separate inductive loads have been connected to form a **delta connection**. This connection receives its name from the fact that a schematic diagram of this connection resembles the Greek letter delta (Δ). In *Figure 25-14*, voltmeters have been connected across the lines and across the phase. Ammeters have been connected in the line and in the phase. **In a delta connection, line voltage and phase voltage are the same**. Notice that both voltmeters indicate a value of 480 V.

$$E_{Line} = E_{Phase}$$

In a delta connection, line voltage and phase voltage are the same.

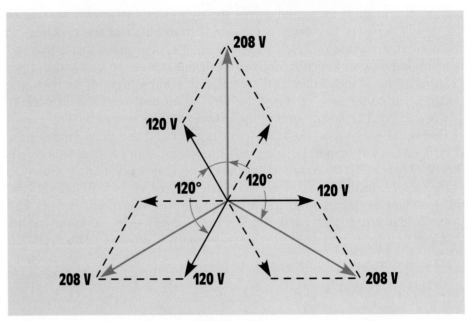

**Figure 25-12** The parallelogram method of adding three-phase vectors

The line current and phase current, however, are different. **The line current of a delta connection is higher than the phase current by a factor of the square root of 3 (1.732).** In the example shown, it is assumed that each of the phase windings has a current flow of 10 A. The current in each of the lines, however, is 17.32 A. The reason for this difference in current is that current flows through different windings at different times in a three-phase circuit. During some periods of times, current will flow between two lines only. At other times, current will flow

**Figure 25-13** Three-phase delta connection

**The line current of a delta connection is higher than the phase current by a factor of the square root of 3 (1.732).**

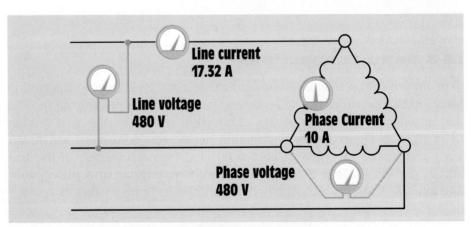

**Figure 25-14** Voltage and current relationships in a delta connection

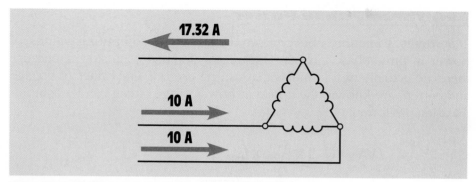

**Figure 25-15**   Division of currents in a delta connection

from two lines to the third *(Figure 25-15)*. The delta connection is similar to a parallel connection because there is always more than one path for current flow. Since these currents are 120° out of phase with each other, vector addition must be used when finding the sum of the currents *(Figure 25-16)*. Formulas for determining the current in a delta connection are

$$I_{Line} = I_{Phase} \times 1.732$$

and

$$I_{Phase} = \frac{I_{Line}}{1.732}$$

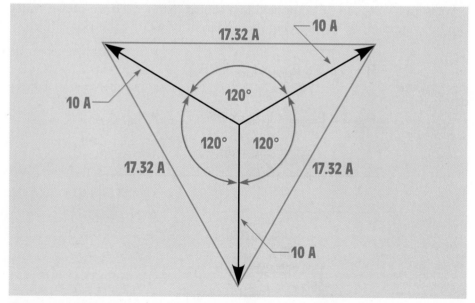

**Figure 25-16**   Vector addition is used to compute the sum of the currents in a delta connection.

### 25-4 Three-Phase Power

Students sometimes become confused when computing values of power in three-phase circuits. One reason for this confusion is because there are actually two formulas that can be used. If *line* values of voltage and current are known, the apparent power of the circuit can be computed using the formula

$$VA = \sqrt{3} \times E_{Line} \times I_{Line}$$

If the *phase* values of voltage and current are known, the apparent power can be computed using the formula

$$VA = 3 \times E_{Phase} \times I_{Phase}$$

Notice that in the first formula, the line values of voltage and current are multiplied by the square root of 3. In the second formula, the phase values of voltage and current are multiplied by 3. The first formula is used more because it is generally more convenient to obtain line values of voltage and current since they can be measured with a voltmeter and clamp-on ammeter.

### 25-5 Watts and VARs

**three-phase watts**

Watts and VARs can be computed in a similar manner. **Three-phase watts** can be computed by multiplying the apparent power by the power factor

$$P = \sqrt{3} \times E_{Line} \times I_{Line} \times PF$$

or

$$P = 3 \times E_{Phase} \times I_{Phase} \times PF$$

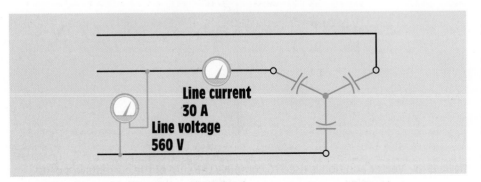

**Line current 30 A**
**Line voltage 560 V**

**Figure 25-17** Pure capacitive three-phase load

(**Note:** When computing the power of a pure resistive load, the voltage and current are in phase with each other and the power factor is 1.)

**Three-phase VARs** can be computed in a similar manner, except that voltage and current values of a pure reactive load are used. For example, a pure capacitive load is shown in *Figure 25-17*. In this example, it is assumed that the line voltage is 480 V and the line current is 30 A. Capacitive VARs can be computed using the formula

$$VARS_C = \sqrt{3} \times E_{Line\ (Capacitive)} \times I_{Line\ (Capacitive)}$$

$$VARS_C = 1.732 \times 560 \times 30$$

$$VARS_C = 29,097.6$$

## 25-6 Three-Phase Circuit Calculations

In the following examples, values of line and phase voltage, line and phase current, and power will be computed for different types of three-phase connections.

A wye-connected three-phase alternator supplies power to a delta-connected resistive load *(Figure 25-18)*. The alternator has a line voltage of 480 V. Each resistor of the delta load has 8 Ω of resistance. Find the following values.

$E_{L(Load)}$ — line voltage of the load

$E_{P(Load)}$ — phase voltage of the load

$I_{P(Load)}$ — phase current of the load

$I_{L(Load)}$ — line current to the load

$I_{L(Alt)}$ — line current delivered by the alternator

$I_{P(Alt)}$ — phase current of the alternator

$E_{P(Alt)}$ — phase voltage of the alternator

P — true power

## Solution

The load is connected directly to the alternator. Therefore, the line voltage supplied by the alternator is the line voltage of the load.

$$E_{L(Load)} = 480\ V$$

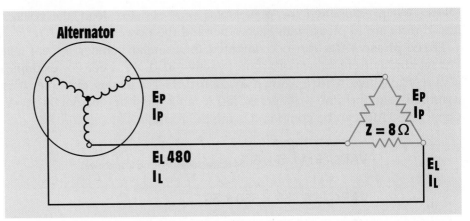

**Figure 25-18** Computing three-phase values using a wye-connected power source and a delta-connected load (example circuit)

The three resistors of the load are connected in a delta connection. In a delta connection, the phase voltage is the same as the line voltage.

$$E_{P(Load)} = E_{L(Load)}$$

$$E_{P(Load)} = 480 \text{ V}$$

Each of the three resistors in the load is one phase of the load. Now that the phase voltage is known (480 V), the amount of phase current can be computed using Ohm's law.

$$I_{P(Load)} = \frac{E_{P(Load)}}{Z}$$

$$I_{P(Load)} = \frac{480}{8}$$

$$I_{P(Load)} = 60 \text{ A}$$

The three load resistors are connected as a delta with 60 A of current flow in each phase. The line current supplying a delta connection must be 1.732 times greater than the phase current.

$$I_{L(Load)} = I_{P(Load)} \times 1.732$$

$$I_{L(Load)} = 60 \times 1.732$$

$$I_{L(Load)} = 103.92 \text{ A}$$

The alternator must supply the line current to the load or loads to which it is connected. In this example, only one load is connected to the alternator. Therefore, the line current of the load will be the same as the line current of the alternator.

$$I_{L(Alt)} = 103.92 \text{ A}$$

The phase windings of the alternator are connected in a wye connection. In a wye connection, the phase current and line current are equal. The phase current of the alternator will, therefore, be the same as the alternator line current.

$$I_{P(Alt)} = 103.92 \text{ A}$$

The phase voltage of a wye connection is less than the line voltage by a factor of the square root of 3. The phase voltage of the alternator will be

$$E_{P(Alt)} = \frac{E_{L(Alt)}}{1.732}$$

$$E_{P(Alt)} = \frac{480}{1.732}$$

$$E_{P(Alt)} = 277.13 \text{ V}$$

In this circuit, the load is pure resistive. The voltage and current are in phase with each other, which produces a unity power factor of 1. The true power in this circuit will be computed using the formula

$$P = 1.732 \times E_{L(Alt)} \times I_{L(Alt)} \times PF$$

$$P = 1.732 \times 480 \times 103.92 \times 1$$

$$P = 86,394.93 \text{ W}$$

A delta-connected alternator is connected to a wye-connected resistive load *(Figure 25-19)*. The alternator produces a line voltage of 240 V and the resistors have a value of 6 Ω each. Find the following values.

**Example 2**

$E_{L(Load)}$ — line voltage of the load

$E_{P(Load)}$ — phase voltage of the load

$I_{P(Load)}$ — phase current of the load

$I_{L(Load)}$ — line current to the load

$I_{L(Alt)}$ — line current delivered by the alternator

$I_{P(Alt)}$ — phase current of the alternator

$E_{P(Alt)}$ — phase voltage of the alternator

P — true power

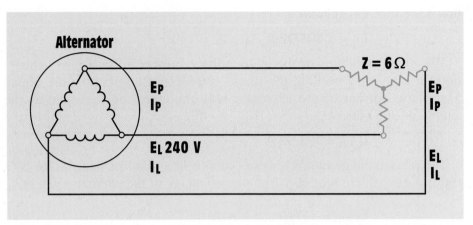

**Figure 25-19** Computing three-phase values using a delta-connected source and a wye-connected load (example circuit)

## Solution

As was the case in the previous example, the load is connected directly to the output of the alternator. The line voltage of the load must, therefore, be the same as the line voltage of the alternator.

$$E_{L(Load)} = 240 \text{ V}$$

The phase voltage of a wye connection is less than the line voltage by a factor of 1.732.

$$E_{P(Load)} = \frac{240}{1.732}$$

$$E_{P(Load)} = 138.57 \text{ V}$$

Each of the three 6-$\Omega$ resistors is one phase of the wye-connected load. Since the phase voltage is 138.57 V, this voltage is applied to each of the three resistors. The amount of phase current can now be determined using Ohm's law.

$$I_{P(Load)} = \frac{E_{P(Load)}}{Z}$$

$$I_{P(Load)} = \frac{138.57}{6}$$

$$I_{P(Load)} = 23.1 \text{ A}$$

The amount of line current needed to supply a wye-connected load is the same as the phase current of the load.

$$I_{L(Load)} = 23.1 \text{ A}$$

Only one load is connected to the alternator. The line current supplied to the load is the same as the line current of the alternator.

$$I_{L(Alt)} = 23.1 \text{ A}$$

The phase windings of the alternator are connected in delta. In a delta connection the phase current is less than the line current by a factor of 1.732.

$$I_{P(Alt)} = \frac{I_{L(Alt)}}{1.732}$$

$$I_{P(Alt)} = \frac{23.1}{1.732}$$

$$I_{P(Alt)} = 13.34 \text{ A}$$

The phase voltage of a delta is the same as the line voltage.

$$E_{P(Alt)} = 240 \text{ V}$$

Since the load in this example is pure resistive, the power factor has a value of unity, or 1. Power will be computed by using the line values of voltage and current.

$$P = 1.732 \times E_L \times I_L \times PF$$

$$P = 1.732 \times 240 \times 23.1 \times 1$$

$$P = 9{,}602.21 \text{ W}$$

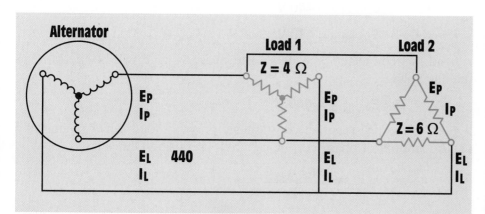

**Figure 25-20** Computing three-phase values using a wye-connected source and two three-phase loads (example circuit)

**Example 3**

The phase windings of an alternator are connected in wye. The alternator produces a line voltage of 440 V and supplies power to two resistive loads. One load contains resistors with a value of 4 Ω each, connected in wye. The second load contains resistors with a value of 6 Ω each, connected in delta *(Figure 25-20)*. Find the following circuit values.

$E_{L(Load\ 2)}$ — line voltage of load 2

$E_{P(Load\ 2)}$ — phase voltage of load 2

$I_{P(Load\ 2)}$ — phase current of load 2

$I_{L(Load\ 2)}$ — line current to load 2

$E_{P(Load\ 1)}$ — phase voltage of load 1

$I_{P(Load\ 1)}$ — phase current of load 1

$I_{L(Load\ 1)}$ — line current to load 1

$I_{L(Alt)}$ — line current delivered by the alternator

$I_{P(Alt)}$ — phase current of the alternator

$E_{P(Alt)}$ — phase voltage of the alternator

P — true power

## Solution

Both loads are connected directly to the output of the alternator. The line voltage for both loads 1 and 2 will be the same as the line voltage of the alternator.

$$E_{L(Load\ 2)} = 440\ V$$

$$E_{L(Load\ 1)} = 440\ V$$

Load 2 is connected as a delta. The phase voltage will be the same as the line voltage.

$$E_{P(Load\ 2)} = 440\ V$$

Each of the resistors that constitutes a phase of load 2 has a value of 6 Ω. The amount of phase current can be found using Ohm's law.

$$I_{P(Load\ 2)} = \frac{E_{P(Load\ 2)}}{Z}$$

$$I_{P(Load\ 2)} = \frac{440}{6}$$

$$I_{P(Load\ 2)} = 73.33\ A$$

The line current supplying a delta-connected load is 1.732 times greater than the phase current. The amount of line current needed for load 2 can be computed by increasing the phase current value by 1.732.

$$I_{L(Load\ 2)} = I_{P(Load\ 2)}\ X\ 1.732$$

$$I_{L(Load\ 2)} = 73.33\ X\ 1.732$$

$$I_{L(Load\ 2)} = 127.01\ A$$

The resistors of load 1 are connected to form a wye. The phase voltage of a wye connection is less than the line voltage by a factor of 1.732.

$$E_{P(Load\ 1)} = \frac{E_{L(Load\ 1)}}{1.732}$$

$$E_{P(Load\ 1)} = \frac{440}{1.732}$$

$$E_{P(Load\ 1)} = 254.04\ V$$

Now that the voltage applied to each of the 4-$\Omega$ resistors is known, the phase current can be computed using Ohm's law.

$$I_{P(Load\ 1)} = \frac{E_{P(Load\ 1)}}{Z}$$

$$I_{P(Load\ 1)} = \frac{254.04}{4}$$

$$I_{P(Load\ 1)} = 63.51\ A$$

The line current supplying a wye-connected load is the same as the phase current. Therefore, the amount of line current needed to supply load 1 is

$$I_{L(Load\ 1)} = 63.51\ A$$

The alternator must supply the line current needed to operate both loads. In this example, both loads are resistive. The total line current supplied by the alternator will be the sum of the line currents of the two loads.

$$I_{L(Alt)} = I_{L(Load\ 1)} + I_{L(Load\ 2)}$$

$$I_{L(Alt)} = 63.51 + 127.01$$

$$I_{L(Alt)} = 190.52\ A$$

Since the phase windings of the alternator in this example are connected in a wye, the phase current will be the same as the line current.

$$I_{P(Alt)} = 190.52 \text{ A}$$

The phase voltage of the alternator will be less than the line voltage by a factor of 1.732.

$$E_{P(Alt)} = \frac{440}{1.732}$$

$$E_{P(Alt)} = 254.04 \text{ V}$$

Both of the loads in this example are resistive and have a unity power factor of 1. The total power in this circuit can be found by using the line voltage and total line current supplied by the alternator.

$$P = 1.732 \times E_L \times I_L \times PF$$

$$P = 1.732 \times 440 \times 190.52 \times 1$$

$$P = 145{,}191.48 \text{ W}$$

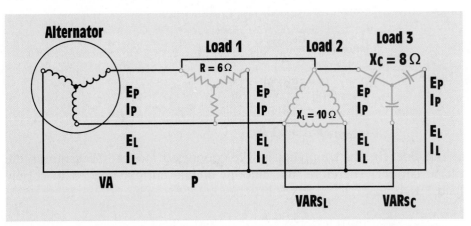

**Figure 25-21** Computing three-phase values with a wye-connected source supplying power to a resistive, inductive, and capacitive load (example circuit)

**Example 4**

A wye-connected three-phase alternator with a line voltage of 560 V supplies power to three different loads *(Figure 25-21)*. The first load is formed by three resistors with a value of 6 Ω each, connected in a wye. The second load comprises three inductors with an inductive reactance of 10 Ω each, connected in delta, and the third load comprises three capacitors with a capacitive reactance of 8 Ω each, connected in wye. Find the following circuit values.

$E_{L(Load\ 3)}$ — line voltage of load 3 (capacitive)

$E_{P(Load\ 3)}$ — phase voltage of load 3 (capacitive)

$I_{P(Load\ 3)}$ — phase current of load 3 (capacitive)

$I_{L(Load\ 3)}$ — line current to load 3 (capacitive)

$E_{L(Load\ 2)}$ — line voltage of load 2 (inductive)

$E_{P(Load\ 2)}$ — phase voltage of load 2 (inductive)

$I_{P(Load\ 2)}$ — phase current of load 2 (inductive)

$I_{L(Load\ 2)}$ — line current to load 2 (inductive)

$E_{L(Load\ 1)}$ — line voltage of load 1 (resistive)

$E_{P(Load\ 1)}$ — phase voltage of load 1 (resistive)

$I_{P(Load\ 1)}$ — phase current of load 1 (resistive)

$I_{L(Load\ 1)}$ — line current to load 1 (resistive)

$I_{L(Alt)}$ — line current delivered by the alternator

$E_{P(Alt)}$ — phase voltage of the alternator

P — true power

$VARS_L$ — reactive power of the inductive load

$VARS_C$ — reactive power of the capacitive load

VA — apparent power

PF — power factor

## Solution

All three loads are connected to the output of the alternator. The line voltage connected to each load is the same as the line voltage of the alternator.

$$E_{L(Load\ 3)} = 560\ V$$

$$E_{L(Load\ 2)} = 560\ V$$

$$E_{L(Load\ 1)} = 560\ V$$

### 25-7 Load 3 Calculations

Load 3 is formed from three capacitors with a capacitive reactance of 8 $\Omega$ each, connected in a wye. Since this load is wye-connected, the phase voltage will be less than the line voltage by a factor of 1.732.

$$E_{P(Load\ 3)} = \frac{E_{L(Load\ 3)}}{1.732}$$

$$E_{P(Load\ 3)} = \frac{560}{1.732}$$

$$E_{P(Load\ 3)} = 323.33\ V$$

Now that the voltage applied to each capacitor is known, the phase current can be computed using Ohm's law.

$$I_{P(Load\ 3)} = \frac{E_{P(Load\ 3)}}{X_C}$$

$$I_{P(Load\ 3)} = \frac{323.33}{8}$$

$$I_{P(Load\ 3)} = 40.42\ A$$

The line current required to supply a wye-connected load is the same as the phase current.

$$I_{L(Load\ 3)} = 40.42\ A$$

The reactive power of load 3 can be found using a formula similar to the formula for computing apparent power. Since load 3 is pure capacitive, the current and voltage are 90° out of phase with each other, and the power factor is zero.

$$VARS_C = 1.732 \times E_{L(Load\ 3)} \times I_{L(Load\ 3)}$$

$$VARS_C = 1.732 \times 560 \times 40.42$$

$$VARS_C = 39,204.17$$

### 25-8 Load 2 Calculations

Load 2 comprises three inductors connected in a delta with an inductive reactance of 10 $\Omega$ each. Since the load is connected in delta, the phase voltage will be same as the line voltage.

$$E_{L(Load\ 2)} = 560\ V$$

The phase current can be computed by using Ohm's law.

$$I_{P(Load\ 2)} = \frac{E_{P(Load\ 2)}}{X_L}$$

$$I_{P(Load\ 2)} = \frac{560}{10}$$

$$I_{P(Load\ 2)} = 56\ A$$

The amount of line current needed to supply a delta-connected load is 1.732 times greater than the phase current of the load.

$$I_{L(Load\ 2)} = I_{P(Load\ 2)} \times 1.732$$

$$I_{L(Load\ 2)} = 56 \times 1.732$$

$$I_{L(Load\ 2)} = 96.99\ A$$

Since load 2 is made up of inductors, the reactive power can be computed using the line values of voltage and current supplied to the load.

$$VARS_L = 1.732 \times E_{L(Load\ 2)} \times I_{L(Load\ 2)}$$

$$VARS_L = 1.732 \times 560 \times 96.99$$

$$VARS_L = 94{,}072.54$$

## 25-9 Load 1 Calculations

Load 1 consists of three resistors with a resistance of 6 Ω each, connected in wye. In a wye connection the phase voltage is less than the line voltage by a factor of 1.732. The phase voltage for load 1 will be the same as the phase voltage for load 3.

$$E_{P(Load\ 1)} = 323.33\ V$$

The amount of phase current can now be computed using the phase voltage and the resistance of each phase.

$$I_{P(Load\ 1)} = \frac{E_{P(Load\ 1)}}{R}$$

$$I_{P(Load\ 1)} = \frac{323.33}{6}$$

$$I_{P(Load\ 1)} = 53.89\ A$$

Since the resistors of load 1 are connected in a wye, the line current will be the same as the phase current.

$$I_{L(Load\ 1)} = 53.89\ A$$

Since load 1 is pure resistive, true power can be computed using the line and phase current values.

$$P = 1.732 \times E_{L(Load\ 1)} \times I_{L(Load\ 1)}$$

$$P = 1.732 \times 560 \times 53.89$$

$$P = 52,267\ W$$

## 25-10 Alternator Calculations

The alternator must supply the line current for each of the loads. In this problem, however, the line currents are out of phase with each other. To find the total line current delivered by the alternator, vector addition must be used. The current flow in load 1 is resistive and in phase with the line voltage. The current flow in load 2 is inductive and lags the line voltage by 90°. The current flow in load 3 is capacitive and leads the line voltage by 90°. A formula similar to the formula used to find total current flow in an RLC parallel circuit can be employed to find the total current delivered by the alternator.

$$I_{L(Alt)} = \sqrt{I_{L(Load\ 1)}^2 + (I_{L(Load\ 2)} - I_{L(Load\ 3)})^2}$$

$$I_{L(Alt)} = \sqrt{53.89^2 + (96.99 - 40.42)^2}$$

$$I_{L(Alt)} = 78.13\ A$$

The apparent power can now be found using the line voltage and current values of the alternator.

$$VA = 1.732 \times E_{L(Alt)} \times I_{L(Alt)}$$

$$VA = 1.732 \times 560 \times 78.13$$

$$VA = 75,779.85$$

The circuit power factor is the ratio of apparent power and true power.

$$PF = \frac{W}{VA}$$

$$PF = \frac{52,267}{75,779.85}$$

$$PF = 69\%$$

## 25-11 Power Factor Correction

**Correcting the power factor** of a three-phase circuit is similar to the procedure used to correct the power factor of a single-phase circuit.

A three-phase motor is connected to a 480-V, 60-Hz line *(Figure 25-22)*. A clamp-on ammeter indicates a running current of 68 A at full load, and a three-phase wattmeter indicates a true power of 40,277 W. Compute the motor power factor first. Then find the amount of capacitance needed to correct the power factor to 95%. Assume that the capacitors used for power factor correction are to be connected in wye, and the capacitor bank is then to be connected in parallel with the motor.

### Solution

First find the amount of apparent power in the circuit.

$$VA = 1.732 \times E_L \times I_L$$
$$VA = 1.732 \times 480 \times 68$$
$$VA = 56{,}532.48$$

The motor power factor can be computed by dividing the true power by the apparent power.

$$PF = \frac{P}{VA}$$
$$PF = \frac{40{,}277}{56{,}532.48}$$
$$PF = 71.2\%$$

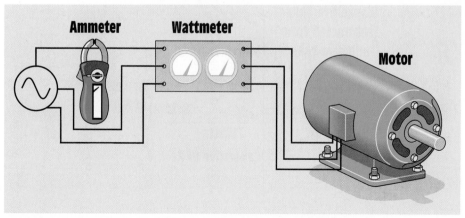

**Figure 25-22**  Determining apparent and true power for a three-phase motor

The inductive VARs in the circuit can be computed using the formula

$$VARS_L = \sqrt{VA^2 - P^2}$$

$$VARS_L = \sqrt{56,532.48^2 - 40,277^2}$$

$$VARS_L = 39,669.69$$

If the power factor is to be corrected to 95%, the apparent power at 95% power factor must be found. This can be done using the formula

$$VA = \frac{P}{PF}$$

$$VA = \frac{40,277}{0.95}$$

$$VA = 42,396.84$$

The amount of inductive VARs needed to produce an apparent power of 42,396.84 VA can be found using the formula

$$VARS_L = \sqrt{VA^2 - P^2}$$

$$VARS_L = \sqrt{42,396.84^2 - 40,277^2}$$

$$VARS_L = 13,238.4$$

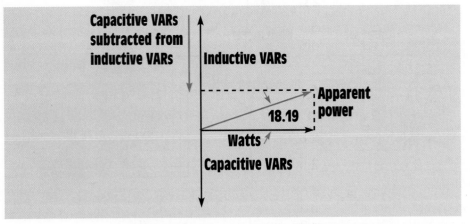

**Figure 25-23**  Vector relationship of powers to correct motor power factor

To correct the power factor to 95%, the inductive VARs must be reduced from 39,669.69 to 13,238.4. This can be done by connecting a bank of capacitors in the circuit that will produce a total of 26,431.29 capacitive VARs (39,669.69 – 13,238.4 = 26,431.29). This amount of capacitive VARs will reduce the inductive VARs to the desired amount *(Figure 25-23)*.

Now that the amount of capacitive VARs needed to correct the power factor is known, the amount of line current supplying the capacitor bank can be computed using the formula

$$I_L = \frac{VARS_C}{E_L \times 1.732}$$

$$I_L = \frac{26,431.29}{480 \times 1.732}$$

$$I_L = 31.79 \text{ A}$$

The capacitive load bank is to be connected in a wye. Therefore, the phase current will be the same as the line current. The phase voltage, however, will be less than the line voltage by a factor of 1.732, or 277.14 V. Ohm's law can be used to find the amount of capacitive reactance needed to produce a phase current of 31.79 A with an applied voltage of 277.14 V.

$$X_C = \frac{E_P}{I_P}$$

$$X_C = \frac{277.14}{31.79}$$

$$X_C = 8.72 \ \Omega$$

The amount of capacitance needed to produce a capacitive reactance of 8.72 Ω can now be computed.

$$C = \frac{1}{2\pi F X_C}$$

$$C = \frac{40}{377 \times 8.72}$$

$$C = 304.2 \ \mu F$$

When a bank of wye-connected capacitors with a value of 304.2 μF each are connected in parallel with the motor, the power factor will be corrected to 95% *(Figure 25-24)*.

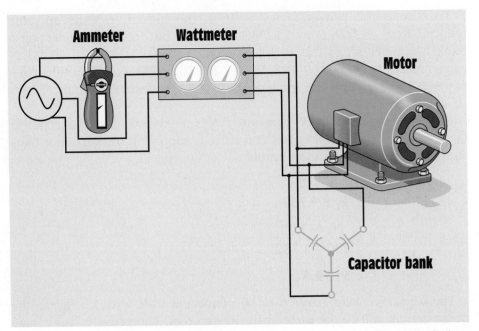

**Figure 25-24** A wye-connected bank of capacitors is used to correct motor power factor.

## Summary

1. The voltages of a three-phase system are 120° out of phase with each other.

2. The two types of three-phase connections are wye and delta.

3. Wye connections are characterized by the fact that one terminal of each of the devices is connected together.

4. In a wye connection, the phase voltage is less than the line voltage by a factor of 1.732. The phase current and line current are the same.

5. In a delta connection, the phase voltage is the same as the line voltage. The phase current is less than the line current by a factor of 1.732.

---

## Review Questions

1. How many degrees out of phase with each other are the voltages of a three-phase system?

2. What are the two main types of three-phase connections?

3. A wye-connected load has a voltage of 480 V applied to it. What is the voltage dropped across each phase?

4. A wye-connected load has a phase current of 25 A. How much current is flowing through the lines supplying the load?

5. A delta connection has a voltage of 560 V connected to it. How much voltage is dropped across each phase?

6. A delta connection has 30 A of current flowing through each phase winding. How much current is flowing through each of the lines supplying power to the load?

7. A three-phase load has a phase voltage of 240 V and a phase current of 18 A. What is the apparent power of this load?

8. If the load in question 7 is connected in a wye, what would be the line voltage and line current supplying the load?

9. An alternator with a line voltage of 2400 V supplies a delta-connected load. The line current supplied to the load is 40 A. Assuming the load is a balanced three-phase load, what is the impedance of each phase?

10. What is the apparent power of the circuit in question 9?

**Practice Problems**

1. Refer to the circuit shown in *Figure 25-18* to answer the following questions. But assume that the alternator has a line voltage of 240 V and the load has an impedance of 12 Ω per phase. Find all the missing values.

$E_{P(A)}$ _____     $E_{P(L)}$ _____

$I_{P(A)}$ _____     $I_{P(L)}$ _____

$E_{L(A)}$ 240     $E_{L(L)}$ _____

$I_{L(A)}$ _____     $I_{L(L)}$ _____

P _____     $Z_{(PHASE)}$ 12 Ω

2. Refer to the circuit shown in *Figure 25-19* to answer the following questions. But assume that the alternator has a line voltage of 4160 V, and the load has an impedance of 60 Ω per phase. Find all the missing values.

$E_{P(A)}$ _____     $E_{P(L)}$ _____

$I_{P(A)}$ _____     $I_{P(L)}$ _____

$E_{L(A)}$ 4160     $E_{L(L)}$ _____

$I_{L(A)}$ _____     $I_{L(L)}$ _____

P _____     $Z_{(PHASE)}$ 60 Ω

3. Refer to the circuit shown in *Figure 25-20* to answer the following questions. But assume that the alternator has a line voltage of 560 V. Load 1 has an impedance of 5 Ω per phase, and load 2 has an impedance of 8 Ω per phase. Find all the missing values.

$E_{P(A)}$ _____     $E_{P(L1)}$ _____     $E_{P(L2)}$ _____

$I_{P(A)}$ _____     $I_{P(L1)}$ _____     $I_{P(L2)}$ _____

$E_{L(A)}$ 560     $E_{L(L1)}$ _____     $E_{L(L2)}$ _____

$I_{L(A)}$ _____     $I_{L(L1)}$ _____     $I_{L(L2)}$ _____

P _____     $Z_{(PHASE)}$ 5 Ω     $Z_{(PHASE)}$ 8 Ω

4. Refer to the circuit shown in *Figure 25-21* to answer the following questions. But assume that the alternator has a line voltage of 480

V. Load 1 has a resistance of 12 Ω per phase. Load 2 has an inductive reactance of 16 Ω per phase, and load 3 has a capacitive reactance of 10 Ω per phase. Find all the missing values.

$E_{P(A)}$ _____    $E_{P(L1)}$ _____    $E_{P(L2)}$ _____    $E_{P(L3)}$ _____

$I_{P(A)}$ _____    $I_{P(L1)}$ _____    $I_{P(L2)}$ _____    $I_{P(L3)}$ _____

$E_{L(A)}$ 480    $E_{L(L1)}$ _____    $E_{L(L2)}$ _____    $E_{L(L3)}$ _____

$I_{L(A)}$ _____    $I_{L(L1)}$ _____    $I_{L(L2)}$ _____    $I_{L(L3)}$ _____

VA _____    $R_{(PHASE)}$ 12 Ω    $X_{L(PHASE)}$ 16 Ω    $X_{C(PHASE)}$ 10 Ω

P _____    $VARS_L$ _____    $VARS_C$ _____

*section*   **10**

*transformers*

One of the
greatest
advantages of
alternating
current over
direct current
is the fact
that AC
current can be
transformed
and DC cannot
be trans-
formed. This
unit discusses
the basic
types of
transformers
and how they
operate.

*Courtesy of Niagara Mohawk Power Corp.*

# Single-Phase Transformers

## Key Terms

Autotransformers

Control transformer

Distribution transformer

Excitation current

Flux leakage

Inrush current

Isolation transformers

Laminated

Neutral conductor

Primary winding

Secondary winding

Step-down transformer

Step-up transformer

Tape wound core

Toroid core

Transformer

Turns ratio

Volts-per-turn ratio

## Outline

**Objectives**

**A**fter studying this unit, you should be able to:

■ Discuss the different types of transformers.

■ Calculate values of voltage, current, and turns for single-phase transformers using formulas.

■ Calculate values of voltage, current, and turns for single-phase transformers using the turns ratio.

■ Connect a transformer and test the voltage output of different windings.

■ Discuss polarity markings on a schematic diagram.

■ Test a transformer to determine the proper polarity marks.

**Preview**

Transformers are among the most common devices found in the electrical field. They range in size from less than one cubic inch to the size of rail cars. Their ratings can range from mVA (milli-volt-amps) to GVA (giga-volt-amps. It is imperative that anyone working in the electrical field have an understanding of transformer types and connections. This unit will present transformers intended for use in single-phase installations. The two main types of voltage transformers, isolation transformers and autotransformers, will be discussed.

### 26-1 Single-Phase Transformers

**trans-former**

A **transformer** is a magnetically operated machine that can change values of voltage, current, and impedance without a change of frequency. Transformers are the most efficient machines known. Their efficiencies commonly range from 90% to 99% at full load. Transformers can be divided into three classifications:

1. Isolation transformer
2. Autotransformer
3. Current transformer (current transformers were discussed in Unit 9).

**All values of a transformer are proportional to its turns ratio**. This does not mean that the exact number of turns of wire on each winding must be known to determine different values of voltage and current for a transformer. What must be known is the *ratio* of turns. For example, assume a transformer has two windings. One winding, the primary, has 1000 turns of wire, and the other, the secondary, has 250 turns of wire *(Figure 26-1)*. The **turns ratio** of this transformer is 4 to 1, or 4:1 (1000/250 = 4). This indicates there are four turns of wire on the primary for every one turn of wire on the secondary.

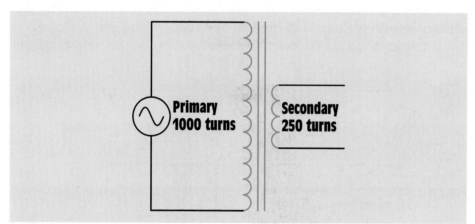

**Figure 26-1** All values of a transformer are proportional to its turns ratio.

## Transformer Formulas

Different formulas can be used to find the values of voltage and current for a transformer. The following is a list of standard formulas, where

$N_P$ = number of turns in the primary

$N_S$ = number of turns in the secondary

$E_P$ = voltage of the primary

$E_S$ = voltage of the secondary

$I_P$ = current in the primary

$I_S$ = current in the secondary

$$\frac{E_P}{E_S} = \frac{N_P}{N_S}$$

$$\frac{E_P}{E_S} = \frac{I_S}{I_P}$$

$$\frac{N_P}{N_S} = \frac{I_S}{I_P}$$

or

$$E_P \times N_S = E_S \times N_P$$

$$E_P \times I_P = E_S \times I_S$$

$$N_P \times I_P = N_S \times I_S$$

**primary winding**

**secondary winding**

The **primary winding** of a transformer is the power input winding. It is the winding that is connected to the incoming power supply. The **secondary winding** is the load winding, or output winding. It is the side of the transformer that is connected to the driven load (Figure 26-2).

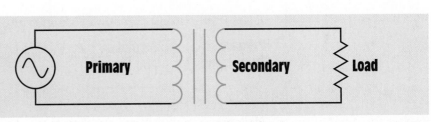

**Figure 26-2** An isolation transformer has its primary and secondary winding electrically separated from each other.

## 26-2 Isolation Transformers

**isolation transformers**

The transformers shown in *Figures 26-1* and *26-2* are **isolation transformers**. This means that the secondary winding is physically and electrically isolated from the primary winding. There is no electrical connection between the primary and secondary winding. This transformer is magnetically coupled, not electrically coupled. This line isolation is often a very desirable characteristic. Since there is no electrical connection between the load and power supply, the transformer becomes a filter between the two. The isolation transformer will greatly reduce any voltage spikes that originate on the supply side before they are transferred to the load side. Some isolation transformers are built with a turns ratio of 1:1. A transformer of this type will have the same input and output voltage and is used for the purpose of isolation only.

The reason that the isolation transformer can greatly reduce any voltage spikes before they reach the secondary is because of the rise time of cur-

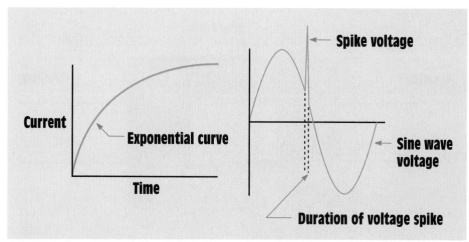

**Figure 26-3** The current through an inductor rises at an exponential rate.

**Figure 26-4** Voltage spikes are generally of very short duration.

rent through an inductor. Recall from Unit 13 that the current in an inductor rises at an exponential rate *(Figure 26-3)*. As the current increases in value, the expanding magnetic field cuts through the conductors of the coil and induces a voltage that is opposed to the applied voltage. The amount of induced voltage is proportional to the rate of change of current. This simply means that the faster current attempts to increase, the greater the opposition to that increase will be. Spike voltages and currents are generally of very short duration, which means that they increase in value very rapidly *(Figure 26-4)*. This rapid change of value causes the opposition to the change to increase just as rapidly. By the time the spike

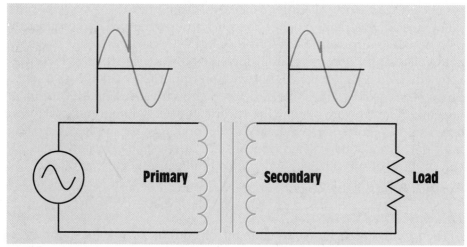

**Figure 26-5** The isolation transformer greatly reduces the voltage spike.

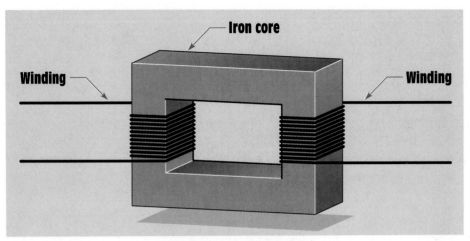

**Figure 26-6** Basic construction of an isolation transformer

has been transferred to the secondary winding of the transformer it has been eliminated or greatly reduced *(Figure 26-5)*.

The basic construction of an isolation transformer is shown in *Figure 26-6*. A metal core is used to provide good magnetic coupling between the two windings. The core is generally made of laminations stacked together. Laminating the core helps reduce power losses caused by eddy current induction.

## Basic Operating Principles

In *Figure 26-7*, one winding of an isolation transformer has been connected to an alternating current supply, and the other winding has been connected to a load. As current increases from zero to its peak positive point, a magnetic field expands outward around the coil. When the current decreases from its peak positive point toward zero, the magnetic field collapses. When the current increases toward its negative peak, the magnetic field again expands, but with an opposite polarity of that previously. The field again collapses when the current decreases from its negative peak toward zero. This continually expanding and collapsing magnetic field cuts the windings of the primary and induces a voltage into it. This induced voltage opposes the applied voltage and limits the current flow of the primary. When a coil induces a voltage into itself, it is known as *self-induction*.

## Excitation Current

There will always be some amount of current flow in the primary of any voltage transformer regardless of type or size even if there is no load

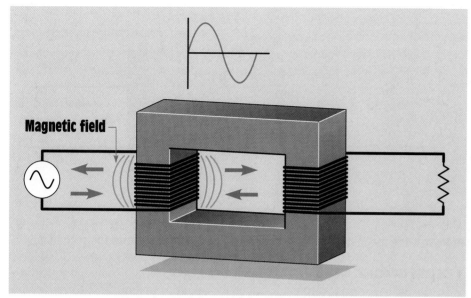

**Figure 26-7** Magnetic field produced by alternating current

connected to the secondary. This current flow is called the **excitation current** of the transformer. The excitation current is the amount of current required to magnetize the core of the transformer. The excitation current remains constant from no load to full load. As a general rule, the excitation current is such a small part of the full load current it is often omitted when making calculations.

## Mutual Induction

Since the secondary windings of an isolation transformer are wound on the same core as the primary, the magnetic field produced by the primary winding cuts the windings of the secondary also *(Figure 26-8)*. This continually changing magnetic field induces a voltage into the secondary winding. The ability of one coil to induce a voltage into another coil is called *mutual induction*. The amount of voltage induced in the secondary is determined by the ratio of the number of turns of wire in the secondary to those in the primary. For example, assume the primary has 240 turns of wire and is connected to 120 VAC. This gives the transformer a **volts-per-turn ratio** of 0.5 (120 V/240 turns = 0.5 volt per turn). Now assume the secondary winding contains 100 turns of wire. Since the transformer has a volts-per-turn ratio of 0.5, the secondary voltage will be 50 V (100 x 0.5 = 50).

**excitation current**

**volts-per-turn ratio**

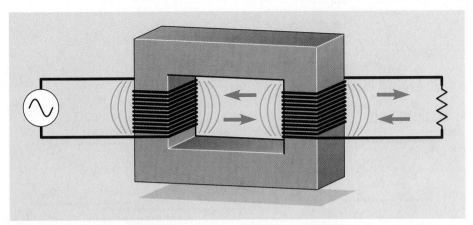

**Figure 26-8** The magnetic field of the primary induces a voltage into the secondary.

### Transformer Calculations

In the following examples, values of voltage, current, and turns for different transformers will be computed.

Assume that the isolation transformer shown in *Figure 26-2* has 240 turns of wire on the primary and 60 turns of wire on the secondary. This is a ratio of 4:1 (240/60 = 4). Now assume that 120 V is connected to the primary winding. What is the voltage of the secondary winding?

$$\frac{E_P}{E_S} = \frac{N_P}{N_S}$$

$$\frac{120}{E_S} = \frac{240}{60}$$

$$240\, E_S = 7200$$

$$E_S = 30\ V$$

The transformer in this example is known as a **step-down transformer** because it has a lower secondary voltage than primary voltage.

Now assume that the load connected to the secondary winding has an impedance of 5 Ω. The next problem is to calculate the current flow in the secondary and primary windings. The current flow of the secondary can be computed using Ohm's law since the voltage and impedance are known.

$$I = \frac{E}{Z}$$

$$I = \frac{30}{5}$$

**step-down trans-former**

$$I = 6 \text{ A}$$

Now that the amount of current flow in the secondary is known, the primary current can be computed using the formula

$$\frac{E_P}{E_S} = \frac{I_S}{I_P}$$

$$\frac{120}{30} = \frac{6}{I_P}$$

$$120\, I_P = 180$$

$$I_P = 1.5 \text{ A}$$

Notice that the primary voltage is higher than the secondary voltage, but the primary current is much less than the secondary current. **A good rule for any type of transformer is that power in must equal power out.** If the primary voltage and current are multiplied together, the product should equal the product of the voltage and current of the secondary.

<table>
<tr><td align="center">Primary</td><td align="center">Secondary</td></tr>
<tr><td align="center">120 x 1.5 = 180 VA</td><td align="center">30 x 6 = 180 VA</td></tr>
</table>

In this example, assume that the primary winding contains 240 turns of wire and the secondary contains 1200 turns of wire. This is a turns ratio of 1:5 (1200/240 = 5). Now assume that 120 V is connected to the primary winding. Compute the voltage output of the secondary winding.

$$\frac{E_P}{E_S} = \frac{N_P}{N_S}$$

$$\frac{120}{E_S} = \frac{240}{1200}$$

$$240\, E_S = 144{,}000$$

$$E_S = 600 \text{ V}$$

Notice that the secondary voltage of this transformer is higher than the primary voltage. This type of transformer is known as a **step-up transformer**.

Now assume that the load connected to the secondary has an impedance of 2400 Ω. Find the amount of current flow in the primary and secondary windings. The current flow in the secondary winding can be computed using Ohm's law.

$$I = \frac{E}{Z}$$

$$I = \frac{600}{2400}$$

$$I = 0.25 \text{ A}$$

Now that the amount of current flow in the secondary is known, the primary current can be computed using the formula

$$\frac{E_P}{E_S} = \frac{I_S}{I_P}$$

$$\frac{120}{600} = \frac{0.25}{I_P}$$

$$120\, I_P = 150$$

$$I_P = 1.25 \text{ A}$$

Notice that the amount of power input equals the amount of power output.

| Primary | Secondary |
|---|---|
| 120 x 1.25 = 150 VA | 600 x 0.25 = 150 VA |

### Calculating Isolation Transformer Values Using the Turns Ratio

As illustrated in the previous examples, transformer values of voltage, current, and turns can be computed using formulas. It is also possible to compute these same values using the turns ratio. To make calculations using the turns ratio, a ratio is established that compares some number to 1, or 1 to some number. For example, assume a transformer has a primary rated at 240 V and a secondary rated at 96 V *(Figure 26-9)*. The turns ratio can be computed by dividing the higher voltage by the lower voltage.

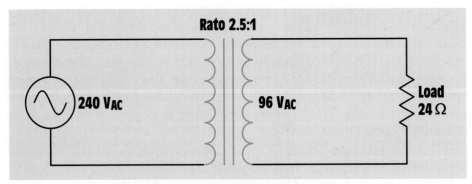

**Rato 2.5:1**

240 V_AC    96 V_AC    Load 24 Ω

**Figure 26-9** Computing transformer values using the turns ratio

$$\text{Ratio} = \frac{240}{96}$$

$$\text{Ratio} = 2.5:1$$

This ratio indicates that there are 2.5 turns of wire in the primary winding for every 1 turn of wire in the secondary. The side of the transformer with the lowest voltage will always have the lowest number (1) of the ratio.

Now assume that a resistance of 24 Ω is connected to the secondary winding. The amount of secondary current can be found using Ohm's law.

$$I_S = \frac{96}{24}$$

$$I_S = 4 \text{ A}$$

The primary current can be found using the turns ratio. Recall that the volt-amps of the primary must equal the volt-amps of the secondary. Since the primary voltage is greater, the primary current will have to be less than the secondary current.

$$I_P = \frac{I_S}{\text{turns ratio}}$$

$$I_P = \frac{4}{2.5}$$

$$I_P = 1.6 \text{ A}$$

To check the answer, find the volt-amps of the primary and secondary.

| Primary | Secondary |
|---|---|
| 240 x 1.6 = 384 VA | 96 x 4 = 384 VA |

Now assume that the secondary winding contains 150 turns of wire. The primary turns can be found by using the turns ratio also. Since the primary voltage is higher than the secondary voltage, the primary must have more turns of wire.

$$N_P = N_S \times \text{turns ratio}$$

$$N_P = 150 \times 2.5$$

$$N_P = 375 \text{ turns}$$

In the next example, assume an isolation transformer has a primary voltage of 120 V and a secondary voltage of 500 V. The secondary has a load impedance of 1200 Ω. The secondary contains 800 turns of wire *(Figure 26-10)*.

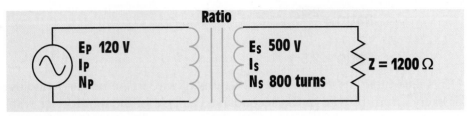

**Figure 26-10** Calculating transformer values

The turns ratio can be found by dividing the higher voltage by the lower voltage.

$$\text{Ratio} = \frac{500}{120}$$

$$\text{Ratio} = 1{:}4.17$$

The secondary current can be found using Ohm's law.

$$I_S = \frac{500}{1200}$$

$$I_S = 0.417 \text{ A}$$

In this example, the primary voltage is lower than the secondary voltage. Therefore, the primary current must be higher.

$$I_P = I_S \times \text{turns ratio}$$

$$I_P = 0.417 \times 4.17$$

$$I_P = 1.74 \text{ A}$$

To check this answer, compute the volt-amps of both windings.

| Primary | Secondary |
|---------|-----------|
| 120 x 1.74 = 208.8 VA | 500 x 0.417 = 208.5 VA |

The slight difference in answers is caused by rounding off values.

Since the primary voltage is less than the secondary voltage, the turns of wire in the primary will be less also.

$$N_P = \frac{N_S}{\text{turns ratio}}$$

$$N_P = \frac{800}{4.17}$$

$$N_P = 192 \text{ turns}$$

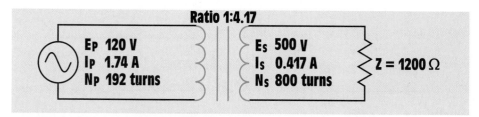

**Figure 26-11** Transformer with completed values

*Figure 26-11* shows the transformer with all completed values.

## Multiple-Tapped Windings

It is not uncommon for isolation transformers to be designed with windings that have more than one set of lead wires connected to the primary or secondary. These are called multiple-tapped windings. The transformer shown in *Figure 26-12* contains a secondary winding rated at 24 V. The primary winding contains several taps, however. One of the primary lead wires is labeled C and is the common for the other leads. The other leads are labeled 120, 208, and 240. This transformer is designed in such a manner that it can be connected to different primary voltages without

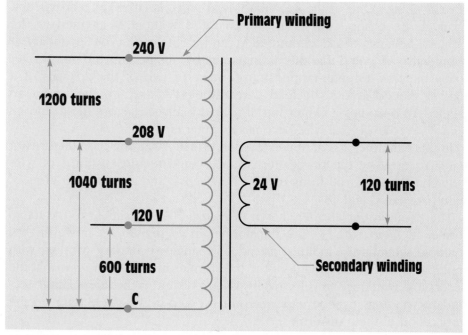

**Figure 26-12** Transformer with multiple-tapped primary winding

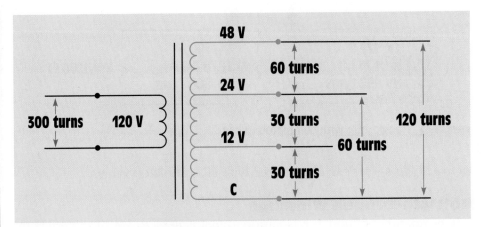

**Figure 26-13** Transformer secondary with multiple taps

changing the value of the secondary voltage. In this example, it is assumed that the secondary winding has a total of 120 turns of wire. To maintain the proper turns ratio, the primary would have 600 turns of wire between C and 120, 1040 turns between C and 208, and 1200 turns between C and 240.

The isolation transformer shown in *Figure 26-13* contains a single primary winding. The secondary winding, however, has been tapped at several points. One of the secondary lead wires is labeled C and is common to the other lead wires. When rated voltage is applied to the primary, voltages of 12, 24, and 48 V can be obtained at the secondary. It should also be noted that this arrangement of taps permits the transformer to be used as a center-tapped transformer for two of the voltages. If a load is placed across the lead wires labeled C and 24, the lead wire labeled 12 becomes a center tap. If a load is placed across the C and 48 lead wires, the 24 lead wire becomes a center tap.

In this example, it is assumed that the primary winding has 300 turns of wire. To produce the proper turns ratio would require 30 turns of wire between C and 12, 60 turns of wire between C and 24, and 120 turns of wire between C and 48.

The isolation transformer shown in *Figure 26-14* is similar to the transformer in *Figure 26-13*. The transformer in *Figure 26-14*, however, has multiple secondary windings instead of a single secondary winding with multiple taps. The advantage of the transformer in *Figure 26-14* is that the secondary windings are electrically isolated from each other. These secondary windings can be either step-up or step-down depending on the application of the transformer.

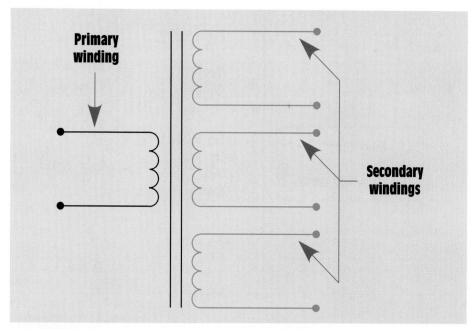

**Figure 26-14**  Transformer with multiple secondary windings

## Computing Values for Isolation Transformers with Multiple Secondaries

When computing the values of an isolation transformer with multiple secondary windings, each secondary must be treated as a different transformer. For example, the transformer in *Figure 26-15* contains one primary winding and three secondary windings. The primary is connected to 120 VAC and contains 300 turns of wire. One secondary has an output voltage of 560 V and a load impedance of 1000 $\Omega$. The second secondary has an output voltage of 208 V and a load impedance of 400 $\Omega$, and the third secondary has an output voltage of 24 V and a load impedance of 6 $\Omega$. The current, turns of wire, and ratio for each secondary and the current of the primary will be found.

The first step will be to compute the turns ratio of the first secondary. The turns ratio can be found by dividing the smaller voltage into the larger.

$$\text{Ratio} = \frac{E_{S1}}{E_P}$$

$$\text{Ratio} = \frac{560}{120}$$

$$\text{Ratio} = 1:4.67$$

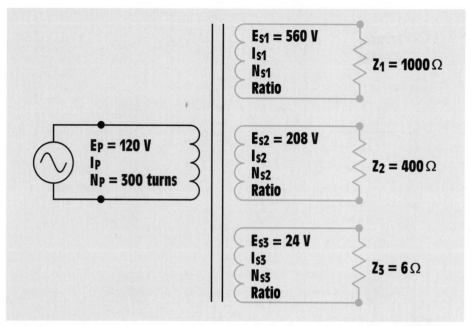

**Figure 26-15** Computing values for a transformer with multiple secondary windings

The current flow in the first secondary can be computed using Ohm's law.

$$I_{S1} = \frac{560}{1000}$$

$$I_{S1} = 0.56 \text{ A}$$

The number of turns of wire in the first secondary winding will be found using the turns ratio. Since this secondary has a higher voltage than the primary, it must have more turns of wire.

$$N_{S1} = N_P \times \text{turns ratio}$$

$$N_{S1} = 300 \times 4.67$$

$$N_{S1} = 1401 \text{ turns}$$

The amount of primary current needed to supply this secondary winding can be found using the turns ratio also. Since the primary has less voltage, it will require more current.

$$I_{P(\text{FIRST SECONDARY})} = I_{S1} \times \text{turns ratio}$$

$$I_{P(\text{FIRST SECONDARY})} = 0.56 \times 4.67$$

$$I_{P(\text{FIRST SECONDARY})} = 2.61 \text{ A}$$

The turns ratio of the second secondary winding will be found by dividing the higher voltage by the lower.

$$\text{Ratio} = \frac{208}{120}$$

$$\text{Ratio} = 1{:}1.73$$

The amount of current flow in this secondary can be determined using Ohm's law.

$$I_{S2} = \frac{208}{400}$$

$$I_{S2} = 0.52 \text{ A}$$

Since the voltage of this secondary is greater than the primary, it will have more turns of wire than the primary. The turns of this secondary will be found using the turns ratio.

$$N_{S2} = N_P \text{ x turns ratio}$$

$$N_{S2} = 300 \text{ x } 1.73$$

$$N_{S2} = 519 \text{ turns}$$

The voltage of the primary is less than this secondary. The primary will, therefore, require a greater amount of current. The amount of current required to operate this secondary will be computed using the turns ratio.

$$I_{P(\text{SECOND SECONDARY})} = I_{S2} \text{ x turns ratio}$$

$$I_{P(\text{SECOND SECONDARY})} = 0.52 \text{ x } 1.732$$

$$I_{P(\text{SECOND SECONDARY})} = 0.9 \text{ A}$$

The turns ratio of the third secondary winding will be computed in the same way as the other two. The larger voltage will be divided by the smaller.

$$\text{Ratio} = \frac{120}{24}$$

$$\text{Ratio} = 5{:}1$$

The primary current will be found using Ohm's law.

$$I_{S3} = \frac{24}{6}$$

$$I_{S3} = 4 \text{ A}$$

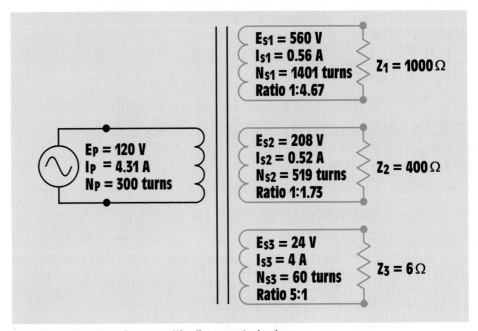

**Figure 26-16**  The transformer with all computed values

The output voltage of the third secondary is less than the primary. The number of turns of wire will, therefore, be less than the primary turns.

$$N_{S3} = \frac{N_P}{\text{turns ratio}}$$

$$N_{S3} = \frac{300}{5}$$

$$N_{S3} = 60 \text{ turns}$$

The primary has a higher voltage than this secondary. The primary current will, therefore, be less by the amount of the turns ratio.

$$I_{P(\text{THIRD SECONDARY})} = \frac{I_{S3}}{\text{turns ratio}}$$

$$I_{P(\text{THIRD SECONDARY})} = \frac{4}{5}$$

$$I_{P(\text{THIRD SECONDARY})} = 0.8 \text{ A}$$

The primary must supply current to each of the three secondary windings. Therefore, the total amount of primary current will be the sum of the currents required to supply each secondary.

$$I_{P(TOTAL)} = I_{P1} + I_{P2} + I_{P3}$$

$$I_{P(TOTAL)} = 2.61 + 0.9 + 0.8$$

$$I_{P(TOTAL)} = 4.31 \text{ A}$$

The transformer with all computed values is shown in *Figure 26-16.*

## Distribution Transformers

A common type of isolation transformer is the **distribution transformer**, *Figure 26-17.* This type of transformer changes the high voltage of power company distribution lines to the common 240/120 V used to supply power to most homes and many businesses. In this example, it is assumed that the primary is connected to a 7200-V line. The secondary is 240 V with a center tap. The center tap is grounded and becomes the **neutral conductor** or common conductor. If voltage is measured across the entire secondary, a voltage of 240 V will be seen. If voltage is measured from either line to the center tap, half of the secondary voltage, or 120 V, will be seen *(Figure 26-18).* The reason is that the voltages between the two

**distribution trans- former**

**neutral conductor**

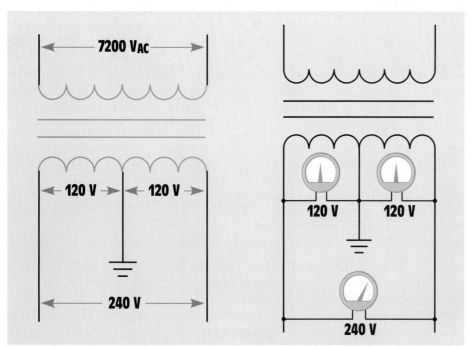

**Figure 26-17**  Distribution transformer

**Figure 26-18**  The voltage from either line to neutral is 120 V. The voltage across the entire secondary winding is 240 V.

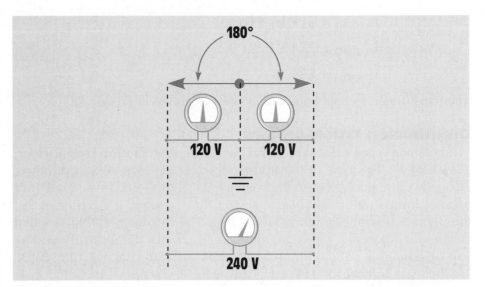

**Figure 26-19** The voltages across the secondary are 180° out of phase with each other.

secondary lines are 180° out of phase with each other. If a vector diagram is drawn to illustrate this condition, it will be seen that the grounded neutral conductor is connected to the axis point of the two voltage vectors *(Figure 26-19)*. Loads that are intended to operate on 240 V, such as water heaters, electric-resistance heating units, and central air conditioners are connected directly across the lines of the secondary *(Figure 26-20)*.

Loads that are intended to operate on 120 V connect from the center tap, or neutral, to one of the secondary lines. The function of the neutral

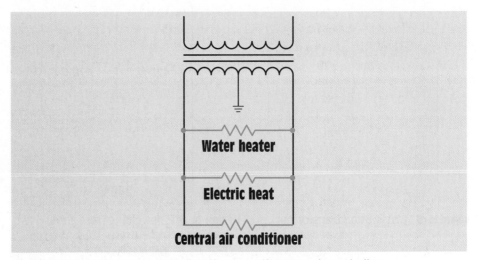

**Figure 26-20** 240-V loads connect directly across the secondary winding.

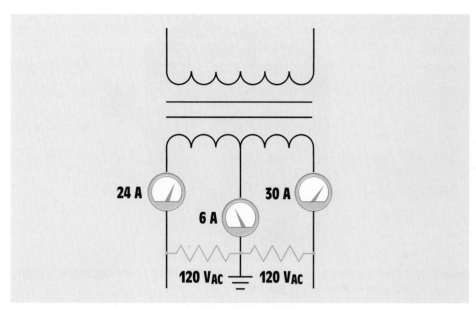

**Figure 26-21**  The neutral carries the sum of the unbalanced load.

is to carry the difference in current between the two secondary lines and maintain a balanced voltage. In *Figure 26-21* one of the secondary lines has a current flow of 30 A and the other has a current flow of 24 A. The neutral conducts the sum of the unbalanced load. In this example, the neutral current will be 6 A (30 − 24 = 6).

## Control Transformers

Another common type of isolation transformer found throughout industry is the **control transformer** *(Figure 26-22)*. The control transformer is used to reduce the line voltage to the value needed to operate control circuits. The most common type of control transformer contains two primary windings and one secondary. The primary windings are generally rated at 240 V each, and the secondary is rated at 120 V. This arrangement provides a 2:1 turns ratio between each of the primary windings and the secondary. For example, assume that each of the primary windings contains 200 turns of wire. The secondary will contain 100 turns of wire.

One of the primary windings in *Figure 26-23* is labeled $H_1$ and $H_2$. The other is labeled $H_3$ and $H_4$. The secondary winding is labeled $X_1$ and $X_2$. If the primary of the transformer is to be connected to 240 V, the two primary windings will be connected in parallel by connecting $H_1$ and $H_3$ together, and $H_2$ and $H_4$ together. When the primary windings are connected in parallel, the same voltage is applied across both windings. This has the same

**control trans-former**

**Figure 26-22** Control transformer with fuse protection added to the secondary winding *(Courtesy of Hevi-Duty Electric.)*

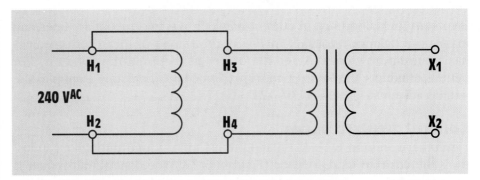

**Figure 26-23** Control transformer connected for 240-V operation

effect as using one primary winding with a total of 200 turns of wire. A turns ratio of 2:1 is maintained, and the secondary voltage will be 120 V.

If the transformer is to be connected to 480 V, the two primary windings will be connected in series by connecting $H_2$ and $H_3$ together *(Figure 26-24)*. The incoming power is connected to $H_1$ and $H_4$. Series-connecting the primary windings has the effect of increasing the number of turns in the primary to 400. This produces a turns ratio of 4:1. When 480 V is connected to the primary the secondary voltage will remain at 120.

The primary leads of a control transformer are generally cross-connected as shown in *Figure 26-25*. This is done so that metal links can be used

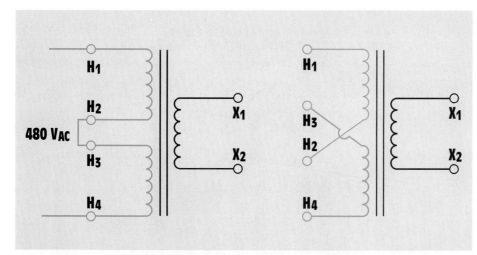

**Figure 26-24** Control transformer connected for 480-V operation

**Figure 26-25** The primary windings of a control transformer are crossed.

to connect the primary for 240- or 480-V operation. If the primary is to be connected for 240-V operation, the metal links will be connected under screws as shown in *Figure 26-26*. Notice that leads $H_1$ and $H_3$ are connected together and leads $H_2$ and $H_4$ are connected together. Compare this connection with the connection shown in *Figure 26-23*.

If the transformer is to be connected for 480-V operation, terminals $H_2$ and $H_3$ are connected as shown in *Figure 26-27*. Compare this connection with the connection shown in *Figure 26-24*.

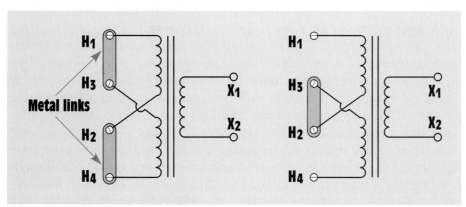

**Figure 26-26** Metal links connect transformer for 240-V operation.

**Figure 26-27** Control transformer connected for 480-V operation

**Figure 26-28** Core of a 600-MVA three-phase transformer *(Courtesy of Houston Lighting and Power.)*

### Transformer Core Types

Several types of cores are used in the construction of transformers. Most cores are made from thin steel punchings **laminated** together to form a solid metal core. The core for a 600-MVA (mega-volt-amp) three-phase transformer is shown in *Figure 26-28*. Laminated cores are preferred because a thin layer of oxide forms on the surface of each lamination and acts as an insulator to reduce the formation of eddy currents inside the core material. The amount of core material needed for a particular transformer is determined by the power rating of the transformer. The amount of core material must be sufficient to prevent saturation at full load. The type and shape of the core generally determine the amount of magnetic coupling between the windings and to some extent the efficiency of the transformer.

The transformer illustrated in *Figure 26-29* is known as a core type transformer. The windings are placed around each end of the core material. As a general rule, the low-voltage winding is placed closest to the core, and the high-voltage winding is placed over the low-voltage winding.

The shell type transformer is constructed in a similar manner to the core type, except that the shell type has a metal core piece through the

**laminated**

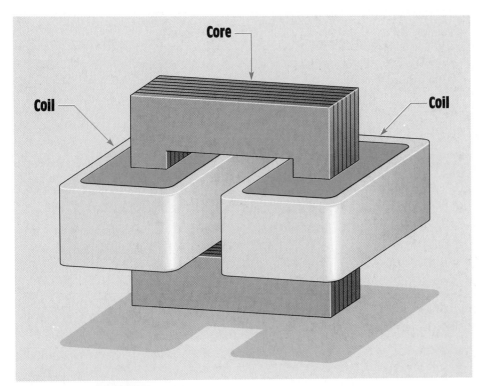

**Figure 26-29**  A core type transformer

middle of the window *(Figure 26-30)*. The primary and secondary windings are wound around the center core piece with the low-voltage winding being closest to the metal core. This arrangement permits the transformer to be surrounded by the core and provides excellent magnetic coupling. When the transformer is in operation, all the magnetic flux must pass through the center core piece. It then divides through the two outer core pieces.

The H type core shown in *Figure 26-31* is similar to the shell type core in that it has an iron core through its center around which the primary and secondary windings are wound. The H core, however, surrounds the windings on four sides instead of two. This extra metal helps reduce stray leakage flux and improve the efficiency of the transformer. The H type core is often found on high-voltage distribution transformers.

The **tape wound core** or **toroid core** *(Figure 26-32)* is constructed by tightly winding one long continuous silicon steel tape into a spiral. The tape may or may not be housed in a plastic container, depending on the application. This type of core does not require steel punchings laminated together. Since the core is one continuous length of metal, **flux leakage**

**tape wound core**

**toroid core**

**flux leakage**

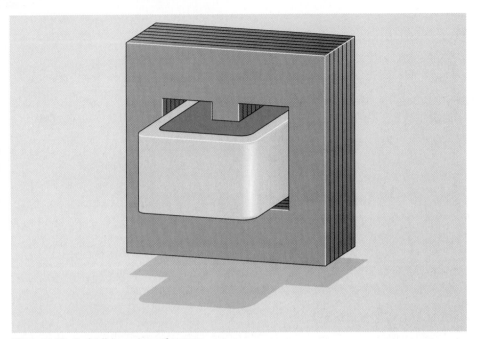

**Figure 26-30**  A shell type transformer

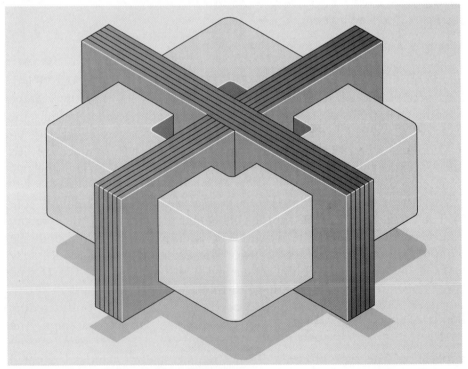

**Figure 26-31**  A transformer with an H type core

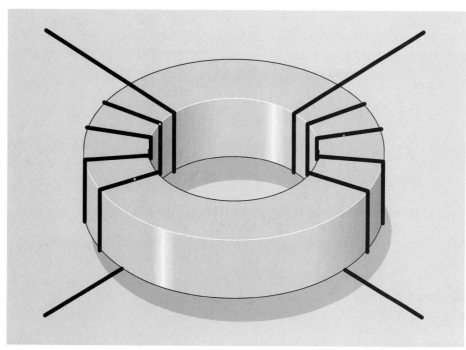

**Figure 26-32**  A toroid transformer

is kept to a minimum. Flux leakage is the amount of magnetic flux lines that do not follow the metal core and are lost to the surrounding air. The tape wound core is one of the most efficient core designs available.

## Transformer Inrush Current

A reactor is an inductor used to add inductance to the circuit. Although transformers and reactors are both inductive devices, there is a great difference in their operating characteristics. Reactors are often connected in series with a low-impedance load to prevent **inrush current** (the amount of current that flows when power is initially applied to the circuit) from becoming excessive *(Figure 26-33)*. Transformers, however, can produce extremely high inrush currents when power is first applied to the primary winding. The type of core used when constructing inductors and transformers is primarily responsible for this difference in characteristics.

## Magnetic Domains

Magnetic materials contain tiny magnetic structures in their molecular material known as *magnetic domains* (refer back to Unit 4). These domains can be affected by outside sources of magnetism. *Figure 26-34*

**inrush current**

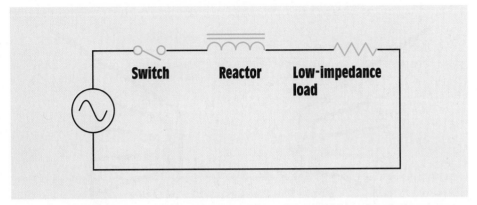

**Figure 26-33** Reactors are used to help prevent inrush current from becoming excessive when power is first turned on.

**Figure 26-34** Magnetic domain in neutral position

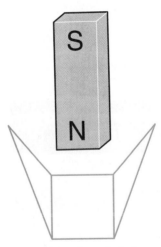

**Figure 26-35** Domain influenced by a north magnetic field

illustrates a magnetic domain that has not been polarized by an outside magnetic source.

Now assume that the north pole of a magnet is placed toward the top of the material that contains the magnetic domains *(Figure 26-35)*. Notice that the structure of the domain has changed to realign the molecules in the direction of the outside magnetic field. If the polarity of the magnetic pole is changed *(Figure 26-36)*, the molecular structure of the domain will change to realign itself with the new magnetic lines of flux. This external influence can be produced by an electromagnet as well as a permanent magnet.

In certain types of cores, the molecular structure of the domain will snap back to its neutral position when the magnetizing force is removed. This type of core is used in the construction of reactors or chokes *(Figure 26-37)*. A core of this type is constructed by separating sections of the steel laminations with an air gap. This air gap breaks the magnetic path through the core material and is responsible for the domains returning to their neutral position once the magnetizing force is removed.

The core construction of a transformer, however, does not contain an air gap. The steel laminations are connected together in such a manner as to produce a very low-reluctance path for the magnetic lines of flux. In this type of core, the domains remain in their set position once the magnetizing force has been removed. This type of core "remembers" where it was last set. This was the principle of operation of the core memory of early computers. It is also the reason that transformers can have extremely high inrush currents when they are first connected to the power line.

The amount of inrush current in the primary of a transformer is limited by three factors:

1. the amount of applied voltage.
2. the resistance of the wire in the primary winding, and
3. the flux change of the magnetic field in the core. The amount of flux change determines the amount of inductive reactance produced in the primary winding when power is applied.

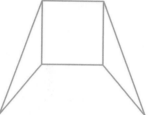

**Figure 26-36**  Domain influenced by a south magnetic field

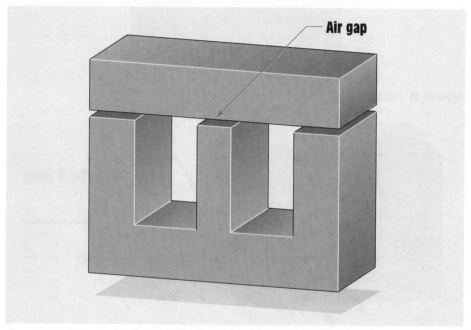

**Figure 26-37**  The core of an inductor contains an air gap.

*Figure 26-38* illustrates a simple isolation type transformer. The alternating current applied to the primary winding produces a magnetic field around the winding. As the current changes in magnitude and direction, the magnetic lines of flux change also. Since the lines of flux in the core are continually changing polarity, the magnetic domains in the core material are changing also. As stated previously, the magnetic domains in the core of a transformer remember their last set position. For this reason, the point on the wave form at which current is disconnected from the primary winding can have a great bearing on the amount of inrush current when the transformer is reconnected to power. For example, assume the power supplying the primary winding is disconnected at the zero crossing point *(Figure 26-39)*. In this instance, the magnetic domains would be set

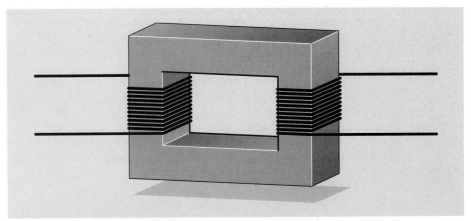

**Figure 26-38** Isolation transformer

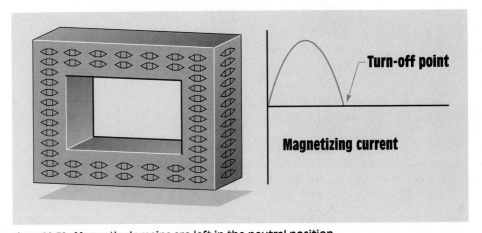

**Figure 26-39** Magnetic domains are left in the neutral position

at the neutral point. When power is restored to the primary winding, the core material can be magnetized by either magnetic polarity. This permits a change of flux, which is the dominant current-limiting factor. In this instance, the amount of inrush current would be relatively low.

If the power supplying current to the primary winding is interrupted at the peak point of the positive or negative half cycle, however, the domains in the core material will be set at that position. *Figure 26-40* illustrates this condition. It is assumed that the current was stopped as it reached its peak positive point. If the power is reconnected to the primary winding during the positive half cycle, only a very small amount of flux change can take place. Since the core material is saturated in the positive direction, the primary winding of the transformer is essentially an air

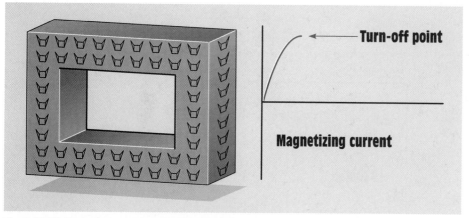

**Figure 26-40**  Domains are set at one end of magnetic polarity.

core inductor, which greatly decreases the inductive characteristics of the winding. The inrush current in this situation would be limited by the resistance of the winding and a very small amount of inductive reactance.

This characteristic of transformers can be demonstrated with a clamp-on ammeter that has a "peak hold" capability. If the ammeter is connected to one of the primary leads and power is switched on and off several times, the amount of inrush current will vary over a wide range.

## 26-3 Autotransformers

**Autotransformers** are one-winding transformers. They use the same winding for both the primary and secondary. The primary winding in *Figure 26-41* is between points B and N and has a voltage of 120 V applied to it. If the turns of wire are counted between points B and N, it can be seen that there are 120 turns of wire. Now assume that the selector switch is set to point D. The load is now connected between points D and N. The secondary of this transformer contains 40 turns of wire. If the amount of voltage applied to the load is to be computed the following formula can be used.

$$\frac{E_P}{E_S} = \frac{N_P}{N_S}$$

$$\frac{120}{E_S} = \frac{120}{40}$$

$$120\,E_S = 4800$$

$$E_S = 40\ V$$

**autotrans-formers**

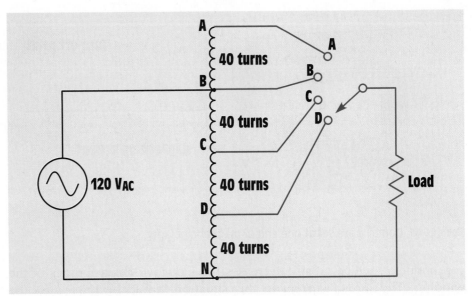

**Figure 26-41** An autotransformer has only one winding used for both the primary and secondary.

Assume that the load connected to the secondary has an impedance of 10 Ω. The amount of current flow in the secondary circuit can be computed using the formula

$$I = \frac{E}{Z}$$

$$I = \frac{40}{10}$$

$$I = 4 \text{ A}$$

The primary current can be computed by using the same formula that was used to compute primary current for an isolation type of transformer.

$$\frac{E_P}{E_S} = \frac{I_S}{I_P}$$

$$\frac{120}{40} = \frac{4}{I_P}$$

$$120 \, I_P = 160$$

$$I_P = 1.333 \text{ A}$$

The amount of power input and output for the autotransformer must be the same, just as they are in an isolation transformer.

| Primary | Secondary |
|---|---|
| 120 x 1.333 = 160 VA | 40 x 4 = 160 VA |

Now assume that the rotary switch is connected to point A. The load is now connected to 160 turns of wire. The voltage applied to the load can be computed by

$$\frac{E_P}{E_S} = \frac{N_P}{N_S}$$

$$\frac{120}{E_S} = \frac{120}{160}$$

$$120\, E_S = 19{,}200$$

$$E_S = 160\ V$$

Notice that the autotransformer, like the isolation transformer, can be either a step-up or step-down transformer.

If the rotary switch shown in *Figure 26-41* were to be removed and replaced with a sliding tap that made contact directly to the transformer winding, the turns ratio could be adjusted continuously. This type of transformer is commonly referred to as a Variac or Powerstat depending

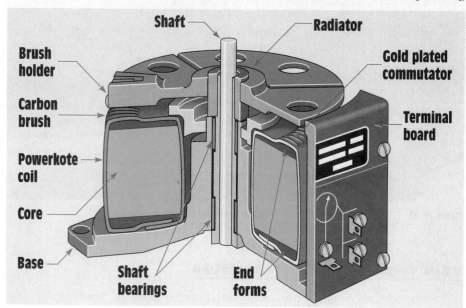

**Figure 26-42**  Cutaway view of a powerstat

on the manufacturer. A cutaway view of a variable autotransformer is shown in *Figure 26-42*. The windings are wrapped around a tape wound toroid core inside a plastic case. The tops of the windings have been milled flat to provide a commutator. A carbon brush makes contact with the windings.

Autotransformers are often used by power companies to provide a small increase or decrease to the line voltage. They help provide voltage regulation to large power lines. A 600-MVA three-phase autotransformer is shown in *Figure 26-43*. This transformer is contained in a housing filled with transformer oil, which acts as a coolant, and prevents moisture from forming in the windings.

The autotransformer does have one disadvantage. Since the load is connected to one side of the power line, there is no line isolation between the incoming power and the load. This can cause problems with certain types of equipment and must be a consideration when designing a power system.

**Figure 26-43**  Three-phase autotransformer *(Courtesy of Magnatek.)*

## 26-4 Transformer Polarities

To understand what is meant by transformer polarity, the voltage produced across a winding must be considered during some point in time. In

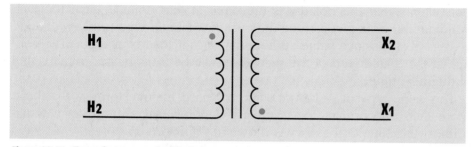

**Figure 26-44** Transformer polarity dots

a 60-Hz AC circuit, the voltage changes polarity 60 times per second. When discussing transformer polarity, it is necessary to consider the relationship between the different windings at the same point in time. It will, therefore, be assumed that this point in time is when the peak positive voltage is being produced across the winding.

## Polarity Markings on Schematics

When a transformer is shown on a schematic diagram it is common practice to indicate the polarity of the transformer windings by placing a dot beside one end of each winding as shown in *Figure 26-44*. These

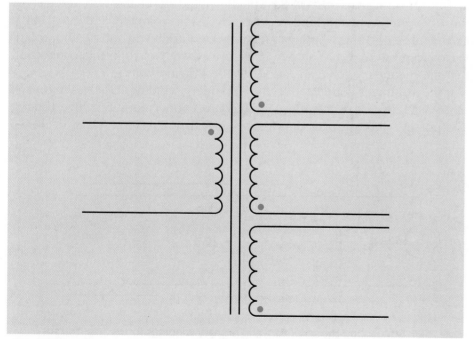

**Figure 26-45**  Polarity marks for multiple secondaries

dots signify that the polarity is the same at that point in time for each winding. For example, assume the voltage applied to the primary winding is at its peak positive value at the terminal indicated by the dot. The voltage at the dotted lead of the secondary will be at its peak positive value at the same time.

This same type of polarity notation is used for transformers that have more than one primary or secondary winding. An example of a transformer with a multisecondary is shown in *Figure 26-45*.

## Additive and Subtractive Polarities

The polarity of transformer windings can be determined by connecting them as an autotransformer and testing for additive or subtractive polarity, often referred to as a boost or buck connection. This is done by connecting one lead of the secondary to one lead of the primary and measuring the voltage across both windings *(Figure 26-46)*. The transformer shown in the example has a primary voltage rating of 120 V and a secondary voltage rating of 24 V. This same circuit has been redrawn in *Figure 26-47* to show the connection more clearly. Notice that the secondary winding has been connected in series with the primary winding. The transformer now contains only one winding and is, therefore, an autotransformer. When 120 V is applied to the primary winding, the voltmeter connected across the secondary will indicate either the *sum* of the two voltages or the *difference* between the two voltages. If this voltmeter indicates 144 V (120 + 24 = 144) the windings are connected additive (boost), and polarity dots can be placed as shown in *Figure 26-48*. Notice in this connection that the secondary voltage is added to the primary voltage.

If the voltmeter connected to the secondary winding indicates a voltage of 96 V (120 − 24 = 96) the windings are connected subtractive (buck), and polarity dots are placed as shown in *Figure 26-49*.

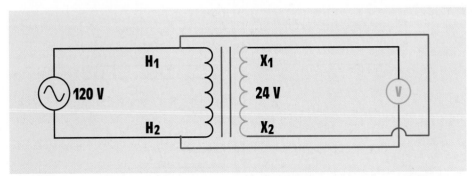

**Figure 26-46** Connecting the secondary and primary windings forms an autotransformer.

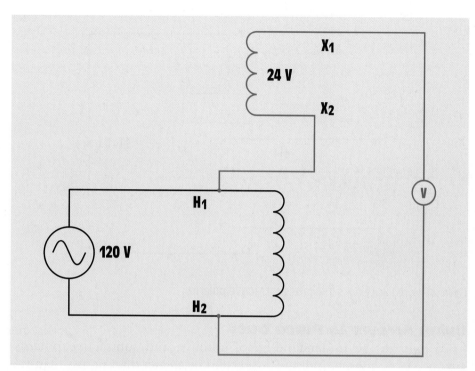

**Figure 26-47**   Redrawing the connection

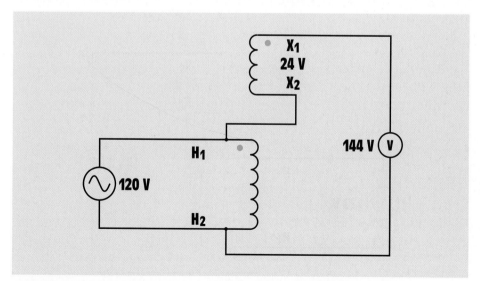

**Figure 26-48**   Placing polarity dots to indicate additive polarity

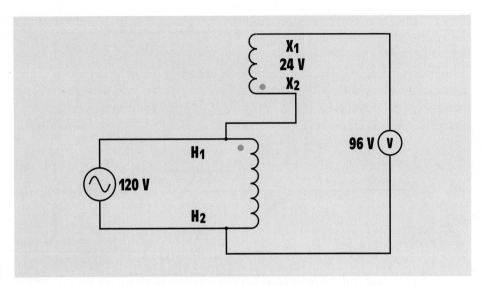

**Figure 26-49** Polarity dots indicate subtractive polarity.

## Using Arrows to Place Dots

To help in the understanding of additive and subtractive polarity, arrows can be used to indicate a direction of greater than or less than values. In *Figure 26-50,* arrows have been added to indicate the direction in which the dot is to be placed. In this example, the transformer is connected additive, or boost, and both of the arrows point in the same direction.

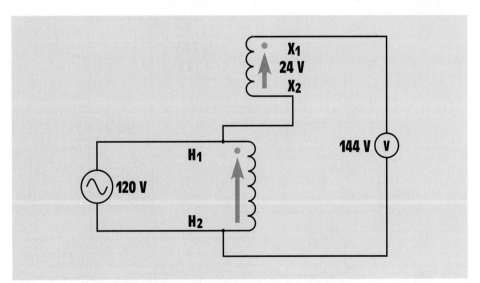

**Figure 26-50** Arrows help indicate the placement of the polarity dots.

Notice that the arrow points to the dot. In *Figure 26-51* it is seen that values of the two arrows add to produce 144 V.

In *Figure 26-52*, arrows have been added to a subtractive, or buck, connection. In this instance, the arrows point in opposite directions, and the voltage of one tries to cancel the voltage of the other. The result is that the smaller value is eliminated, and the larger value is reduced as shown in *Figure 26-53*.

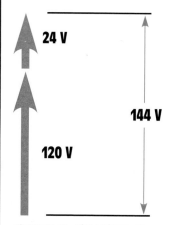

**Figure 26-51** The values of the arrows add to indicate additive polarity (boost connection).

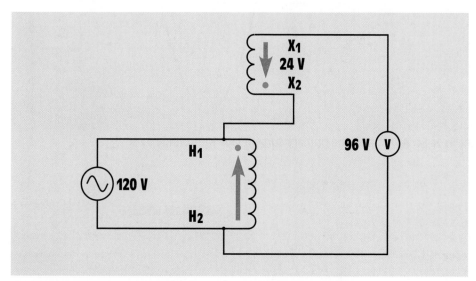

**Figure 26-52** The arrows help indicate subtractive polarity.

## 26-5 Voltage and Current Relationships in a Transformer

When the primary of a transformer is connected to power but there is no load connected to the secondary, current is limited by the inductive reactance of the primary. At this time, the transformer is essentially an inductor, and the excitation current is lagging the applied voltage by 90° *(Figure 26-54)*. The primary current induces a voltage in the secondary. This induced voltage is proportional to the rate of change of current. The secondary voltage will be maximum during the periods that the primary current is changing the most (0°, 180°, and 360°), and it will be zero when the primary current is not changing (90° and 270°). A plot of the primary current and secondary voltage shows that the secondary voltage lags the primary current by 90° *(Figure 26-55)*. Since the secondary voltage

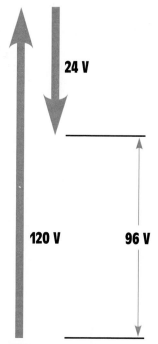

**Figure 26-53** The values of the arrows subtract (buck connection).

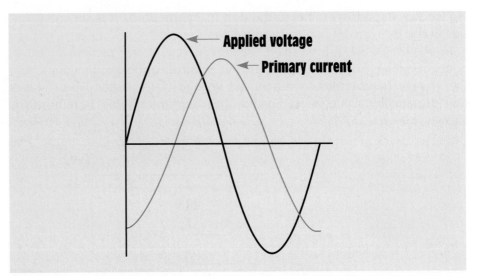

**Figure 26-54** At no load, the primary current lags the voltage by 90°.

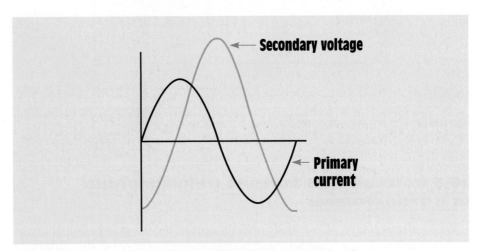

**Figure 26-55** The secondary voltage lags the primary current by 90°.

lags the primary current by 90° and the applied voltage leads the primary current by 90°, the secondary voltage is 180° out of phase with the applied voltage and in phase with the induced voltage in the primary.

## Adding Load to the Secondary

When a load is connected to the secondary, current begins to flow. Because the transformer is an inductive device, the secondary current lags the secondary voltage by 90°. Since the secondary voltage lags the primary current by 90°, the secondary current is 180° out of phase with the primary current *(Figure 26-56)*.

The current of the secondary induces a counter-voltage in the secondary windings that is in opposition to the counter-voltage induced in the primary. The counter-voltage of the secondary weakens the counter-voltage of the primary and permits more primary current to flow. As secondary current increases, primary current increases proportionally.

Since the secondary current causes a decrease in the counter-voltage produced in the primary, the current of the primary is limited less by inductive reactance and more by the resistance of the windings as load is added to the secondary. If a wattmeter were connected to the primary, you would see that the true power would increase as load was added to the secondary.

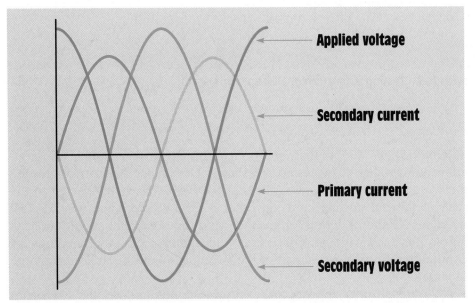

**Figure 26-56**  Voltage and current relationships of the primary and secondary windings

## 26-6 Testing the Transformer

Several tests can be made to determine the condition of the transformer. A simple test for grounds, shorts, or opens can be made with an ohmmeter *(Figure 26-57)*. Ohmmeter A is connected to one lead of the primary and one lead of the secondary. This test checks for shorted windings between the primary and secondary. The ohmmeter should indicate

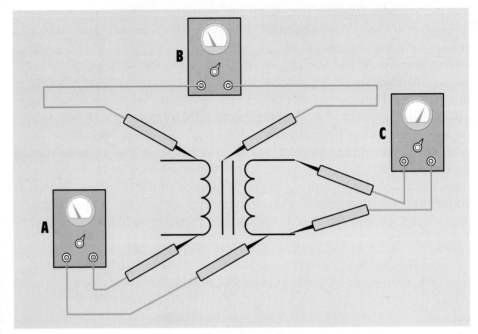

**Figure 26-57** Testing a transformer with an ohmmeter

infinity. If there is more than one primary or secondary winding, all isolated windings should be tested for shorts. Ohmmeter B illustrates testing the windings for grounds. One lead of the ohmmeter is connected to the case of the transformer, and the other is connected to the winding. All windings should be tested for grounds, and the ohmmeter should indicate infinity for each winding. Ohmmeter C illustrates testing the windings for continuity. The wire resistance of the winding should be indicated by the ohmmeter.

If the transformer appears to be in good condition after the ohmmeter test, it should then be tested for shorts and grounds with a megohmmeter. A MEGGER® will reveal problems of insulation breakdown that an ohmmeter will not. Large oil-filled transformers should have the condition of the dielectric oil tested at periodic intervals. This test involves taking a sample of the oil and performing certain tests for dielectric strength and contamination.

## 26-7 Transformer Ratings

Most transformers contain a nameplate that lists information concerning the transformer. The information listed is generally determined by the size, type, and manufacturer. Almost all nameplates will list the primary voltage, secondary voltage, and KVA (kilo-volt-amps) rating. Transformers are rated in kilo-volt-amps and not kilowatts because the true power is determined by the power factor of the load. Other information that may or may not be listed is frequency, temperature rise in C°, impedance, type of insulating oil, gallons of insulating oil, serial number, type number, model number, and whether the transformer is single-phase or three-phase.

## 26-8 Determining Maximum Current

The nameplate does not list the current rating of the windings. Since power input must equal power output, the current rating for a winding can be determined by dividing the KVA rating by the winding voltage. For example, assume a transformer has a KVA rating of 0.5 KVA, a primary voltage of 480 V, and a secondary voltage of 120 V. To determine the maximum current that can be supplied by the secondary, divide the KVA rating by the secondary voltage.

$$I_S = \frac{KVA}{E_S}$$

$$I_S = \frac{500}{120}$$

$$I_S = 4.16 \text{ A}$$

The primary current can be computed in the same way.

$$I_P = \frac{KVA}{E_P}$$

$$I_P = \frac{500}{480}$$

$$I_P = 1.04 \text{ A}$$

Transformers with multiple secondary windings will generally have the current rating listed with the voltage rating.

## Summary

1. All values of voltage, current, and impedance in a transformer are proportional to the turns ratio.

2. Transformers can change values of voltage, current, and impedance, but cannot change the frequency.

3. The primary winding of a transformer is connected to the power line.

4. The secondary winding is connected to the load.

5. A transformer that has a lower secondary voltage than primary voltage is a step-down transformer.

6. A transformer that has a higher secondary voltage than primary voltage is a step-up transformer.

7. An isolation transformer has its primary and secondary windings electrically and mechanically separated from each other.

8. When a coil induces a voltage into itself, it is known as self-induction.

9. When a coil induces a voltage into another coil, it is known as mutual induction.

10. Transformers can have very high inrush current when first connected to the power line because of the magnetic domains in the core material.

11. Inductors provide an air gap in their core material that causes the magnetic domains to reset to a neutral position.

12. Autotransformers have only one winding, which is used as both the primary and secondary.

13. Autotransformers have a disadvantage in that they have no line isolation between the primary and secondary winding.

14. Isolation transformers help filter voltage and current spikes between the primary and secondary side.

15. Polarity dots are often added to schematic diagrams to indicate transformer polarity.

16. Transformers can be connected as additive or subtractive polarity.

## Review Questions

1. What is a transformer?

2. What are common efficiencies for transformers?

3. What is an isolation transformer?

4. All values of a transformer are proportional to its

   _____ _____ .

5. What is an autotransformer?

6. What is a disadvantage of an autotransformer?

7. Explain the difference between a step-up and a step-down transformer.

8. A transformer has a primary voltage of 240 V and a secondary voltage of 48 V. What is the turns ratio of this transformer?

9. A transformer has an output of 750 VA. The primary voltage is 120 V. What is the primary current?

10. A transformer has a turns ratio of 1:6. The primary current is 18 A. What is the secondary current?

11. What do the dots shown beside the terminal leads of a transformer represent on a schematic?

12. A transformer has a primary voltage rating of 240 V and a secondary voltage rating of 80 V. If the windings were connected subtractive, what voltage would appear across the entire connection?

13. If the windings of the transformer in question 12 were to be connected additive, what voltage would appear across the entire winding?

14. The primary leads of a transformer are labeled 1 and 2. The secondary leads are labeled 3 and 4. If polarity dots are placed beside leads 1 and 4, which secondary lead would be connected to terminal 2 to make the connection additive?

**Practice Problems**

Refer to *Figure 26-58* to answer the following questions. Find all the missing values.

1.

| $E_P$ 120 | $E_S$ 24 |
|---|---|
| $I_P$ _____ | $I_S$ _____ |
| $N_P$ 300 | $N_S$ _____ |
| Ratio _____ | $Z = 3\ \Omega$ |

2.

| $E_P$ 240 | $E_S$ 320 |
|---|---|
| $I_P$ _____ | $I_S$ _____ |
| $N_P$ _____ | $N_S$ 280 |
| Ratio _____ | $Z = 500\ \Omega$ |

3.

| $E_P$ _____ | $E_S$ 160 |
|---|---|
| $I_P$ _____ | $I_S$ _____ |
| $N_P$ _____ | $N_S$ 80 |
| Ratio 1:2.5 | $Z = 12\ \Omega$ |

4.

| $E_P$ 48 | $E_S$ 240 |
|---|---|
| $I_P$ _____ | $I_S$ _____ |
| $N_P$ 220 | $N_S$ _____ |
| Ratio _____ | $Z = 360\ \Omega$ |

5.

| $E_P$ _____ | $E_S$ _____ |
|---|---|
| $I_P$ 16.5 | $I_S$ 3.25 |
| $N_P$ _____ | $N_S$ 450 |
| Ratio _____ | $Z = 56\ \Omega$ |

6.

| $E_P$ 480 | $E_S$ _____ |
|---|---|
| $I_P$ _____ | $I_S$ _____ |
| $N_P$ 275 | $N_S$ 525 |
| Ratio _____ | $Z = 1.2\ k\Omega$ |

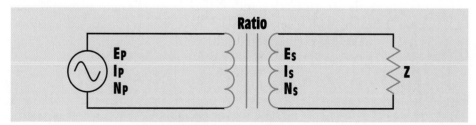

**Figure 26-58** Isolation transformer practice problems

Refer to *Figure 26-59* to answer the following questions. Find all the missing values .

7.

| $E_P$ 208 | $E_{S1}$ 320 | $E_{S2}$ 120 | $E_{S3}$ 24 |
|---|---|---|---|
| $I_P$ _____ | $I_{S1}$ _____ | $I_{S2}$ _____ | $I_{S3}$ _____ |
| $N_P$ 800 | $N_{S1}$ _____ | $N_{S2}$ _____ | $N_{S3}$ _____ |
| | Ratio 1: | Ratio 2: | Ratio 3: |
| | $R_1$ 12 k$\Omega$ | $R_2$ 6 $\Omega$ | $R_3$ 8 $\Omega$ |

8.

| $E_P$ 277 | $E_{S1}$ 480 | $E_{S2}$ 208 | $E_{S3}$ 120 |
|---|---|---|---|
| $I_P$ _____ | $I_{S1}$ _____ | $I_{S2}$ _____ | $I_{S3}$ _____ |
| $N_P$ 350 | $N_{S1}$ _____ | $N_{S2}$ _____ | $N_{S3}$ _____ |
| | Ratio 1: | Ratio 2: | Ratio 3: |
| | $R_1$ 200 $\Omega$ | $R_2$ 60 $\Omega$ | $R_3$ 24 $\Omega$ |

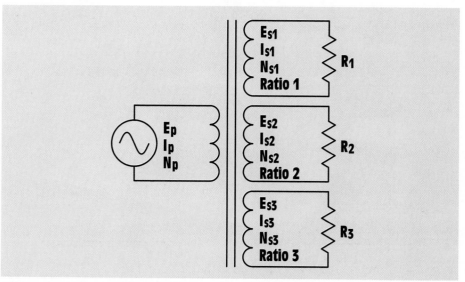

**Figure 26-59** Single-phase transformer with multi-secondaries

**The basic concepts of three-phase transformers are presented in this unit. Example problems are used to illustrate how values of voltage and current can be determined for different three-phase transformer connections.**

*Courtesy of Magnatek.*

# Three-Phase Transformers

## Outline

## Key Terms

Closing a delta

Delta-wye

Dielectric oil

High leg

Neutral conductor

One-line diagram

Open delta

Orange wire

Single-phase loads

Tagging

Three-phase bank

Wye-delta

**Objectives**

*After studying this unit, you should be able to:*

- Discuss the operation of three-phase transformers.
- Connect three single-phase transformers to form a three-phase bank.
- Calculate voltage and current values for a three-phase transformer connection.
- Connect two single-phase transformers to form a three-phase open delta connection.
- Discuss the characteristics of an open delta connection.
- Discuss different types of three-phase transformer connections and how they are used to supply single-phase loads.
- Calculate values of voltage and current for a three-phase transformer used to supply both three-phase and single-phase loads.

**Preview**

Three-phase transformers are used throughout industry to change values of three-phase voltage and current. Since three-phase power is the most common way in which power is produced, transmitted, and used, an understanding of how three-phase transformer connections are made is essential. This unit will discuss different types of three-phase transformer connections, and present examples of how values of voltage and current for these connections are computed.

## 27-1 Three-Phase Transformers

A three-phase transformer is constructed by winding three single-phase transformers on a single core *(Figure 27-1)*. A photograph of a three-phase transformer is shown in *Figure 27-2*. The transformer is shown before it is mounted in an enclosure, which will be filled with a **dielectric oil**. The dielectric oil performs several functions. Since it is a dielectric, it provides electrical insulation between the windings and the case. It is also used to help provide cooling and to prevent the formation of moisture, which can deteriorate the winding insulation.

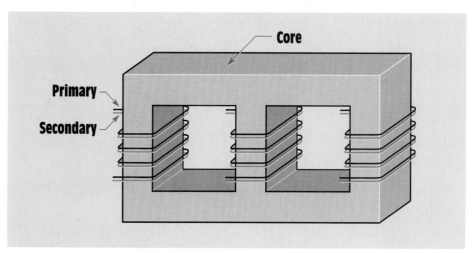

**Figure 27-1**  Basic construction of a three-phase transformer

## Three-Phase Transformer Connections

Three-phase transformers are connected in delta or wye configurations. A **wye-delta** transformer, for example, has its primary winding connected in a wye and its secondary winding connected in a delta *(Figure 27-3)*. A **delta-wye** transformer would have its primary winding connected in delta and its secondary connected in wye *(Figure 27-4)*.

## Connecting Single-Phase Transformers into a Three-Phase Bank

If three-phase transformation is needed and a three-phase transformer of the proper size and turns ratio is not available, three single-phase transformers can be connected to form a **three-phase bank**. When three single-phase transformers are used to make a three-phase transformer

**Figure 27-2** Three-phase transformer *(Courtesy of Magnatek.)*

bank, their primary and secondary windings are connected in a wye or delta connection. The three transformer windings in *Figure 27-5* are labeled A, B, and C. One end of each primary lead is labeled $H_1$, and the other end is labeled $H_2$. One end of each secondary lead is labeled $X_1$, and the other end is labeled $X_2$.

*Figure 27-6* shows three single-phase transformers labeled A, B, and C. The primary leads of each transformer are labeled $H_1$ and $H_2$, and the secondary leads are labeled $X_1$ and $X_2$. The schematic diagram of *Figure 27-5* will be used to connect the three single-phase transformers into a three-phase wye-delta connection as shown in *Figure 27-7*.

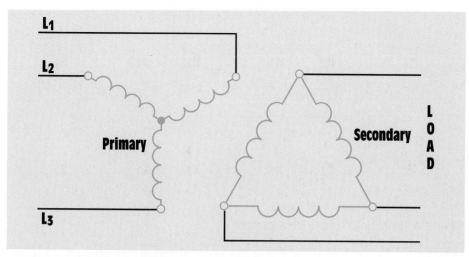

**Figure 27-3** Wye-delta connected three-phase transformer

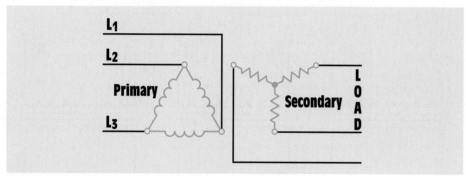

**Figure 27-4** Delta-wye connected three-phase transformer

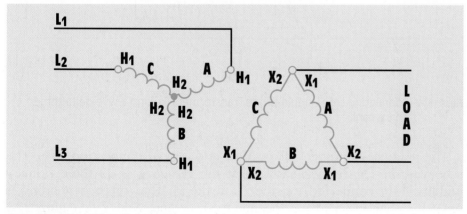

**Figure 27-5** Identifying the windings

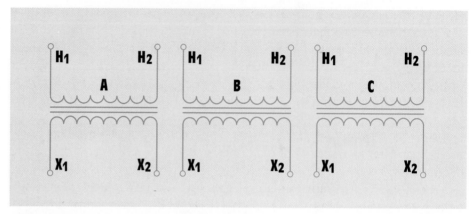

**Figure 27-6** Three single-phase transformers

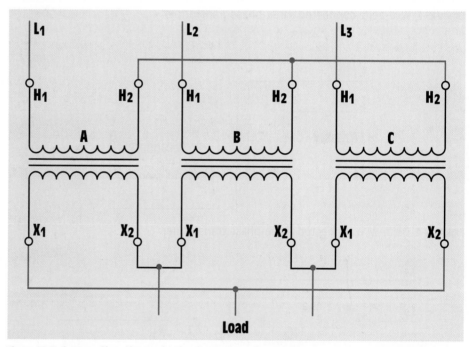

**Figure 27-7** Connecting three single-phase transformers to form a wye-delta three-phase bank

The primary winding will be tied into a wye connection first. The schematic in *Figure 27-5* shows that the $H_2$ leads of all three primary windings are connected together, and the $H_1$ lead of each winding is open for connection to the incoming power line. Notice in *Figure 27-7* that the $H_2$ leads of the primary windings are connected together, and the

$H_1$ lead of each winding has been connected to the incoming power line.

*Figure 27-5* shows that the $X_1$ lead of transformer A is connected to the $X_2$ lead of transformer C. Notice that this same connection has been made in *Figure 27-7*. The $X_1$ lead of transformer B is connected to the $X_2$ lead of transformer A, and the $X_1$ lead of transformer C is connected to the $X_2$ lead of transformer B. The load is connected to the points of the delta connection.

Although *Figure 27-5* illustrates the proper schematic symbology for a three-phase transformer connection, some electrical schematics and wiring diagrams do not illustrate three-phase transformer connections in this manner. One type of diagram, called the **one-line diagram**, would illustrate a delta-wye connection as shown in *Figure 27-8*. These diagrams are generally used to show the main power distribution system of a large industrial plant. The one-line diagram in *Figure 27-9* shows the main power to the plant and the transformation of voltages to different subfeeders. Notice that each transformer shows whether the primary and secondary are connected as a wye or delta and the secondary voltage of the subfeeder.

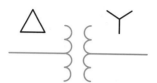

**Figure 27-8** One-line diagram symbol used to represent a delta-wye three-phase transformer connection

**one-line diagram**

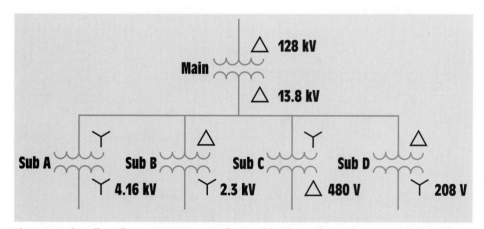

**Figure 27-9** One-line diagrams are generally used to show the main power distribution of a plant.

## 27-2 Closing a Delta

When **closing a delta**, connections should be checked for proper polarity before making the final connection and applying power. If the phase winding of one transformer is reversed, an extremely high current will flow when power is applied. Proper phasing can be checked with a

**closing a delta**

voltmeter as shown in *Figure 27-10*. If power is applied to the transformer bank before the delta connection is closed, the voltmeter should indicate 0 V. If one phase winding has been reversed, however, the voltmeter will indicate double the amount of voltage. For example, assume that the output voltage of a delta secondary is 240 V. If the voltage is checked before the delta is closed, the voltmeter should indicate a voltage of 0 V if all windings have been phased properly. If one winding has been reversed, however, the voltmeter will indicate a voltage of 480 V (240 + 240). This test will confirm whether a phase winding has been reversed, but it will not indicate whether the reversed winding is located in the primary or secondary. If either primary or secondary windings have been reversed, the voltmeter will indicate double the output voltage.

It should be noted, however, that a voltmeter is a high-impedance device. It is not unusual for a voltmeter to indicate some amount of voltage before the delta is closed, especially if the primary has been connected as a wye and the secondary as a delta. When this is the case, however, the voltmeter will generally indicate close to the normal output voltage if the connection is correct and double the output voltage if the connection is incorrect.

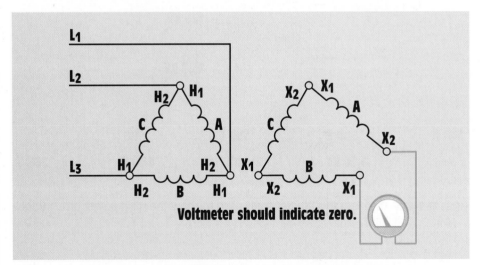

**Figure 27-10** Testing for proper transformer polarity before closing the delta

## 27-3 Three-Phase Transformer Calculations

To compute the values of voltage and current for three-phase transformers, the formulas used for making transformer calculations and three-phase calculations must be followed. Another very important rule is that **only phase values of voltage and current can be used when computing transformer values**. Refer to transformer A in *Figure 27-6*. All transformation of voltage and current takes place between the primary and secondary windings. Since these windings form the phase values of the three-phase connection, only phase and not line values can be used when calculating transformed voltages and currents.

A three-phase transformer connection is shown in *Figure 27-11*. Three single-phase transformers have been connected to form a wye-delta bank. The primary is connected to a three-phase line of 13,800 V, and the secondary voltage is 480 V. A three-phase resistive load with an impedance of 2.77 Ω per phase is connected to the secondary of the transformer. Compute the following values for this circuit.

| | | |
|---|---|---|
| $E_{P(PRIMARY)}$ | — | phase voltage of the primary |
| $E_{P(SECONDARY)}$ | — | phase voltage of the secondary |
| Ratio | — | turns ratio of the transformer |
| $E_{P(LOAD)}$ | — | phase voltage of the load bank |
| $I_{P(LOAD)}$ | — | phase current of the load bank |
| $I_{L(SECONDARY)}$ | — | secondary line current |
| $I_{P(SECONDARY)}$ | — | phase current of the secondary |
| $I_{P(PRIMARY)}$ | — | phase current of the primary |
| $I_{L(PRIMARY)}$ | — | line current of the primary |

## Solution

The primary windings of the three single-phase transformers have been connected to form a wye connection. In a wye connection, the phase

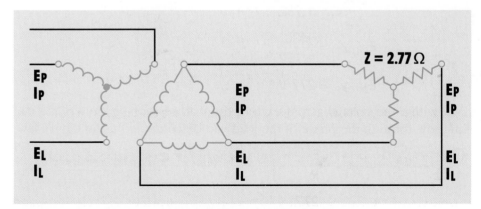

**Figure 27-11**   Example circuit 1 three-phase transformer calculation

**Example 1**

voltage is less than the line voltage by a factor of 1.732 (the square root of 3). Therefore, the phase value of the primary voltage can be computed using the formula

$$E_{P(PRIMARY)} = \frac{E_L}{1.732}$$

$$E_{P(PRIMARY)} = \frac{13,800}{1.732}$$

$$E_{P(PRIMARY)} = 7967.67 \text{ V}$$

The secondary windings are connected as a delta. In a delta connection, the phase voltage and line voltage are the same.

$$E_{P(SECONDARY)} = E_{L(SECONDARY)}$$

$$E_{P(SECONDARY)} = 480 \text{ V}$$

The turns ratio can be computed by comparing the phase voltage of the primary with the phase voltage of the secondary.

$$\text{Ratio} = \frac{\text{primary voltage}}{\text{secondary voltage}}$$

$$\text{Ratio} = \frac{7967.67}{480}$$

$$\text{Ratio} = 16.6:1$$

The load bank is connected in a wye connection. The voltage across the phase of the load bank will be less than the line voltage by a factor of 1.732.

$$E_{P(LOAD)} = \frac{E_{L(LOAD)}}{1.732}$$

$$E_{P(LOAD)} = \frac{480}{1.732}$$

$$E_{P(LOAD)} = 277 \text{ V}$$

Now that the voltage across each of the load resistors is known, the current flow through the phase of the load can be computed using Ohm's law.

$$I_{P(LOAD)} = \frac{E}{R}$$

$$I_{P(LOAD)} = \frac{277}{2.77}$$

$$I_{P(LOAD)} = 100 \text{ A}$$

Since the load is connected as a wye connection, the line current will be the same as the phase current.

$$I_{L(SECONDARY)} = 100 \text{ A}$$

The secondary of the transformer bank is connected as a delta. The phase current of the delta is less than the line current by a factor of 1.732.

$$I_{P(SECONDARY)} = \frac{I_L}{1.732}$$

$$I_{P(SECONDARY)} = \frac{100}{1.732}$$

$$I_{P(SECONDARY)} = 57.74 \text{ A}$$

The amount of current flow through the primary can be computed using the turns ratio. Since the primary has a higher voltage than the secondary, it will have a lower current. (Volts times amps input must equal volts times amps output.)

$$\text{primary current} = \frac{\text{secondary current}}{\text{turns ratio}}$$

$$I_{P(PRIMARY)} = \frac{57.74}{16.6}$$

$$I_{P(PRIMARY)} = 3.48 \text{ A}$$

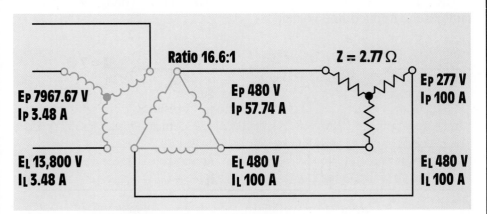

**Figure 27-12**   Example circuit 1 with all missing values

Since all transformed values of voltage and current take place across the phases, the primary has a phase current of 3.48 A. In a wye connection, the phase current is the same as the line current.

$$I_{L(PRIMARY)} = 3.48 \text{ A}$$

The transformer connection with all computed values is shown in *Figure 27-12*.

**Example 2**

A three-phase transformer is connected in a delta-delta configuration *(Figure 27-13)*. The load is connected as a wye, and each phase has an impedance of 7 Ω. The primary is connected to a line voltage of 4160 V and the secondary line voltage is 440 V. Find the following values.

| | |
|---|---|
| $E_{P(PRIMARY)}$ | — phase voltage of the primary |
| $E_{P(SECONDARY)}$ | — phase voltage of the secondary |
| **Ratio** | — turns ratio of the transformer |
| $E_{L(LOAD)}$ | — line voltage of the load |
| $E_{P(LOAD)}$ | — phase voltage of the load bank |
| $I_{P(LOAD)}$ | — phase current of the load bank |
| $I_{L(LOAD)}$ | — line current of the load |
| $I_{L(SECONDARY)}$ | — secondary line current |
| $I_{P(SECONDARY)}$ | — phase current of the secondary |
| $I_{P(PRIMARY)}$ | — phase current of the primary |
| $I_{L(PRIMARY)}$ | — line current of the primary |

### Solution

The primary is connected as a delta. The phase voltage will be the same as the applied line voltage.

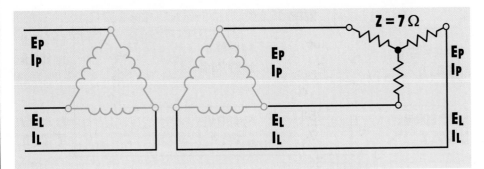

**Figure 27-13** Example circuit 2 three-phase transformer calculation

$$E_{P(PRIMARY)} = E_{L(PRIMARY)}$$

$$E_{P(PRIMARY)} = 4160 \text{ V}$$

The secondary of the transformer is connected as a delta also. Therefore, the phase voltage of the secondary will be the same as the line voltage of the secondary.

$$E_{P(SECONDARY)} = 440 \text{ V}$$

All transformer values must be computed using phase values of voltage and current. The turns ratio can be found by dividing the phase voltage of the primary by the phase voltage of the secondary.

$$\text{Ratio} = \frac{E_{P(PRIMARY)}}{E_{P(SECONDARY)}}$$

$$\text{Ratio} = \frac{4160}{440}$$

$$\text{Ratio} = 9.45:1$$

The load is connected directly to the output of the secondary. The line voltage applied to the load must, therefore, be the same as the line voltage of the secondary.

$$E_{L(LOAD)} = 440 \text{ V}$$

The load is connected in a wye. The voltage applied across each phase will be less than the line voltage by a factor of 1.732.

$$E_{P(LOAD)} = \frac{E_{L(LOAD)}}{1.732}$$

$$E_{P(LOAD)} = \frac{440}{1.732}$$

$$E_{P(LOAD)} = 254 \text{ V}$$

The phase current of the load can be computed using Ohm's law.

$$I_{P(LOAD)} = \frac{E_{P(LOAD)}}{Z}$$

$$I_{P(LOAD)} = \frac{254}{7}$$

$$I_{P(LOAD)} = 36.29 \text{ A}$$

The amount of line current supplying a wye-connected load will be the same as the phase current of the load.

$$I_{L(LOAD)} = 36.29 \text{ A}$$

Since the secondary of the transformer is supplying current to only one load, the line current of the secondary will be the same as the line current of the load.

$$I_{L(SECONDARY)} = 36.29 \text{ A}$$

The phase current in a delta connection is less than the line current by a factor of 1.732.

$$I_{P(SECONDARY)} = \frac{I_{L(SECONDARY)}}{1.732}$$

$$I_{P(SECONDARY)} = \frac{36.29}{1.732}$$

$$I_{P(SECONDARY)} = 20.95 \text{ A}$$

The phase current of the transformer primary can now be computed using the phase current of the secondary and the turns ratio.

$$I_{P(PRIMARY)} = \frac{I_{P(SECONDARY)}}{\text{turns ratio}}$$

$$I_{P(PRIMARY)} = \frac{20.95}{9.45}$$

$$I_{P(PRIMARY)} = 2.27 \text{A}$$

In this example, the primary of the transformer is connected as a delta. The line current supplying the transformer will be higher than the phase current by a factor of 1.732.

$$I_{L(PRIMARY)} = I_{P(PRIMARY)} \times 1.732$$

$$I_{L(PRIMARY)} = 2.27 \times 1.732$$

$$I_{L(PRIMARY)} = 3.93 \text{ A}$$

The circuit with all computed values is shown in *Figure 27-14*.

## 27-4 Open Delta Connection

The **open delta** transformer connection can be made with only two transformers instead of three *(Figure 27-15)*. This connection is often

**open delta**

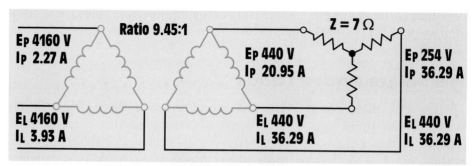

**Figure 27-14**   Example circuit 2 with all missing values

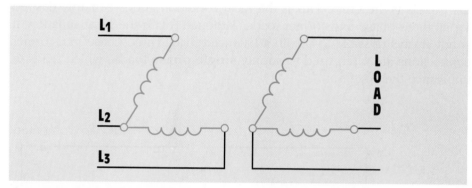

**Figure 27-15**   Open delta connection

used when the amount of three-phase power needed is not excessive, such as in a small business. It should be noted that the output power of an open delta connection is only 87% of the rated power of the two transformers. For example, assume two transformers, each having a capacity of 25 kVA (kilo-volt-amperes), are connected in an open delta connection. The total output power of this connection is 43.5 kVA (50 kVA x 0.87 = 43.5 kVA).

Another figure given for this calculation is 58%. This percentage assumes a closed delta bank containing three transformers. If three 25-kVA transformers were connected to form a closed delta connection, the total output power would be 75 kVA (3 x 25 kVA = 75 kVA). If one of these transformers were removed and the transformer bank operated as an open delta connection, the output power would be reduced to 58% of its original capacity of 75 kVA. The output capacity of the open delta bank is 43.5 kVA (75 kVA x 0.58 = 43.5 kVA).

The voltage and current values of an open delta connection are computed in the same manner as a standard delta-delta connection when three transformers are employed. The voltage and current rules for a delta connection must be used when determining line and phase values of voltage and current.

### 27-5 Single-Phase Loads

When true three-phase loads are connected to a three-phase transformer bank, there are no problems in balancing the currents and voltages of the individual phases. *Figure 27-16* illustrates this condition. In this circuit, a delta-wye three-phase transformer bank is supplying power to a wye-connected three-phase load in which the impedances of the three phases are the same. Notice that the amount of current flow in the phases is the same. This is the ideal condition and is certainly desired for all three-phase transformer loads. Although this is the ideal situation, it is not always possible to obtain a balanced load. Three-phase transformer connections are often used to supply **single-phase loads**, which tends to unbalance the system.

**single-phase loads**

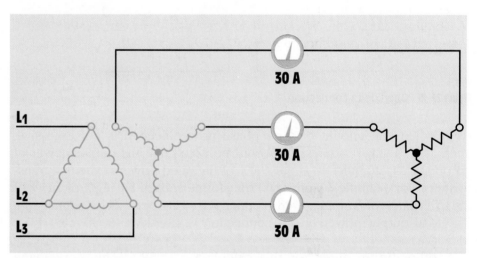

**Figure 27-16**   Three-phase transformer connected to a balanced three-phase load

### Open Delta Connection Supplying a Single-Phase Load

The type of three-phase transformer connection used is generally determined by the amount of power needed. When a transformer bank must supply both three-phase and single-phase loads, the utility company will often provide an open delta connection with one transformer center-tapped as shown in *Figure 27-17*. In this connection, it is assumed that the amount of three-phase power needed is 20 kVA, and the amount of single-phase power needed is 30 kVA. Notice that the transformer that has been center-tapped must supply power to both the three-phase and single-phase loads. Since this is an open delta connection, the transformer bank can be loaded to only 87% of its full capacity when supplying a three-phase load. The rating of the three-phase transformer bank must therefore be 23 kVA (20 kVA/0.87 = 23 kVA). Since the rating of the two transformers can be added to obtain a total output power rating, one transformer is rated at only half the total amount of power needed, or 12 kVA (23 kVA/2 = 11.5 kVA). The transformer that is used to supply power to the three-phase load only will be rated at 12 kVA. The transformer that has been center-tapped must supply power to both the single-phase and three-phase load. Its capacity will, therefore, be 42 kVA (12 kVA + 30 kVA). A 45-kVA transformer will be used.

## Voltage Values

The connection shown in *Figure 27-17* has a line-to-line voltage of 240 V. The three voltmeters $V_1$, $V_2$, and $V_3$ have all been connected across the three-phase lines and should indicate 240 V each. Voltmeters $V_4$ and $V_5$

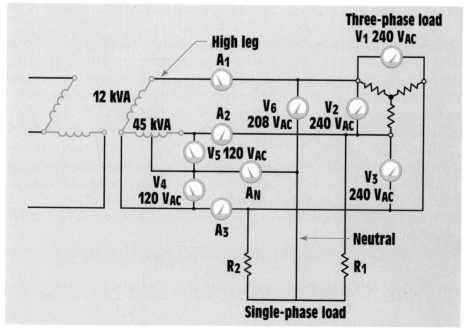

**Figure 27-17**   Three-phase open delta transformer supplying both three-phase and single-phase loads

have been connected between the two lines of the larger transformer and its center tap. These two voltmeters will indicate a voltage of 120 V each. Notice that it is these two lines and the center tap that are used to supply the single-phase power needed. The center tap of the larger transformer is used as a neutral conductor for the single-phase loads. Voltmeter $V_6$ has been connected between the center tap of the larger transformer and the line of the smaller transformer. This line is known as a **high leg**, because the voltage between this line and the neutral conductor will be higher than the voltage between the neutral and either of the other two conductors. The high-leg voltage can be computed by increasing the single-phase center-tapped voltage value by 1.732. In this case, the high-leg voltage will be 208 V (120 x 1.732 = 208). When this type of connection is employed, the *National Electrical Code* requires that the high leg be identified by connecting it to an **orange wire** or by **tagging** it at any point that it enters an enclosure with the neutral conductor.

**high leg**

**orange wire**

**tagging**

### Load Conditions

In the first load condition, it will be assumed that only the three-phase load is in operation and none of the single-phase load is operating. If the three-phase load is operating at maximum capacity, ammeters $A_1$, $A_2$, and $A_3$ will indicate a current flow of 48.1 A each (20 kVA/240 V x 1.732 = 48.1 A). Notice that only when the three-phase load is in operation is the current on each line balanced.

Now assume that none of the three-phase load is in operation and only the single-phase load is operating. If all the single-phase load is operating at maximum capacity, ammeters $A_2$ and $A_3$ will each indicate a value of 125 A (30 kVA/240 V = 125 A). Ammeter $A_1$ will indicate a current flow of 0 A because all the load is connected between the other two lines of the transformer connection. Ammeter $A_N$ will also indicate a value of 0 A. Ammeter $A_N$ is connected in the neutral conductor, and the neutral conductor carries the sum of the unbalanced load between the two phase conductors. Another way of stating this is to say that the neutral conductor carries the difference between the two line currents. Since both of these conductors are now carrying the same amount of current, the difference between them is 0 A.

Now assume that one side of the single-phase load, resistor $R_2$, has been opened and no current flows through it. If the other line maintains a current flow of 125 A, the neutral conductor will have a current flow of 125 A also (125 − 0 = 125).

Now assume that resistor $R_2$ has a value that will permit a current flow of 50 A on that phase. The neutral current will now be 75 A (125 − 50 = 75). Since the neutral conductor carries the sum of the unbalanced load, the neutral conductor never needs to be larger than the largest line conductor.

It will now be assumed that both three-phase and single-phase loads are operating at the same time. If the three-phase load is operating at maximum capacity, and the single-phase load is operating in such a manner that 125 A flow through resistor $R_1$ and 50 A flow through resistor $R_2$, the ammeters will indicate the following values:

$$A_1 = 48.1 \text{ A}$$

$$A_2 = 173.1 \text{ A} \ (48.1 + 125 = 173.1)$$

$$A_3 = 98.1 \text{ A} \ (48.1 + 50 = 98.1)$$

$$A_N = 75 \text{ A} \ (125 - 50 = 75)$$

Notice that the smaller of the two transformers is supplying current to only the three-phase load, but the larger transformer must supply current for both the single-phase and three-phase loads.

Although the circuit shown in *Figure 27-17* is the most common method of connecting both three-phase and single-phase loads to an open delta transformer bank, it is possible to use the high leg to supply power to a single-phase load also. The circuit shown in *Figure 27-18* is a circuit of this type. Resistors $R_1$ and $R_2$ are connected to the lines of the transformer that has been center-tapped, and resistor $R_3$ is connected to the line of the other transformer. If the line-to-line voltage is 240 V, voltmeters $V_1$ and $V_2$ will each indicate a value of 120 V across resistors $R_1$ and $R_2$. Voltmeter $V_3$, however, will indicate that a voltage of 208 V is applied across resistor $R_3$.

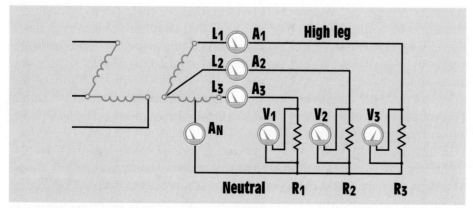

**Figure 27-18**   High leg supplies a single-phase load.

**neutral conductor**

## Calculating Neutral Current

The amount of current flow in the **neutral conductor** will still be the sum of the unbalanced load between lines $L_2$ and $L_3$, with the addition of the current flow in the high leg, $L_1$. To determine the amount of neutral current, use the formula

$$A_N = A_1 + (A_2 - A_3)$$

For example, assume line $L_1$ has a current flow of 100 A, line $L_2$ has a current flow of 75 A, and line $L_3$ has a current flow of 50 A. The amount of current flow in the neutral conductor would be

$$A_N = A_1 + (A_2 - A_3)$$
$$A_N = 100 + (75 - 50)$$
$$A_N = 100 + 25$$
$$A_N = 125 \ A$$

In this circuit, it is possible for the neutral conductor to carry more current than any of the three-phase lines. This circuit is more of an example of why the *National Electrical Code* requires a high leg to be identified than it is a practical working circuit. It is seldom that the high-leg side of this type of connection will be connected to the neutral conductor.

## 27-6 Closed Delta with Center Tap

Another three-phase transformer configuration used to supply power to single-phase and three-phase loads is shown in *Figure 27-19*. This circuit is virtually identical to the circuit shown in *Figure 27-17*, with the exception that a third transformer has been added to close the delta. Closing the delta permits more power to be supplied for the operation of three-phase loads. In this circuit, it is assumed that the three-phase load has a power requirement of 75 kVA, and the single-phase load requires an additional 50 kVA. Three 25-kVA transformers could be used to supply the three-phase power needed (25 kVA x 3 = 75 kVA). The addition of the single-phase load, however, requires one of the transformers to be larger. This transformer must supply both the three-phase and single-phase load, which requires it to have a rating of 75 kVA (25 kVA + 50 kVA = 75 kVA).

In this circuit, the primary is connected in a delta configuration. Since the secondary side of the transformer bank is a delta connection, either a wye or a delta primary could have been used. This, however, will not be true of all three-phase transformer connections supplying single-phase loads.

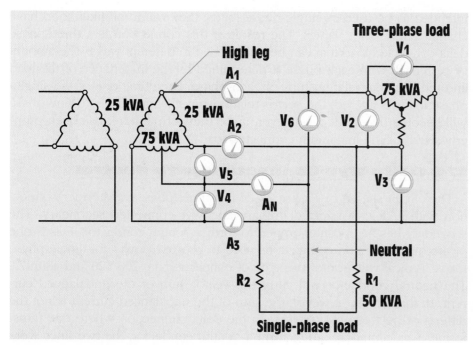

**Figure 27-19**  Closed delta connection with high leg

## 27-7 Closed Delta without Center Tap

In the circuit shown in *Figure 27-20,* the transformer bank has been connected in a wye-delta configuration. Notice that there is no transformer secondary with a center-tapped winding. In this circuit, there is no neutral conductor. The three loads have been connected directly across the three-phase lines. Since these three loads are connected directly across the lines, they form a delta-connected load. If these three loads are

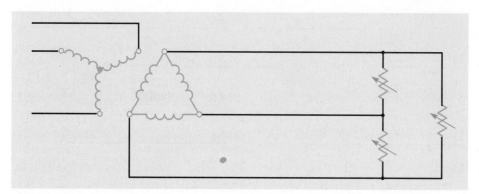

**Figure 27-20**  Single-phase loads supplied by a wye-delta transformer connection

intended to be used as single-phase loads, they will in all likelihood have changing resistance values. The result of this connection is a three-phase delta-connected load that can be unbalanced in different ways. The amount of current flow in each phase is determined by the impedance of the load and the vectorial relationships of each phase. Each time one of the single-phase loads is altered, the vector relationship changes also. No one phase will become overloaded, however, if the transformer bank has been properly sized for the maximum connected load.

## 27-8 Delta-Wye Connection with Neutral

The circuit shown in *Figure 27-21* is a three-phase transformer connection with a delta-connected primary and wye-connected secondary. The secondary has been center-tapped to form a neutral conductor. This is one of the most common connections used to provide power for single-phase loads. Typical voltages for this type of connection are 208/120 and 480/277. The neutral conductor will carry the vector sum of the unbalanced current. In this circuit, however, the sum of the unbalanced current is not the difference between two phases. In the delta connection where one transformer was center-tapped to form a neutral conductor, the two lines were 180° out of phase when compared with the center tap. In the wye connection, the lines will be 120° out of phase. When all three lines are carrying the same amount of amperage, the neutral current will be zero.

A wye-connected secondary with center tap can, under the right conditions, experience extreme unbalance problems. **If this transformer connection is powered by a three-phase three-wire system, the primary winding must be connected in a delta configuration.** If the primary is connected as a wye connection, the circuit will become exceedingly unbalanced when load is added to the circuit. Connecting the

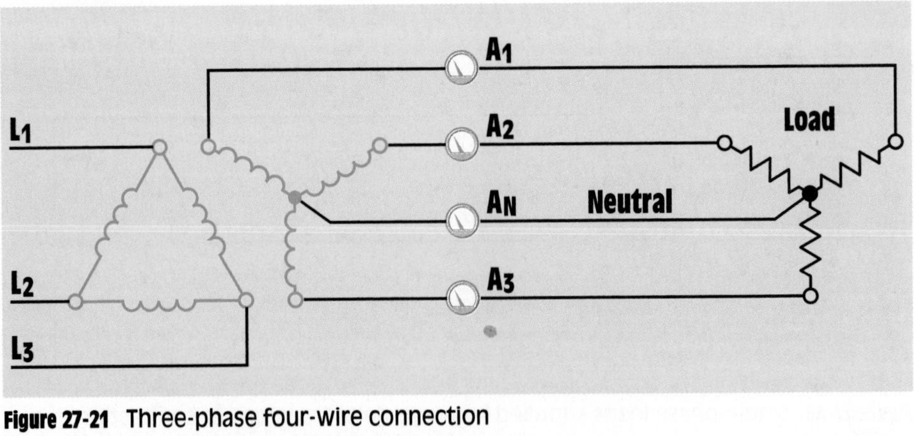

**If this transformer connection is powered by a three-phase three-wire system, the primary winding must be connected in a delta configuration.**

**Figure 27-21**  Three-phase four-wire connection

center tap of the primary to the center tap of the secondary will not solve the unbalance problem if a wye primary is used on a three-wire system.

If the incoming power is a three-phase four-wire system as shown in *Figure 27-22*, however, a wye-connected primary can be used without problem. The neutral conductor connected to the center tap of the primary prevents the unbalance problems. It is a common practice with this type of connection to tie the neutral conductor of both primary and secondary together as shown. When this is done, however, line isolation between the primary and secondary windings is lost.

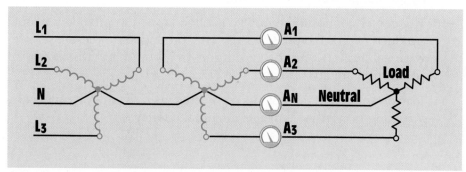

**Figure 27-22** Neutral conductor is supplied by the incoming power.

## Summary

1. Three-phase transformers are constructed by winding three separate transformers on the same core material.

2. Single-phase transformers can be used as a three-phase transformer bank by connecting their primary and secondary windings as either wyes or deltas.

3. When computing three-phase transformer values, the rules for three-phase circuits must be followed as well as the rules for transformers.

4. Phase values of voltage and current must be used when computing the values associated with the transformer.

5. The total power output of a three-phase transformer bank will be the sum of the rating of the three transformers.

6. An open delta connection can be made with the use of only two transformers.

7. When an open delta connection is used, the total output power is 87% of the sum of the power rating of the two transformers.

8. It is common practice to center-tap one of the transformers in a delta connection to provide power for single-phase loads. When this is done, the remaining phase connection becomes a high leg.

9. The *National Electrical Code* requires that a high leg be identified by an orange wire or by tagging.

10. The center connection of a wye is often tapped to provide a neutral conductor for three-phase loads. This produces a three-phase four-wire system. Common voltages produced by this type of connection are 208/120 and 480/277.

11. Transformers should not be connected as a wye-wye unless the incoming power line contains a neutral conductor.

## Review Questions

1. How many transformers are needed to make an open delta connection?

2. Two transformers rated at 100 kVA each are connected in an open delta connection. What is the total output power that can be supplied by this bank?

3. How does the *National Electrical Code* specify that the high leg of a four-wire delta connection be marked?

4. An open delta three-phase transformer system has one transformer center-tapped to provide a neutral for single-phase voltages. If the voltage from line to center tap is 277 V, what is the high-leg voltage?

5. If a single-phase load is connected across the two line conductors and neutral of the above transformer, and one line has a current of 80 A and the other line has a current of 68 A, how much current is flowing in the neutral conductor?

6. A three-phase transformer connection has a delta-connected secondary and one of the transformers has been center-tapped to form a

neutral conductor. The phase-to-neutral value of the center-tapped secondary winding is 120 V. If the high leg is connected to a single-phase load, how much voltage will be applied to that load?

7. A three-phase transformer connection has a delta-connected primary and a wye-connected secondary. The center tap of the wye is used as a neutral conductor. If the line-to-line voltage is 480 V, what is the voltage between any one phase conductor and the neutral conductor?

8. A three-phase transformer bank has the secondary connected in a wye configuration. The center tap is used as a neutral conductor. If the voltage across any phase conductor and neutral is 120 V, how much voltage would be applied to a three-phase load connected to the secondary of this transformer bank?

9. A three-phase transformer bank has the primary and secondary windings connected in a wye configuration. The secondary center tap is being used as a neutral to supply single-phase loads. Will connecting the center tap connection of the secondary to the center tap connection of the primary permit the secondary voltage to stay in balance when a single-phase load is added to the secondary?

10. Referring to the transformer connection in question 9, if the center tap of the primary is connected to a neutral conductor on the incoming power, will it permit the secondary voltages to be balanced when single-phase loads are added?

**Practice Problems**

Refer to the transformer shown in *Figure 27-11* and find all the missing values.

1.

| Primary | Secondary | Load |
|---------|-----------|------|
| $E_P$ _____ | $E_P$ _____ | $E_P$ _____ |
| $I_P$ _____ | $I_P$ _____ | $I_P$ _____ |

| | | |
|---|---|---|
| $E_L$ 4160 | $E_L$ 440 | $E_L$ _____ |
| $I_L$ _____ | $I_L$ _____ | $I_L$ _____ |
| Ratio | $Z = 3.5\ \Omega$ | |

2.

| Primary | Secondary | Load |
|---|---|---|
| $E_P$ _____ | $E_P$ _____ | $E_P$ _____ |
| $I_P$ _____ | $I_P$ _____ | $I_P$ _____ |
| $E_L$ 7200 | $E_L$ 240 | $E_L$ _____ |
| $I_L$ _____ | $I_L$ _____ | $I_L$ _____ |
| Ratio | $Z = 4\ \Omega$ | |

Refer to the transformer connection shown in *Figure 27-23* and fill in the missing values.

3.

| Primary | Secondary | Load |
|---|---|---|
| $E_P$ _____ | $E_P$ _____ | $E_P$ _____ |
| $I_P$ _____ | $I_P$ _____ | $I_P$ _____ |
| $E_L$ 13,800 | $E_L$ 480 | $E_L$ _____ |
| $I_L$ _____ | $I_L$ _____ | $I_L$ _____ |
| Ratio | $Z = 2.5\ \Omega$ | |

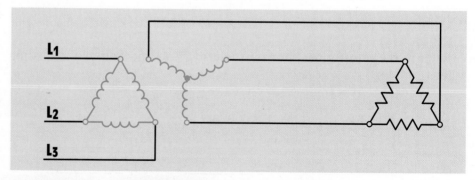

**Figure 27-23** Practice problems circuit

4.

| Primary | Secondary | Load |
|---------|-----------|------|
| $E_P$ _____ | $E_P$ _____ | $E_P$ _____ |
| $I_P$ _____ | $I_P$ _____ | $I_P$ _____ |
| $E_L$ 23,000 | $E_L$ 208 | $E_L$ _____ |
| $I_L$ _____ | $I_L$ _____ | $I_L$ _____ |
| Ratio | $Z = 3\ \Omega$ | |

**11**

# direct current machines

A generator is a device that converts mechanical energy into electrical energy. This unit discusses the different types of machines used to produce direct current.

*Courtesy of Miller Electric Manufacturing Co.*

# Direct Current Generators

## Key Terms

Armature

Armature reaction

Brushes

Commutator

Compound generators

Compounding

Countertorque

Cumulative compound

Differential compound

Field excitation current

Frogleg wound armatures

Generator

Lap wound armatures

Left-hand generator rule

Pole pieces

Series field diverter

Series field windings

Series generator

Shunt field windings

Shunt generators

Wave wound armatures

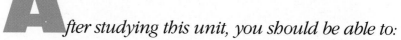

**A**fter studying this unit, you should be able to:

- Discuss the theory of operation of direct current generators.
- List the factors that determine the amount of output voltage produced by a generator.
- List the three major types of direct current generators.
- List different types of armature windings.
- Describe the differences between series and shunt field windings.
- Discuss the operating differences between different types of generators.
- Draw schematic diagrams for different types of direct current generators.
- Set the brushes to the neutral plane position on the commutator of a DC machine.

**Preview**

    Although most of the electric power generated throughout the world is alternating current, direct current is used for some applications. Many industrial plants use direct current generators to produce the power needed to operate large direct current motors. Direct current motors have characteristics that make them superior to alternating current motors for certain applications. Direct current generators and motors are also used in diesel locomotives. The diesel engine in most locomotives is used to operate a large direct current generator. The generator is used to provide power to direct current motors connected to the wheels.

## 28-1 What Is a Generator?

A **generator** is a device that converts mechanical energy into electrical energy. Direct current generators operate on the principle of magnetic induction. In Unit 13 it was shown that a voltage is induced in a conductor when it cuts magnetic lines of flux *(Figure 28-1)*. In this example, the ends of the wire loop have been connected to two sliprings mounted on the shaft. Brushes are used to carry the current from the loop to the outside circuit.

In *Figure 28-2*, an end view of the shaft and wire loop is shown. At this particular instant, the loop of wire is parallel to the magnetic lines of flux, and no cutting action is taking place. Since the lines of flux are not being cut by the loop, no voltage is induced in the loop.

In *Figure 28-3*, the shaft has been turned 90° clockwise. The loop of wire cuts through the magnetic lines of flux, and a voltage is induced in the loop. When the loop is rotated 90°, it is cutting the maximum number of lines of flux per second and the voltage reaches its maximum, or peak, value.

After another 90° of rotation *(Figure 28-4)*, the loop has completed 180° of rotation and is again parallel to the lines of flux. As the loop was turned, the voltage decreased until it again reached zero.

**generator**

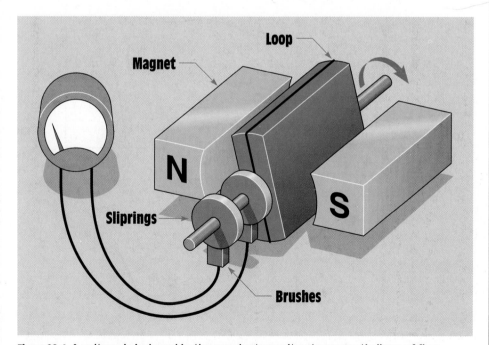

**Loop**

**Magnet**

**N**

**S**

**Sliprings**

**Brushes**

**Figure 28-1** A voltage is induced in the conductor as it cuts magnetic lines of flux.

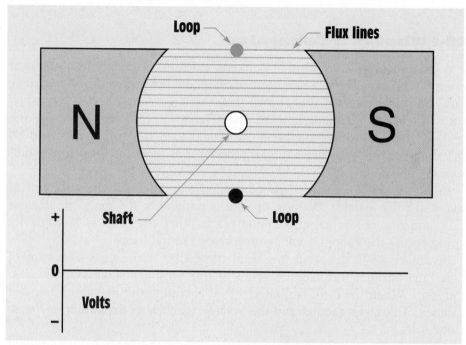

**Figure 28-2** The loop is parallel to the lines of flux and no cutting action is taking place.

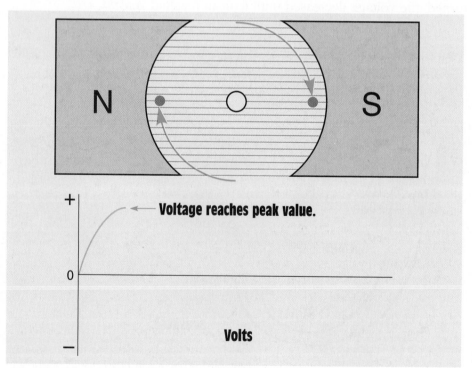

**Figure 28-3** Induced voltage after 90° of rotation

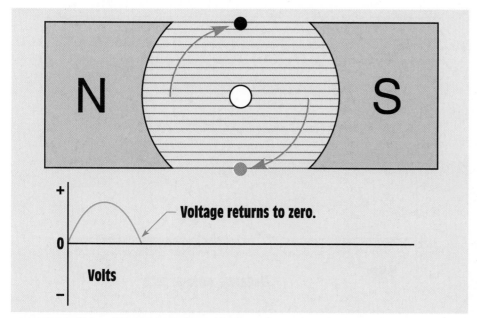

**Figure 28-4** Induced voltage after 180° of rotation

As the loop continues to turn, the conductors again cut the lines of magnetic flux *(Figure 28-5)*. This time, however, the conductor that previously cut through the flux lines of the south magnetic field is cutting the lines of the north magnetic field, and the conductor that previously cut the lines of the north magnetic field is cutting the lines of the south field. Since the conductors are cutting the flux lines of opposite magnetic polarity, the polarity of voltage reverses. After 270° of rotation, the loop has rotated to the position shown, and the maximum amount of voltage in the negative direction is being produced.

After another 90° of rotation, the loop has completed one rotation of 360° and returned to its starting position *(Figure 28-6)*. The voltage decreased from its negative peak back to zero. Notice that the voltage produced in the **armature** (the rotating member of the machine), alternates polarity. **The voltage produced in all rotating armatures is alternating voltage**.

Since direct-current generators must produce DC current instead of AC current, some device must be used to change the alternating voltage produced in the armature windings into direct voltage before it leaves the generator. This job is performed by the **commutator**. The commutator is constructed from a copper ring split into segments with insulating material between the segments *(Figure 28-7)*. Brushes riding against the commutator segments carry the power to the outside circuit.

**armature**

The voltage produced in all rotating armatures is alternating voltage.

**commu- tator**

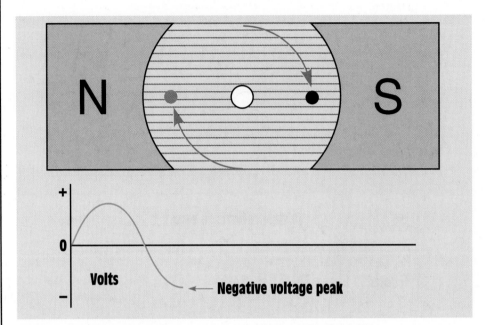

**Figure 28-5** The negative voltage peak is reached after 270° of rotation.

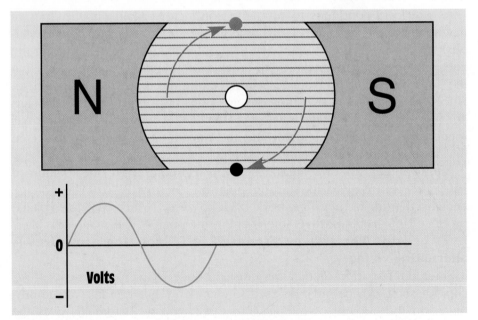

**Figure 28-6** Voltage produced after 360° of rotation

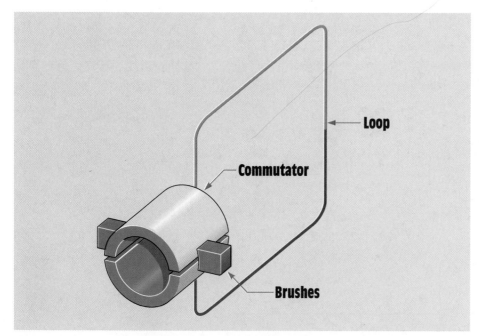

**Figure 28-7** The commutator is used to convert the AC voltage produced in the armature into DC voltage.

In *Figure 28-8*, the loop has been placed between the poles of two magnets. At this point in time, the loop is parallel to the magnetic lines of flux, and no voltage is induced in the loop. Note that the brushes make contact with both of the commutator segments at this time. The position at which the windings are parallel to the lines of flux and there is no induced voltage is called the neutral plane. The brushes should be set to make contact between commutator segments when the armature windings are in the neutral plane position.

As the loop rotates, the conductors begin to cut through the magnetic lines of flux. The conductor cutting through the south magnetic field is connected to the positive brush, and the conductor cutting through the north magnetic field is connected to the negative brush *(Figure 28-9)*. Since the loop is cutting lines of flux, a voltage is induced into the loop. After 90° of rotation, the voltage reaches its most positive point.

As the loop continues to rotate, the voltage decreases to zero. After 180° of rotation, the conductors are again parallel to the lines of flux, and no voltage is induced in the loop. Note that the brushes again make contact with both segments of the commutator at the time when there is no induced voltage in the conductors *(Figure 28-10)*.

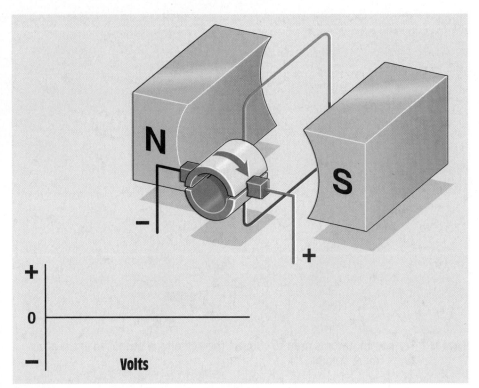

**Figure 28-8** The loop is parallel to the lines of flux.

During the next 90° of rotation, the conductors again cut through the magnetic lines of flux. This time, however, the conductor that previously cut through the south magnetic field is now cutting the flux lines of the north field, and the conductor that previously cut the lines of flux of the north magnetic field is cutting the lines of flux of the south magnetic field *(Figure 28-11)*. Since these conductors are cutting the lines of flux of opposite magnetic polarities, the polarity of induced voltage is different for each of the conductors. The commutator, however, maintains the correct polarity to each brush. The conductor cutting through the north magnetic field will always be connected to the negative brush, and the conductor cutting through the south field will always be connected to the positive brush. Since the polarity at the brushes has remained constant, the voltage will increase to its peak value in the same direction.

As the loop continues to rotate *(Figure 28-12)*, the induced voltage again decreases to zero when the conductors become parallel to the magnetic lines of flux. Notice that during this 360° rotation of the loop the polarity of voltage remained the same for both halves of the wave form.

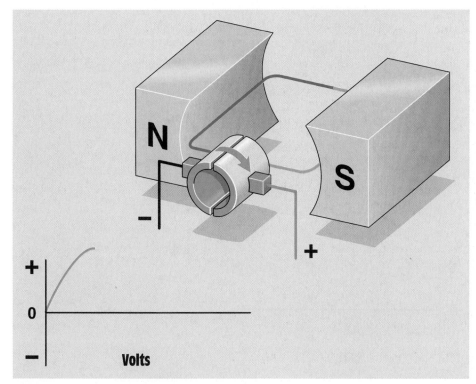

**Figure 28-9** The loop has rotated 90°.

This is called rectified DC voltage. The voltage is pulsating or fluctuating. It does turn on and off, but it never reverses polarity. Since the polarity for each brush remains constant, the output voltage is DC.

To increase the amount of output voltage, it is common practice to increase the number of turns of wire for each loop *(Figure 28-13)*. If a loop contains 20 turns of wire, the induced voltage will be 20 times greater than that for a single-loop conductor. The reason for this is that each loop is connected in series with the other loops. Since the loops form a series path, the voltage induced in the loops will add. In this example, if each loop has an induced voltage of 2 V, the total voltage for this winding would be 40 V (2 V x 20 loops = 40 V).

It is also common practice to use more than one loop of wire *(Figure 28-14)*. When more than one loop is used, the average output voltage is higher and there is less pulsation of the rectified voltage. This pulsation is called *ripple*.

The loops are generally placed in slots of an iron core *(Figure 28-15)*. The iron acts as a magnetic conductor by providing a low-reluctance path

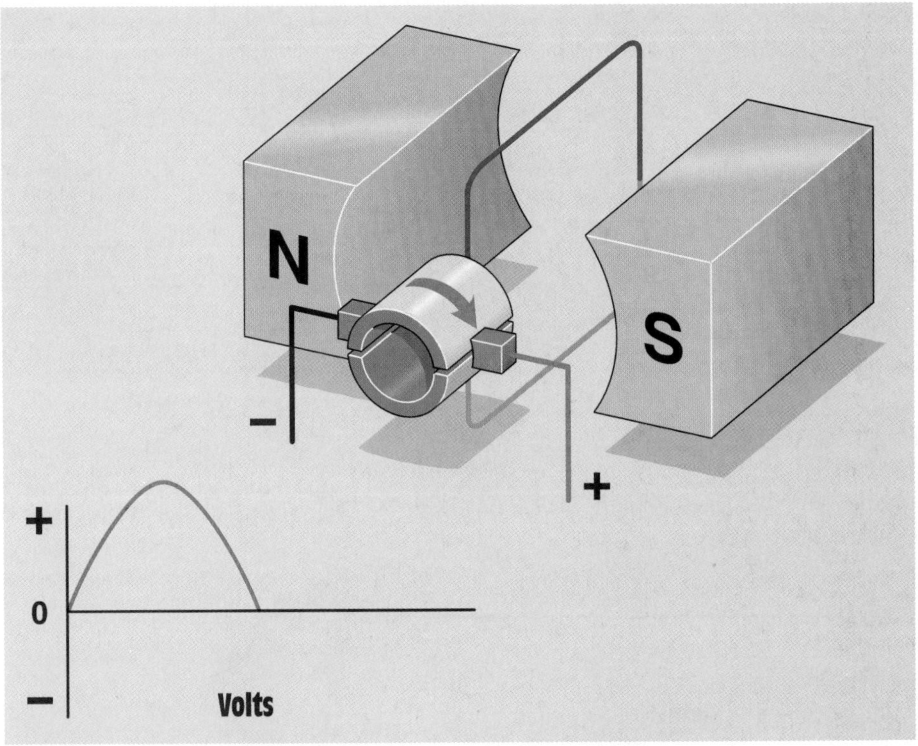


**Figure 28-10** The loop has rotated 180°.

for magnetic lines of flux to increase the inductance of the loops and provide a higher induced voltage. The commutator is connected to the slotted iron core. The entire assembly of iron core, commutator, and windings is called the armature *(Figure 28-16)*. The windings of armatures are connected in different ways depending on the requirements of the machine. The three basic types of armature windings are the *lap, wave,* and *frogleg.*

**lap wound armatures**

**Lap wound armatures are used in machines designed for low voltage and high current.**

## 28-2 Armature Windings

### Lap Wound Armatures

***Lap wound armatures* are used in machines designed for low voltage and high current.** These armatures are generally constructed with large wire because of high current. A good example of where lap wound armatures are used is in the starter motor of almost all automobiles. One characteristic of machines that use a lap wound armature is that they will have as many pairs of brushes as there are pairs of poles. The windings of a lap wound armature are connected in parallel *(Figure 28-17)*. This permits the current capacity of each winding to be added

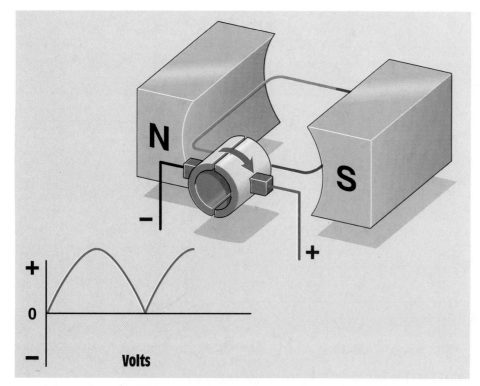

**Figure 28-11**   The commutator maintains the proper polarity.

and provides a higher operating current. Lap wound armatures have as many parallel paths through the armature as there are pole pieces.

## Wave Wound Armatures

*Wave wound armatures* **are used in machines designed for high voltage and low current**. These armatures have their windings connected in series as shown in *Figure 28-18*. When the windings are connected in series, the voltage of each winding adds, but the current capacity remains the same. A good of example of where wave wound armatures are used is in the small generator in hand-cranked megohmmeters. Wave wound armatures never contain more than two parallel paths for current flow regardless of the number of pole pieces, and they never contain more than one set of brushes (a set being one brush or group of brushes for positive and one brush or group of brushes for negative).

## Frogleg Wound Armatures

*Frogleg wound armatures* **are probably the most used. These armatures are used in machines designed for use with moderate**

**wave wound armatures**

Wave wound armatures are used in machines designed for high voltage and low current.

**frogleg wound armatures**

Frogleg wound armatures are used in machines designed for use with moderate current and moderate voltage.

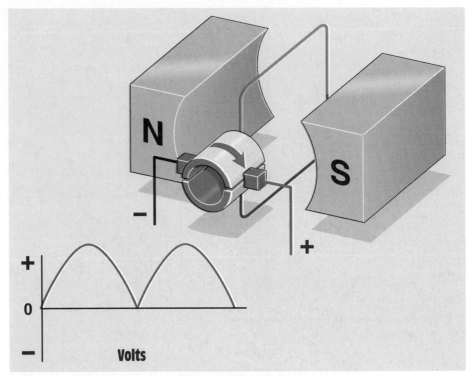

**Figure 28-12** The loop completes one complete rotation.

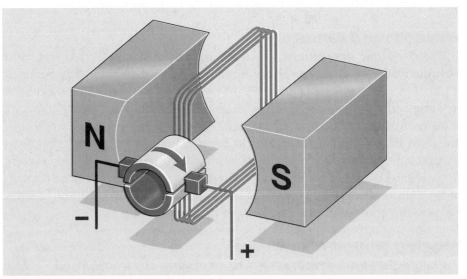

**Figure 28-13** Increasing the number of turns increases the output voltage.

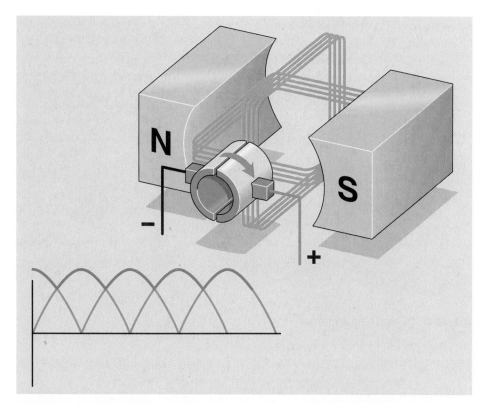

**Figure 28-14**  Increasing the number of loops produces a smoother output voltage.

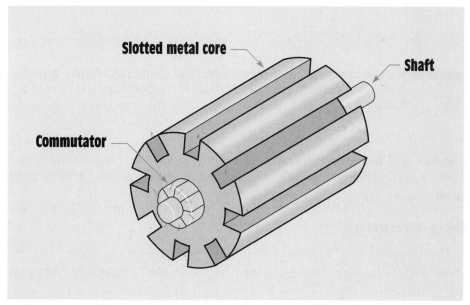

**Figure 28-15**  The loops of wire are wound around slots in a metal core.

**Figure 28-16** DC machine armature

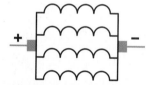

**Figure 28-17** Lap wound armatures have their windings connected in parallel. They are used in machines intended for high-current and low-voltage operation.

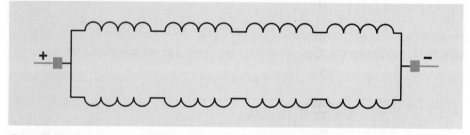

**Figure 28-18** Wave wound armatures have their windings connected in series. Wave windings are used in machines intended for high-voltage, low-current operation.

**current and moderate voltage.** The windings of a frogleg wound armature are connected in series-parallel as shown in *Figure 28-19*. Most large DC machines use frogleg wound armatures.

### 28-3 Brushes

The **brushes** ride against the commutator segments and are used to connect the armature to the external circuit of the DC machine. Brushes are made from a material that is softer than the copper bars of the commutator. This permits the brushes, which are easy to replace, to wear

**brushes**

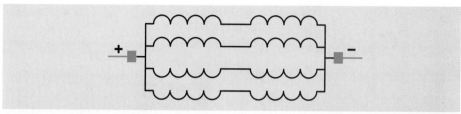

**Figure 28-19** Frogleg wound armatures are connected in series-parallel. These windings are generally used in machines intended for medium voltage and current operation.

instead of the commutator. The brush leads are generally marked $A_1$ and $A_2$ and are referred to as the armature leads.

## 28-4 Pole Pieces

The **pole pieces** are located inside the housing of the DC machine (*Figure 28-20*). The pole pieces provide the magnetic field necessary for the operation of the machine. They are constructed of some type of good magnetic conductive material such as soft iron or silicon steel. Some DC generators use permanent magnets to provide the magnetic field instead of electromagnets. These machines are generally small and rated about one horsepower or less. A DC generator that uses permanent magnets as its field is referred to as a *magneto*.

## 28-5 Field Windings

Most DC machines use wound electromagnets to provide the magnetic field. Two types of field windings are used. One is the series field and the other is the shunt field. **Series field windings** are made with relatively few turns of very large wire and have a very low resistance. They are so named because they are connected in series with the armature. The terminal leads of the series field are labeled $S_1$ and $S_2$. It is not uncommon to find the series field of large-horsepower machines wound with square or rectangular wire (*Figure 28-21*). The use of square wire permits the windings to be laid closer together, which increases the number of turns that can be wound in a particular space. Additionally, smaller square and rectangular wire can be used to yield the same surface area of larger round wire (*Figure 28-22*).

**Shunt field windings** are made with many turns of small wire. Since the shunt field is constructed with relatively small wire, it has a much higher resistance than the series field. The shunt field is intended to be connected in parallel with, or shunt, the armature. The resistance of the

**pole pieces**

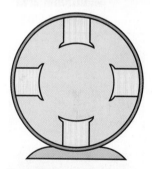

**Figure 28-20** Pole pieces are constructed of soft iron and placed on the inside of the housing.

**series field windings**

**shunt field windings**

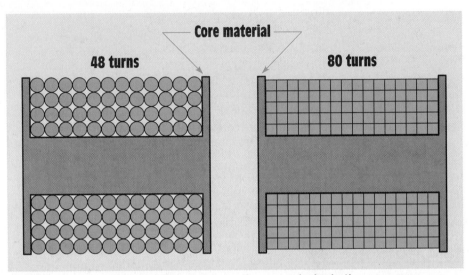

**Figure 28-21** Square wire permits more turns than round wire in the same area.

shunt field must be high because its resistance is used to limit current flow through the field. The shunt field is often referred to as the "field," and its terminal leads are labeled $F_1$ and $F_2$.

When a DC machine uses both series and shunt fields, each pole piece

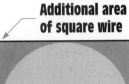

**Figure 28-22** A square wire of equal size contains more surface area than round wire.

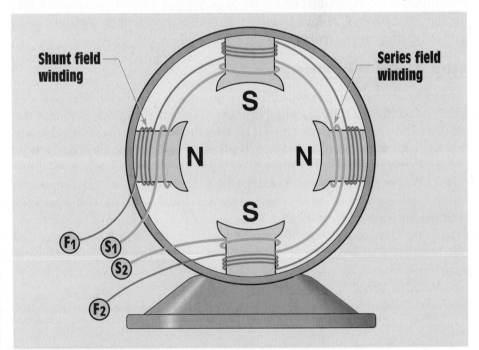

**Figure 28-23** Both series and shunt field windings are contained on each pole piece.

will contain both windings *(Figure 28-23)*. The windings are wound on the pole pieces in such a manner that when current flows through the winding it will produce alternate magnetic polarities. In the illustration shown in *Figure 28-23*, two pole pieces will form north magnetic polarities and two will form south magnetic polarities. A DC machine with two field poles and one interpole is shown in *Figure 28-24*. Interpoles will be discussed later in this unit.

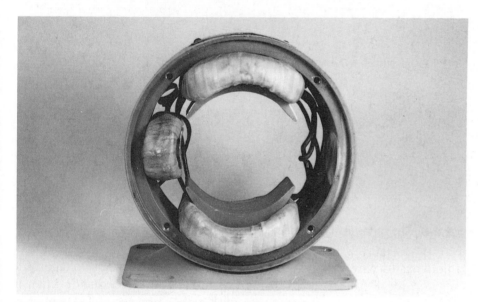

**Figure 28-24** A two-pole DC machine with one interpole

## 28-6 Series Generators

There are three basic types of direct current generators: the series, shunt, and compound. The type is determined by the arrangement and connection of field coils. The **series generator** contains only a series field connected in series with the armature *(Figure 28-25)*. A schematic diagram used to represent a series-connected DC machine is shown in *Figure 28-26*. The series generator must be *self-excited*, which means that the pole pieces contain some amount of residual magnetism. This residual magnetism produces an initial output voltage that permits current to flow through the field if a load is connected to the generator. The amount of output voltage produced by the generator is proportional to three factors:

1. the number of turns of wire in the armature,
2. the strength of the magnetic field of the pole pieces, and
3. the speed of the cutting action (speed of rotation).

**series generator**

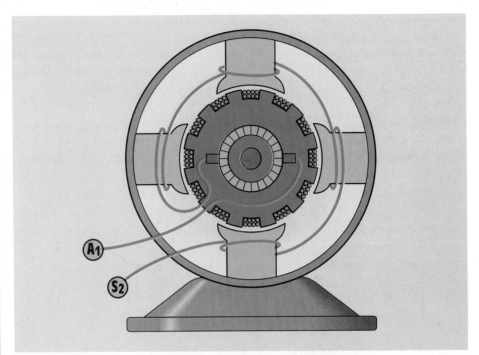

**Figure 28-25** The series field is connected in series with the armature.

To understand why these three factors determine the output voltage of a generator, recall from Unit 13 that 1 V is induced in a conductor when it cuts magnetic lines of flux at a rate of 1 weber per second (1 weber = 100,000,000 lines of flux). When conductors are wound into a loop, each turn acts like a separate conductor. Since the turns are connected in series, the voltage induced into each conductor will add. If one conductor has an induced voltage of 0.5 V and there are 20 turns, the total induced voltage would be 10 V.

The second factor is the strength of the magnetic field. Flux density is a measure of the strength of a magnetic field. If the number of turns of wire in the armature remains constant and the speed remains constant, the out-

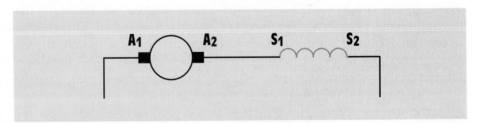

**Figure 28-26** Schematic drawing of a series generator

put voltage can be controlled by the number of flux lines produced by the field poles. Increasing the lines of flux will increase the number of flux lines cut per second and, therefore, the output voltage. The magnetic field strength can be increased until the iron of the pole pieces reaches saturation.

Induced voltage is proportional to the number of flux lines cut per second. If the strength of the magnetic field remains constant and the number of turns of wire in the armature remains constant, the output voltage will be determined by the speed at which the conductors cut the flux lines. Increasing the speed of the armature will increase the speed of the cutting action, which will increase the output voltage. Likewise, decreasing the speed of the armature will decrease the output voltage.

## Connecting Load to the Series Generator

When a load is connected to the output of a series generator, the initial voltage produced by the residual magnetism of the pole pieces produces a current flow through the load *(Figure 28-27)*. Since the series field is connected in series with the armature, the current flowing through the armature and load must also flow through the series field. This causes the magnetism of the pole pieces to become stronger and produce more magnetic lines of flux. When the strength of the magnetic pole pieces increases, the output voltage increases also.

If another load is added *(Figure 28-28)*, more current flows, and the pole pieces produce more magnetic lines of flux, which again increases the output voltage. Each time a load is added to the series generator, its output voltage increases. This increase of voltage will continue until the iron in the pole pieces and armature becomes saturated. At that point, an increase of load will result in a decrease of output voltage *(Figure 28-29)*.

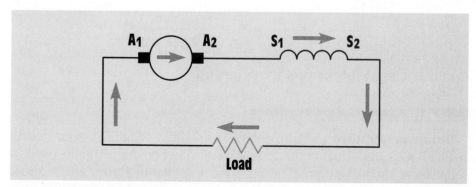

**Figure 28-27** Residual magnetism produces an initial voltage used to provide a current flow through the load.

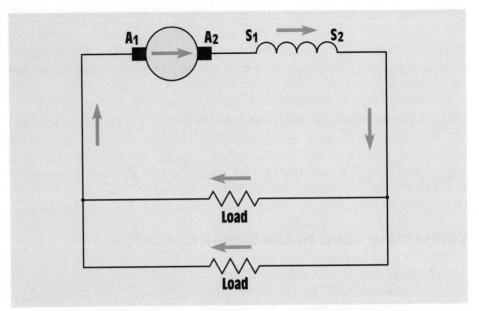

**Figure 28-28** If more load is added, current flow and output voltage increase.

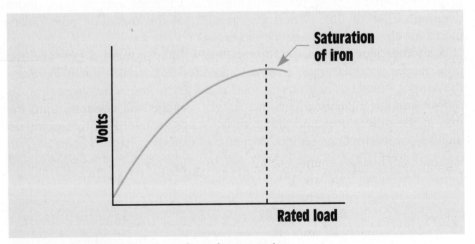

**Figure 28-29** Characteristic curve of a series generator

## 28-7 Shunt Generators

**Shunt generators** contain only a shunt field winding connected in parallel with the armature *(Figure 28-30)*. A schematic diagram used to represent a shunt-connected DC machine is shown in *Figure 28-31*. Shunt generators can be either self-excited or separately excited. Self-excited shunt generators are similar to self-excited series generators in that resid-

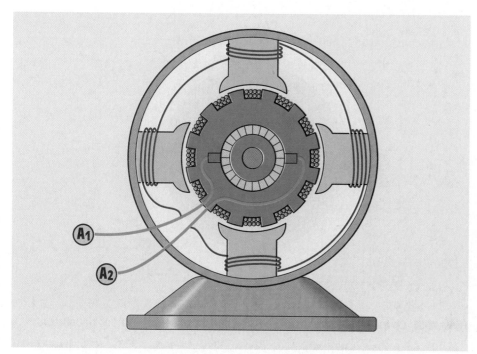

**Figure 28-30** Shunt field windings are connected in parallel with the armature.

ual magnetism in the pole pieces is used to produce an initial output voltage. In the case of a shunt generator, however, the initial voltage is used to produce a current flow through the shunt field. This current increases the magnetic field strength of the pole pieces, which produces a higher output voltage *(Figure 28-32)*. This buildup of voltage continues until a maximum value, determined by the speed of rotation, the turns of wire in the armature, and the turns of wire on the pole pieces, is reached.

Another difference between the series generator and the self-excited shunt generator is that the series generator must be connected to a load before voltage can increase. The load is required to form a complete path for current to flow through the armature and series field *(Figure 28-27)*. In a self-excited shunt generator, the shunt field winding provides a complete circuit across the armature, permitting the full output voltage to be obtained before a load is connected to the generator.

*Separately excited generators* have their fields connected to an external source of direct current *(Figure 28-33)*. The advantages of the separately excited machine are that it gives better control of the output voltage and that its voltage drop is less when load is added. The characteristic curves

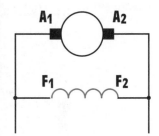

**Figure 28-31** Schematic drawing of a shunt generator

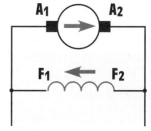

**Figure 28-32** Residual magnetism in the pole pieces produces an initial voltage, which causes current to flow through the shunt field increasing field flux.

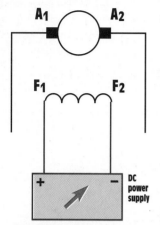

**Figure 28-33** Separately excited shunt generators must have an external power source to provide excitation current for the shunt field.

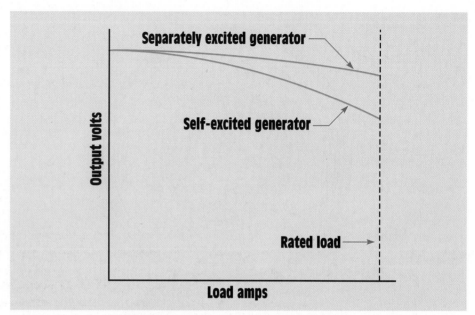

**Figure 28-34** Characteristic curves of self- and separately excited shunt generators

of both self-excited and separately excited shunt generators are shown in *Figure 28-34.*

The self-excited generator exhibits a greater drop in voltage when load is added because the armature voltage is used to produce the current flow in the shunt field. Each time the voltage decreases, the current flow through the field decreases, causing a decrease in the amount of magnetic flux lines in the pole pieces. This decrease of flux in the pole pieces causes a further decrease of output voltage. The separately excited machine does not have this problem because the field flux is held constant by the external power source.

## Field Excitation Current

Regardless of which type of shunt generator is used, the amount of output voltage is generally controlled by the amount of **field excitation current**. Field excitation current is the DC current that flows through the shunt field winding. This current is used to turn the iron pole pieces into electromagnets. Since one of the factors that determines the output voltage of a DC generator is the strength of the magnetic field, the output voltage can be controlled by the amount of current flow through the field coils. A simple method of controlling the output voltage is by the use of a shunt field rheostat. The shunt field rheostat is connected in series with the shunt field winding *(Figure 28-35)*. By adding or removing resistance

field
excitation
current

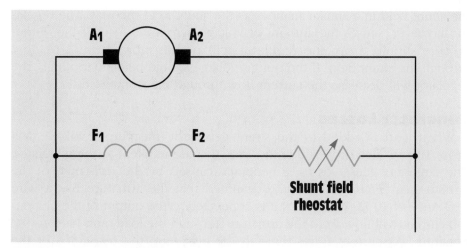

**Figure 28-35**  The shunt field rheostat is used to control the output voltage.

connected in series with the shunt field winding, the amount of current flow through the field can be controlled. This in turn controls the strength of the magnetic field of the pole pieces.

When it is important that the output voltage remain constant regardless of load, an electronic voltage regulator can be used to adjust the shunt field current *(Figure 28-36)*. The voltage regulator connects in series with

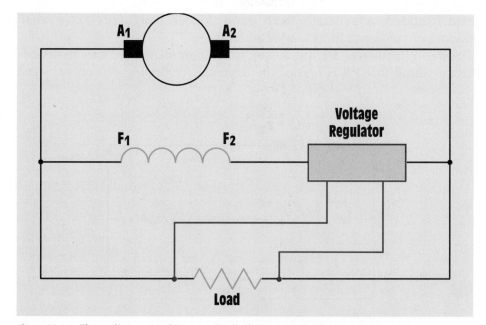

**Figure 28-36**  The voltage regulator controls the amount of shunt field current.

**voltage regulator**

A device that maintains a constant output voltage from a source, adjusting for variations in load resistance and the input voltage.

the shunt field in a similar manner as the shunt field rheostat. The regulator, however, senses the amount of voltage across the load. If the output voltage should drop, the regulator will permit more current to flow through the shunt field. If the output voltage should become too high, the regulator will decrease the current flow through the shunt field.

### Generator Losses

When load is added to the shunt generator, the output voltage will drop. This voltage drop is due to losses that are inherent to the generator. The largest of these losses is generally caused by the resistance of the armature. In *Figure 28-37*, it is assumed that the armature has a wire resistance of 10 $\Omega$. When a load is connected to the output of the generator, current will flow from the armature, through the load, and back to the armature. As current flows through the armature, the resistance of the wire causes a voltage drop. Assume that the armature has a current flow of 2 A. If the resistance of the armature is 10 $\Omega$, it will require 20 V to push the current through the resistance of the armature.

Now assume that the armature has a resistance of 2 $\Omega$. The same 2 A of current flow requires only 4 V to push the current through the armature resistance. A low-resistance armature is generally a very desirable characteristic for DC machines. **In the case of a generator, the voltage regulation is determined by the resistance of the armature**. Voltage regulation is measured by the amount that output voltage will drop as load is added. A generator with good voltage regulation has a small amount of voltage drop as load is added.

Some other losses are $I^2R$ losses, eddy current losses, and hysteresis losses. Recall that $I^2R$ is one of the formulas for finding power, or watts.

> In the case of a generator, the voltage regulation is determined by the resistance of the armature.

**Figure 28-37** Armature resistance causes a drop in output voltage.

In the case of a DC machine, it describes the power loss associated with heat due to the resistance of the wire in both the armature and field windings.

Eddy currents are currents that are induced into the metal core material by the changing magnetic field as the armature spins through the flux lines of the pole pieces. Eddy currents are so named because they circulate around inside the metal in a manner similar to the swirling eddies in a river *(Figure 28-38)*. These swirling currents produce heat, which is a power loss. Many machines are constructed with laminated pole pieces and armature cores to help reduce eddy currents. The surface of each lamination forms a layer of iron oxide, which acts as an insulator to help prevent the formation of eddy currents.

Hysteresis losses are losses due to molecular friction. As discussed previously, alternating current is produced inside the armature. This reversal of the direction of current flow causes the molecules of iron in the core to realign themselves each time the current changes direction. The molecules of iron are continually rubbing against each other as they realign magnetically. The friction of the molecules rubbing together causes heat, which is a power loss. Hysteresis loss is proportional to the speed of rotation of the armature. The faster the armature rotates, the more current reversals there are per second, and the more heat is produced because of friction.

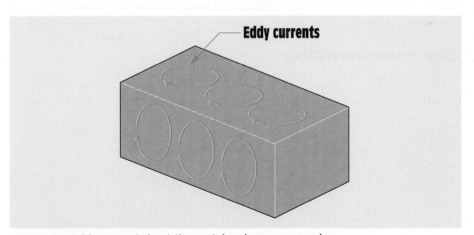

**Eddy currents**

**.Figure 28-38**  Eddy currents heat the metal and cause power loss.

## 28-8 Compound Generators

**Compound generators** contain both series and shunt fields. Most large DC machines are compound wound. The series and shunt fields can be connected in two ways. One connection is called long shunt *(Figure*

**compound generators**

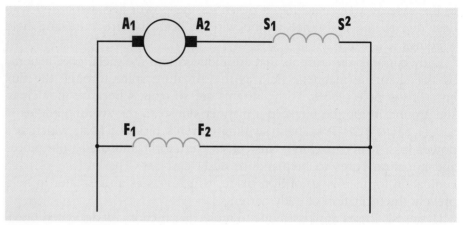

**Figure 28-39** Schematic drawing of a long shunt compound generator

28-39). The long shunt connection has the shunt field connected in parallel with both the armature and series field. This is the most used of the two connections.

The second connection is called short shunt *(Figure 28-40)*. The short shunt connection has the shunt field connected in parallel with the armature. The series field is connected in series with the armature. This is a very common connection for DC generators that must be operated in parallel with each other.

## 28-9 Compounding

**compound-ing**

The relationship of the strengths of the two fields in a generator determines the amount of **compounding** for the machine. A machine is

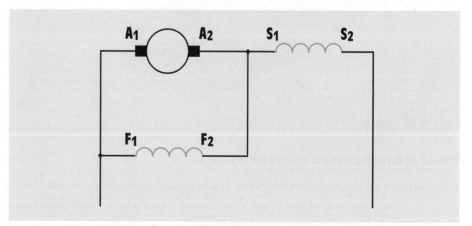

**Figure 28-40** Schematic drawing of a short shunt compound generator

overcompounded when the series field has too much control and the output voltage increases each time a load is added to the generator. Basically, the generator begins to take on the characteristics of a series generator. *Overcompounding* is characterized by the fact that the output voltage at full load will be greater than the output voltage at no load *(Figure 28-41)*.

When the generator is flat-compounded, the output voltage will be the same at full load as it is at no load. *Flat compounding* is accomplished by permitting the series field to increase the output voltage by an amount that is equal to the losses of the generator.

If the series field is too weak, however, the generator will become *undercompounded*. This condition is characterized by the fact that the output voltage will be less at full load than it is at no load. When a generator is undercompounded it has characteristics similar to those of a shunt generator.

## Controlling Compounding

Most DC machines are constructed in such a manner that they are overcompounded if no control is used. This permits the series field strength to

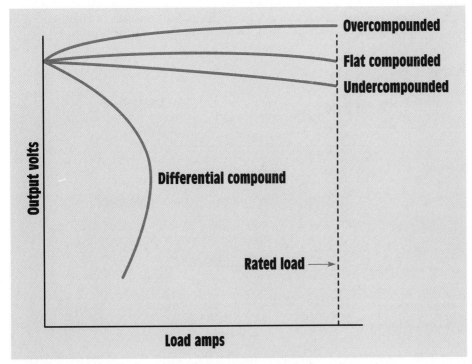

**Figure 28-41** Characteristic curves of compound generators

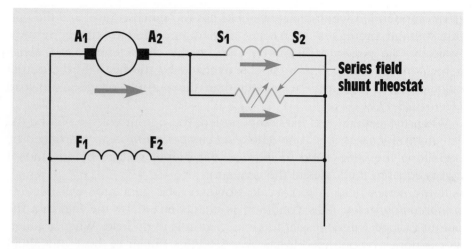

**Figure 28-42**  The series field shunt rheostat controls the amount of compounding.

**series field
diverter**

be weakened and thereby control the amount of compounding. The amount of compounding is controlled by connecting a low-value variable resistor in parallel with the series field *(Figure 28-42)*. This resistor is known as the series field shunt rheostat, or the **series field diverter**. The rheostat permits part of the current that normally flows through the series field to flow through the resistor. This reduces the amount of magnetic flux produced by the series field, which reduces the amount of compounding.

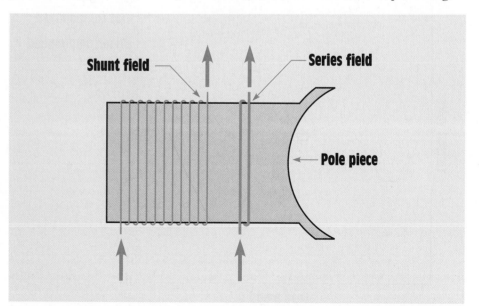

**Figure 28-43**  In a cumulative-compound machine the current flows in the same direction through both the series and shunt field.

## Cumulative and Differential Compounding

Direct current generators are generally connected in such a manner that they are a **cumulative compound**. This means that the shunt and series fields are connected in such a manner that when current flows through them they aid each other in the production of magnetism *(Figure 28-43)*. In the example shown, each of the field windings would produce the same magnetic polarity for the pole piece.

A **differential compound** generator has its fields connected in such a manner that they oppose each other in the production of magnetism *(Figure 28-44)*. In this example, the shunt and series fields are attempting to produce opposite magnetic polarities for the same pole piece. This results in the magnetic field becoming weaker as current flow through the series field increases. Although there are some applications for a differential-compound machine, they are very limited.

**cumulative compound**

**differential compound**

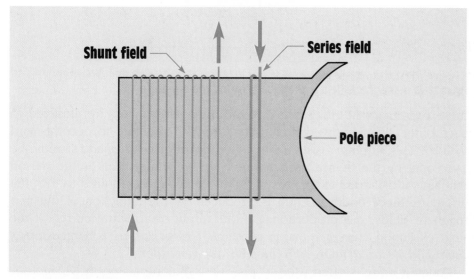

**Figure 28-44** In a differential-compound machine the current flows through the shunt field in a direction opposite that of the current flow through the series field.

## 28-10 Countertorque

When a load is connected to the output of a generator, current flows from the armature, through the load, and back to the armature. As current flows through the armature, a magnetic field is produced around the armature *(Figure 28-45)*. In accord with Lenz's law, the magnetic field of the armature will be opposite in polarity to that of the pole pieces. Since

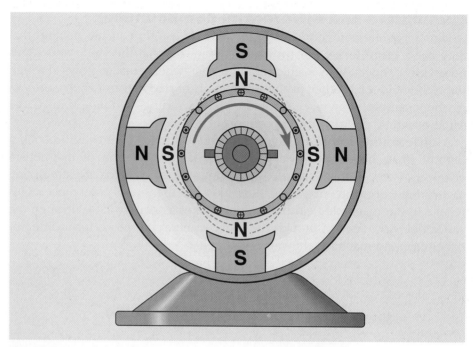

**Figure 28-45**  A magnetic field is produced around the armature.

counter-
torque

these two magnetic fields are opposite in polarity, they are attracted to each other. This magnetic attraction causes the armature to become hard to turn. This turning resistance is called **countertorque**, and it must be overcome by the device used to drive the generator. This is the reason that as load is added to the generator, more power is required to turn the armature. Since countertorque is produced by the attraction of the two magnetic fields, it is proportional to the output or armature current if the field excitation current remains constant. *Countertorque is a measure of the useful electrical energy produced by the generator.*

Countertorque is often used to provide a braking action in DC motors. If the field excitation current remains turned on, the motor can be converted into a generator very quickly by disconnecting the armature from its source of power and reconnecting it to a load resistance. The armature now supplies current to the load resistance. The countertorque developed by the generator action causes the armature to decrease in speed. When this type of braking action is used, it is referred to as *dynamic braking* or *regenerative braking.*

## 28-11 Armature Reaction

**Armature reaction** is the twisting or bending of the magnetic lines of flux of the pole pieces. It is caused by the magnetic field produced

armature
reaction

around the armature as it supplies current to the load *(Figure 28-46)*. This distortion of the main magnetic field causes the position of the neutral plane to change position. When the neutral plane changes, the brushes no longer make contact between commutator segments at a time when no voltage is induced in the armature. This results in power loss and arcing and sparking at the brushes, which can cause overheating and damage to both the commutator and brushes. The amount of armature reaction is proportional to armature current.

## Correcting Armature Reaction

Armature reaction can be corrected in several ways. One method is to rotate the brushes an equal amount to the shift of the neutral plane *(Figure 28-47)*. This method would only be satisfactory, however, if the generator delivered a constant current. Since the distortion of the main magnetic field is proportional to armature current, the brushes would have to be adjusted each time the load current changed. In the case of a generator, the brushes would be rotated in the direction of rotation of the armature. In the case of a motor, the brushes would be rotated in a direction opposite that of armature rotation.

Another method that is used often is to insert small pole pieces, called interpoles or commutating poles, between the main field poles *(Figure 28-24)*. The interpoles are sometimes referred to as the commutating winding

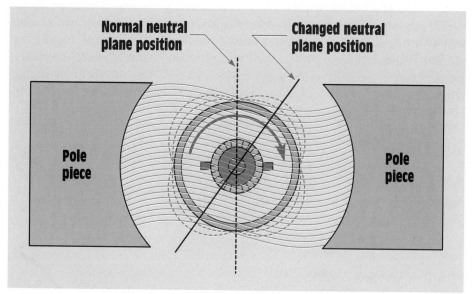

**Figure 28-46** Armature reaction changes the position of the neutral plane.

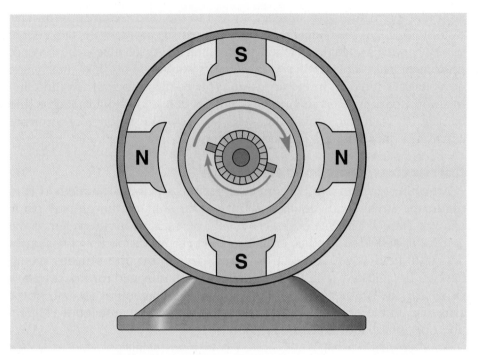

**Figure 28-47** In a generator, the brushes are rotated in the direction of armature rotation to correct armature reaction.

because they are wound with a few turns of large wire similar to the series field winding. The interpoles are connected in series with the armature, which permits their strength to increase with an increase of armature current *(Figure 28-48)*. Interpole connections are often made inside the housing of the machine. When the interpole connection is made internally, the $A_2$ lead is actually connected to one end of the interpole winding. When the interpole windings are brought out of the machine separately,

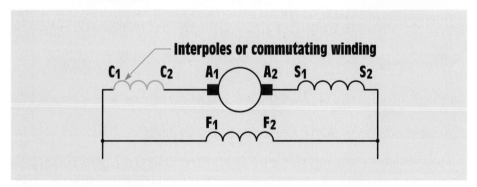

**Figure 28-48** Interpoles are connected in series with the armature.

they are generally labeled $C_1$ and $C_2$, which stands for commutating field. It is not unusual, however, to find them labeled $S_3$ and $S_4$.

In a generator, the magnetic field of the armature tends to bend the main magnetic field upward as shown in *Figure 28-46*. In a motor, the armature field bends the main field downward. The function of the interpoles is to restore the field to its normal condition. When a DC machine is used as a generator, the interpoles will have the same polarity as the main field pole directly ahead of them (ahead in the sense of the direction of rotation of the armature) *(Figure 28-49)*. When a DC machine is used as a motor, the interpoles will have the same polarity as the pole piece behind them in the sense of direction of rotation of the armature.

Interpoles do have one disadvantage. They restore the field only in their immediate area and are not able to overcome all the field distortion. Large DC generators use another set of windings called compensating windings to help restore the main magnetic field. Compensating windings are made by placing a few large wires in the face of the pole piece parallel to the armature windings *(Figure 28-50)*. The compensating winding is connected in series with the armature so that its strength increases with an increase of output current.

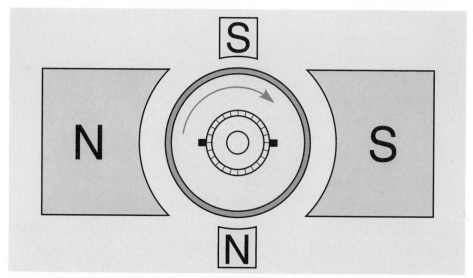

**Figure 28-49** Interpoles must have the same polarity as the pole piece directly ahead of them.

## 28-12 Setting the Neutral Plane

Most DC machines are designed in such a manner that the position of the brushes on the commutator can be set or adjusted. An exposed view

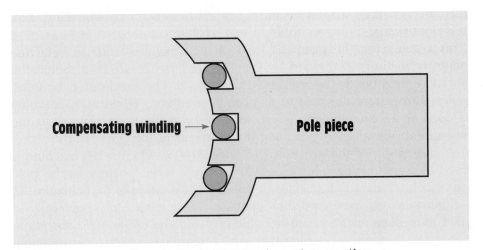

**Figure 28-50** Compensating winding helps correct armature reaction.

of the brushes and brush yoke of a direct current machine is shown in *Figure 28-51*. The simplest method of setting the brushes to the neutral plane position is to connect an AC voltmeter across the shunt field leads.

**Figure 28-51** Exposed view of a direct current machine *(Photograph courtesy of GE Motors – DM&G.)*

Low-voltage alternating current is then applied to the armature *(Figure 28-52)*. The armature acts like the primary of a transformer, and the shunt field acts like the secondary. If the brushes are not set at the neutral plane position, the changing magnetic field of the armature will induce a voltage into the shunt field. The brush position can be set by observing the action of the AC voltmeter. If the brush yoke is loosened to permit the brushes to be moved back and forth on the commutator, the voltmeter pointer will move up and down the scale. The brushes are set to the neutral plane position when the voltmeter is at its lowest possible reading.

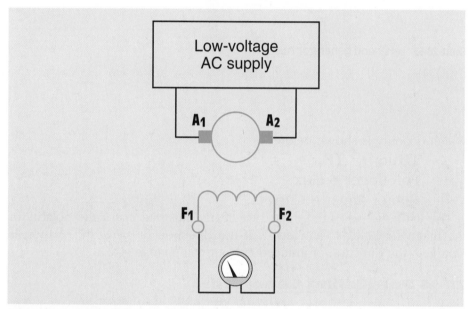

**Figure 28-52**  Setting the brushes at the neutral plane

## 28-13 Fleming's Left-Hand Generator Rule

Fleming's **left-hand generator rule** can be used to determine the relationship of the motion of the conductor in a magnetic field to the direction of the induced current. To use the left-hand rule, place the thumb, forefinger, and center finger at right angles to each other as shown in *Figure 28-53*. **The forefinger points in the direction of the field flux**, assuming that magnetic lines of force are in a direction of north to south. **The thumb points in the direction of thrust**, or movement of the conductor, **and the center finger shows the direction of the current induced into the armature**. An easy method of remembering which finger represents

**left-hand generator rule**

The forefinger points in the direction of the field flux. The thumb points in the direction of thrust, and the center finger shows the direction of the current induced into the armature.

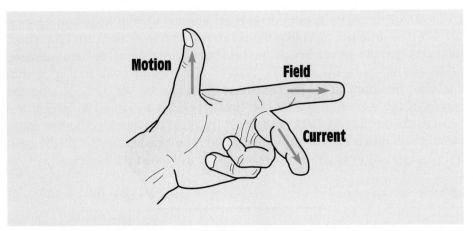

**Figure 28-53** Left-hand generator rule

which quantity is shown below.

**TH**umb = **TH**rust

**F**orefinger = **F**lux

**C**enter finger = **C**urrent

The left-hand rule can be used to clearly illustrate that if the polarity of the magnetic field is changed or if the direction of armature rotation is changed, the direction of induced current will change also.

## 28-14 Paralleling Generators

There may be occasions when one direct current generator cannot supply enough current to operate the connected load. In such a case, another generator is connected in parallel with the first. Direct current generators should never be connected in parallel without an equalizer connection *(Figure 28-54)*. The equalizing connection is used to connect the series fields of the two machines in parallel with each other. This arrangement prevents one machine from taking the other over as a motor.

Assume that two generators are to be connected in parallel, and the equalizing connection has not been made. Unless both machines are operating with identical field excitation when they are connected in parallel, the machine with the greatest excitation will take the entire load and begin operating the other machine as a motor. The series field of the machine that accepted the load would be strengthened, and the series field of the machine that gave up the load would be weakened. The

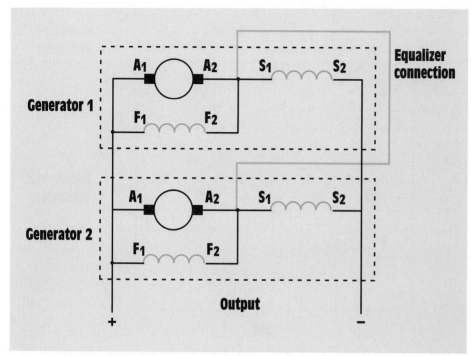

**Figure 28-54**  The equalizer connection is used to connect the series fields in parallel with each other.

machine with the stronger series field will take even greater load and the machine with the weaker series field will reduce load even further.

The generator that begins motoring will have the current flow through its series field reversed, which will cause it to operate as a differential-compound motor *(Figure 28-55)*. If the motoring generator is not removed from the line, the magnetic field strength of the series field will become greater than the field strength of the shunt field. This will cause the polarity of the residual magnetism in the pole pieces to reverse. This is often referred to as flashing the field. Flashing the field results in the polarity of the output voltage being reversed when the machine is restarted as a generator. The equalizer connection prevents field reversal even if the generator becomes a motor.

The resistance of the equalizer cable should not exceed 20% of the resistance of the series field winding of the smallest paralleled generator. This will ensure that the current flow provided to the series fields will divide in the approximate inverse ratio of the respective series field winding.

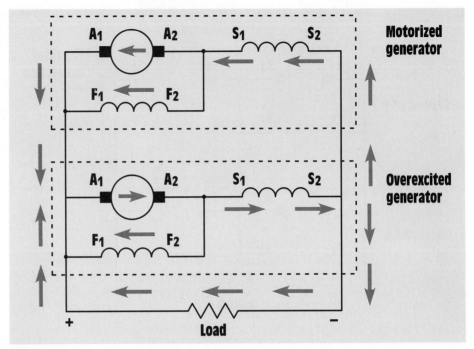

**Figure 28-55** One generator takes all the load, and the other becomes a motor.

## Summary

1. A generator is a machine that converts mechanical energy into electrical energy.

2. Generators operate on the principle of magnetic induction.

3. Alternating current is produced in all rotating armatures.

4. The commutator changes the alternating current produced in the armature into direct current.

5. The brushes are used to make contact with the commutator and to carry the output current to the outside circuit.

6. The position at which there is no induced voltage in the armature is called the neutral plane.

7. Lap wound armatures are used in machines designed for low-voltage and high-current operation.

8. Wave wound armatures are used in machines designed for high-voltage and low-current operation.

9. Frogleg wound armatures are the most used and are intended for machines designed for moderate voltages and current.

10. The loops of wire, iron core, and commutator are made as one unit and are referred to as the armature.

11. The armature connection is marked $A_1$ and $A_2$.

12. Series field windings are made with a few turns of large wire and have a very low resistance.

13. Series field windings are connected in series with the armature.

14. Series field windings are marked $S_1$ and $S_2$.

15. Shunt field windings are made with many turns of small wire and have a high resistance.

16. Shunt field windings are connected in parallel with the armature.

17. The shunt field windings are marked $F_1$ and $F_2$.

18. Three factors that determine the voltage produced by a generator are:
    A. The number of turns of wire in the armature.
    B. The strength of the magnetic field of the pole pieces.
    C. The speed of the armature.

19. Series generators increase their output voltage as load is added.

20. Shunt generators decrease their output voltage as load is added.

21. The voltage regulation of a DC generator is proportional to the resistance of the armature.

22. Compound generators contain both series and shunt field windings.

23. A long shunt compound generator has the shunt field connected in parallel with both the armature and series field.

24. A short shunt compound generator has the shunt field connected in parallel with the armature, but in series with the series field.

25. When a generator is overcompounded, the output voltage will be higher at full load than it is at no load.

26. When a generator is flat-compounded, the output voltage will be the same at full load and no load.

27. When a generator is undercompounded, the output voltage will be less at full load than it is at no load.

28. Cumulative-compound generators have their series and shunt fields connected in such a manner that they aid each other in the production of magnetism.

29. Differential-compound generators have their series and shunt field winding connected in such a manner that they oppose each other in the production of magnetism.

30. Armature reaction is the twisting or bending of the main magnetic field.

31. Armature reaction is caused by the interaction of the magnetic field produced in the armature.

32. Armature reaction is proportional to armature current.

33. Interpoles are small pole pieces connected between the main field poles used to help correct armature reaction.

34. Interpoles are connected in series with the armature.

35. Interpoles used in a generator must have the same polarity as the main field pole directly ahead of them in the sense of rotation of the armature.

36. Interpoles used in a motor must have the same polarity as the main field pole directly behind them in the sense of rotation of the armature.

37. Interpole leads are sometimes marked $C_1$ and $C_2$ or $S_3$ and $S_4$.

38. Interpole leads are not always brought out of the machine.

39. The neutral plane can be set by connecting an AC voltmeter to the shunt field and a source of low-voltage AC to the armature. The brushes are then adjusted until the voltmeter indicates the lowest possible voltage.

40. When a generator supplies current to a load, countertorque is produced, which makes the armature harder to turn.

41. Countertorque is proportional to the armature current if the field excitation current remains constant.

42. Countertorque is a measure of the useful electrical energy produced by the generator.

## Review Questions

1. What is a generator?

2. What type of voltage is produced in all rotating armatures?

3. What are the three types of armature winding?

4. What type of armature winding would be used for a machine intended for high-voltage, low-current operation?

5. What are interpoles, and what is their purpose?

6. How are interpoles connected in relation to the armature?

7. What type of field winding is made with many turns of small wire?

8. How is the series field connected in relation to the armature?

9. How is the shunt field connected in relation to the armature?

10. What is armature reaction?

11. What is armature reaction proportional to?

12. What are eddy currents?

13. What condition characterizes overcompounding?

14. What is the function of the shunt field rheostat?

15. What is used to control the amount of compounding for a generator?

16. Explain the difference between cumulative- and differential-compounded connections.

17. What three factors determine the amount of output voltage for a DC generator?

18. What is the voltage regulation of a DC generator proportional to?

19. What is countertorque proportional to?

20. What is countertorque a measure of?

**A motor is a device that converts electrical energy into mechanical energy. This unit presents different types of motors used to change direct current into mechanical energy.**

*Courtesy of National Railway Passenger Corp.*

# Direct Current Motors

## Outline

## Key Terms

Brushless DC motors

Compound motor

Constant speed motors

Counter-EMF (CEMF) (back-EMF)

Cumulative-compounded motors

Differential-compounded motors

Field loss relay

Motor

Permanent magnet motors

Printed circuit motor

Pulse-width modulation

Series motor

ServoDisk® motor

Servomotors

Shunt motor

Speed regulation

Torque

**Objectives**

*After studying this unit, you should be able to:*

■ Discuss the principle of operation of direct current motors.

■ Discuss different types of DC motors.

■ Draw schematic diagrams of different types of DC motors.

■ Be able to connect a DC motor for a particular direction of rotation.

■ Discuss counter-EMF.

■ Describe methods for controlling the speed of direct current motors.

**Preview**

Direct current motors are used throughout industry in applications where variable speed is desirable. The speed-torque characteristic of direct current motors makes them desirable for many uses. The automotive industry uses DC motors to start internal combustion engines and to operate blower fans, power seats, and other devices where a small motor is needed.

### 29-1 DC Motor Principles

**motor**

Direct current motors operate on the principle of repulsion and attraction of magnetism. Motors use the same types of armature windings and field windings as the generators discussed in Unit 28. A motor, however, performs the opposite function of a generator. A **motor** is a device used to convert electrical energy into mechanical energy. Direct current motors were, in fact, the first electric motors to be invented. For many years, it was not believed possible to make a motor that could operate using alternating current.

DC motors are the same basic machines as DC generators. In the case of a generator, some device is used to turn the shaft of the armature, and the power produced by the turning armature is supplied to a load. In the case of a motor, power connected to the armature causes it to turn. To understand why the armature turns when current is applied to it, refer to the simple one-loop armature in *Figure 29-1*. Electrons enter the loop through the negative brush, flow around the loop, and exit through the positive brush. As current flows through the loop, a magnetic field is created around the loop. An end view, illustrating the pole pieces and the two conductors of a single loop, is shown in *Figure 29-2*. The X indicates electrons moving away from the observer like the back of an arrow moving away. The dot represents electrons moving toward the observer like the point of an approaching arrow. The left-hand rule for magnetism can be used to check the direction of the magnetic field around the conductors.

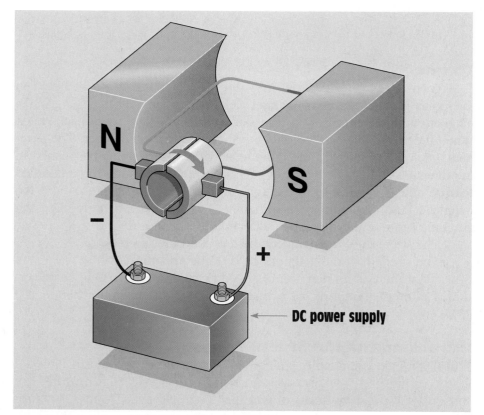

**Figure 29-1** Direct current is supplied to the loop.

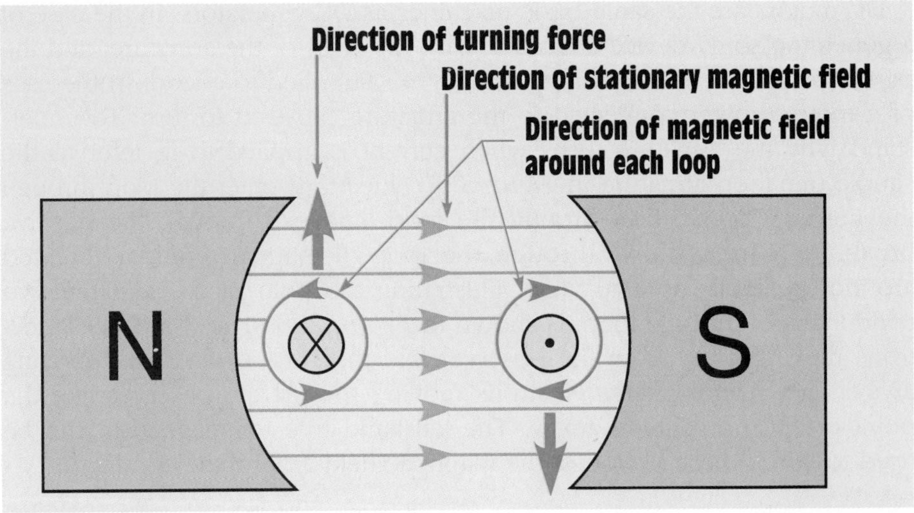

**Figure 29-2** Flux lines in the same direction repel each other, and flux lines in the opposite direction attract each other.

## Torque

Magnetic lines of force flow in a direction of north to south between the poles of the stationary magnet (left to right in *Figure 29-2*). When magnetic lines of flux flow in the same direction, they repel each other. When they flow in opposite directions, they attract each other. The magnetic lines of flux around the conductors cause the loop to be pushed in the direction shown by the arrows. This pushing or turning force is called **torque** and is created by the magnetic field of the pole pieces and the magnetic field of the loop or armature. Two factors that determine the amount of torque produced by a direct current motor are

    1. the strength of the magnetic field of the pole pieces and

    2. the strength of the magnetic field of the armature.

Notice that there is no mention of speed or cutting action. One characteristic of a direct current motor is that it can develop maximum torque at 0 RPM.

## Increasing the Number of Loops

In the previous example, a single-loop armature was used to illustrate the operating principle of a DC motor. In actual practice, armatures are constructed with many turns of wire per loop and many loops. This provides a strong continuous turning force for the armature *(Figure 29-3)*.

**torque**

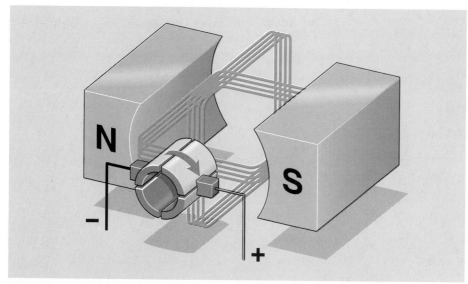

**Figure 29-3** Increasing the number of loops and turns increases the torque.

## The Commutator

When a DC machine is used as a generator, the commutator performs the function of a mechanical rectifier to change the alternating current produced in the armature into direct current before it exits the machine through the brushes. When a DC machine is used as a motor, the commutator performs the function of a rotary switch and maintains the correct direction of current flow through the armature windings. In order for the motor to develop a turning force, the magnetic field polarity of the armature must remain constant in relation to the polarity of the pole pieces. The commutator forces the direction of current flow to remain constant through certain sections of the armature as it rotates. The brushes are used to provide power to the armature from an external power source *(Figure 29-4).*

## 29-2 Shunt Motors

There are three basic types of direct current motors: the shunt, series, and compound. The direct current **shunt motor** has the shunt field connected in parallel with the armature *(Figure 29-5).* This permits an external power source to supply current to the shunt field and maintain a constant magnetic field. The shunt motor has very good speed characteristics. The full-load speed will generally remain within 10% of the no-load speed. Shunt motors are often referred to as **constant speed motors**.

**shunt motor**

**constant speed motors**

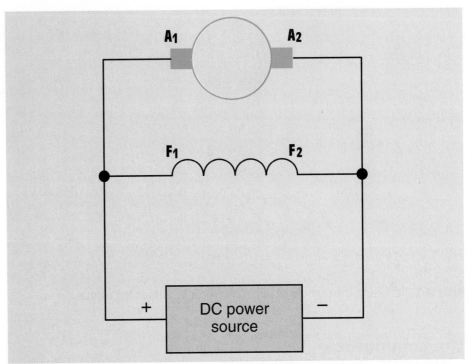

**Figure 29-4** Brushes provide power connection from an outside source to the armature.

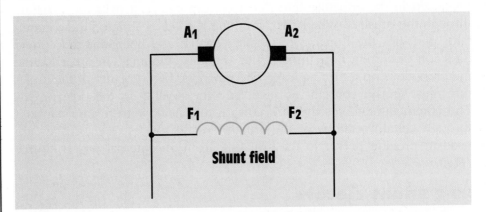

**Figure 29-5** The shunt motor has the shunt field connected in parallel with the armature.

Characteristic curves for shunt, series, and compound motors are shown in *Figure 29-6*. Note that the shunt motor maintains the most constant speed as load is added and armature current increases.

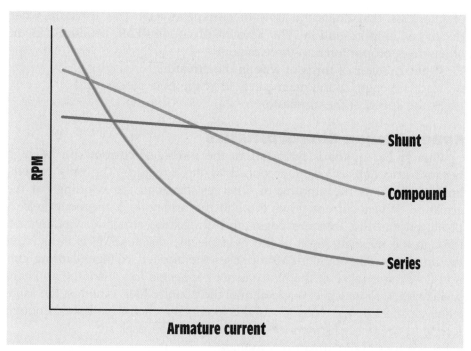

**Figure 29-6** Characteristic speed curves for different direct current motors

## Counter-EMF

When the windings of the armature spin through the magnetic field produced by the pole pieces, a voltage is induced into the armature. This induced voltage is opposite in polarity to the applied voltage and is known as **counter-EMF** (CEMF) or **back-EMF** *(Figure 29-7)*. It is the

counter-
EMF
(back-EMF)

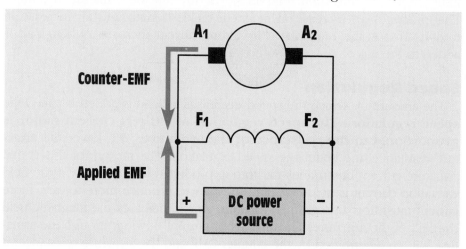

**Figure 29-7** Counter-EMF limits the flow of current through the armature.

counter-EMF that limits the flow of current through the armature when the motor is in operation. The amount of counter-EMF produced in the armature is proportional to three factors:

1. the number of turns of wire in the armature,
2. the strength of the magnetic field of the pole pieces, and
3. the speed of the armature.

## Speed-Torque Characteristics

When a DC motor is first started, the inrush of current can be high because no CEMF is being produced by the armature. The only current-limiting factor is the amount of wire resistance in the windings of the armature. When current flows through the armature, a magnetic field is produced and the armature begins to turn. As the armature windings cut through the magnetic field of the pole pieces, counter-EMF is induced in the armature. The counter-EMF opposes the applied voltage, causing current flow to decrease. If the motor is not connected to a load, the armature will continue to increase in speed until the counter-EMF is almost the same value as the applied voltage. At this point, the motor produces enough torque to overcome its own losses. Some of these losses are:

1. $I^2R$ loss in the armature windings,
2. windage loss,
3. bearing friction, and
4. brush friction.

When a load is added to the motor, the torque will not be sufficient to support the load at the speed at which the armature is turning. The armature will, therefore, slow down. When the armature slows down, counter-EMF is reduced and more current flows through the armature windings. This produces an increase in magnetic field strength and an increase in torque. This is the reason that armature current increases when load is added to the motor.

## Speed Regulation

The amount by which the speed decreases as load is added is called the **speed regulation. The speed regulation of a direct current motor is proportional to the resistance of the armature.** The lower the armature resistance, the better the speed regulation. The reason for this is that armature current determines the torque produced by the motor if the field excitation current is held constant. In order to produce more torque, more current must flow though the armature, which increases the magnetic field strength of the armature. The amount of current flowing through the armature will be determined by the counter-EMF and the armature resistance. If the field excitation current is constant, the amount of counter-EMF will be

**speed regulation**

**The speed regulation of a direct current motor is proportional to the resistance of the armature.**

proportional to the speed of the armature. The faster the armature turns, the higher the counter-EMF. When the speed of the armature decreases, the counter-EMF decreases also.

Assume an armature has a resistance of 6 Ω. Now assume that when load is added to the motor, an additional 3 A of armature current will be required to produce the torque necessary to overcome the added load. In this example, a voltage of 18 V will be required to increase the armature current by 3 A (3 A x 6 Ω = 18 V). This means that the speed of the armature must drop enough so that the counter-EMF is 18 V less than it was before. The reduction in counter-EMF permits the applied voltage to push more current through the resistance of the armature.

Now assume that the armature has a resistance of 1 Ω. If a load is added that requires an additional 3 A of armature current, the speed of the armature must drop enough to permit a 3-V reduction in counter-EMF (3 A x 1 Ω = 3 V). The armature does not have to reduce speed as much to cause a 3-V reduction in counter-EMF as it does for a reduction of 18 V.

## 29-3 Series Motors

The operating characteristics of the direct current **series motor** are very different from those of the shunt motor. The reason is that the series motor has only a series field connected in series with the armature *(Figure 29-8)*. The armature current, therefore, flows through the series field. The speed of the series motor is controlled by the amount of load connected to the motor. When load is increased, the speed of the motor will decrease. This causes a reduction in the amount of counter-EMF produced in the armature and an increase in armature and series field current. Since the current increases in both the armature and series field,

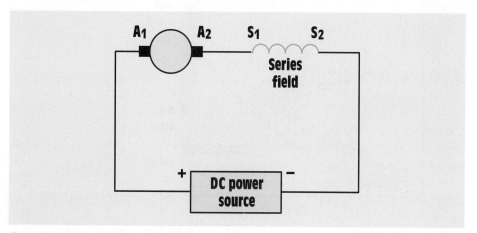

**Figure 29-8** Series motor connection

the torque will increase by the square of the current. In other words, if the current doubles, the torque will increase four times. Characteristic curves showing the relationship of torque and armature current for the three main types of DC motors are shown in *Figure 29-9.* Notice that the series motor produces the most torque of the three motors.

## Series Motor Speed Characteristics

Series motors have no natural speed limit and should, therefore, never be operated in a no-load condition. Large series motors that suddenly lose their load will race to speeds that will destroy the motor. Series motors operating at no load have been known to develop such an extremely high RPM that centrifugal force slings both the windings out of the slots in the armature and the copper bars out of the commutator. For this reason, series motors should be coupled directly to a load. Belts or chains should never be used to connect a series motor to a load.

Series motors have the ability to develop extremely high starting torques. An average of about 450% of full torque is common. These motors are generally used for applications that require a high starting torque, such as the starter motor on an automobile, cranes and hoists, and electric buses and street cars.

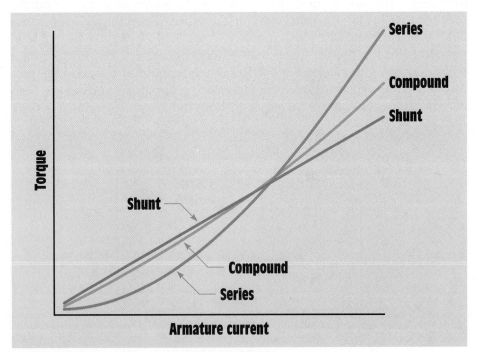

**Figure 29-9** Torque curves of direct current motors

## 29-4 Compound Motors

The **compound motor** uses both a series field and a shunt field. This motor is used to combine the operating characteristics of both the series and shunt motor. The series field of the compound motor permits the motor to develop high torque, and the shunt field permits speed control and regulation. The compound motor is used more than any other type of direct current motor in industry. The compound motor will not develop as much torque as the series motor, but it will develop more than the shunt motor. The speed regulation of a compound motor is not as good as a shunt motor, but it is much better than a series motor *(Figure 29-6)*.

Compound motors can be connected as short shunt or long shunt just as compound generators can *(Figure 29-10)*. The long shunt connection is more common because it has superior speed regulation. Compound motors can also be **cumulative-compounded motors** or **differential-compounded motors.** Although there are some applications for differential-compounded motors, it is generally a connection to be avoided. If a generator is inadvertently connected as a differential-compound machine, the greatest consequence will be that the output voltage drops rapidly as load is added. This is not the case with a motor, however. When a motor is connected differential-compound, the shunt field will determine the direction of rotation of the motor at no load or light load. When load is added to the motor, the series field will become stronger. If enough load is added, the magnetic

**compound motor**

**cumulative-com-pounded motors**

**differen-tial-com-pounded motors**

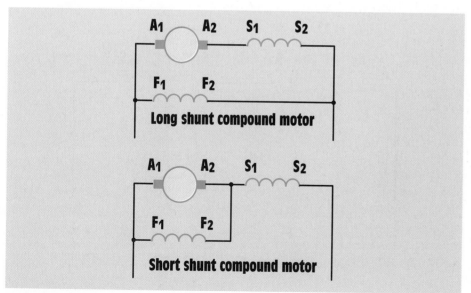

**Figure 29-10** Compound motor connections

field of the pole pieces will reverse polarity, and the motor will suddenly stop, reverse direction, and begin operating like a series motor. This can damage the motor and the equipment to which the motor is connected.

If a compound motor is to be connected the following steps can be followed to prevent the motor from being accidentally connected differential-compound.

1. Disconnect the motor from the load.
2. Connect the series field and armature windings together to form a series motor connection, leaving the shunt field disconnected. Connect the motor to the power source *(Figure 29-11)*.
3. Turn the power on momentarily to determine the direction of rotation. This application of power must be of very short duration because the motor is now being operated as a series motor with no load. The idea is to "bump" the motor just to check for direction of rotation. If the motor turns in the opposite direction than desired, reverse the connection of the armature leads. This will reverse the direction of rotation of the motor.
4. Connect the shunt field leads to the incoming power *(Figure 29-12)*. Again turn on the power and check the direction of rotation. If the motor operates in the desired direction, it is connected cumulative-compound. If the motor turns in the opposite direction, it is connected differential-compound, and the shunt field leads should be reversed.

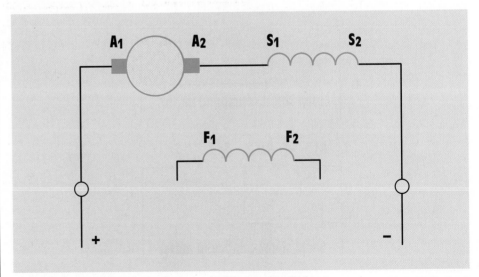

**Figure 29-11** The motor is first connected as a series motor.

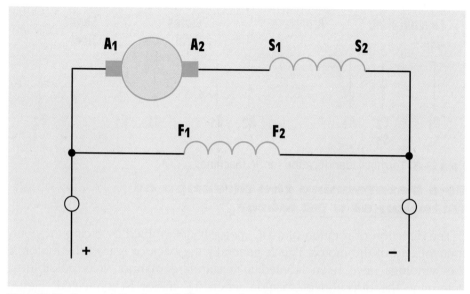

**Figure 29-12**   Connect the shunt field and again test for direction of rotation.

This test can be used to check for a differential- or cumulative-compound connection because the shunt field controls the direction of rotation at no load. If the motor operates in the same direction as both a series and compound motor, the magnetic polarity of the pole pieces must be the same for both connections. This indicates that both the series and shunt field windings must be producing the same magnetic polarity and are, therefore, connected cumulative-compound.

## 29-5 Terminal Identification for DC Motors

The terminal leads of DC machines are labeled so they can be identified when they are brought outside the motor housing to the terminal box. DC motors have the same terminal identification as that used for DC generators. *Figure 29-13* illustrates this standard identification. Terminals $A_1$ and $A_2$ are connected to the armature through the brushes. The ends of the series field are identified with $S_1$ and $S_2$, and the shunt field leads are marked $F_1$ and $F_2$. Some DC machines will provide access to another set of windings called the commutating field or interpoles. The ends of this winding will be labeled $C_1$ and $C_2$ or $S_3$ and $S_4$. It is common practice to provide access to the interpole winding on machines designed to be used as motors or generators.

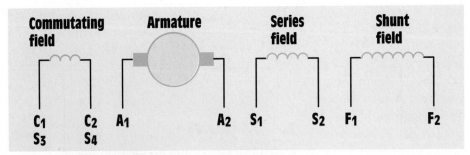

**Figure 29-13** Terminal identification for DC machines

## 29-6 Determining the Direction of Rotation of a DC Motor

The direction of rotation of a DC motor is determined by facing the commutator end of the motor. This is generally the back or rear of the motor. If the windings have been labeled in a standard manner, it is possible to determine the direction of rotation when the motor is connected. *Figure 29-14* illustrates the standard connections for a series motor. The standard

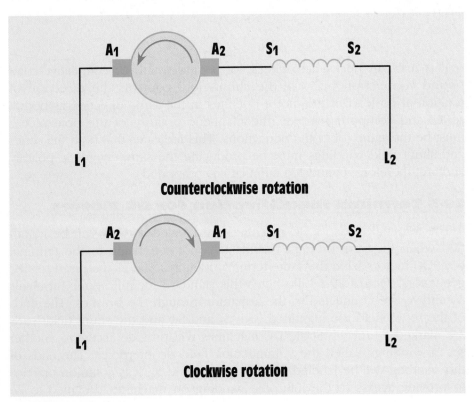

**Figure 29-14** Series motors

connections for a shunt motor are illustrated in *Figure 29-15,* and the standard connections for a compound motor are shown in *Figure 29-16.*

The direction of rotation of a DC motor can be reversed by changing the connections of the armature leads or the field leads. It is common practice to change the connection of the armature leads. This is done to prevent changing a cumulative-compound motor into a differential-compound motor.

Although it is standard practice to change the connection of the armature leads to reverse the direction of rotation, it is not uncommon to reverse the rotation of small shunt motors by changing the connection of the field leads. If a motor contains only a shunt field, there is no danger of changing the motor from a cumulative- to a differential-compound motor. The shunt field leads are often changed on small motors because the amount of current flow through the field is much less than the current flow through the armature. This permits a small double-pole double-throw switch to be used as a control for reversing the direction of rotation *(Figure 29-17).*

Large compound motors often use a control circuit similar to the one shown in *Figure 29-18* for reversing the direction of rotation. This control

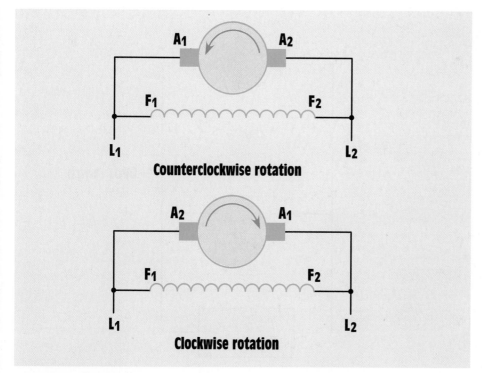

Counterclockwise rotation

Clockwise rotation

**Figure 29-15**  Shunt motors

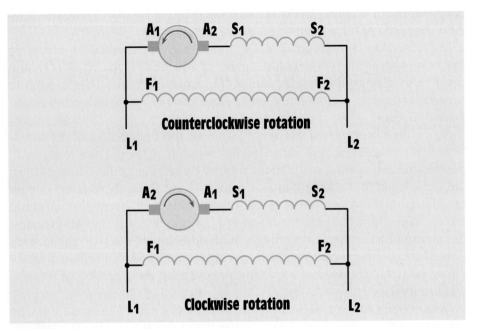

**Figure 29-16** Compound motors

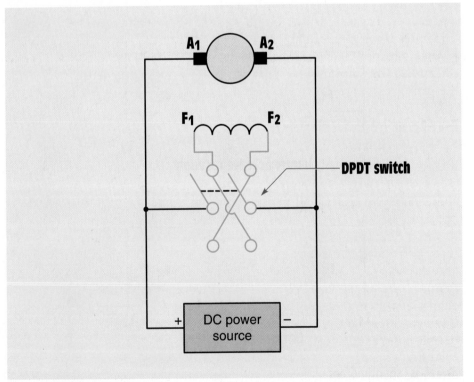

**Figure 29-17** Reversing the rotation of a shunt motor by changing the shunt field leads

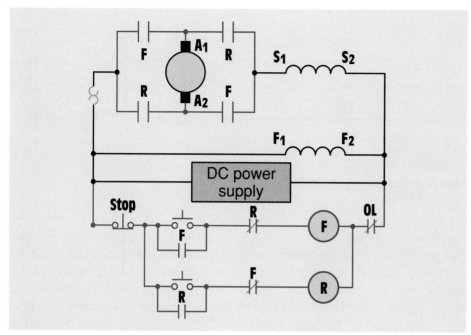

**Figure 29-18** Forward-reverse control for a compound motor

circuit uses magnetic contactors to reverse the flow of current through the armature. If the circuit is traced, it will be seen that when the forward or reverse direction is chosen, only the current through the armature changes direction. The current flow through the shunt and series fields remains the same.

## 29-7 Speed Control

There are several ways of controlling the speed of direct current motors. The method employed is generally dictated by the requirements of the load. When full voltage is connected to both the armature and shunt field, the motor will operate at its base speed. If the motor is to be operated at below base speed or under speed, full voltage is maintained to the shunt field and the amount of armature current is reduced. The reduction of armature current causes the motor to produce less torque, and the speed decreases. One method of reducing armature current is to connect resistance in series with the armature *(Figure 29-19)*. In the example shown, three resistors are connected in series with the armature. Contacts $S_1$, $S_2$, and $S_3$ can be used to shunt out steps of resistance and permit more armature current to flow.

Although adding resistance in the armature circuit does permit the motor to be underspeeded, it has several disadvantages. As current flows

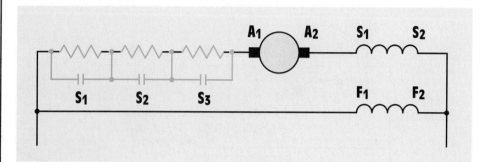

**Figure 29-19** Resistors limit armature current.

through the resistors, they waste power in the form of heat. Also, the speed of the motor can only be controlled in steps. There is no smooth increase or decrease in speed. Most large direct current motors use an electronic controller to supply variable voltage to the armature circuit separately from the field *(Figure 29-20)*. This permits continuous adjustment of the speed from zero to full RPM. Electronic power supplies can also sense the speed of the motor and maintain a constant speed as load is changed. Most of these power supplies are current-limited. The maximum current output can be set to a value that will not permit the motor to be harmed if it stalls or if the load becomes too great.

A direct current motor can be overspeeded by connecting full voltage to the armature and reducing the current flow through the shunt field. In *Figure 29-21,* a shunt field rheostat has been connected in series with the shunt field. As resistance is added to the shunt field, field current decreas-

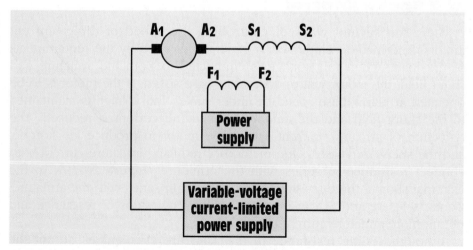

**Figure 29-20** Armature current is controlled by a variable-voltage power supply.

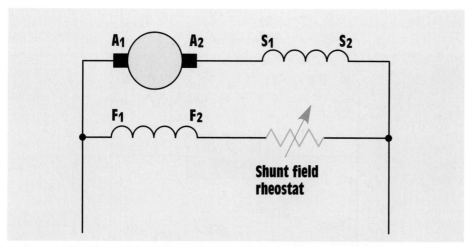

**Figure 29-21** The shunt field rheostat can be used to overspeed the motor.

es, which causes a decrease in the flux density of the pole pieces. This decrease in flux density produces less counter-EMF in the armature, which permits more armature current to flow. The increased armature current causes an increase in the magnetic field strength of the armature. This increased magnetic field strength of the armature produces a net gain in torque, which causes the motor speed to increase.

## 29-8 The Field Loss Relay

Most large compound direct current motors have a protective device connected in series with the shunt field called the **field loss relay** or shunt field relay *(Figure 29-22)*. The function of the field loss relay is to disconnect power to the armature if current flow through the shunt field should decrease below a certain level. If the shunt field current stopped completely, the compound motor would become a series motor and would increase rapidly in speed. This could cause damage to both the motor and the load.

Many large DC compound motors intended to operate in an overspeed condition will actually contain two separate shunt fields *(Figure 29-23)*. One shunt field is connected to a fixed voltage and maintains a constant field to provide an upper limit to motor speed. This shunt field will be connected to the field loss, or shunt field relay. The second shunt field is connected to a source of variable voltage. This shunt field is used to increase speed above the base speed. For this type of motor, base speed is achieved by applying full voltage to the armature and both shunt fields.

**field loss relay**

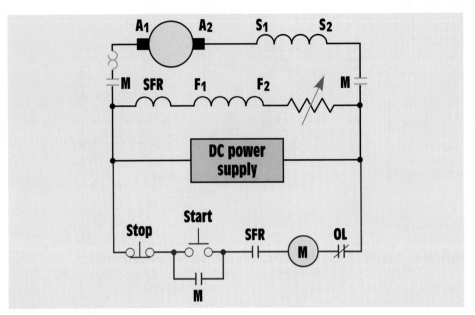

**Figure 29-22** The shunt field relay disconnects power to the armature if shunt field current stops.

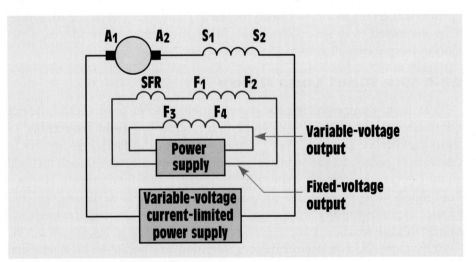

**Figure 29-23** One shunt field is used to provide a stable speed; the other shunt field provides overspeed control.

It should be noted that most large DC motors have voltage applied to the shunt field at all times, even when the motor is not in operation. The resistance of the winding produces heat, which is used to prevent any formation of moisture inside the motor.

## 29-9 Horsepower

When James Watt first began to try to market steam engines, he found that he needed a way to compare them to the horses they were to replace. After conducting experiments, Watt found that the average horse could do work at a rate of 550 ft-lb/s. This became the basic horsepower measurement. Horsepower can also be expressed in the basic electrical unit for power, which is the watt.

$$1 \text{ horsepower} = 746 \text{ W}$$

Once horsepower has been converted to a basic unit of power, it can be converted to other power units such as:

$$1 \text{ W} = 3.42 \text{ BTUs per hour}$$

$$1055 \text{ W} = 1 \text{ BTU per second}$$

$$4.19 \text{ W} = 1 \text{ calorie per second}$$

$$1.36 \text{ W} = 1 \text{ ft-lb per second}$$

In order to determine the horsepower output of a motor, the rate at which the motor is doing work must be known. The following formula can be used to determine the horsepower output of a motor.

$$hp = \frac{(1.59) \, (torque) \, (RPM)}{100,000}$$

where

$$hp = \text{horsepower}$$

$$1.59 = \text{a constant}$$

$$torque = \text{torque in lb-in.}$$

$$RPM = \text{speed}$$

$$100,000 = \text{a constant}$$

How much horsepower is being produced by a motor turning a load of 350 lb-in. at a speed of 1375 RPM?

**Example 1**

### Solution

$$hp = \frac{1.59 \times 350 \times 1375}{100,000}$$

$$hp = 7.65$$

Once the output horsepower is known, it is possible to determine the efficiency of the motor by using the formula

$$\text{Eff.} = \frac{\text{power out}}{\text{power in}} \times 100$$

**Example 2**

A direct current motor is connected to a 120-V DC line and has a current draw of 1.3 A. The motor is operating a load that requires 8 lb-in. of torque and is turning at a speed of 1250 RPM. What is the efficiency of the motor?

## Solution

The first step is to determine the horsepower output of the motor.

$$\text{hp} = \frac{1.59 \times 8 \times 1250}{100,000}$$

$$\text{hp} = 0.159$$

Now that the output horsepower is known, horsepower can be changed into watts using the formula

$$1 \text{ hp} = 746 \text{ W}$$

$$0.159 \times 746 = 118.6 \text{ W}$$

The amount of input power can be found by using the formula

$$\text{watts} = \text{volts} \times \text{amps}$$

$$\text{watts} = 120 \times 1.3$$

$$\text{watts} = 156$$

The efficiency of the motor can be found by using the formula

$$\text{Eff.} = \frac{\text{power out}}{\text{power in}} \times 100$$

The answer is multiplied by 100 to change it to a percent.

$$\text{Eff.} = \frac{118.6}{156} \times 100$$

$$\text{Eff.} = 76\%$$

Torque is often measured in pound-feet instead of pound-inches. Another formula often used to determine horsepower when the torque is measured in pound-feet is

$$\text{hp} = \frac{(2\pi) \text{ (torque) (RPM)}}{33,000}$$

where

hp      =  horsepower

$\pi$      =  3.1416

torque  =  lb-ft

RPM    =  speed in revolutions per minute

33,000  =  a constant

## 29-10 Brushless DC Motors

**Brushless DC motors** do not contain a wound armature, commutator, or brushes. The armature or rotor (rotating member) contains permanent magnets. The rotor is surrounded by fixed stator windings *(Figure 29-24)*. The stationary armature or stator winding is generally three-phase, but some motors are designed to operate on four-phase or two-phase power.

**brushless DC motors**

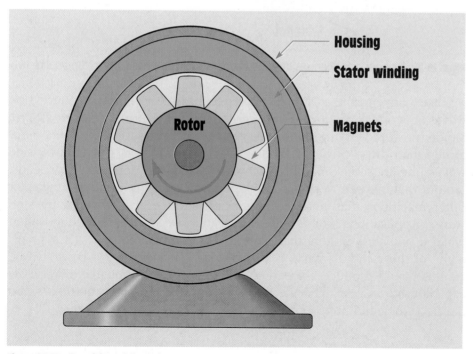

Figure 29-24  Brushless DC motor

Two-phase stator windings are commonly used for motors intended to operate small fans.

The phases are provided by a converter that changes the direct current into alternating current *(Figure 29-25)*. Alternating current is used to create a rotating magnetic field inside the stator of the motor. This rotating magnetic field attracts the permanent magnets of the rotor and causes the rotor to turn in the same direction as the rotating field. Rotating magnetic fields will be discussed in Unit 31. The speed is determined by the number of stator poles per phase and the frequency of the AC voltage.

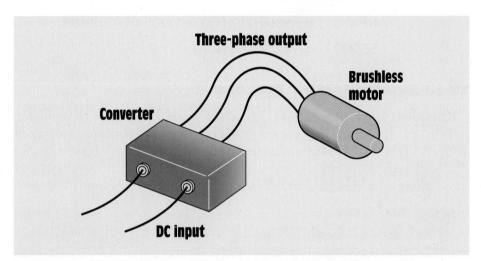

**Figure 29-25**  A converter changes direct current into three-phase alternating current.

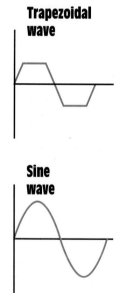

**Figure 29-26** Most converters produce a trapezoidal wave, but some produce a sine wave.

Some converters that supply power to brushless motors produce sine waves, but most produce a trapezoidal AC wave form *(Figure 29-26)*. Motors powered by the trapezoidal wave form produce about 10% more torque than those powered by sine waves. But motors powered by sine waves operate more smoothly and with less torque ripple at low speeds. Motors that require very smooth operation such as grinders, mills, or other machines that produce a finished surface are powered by sine waves. Applications where high torque and low speed are required often employ brushless motors that have a very high number of stator poles. Some of these motors have as many as 64 stator poles per phase. At 60 Hz this would produce a speed of 112.5 RPM. The formula shown below can be used to determine the speed of the rotating magnetic field when the frequency and number of stator poles per phase are known.

$$S = \frac{120 \times F}{P}$$

where

S   =   speed in revolutions per minute (RPM)

120  =  a constant

F   =  frequency in hertz

P   =  number of poles per phase

The speed of the rotating magnetic field for the motor described above would be

$$S = \frac{120 \times 60}{64}$$

$$S = 112.5 \text{ RPM}$$

The use of this many poles to provide low speed and high torque is often referred to as magnetic gearing, because it eliminates the need for mechanical gears and other speed-reducing equipment. This in turn eliminates friction and backlash associated with mechanical gears. Motors that use a high pole count for low-speed high-torque applications are often called ring motors or ring torquers.

## Inside-Out Motors

Another type of brushless DC motor uses a rotor shaped like a hollow cylinder or cup. The stator windings are wound inside the rotor (*Figure 29-27*). The permanent magnet rotor is mounted on the outside of the stator windings. The size and shape of the rotor cause it to act like a flywheel, which gives these motors a large amount of inertia. **Motors with a large amount of inertia exhibit superior speed regulation characteristics.** Inside-out motors are often used to drive the hard disks in computers, to operate tape cartridge drives, to provide power for robots, and to operate fans and blowers in high-speed air-conditioning systems.

## Differences between Brush Type and Brushless Motors

Since brushless motors do not have a commutator or brushes, they are generally smaller and cost less than brush type motors. The added expense of the converter, however, makes the cost about the same as a brush type motor. Brushless motors do dissipate heat more quickly because the stator windings can dissipate heat faster than a wound armature. They are more efficient than brush type motors, require less maintenance, and as a general rule have less down time.

**Motors with a large amount of inertia exhibit superior speed regulation characteristics.**

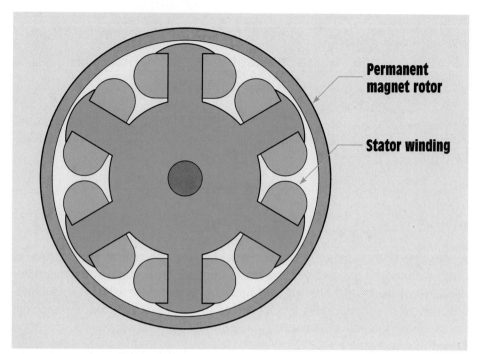

**Figure 29-27**  Inside-out brushless DC motor

## 29-11 Converters

As stated previously, converters are used to change the direct current into multiphase alternating current to produce a rotating magnetic field in the stationary armature or stator of the brushless DC motor. An example circuit for a DC-to-three-phase converter is shown in *Figure 29-28*. In this example, transistors $Q_1$ through $Q_6$ are used as switches. By turning the transistors on or off in the proper sequence, current can be routed through the stator windings in such a manner that it produces three alternating currents 120° out of phase with each other. The commutation sequence provides an action similar to that of the commutator in a brush type motor. A chart illustrating the firing order of the transistors to produce this commutation sequence is shown in *Figure 29-29*. The current flow through the three stator windings is illustrated in the same chart.

## 29-12 Permanent Magnet Motors

**permanent magnet motors**

**Permanent magnet motors** contain a wound armature and brushes like a conventional direct current motor. The pole pieces, however, are permanent magnets. This eliminates the need for shunt or series field windings *(Figure 29-30)*. Permanent magnet motors have a higher efficiency than

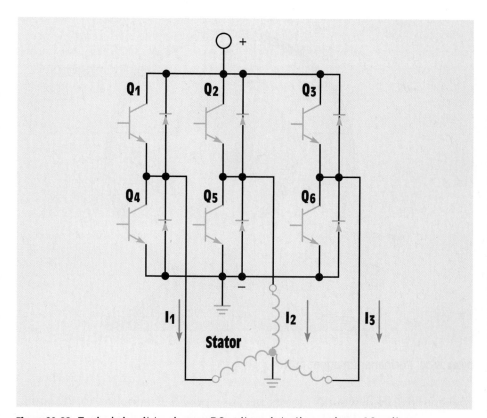

**Figure 29-28** Typical circuit to change DC voltage into three-phase AC voltage

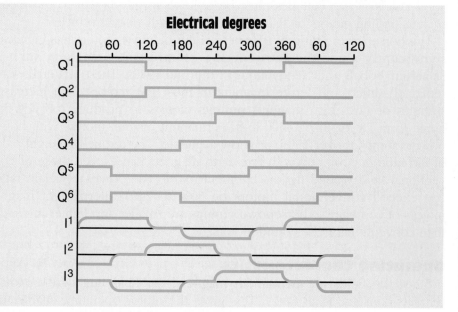

**Figure 29-29** Converter commutation sequence

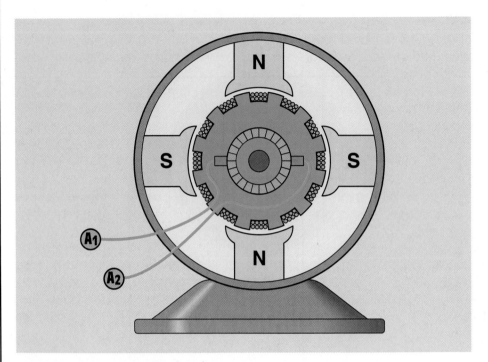

**Figure 29-30** Permanent magnet motor

conventional field wound motors because power is supplied to the arma-ture circuit only. These motors have been popular for many years in applications where batteries must be used to supply power to the motor, such as trolling motors on fishing boats and small electric vehicles.

The horsepower rating of PM (permanent magnet) motors has increased significantly since the introduction of rare-earth magnets such as Samarium-cobalt and Neodymium in the mid-1970s. These materials have replaced Alnico and ferrite magnets in most PM motors. The increased strength of rare-earth magnets permits motors to produce torques that range from 7.0 oz-in. to 4500 lb-ft. Permanent magnet motors with horse-power ratings of over 15 hp are now available. The torque-to-weight ratios of PM motors equipped with rare-earth magnets can exceed those of con-ventional field wound motors by 40% to 90%. The power-to-weight ratios can exceed conventional motors by 50% to 200%. Permanent magnet motors of comparable horsepower ratings are smaller and lighter in weight than conventional field wound motors.

## Operating Characteristics

Since the fields are permanent magnets, the field flux of PM motors remains constant at all times. This gives the motor operating characteristics very similar to those of conventional separately excited shunt motors. The speed can be controlled by the amount of voltage applied to the armature, and the direction of rotation is reversed by reversing the polarity of voltage applied to the armature leads *(Figure 29-31).*

## DC Servomotors

Small permanent magnet motors are used as **servomotors**. These motors have small, lightweight armatures that contain very little inertia. This permits servomotors to be operated at high speed and then stopped or reversed very quickly. Servomotors generally contain from two to six poles. They are used to operate tape drives on computers and to power the spindles on numerically controlled (NC) machines such as milling machines and lathes.

## DC ServoDisc Motors

Another type of direct current servomotor that is totally different in design is the **ServoDisc® motor**. This motor uses permanent magnets to provide a constant magnetic field like conventional servomotors, but the

servo-
motors

ServoDisc®
motor

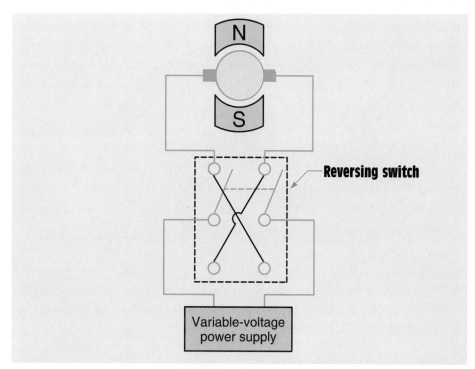

**Figure 29-31** PM motors can be reversed by reversing the polarity to the armature and the speed can be controlled by variable voltage.

design of the armature is completely different. In a conventional servomotor, the armature is constructed in the same way as other DC motors by cutting slots in an iron core and winding wire through the slots. A commutator is then connected to one end to provide commutation, which switches the current path as the armature turns. This maintains a constant magnetic polarity for different sections of the armature. In the conventional servomotor, the permanent magnets are mounted on the motor housing in such a manner that they create a radial magnetic field that is perpendicular to the windings of the armature *(Figure 29-32)*. The armature core is made of iron because it must conduct the magnetic lines of flux between the two pole pieces.

The armature of the ServoDisc motor does not contain any iron. It is made of two to four layers of copper conductors formed into a thin disc. The conductors are "printed" on a fiberglass material in much the same way as a printed circuit. For this reason, the ServoDisc motor is often called a **printed circuit motor**. Since the disc armature is very thin, it permits the permanent magnets to be mounted on either side of the disc and parallel to the shaft of the motor *(Figure 29-33)*. Since the disc is very thin, the air gap between the two magnets is small.

**printed circuit motor**

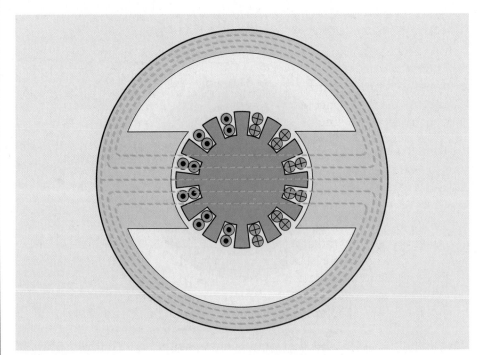

**Figure 29-32** Magnetic lines of flux must travel a great distance between the poles of the permanent magnets.

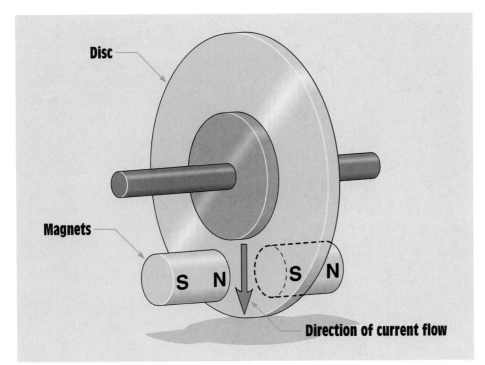

**Figure 29-33**  Basic construction of a ServoDisc motor

Torque is produced when current flowing though the copper conductors of the disc produces a magnetic field that reacts with the magnetic field of the permanent magnets. The permanent magnet pairs are arranged around the circumference of the motor housing in such a manner that they provide alternate magnetic fields *(Figure 29-34)*. The conductors on each side of the armature, upper and lower, are arranged in such a manner that when current flows through them, a force tangent to the pole piece is produced. This tangential force is the vector sum of the forces produced by the upper and lower conductors *(Figure 29-35)*. The ServoDisc motor produces a relatively strong torque for its size and weight.

The conductors of the armature are so arranged on the fiberglass disc that they form a commutator on one side of the armature. Brushes riding against this commutator supply direct current to the armature conductors. Since the armature contains no iron, it has almost no inductance. This greatly reduces any arcing at the brushes, which results in extremely long brush life. A typical ServoDisc motor is shown in *Figure 29-36A and B.*

## Characteristics of ServoDisc Motors

The unique construction of the disc type servomotor gives it some operating characteristics that are different from those of other types of

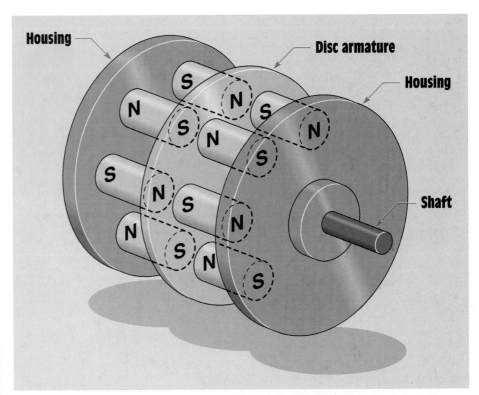

**Figure 29-34** The permanent magnets are arranged to produce alternate magnetic polarities.

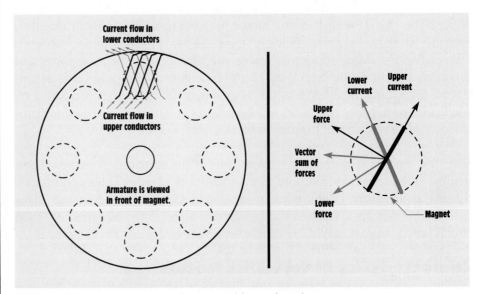

**Figure 29-35** A force tangent to the magnet is produced.

**Figure 29-36**  DC ServoDisc motors *(Courtesy of PMI Motion Technologies.)*

permanent magnetic DC motors. Permanent magnet direct current motors with iron armatures have a problem with cogging at low speeds. Cogging is caused by the reaction of the iron armature with the field of the permanent magnets. If the armature is turned by hand, it will jump, or "cog," from one position to another. As the armature is turned, the magnetic flux lines of the pole pieces pass through the core of the armature *(Figure 29-32)*. Since the armature is slotted, certain areas will offer a path of lower reluctance than others. These areas are more strongly attracted to the pole pieces, which causes the armature to jump from one position to the next. Since the armature of the disc motor contains no iron, there is no cogging action. This permits very smooth operation at low speeds.

Another characteristic of disc motors is extremely fast acceleration. The thin, low-inertia disc armature permits an exceptional torque-inertia ratio. A typical ServoDisc motor can accelerate from 0 to 3000 RPM in about 60° of rotation. In other words, the motor can accelerate from 0 to 3000 RPM in one sixth of a revolution. The low-inertia armature also permits rapid stops and reversals. Disc servomotors can operate at speeds over 4000 RPM.

The speed of the disc servomotor can be varied by changing the amount of voltage supplied to the armature. The voltage is generally varied using **pulse-width modulation**. Most amplifiers for the disc servomotor produce a pulsating DC voltage at a frequency of about 20 kHz. The average voltage supplied to the armature is determined by the length of time the voltage is turned on as compared with the time it is turned off (pulse-

**pulse-width modulation**

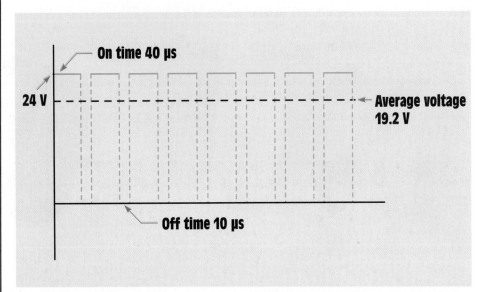

**Figure 29-37**  The amount of average voltage is determined by the peak voltage and the amount of time it is turned on or off.

width). At a frequency of 20 kHz, the pulses have a width of 50 μs (microseconds) (pulse-width = 1/frequency). Assume that the pulses have a peak voltage of 24 V and are turned on for 40 μs and off for 10 μs during each pulse *(Figure 29-37)*. In this example, 24 V is supplied to the motor for 80% of the time, producing an average voltage of 19.2 V (24 x

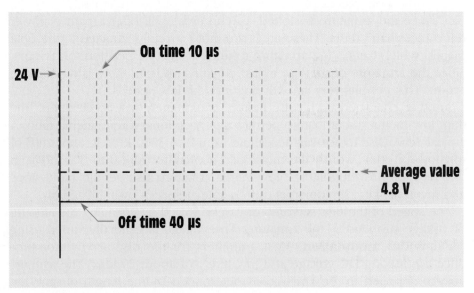

**Figure 29-38**  Reducing the on time reduces the average voltage.

0.80 = 19.2). Now assume that the pulse-width is changed so that the on time is 10 µs and the off time is 40 µs *(Figure 29-38)*. A voltage of 24 V is now supplied to the motor 20% of the time, resulting in an average value of 4.8 V (24 x 0.20 = 4.8).

Because of the basic design of the motor, the ServoDisc motor can be constructed in a very small, thin case *(Figure 29-39)*. This makes it very useful in some applications where other types of servomotors could not be used.

**Figure 29-39**  The DC ServoDisc motor can be constructed with small, thin cases. *(Courtesy of PMI Motion Technologies.)*

## 29-13 The Right-Hand Motor Rule

In Unit 28 it was shown that it is possible to determine the direction of current flow through the armature of a generator using the fingers of the left hand when the polarity of the field poles and the direction of rotation of the armature are known. Similarly, the fingers of the right hand can be used to determine the direction of rotation of the armature when the magnetic field polarity of the pole pieces and the direction of current flow through the armature are known *(Figure 29-40)*. The thumb indicates the direction of thrust or movement of the armature. The forefinger indicates the direction of the field flux assuming that flux lines are in a direction of north to south, and the center finger indicates the direction of current

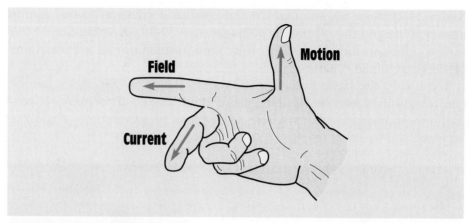

**Figure 29-40** Right-hand motor rule

flow through the armature. A simple method of remembering which finger represents which quantity is:

| | | |
|---|---|---|
| **THumb** | = | **THrust** (direction of armature rotation) |
| **Forefinger** | = | **Field** (direction of magnetic field) |
| **Center finger** | = | **Current** (direction of armature current) |

## Summary

1. A motor is a machine that converts electrical energy into mechanical energy.

2. Direct current motors operate on the principle of attraction and repulsion of magnetism.

3. Two factors that determine the torque produced by a motor are:
   A. The strength of the magnetic field of the pole pieces.
   B. The strength of the magnetic field of the armature.

4. Torque is proportional to armature current if field excitation current remains constant.

5. Current flow through the armature is limited by counter-EMF and armature resistance.

6. Three factors that determine the amount of counter-EMF produced by a motor are:
   A. The strength of the magnetic field of the pole pieces.
   B. The number of turns of wire in the armature.
   C. The speed of the armature.

7. Three basic types of DC motors are the series, shunt, and compound.

8. Shunt motors are sometimes known as constant speed motors.

9. Series motors must never be operated under a no-load condition.

10. Series motors can develop extremely high starting torque.

11. Compound motors contain both series and shunt field windings.

12. When full voltage is applied to both the armature and shunt field, the motor will operate at base speed.

13. When full voltage is applied to the field and reduced voltage is applied to the armature, the motor will operate below base speed.

14. When full voltage is applied to the armature and reduced voltage is applied to the shunt field, the motor will operate above base speed.

15. The direction of rotation of a direct current motor can be changed by reversing the connection of either the armature or the field leads.

16. It is common practice to reverse the connection of the armature leads to prevent changing a compound motor from a cumulative- to a differential-compound connection.

17. The shunt field relay is used to disconnect power to the armature if shunt field current drops below a certain level.

18. Brushless DC motors do not contain a wound armature, commutator, or brushes.

19. The stator windings for brushless motors are generally three-phase, but can be four-phase or two-phase.

20. Brushless DC motors require a converter to change the direct current into multiphase alternating current to operate the motor.

21. Brushless motors generally require less maintenance and have less down time than conventional DC motors.

22. Permanent magnet motors use permanent magnets as the pole pieces and do not require series or shunt field windings.

23. Permanent magnet motors are more efficient than conventional DC motors.

24. The operating characteristics of a PM motor are similar to those of a shunt motor with external excitation.

25. Servomotors have lightweight armatures with low inertia.

26. Disc servomotors contain an armature that is made of several layers of copper conductors.

27. Since there is no iron in the armature of a disc servomotor, it does not have a problem with cogging.

28. Disc servomotors can be accelerated very rapidly.

## Review Questions

1. What is the principle of operation of a direct current motor?

2. What is a motor?

3. What is the function of the commutator in a direct current motor?

4. What two factors determine the amount of torque developed by a DC motor?

5. What type of motor is known as a constant speed motor?

6. What is counter-EMF?

7. What three factors determine the amount of counter-EMF produced in the armature?

8. What limits the amount of current flow through the armature when power is first applied to the motor?

9. What factor determines the speed regulation of a DC motor?

10. What type of motor should never be operated at no load?

11. In general, what type of compound motor connection should be avoided?

12. What is the most common way of changing the direction of rotation of a compound motor?

13. How can a DC motor be made to operate at its base speed?

14. How can a DC motor be made to operate above its base speed?

15. What device is used to disconnect power to the armature if the shunt field current drops below a certain level?

16. Why do many industries leave power connected to the shunt field at all times even when the motor is not operating?

17. Who was the first person to establish a measurement for horsepower?

18. One horsepower is equal to how many watts?

19. A motor is operating a load that requires a torque of 750 lb-in. and is turning at a speed of 1575 RPM. How much horsepower is the motor producing?

20. The motor is question 19 is connected to a 250-V DC line and has a current draw of 80 A. What is the efficiency of this motor?

*section*

# alternating current machines

Courtesy of New York Power Authority.

# Three-Phase Alternators

## Outline

## Key Terms

Alternators

Brushless exciter

Cooling

Excitation current

Field discharge resistor

Frequency

Hydrogen

Parallel alternators

Phase rotation

Revolving armature

Revolving field

Rotor

Sliprings

Stator

Synchroscope

**A**fter studying this unit, you should be able to:

- Discuss the operation of a three-phase alternator.
- Explain the effect of speed of rotation on frequency.
- Explain the effect of field excitation on output voltage.
- Connect a three-phase alternator and make measurements using test instruments.

Most of the electrical power in the world today is produced by alternating current generators or alternators. Electrical power companies use alternators rated in gigawatts (1 gigawatt = 1,000,000,000 W) to produce the power used throughout the United States and Canada. The entire North American continent is powered by alternating current generators connected together in parallel. These alternators are powered by steam turbines. The turbines, called prime movers, are powered by oil, coal, natural gas, or nuclear energy.

### 30-1 Three-Phase Alternators

**Alternators** operate on the same principle of electromagnetic induction as direct current generators, but they have no commutator to change the alternating current produced in the armature into direct current. There are two basic types of alternators: the revolving armature type and the revolving field type. Although there are some single-phase alternators that are used as portable power units for emergency home use or to operate power tools in a remote location, most alternators are three-phase.

### Revolving Armature Type Alternators

The **revolving armature** type alternator is the least used of the two basic types. This alternator uses an armature similar to that of a direct current machine with the exception that the loops of wire are connected to **sliprings** instead of to a commutator *(Figure 30-1)*. There are three separate windings, which are connected in either delta or wye. The armature windings are rotated inside a magnetic field *(Figure 30-2)*. Power is carried to the outside circuit via brushes riding against the sliprings. This alternator is the least used because it is very limited in the amount of output voltage and kilo-volt-amp (kVA) capacity it can develop.

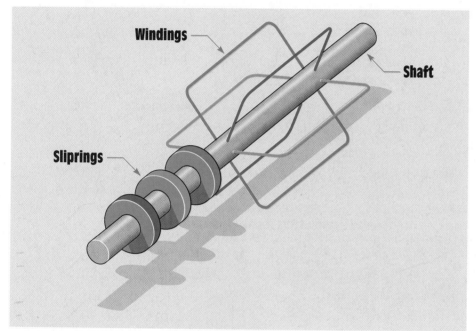

**Figure 30-1** Basic design of a three-phase armature

## Revolving Field Type Alternators

The **revolving field** type alternator uses a stationary armature called the **stator** and a rotating magnetic field. This design permits higher voltage and kVA ratings because the outside circuit is connected directly to the stator and is not routed through sliprings and brushes. This type of alternator is constructed by placing three sets of windings 120° apart *(Figure 30-3)*. In *Figure 30-3*, the winding of phase 1 winds around the top center pole piece. It then proceeds 180° around the stator and winds around the opposite pole piece in the opposite direction. The second phase winding winds around the top pole piece directly to the left of the top center pole piece. The second phase winding is wound in an opposite direction to the first. It then proceeds 180° around the stator housing

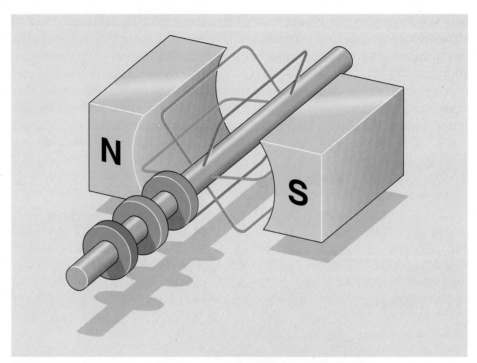

**Figure 30-2** The armature conductors rotate inside a magnetic field.

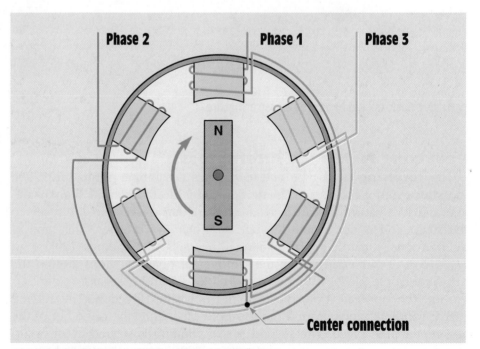

**Figure 30-3** Basic design of a three-phase alternator

and winds around the opposite pole piece in the opposite direction. The finish end of phase 2 connects to the finish end of phase 1. The start end of phase 3 winds around the top pole piece to the right of the top center pole piece. This winding is wound in an opposite direction than phase 1 also. The winding then proceeds 180° around the stator frame to its opposite pole piece and winds around it in an opposite direction. The finish end of phase 3 is then connected to the finish ends of phases 1 and 2. This forms a wye connection for the stator winding. When the magnet is rotated, voltage will be induced in the three windings. Since these windings are spaced 120° apart, the induced voltages will be 120° out of phase with each other *(Figure 30-4)*.

The stator shown in *Figure 30-3* is drawn in a manner to aid in understanding how the three phase windings are arranged and connected. In actual practice, the stator windings are placed in a smooth cylindrical core without projecting pole pieces *(Figure 30-5)*. This design provides a better path for magnetic lines of flux and increases the efficiency of the alternator.

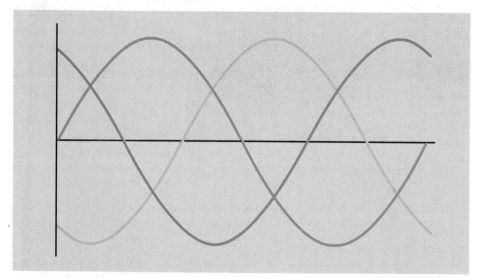

**Figure 30-4** The alternator produces three sine wave voltages 120° out of phase with each other.

## 30-2 The Rotor

The **rotor** is the rotating member of the machine. It provides the magnetism needed to induce voltage into the stator windings. The magnets of the rotor are electromagnets and require some source of external direct

**rotor**

**Figure 30-5** Wound stator *(Courtesy of Magnatek.)*

current to excite the alternator. This direct current is known as **excitation current**. The alternator cannot produce an output voltage until the rotor has been excited. Some alternators use sliprings and brushes to provide the excitation current to the rotor *(Figure 30-6)*. A good example of this type of rotor can be found in the alternator of most automobiles. The DC excitation current can be varied in order to change the strength of the magnetic field. A rotor with salient (projecting) poles is shown in *Figure 30-7*.

## 30-3 The Brushless Exciter

Most large alternators use an exciter that contains no brushes. This is accomplished by adding a separate small alternator of the armature type

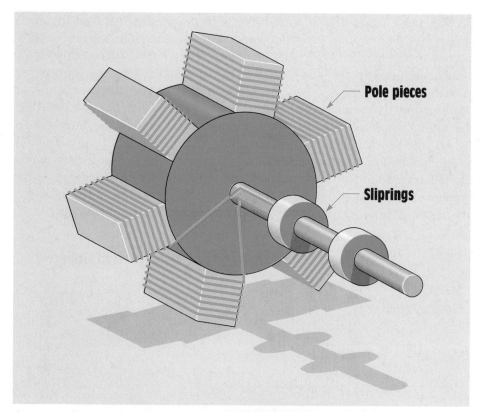

**Figure 30-6** The rotor contains pole pieces that become electromagnets.

**Figure 30-7** Rotor of the salient pole type (*From Duff and Herman,* Alternating Current Fundamentals, *4th ed., copyright 1991 by Delmar Publishers Inc.)*

on the same shaft of the rotor of the larger alternator. The armature rotates between wound electromagnets. The DC excitation current is connected to the wound stationary magnets *(Figure 30-8)*. The amount of voltage induced in the rotor can be varied by changing the amount of excitation current supplied to the electromagnets. The output voltage of the armature is connected to a three-phase bridge rectifier mounted on the rotor shaft *(Figure 30-9)*. The bridge rectifier converts the three-phase AC voltage produced in the armature into DC voltage before it is applied to the main rotor windings. Since the armature, rectifier, and rotor winding are connected to the main rotor shaft, they all rotate together and no brushes or sliprings are needed to provide excitation current for the large alternator. A photograph of the **brushless exciter** assembly is shown in

**brushless exciter**

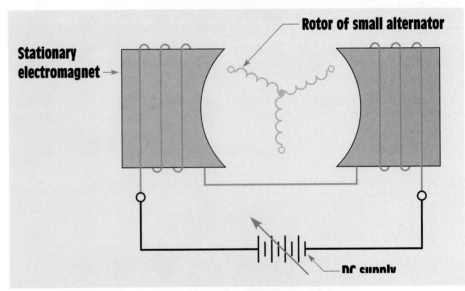

**Figure 30-8** The brushless exciter uses stationary electromagnets.

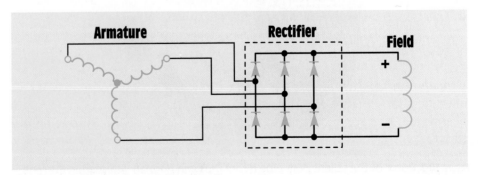

**Figure 30-9** Basic brushless exciter circuit

**Figure 30-10**  Brushless exciter assembly *(Courtesy of Magnatek.)*

*Figure 30-10*. The field winding is placed in slots cut in the core material of the rotor *(Figure 30-11)*.

## 30-4 Alternator Cooling

There are two main methods of **cooling** alternators. Alternators of small kVA rating are generally air-cooled. Open spaces are left in the stator windings, and slots are often provided in the core material for the passage of air. Air-cooled alternators have a fan attached to one end of the shaft that circulates air through the entire assembly.

Large-capacity alternators are often enclosed and operate in a **hydrogen** atmosphere. There are several advantages in using hydrogen. Hydrogen is less dense than air at the same pressure. The lower density reduces the windage loss of the spinning rotor. A second advantage in operating an alternator in a hydrogen atmosphere is that hydrogen has the ability to absorb and remove heat much faster than air. At a pressure of one atmosphere, hydrogen has a specific heat of approximately 3.42. The specific heat of air at a pressure of one atmosphere is approximately 0.238. This means that hydrogen has the ability to absorb approximately 13.8 times

**cooling**

**hydrogen**

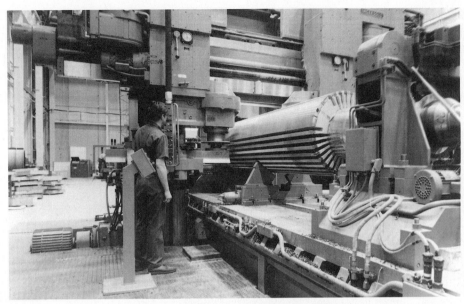

**Figure 30-11** Two-pole rotor slotting *(Courtesy of Houston Lighting and Power.)*

more heat than air. A cutaway drawing of an alternator intended to operate in a hydrogen atmosphere is shown in *Figure 30-12.*

### 30-5 Frequency

**frequency**

The output **frequency** of an alternator is determined by two factors:
1. the number of stator poles and
2. the speed of rotation of the rotor.

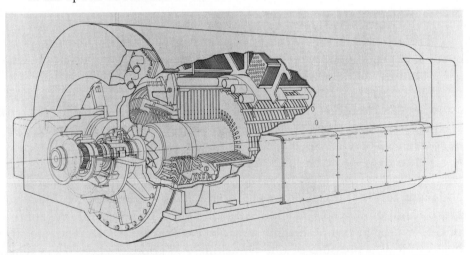

**Figure 30-12** Two-pole turbine-driven hydrogen-cooled alternator *(Courtesy of Houston Lighting and Power.)*

Since the number of stator poles is constant for a particular machine, the output frequency is controlled by adjusting the speed of the rotor. The chart below shows the speed of rotation needed to produce 60 Hz for alternators with different numbers of poles.

| RPM | Stator Poles |
|------|--------------|
| 3600 | 2 |
| 1800 | 4 |
| 1200 | 6 |
| 900 | 8 |

The formula shown below can also be used to determine the frequency when the poles and RPM (revolutions per minute) are known.

$$F = \frac{PS}{120}$$

where

F = frequency in hertz

P = number of pairs of poles (one north and one south)

S = speed in RPM

120 = a constant

What is the output frequency of an alternator that contains six poles per phase and is turning at a speed of 1000 RPM?

Example 1

**Solution**

$$F = \frac{6 \times 1000}{120}$$

F = 50 Hz

## 30-6 Output Voltage

Three factors determine the amount of output voltage of an alternator:
1. the length of the armature or stator conductors (number of turns),

2. the strength of the magnetic field of the rotor, and

3. the speed of rotation of the rotor.

The formula shown below can be used to compute the amount of voltage induced in the stator winding.

$$E \ = \ \frac{BLv}{10^8}$$

where

$10^8$ = flux lines equal to 1 weber

E   = induced voltage

B   = flux density in gauss

L   = length of the conductor

v   = velocity

**One of the factors that determines the amount of induced voltage is the length of the conductor. This factor is often stated as number of turns of wire in the stator because the voltage induced in each turn adds.** Increasing the number of turns of wire has the same effect as increasing the length of one conductor.

### Controlling Output Voltage

The number of turns of wire in the stator cannot be changed in a particular machine without rewinding the stator, and the speed of rotation is generally maintained at a certain level to provide a constant output frequency. Therefore, the output voltage is controlled by increasing or decreasing the strength of the magnetic field of the rotor. The magnetic field strength can be controlled by controlling the DC excitation current to the rotor.

### 30-7 Paralleling Alternators

Since one alternator cannot produce all the power that is required, it often becomes necessary to use more than one machine. When more than one alternator is to be used, they are connected in parallel with each other. Several conditions must be met before **parallel alternators** can be used:

1. The phases must be connected in such a manner that the phase rotation of all the machines is the same.

2. Phases A, B, and C of one machine must be in sequence with phases A, B, and C of the other machine. For example, phase A of

One of the factors that determines the amount of induced voltage is the length of the conductor. This factor is often stated as number of turns of wire in the stator because the voltage induced in each turn adds.

parallel alternators

alternator 1 must reach its positive peak value of voltage at the same time phase A of alternator 2 does *(Figure 30-13)*.

3. The output voltage of the two alternators should be the same.

## Determining Phase Rotation

The most common method of detecting when the **phase rotation** (the direction of magnetic field rotation) of one alternator is matched to the phase rotation of the other is with the use of three lights *(Figure 30-14)*. In *Figure 30-14*, the two alternators that are to be paralleled are connected together through a synchronizing switch. A set of lamps acts as a resistive load between the two machines when the switch contacts are in the open position. The voltage developed across the lamps is proportional to the difference in voltage between the two alternators. The lamps are used to indicate two conditions.

1. The lamps indicate when the phase rotation of one machine is matched to the phase rotation of the other. When both alternators are operating, both are producing a voltage. The lamps will blink on and off when the phase rotation of one machine is not synchronized to the phase rotation of the other machine. If all three lamps blink on and off at the same time, or in unison, the phase rotation of alternator 1 is correctly matched to the phase rotation of alternator 2. If the lamps blink on and off, but not in unison, the phase rotation between the two machines is not correctly matched, and two lines of alternator 2 should be switched.

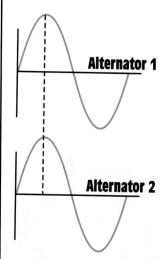

**phase rotation**

**Figure 30-13**  The voltages of both alternators must be in phase with each other.

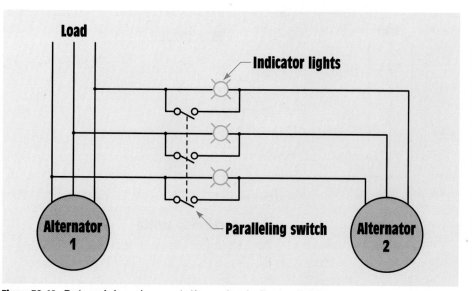

**Figure 30-14**  Determining phase rotation using indicator lights

2. The lamps also indicate when the phase of one machine is synchronized with the phase of the other machine. If the positive peak of alternator 1 does not occur at the same time as the positive peak of alternator 2, there will be a potential between the two machines. This will permit the lamps to glow. The brightness of the lamps indicates how far out of synchronism the two machines are. When the peak voltages of the two alternators occur at the same time, there is no potential difference between them. The lamps should be off at this time. The synchronizing switch should never be closed when the lamps are glowing.

## The Synchroscope

Another instrument often used for paralleling two alternators is the **synchroscope** *(Figure 30-15)*. The synchroscope measures the difference in voltage and frequency of the two alternators. The pointer of the synchroscope is free to rotate in a 360° arc. The alternator already connected to the load is considered to be the base machine. The synchroscope will indicate if the frequency of the alternator to be parallel to the base machine is fast or slow. When the voltages of the two alternators are in phase, the pointer will cover the shaded area on the face of the meter. When the two alternators are synchronized, the paralleling switch is closed.

If a synchroscope is not available, the two alternators can be paralleled using three lamps as described earlier. If the three-lamp method is used,

**Figure 30-15** Synchroscope

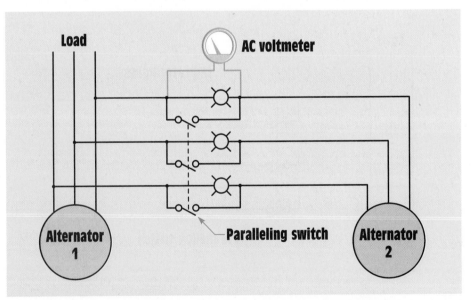

**Figure 30-16** AC voltmeter indicates when the two alternators are in phase.

an AC voltmeter connected across the same phase of each machine will indicate when the potential difference between the two machines is zero *(Figure 30-16)*. That is the point at which the paralleling switch should be closed.

## 30-8 Sharing the Load

After the alternators have been paralleled, the power input to alternator 2 must be increased to permit it to share part of the load. For example, if the alternator is being driven by a steam turbine, the power of the turbine would have to be increased. When this is done, the power to the load will remain constant. The power output of the base alternator will decrease, and the power output of the second alternator will increase.

## 30-9 Field Discharge Protection

When the DC excitation current is disconnected, the collapsing magnetic field can induce a high voltage in the rotor winding. This voltage can be high enough to arc contacts and damage the rotor winding or other circuit components. One method of preventing the induced voltage from becoming excessive is with the use of a **field discharge resistor**. A special double-pole, single-throw switch with a separate blade is used to connect the resistor to the field before the switch contacts open. When the switch is closed and direct current is connected to the field, the circuit connecting the resistor to the field is open *(Figure 30-17)*. When the switch is opened, the special blade connects the resistor to the field before the main contacts open *(Figure 30-18)*.

**field discharge resistor**

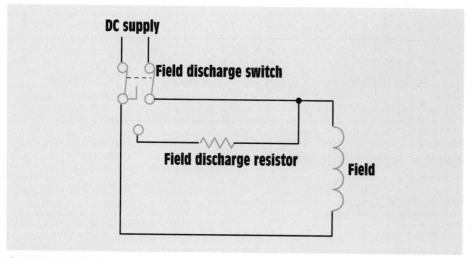

**Figure 30-17**  Switch in closed position

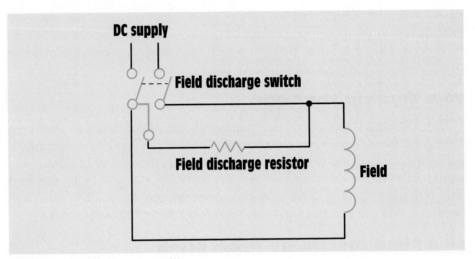

**Figure 30-18** Switch in open position

Another method of preventing the high voltage discharge is to connect a diode in parallel with the field *(Figure 30-19)*. The diode is connected in such a manner that when excitation current is flowing, the diode is reverse-biased and no current flows through the diode.

When the switch opens and the magnetic field collapses, the induced voltage is opposite in polarity to the applied voltage *(Figure 30-20)*. The diode in now forward-biased, permitting current to flow through the diode. The energy contained in the magnetic field is dissipated in the form of heat by the diode and field winding.

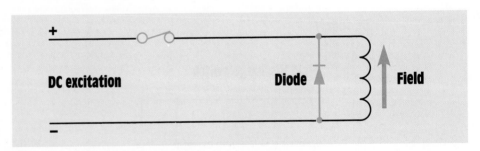

**Figure 30-19** Normal current flow

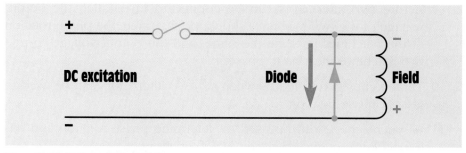

**Figure 30-20** Induced current flow

---

## Summary

1. There are two basic types of three-phase alternators: the rotating armature type and the rotating field type.

2. The rotating armature type is the least used because of its limited voltage and power rating.

3. The rotor of the rotating field type alternator contains electromagnets.

4. Direct current must be supplied to the field before the alternator can produce an output voltage.

5. The direct current supplied to the field is called excitation current.

6. The output frequency of an alternator is determined by the number of stator poles and the speed of rotation.

7. Three factors that determine the output voltage of an alternator are:
   A. The length of the conductor of the armature or stator winding.
   B. The strength of the magnetic field of the rotor.
   C. The speed of the rotor.

8. The output voltage is controlled by the amount of DC excitation current.

9. Before two alternators can be connected in parallel, the output voltage of the two machines should be the same, the phase rotation of the machines must be the same, and the output voltages of the two machines must be in phase.

10. Three lamps connected between the two alternators can be used to test for phase rotation.

11. A synchroscope can be used to determine phase rotation and difference of frequency between two alternators.

12. Two devices used to prevent a high voltage being induced in the rotor when the DC excitation current is stopped are a field discharge resistor and a diode.

13. Many large alternators use a brushless exciter to supply direct current to the rotor winding.

## Review Questions

1. What conditions must be met before two alternators can be paralleled together?

2. How can the phase rotation of one alternator be changed in relationship to the other alternator?

3. What is the function of the synchronizing lamps?

4. What is a synchroscope?

5. Assume that alternator A is supplying power to a load and that alternator B is to be paralleled to A. After the paralleling has been completed, what must be done to permit alternator B to share the load with alternator A?

6. What two factors determine the output frequency of an alternator?

7. At what speed must a six-pole alternator turn to produce 60 Hz?

8. What three factors determine the output voltage of an alternator?

9. What are sliprings used for on a rotating field type alternator?

10. Is the rotor excitation current AC or DC?

11. When a brushless exciter is used, what converts the alternating current produced in the armature winding into direct current before it is supplied to the field winding?

12. What two devices are used to eliminate the induced voltage produced in the rotor when the field excitation current is stopped?

**Most of the motors used throughout industry operate directly on three-phase power. This unit presents the three basic types of three-phase motors and their operating characteristics.**

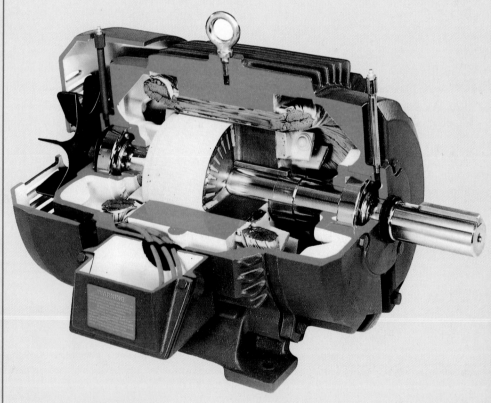

*Courtesy of Reliance Electric.*

# Three-Phase Motors

## Key Terms

Amortisseur winding

Code letter

Consequent pole

Differential selsyn

Direction of rotation

Dual-voltage motors

Percent slip

Phase rotation meter

Rotating magnetic field

Rotating transformers

Rotor frequency

Selsyn motors

Single-phasing

Squirrel cage

Synchronous

Synchronous condenser

Synchronous speed

Torque

Wound rotor

Objectives

Preview

**A**fter studying this unit, you should be able to:

- Discuss the basic operating principles of three-phase motors.
- List factors that produce a rotating magnetic field.
- List different types of three-phase motors.
- Discuss the operating principles of squirrel-cage motors.
- Connect dual voltage motors for proper operation on the desired voltage.
- Discuss the operation of consequent pole motors.
- Discuss the operation of wound rotor motors.
- Discuss the operation of synchronous motors.
- Determine the direction of rotation of a three-phase motor using a phase rotation meter.

Three-phase motors are used throughout the United States and Canada as the prime mover for industry. These motors convert the three-phase alternating current into mechanical energy to operate all types of machinery. Three-phase motors are smaller, lighter, and have higher efficiencies per horsepower than single-phase motors. They are extremely rugged and require very little maintenance. Many of these motors are operated 24 hours a day, seven days a week for many years without problem.

### 31-1 Three-Phase Motors

There are three basic types of three-phase motors:
1. the squirrel-cage induction motor,
2. the wound rotor induction motor, and
3. the synchronous motor.

All three motors operate on the same principle, and they all use the same basic design for the stator windings. The difference between them is the type of rotor used. Two of the three motors are induction motors and operate on the principle of electromagnetic induction in a manner similar to that of transformers. In fact, AC induction motors were patented as **rotating transformers** by Nikola Tesla. The stator winding of a motor is often referred to as the motor primary, and the rotor is referred to as the motor secondary.

## 31-2 The Rotating Magnetic Field

The principle of operation for all three-phase motors is the **rotating magnetic field**. Three factors cause the magnetic field to rotate:

1. the fact that the voltages in a three-phase system are 120° out of phase with each other,

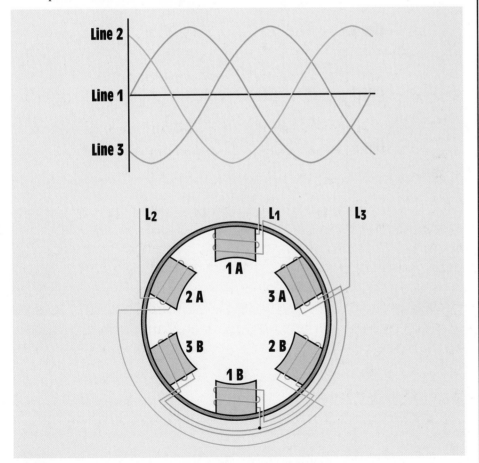

**Figure 31-1** Three-phase stator and three sine wave voltages

rotating trans-formers

rotating magnetic field

2. the fact that the three voltages change polarity at regular intervals, and

3. the arrangement of the stator windings around the inside of the motor.

*Figure 31-1* shows three AC sine waves 120° out of phase with each other and the stator winding of a three-phase motor. The stator illustrates a two-pole three-phase motor. Two-pole means that there are two poles per phase. AC motors do not generally have actual pole pieces as shown in *Figure 31-1*, but they will be used here to aid in understanding how the rotating magnetic field is created in a three-phase motor. Notice that pole pieces 1A and 1B are located opposite each other. The same is true for poles 2A and 2B, and for 3A and 3B. Pole pieces 1A and 1B are wound in such a manner that when current flows through the winding they will develop opposite magnetic polarities. This is true also for poles 2A and

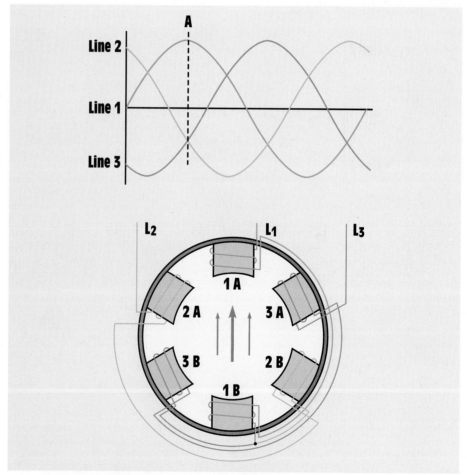

**Figure 31-2** The magnetic field is concentrated between poles 1A and 1B.

2B and for 3A and 3B. The windings of poles 2A and 3A are wound in the same direction in relation to each other, but in an opposite direction to that of pole 1A. The start end of the winding for poles 1A and 1B is connected to line 1, the start end of the winding for poles 2A and 2B is connected to line 2, and the start end of the winding for poles 3A and 3B is connected to line 3. The finish end of all three windings is joined to form a wye connection for the stator.

To understand how the magnetic field rotates around the inside of the stator refer to *Figure 31-2*. A dashed line labeled A has been drawn through the three sine waves of the three-phase system. This line is used to illustrate the condition of the three voltages at this point in time. The arrows drawn inside the motor indicate the greatest concentration of magnetic lines of flux at this point in time. Assume that the arrow is pointing in the direction of the north magnetic polarity. At this point in time, line 1

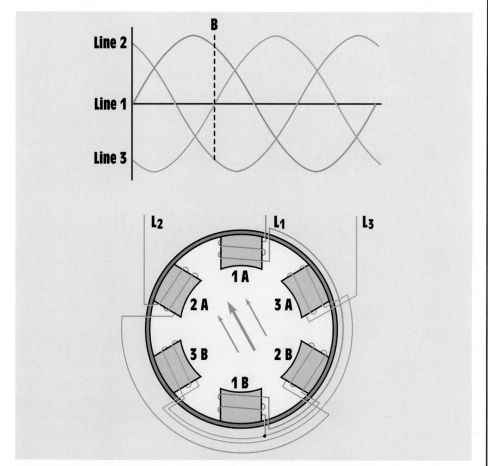

**Figure 31-3** The magnetic field is concentrated between the poles of phases 1 and 2.

has reached its maximum peak voltage in the positive direction, and lines 2 and 3 are less than maximum and in the negative direction. The magnetic field is concentrated between poles 1A and 1B.

In *Figure 31-3*, line B is drawn at a point in time when the voltage of line 3 is zero and the voltages of lines 1 and 2 are less than maximum but opposite in polarity. The magnetic field is now concentrated between the pole pieces of phases 1 and 2.

In *Figure 31-4*, line C is drawn during a period of time that line 2 has reached its maximum negative value and lines 1 and 3 are both less than maximum and have a positive polarity. At this point in time, the magnetic field is concentrated between poles 2A and 2B.

Line D in *Figure 31-5* indicates a point in time when line 1 is zero and lines 2 and 3 are less than maximum and opposite in polarity. The magnetic field is now concentrated between the poles of phases 2 and 3.

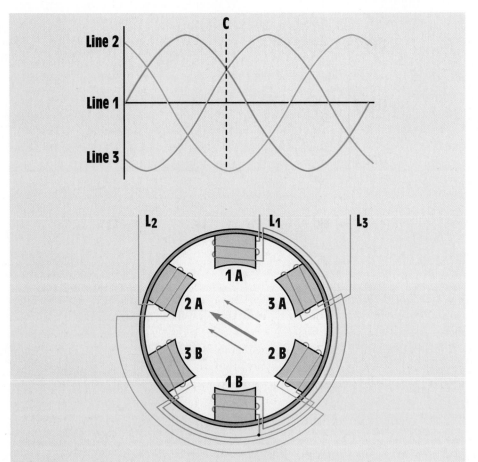

**Figure 31-4** The magnetic field is concentrated between poles 2A and 2B.

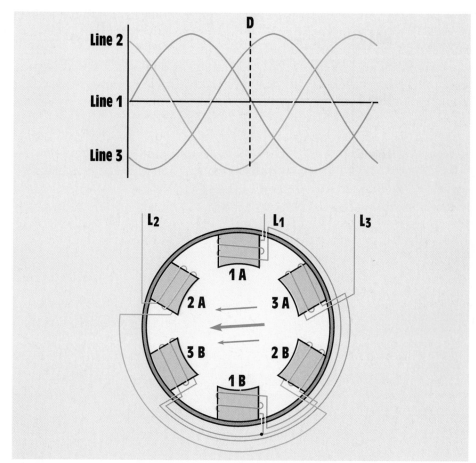

**Figure 31-5** The magnetic field is concentrated between phases 2 and 3.

In *Figure 31-6*, line E is drawn to indicate a point in time when line 3 has reached its peak positive point and lines 1 and 2 are less than maximum and negative. The magnetic field is now concentrated between poles 3A and 3B.

Line F in *Figure 31-7* indicates a point in time when line 2 is zero and lines 1 and 3 are less than maximum but have opposite polarities. The magnetic field is now concentrated between the poles of phases 1 and 3.

In *Figure 31-8*, line G indicates a point in time when line 1 has reached its maximum negative value and lines 2 and 3 are less than maximum and have a positive polarity. The magnetic field is again concentrated between poles 1A and 1B. At this point in time, however, pole 1B has a north magnetic polarity instead of pole 1A.

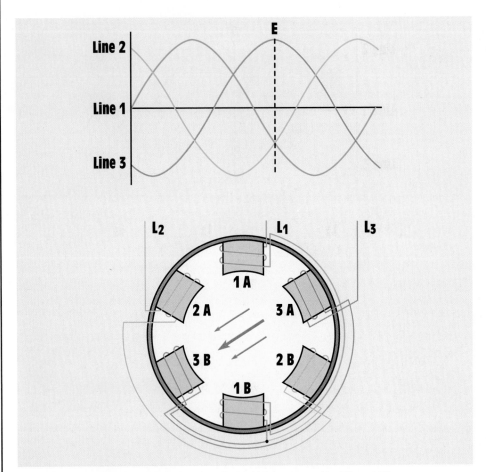

**Figure 31-6** The magnetic field is concentrated between poles 3A and 3B.

The magnetic field has rotated 180° during one half cycle. If the position of the magnetic field is traced for a period of one complete cycle, it will be seen that the magnetic field will rotate 360°.

### Synchronous Speed

synchro-
nous speed

The speed at which the magnetic field rotates is called the **synchronous speed**. Two factors that determine the synchronous speed of the rotating magnetic field are

    1. the number of stator poles (per phase) and
    2. the frequency of the applied voltage.

The chart below shows the synchronous speed at 60 Hz for different numbers of stator poles.

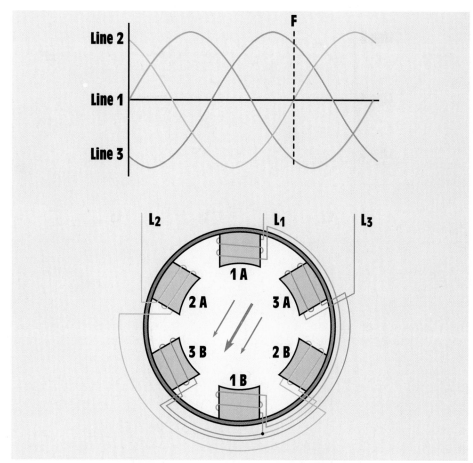

**Figure 31-7** The magnetic field is concentrated between phases 1 and 3.

| RPM | Stator Poles |
|------|------|
| 3600 | 2 |
| 1800 | 4 |
| 1200 | 6 |
| 900 | 8 |

The stator winding of a three-phase motor is shown in *Figure 31-9*. The synchronous speed can be calculated using the formula

$$S = \frac{120\ F}{P}$$

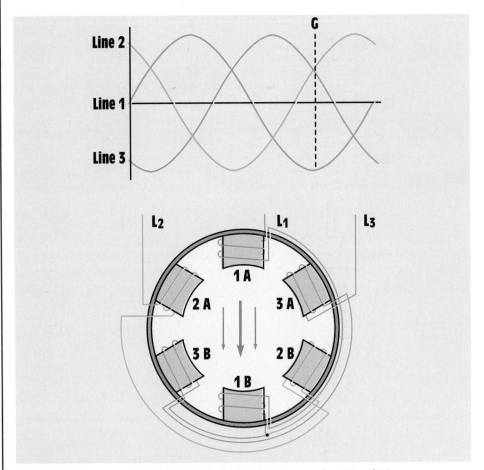

**Figure 31-8** The magnetic field is concentrated between poles 1B and 1A.

**Figure 31-9** Stator of a three-phase motor *(From Duff and Herman,* Alternating Current Fundamentals, *4th ed., copyright 1991 by Delmar Publishers Inc.)*

where

$$S = \text{speed in RPM}$$

$$F = \text{frequency in Hz}$$

$$P = \text{number of stator poles (per phase)}$$

What is the synchronous speed of a four-pole motor connected to 50 Hz?

Example 1

### Solution

$$S = \frac{120 \times 50}{4}$$

$$S = 1500 \text{ RPM}$$

What frequency should be applied to a six-pole motor to produce a synchronous speed of 400 RPM?

Example 2

### Solution

First change the base formula to find frequency. Once that is done, known values can be substituted in the formula.

$$F = \frac{PS}{120}$$

$$F = \frac{6 \times 400}{120}$$

$$F = 20 \text{ Hz}$$

## Determining the Direction of Rotation for Three-Phase Motors

On many types of machinery, the direction of rotation of the motor is critical. **The direction of rotation of any three-phase motor can be changed by reversing two of its stator leads**. This causes the direction of the rotating magnetic field to reverse. When a motor is connected to a machine that will not be damaged when its direction of rotation is reversed, power can be momentarily applied to the motor to observe its direction of rotation. If the rotation is incorrect, any two line leads can be interchanged to reverse the motor's rotation.

When a motor is to be connected to a machine that can be damaged by incorrect rotation, however, the direction of rotation must be determined before the motor is connected to its load. The **direction of rotation** can

**The direction of rotation of any three-phase motor can be changed by reversing two of its stator leads.**

**direction of rotation**

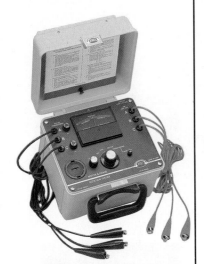

**Figure 31-10** Phase rotation meter *(Courtesy of Biddle Instruments.)*

phase
rotation
meter

be determined in two basic ways. One way is to make an electrical connection to the motor before it is mechanically connected to the load. The direction of rotation can then be tested by momentarily applying power to the motor before it is coupled to the load.

There may be occasions when it is not practical or is very inconvenient to apply power to the motor before it is connected to the load. In such a case a **phase rotation meter** can be used *(Figure 31-10)*. The phase rotation meter compares the phase rotation of two different three-phase connections. The meter contains six terminal leads. Three of the leads are connected to one side of the meter and are labeled MOTOR. These three motor leads are labeled A, B, or C. The LINE leads are located on the other side of the meter, and are labeled A, B, or C.

To determine the direction of rotation of the motor, first zero the meter by following the instructions provided by the manufacturer. Then set the meter selector switch to MOTOR, and connect the three MOTOR leads of the meter to the "T" leads of the motor as shown in *Figure 31-11*. The phase rotation meter contains a zero-center voltmeter. One side of the voltmeter is labeled INCORRECT, and the other side is labeled CORRECT. While observing the zero-center voltmeter, manually turn the motor shaft

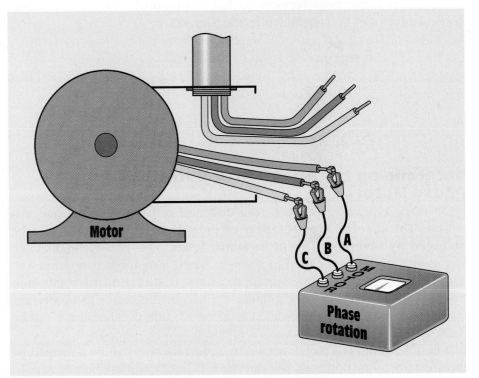

**Figure 31-11** Connecting the phase rotation meter to the motor

in the direction of desired rotation. The zero-center voltmeter will immediately swing in the CORRECT or INCORRECT direction. When the motor shaft stops turning, the needle may swing in the opposite direction. It is the *first* indication of the voltmeter that is to be used.

If the voltmeter needle indicated CORRECT, label the motor T leads A, B, or C to correspond with the MOTOR leads from the phase rotation meter. If the voltmeter needle indicated INCORRECT, change any two of the MOTOR leads from the phase rotation meter and again turn the motor shaft. The voltmeter needle should now indicate CORRECT. The motor T leads can now be labeled to correspond with the MOTOR leads from the phase rotation meter.

After the motor T leads have been labeled A, B, or C to correspond with the leads of the phase rotation meter, the rotation of the line supplying power to the motor must be determined. Set the selector switch on the phase rotation meter to the LINE position. After making certain the power has been turned off, connect the three LINE leads of the phase rotation meter to the incoming power line *(Figure 31-12)*. Turn on the power and observe the zero-center voltmeter. If the meter is pointing in the CORRECT direction, turn off the power and label the line leads A, B, or C to correspond with the LINE leads of the phase rotation meter.

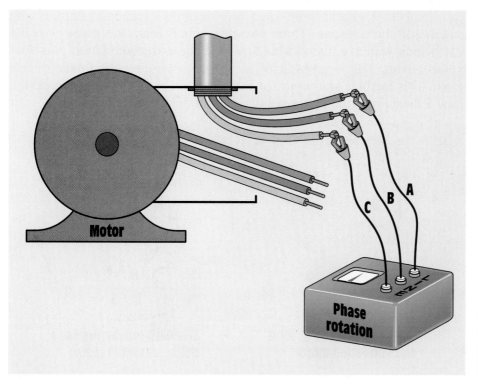

**Figure 31-12**   Connecting the phase rotation meter to the line

If the voltmeter is pointing in the INCORRECT direction, turn off the power and change any two of the leads from the phase rotation meter. When the power is turned on, the voltmeter should point in the CORRECT direction. Turn off the power and label the line leads A, B, or C to correspond with the leads from the phase rotation meter.

Now that the motor T leads and the incoming power leads have been labeled, connect the line lead labeled A to the T lead labeled A, the line lead labeled B to the T lead labeled B, and the line lead labeled C to the T lead labeled C. When power is connected to the motor, it will operate in the proper direction.

## 31-3 Connecting Dual-Voltage Three-Phase Motors

Many of the three-phase motors used in industry are designed to be operated on two voltages, such as 240 V and 480 V. Motors of this type, called **dual-voltage motors**, contain two sets of windings per phase. Most dual-voltage motors bring out nine T leads at the terminal box. There is a standard method used to number these leads as shown in *Figure 31-13*. Starting with terminal 1, the leads are numbered in a decreasing spiral as shown. Another method of determining the proper lead numbers is to add three to each terminal. For example, starting with lead 1, add three to one. Three plus one equals four. The phase winding that begins with 1 ends with 4. Now add three to four. Three plus four equals seven. The beginning of the second winding for phase one is seven. This method will work for the windings of all phases. If in doubt, draw a diagram of the phase windings and number them in a spiral.

**dual-voltage motors**

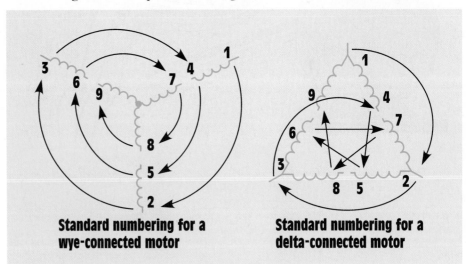

**Standard numbering for a wye-connected motor**

**Standard numbering for a delta-connected motor**

**Figure 31-13** Standard numbering for three-phase motors

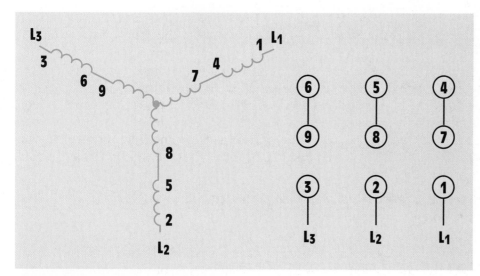

**Figure 31-14**   High-voltage wye connection

## High-Voltage Connections

Three-phase motors can be constructed to operate in either wye or delta. If a motor is to be connected to high voltage, the phase windings will be connected in series. In *Figure 31-14*, a schematic diagram and terminal connection chart for high voltage are shown for a wye-connected motor. In *Figure 31-15*, a schematic diagram and terminal connection chart for high voltage are shown for a delta-connected motor. Notice that in both cases the windings are connected in series.

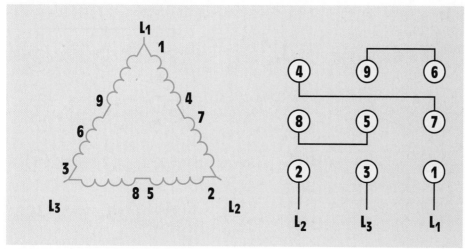

**Figure 31-15**   High-voltage delta connection

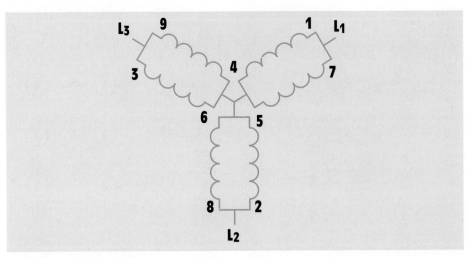

**Figure 31-16** Stator windings connected in parallel

## Low-Voltage Connections

When a motor is to be connected for low-voltage operation, the phase windings must be connected in parallel. *Figure 31-16* shows the basic schematic diagram for a wye-connected motor with parallel phase windings. In actual practice, however, it is not possible to make this exact connection with a nine-lead motor. The schematic shows that terminal 4 connects to the other end of the phase winding that starts with terminal 7. Terminal 5 connects to the other end of winding 8, and terminal 6 connects to the other end of winding 9. In actual motor construction, the

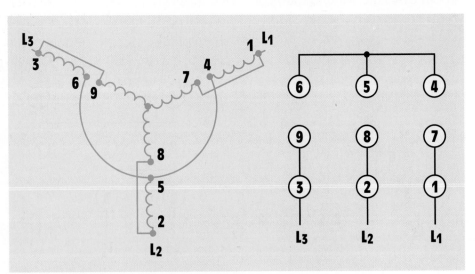

**Figure 31-17** Low-voltage wye connection

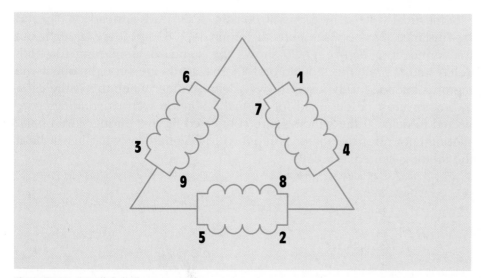

**Figure 31-18**  Parallel delta connection

opposite ends of windings 7, 8, and 9 are connected together inside the motor and are not brought outside the motor case. The problem is solved, however, by forming a second wye connection by connecting terminals 4, 5, and 6 together as shown in *Figure 31-17*.

The phase windings of a delta-connected motor must also be connected in parallel for use on low voltage. A schematic for this connection is shown in *Figure 31-18*. A connection diagram and terminal connection chart for this hook-up are shown in *Figure 31-19*.

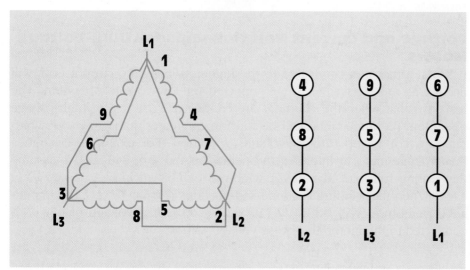

**Figure 31-19**  Low-voltage delta connection

Some dual-voltage motors will contain 12 T leads instead of nine. In this instance, the opposite ends of terminals 7, 8, and 9 are brought out for connection. *Figure 31-20* shows the standard numbering for both delta- and wye-connected motors. Twelve leads are brought out if the motor is intended to be used for wye-delta starting. When this is the case, the motor must be designed for normal operation with its windings connected in delta. If the windings are connected in wye during starting, the starting current of the motor is reduced to one-third of what it will be if the motor is started as a delta.

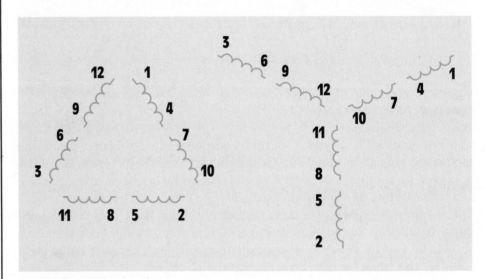

**Figure 31-20** A 12-lead motor

## Voltage and Current Relationships for Dual-Voltage Motors

When a motor is connected to the higher voltage, the current flow will be half as much as when it is connected for low-voltage operation. The reason is that when the windings are connected in series for high-voltage operation, the impedance will be four times greater than when the windings are connected for low-voltage operation. For example, assume a dual-voltage motor is intended to operate on 480 V or 240 V. Also assume that during full load, the motor windings exhibit an impedance of 10 $\Omega$ each. When the winding is connected in series *(Figure 31-21)*, the impedance per phase will be 20 $\Omega$ (10 + 10 = 20). If a voltage of 480 V is connected to the motor, the phase voltage will be

$$E_{PHASE} = \frac{E_{LINE}}{1.732}$$

$$E_{PHASE} = \frac{480}{1.732}$$

$$E_{PHASE} = 277 \text{ V}$$

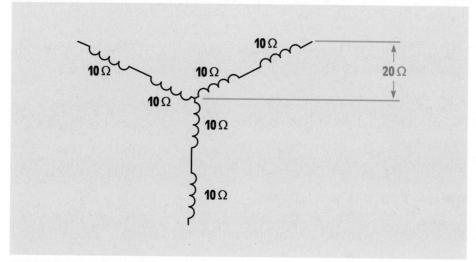

**Figure 31-21**  Impedance adds in series.

The amount of current flow through the phase can be computed using Ohm's law.

$$I = \frac{E}{Z}$$

$$I = \frac{277}{20}$$

$$I = 13.85 \text{ A}$$

If the stator windings are connected in parallel, the total impedance will be found by adding the reciprocals of the impedances of the windings *(Figure 31-22)*.

$$Z_T = \frac{1}{\dfrac{1}{Z_1} + \dfrac{1}{Z_2}}$$

$$Z_T = 5 \text{ }\Omega$$

If a voltage of 240 V is connected to the motor, the voltage applied across each phase will be 138.6 V (240/1.732 = 138.6). The amount of

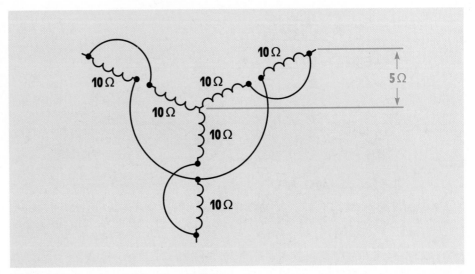

**Figure 31-22** Impedance is less in parallel.

phase current can now be computed using Ohm's law.

$$I = \frac{E}{Z}$$

$$I = \frac{138.6}{5}$$

$$I = 27.12 \text{ A}$$

## 31-4 Squirrel-Cage Induction Motors

**squirrel cage**

The **squirrel-cage** induction motor receives its name from the type of rotor used in the motor. A squirrel-cage rotor is made by connecting bars to two end rings. If the metal laminations were removed from the rotor, the result would look very similar to a squirrel cage *(Figure 31-23)*. A squirrel cage is a cylindrical device constructed of heavy wire. A shaft placed through the center of the cage permits the cage to spin around the shaft. A squirrel cage is placed inside the cage of small pets such as squirrels and hamsters to permit them to exercise by running inside of the squirrel cage. A squirrel-cage rotor is shown in *Figure 31-24*.

### Principles of Operation

The squirrel-cage motor is an induction motor. That means that the current flow in the rotor is produced by induced voltage from the rotating magnetic field of the stator. In *Figure 31-25*, a squirrel-cage rotor is shown

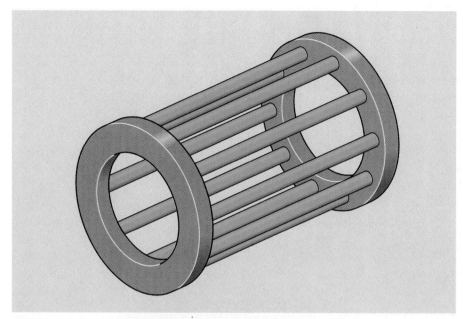

**Figure 31-23**   Basic squirrel-cage rotor without laminations

**Figure 31-24**   Squirrel-cage rotor *(Courtesy of Electric Machinery Corp.)*

inside the stator of a three-phase motor. It will be assumed that the motor shown in *Figure 31-25* contains four poles per phase, which produces a rotating magnetic field with a synchronous speed of 1800 RPM when the stator is connected to a 60-Hz line. When power is first connected to the stator, the rotor is not turning. The magnetic field of the stator cuts the rotor bars at a rate of 1800 RPM. This cutting action induces a voltage into the rotor bars. This induced voltage will be the same frequency as the voltage applied to the stator. The amount of induced voltage is determined by three factors:

1. the strength of the magnetic field of the stator,
2. the number of turns of wire cut by the magnetic field (In the case of a

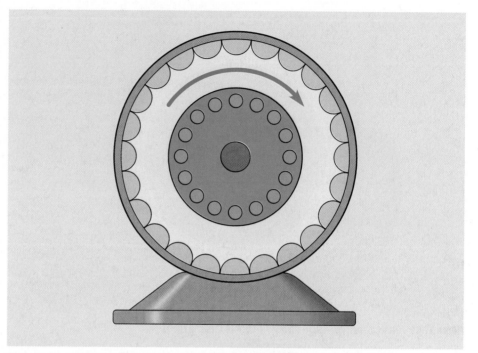

**Figure 31-25** Voltage is induced in the rotor by the rotating magnetic field.

    squirrel-cage rotor this will be the number of bars in the rotor.) and
3. the speed of the cutting action.

Since the rotor is stationary at this time, maximum voltage is induced into the rotor. The induced voltage causes current to flow through the rotor bars. As current flows through the rotor, a magnetic field is produced around each bar *(Figure 31-26)*.

The magnetic field of the rotor is attracted to the magnetic field of the stator, and the rotor begins to turn in the same direction as the rotating magnetic field.

As the speed of the rotor increases, the rotating magnetic field cuts the rotor bars at a slower rate. For example, assume the rotor has accelerated to a speed of 600 RPM. The synchronous speed of the rotating magnetic field is 1800 RPM. Therefore, the rotor bars are being cut at a rate of 1200 RPM (1800 RPM – 600 RPM = 1200 RPM). Since the rotor bars are being cut at a slower rate, less voltage is induced in the rotor, reducing rotor current. When the rotor current decreases, the stator current decreases also.

As the rotor continues to accelerate, the rotating magnetic field cuts the rotor bars at a decreasing rate. This reduces the amount of induced voltage, and, therefore, the amount of rotor current. If the motor is operating

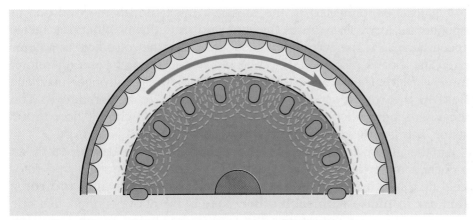

**Figure 31-26**   A magnetic field is produced around each rotor bar.

without a load, the rotor will continue to accelerate until it reaches a speed close to that of the rotating magnetic field.

## Torque

The amount of **torque** produced by an AC induction motor is determined by three factors:

1. the strength of the magnetic field of the stator,
2. the strength of the magnetic field of the rotor,
3. the phase angle difference between rotor and stator fields.

$$T = K_T \times \varphi_S \times I_R \times COS\ \theta_R$$

where

$T$ = torque in lb-ft

$K_T$ = torque constant

$\varphi_S$ = stator flux (constant at all speeds)

$I_R$ = rotor current

$COS\ \theta_R$ = rotor power factor

Notice that one of the factors that determines the amount of torque produced by an induction motor is the strength of the magnetic field of the rotor. **An induction motor can never reach synchronous speed.** If the rotor were to turn at the same speed as the rotating magnetic field, there would be no induced voltage in the rotor, and consequently no rotor current. Without rotor current there could be no magnetic field developed by the rotor and, therefore, no torque or turning force. A motor operating at no load will accelerate until the torque developed is proportional to the windage and bearing friction losses.

torque

An induction motor can never reach synchronous speed.

If a load is connected to the motor, it must furnish more torque to operate the load. This causes the motor to slow down. When the motor speed decreases, the rotating magnetic field cuts the rotor bars at a faster rate. This causes more voltage to be induced in the rotor and, therefore, more current. The increased current flow produces a stronger magnetic field in the rotor, which causes more torque to be produced. The increased current flow in the rotor causes increased current flow in the stator. This is why motor current will increase as load is added.

Another factor that determines the amount of torque developed by an induction motor is the phase angle difference between stator and rotor field flux. **Maximum torque is developed when the stator and rotor flux are in phase with each other.** Note in the above formula that one of the factors that determines the torque developed by an induction motor is the cosine of the rotor power factor. The cosine function reaches its maximum value of 1 when the phase angle is 0 (COS 0° = 1).

### Starting Characteristics

When a squirrel-cage motor is first started, it will have a current draw several times greater than its normal running current. The actual amount of starting current is determined by the type of rotor bars, the horsepower rating of the motor, and the applied voltage. The type of rotor bars is indicated by the code letter found on the nameplate of a squirrel-cage motor. Table 430-7b of the *National Electrical Code* can be used to compute the locked rotor current (starting current) of a squirrel-cage motor when the applied voltage, horsepower, and code letter are known.

An 800-hp, three-phase squirrel-cage motor is connected to 2300 V. The motor has a code letter of J. What is the starting current of this motor?

### Solution

Table 430-7b of the *National Electrical Code* gives a value of 7.1 to 7.99 kilo-volt-amperes per horsepower as the locked rotor current of a motor with a code letter J *(Figure 31-27)*. An average value of 7.5 will be used for this calculation. The apparent power can be computed by multiplying the 7.5 times the horsepower rating of the motor.

$$kVA = 7.5 \times 800$$

$$kVA = 6000$$

The line current supplying the motor can now be computed using the formula

$$I_{(LINE)} = \frac{VA}{E_{(LINE)} \times 1.732}$$

Maximum torque is developed when the stator and rotor flux are in phase with each other.

**Example 3**

$$I_{(LINE)} = \frac{6,000,000}{2300 \times 1.732}$$

$$I_{(LINE)} = 1506.175 \text{ A}$$

**Table 430-7(b).   Locked-Rotor Indicating Code Letters**

| Code Letter | | Kilovolt-Amperes per Horsepower with Locked Rotor |
|---|---|---|
| A | | 0 — 3.14 |
| B | | 3.15 — 3.54 |
| C | | 3.55 — 3.99 |
| D | | 4.0 — 4.49 |
| E | | 4.5 — 4.99 |
| F | | 5.0 — 5.59 |
| G | | 5.6 — 6.29 |
| H | | 6.3 — 7.09 |
| J | | 7.1 — 7.99 |
| K | | 8.0 — 8.99 |
| L | | 9.0 — 9.99 |
| M | | 10.0 — 11.19 |
| N | | 11.2 — 12.49 |
| P | | 12.5 — 13.99 |
| R | | 14.0 — 15.99 |
| S | | 16.0 — 17.99 |
| T | | 18.0 — 19.99 |
| U | | 20.0 — 22.39 |
| V | | 22.4 — and up |

**Figure 31-27**   Table 430-7B *(Reprinted with permission from NFPA 70-1990, the National Electrical Code®, Copyright© 1989, National Fire Protection Association, Quincy, MA 02269. This reprinted material is not the complete and official position of the National Fire Protection Association, on the referenced subject which is represented only by the standard in its entirety.)*

This large starting current is caused by the fact that the rotor is not turning when power is first applied to the stator. Since the rotor is not turning, the squirrel-cage bars are cut by the rotating magnetic field at a fast rate. Remember that one of the factors that determines the amount of induced voltage is speed of the cutting action. This high induced voltage causes a large amount of current to flow in the rotor. The large current flow in the rotor causes a large amount of current flow in the stator. Since a large amount of current flows in both the stator and rotor, a strong magnetic field is established in both.

It would first appear that the starting torque of a squirrel-cage motor is high because the magnetic fields, of both the stator and rotor are strong at this point. Recall that the third factor for determining the torque developed by an induction motor is the difference in phase angle between stator flux and rotor flux. Since the rotor is being cut at a high rate of speed by the rotating stator field, the bars in the squirrel-cage rotor appear to be very

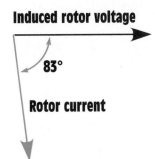

**Induced rotor voltage**

83°

**Rotor current**

**Figure 31-28** Rotor current is almost 90° out of phase with the induced voltage at the moment of starting.

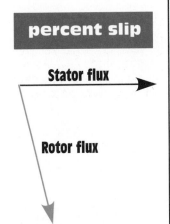

**percent slip**

**Stator flux**

**Rotor flux**

**Figure 31-29** Rotor flux lags the stator flux by a large amount during starting.

**rotor frequency**

inductive at this point because of the high frequency of the induced voltage. This causes the phase angle difference between the induced voltage in the rotor and rotor current to be almost 90° out of phase with each other, producing a lagging power factor for the rotor *(Figure 31-28)*. This causes the rotor flux to lag the stator flux by a large amount, and consequently a relatively weak starting torque, compared with other types of three-phase motors, is developed *(Figure 31-29)*.

## Percent Slip

The speed performance of an induction motor is measured in **percent slip.** The percent slip can be determined by subtracting the synchronous speed from the speed of the rotor. For example, assume an induction motor has a synchronous speed of 1800 RPM and at full load the rotor turns at a speed of 1725 RPM. The difference between the two speeds is 75 RPM (1800 − 1725 = 75). The percent slip can be determined using the formula

$$\text{percent slip} = \frac{\text{synchronous speed} - \text{rotor speed}}{\text{synchronous speed}} \times 100$$

$$\text{percent slip} = \frac{75}{1800} \times 100$$

$$\text{percent slip} = 4.16\%$$

A rotor slip of 2% to 5% is common for most squirrel-cage induction motors. The amount of slip for a particular motor is greatly affected by the type of rotor bars used in the construction of the rotor. Squirrel-cage motors are considered to be constant speed motors because there is a small difference between no-load speed and full-load speed.

## Rotor Frequency

In the previous example, the rotor slips behind the rotating magnetic field by 75 RPM. This means that at full load, the bars of the rotor are being cut by magnetic lines of flux at a rate of 75 RPM. Therefore, the voltage being induced in the rotor at this point in time is at a much lower frequency than when the motor was started. The **rotor frequency** can be determined using the formula

$$F = \frac{P \times S_R}{120}$$

where

$$F = \text{frequency in Hz}$$

$$P = \text{number of stator poles}$$

$$S_R = \text{rotor slip in RPM}$$

$$F = \frac{4 \times 75}{120}$$

$$F = 2.5\ \text{Hz}$$

Because the frequency of the current in the rotor decreases as the rotor approaches synchronous speed, the rotor bars become less inductive. The current flow through the rotor becomes limited more by the resistance of the bars and less by inductive reactance. The current flow in the rotor becomes more in phase with the induced voltage, which causes less phase angle shift between stator and rotor flux *(Figure 31-30)*. This is the reason that squirrel-cage motors generally have a relatively poor starting torque compared with other types of three-phase motors but a good running torque.

## Reduced Voltage Starting

Because many squirrel-cage motors require a large amount of starting current, it is sometimes necessary to reduce the voltage during the starting period. When the voltage is reduced, the starting torque is reduced also. If the applied voltage is reduced to 50% of its normal value, the magnetic fields of both the stator and rotor are reduced to 50% of normal. The 50% reduction of the magnetic fields causes the starting torque to be reduced to 25% of normal. A chart showing a typical torque curve for a squirrel-cage motor is shown in *Figure 31-31.*

The torque formula given earlier can be used to show why this large reduction of torque occurs. Both the stator flux, $\varphi_S$, and the rotor current, $I_R$, are reduced to half their normal value. The product of these two values, torque, is reduced to one-fourth. The torque varies as the square of the applied voltage for any given value of slip.

## Code Letters

Squirrel-cage rotors are not all the same. Rotors are made with different types of bars. The type of rotor bars used in the construction of the rotor determines the operating characteristics of the motor. AC squirrel-cage motors are given a **code letter** on their nameplate. The code letter indicates the type of bars used in the rotor. *Figure 31-32* shows a rotor with type A bars. A type A rotor has the highest resistance of any squirrel-cage rotor. This means that the starting torque will be high, since the rotor current is closer to being in phase with the induced voltage than on any other type of rotor. Also, the high resistance of the rotor bars limits the

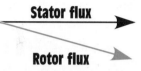

**Stator flux**

**Rotor flux**

**Figure 31-30**  Rotor and stator flux become more in phase with each other as motor speed increases.

**code letter**

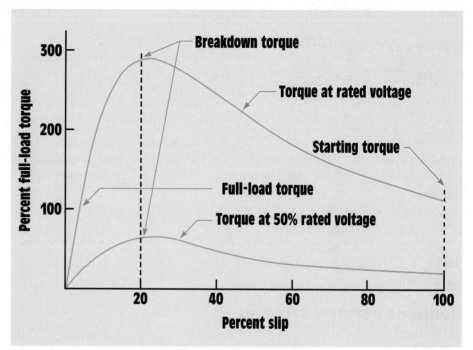

**Figure 31-31** Typical torque curves for a squirrel-cage motor

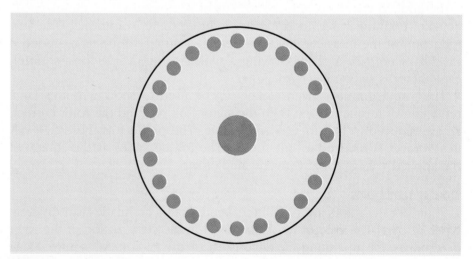

**Figure 31-32** Type A rotor

amount of current flow in the rotor when starting. This produces a low starting current for the motor. A rotor with type A bars has very poor running characteristics, however. Since the bars are resistive, a large amount of voltage will have to be induced into the rotor to produce an increase

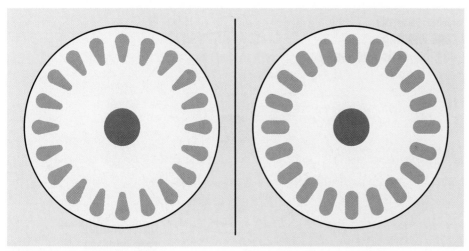

**Figure 31-33** Type B – E rotor    **Figure 31-34** Type F – V rotor

in rotor current and, therefore, an increase in the rotor magnetic field. This means that when load is added to the motor, the rotor must slow down a great amount to produce enough current in the rotor to increase the torque. Motors with type A rotors have the highest percent slip of any squirrel-cage motor. Motors with type A rotors are generally used in applications where starting is a problem, such as a motor that must accelerate a large flywheel from 0 RPM to its full speed. Flywheels can have a very large amount of inertia, which may require several minutes to accelerate them to their running speed when they are started.

*Figure 31-33* shows a rotor with bars similar to those found in rotors with code letters B through E. These rotor bars have lower resistance than the type A rotor. Rotors of this type have fair starting torque, low starting current, and fair speed regulation.

*Figure 31-34* shows a rotor with bars similar to those found in rotors with code letters F through V. This rotor has low starting torque, high starting current, and good running torque. Motors containing rotors of this type generally have very good speed regulation and low percent slip.

## The Double Squirrel-Cage Rotor

Some motors use a rotor that contains two sets of squirrel-cage windings *(Figure 31-35)*. The outer winding consists of bars with a relatively high resistance located close to the top of the iron core. Since these bars are located close to the surface they have a relatively low reactance. The inner winding consists of bars with a large cross-sectional area, which gives them a low resistance. The inner winding is placed deeper in the core material, which causes it to have a much higher reactance.

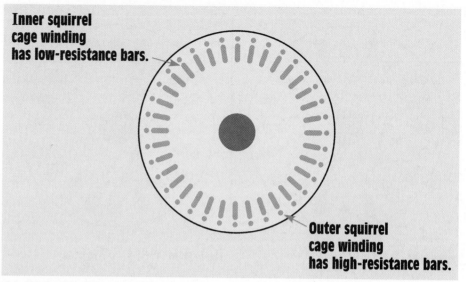

Inner squirrel cage winding has low-resistance bars.

Outer squirrel cage winding has high-resistance bars.

**Figure 31-35**  Double squirrel-cage rotor

When the double squirrel-cage motor is started, the rotor frequency is high. Since the inner winding is inductive, its impedance will be high compared with the resistance of the outer winding. During this period of time, most of the rotor current flows through the outer winding. The resistance of the outer winding limits the current flow through the rotor, which limits the starting current to a relatively low value. Since the current is close to being in phase with the induced voltage, the rotor flux and stator flux are close to being in phase with each other, and a strong starting torque is developed. The starting torque of a double squirrel-cage motor can be as high as 250% of rated full-load torque.

When the rotor reaches its full load speed, rotor frequency decreases to 2 or 3 Hz. The inductive reactance of the inner winding has now decreased to a low value. Most of the rotor current now flows through the low-resistance inner winding. This type of motor has good running torque and excellent speed regulation.

## Power Factor of a Squirrel-Cage Induction Motor

At no load, most of the current is used to magnetize the stator and rotor. Since most of the current is magnetizing current, it is inductive and lags the applied voltage by close to 90°. A very small resistive component is present, caused mostly by the resistance of the wire in the stator and the power needed to overcome bearing friction and windage loss. At no load, the motor appears to be a resistive-inductive series circuit with a large inductive component as compared with resistance *(Figure 31-36)*. A power factor of about 10% is common for a squirrel-cage motor at no load.

As load is added, electrical energy is converted into mechanical energy, and the in-phase component of current increases. The circuit now appears to contain more resistance than inductance *(Figure 31-37)*. This causes the phase angle between applied voltage and motor current to decrease, causing the power factor to increase. In practice, the power factor of an induction motor at full load will be from about 85% to 90% lagging.

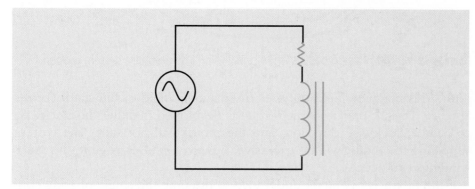

**Figure 31-36**   At no load the motor appears to have a large amount of inductance and a very small resistance.

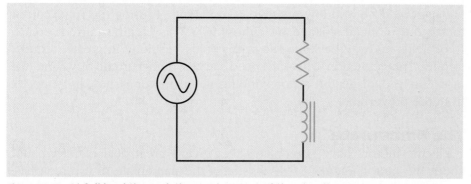

**Figure 31-37**   At full load the resistive component of the circuit appears to be greater than the inductive component.

## Single-Phasing

Three lines supply power to a three-phase motor. If one of these lines should open, the motor will be connected to single-phase power *(Figure 31-38)*. This condition is known as **single-phasing.**

If the motor is not running and single-phase power is applied to the motor, the induced voltage in the rotor sets up a magnetic field in the

**single-phasing**

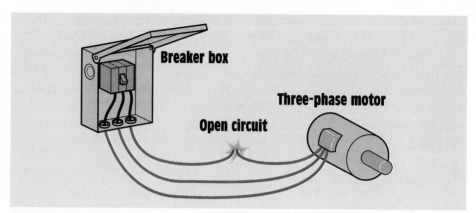

**Figure 31-38** Single-phasing occurs when one line of a three-phase system is open.

rotor. This magnetic field opposes the magnetic field of the stator (Lenz's law). As a result practically no torque is developed in either the clockwise or counterclockwise direction, and the motor will not start. The current supplying the motor will be excessive, however, and damage to the stator windings can occur.

If the motor is operating under load at the time the single-phasing condition occurs, the rotor will continue to turn at a reduced speed. The moving bars of the rotor cut the stator field flux, which continues to induce voltage and current in the bars. Due to reduced speed, the rotor has high reactive and low resistive components, causing the rotor current to lag the induced voltage by almost 90°. This lagging current creates rotor fields midway between the stator poles, resulting in greatly reduced torque. The reduction in rotor speed causes high current flow and will most likely damage the stator winding if the motor is not disconnected from the power line.

## The Nameplate

Electric motors have nameplates that give a great deal of information about the motor. *Figure 31-39* illustrates the nameplate of a three-phase squirrel-cage induction motor. The nameplate shows that the motor is 10 hp, it is a three-phase motor, and operates on 240 or 480 V. The full-load running current of the motor is 28 A when operated on 240 V and 14 A when operated on 480 V. The motor is designed to be operated on a 60-Hz AC voltage, and has a full-load speed of 1745 RPM. This speed indicates that the motor has four poles per phase. Since the full-load speed is 1745 RPM, the synchronous speed would be 1800 RPM. The motor contains a type J squirrel-cage rotor, and has a service factor of 1.25. The service factor is used to determine the amperage rating of the overload

| Manufacturer | |
|---|---|
| HP<br>10 | Phase<br>3 |
| Volts<br>240/480 | Amps<br>28/14 |
| Hz<br>60 | Fl speed<br>1745 RPM |
| Code<br>J | SF<br>1.25 |
| Frame<br>XXXX | Model No.<br>XXXX |

**Figure 31-39**   Motor nameplate

protection for the motor. Some motors indicate a marked temperature rise in Celsius degrees instead of a service factor. The frame number indicates the type of mounting the motor has. *Figure 31-40* shows the schematic symbol used to represent a three-phase squirrel-cage motor.

## Consequent Pole Squirrel-Cage Motors

**Consequent pole** squirrel-cage motors permit the synchronous speed to be changed by changing the number of stator poles. If the number of poles is doubled, the synchronous speed will be reduced by one-half. A two-pole motor has a synchronous speed of 3600 RPM when operated at 60 Hz. If the number of poles is doubled to four, the synchronous speed becomes 1800 RPM. The number of stator poles can be changed by changing the direction of current flow through alternate pairs of poles.

*Figure 31-41* illustrates this concept. In *Figure 31-41A,* two coils are connected in such a manner that current flows through them in the same direction. Both poles will produce the same magnetic polarity and are essentially one pole. In *Figure 31-41B,* the coils have been reconnected in such a manner that current flows through them in opposite directions. The coils now produce the opposite magnetic polarities, and are essentially two different poles.

Consequent pole motors with one stator winding bring out six leads labeled $T_1$ through $T_6$. Depending on the application, the windings will be connected as a series delta or a parallel wye. If it is intended that the motor maintain the same horsepower rating for both high and low speed,

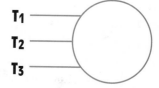

**Figure 31-40**   Schematic symbol of a three-phase squirrel-cage induction motor

**consequent pole**

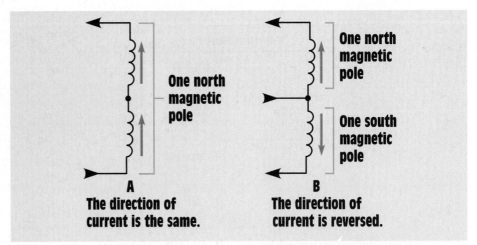

**Figure 31-41** The number of poles can be changed by reversing the current flow through alternate poles.

the high-speed connection will be a series delta *(Figure 31-42)*. The low-speed connection will be a parallel wye *(Figure 31-43)*.

If it is intended that the motor maintain constant torque for both low and high speeds, the series delta connection will provide low speed, and the parallel wye will provide high speed.

Since the speed range of a consequent pole motor is limited to a 1:2

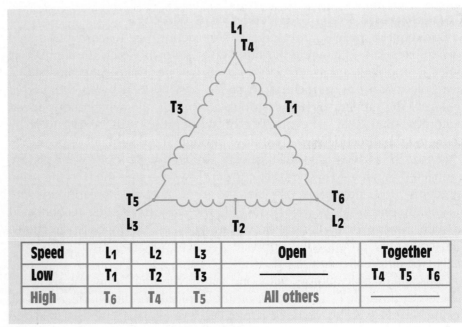

| Speed | L₁ | L₂ | L₃ | Open | Together |
|-------|-----|-----|-----|------------|---------------|
| Low | T₁ | T₂ | T₃ | ————— | T₄ T₅ T₆ |
| High | T₆ | T₄ | T₅ | All others | ————— |

**Figure 31-42** High-speed series delta connection

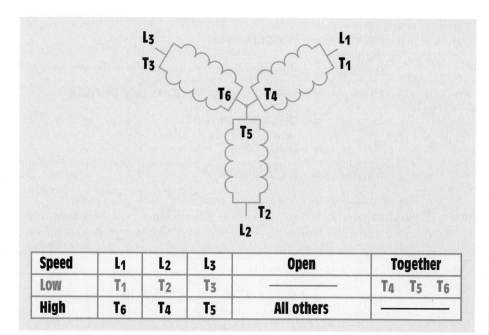

| Speed | L₁ | L₂ | L₃ | Open | Together |
|---|---|---|---|---|---|
| Low | T₁ | T₂ | T₃ | —————— | T₄ T₅ T₆ |
| High | T₆ | T₄ | T₅ | All others | —————— |

**Figure 31-43**  Low-speed parallel wye connection

ratio, motors intended to operate at more than two speeds contain more than one stator winding. A consequent pole motor with three speeds, for example, will have one stator winding for one speed only and a second winding with taps. The tapped winding may provide synchronous speeds of 1800 and 900 RPM, and the separate winding may provide a speed of 1200 RPM. Consequent pole motors with four speeds contain two separate stator windings with taps. If the second stator winding of the motor in this example were to be tapped, the motor would provide synchronous speeds of 1800, 1200, 900, and 600 RPM.

## Motor Calculations

In the following example, output horsepower and motor efficiency will be computed. It is assumed that a 1/2-hp squirrel-cage motor is connected to a load. A wattmeter is connected to the motor, and the load torque measurement is calibrated in pound-inches. The motor is operating at a speed of 1725 RPM and producing a torque of 16 lb/in. The wattmeter is indicating an input power of 500 W.

The actual amount of horsepower being produced by the motor can be calculated by using the formula

$$hp = \frac{6.28 \times RPM \times L \times P}{33{,}000}$$

where

| | | |
|---|---|---|
| hp | = | horsepower |
| 6.28 | = | a constant |
| RPM | = | speed in revolutions per minute |
| L | = | distance in feet |
| P | = | pounds |
| 33,000 | = | a constant |

Since the formula uses feet for the distance, and the torque of the motor is rated in pound-inches, L will be changed to 1/12 of a foot, or 1 in. To simplify the calculation, the fraction 1/12 will be changed into its decimal equivalent (0.08333). To compute the output horsepower, substitute the known values in the formula.

$$hp = \frac{6.28 \times 1725 \times 0.08333 \times 16}{33,000}$$

$$hp = \frac{14,444}{33,000}$$

$$hp = 0.438$$

One horsepower is equal to 746 W. The output power of the motor can be computed by multiplying the output horsepower by 746.

$$power\ out = 746 \times 0.438$$

$$power\ out = 326.5\ W$$

The efficiency of the motor can be computed by using the formula

$$Eff. = \frac{power\ out}{power\ in} \times 100$$

$$Eff. = \frac{326.5}{500} \times 100$$

$$Eff. = 0.653 \times 100$$

$$Eff. = 65.3\%$$

## 31-5 Wound Rotor Induction Motors

**wound rotor**

The **wound rotor** induction motor is very popular in industry because of its high starting torque and low starting current. The stator winding of the wound rotor motor is the same as the squirrel-cage motor. The difference between the two motors lies in the construction of the rotor. Recall that the squirrel-cage rotor is constructed of bars connected together at each end by a shorting ring as shown in *Figure 31-23*.

The rotor of a wound rotor motor is constructed by winding three separate coils on the rotor 120° apart. The rotor will contain as many poles per phase as the stator winding. These coils are then connected to three sliprings located on the rotor shaft as shown in *Figure 31-44*. Brushes, connected to the sliprings, provide external connection to the rotor. This permits the rotor circuit to be connected to a set of resistors as shown in *Figure 31-45*.

The stator terminal connections are generally labeled $T_1$, $T_2$, and $T_3$. The rotor connections are commonly labeled $M_1$, $M_2$, and $M_3$. The $M_2$ lead is generally connected to the middle slip ring, and the $M_3$ lead is connected close to the rotor windings. The direction of rotation for the wound rotor motor is reversed by changing any two stator leads. Changing the M leads will have no effect on the direction of rotation. The schematic symbol for a wound rotor motor is shown in *Figure 31-46*.

**Figure 31-44**  Rotor of a wound rotor induction motor *(From Duff and Herman, Alternating Current Fundamentals, 4th ed., copyright 1991 by Delmar Publishers Inc.)*

## Principles of Operation

When power is applied to the stator winding, a rotating magnetic field is created in the motor. This magnetic field cuts through the windings of the rotor and induces a voltage into them. The amount of current flow in the rotor is determined by the amount of induced voltage and the total

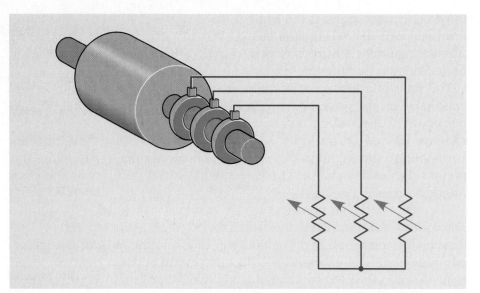

**Figure 31-45** The rotor of a wound rotor motor is connected to external resistors.

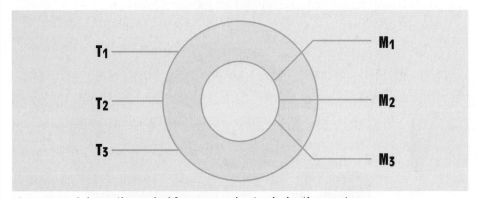

**Figure 31-46** Schematic symbol for a wound rotor induction motor

impedance of the rotor circuit (I = E/Z). The rotor impedance is a combination of inductive reactance created in the rotor windings and the external resistance. The impedance could be calculated using the formula for resistance and inductive reactance connected in series.

$$Z = \sqrt{R^2 + X_L^2}$$

As the rotor speed increases, the frequency of the induced voltage will decrease just as it does in the squirrel-cage motor. The reduction in frequency causes the rotor circuit to become more resistive and less inductive, decreasing the phase angle between induced voltage and rotor current.

When current flows through the rotor, a magnetic field is produced. This magnetic field is attracted to the rotating magnetic field of the stator. As the rotor speed increases, the induced voltage decreases because of less cutting action between the rotor windings and rotating magnetic field. The decrease in induced voltage produces less current flow in the rotor and, therefore, less torque. If the rotor circuit resistance is reduced, more current can flow, which will increase motor torque, and the rotor will increase in speed. This action continues until all external resistance has been removed from the rotor circuit by shorting the M leads together and the motor is operating at maximum speed. At this point, the wound rotor motor is operating in the same manner as a squirrel-cage motor.

## Starting Characteristics of a Wound Rotor Motor

Although the overall starting torque of a wound rotor motor is less than that of an equivalent horsepower squirrel-cage motor, due to reduced current in both the rotor and stator during the starting period, the starting torque will be higher per amp than in a squirrel-cage motor. The starting current is less because resistance is connected in the rotor circuit during starting. This resistance limits the amount of current that can flow in the rotor circuit. Since the stator current is proportional to rotor current because of transformer action, the stator current is less also. The starting torque is higher per amp than that of a squirrel-cage motor because of the resistance in the rotor circuit. Recall that one of the factors that determines motor torque is the phase angle difference between stator flux and rotor flux. Since resistance is connected in the rotor circuit, stator and rotor flux are close to being in phase with each other producing a high starting torque for the wound rotor induction motor. The wound rotor motor will generally exhibit a higher starting torque than an equivalent size squirrel-cage motor connected to a reduced voltage starter.

If an attempt is made to start the motor with no circuit connected to the rotor, the motor cannot start. If no resistance is connected to the rotor circuit, there can be no current flow and consequently no magnetic field can be developed in the rotor.

## Speed Control

The speed of a wound rotor motor can be controlled by permitting resistance to remain in the rotor circuit during operation. When this is done, the rotor and stator current is limited, which reduces the strength of both magnetic fields. The reduced magnetic field strength permits the rotor to slip behind the rotating magnetic field of the stator. The resistors of speed controllers must have higher power ratings than the resistors of starters because they operate for extended periods of time.

The operating characteristics of a wound rotor motor with the sliprings shorted are almost identical to those of a squirrel-cage motor. The percent slip, power factor, and efficiency are very similar for motors of equal horsepower rating.

### 31-6 Synchronous Motors

The three-phase **synchronous** motor has several characteristics that separate it from the other types of three-phase motors. Some of these characteristics are:

1. The synchronous motor is not an induction motor. It does not depend on induced current in the rotor to produce a torque.
2. It will operate at a constant speed from full load to no load.
3. The synchronous motor must have DC excitation to operate.
4. It will operate at the speed of the rotating magnetic field (synchronous speed).
5. It has the ability to correct its own power factor and the power factor of other devices connected to the same line.

### Rotor Construction

The synchronous motor has the same type of stator windings as the other two three-phase motors. The rotor of a synchronous motor has windings similar to the rotor of an alternator *(see Figure 30-6)*. Wound pole pieces become electromagnets when direct current is applied to them. The excitation current can be applied to the rotor through two sliprings located on the rotor shaft or by a brushless exciter. The brushless exciter for a synchronous motor is the same as that used for the alternator discussed in Unit 30.

### Starting A Synchronous Motor

The rotor of a synchronous motor also contains a set of squirrel-cage bars similar to those found in a type A rotor. This set of squirrel-cage bars is used to start the motor and is known as the **amortisseur winding** *(Figure 31-47)*. When power is first connected to the stator, the rotating magnetic field cuts through the squirrel-cage bars. The cutting action of the field induces a current into the squirrel-cage winding. The current flow through the amortisseur winding produces a rotor magnetic field that is attracted to the rotating magnetic field of the stator. This causes the rotor to begin turning in the direction of rotation of the stator field. When the rotor has accelerated to a speed that is close to the synchronous speed of the field, direct current is connected to the rotor through the sliprings on the rotor shaft or by a brushless exciter *(Figure 31-48)*.

**Figure 31-47** Synchronous motor rotor with amortisseur winding *(Courtesy of Electric Machinery Manufacturing Corp.)*

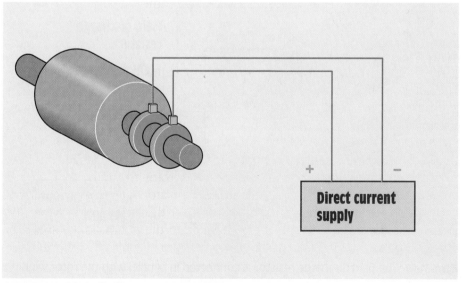

**Direct current supply**

**Figure 31-48** DC excitation current supplied through sliprings

When DC current is applied to the rotor, the windings of the rotor become electromagnets. The electromagnetic field of the rotor locks in step with the rotating magnetic field of the stator. The rotor will now turn at the same speed as the rotating magnetic field. When the rotor turns at the synchronous speed of the field, there is no more cutting action between the stator field and the amortisseur winding. This causes the current flow in the amortisseur winding to cease.

Notice that the synchronous motor starts as a squirrel-cage induction motor. Since the rotor uses bars that are similar to those used in a type A rotor, they have a relatively high resistance, which gives the motor good starting torque and low starting current. **A synchronous motor must never be started with DC current connected to the rotor.** If DC current is applied to the rotor, the field poles of the rotor become electromagnets. When the stator is energized, the rotating magnetic field begins turning at synchronous speed. The electromagnets are alternately attracted and repelled by the stator field. As a result, the rotor does not turn. The rotor and power supply can be damaged by high induced voltages, however.

### The Field Discharge Resistor

When the stator winding is first energized, the rotating magnetic field cuts through the rotor winding at a fast rate of speed. This causes a large

**A synchronous motor must never be started with DC current connected to the rotor.**

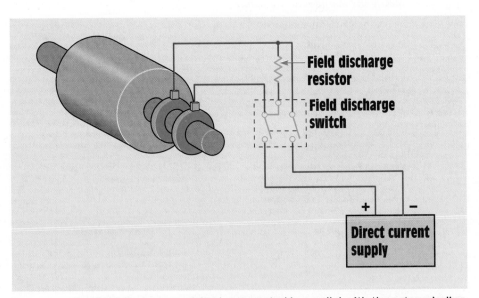

**Field discharge resistor**

**Field discharge switch**

**Direct current supply**

**Figure 31-49**  The field discharge resistor is connected in parallel with the rotor winding during starting.

amount of voltage to be induced into the winding of the rotor. To prevent this from becoming excessive, a resistor is connected across the winding. This resistor is known as the field discharge resistor *(Figure 31-49)*. It also helps to reduce the voltage induced into the rotor by the collapsing magnetic field when the DC current is disconnected from the rotor. The field discharge resistor is connected in parallel with the rotor winding during starting. If the motor is manually started a field discharge switch is used to connect the excitation current to the rotor. If the motor is automatically started, a special type of relay is used to connect excitation current to the rotor and disconnect the field discharge resistor.

## Constant Speed Operation

Although the synchronous motor starts as an induction motor, it does not operate as one. After the amortisseur winding has been used to accelerate the rotor to about 95% of the speed of the rotating magnetic field, direct current is connected to the rotor, and the electromagnets lock in step with the rotating field. Notice that the synchronous motor does not depend on induced voltage from the stator field to produce a magnetic field in the rotor. The magnetic field of the rotor is produced by external DC current applied to the rotor. This is the reason that the synchronous motor has the ability to operate at the speed of the rotating magnetic field.

As load is added to the motor, the magnetic field of the rotor remains locked with the rotating magnetic field of the stator, and the rotor continues to turn at the same speed. The added load, however, causes the magnetic fields of the rotor and stator to become stressed *(Figure 31-50)*. The action is similar to connecting the north and south ends of two magnets together and then trying to pull them apart. If the force being used to pull the magnets apart becomes greater than the strength of the magnetic attraction, the magnetic coupling will be broken and the magnets can be separated. The same is true for the synchronous motor. If the load on the motor becomes too great, the rotor will be pulled out of sync with the rotating magnetic field. The amount of torque necessary to cause this condition is called the *pullout torque*. The pullout torque for most synchronous motors will range from 150% to 200% of rated full-load torque. If pullout torque is reached, the motor must be stopped and restarted.

## The Power Supply

The DC power supply of a synchronous motor can be provided by several methods. The most common of these methods is either a small DC generator mounted to the shaft of the motor or an electronic power supply that converts the AC line voltage into DC voltage.

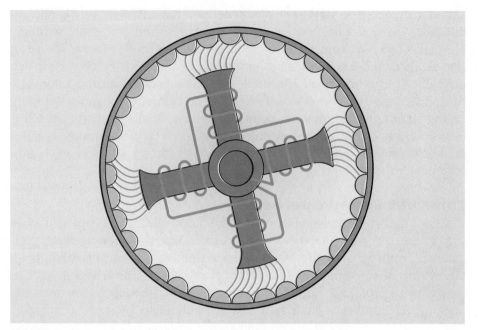

**Figure 31-50**  The magnetic field becomes stressed as load is added.

## Power Factor Correction

The synchronous motor has the ability to correct its own power factor and the power factor of other devices connected to the same line. The amount of power factor correction is controlled by the amount of excitation current in the rotor. If the rotor of a synchronous motor is underexcited, the motor will have a lagging power factor like a common induction motor. As rotor excitation current is increased, the synchronous motor appears to be more capacitive. When the excitation current reaches a point that the power factor of the motor is at unity or 100%, it is at the normal excitation level. At this point, the current supplying the motor will drop to its lowest value.

If the excitation current is increased above the normal level, the motor will have a leading power factor and appear as a capacitive load. When the rotor is overexcited, the current supplying the motor will increase due to the change in power factor. The power factor at this point, however, is leading and not lagging. Since capacitance has now been added to the line, it will correct the lagging power factor of other inductive devices connected to the same line. Changes in the amount of excitation current will not affect the speed of the motor.

## Interaction of the Direct and Alternating Current Fields

*Figure 31-51* illustrates how the magnetic flux of the AC field aids or opposes the DC field. In this example, it is assumed that the DC field is held stationary and the rotating armature is connected to the AC source. Although most synchronous motors have a stationary AC field and a rotating DC field, the principle of operation is the same. When the DC excitation current is less than the amount required for normal excitation, the AC current must supply some portion of the magnetizing current to aid the weak DC current *(Figure 31-51A)*. This portion of magnetizing current lags the applied voltage by 90°. The current wave form shown in *Figure 31-51A* depicts only the portion of magnetizing current that is out of phase with the voltage. The remaining part of the AC current is used to produce the torque necessary to operate the load. The synchronous motor will have a lagging power factor at this time.

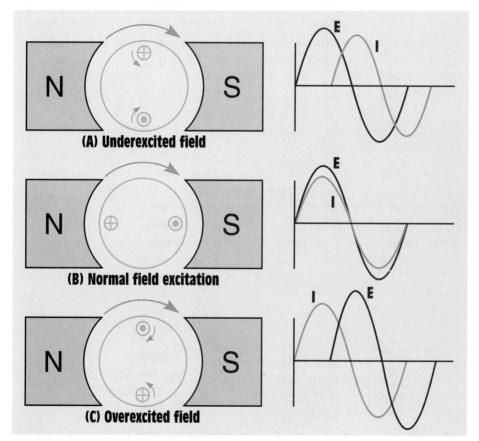

**(A) Underexcited field**

**(B) Normal field excitation**

**(C) Overexcited field**

**Figure 31-51**   Field excitation in a synchronous motor

In *Figure 31-51B*, the DC excitation current has been increased to the normal excitation value. All of the AC current is now used to produce the torque necessary to operate the load. Since the AC current no longer supplies any of the magnetizing current, it is in phase with the voltage and the motor power factor is at unity or 100%. The amount of AC current supplied to the motor will be at its lowest value during this period.

In *Figure 31-51C* the DC excitation current is greater than that needed for normal excitation. The AC current now supplies a demagnetizing component of current. The portion of AC current used to demagnetize the overexcited DC field will lead the applied voltage by 90°. The current wave form shown in *Figure 31-51C* illustrates only the portion of AC current used to demagnetize the DC field and does not take into account the amount of AC current used to produce torque for the load. The synchronous motor now has a leading power factor.

## Synchronous Motor Applications

Synchronous motors are very popular in industry, especially in the large horsepower ratings (motors up to 5000 hp are not uncommon). They have a low starting current per horsepower and a high starting torque. They operate at a constant speed from no load to full load and maintain maximum efficiency. Synchronous motors are used to operate DC generators, fans, blowers, pumps, and centrifuges. They correct their own power factor and can correct the power factor of other inductive loads connected to the same feeder *(Figure 31-52)*. Synchronous motors are sometimes operated at no load and are used for power factor correc-

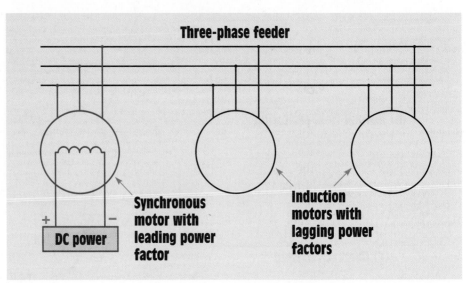

**Three-phase feeder**

**Synchronous motor with leading power factor**

**Induction motors with lagging power factors**

**DC power**

**+** **−**

**Figure 31-52** Synchronous motor used to correct the power factor of other motors

tion only. When this is done, the motor is referred to as a **synchronous condenser**.

## Advantages of the Synchronous Condenser

The advantage of using a synchronous condenser over a bank of capacitors for power factor correction is that the amount of correction is easily controlled. When a bank of capacitors is used for correcting power factor, capacitors must be added to or removed from the bank if a change in the amount of correction is needed. When a synchronous condenser is used, only the excitation current must be changed to cause an alteration of power factor. The schematic symbol for a synchronous motor is shown in *Figure 31-53*.

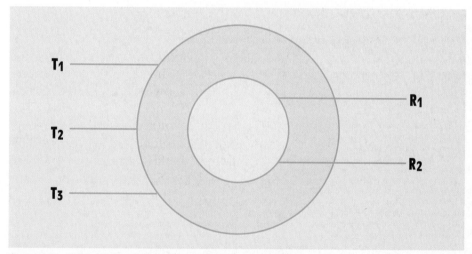

**Figure 31-53**   Schematic symbol for a synchronous motor

## 31-7 Selsyn Motors

The word *selsyn* is a contraction derived from *self-synchronous*. **Selsyn motors** are used to provide position control and angular feedback information in industrial applications. Although selsyn motors are actually operated on single-phase alternating current, they do contain three-phase windings *(Figure 31-54)*. The schematic symbol for a selsyn motor is shown in *Figure 31-55*. This symbol is very similar to the symbol used to represent a three-phase synchronous motor. The stator windings are labeled $S_1$, $S_2$, and $S_3$. The rotor leads are labeled $R_1$ and $R_2$. The rotor leads are connected to the rotor winding by means of sliprings and brushes.

When selsyn motors are employed, at least two are used together. One motor is referred to as the transmitter and the other is called the receiver. It

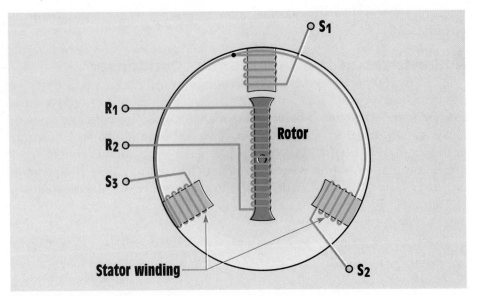

**Figure 31-54**  Selsyn motor

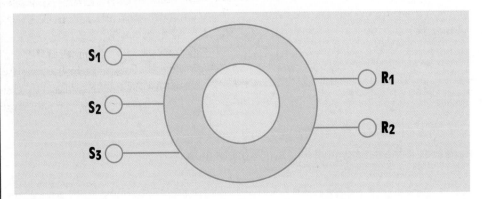

**Figure 31-55**  Schematic symbol for a selsyn motor

makes no difference which motor acts as the transmitter and which acts as the receiver. Connection is made by connecting $S_1$ of the transmitter to $S_1$ of the receiver, $S_2$ of the transmitter to $S_2$ of the receiver, and $S_3$ of the transmitter to $S_3$ of the receiver. The rotor leads of each motor are connected to a source of single-phase alternating current *(Figure 31-56)*. If the stator winding leads of the two selsyn motors are connected improperly, the receiver will rotate in a direction opposite that of the transmitter. If the rotor leads are connected improperly, the rotor of the transmitter and the rotor of the receiver will have an angle difference of 180°.

## Selsyn Motor Operation

Selsyn motors actually operate as transformers. The rotor winding is the primary and the stator winding is the secondary. In *Figure 31-56*, the rotor of the transmitter is in line with stator winding $S_1$. Since the rotor is connected to a source of alternating current, an alternating magnetic field exists in the rotor. This alternating magnetic field will induce a voltage into the windings of the stator. Since the rotors of both motors are connected to the same source of alternating current, magnetic fields of identical strength and polarity exist in both motors.

Since the rotor of the transmitter is in line with stator winding $S_1$, maximum voltage and current are being induced in stator $S_1$, and less than maximum voltage and current are being induced in stator windings $S_2$ and $S_3$. Since the stator windings of the receiver are connected to the stator windings of the transmitter, the same current will flow through the receiver, producing a magnetic field in the receiver. This magnetic field will attract or repel the magnetic field of the rotor depending on the relative polarity of the two fields. When the rotor of the receiver is in the same position as the rotor of the transmitter, an equal amount of voltage will be induced in the stator windings of the receiver, causing stator winding current to become zero.

If the rotor of the transmitter is turned to a different position, the magnetic field of the stator will change, resulting in a change of the magnetic field in the stator of the receiver. This will cause the rotor of the receiver to rotate to a new position, where the two stator magnetic fields again cancel each other. Each time the rotor position of the transmitter is changed, the rotor of the receiver will change the same amount.

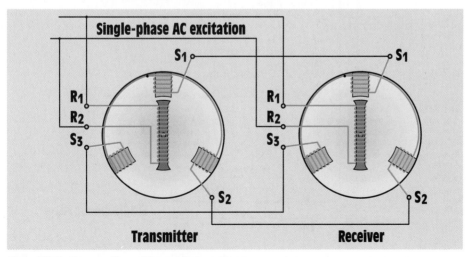

**Figure 31-56** Connection of two selsyn motors

differential
selsyn

### The Differential Selsyn

The **differential selsyn** is used to produce the algebraic sum of the rotation of two other selsyn units. Differential selsyns are constructed in a manner different from other selsyn motors. The differential selsyn contains three rotor windings connected in wye as well as three stator windings connected in wye. The rotor windings are brought out through three sliprings and brushes in a manner very similar to a wound rotor induction motor. The differential selsyn is not connected to a source of power. Power must be provided by one of the other selsyn motors connected to it *(Figure 31-57)*.

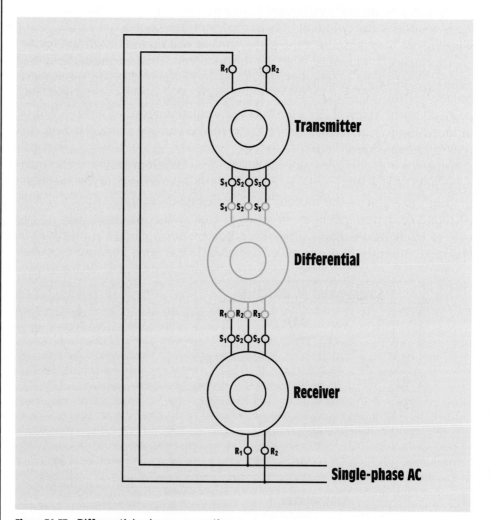

**Figure 31-57** Differential selsyn connection

If any one of the selsyn units is held in place and a second unit is turned, the third will turn by the same amount. If any two of the selsyn units are turned at the same time, the third will turn an amount equal to the sum of the angle of rotation of the other two.

## Summary

1. Three basic types of three-phase motors are:
   A. Squirrel-cage induction motor.
   B. Wound rotor induction motor.
   C. Synchronous motor.

2. All three-phase motors operate on the principle of a rotating magnetic field.

3. Three factors that cause a magnetic field to rotate are:
   A. The fact that the voltages of a three-phase system are 120° out of phase with each other.
   B. The fact that voltages change polarity at regular intervals.
   C. The arrangement of the stator windings.

4. The speed of the rotating magnetic field is called the synchronous speed.

5. Two factors that determine the synchronous speed are:
   A. Number of stator poles per phase.
   B. Frequency of the applied voltage.

6. The direction of rotation of any three-phase motor can be changed by reversing the connection of any two stator leads.

7. The direction of rotation of a three-phase motor can be determined with a phase rotation meter before power is applied to the motor.

8. Dual-voltage motors will have 9 or 12 leads brought out at the terminal connection box.

9. Dual-voltage motors intended for high-voltage connection have their phase windings connected in series.

10. Dual-voltage motors intended for low-voltage connection have their phase windings connected in parallel.

11. Motors that bring out 12 leads are generally intended for wye-delta starting.

12. Three factors that determine the torque produced by an induction motor are:
    A. The strength of the magnetic field of the stator.
    B. The strength of the magnetic field of the rotor.
    C. The phase angle difference between rotor and stator flux.

13. Maximum torque is developed when stator and rotor flux are in phase with each other.

14. The code letter on the nameplate of a squirrel-cage motor indicates the type of rotor bars used in the construction of the rotor.

15. The type A rotor has the lowest starting current, highest starting torque, and poorest speed regulation of any type squirrel-cage rotor.

16. The double squirrel-cage rotor contains two sets of squirrel-cage windings in the same rotor.

17. Consequent pole squirrel-cage motors change speed by changing the number of stator poles.

18. Wound rotor induction motors have wound rotors that contain three-phase windings.

19. Wound rotor motors have three sliprings on the rotor shaft to provide external connection to the rotor.

20. Wound rotor motors have higher starting torque and lower starting current than squirrel-cage motors of the same horsepower.

21. The speed of a wound rotor motor can be controlled by permitting resistance to remain in the rotor circuit during operation.

22. Synchronous motors operate at synchronous speed.

23. Synchronous motors operate at a constant speed from no load to full load.

24. When load is connected to a synchronous motor, stress develops between the magnetic fields of the rotor and stator.

25. Synchronous motors must have DC excitation from an external source.

26. DC excitation is provided to some synchronous motors through two sliprings located on the rotor shaft, and other motors use a brushless exciter.

27. Synchronous motors have the ability to produce a leading power factor by overexcitation of the DC current supplied to the rotor.

28. Synchronous motors have a set of type A squirrel-cage bars used for starting. This squirrel-cage winding is called the amortisseur winding.

29. A field discharge resistor is connected across the rotor winding during starting to prevent high voltage in the rotor due to induction.

30. Changing the DC excitation current does not affect the speed of the motor.

31. Selsyn motors are used to provide position control and angular feedback information.

32. Although selsyn motors contain three-phase windings, they operate on single-phase AC.

33. A differential selsyn unit can be used to determine the algebraic sum of the rotation of two other selsyn units.

## Review Questions

1. What are the three basic types of three-phase motors?

2. What is the principle of operation of all three-phase motors?

3. What is synchronous speed?

4. What two factors determine synchronous speed?

5. Name three factors that cause the magnetic field to rotate.

6. Name three factors that determine the torque produced by an induction motor.

7. Is the synchronous motor an induction motor?

8. What is the amortisseur winding?

9. Why must a synchronous motor never be started when DC excitation current is applied to the rotor?

10. Name three characteristics that make the synchronous motor different from an induction motor.

11. What is the function of the field discharge resistor?

12. Why can an induction motor never operate at synchronous speed?

13. A squirrel-cage induction motor is operating at 1175 RPM and producing a torque of 22 lb-ft. What is the horsepower output of the motor?

14. A wattmeter measures the input power of the motor in problem 13 to be 5650 W. What is the efficiency of the motor?

15. What is the difference between a squirrel-cage motor and a wound rotor motor?

16. What is the advantage of the wound rotor motor over the squirrel-cage motor?

17. Name three factors that determine the amount of voltage induced in the rotor of a wound rotor motor.

18. Why will the rotor of a wound rotor motor not turn if the rotor circuit is left open with no resistance connected to it?

19. Why is the starting torque of a wound rotor motor higher than that of a squirrel-cage motor although the starting current is less?

20. When is a synchronous motor a synchronous condenser?

21. What determines when a synchronous motor is at normal excitation?

22. How can a synchronous motor be made to have a leading power factor?

23. Is the excitation current of a synchronous motor AC or DC?

24. How is the speed of a consequent pole squirrel-cage motor changed?

25. A three-phase squirrel-cage motor is connected to a 60-Hz line. The full-load speed is 870 RPM. How many poles per phase does the stator have?

Single-phase motors are used in locations where three-phase power is not generally available, such as homes and small office buildings. This unit presents the different types of single-phase motors and their operating characteristics.

Courtesy of Crest Industries, Inc.

# Single-Phase Motors

## Outline

## Key Terms

Centrifugal switch

Compensating winding

Conductive compensation

Consequent pole motor

Holtz motor

Inductive compensation

Multispeed motors

Neutral plane

Repulsion motor

Run winding

Shaded-pole induction motor

Shading coil

Split-phase motors

Start winding

Stepping motors

Synchronous motors

Synchronous speed

Two-phase

Universal motor

Warren motor

**Objectives**

*fter studying this unit, you should be able to:*

- List the different types of split-phase motors.

- Discuss the operation of split-phase motors.

- Reverse the direction of rotation of a split-phase motor.

- Discuss the operation of multispeed split-phase motors.

- Discuss the operation of shaded-pole type motors.

- Discuss the operation of repulsion type motors.

- Discuss the operation of stepping motors.

- Discuss the operation of universal motors.

**Preview**

Although most of the large motors used in industry are three-phase, at times single-phase motors must be used. Single-phase motors are used almost exclusively to operate home appliances such as air conditioners, refrigerators, well pumps, and fans. They are generally designed to operate on 120 V or 240 V. They range in size from fractional horsepower to several horsepower, depending on the application.

### 32-1 Single-Phase Motors

In Unit 31, it was stated that there are three basic types of three-phase motors and that all operate on the principle of a rotating magnetic field. While that is true for three-phase motors, it is not true for single-phase motors. There are not only many different types of single-phase motors, but they have different operating principles.

## 32-2 Split-Phase Motors

Split-phase motors fall into three general classifications:
1. the resistance-start induction-run motor,
2. the capacitor-start induction-run motor, and
3. the capacitor-start capacitor-run motor.

Although all these motors have different operating characteristics, they are similar in construction and use the same operating principle. **Split-phase motors** receive their name from the manner in which they operate. Like three-phase motors, split-phase motors operate on the principle of a rotating magnetic field. A rotating magnetic field, however, cannot be produced with only one phase. Split-phase motors, therefore, split the current flow through two separate windings to simulate a two-phase power system. A rotating magnetic field can be produced with a two-phase system.

### The Two-Phase System

In some parts of the world two-phase power is produced. A **two-phase** system is produced by having an alternator with two sets of coils wound 90° apart *(Figure 32-1)*. The voltages of a two phase system are, therefore, 90° out of phase with each other. These two out-of-phase voltages can be used to produce a rotating magnetic field in a manner similar to that of producing a rotating magnetic field with the voltages of a three-phase system. Since there have to be two voltages or currents out of phase with each other to produce a rotating magnetic field, split-phase

**split-phase motors**

**two-phase**

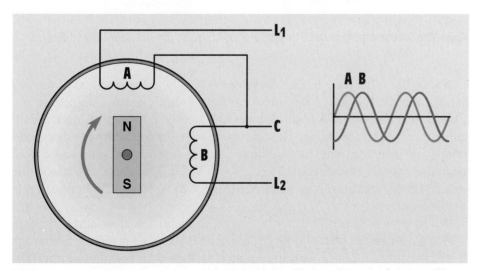

**Figure 32-1** A two-phase alternator produces voltages that are 90° out of phase with each other.

motors use two separate windings to create a phase difference between the currents in each of these windings. These motors literally split one phase and produce a second phase, hence the name split-phase motor.

## Stator Windings

The stator of a split-phase motor contains two separate windings, the **start winding** and the **run winding**. The start winding is made of small wire and is placed near the top of the stator core. The run winding is made of relatively large wire and is placed in the bottom of the stator core. *Figure 32-2* shows a photograph of two split-phase stators. The stator on the left is used for a resistance-start induction-run motor, or a capacitor-start induction-run motor. The stator on the right is used for a capacitor-start capacitor-run motor. Both stators contain four poles, and the start winding is placed at a 90° angle from the run winding.

**start winding**

**run winding**

**Figure 32-2** Stator windings used in single-phase motors *(Courtesy of Bodine Electric Co.)*

Notice the difference in size and position of the two windings of the stator shown on the left. The start winding is made from small wire and placed near the top of the stator core. This causes it to have a higher resistance than the run winding. The start winding is located between the poles of the run winding. The run winding is made with larger wire and placed near the bottom of the core. This gives it higher inductive reactance and less resistance than the start winding. These two windings are connected in parallel with each other *(Figure 32-3)*.

When power is applied to the stator, current will flow through both windings. Since the start winding is more resistive, the current flow through it will be more in phase with the applied voltage than will the current flow through the run winding. The current flow through the run winding will lag the applied voltage due to inductive reactance. These two out-of-phase

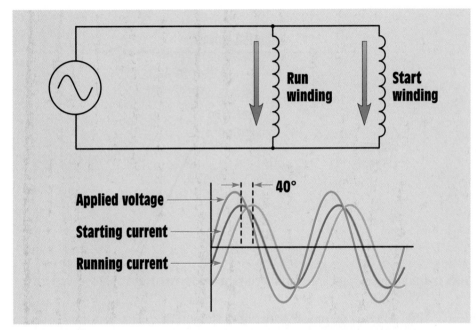

**Figure 32-3** The start and run windings are connected in parallel with each other.

two out-of-phase currents are used to create a rotating magnetic field in the stator. The speed of this rotating magnetic field is called **synchronous speed** and is determined by the same two factors that determined the synchronous speed for a three-phase motor:

1. number of stator poles per phase, and
2. frequency of the applied voltage.

## 32-3 Resistance-Start Induction-Run Motors

The resistance-start induction-run motor receives its name from the fact that the out-of-phase condition between start and run winding current is caused by the start winding being more resistive than the run winding. The amount of starting torque produced by a split-phase motor is determined by three factors:

1. the strength of the magnetic field of the stator,
2. the strength of the magnetic field of the rotor, and
3. the phase angle difference between current in the start winding and current in the run winding. (Maximum torque is produced when these two currents are 90° out of phase with each other.)

Although these two currents are out of phase with each other, they are not 90° out of phase. The run winding is more inductive than the start winding, but it does have some resistance, which prevents the current

**synchronous speed**

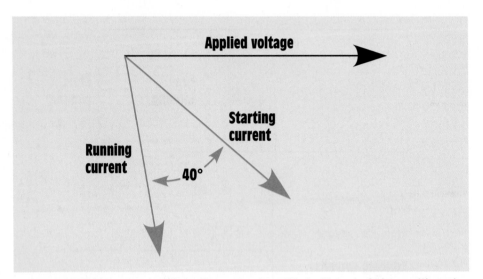

**Figure 32-4** Running current and starting current are 35° to 40° out of phase with each other.

from being 90° out of phase with the voltage. The start winding is more resistive than the run winding, but it does have some inductive reactance, preventing the current from being in phase with the applied voltage. Therefore, a phase angle difference of 35° to 40° is produced between these two currents, resulting in a rather poor starting torque *(Figure 32-4)*.

## Disconnecting the Start Winding

A stator rotating magnetic field is necessary only to start the rotor turning. Once the rotor has accelerated to approximately 75% of rated speed, the start winding can be disconnected from the circuit and the motor will continue to operate with only the run winding energized. Motors that are

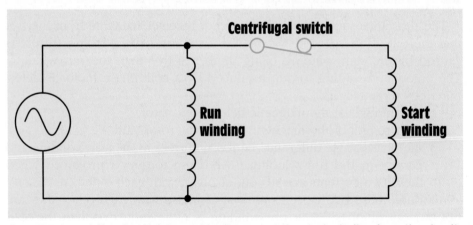

**Figure 32-5** A centrifugal switch is used to disconnect the start winding from the circuit.

**centrifugal switch**

not hermetically sealed (most refrigeration and air-conditioning compressors are hermetically sealed) use a **centrifugal switch** to disconnect the start windings from the circuit. The contacts of the centrifugal switch are connected in series with the start winding *(Figure 32-5)*. The centrifugal switch contains a set of spring-loaded weights. When the shaft is not turning, the springs hold a fiber washer in contact with the movable contact of the switch *(Figure 32-6)*. The fiber washer causes the movable contact to complete a circuit with a stationary contact.

When the rotor accelerates to about 75% of rated speed, centrifugal force causes the weights to overcome the force of the springs. The fiber washer retracts and permits the contacts to open and disconnect the start winding from the circuit *(Figure 32-7)*. The start winding of this type motor is intended to be energized only during the period of time that the motor is actually starting. If the start winding is not disconnected, it will be damaged by excessive current flow.

**Figure 32-6** The centrifugal switch is closed when the rotor is not turning.

## Relationship of Stator and Rotor Fields

The split-phase motor contains a squirrel-cage rotor very similar to those used with three-phase squirrel-cage motors *(Figure 32-8)*. When power is connected to the stator windings, the rotating magnetic field induces a voltage into the bars of the squirrel-cage rotor. The induced voltage causes current to flow in the rotor, and a magnetic field is produced around the rotor bars. The magnetic field of the rotor is attracted to the stator field, and the rotor begins to turn in the direction of the rotating

**Figure 32-7** The contact opens when the rotor reaches about 75% of rated speed.

**Figure 32-8** Squirrel-cage rotor used in a split-phase motor *(Courtesy of Bodine Electric Co.)*

magnetic field. After the centrifugal switch opens, only the run winding induces voltage into the rotor. This induced voltage is in phase with the stator current. The inductive reactance of the rotor is high, causing the rotor current to be almost 90° out of phase with the induced voltage. This causes the pulsating magnetic field of the rotor to lag the pulsating magnetic field of the stator by 90°. Magnetic poles, located midway between the stator poles, are created in the rotor *(Figure 32-9)*. These two pulsating magnetic fields produce a rotating magnetic field of their own, and the rotor continues to rotate.

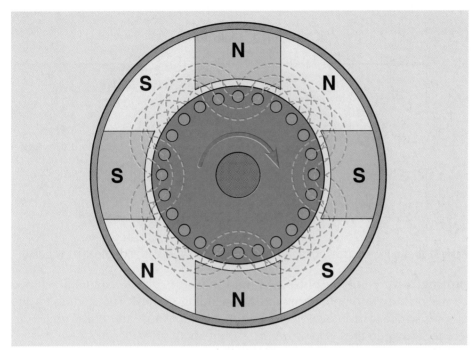

**Figure 32-9** A rotating magnetic field is produced by the stator and rotor flux.

## Direction of Rotation

The direction of rotation for the motor is determined by the direction of rotation of the rotating magnetic field created by the run and start windings when the motor is first started. The direction of motor rotation can be changed by reversing the connection of either the start winding or the run winding but not both. If the start winding is disconnected, the motor can be operated in either direction by manually turning the rotor shaft in the desired direction of rotation.

## 32-4 Capacitor-Start Induction-Run Motors

The capacitor-start induction-run motor is very similar in construction and operation to the resistance-start induction-run motor. The capacitor-start induction-run motor, however, has an AC electrolytic capacitor connected in series with the centrifugal switch and start winding *(Figure 32-10)*. Although the running characteristics of the capacitor-start induction-run motor and the resistance-start induction-run motor are identical, the starting characteristics are not. The capacitor-start induction-run motor produces a starting torque that is substantially higher than that of the resistance-start induction-run motor. Recall that one of the factors that determines the

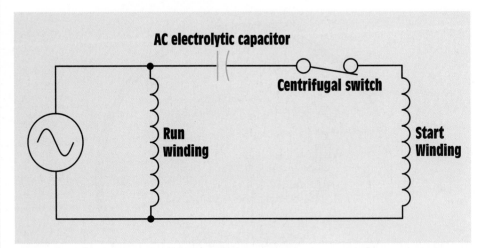

**Figure 32-10** An AC electrolytic capacitor is connected in series with the start winding.

starting torque for a split-phase motor is the phase angle difference between start winding current and run winding current. The starting torque of a resistance-start induction-run motor is low because the phase angle difference between these two currents is only about 40° *(Figure 32-4)*.

When a capacitor of the proper size is connected in series with the start winding, it causes the start winding current to lead the applied voltage. This leading current produces a 90° phase shift between run winding current and start winding current *(Figure 32-11)*. Maximum starting torque is developed at this point.

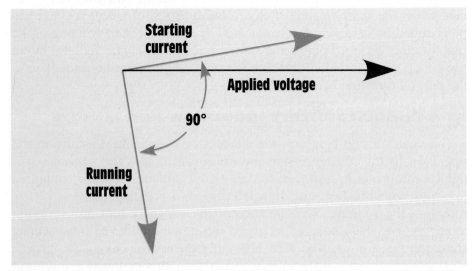

**Figure 32-11** Run winding current and start winding current are 90° out of phase with each other.

**Figure 32-12**   Capacitor-start induction-run motor *(From Duff and Herman,* Alternating Current Fundamentals, *4th ed., copyright 1991 by Delmar Publishers Inc.)*

**Although the capacitor-start induction-run motor has a high starting torque, the motor should not be started more than about eight times per hour.** Frequent starting can damage the start capacitor due to overheating. If the capacitor must be replaced, care should be taken to use a capacitor of the correct microfarad rating. If a capacitor with too little capacitance is used, the starting current will be less than 90° out of phase with the running current, and the starting torque will be reduced. If the capacitance value is too great, the starting current will be more than 90° out of phase with the running current, and the starting torque will again be reduced. A capacitor-start induction-run motor is shown in *Figure 32-12*.

## 32-5 Dual-Voltage Split-Phase Motors

Many split-phase motors are designed for operation on 120 or 240 V. *Figure 32-13* shows the schematic diagram of a split-phase motor designed for dual-voltage operation. This particular motor contains two run windings and two start windings. The lead numbers for single-phase motors are numbered in a standard manner. One of the run windings has lead numbers of $T_1$ and $T_2$. The other run winding has its leads numbered $T_3$ and $T_4$. This particular motor uses two different sets of start winding leads. One set is labeled $T_5$ and $T_6$, and the other set is labeled $T_7$ and $T_8$.

If the motor is to be connected for high-voltage operation, the run windings and start windings will be connected in series as shown in

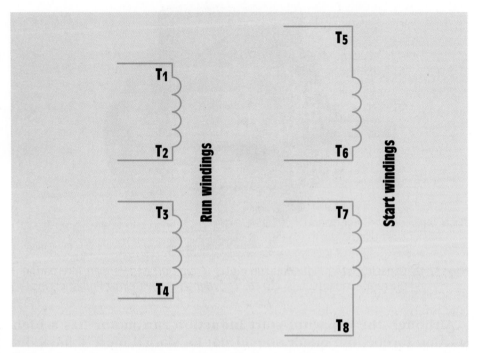

**Figure 32-13** Dual-voltage windings for a split-phase motor

*Figure 32-14.* The start windings are then connected in parallel with the run windings. If the opposite direction of rotation is desired, $T_5$ and $T_8$ will be changed.

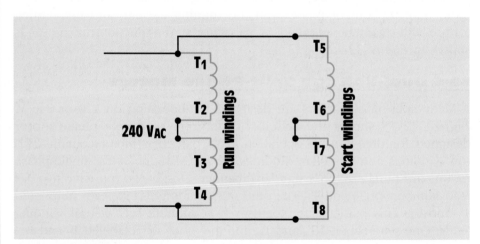

**Figure 32-14** High-voltage connection for a split-phase motor with two run and two start windings

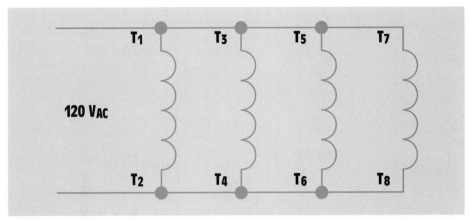

**Figure 32-15**  Low-voltage connection for a split-phase motor with two run and two start windings

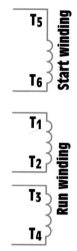

**Figure 32-16**  Dual-voltage motor with one start winding labeled $T_5$ and $T_6$

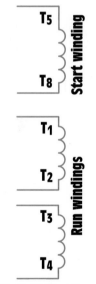

**Figure 32-17**  Dual-voltage motor with one start winding labeled $T_5$ and $T_8$

For low-voltage operation, the windings must be connected in parallel as shown in *Figure 32-15*. This connection is made by first connecting the run windings in parallel by hooking $T_1$ and $T_3$ together and $T_2$ and $T_4$ together. The start windings are paralleled by connecting $T_5$ and $T_7$ together and $T_6$ and $T_8$ together. The start windings are then connected in parallel with the run windings. If the opposite direction of rotation is desired, $T_5$ and $T_6$ should be reversed along with $T_7$ and $T_8$.

Not all dual-voltage single-phase motors contain two sets of start windings. *Figure 32-16* shows the schematic diagram of a motor that contains two sets of run windings and only one start winding. In this illustration, the start winding is labeled $T_5$ and $T_6$. Some motors, however, identify the start winding by labeling it $T_5$ and $T_8$ as shown in *Figure 32-17*.

Regardless of which method is used to label the terminal leads of the start winding, the connection will be the same. If the motor is to be connected for high-voltage operation, the run windings will be connected in series and the start winding will be connected in parallel with one of the run windings, as shown in *Figure 32-18*. In this type of motor, each winding is rated at 120 V. If the run windings are connected in series across 240 V, each winding will have a voltage drop of 120 V. By connecting the start winding in parallel across only one run winding, it will receive only 120 V when power is applied to the motor. If the opposite direction of rotation is desired, $T_5$ and $T_8$ should be changed.

If the motor is to be operated on low voltage, the windings are connected in parallel as shown in *Figure 32-19*. Since all windings are connected in parallel, each will receive 120 V when power is applied to the motor.

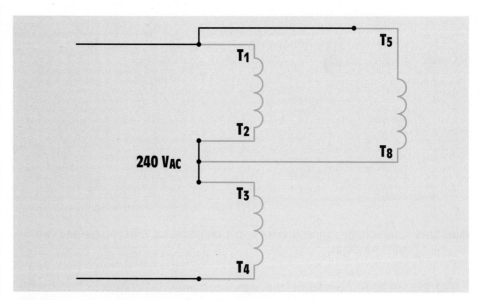

**Figure 32-18** High-voltage connection with one start winding

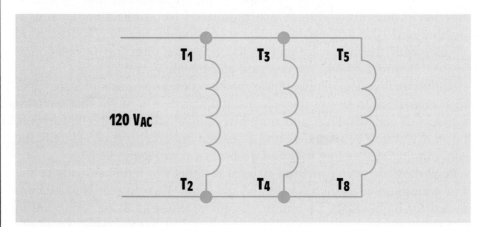

**Figure 32-19** Low-voltage connection for a split-phase motor with one start winding

## 32-6 Determining the Direction of Rotation for Split-Phase Motors

The direction of rotation of a single-phase motor can generally be determined when the motor is connected. The direction of rotation is determined by facing the back or rear of the motor. *Figure 32-20* shows a connection diagram for rotation. If clockwise rotation is desired, $T_5$ should be connected to $T_1$. If counterclockwise rotation is desired, $T_8$ (or $T_6$) should be connected to $T_1$. This connection diagram assumes that the

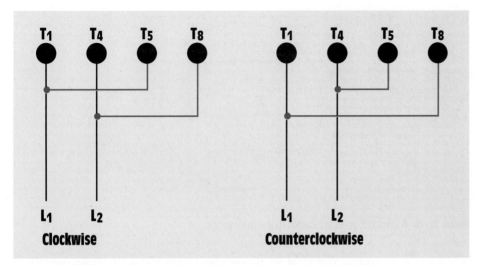

**Figure 32-20**   Determining direction of rotation for a split-phase motor

motor contains two sets of run and two sets of start windings. The type of motor used will determine the actual connection. For example, *Figure 32-18* shows the connection of a motor with two run windings and only one start winding. If this motor were to be connected for clockwise rotation, terminal $T_5$ would have to be connected to $T_1$, and terminal $T_8$ would have to be connected to $T_2$ and $T_3$. If counterclockwise rotation is desired, terminal $T_8$ would have to be connected to $T_1$, and terminal $T_5$ would have to be connected to $T_2$ and $T_3$.

## 32-7 Capacitor-Start Capacitor-Run Motors

Although the capacitor-start capacitor-run motor is a split-phase motor, it operates on a different principle than the resistance-start induction-run motor or the capacitor-start induction-run motor. The capacitor-start capacitor-run motor is designed in such a manner that its start winding remains energized at all times. A capacitor is connected in series with the winding to provide a continuous leading current in the start winding *(Figure 32-21)*. Since the start winding remains energized at all times, no centrifugal switch is needed to disconnect the start winding as the motor approaches full speed. The capacitor used in this type of motor will generally be of the oil-filled type since it is intended for continuous use. An exception to this general rule are small fractional-horsepower motors used in reversible ceiling fans. These fans have a low current draw and use an AC electrolytic capacitor to help save space.

The capacitor-start capacitor-run motor actually operates on the principle of a rotating magnetic field in the stator. Since both run and start

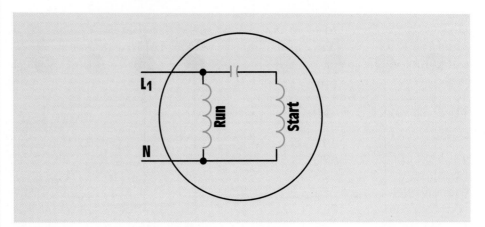

**Figure 32-21** A capacitor-start capacitor-run motor

windings remain energized at all times, the stator magnetic field continues to rotate and the motor operates as a two-phase motor. This motor has excellent starting and running torque. It is quiet in operation and has a high efficiency. Since the capacitor remains connected in the circuit at all times, the motor power factor is close to unity.

Although the capacitor-start capacitor-run motor does not require a centrifugal switch to disconnect the capacitor from the start winding, some motors use a second capacitor during the starting period to help improve starting torque *(Figure 32-22)*. A good example of this can be found on the compressor of a central air conditioning unit designed for operation on single-phase power. If the motor is not hermetically sealed, a centrifu-

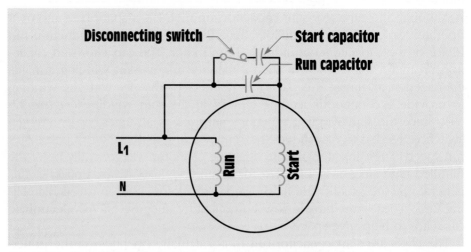

**Figure 32-22** Capacitor-start capacitor-run motor with additional starting capacitor

gal switch will be used to disconnect the start capacitor from the circuit when the motor reaches approximately 75% of rated speed. Hermetically sealed motors, however, must use some type of external switch to disconnect the start capacitor from the circuit.

## 32-8 Shaded-Pole Induction Motors

The **shaded-pole induction motor** is popular because of its simplicity and long life. This motor contains no start windings or centrifugal switch. It contains a squirrel-cage rotor and operates on the principle of a rotating magnetic field. The rotating magnetic field is created by a **shading coil** wound on one side of each pole piece. Shaded-pole motors are generally fractional-horsepower motors and are used for low-torque applications such as operating fans and blowers.

### The Shading Coil

The shading coil is wound around one end of the pole piece *(Figure 32-23)*. The shading coil is actually a large loop of copper wire or a copper band. The two ends are connected to form a complete circuit. The shading coil acts in the same manner as a transformer with a shorted secondary winding. When the current of the AC wave form increases from zero toward its positive peak, a magnetic field is created in the pole piece. As magnetic lines of flux cut through the shading coil, a voltage is induced in the coil. Since the coil is a low-resistance short circuit, a large amount of current flows in the loop. This current causes an opposition to the change of magnetic flux *(Figure 32-24)*. As long as voltage is induced into the shading coil, there will be an opposition to the change of magnetic flux.

**shaded-pole induction motor**

**shading coil**

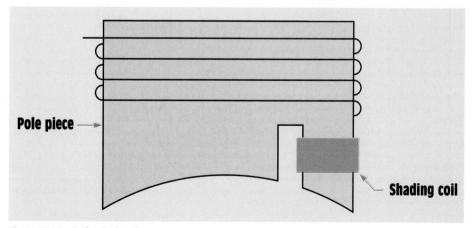

**Pole piece** →

**Shading coil**

**Figure 32-23**   A shaded pole

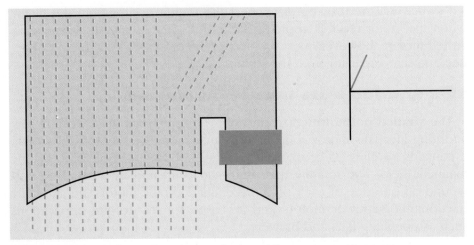

**Figure 32-24** The shading coil opposes a change of flux as current increases.

When the AC current reaches its peak value, it is no longer changing, and no voltage is being induced into the shading coil. Since there is no current flow in the shading coil, there is no opposition to the magnetic flux. The magnetic flux of the pole piece is now uniform across the pole face *(Figure 32-25)*.

When the AC current begins to decrease from its peak value back toward zero, the magnetic field of the pole piece begins to collapse. A voltage is again induced into the shading coil. This induced voltage creates a current that opposes the change of magnetic flux *(Figure 32-26)*. This causes the magnetic flux to be concentrated in the shaded section of the pole piece.

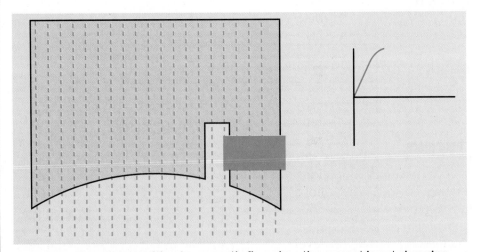

**Figure 32-25** There is opposition to magnetic flux when the current is not changing.

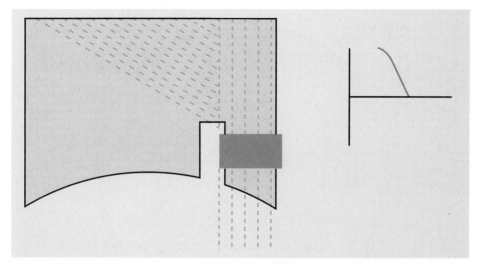

**Figure 32-26**  The shading coil opposes a change of flux when the current decreases.

When the AC current passes through zero and begins to increase in the negative direction, the same set of events happens, except that the polarity of the magnetic field is reversed. If these events were to be viewed in rapid order, the magnetic field would be seen to rotate across the face of the pole piece.

## Speed

The speed of the shaded-pole induction motor is determined by the same factors that determine the synchronous speed of other induction motors: frequency and number of stator poles. Shaded-pole motors are commonly wound as four- or six-pole motors. *Figure 32-27* shows a drawing of a four-pole shaded-pole induction motor.

## General Operating Characteristics

The shaded-pole motor contains a standard squirrel-cage rotor. The amount of torque produced is determined by the strength of the magnetic field of the stator, the strength of the magnetic field of the rotor, and the phase angle difference between rotor and stator flux. The shaded-pole induction motor has low starting and running torque.

The direction of rotation is determined by the direction in which the rotating magnetic field moves across the pole face. The rotor will turn in the direction shown by the arrow in *Figure 32-27*. The direction can be changed by removing the stator winding and turning it around. This is not a common practice, however. As a general rule the shaded-pole induction

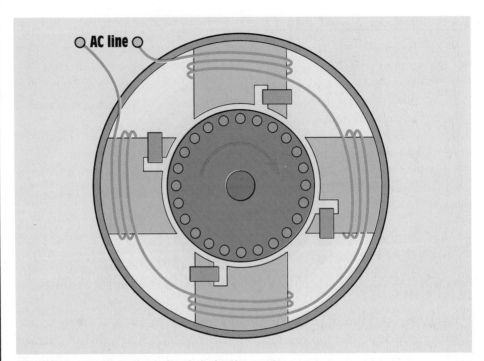

**Figure 32-27** Four-pole shaded-pole induction motor

motor is considered to be nonreversible. *Figure 32-28* shows a photograph of the stator winding and rotor of a shaded-pole induction motor.

**Figure 32-28** Stator winding and rotor of a shaded-pole induction motor *(Courtesy of A. O. Smith)*

## 32-9 Multispeed Motors

There are two basic types of **multispeed** single-phase **motors**. One is the consequent pole type and the other is a specially wound capacitor-start capacitor-run motor or shaded-pole induction motor. The **consequent pole** single-phase **motor** operates in the same basic way as the three-phase consequent pole discussed in Unit 31. The speed is changed by reversing the current flow through alternate poles and increasing or decreasing the total number of stator poles. The consequent pole motor is used where high running torque must be maintained at different speeds. A good example of where this type of motor is used is in two-speed compressors for central air conditioning units.

### Multispeed Fan Motors

Multispeed fan motors have been used for many years. These motors are generally wound for two to five steps of speed and operate fans and squirrel-cage blowers. A schematic drawing of a three-speed motor is shown in *Figure 32-29*. Notice that the run winding has been tapped to produce low, medium, and high speed. The start winding is connected in parallel with the run winding section. The other end of the start winding lead is connected to an external oil-filled capacitor. This motor obtains a change of speed by inserting inductance in series with the run winding. The actual run winding for this motor is between the terminals marked

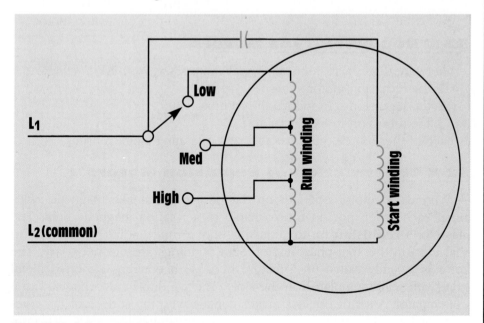

**Figure 32-29**  A three-speed motor

multispeed motors

consequent pole motor

*high* and *common*. The winding shown between *high* and *medium* is connected in series with the main run winding. When the rotary switch is connected to the medium speed position, the inductive reactance of this coil limits the amount of current flow through the run winding. When the current of the run winding is reduced, the strength of the magnetic field of the run winding is reduced and the motor produces less torque. This causes a greater amount of slip and the motor speed to decrease.

If the rotary switch is changed to the *low* position, more inductance is inserted in series with the run winding. This causes less current to flow through the run winding and another reduction in torque. When the torque is reduced the motor speed decreases again.

Common speeds for a four-pole motor of this type are 1625, 1500, and 1350 RPM. Notice that this motor does not have wide ranges between speeds as would be the case with a consequent pole motor. Most induction motors would overheat and damage the motor winding if the speed were reduced to this extent. This type of motor, however, has much higher impedance windings than most motors. The run windings of most split-phase motors have a wire resistance of 1–4 $\Omega$. This motor will generally have a resistance of 10–15 $\Omega$ in its run winding. It is the high impedance of the windings that permits the motor to be operated in this manner without damage.

Since this motor is designed to slow down when load is added, it is not used to operate high-torque loads. This type of motor is generally used to operate only low-torque loads such as fans and blowers.

### 32-10 Repulsion Type Motors

There are three basic repulsion types motors:
1. the repulsion motor,
2. the repulsion-start induction-run motor, and
3. the repulsion-induction motor.

Each of these three types has different operating characteristics.

### 32-11 Construction of Repulsion Motors

**repulsion motor**

A repulsion motor operates on the principle that like magnetic poles repel each other, not on the principle of a rotating magnetic field. The stator of a **repulsion motor** contains only a run winding very similar to that used in the split-phase motor. Start windings are not necessary. The rotor is actually called an armature because it contains a slotted metal core with windings placed in the slots. The windings are connected to a commutator. A set of brushes makes contact with the surface of the commutator bars. The entire assembly looks very much like a DC armature and brush assembly. One difference, however, is that the brushes of the repulsion motor are shorted together. Their function is to provide a current path through certain parts of the armature, not to provide power to the armature from an external source.

## Operation

Although the repulsion motor does not operate on the principle of a rotating magnetic field, it is an induction motor. When AC power is connected to the stator winding, a magnetic field with alternating polarities is produced in the poles. This alternating field induces a voltage into the windings of the armature. When the brushes are placed in the proper position, current flows through the armature windings, producing a magnetic field of the same polarity in the armature. The armature magnetic field is repelled by the stator magnetic field, causing the armature to rotate. Repulsion motors will contain the same number of brushes as there are stator poles. Repulsion motors are commonly wound for four, six, or eight poles.

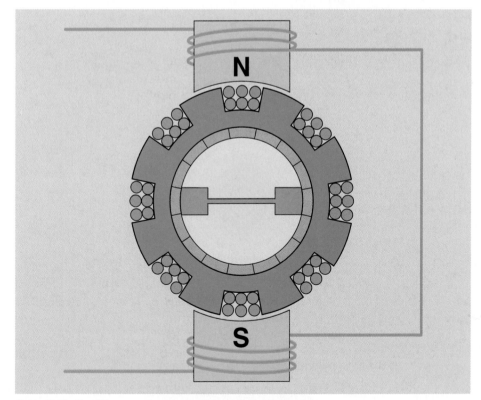

**Figure 32-30**  Brushes are placed at a 90° angle to the poles.

## Brush Position

The position of the brushes is very important. Maximum torque is developed when the brushes are placed 15° on either side of the pole pieces. *Figure 32-30* shows the effect of having the brushes placed at a 90° angle to the pole pieces. When the brushes are in this position, a circuit is completed between the coils located at a right angle to the poles. In this position, there is no induced voltage in the armature windings, and no torque is produced by the motor.

In *Figure 32-31*, the brushes have been moved to a position so that they are in line with the pole pieces. In this position, a large amount of current flows through the coils directly under the pole pieces. This current produces a magnetic field of the same polarity as the pole piece. Since the magnetic field produced in the armature is at a 0° angle to the magnetic field of the pole piece, no twisting or turning force is developed and the armature does not turn.

In *Figure 32-32*, the brushes have been shifted in a clockwise direction so that they are located 15° from the pole piece. The induced voltage in the armature winding produces a magnetic field of the same polarity as

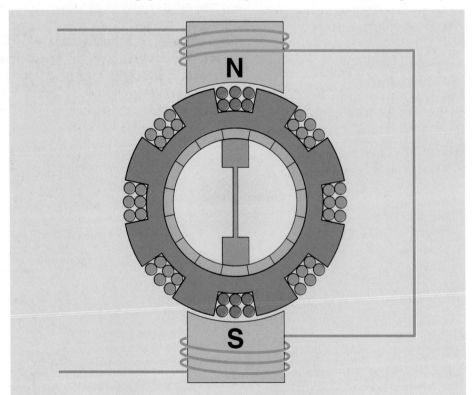

**Figure 32-31** The brushes are set at a 0° angle to the pole pieces.

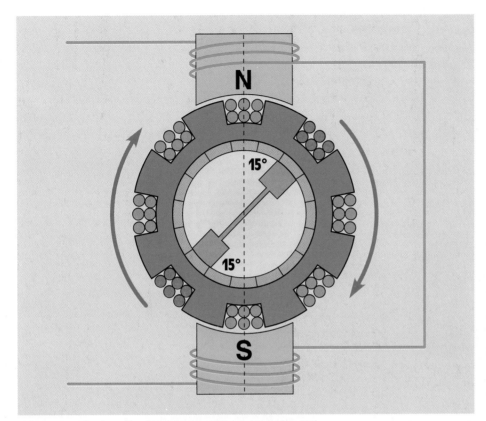

**Figure 32-32** The brushes have been shifted clockwise 15°.

the pole piece. The magnetic field of the armature is repelled by the magnetic field of the pole piece and the armature turns in the clockwise direction.

In *Figure 32-33*, the brushes have been shifted counterclockwise to a position 15° from the center of the pole piece. The magnetic field developed in the armature again repels the magnetic field of the pole piece, and the armature turns in the counterclockwise direction.

The direction of armature rotation is determined by the setting of the brushes. The direction of rotation for any type of repulsion motor is changed by setting the brushes 15° on either side of the pole pieces. Repulsion type motors have the highest starting torque of any single-phase motor. The speed of a repulsion motor, not to be confused with the repulsion-start induction-run motor or the repulsion-induction motor, can be varied by changing the AC voltage supplying power for the motor. The repulsion motor has excellent starting and running torque but can exhibit unstable speed characteristics. The repulsion motor can race to very high speed if operated with no mechanical load connected to the shaft.

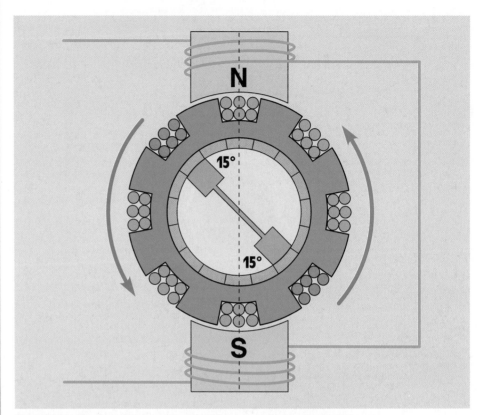

**Figure 32-33** The brushes have been shifted counterclockwise 15°.

## 32-12 Repulsion-Start Induction-Run Motors

The repulsion-start induction-run motor starts as a repulsion motor, but runs like a squirrel-cage motor. There are two types of repulsion-start induction run motors:

1. the brush-riding type and
2. the brush-lifting type.

The brush-riding type uses an axial commutator *(Figure 32-34)*. The brushes ride against the commutator segments at all times when the motor is in operation. After the motor has accelerated to approximately 75% of its full-load speed, centrifugal force causes copper segments of a short-circuiting ring to overcome the force of a spring (Figure 32-35). The segments sling out and make contact with the segments of the commutator. This effectively short-circuits all the commutator segments together, and the motor operates in the same manner as a squirrel-cage motor.

The brush-lifting type motor uses a radial commutator *(Figure 32-36)*. Weights are mounted at the front of the armature. When the motor reach-

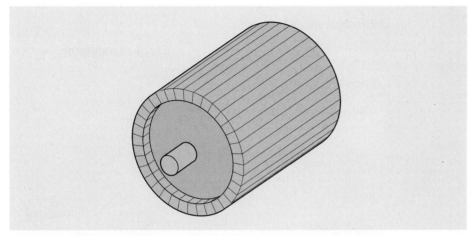

**Figure 32-34** Axial commutator

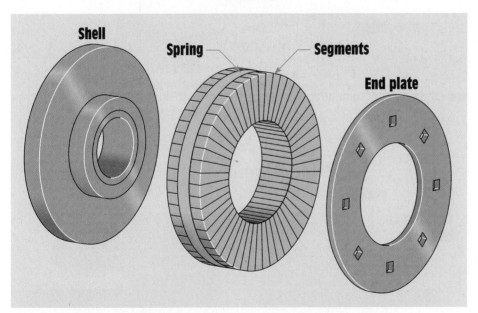

**Figure 32-35** Short-circuiting ring for brush-riding type repulsion-start induction-run motor

es about 75% of full speed, these weights swing outward due to centrifugal force and cause two push rods to act against a spring barrel and short-circuiting necklace. The weights overcome the force of the spring and cause the entire spring barrel and brush holder assembly to move toward the back of the motor (Figure 32-37). The motor is so designed that the short-circuiting necklace will short-circuit the commutator bars before the brushes lift off the surface of the radial commutator. The motor

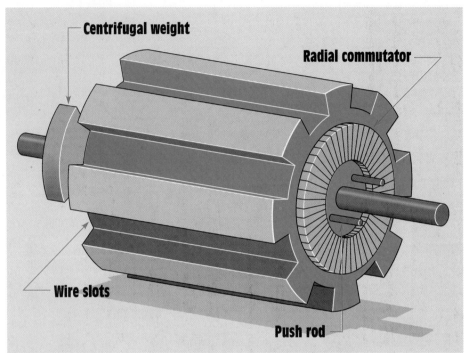

**Figure 32-36**   A radial commutator is used with the brush-lifting type motor.

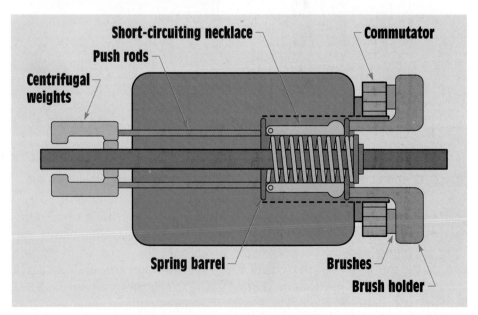

**Figure 32-37**   Brush-lifting type repulsion-start induction-run motor

will now operate as a squirrel-cage induction motor. The brush-lifting motor has several advantages over the brush-riding motor. Since the brushes lift away from the commutator surface during operation, wear on both the commutator and brushes is greatly reduced. Also, the motor does not have to overcome the friction of the brushes riding against the commutator surface during operation. As a result the brush-lifting motor is quieter in operation.

## 32-13 Repulsion-Induction Motors

The repulsion-induction motor is basically the same as the repulsion motor except that a set of squirrel-cage windings are added to the armature *(Figure 32-38)*. This type of motor contains no centrifugal mechanism or short-circuiting device. The brushes ride against the commutator at all times. The repulsion-induction motor has very high starting torque because it starts as a repulsion motor. The squirrel-cage winding, however, gives it much better speed characteristics than a standard repulsion motor. This motor has very good speed regulation between no load and full load. Its running characteristics are similar to a DC compound motor. The schematic symbol for a repulsion motor is shown in *Figure 32-39.*

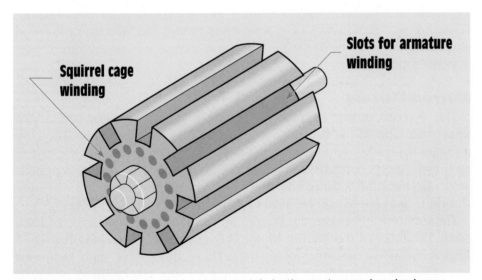

**Squirrel cage winding**

**Slots for armature winding**

**Figure 32-38**   Repulsion-induction motors contain both armature and squirrel-cage windings.

## 32-14 Single-Phase Synchronous Motors

Single-phase **synchronous motors** are small and develop only fractional horsepower. They operate on the principle of a rotating magnetic

**synchro-nous motors**

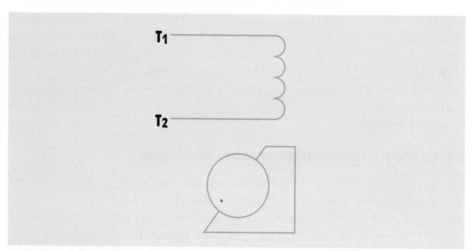

**Figure 32-39** Schematic symbol for a repulsion motor

field developed by a shaded-pole stator. Although they will operate at synchronous speed, they do not require DC excitation current. They are used in applications where constant speed is required such as clock motors, timers, and recording instruments. They also are used as the driving force for small fans because they are small and inexpensive to manufacture. There are two basic types of synchronous motor: the Warren, or General Electric motor, and the Holtz motor. These motors are also referred to as hysteresis motors.

### Warren Motors

The **Warren motor** is constructed with a laminated stator core and a single coil. The coil is generally wound for 120-V AC operation. The core contains two poles, which are divided into two sections each. One-half of each pole piece contains a shading coil to produce a rotating magnetic field *(Figure 32-40)*. Since the stator is divided into two poles, the synchronous field speed will be 3600 RPM when connected to 60 Hz.

The difference between the Warren and Holtz motor is the type of rotor used. The rotor of the Warren motor is constructed by stacking hardened steel laminations onto the rotor shaft. These disks have high hysteresis loss. The laminations form two crossbars for the rotor. When power is connected to the motor, the rotating magnetic field induces a voltage into the rotor, and a strong starting torque is developed, causing the rotor to accelerate to near-synchronous speed. Once the motor has accelerated to near-synchronous speed, the flux of the rotating magnetic field will follow the path of minimum reluctance (magnetic resistance) through the two crossbars. This causes the rotor to lock in step with the rotating magnetic

**Warren motor**

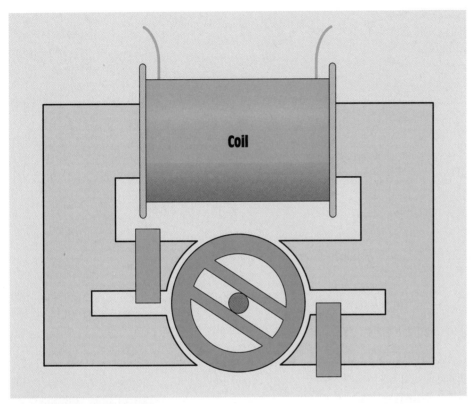

**Figure 32-40**  A warren motor

field, and the motor operates at 3600 RPM. These motors are often used with small geartrains to reduce the speed to the desired level.

## Holtz Motors

The **Holtz motor** uses a different type of rotor *(Figure 32-41)*. This rotor is cut in such a manner that six slots are formed. These slots form six salient (projecting or jutting) poles for the rotor. A squirrel-cage winding is constructed by inserting a metal bar at the bottom of each slot. When power is connected to the motor, the squirrel-cage winding provides the torque necessary to start the rotor turning. When the rotor approaches synchronous speed, the salient poles will lock in step with the field poles each half cycle. This produces a rotor speed of 1200 RPM (one-third of synchronous speed) for the motor.

## 32-15 Stepping Motors

**Stepping motors** are devices that convert electrical impulses into mechanical movement. Stepping motors differ from other types of DC or

**Holtz motor**

**Stepping motors**

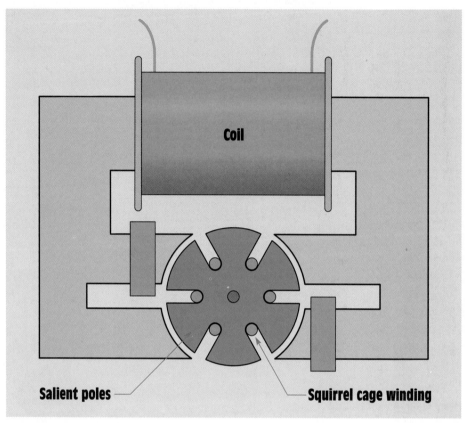

**Figure 32-41** A Holtz motor

AC motors in that their output shaft moves through a specific angular rotation each time the motor receives a pulse. The stepping motor allows a load to be controlled as to speed, distance, or position. These motors are very accurate in their control performance. There is generally less than 5% error per angle of rotation, and this error is not cumulative regardless of the number of rotations. Stepping motors are operated on DC power, but can be used as a two-phase synchronous motor when connected to AC power.

## Theory of Operation

Stepping motors operate on the theory that like magnetic poles repel and unlike magnetic poles attract. Consider the circuit shown in *Figure 32-42*. In this illustration, the rotor is a permanent magnet and the stator windings consist of two electromagnets. If current flows through the winding of stator pole A in such a direction that it creates a north magnet-

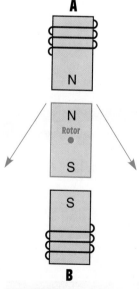

**Figure 32-42** The rotor could turn in either direction.

ic pole and through B in such a direction that it creates a south magnetic pole, it would be impossible to determine the direction of rotation. In this condition, the rotor could turn in either direction.

Now consider the circuit shown in *Figure 32-43*. In this circuit, the motor contains four stator poles instead of two. The direction of current flow through stator pole A is still in such a direction as to produce a north magnetic field, and the current flow through pole B produces a south magnetic field. The current flow through stator pole C, however, produces a south magnetic field and the current flow through pole D produces a north magnetic field. In this illustration, there is no doubt as to the direction or angle of rotation. In this example, the rotor shaft will turn 90° in a counterclockwise direction.

*Figure 32-44* shows yet another condition. In this example, the current flow through poles A and C is in such a direction as to form a north magnetic pole, and the direction of current flow through poles B and D forms south magnetic poles. In this illustration, the permanent magnetic rotor has rotated to a position between the actual pole pieces.

To allow for better stepping resolution, most stepping motors have eight stator poles, and the pole pieces and rotor have teeth machined into them as shown in *Figure 32-45*. In actual practice the number of teeth machined in the stator and rotor determines the angular rotation achieved

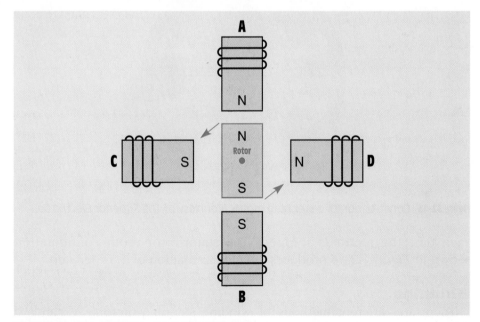

**Figure 32-43** The direction of rotation is known.

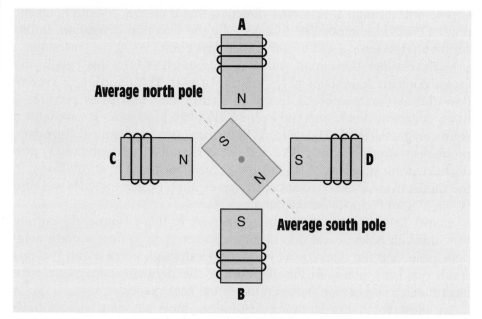

**Figure 32-44** The magnet aligns with the average magnetic pole.

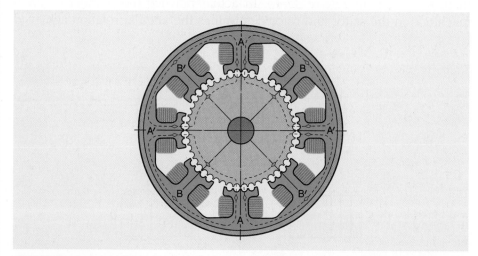

**Figure 32-45** Construction of a stepping motor *(Courtesy of The Superior Electric Co.)*

each time the motor is stepped. The stator-rotor tooth configuration shown in *Figure 32-45* produces an angular rotation of 1.8° per step.

## Windings

There are different methods of winding stepping motors. A standard three-lead motor is shown in *Figure 32-46*. The common terminal of the

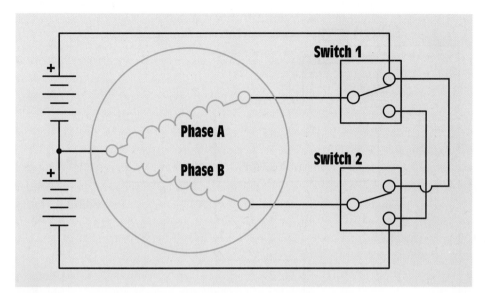

**Figure 32-46**  A standard three-lead motor

two windings is connected to ground of an above- and belowground power supply. Terminal 1 is connected to the common of a single-pole double-throw switch (switch 1) and terminal 3 is connected to the common of another single-pole double-throw switch (switch 2). One of the stationary contacts of each switch is connected to the positive, or aboveground, voltage, and the other stationary contact is connected to the negative, or belowground, voltage. The polarity of each winding is determined by the position setting of its control switch.

Stepping motors can also be wound bifilar as shown in *Figure 32-47*. The term *bifilar* means that two windings are wound together. This is similar to a transformer winding with a center-tap lead. Bifilar stepping motors have twice as many windings as the three-lead type, which makes it necessary to use smaller wire in the windings. This results in higher wire resistance in the winding, producing a better inductive-resistive (LR) time constant for the bifilar wound motor. The increased LR time constant results in better motor performance. The use of a bifilar stepping motor also simplifies the drive circuitry requirements. Notice that the bifilar motor does not require an above- and belowground power supply. As a general rule, the power supply voltage should be about five times greater than the motor voltage. A current-limiting resistance is used in the common lead of the motor. This current-limiting resistor also helps to improve the LR time constant.

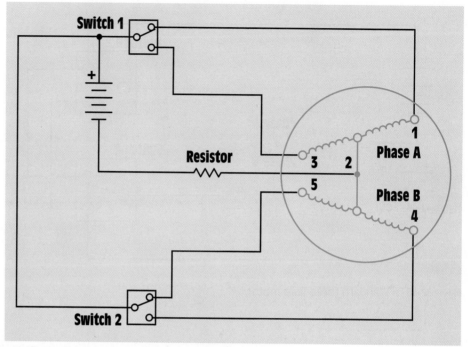

**Figure 32-47** Bilfilar-wound stepping motor

### Four-Step Switching (Full-Stepping)

The switching arrangement shown in *Figure 32-47* can be used for a four-step switching sequence (full-stepping). Each time one of the switches changes position the rotor will advance one-fourth of a tooth. After four steps, the rotor has turned the angular rotation of one "full" tooth. If the rotor and stator have 50 teeth, it will require 200 steps for the motor to rotate one full revolution. This corresponds to an angular rotation of 1.8° per step (360°/200 steps = 1.8° per step). The chart shown in *Figure 32-48* illustrates the switch positions for each step.

### Eight-Step Switching (Half-Stepping)

*Figure 32-49* illustrates the connections for an eight-step switching sequence (half-stepping). In this arrangement, the center-tap leads for phases A and B are connected through their own separate current-limiting resistors back to the negative of the power supply. This circuit contains four separate single-pole switches instead of two switches. The advantage of this arrangement is that each step causes the motor to rotate one-eighth of a tooth instead of one-fourth of a tooth. The motor now requires 400 steps to produce one revolution, which produces an angular rotation of 0.9° per step. This results in better stepping resolution and

| Step | Switch 1 | Switch 2 |
|:----:|:--------:|:--------:|
| 1 | 1 | 5 |
| 2 | 1 | 4 |
| 3 | 3 | 4 |
| 4 | 3 | 5 |
| 1 | 1 | 5 |

**Figure 32-48**  Four-step switching sequence

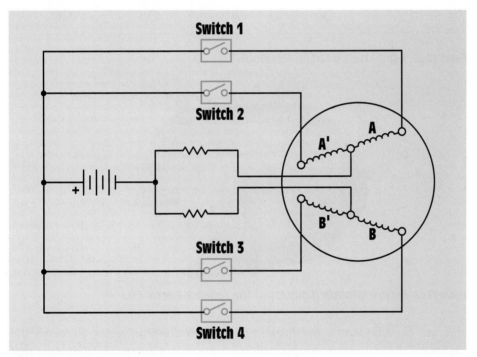

**Figure 32-49**  Eight-step switching

greater speed capability. The chart in *Figure 32-50* illustrates the switch position for each step. A stepping motor is shown in *Figure 32-51*.

## AC Operation

Stepping motors can be operated on AC voltage. In this mode of operation, they become two-phase, AC, synchronous, constant speed motors

| Step | Switch 1 | Switch 2 | Switch 3 | Switch 4 |
|------|----------|----------|----------|----------|
| 1 | On | Off | On | Off |
| 2 | On | Off | Off | Off |
| 3 | On | Off | Off | On |
| 4 | Off | Off | Off | On |
| 5 | Off | On | Off | On |
| 6 | Off | On | Off | Off |
| 7 | Off | On | On | Off |
| 8 | Off | Off | On | Off |
| 1 | On | Off | On | Off |

**Figure 32-50** Eight-step switching sequence

**Figure 32-51** Stepping motor *(Courtesy of The Superior Electric Co.)*

and are classified as a *permanent magnet induction motor.* Refer to the exploded diagram of a stepping motor shown in *Figure 32-52.* Notice that this motor has no brushes, sliprings, commutator, gears, or belts. Bearings maintain a constant air gap between the permanent magnet rotor and the stator windings. A typical eight stator pole stepping motor will have a synchronous speed of 72 RPM when connected to a 60-Hz, two-phase AC power line.

A resistive-capacitive network can be used to provide the 90° phase shift needed to change single-phase AC into two-phase AC. A simple forward-off-reverse switch can be added to provide directional control. A

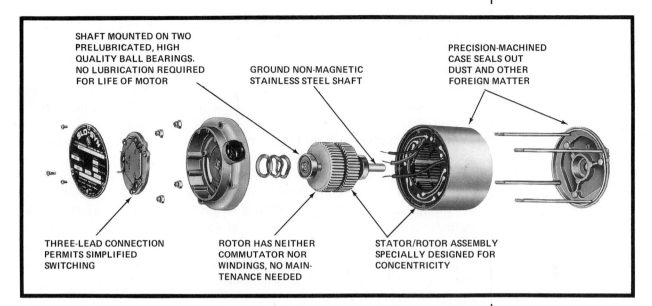

SHAFT MOUNTED ON TWO PRELUBRICATED, HIGH QUALITY BALL BEARINGS. NO LUBRICATION REQUIRED FOR LIFE OF MOTOR

GROUND NON-MAGNETIC STAINLESS STEEL SHAFT

PRECISION-MACHINED CASE SEALS OUT DUST AND OTHER FOREIGN MATTER

THREE-LEAD CONNECTION PERMITS SIMPLIFIED SWITCHING

ROTOR HAS NEITHER COMMUTATOR NOR WINDINGS, NO MAIN-TENANCE NEEDED

STATOR/ROTOR ASSEMBLY SPECIALLY DESIGNED FOR CONCENTRICITY

**Figure 32-52**  Exploded diagram of a stepping motor *(Courtesy of The Superior Electric Co.)*

sample circuit of this type is shown in *Figure 32-53*. The correct values of resistance and capacitance are necessary for proper operation. Incorrect values can result in random direction of rotation when the motor is started, change of direction when the load is varied, erratic and unstable operation, and the motor may fail to start. The correct values of resistance and capacitance will be different with different stepping motors. The manufacturer's recommendations should be followed for the particular type of stepping motor used.

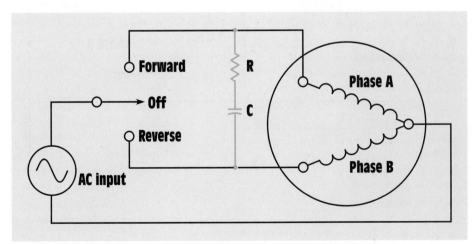

**Figure 32-53**  Phase-shift circuit converts single-phase into two-phase.

## Stepping Motor Characteristics

When stepping motors are used as two-phase synchronous motors they can start, stop, or reverse direction of rotation virtually instantly. The motor will start within about 1-1/2 cycles of the applied voltage and will stop within 5-25 ms. The motor can maintain a stalled condition without harm to the motor. Since the rotor is a permanent magnet, there is no induced current in the rotor. There is no high inrush of current when the motor is started. The starting and running currents are the same. This simplifies the power requirements of the circuit used to supply the motor. Due to the permanent magnetic structure of the rotor, the motor does provide holding torque when turned off. If more holding torque is needed, DC voltage can be applied to one or both windings when the motor is turned off. An example circuit of this type is shown in *Figure 32-54*. If DC is applied to one winding, the holding torque will be approximately 20% greater than the rated torque of the motor. If DC is applied to both windings, the holding torque will be about 1-1/2 times greater than the rated torque.

### 32-16 Universal Motors

The **universal motor** is often referred to as an AC series motor. This motor is very similar to a DC series motor in its construction in that it con-

**universal motor**

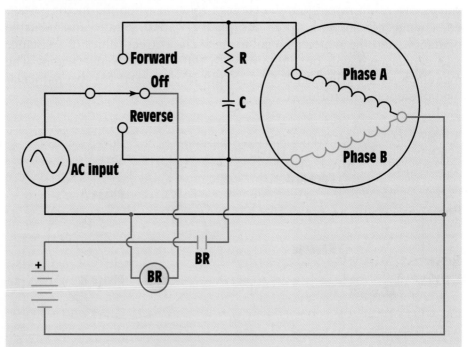

**Figure 32-54**  Applying DC voltage to increase holding torque

tains a wound armature and brushes *(Figure 32-55)*. The universal motor, however, has the addition of a **compensating winding.** If a DC series motor were connected to alternating current, the motor would operate poorly for several reasons. The armature windings would have a large amount of inductive reactance when connected to alternating current. Another reason for poor operation is that the field poles of most DC machines contain solid metal pole pieces. If the field were connected to AC, a large amount of power would be lost to eddy current induction in the pole pieces. Universal motors contain a laminated core to help prevent this problem. The compensating winding is wound around the stator and functions to counteract the inductive reactance in the armature winding.

The universal motor is so named because it can be operated on AC or DC voltage. When the motor is operated on direct current, the compensating winding is connected in series with the series field winding *(Figure 32-56)*.

**compensating winding**

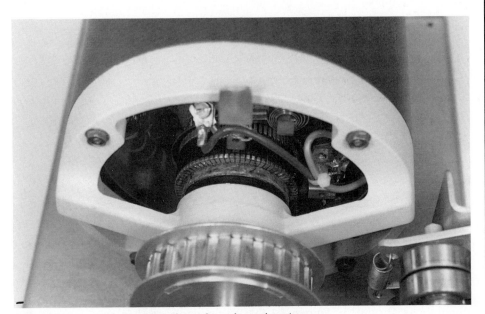

**Figure 32-55** Armature and brushes of a universal motor

## Connecting the Compensating Winding for AC Current

When the universal motor is operated with AC power, the compensating winding can be connected in two ways. If it is connected in series with the armature as shown in *Figure 32-57* it is known as **conductive compensation.**

**conductive compensation**

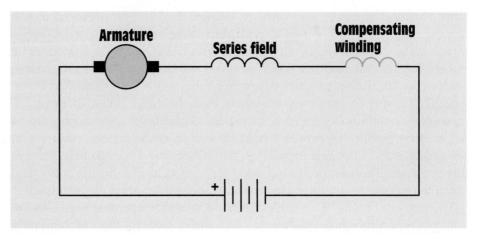

**Figure 32-56** The compensating winding is connected in series with the series field winding.

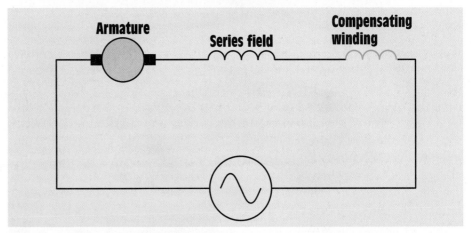

**Figure 32-57** Conductive compensation

The compensating winding can also be connected by shorting its leads together as shown in *Figure 32-58.* When connected in this manner, the winding acts like a shorted secondary winding of a transformer. Induced current permits the winding to operate when connected in this manner. This connection is known as **inductive compensation.** Inductive compensation cannot be used when the motor is connected to direct current.

### The Neutral Plane

Since the universal motor contains a wound armature, commutator, and brushes, the brushes should be set at the **neutral plane** position. This can be done in the universal motor in a manner similar to that of setting the

**inductive compensation**

**neutral plane**

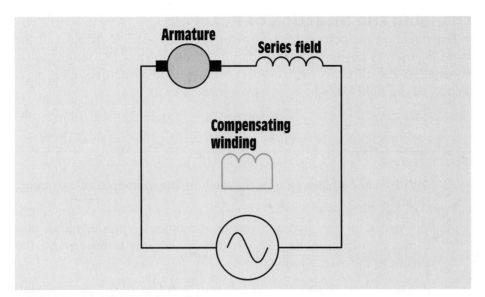

**Figure 32-58**  Inductive compensation

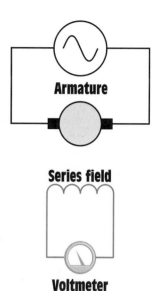

**Figure 32-59**  Using the series field to set the brushes at the neutral plane position

neutral plane of a DC machine. When setting the brushes to the neutral plane position in a universal motor, either the series or compensating winding can be used. To set the brushes to the neutral plane position using the series winding *(Figure 32-59)* alternating current is connected to the armature leads. A voltmeter is connected to the series winding. Voltage is then applied to the armature. The brush position is then moved until the voltmeter connected to the series field reaches a null position. (The null position is reached when the voltmeter reaches its lowest point.)

If the compensating winding is used to set the neutral plane, alternating current is again connected to the armature and a voltmeter is connected to the compensating winding *(Figure 32-60)*. Alternating current is then applied to the armature. The brushes are then moved until the voltmeter indicates its highest or peak voltage.

## Speed Regulation

The speed regulation of the universal motor is very poor. Since this motor is a series motor, it has the same poor speed regulation as a DC series motor. If the universal motor is connected to a light load or no load, its speed is almost unlimited. It is not unusual for this motor to be operated at several thousand revolutions per minute. Universal motors are used in a number of portable appliances where high horsepower and light weight are needed, such as drill motors, skill saws, and vacuum cleaners. The universal motor is able to produce a high horsepower for its size and weight because of its high operating speed.

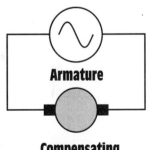

**Figure 32-60**  Using the compensating winding to set the brushes to the neutral plane position

## Changing the Direction of Rotation

The direction of rotation of the universal motor can be changed in the same manner as changing the direction of rotation of a DC series motor. To change the direction of rotation, change the armature leads with respect to the field leads.

## Summary

1. Not all single-phase motors operate on the principle of a rotating magnetic field.

2. Split-phase motors start as two-phase motors by producing an out-of-phase condition for the current in the run winding and the current in the start winding.

3. The resistance of the wire in the start winding of a resistance-start induction-run motor is used to produce a phase angle difference between the current in the start winding and the current in the run winding.

4. The capacitor-start induction-run motor uses an AC electrolytic capacitor to increase the phase angle difference between starting and running current. This causes an increase in starting torque.

5. Maximum starting torque for a split-phase motor is developed when the start winding current and run winding current are 90° out of phase with each other.

6. Most resistance-start induction run motors and capacitor-start induction-run motors use a centrifugal switch to disconnect the start windings when the motor reaches approximately 75% of full-load speed.

7. The capacitor-start capacitor-run motor operates like a two-phase motor because both the start and run windings remain energized during motor operation.

8. Most capacitor-start capacitor-run motors use an AC oil-filled capacitor connected in series with the start winding.

9. The capacitor of the capacitor-start capacitor-run motor does help to correct the power factor.

10. Shaded-pole induction motors operate on the principle of a rotating magnetic field.

11. The rotating magnetic field of a shaded-pole induction motor is produced by placing shading loops or coils on one side of the pole piece.

12. The synchronous field speed of a single-phase motor is determined by the number of stator poles and the frequency of the applied voltage.

13. Consequent pole motors are used when a change of motor speed is desired and high torque must be maintained.

14. Multispeed fan motors are constructed by connecting windings in series with the main run winding.

15. Multispeed fan motors have high-impedance stator windings to prevent them from overheating when their speed is reduced.

16. There are three basic repulsion type motors: the repulsion motor, the repulsion-start induction-run motor, and the repulsion-induction motor.

17. Repulsion motors have the highest starting torque of any single-phase motor.

18. The direction of rotation of repulsion motors is changed by setting the brushes 15° on either side of the pole pieces.

19. The direction of rotation for split-phase motors is changed by reversing the start winding in relation to the run winding.

20. Shaded-pole motors are generally considered to be nonreversible.

21. There are two types of repulsion-start induction-run motors: the brush-riding type and the brush-lifting type.

22. The brush-riding type of motor uses an axial commutator and a short-circuiting device, which short-circuits the commutator segments when the motor reaches approximately 75% of full-load speed.

23. The brush-lifting type of repulsion-start induction-run motor uses a radial commutator. A centrifugal device causes the brushes to move away from the commutator and a short-circuiting necklace to short-circuit the commutator when the motor reaches about 75% of full-load speed.

24. The repulsion-induction motor contains both a wound armature and squirrel-cage windings.

25. There are two types of single-phase synchronous motor: the Warren and the Holtz.

26. Single-phase synchronous motors are sometimes called hysteresis motors.

27. The Warren motor operates at a speed of 3600 RPM.

28. The Holtz motor operates at a speed of 1200 RPM.

29. Stepping motors generally operate on direct current and are used to produce angular movements in steps.

30. Stepping motors are generally used for position control.

31. Stepping motors can be used as synchronous motors when connected to two-phase alternating current.

32. Stepping motors operate at a speed of 72 RPM when connected to 60-Hz power.

33. Stepping motors can produce a holding torque when direct current is connected to their windings.

34. Universal motors operate on direct or alternating current.

35. Universal motors contain a wound armature and brushes.

36. Universal motors are also called AC series motors.

37. Universal motors have a compensating winding that helps overcome inductive reactance.

38. The direction of rotation for a universal motor can be changed by reversing the armature leads with respect to the field leads.

## Review Questions

1. What are the three basic types of split-phase motors?

2. The voltages of a two-phase system are how many degrees out of phase with each other?

3. How are the start and run windings of a split-phase motor connected in relation to each other?

4. In order to produce maximum starting torque in a split-phase motor, how many degrees out of phase should the start and run winding currents be with each other?

5. What is the advantage of the capacitor-start induction-run motor over the resistance-start induction-run motor?

6. On the average, how many degrees out of phase with each other are the start and run winding currents in a resistance-start induction-run motor?

7. What device is used to disconnect the start windings for the circuit in most nonhermetically sealed capacitor-start induction-run motors?

8. Why does a split-phase motor continue to operate after the start windings have been disconnected from the circuit?

9. How can the direction of rotation of a split-phase motor be reversed?

10. If a dual-voltage split-phase motor is to be operated on high voltage, how are the run windings connected in relation to each other?

11. When determining the direction of rotation for a split-phase motor, should you face the motor from the front or from the rear?

12. What type of split-phase motor does not generally contain a centrifugal switch?

13. What type of single-phase motor develops the highest starting torque?

14. What is the principle of operation of a repulsion motor?

15. What type of commutator is used with a brush-lifting type repulsion-start induction-run motor?

16. When a repulsion-start induction-run motor reaches about 75% of rated full-load speed, it stops operating as a repulsion motor and starts operating as a squirrel-cage motor. What must be done to cause the motor to begin operating as a squirrel-cage motor?

17. What is the principle of operation of a capacitor-start capacitor-run motor?

18. What causes the magnetic field to rotate in a shaded-pole induction motor?

19. How can the direction of rotation of a shaded-pole induction motor be changed?

20. How is the speed of a consequent pole motor changed?

21. Why can a multispeed fan motor be operated at lower speed than most induction motors without harm to the motor windings?

22. What is the speed of operation of the Warren motor?

23. What is the speed of operation of the Holtz motor?

24. Explain the difference in operation between a stepping motor and a common DC motor.

25. What is the principle of operation of a stepping motor?

26. What does the term *bifilar* mean?

27. Why do stepping motors have teeth machined in the stator poles and rotor?

28. When a stepping motor is connected to AC power, how many phases must be applied to the motor?

29. What is the synchronous speed of an eight-pole stepping motor when connected to a two-phase 60-Hz AC line?

30. How can the holding torque of a stepping motor be increased?

31. Why is the AC series motor often referred to as a universal motor?

32. What is the function of the compensating winding?

33. How is the direction of rotation of the universal motor reversed?

34. When the motor is connected to DC voltage, how must the compensating winding be connected?

35. Explain how to set the neutral plane position of the brushes using the series field.

36. Explain how to set the neutral plane position using the compensating winding.

# Trigonometric Functions— Natural Sines and Cosines

NOTE: For cosines, use right-hand column of degrees and lower line of tenths.

| Deg. | °0.0 | °0.1 | °0.2 | °0.3 | °0.4 | °0.5 | °0.6 | °0.7 | °0.8 | °0.9 | °1.0 | |
|------|------|------|------|------|------|------|------|------|------|------|------|----|
| 0° | 0.0000 | 0.0017 | 0.0035 | 0.0052 | 0.0070 | 0.0087 | 0.0105 | 0.0122 | 0.0140 | 0.0157 | 0.0175 | 89 |
| 1 | 0.0175 | 0.0192 | 0.0209 | 0.0227 | 0.0244 | 0.0262 | 0.0279 | 0.0297 | 0.0314 | 0.0332 | 0.0349 | 88 |
| 2 | 0.0349 | 0.0366 | 0.0384 | 0.0401 | 0.0419 | 0.0436 | 0.0454 | 0.0471 | 0.0488 | 0.0506 | 0.0523 | 87 |
| 3 | 0.0523 | 0.0541 | 0.0558 | 0.0576 | 0.0593 | 0.0610 | 0.0628 | 0.0645 | 0.0663 | 0.0680 | 0.0698 | 86 |
| 4 | 0.0698 | 0.0715 | 0.0732 | 0.0750 | 0.0767 | 0.0785 | 0.0802 | 0.0819 | 0.0837 | 0.0854 | 0.0872 | 85 |
| 5 | 0.0872 | 0.0889 | 0.0906 | 0.0924 | 0.0941 | 0.0958 | 0.0976 | 0.0993 | 0.1011 | 0.1028 | 0.1045 | 84 |
| 6 | 0.1045 | 0.1063 | 0.1080 | 0.1097 | 0.1115 | 0.1132 | 0.1149 | 0.1167 | 0.1184 | 0.1201 | 0.1219 | 83 |
| 7 | 0.1219 | 0.1236 | 0.1253 | 0.1271 | 0.1288 | 0.1305 | 0.1323 | 0.1340 | 0.1357 | 0.1374 | 0.1392 | 82 |
| 8 | 0.1392 | 0.1409 | 0.1426 | 0.1444 | 0.1461 | 0.1478 | 0.1495 | 0.1513 | 0.1530 | 0.1547 | 0.1564 | 81 |
| 9 | 0.1564 | 0.1582 | 0.1599 | 0.1616 | 0.1633 | 0.1650 | 0.1668 | 0.1685 | 0.1702 | 0.1719 | 0.1736 | 80° |
| 10° | 0.1736 | 0.1754 | 0.1771 | 0.1788 | 0.1805 | 0.1822 | 0.1840 | 0.1857 | 0.1874 | 0.1891 | 0.1908 | 79 |
| 11 | 0.1908 | 0.1925 | 0.1942 | 0.1959 | 0.1977 | 0.1994 | 0.2011 | 0.2028 | 0.2045 | 0.2062 | 0.2079 | 78 |
| 12 | 0.2079 | 0.2096 | 0.2113 | 0.2130 | 0.2147 | 0.2164 | 0.2181 | 0.2198 | 0.2215 | 0.2232 | 0.2250 | 77 |
| 13 | 0.2250 | 0.2267 | 0.2284 | 0.2300 | 0.2317 | 0.2334 | 0.2351 | 0.2368 | 0.2385 | 0.2402 | 0.2419 | 76 |
| 14 | 0.2419 | 0.2436 | 0.2453 | 0.2470 | 0.2487 | 0.2504 | 0.2521 | 0.2538 | 0.2554 | 0.2571 | 0.2588 | 75 |
| 15 | 0.2588 | 0.2605 | 0.2622 | 0.2639 | 0.2656 | 0.2672 | 0.2689 | 0.2706 | 0.2723 | 0.2740 | 0.2756 | 74 |
| 16 | 0.2756 | 0.2773 | 0.2790 | 0.2807 | 0.2823 | 0.2840 | 0.2857 | 0.2874 | 0.2890 | 0.2907 | 0.2924 | 73 |
| 17 | 0.2924 | 0.2940 | 0.2957 | 0.2974 | 0.2990 | 0.3007 | 0.3024 | 0.3040 | 0.3057 | 0.3074 | 0.3090 | 72 |
| 18 | 0.3090 | 0.3107 | 0.3123 | 0.3140 | 0.3156 | 0.3173 | 0.3190 | 0.3206 | 0.3223 | 0.3239 | 0.3256 | 71 |
| 19 | 0.3256 | 0.3272 | 0.3289 | 0.3305 | 0.3322 | 0.3338 | 0.3355 | 0.3371 | 0.3387 | 0.3404 | 0.3420 | 70° |
| 20° | 0.3420 | 0.3437 | 0.3453 | 0.3469 | 0.3486 | 0.3502 | 0.3518 | 0.3535 | 0.3551 | 0.3567 | 0.3584 | 69 |
| 21 | 0.3584 | 0.3600 | 0.3616 | 0.3633 | 0.3649 | 0.3665 | 0.3681 | 0.3697 | 0.3714 | 0.3730 | 0.3746 | 68 |
| 22 | 0.3746 | 0.3762 | 0.3778 | 0.3795 | 0.3811 | 0.3827 | 0.3843 | 0.3859 | 0.3875 | 0.3891 | 0.3907 | 67 |
| 23 | 0.3907 | 0.3923 | 0.3939 | 0.3955 | 0.3971 | 0.3987 | 0.4003 | 0.4019 | 0.4035 | 0.4051 | 0.4067 | 66 |
| 24 | 0.4067 | 0.4083 | 0.4099 | 0.4115 | 0.4131 | 0.4147 | 0.4163 | 0.4179 | 0.4195 | 0.4210 | 0.4226 | 65 |
| 25 | 0.4226 | 0.4242 | 0.4258 | 0.4274 | 0.4289 | 0.4305 | 0.4321 | 0.4337 | 0.4352 | 0.4368 | 0.4384 | 64 |
| 26 | 0.4384 | 0.4399 | 0.4415 | 0.4431 | 0.4446 | 0.4462 | 0.4478 | 0.4493 | 0.4509 | 0.4524 | 0.4540 | 63 |
| 27 | 0.4540 | 0.4555 | 0.4571 | 0.4586 | 0.4602 | 0.4617 | 0.4633 | 0.4648 | 0.4664 | 0.4679 | 0.4695 | 62 |
| 28 | 0.4695 | 0.4710 | 0.4726 | 0.4741 | 0.4756 | 0.4772 | 0.4787 | 0.4802 | 0.4818 | 0.4833 | 0.4848 | 61 |
| 29 | 0.4848 | 0.4863 | 0.4879 | 0.4894 | 0.4909 | 0.4924 | 0.4939 | 0.4955 | 0.4970 | 0.4985 | 0.5000 | 60° |
| 30° | 0.5000 | 0.5015 | 0.5030 | 0.5045 | 0.5060 | 0.5075 | 0.5090 | 0.5105 | 0.5120 | 0.5135 | 0.5150 | 59 |
| 31 | 0.5150 | 0.5165 | 0.5180 | 0.5195 | 0.5210 | 0.5225 | 0.5240 | 0.5255 | 0.5270 | 0.5284 | 0.5299 | 58 |
| 32 | 0.5299 | 0.5314 | 0.5329 | 0.5344 | 0.5358 | 0.5373 | 0.5388 | 0.5402 | 0.5417 | 0.5432 | 0.5446 | 57 |
| 33 | 0.5446 | 0.5461 | 0.5476 | 0.5490 | 0.5505 | 0.5519 | 0.5534 | 0.5548 | 0.5563 | 0.5577 | 0.5592 | 56 |
| 34 | 0.5592 | 0.5606 | 0.5621 | 0.5635 | 0.5650 | 0.5664 | 0.5678 | 0.5693 | 0.5707 | 0.5721 | 0.5736 | 55 |
| 35 | 0.5736 | 0.5750 | 0.5764 | 0.5779 | 0.5793 | 0.5807 | 0.5821 | 0.5835 | 0.5850 | 0.5864 | 0.5878 | 54 |
| 36 | 0.5878 | 0.5892 | 0.5906 | 0.5920 | 0.5934 | 0.5948 | 0.5962 | 0.5976 | 0.5990 | 0.6004 | 0.6018 | 53 |
| 37 | 0.6018 | 0.6032 | 0.6046 | 0.6060 | 0.6074 | 0.6088 | 0.6101 | 0.6115 | 0.6129 | 0.6143 | 0.6157 | 52 |
| 38 | 0.6157 | 0.6170 | 0.6184 | 0.6198 | 0.6211 | 0.6225 | 0.6239 | 0.6252 | 0.6266 | 0.6280 | 0.6293 | 51 |
| 39 | 0.6293 | 0.6307 | 0.6320 | 0.6334 | 0.6347 | 0.6361 | 0.6374 | 0.6388 | 0.6401 | 0.6414 | 0.6428 | 50° |
| 40° | 0.6428 | 0.6441 | 0.6455 | 0.6468 | 0.6481 | 0.6494 | 0.6508 | 0.6521 | 0.6534 | 0.6547 | 0.6561 | 49 |
| 41 | 0.6561 | 0.6574 | 0.6587 | 0.6600 | 0.6613 | 0.6626 | 0.6639 | 0.6652 | 0.6665 | 0.6678 | 0.6691 | 48 |
| 42 | 0.6691 | 0.6704 | 0.6717 | 0.6730 | 0.6743 | 0.6756 | 0.6769 | 0.6782 | 0.6794 | 0.6807 | 0.6820 | 47 |
| 43 | 0.6820 | 0.6833 | 0.6845 | 0.6858 | 0.6871 | 0.6884 | 0.6896 | 0.6909 | 0.6921 | 0.6934 | 0.6947 | 46 |
| 44 | 0.6947 | 0.6959 | 0.6972 | 0.6984 | 0.6997 | 0.7009 | 0.7022 | 0.7034 | 0.7046 | 0.7059 | 0.7071 | 45 |
| | °1.0 | °0.9 | °0.8 | °0.7 | °0.6 | °0.5 | °0.4 | °0.3 | °0.2 | °0.1 | °0.0 | Deg. |

NOTE: For cosines, use right-hand column of degrees and lower line of tenths.

| Deg. | °0.0 | °0.1 | °0.2 | °0.3 | °0.4 | °0.5 | °0.6 | °0.7 | °0.8 | °0.9 | °1.0 | |
|---|---|---|---|---|---|---|---|---|---|---|---|---|
| 45 | 0.7071 | 0.7083 | 0.7096 | 0.7108 | 0.7120 | 0.7133 | 0.7145 | 0.7157 | 0.7169 | 0.7181 | 0.7193 | 44 |
| 46 | 0.7193 | 0.7206 | 0.7218 | 0.7230 | 0.7242 | 0.7254 | 0.7266 | 0.7278 | 0.7290 | 0.7302 | 0.7314 | 43 |
| 47 | 0.7314 | 0.7325 | 0.7337 | 0.7349 | 0.7361 | 0.7373 | 0.7385 | 0.7396 | 0.7408 | 0.7420 | 0.7431 | 42 |
| 48 | 0.7431 | 0.7443 | 0.7455 | 0.7466 | 0.7478 | 0.7490 | 0.7501 | 0.7513 | 0.7524 | 0.7536 | 0.7547 | 41 |
| 49 | 0.7547 | 0.7559 | 0.7570 | 0.7581 | 0.7593 | 0.7604 | 0.7615 | 0.7627 | 0.7638 | 0.7649 | 0.7660 | 40° |
| 50° | 0.7660 | 0.7672 | 0.7683 | 0.7694 | 0.7705 | 0.7716 | 9.7727 | 0.7738 | 0.7749 | 0.7760 | 0.7771 | 39 |
| 51 | 0.7771 | 0.7782 | 0.7793 | 0.7804 | 0.7815 | 0.7826 | 0.7837 | 0.7848 | 0.7859 | 0.7869 | 0.7880 | 38 |
| 52 | 0.7880 | 0.7891 | 0.7902 | 0.7912 | 0.7923 | 0.7934 | 0.7944 | 0.7955 | 0.7965 | 0.7976 | 0.7986 | 37 |
| 53 | 0.7986 | 0.7997 | 0.8007 | 0.8018 | 0.8028 | 0.8039 | 0.8049 | 0.8059 | 0.8070 | 0.8080 | 0.8090 | 36 |
| 54 | 0.8090 | 0.8100 | 0.8111 | 0.8121 | 0.8131 | 0.8141 | 0.8151 | 0.8161 | 0.8171 | 0.8181 | 0.8192 | 35 |
| 55 | 0.8192 | 0.8202 | 0.8211 | 0.8221 | 0.8231 | 0.8241 | 0.8251 | 0.8261 | 0.8271 | 0.8281 | 0.8290 | 34 |
| 56 | 0.8290 | 0.8300 | 0.8310 | 0.8320 | 0.8329 | 0.8339 | 0.8348 | 0.8358 | 0.8368 | 0.8377 | 0.8387 | 33 |
| 57 | 0.8387 | 0.8396 | 0.8406 | 0.8415 | 0.8425 | 0.8434 | 0.8443 | 0.8453 | 0.8462 | 0.8471 | 0.8480 | 32 |
| 58 | 0.8480 | 0.8490 | 0.8499 | 0.8508 | 0.8517 | 0.8526 | 0.8536 | 0.8545 | 0.8554 | 0.8563 | 0.8572 | 31 |
| 59 | 0.8572 | 0.8581 | 0.8590 | 0.8599 | 0.8607 | 0.8616 | 0.8625 | 0.8634 | 0.8643 | 0.8652 | 0.8660 | 30° |
| 60° | 0.8660 | 0.8669 | 0.8678 | 0.8686 | 0.8695 | 0.8704 | 0.8712 | 0.8721 | 0.8729 | 0.8738 | 0.8746 | 29 |
| 61 | 0.8746 | 0.8755 | 0.8763 | 0.8771 | 0.8780 | 0.8788 | 0.8796 | 0.8805 | 0.8813 | 0.8821 | 0.8829 | 28 |
| 62 | 0.8829 | 0.8838 | 0.8846 | 0.8854 | 0.8862 | 0.8870 | 0.8878 | 0.8886 | 0.8894 | 0.8902 | 0.8910 | 27 |
| 63 | 0.8910 | 0.8918 | 0.8926 | 0.8934 | 0.8942 | 0.8949 | 0.8957 | 0.8965 | 0.8973 | 0.8980 | 0.8988 | 26 |
| 64 | 0.8988 | 0.8996 | 0.9003 | 0.9011 | 0.9018 | 0.9026 | 0.9033 | 0.9041 | 0.9048 | 0.9056 | 0.9063 | 25 |
| 65 | 0.9063 | 0.9070 | 0.9078 | 0.9085 | 0.9092 | 0.9100 | 0.9107 | 0.9114 | 0.9121 | 0.9128 | 0.9135 | 24 |
| 66 | 0.9135 | 0.9143 | 0.9150 | 0.9157 | 0.9164 | 0.9171 | 0.9178 | 0.9184 | 0.9191 | 0.9198 | 0.9205 | 23 |
| 67 | 0.9205 | 0.9212 | 0.9219 | 0.9225 | 0.9232 | 0.9239 | 0.9245 | 0.9252 | 0.9259 | 0.9265 | 0.9272 | 22 |
| 68 | 0.9272 | 0.9278 | 0.9285 | 0.9291 | 0.9298 | 0.9304 | 0.9311 | 0.9317 | 0.9323 | 0.9330 | 0.9336 | 21 |
| 69 | 0.9336 | 0.9342 | 0.9348 | 0.9354 | 0.9361 | 0.9367 | 0.9373 | 0.9379 | 0.9385 | 0.9391 | 0.9397 | 20° |
| 70° | 0.9397 | 0.9403 | 0.9409 | 0.9415 | 0.9421 | 0.9426 | 0.9432 | 0.9438 | 0.9444 | 0.9449 | 0.9455 | 19 |
| 71 | 0.9455 | 0.9461 | 0.9466 | 0.9472 | 0.9478 | 0.9483 | 0.9489 | 0.9494 | 0.9500 | 0.9505 | 0.9511 | 18 |
| 72 | 0.9511 | 0.9516 | 0.9521 | 0.9527 | 0.9532 | 0.9537 | 0.9542 | 0.9548 | 0.9553 | 0.9558 | 0.9563 | 17 |
| 73 | 0.9563 | 0.9568 | 0.9573 | 0.9578 | 0.9583 | 0.9588 | 0.9593 | 0.9598 | 0.9603 | 0.9608 | 0.9613 | 16 |
| 74 | 0.9613 | 0.9617 | 0.9622 | 0.9627 | 0.9632 | 0.9636 | 0.9641 | 0.9646 | 0.9650 | 0.9655 | 0.9659 | 15 |
| 75 | 0.9659 | 0.9664 | 0.9668 | 0.9673 | 0.9677 | 0.9681 | 0.9686 | 0.9690 | 0.9694 | 0.9699 | 0.9703 | 14 |
| 76 | 0.9703 | 0.9707 | 0.9711 | 0.9715 | 0.9720 | 0.9724 | 0.9728 | 0.9732 | 0.9736 | 0.9740 | 0.9744 | 13 |
| 77 | 0.9744 | 0.9748 | 0.9751 | 0.9755 | 0.9759 | 0.9763 | 0.9767 | 0.9770 | 0.9774 | 0.9778 | 0.9781 | 12 |
| 78 | 0.9781 | 0.9785 | 0.9789 | 0.9792 | 0.9796 | 0.9799 | 0.9803 | 0.9806 | 0.9810 | 0.9813 | 0.9816 | 11 |
| 79 | 0.9816 | 0.9820 | 0.9823 | 0.9826 | 0.9829 | 0.9833 | 0.9836 | 0.9839 | 0.9842 | 0.9845 | 0.9848 | 10° |
| 80° | 0.9848 | 0.9851 | 0.9854 | 0.9857 | 0.9860 | 0.9863 | 0.9866 | 0.9869 | 0.9871 | 0.9874 | 0.9877 | 9 |
| 81 | 0.9877 | 0.9880 | 0.9882 | 0.9885 | 0.9888 | 0.9890 | 0.9893 | 0.9895 | 0.9898 | 0.9900 | 0.9903 | 8 |
| 82 | 0.9903 | 0.9905 | 0.9907 | 0.9910 | 0.9912 | 0.9914 | 0.9917 | 0.9919 | 0.9921 | 0.9923 | 0.9925 | 7 |
| 83 | 0.9925 | 0.9928 | 0.9930 | 0.9932 | 0.9934 | 0.9936 | 0.9938 | 0.9940 | 0.9942 | 0.9943 | 0.9945 | 6 |
| 84 | 0.9945 | 0.9947 | 0.9949 | 0.9951 | 0.9952 | 0.9954 | 0.9956 | 0.9957 | 0.9959 | 0.9960 | 0.9962 | 5 |
| 85 | 0.9962 | 0.9963 | 0.9965 | 0.9966 | 0.9968 | 0.9969 | 0.9971 | 0.9972 | 0.9973 | 0.9974 | 0.9976 | 4 |
| 86 | 0.9976 | 0.9977 | 0.9978 | 0.9979 | 0.9980 | 0.9981 | 0.9982 | 0.9983 | 0.9984 | 0.9985 | 0.9986 | 3 |
| 87 | 0.9986 | 0.9987 | 0.9988 | 0.9989 | 0.9990 | 0.9990 | 0.9991 | 0.9992 | 0.9993 | 0.9993 | 0.9994 | 2 |
| 88 | 0.9994 | 0.9995 | 0.9995 | 0.9996 | 0.9996 | 0.9997 | 0.9997 | 0.9997 | 0.9998 | 0.9998 | 0.9998 | 1 |
| 89 | 0.9998 | 0.9999 | 0.9999 | 0.9999 | 0.9999 | 1.0000 | 1.0000 | 1.0000 | 1.0000 | 1.0000 | 1.0000 | 0° |
| | °1.0 | °0.9 | °0.8 | °0.7 | °0.6 | °0.5 | °0.4 | °0.3 | °0.2 | °0.1 | °0.0 | Deg. |

# Trigonometric Functions— Natural Tangents and Cotangents

NOTE: For cotangents, use right-hand column of degrees and lower line of tenths.

| Deg. | °0.0 | °0.1 | °0.2 | °0.3 | °0.4 | °0.5 | °0.6 | °0.7 | °0.8 | °0.9 | °1.0 | |
|------|------|------|------|------|------|------|------|------|------|------|------|------|
| 0° | 0.0000 | 0.0017 | 0.0035 | 0.0052 | 0.0070 | 0.0087 | 0.0105 | 0.0122 | 0.0140 | 0.0157 | 0.0175 | 89 |
| 1 | 0.0175 | 0.0192 | 0.0209 | 0.0227 | 0.0244 | 0.0262 | 0.0279 | 0.0297 | 0.0314 | 0.0332 | 0.0349 | 88 |
| 2 | 0.0349 | 0.0367 | 0.0384 | 0.0402 | 0.0419 | 0.0437 | 0.0454 | 0.0472 | 0.0489 | 0.0507 | 0.0524 | 87 |
| 3 | 0.0524 | 0.0542 | 0.0559 | 0.0577 | 0.0594 | 0.0612 | 0.0629 | 0.0647 | 0.0664 | 0.0682 | 0.0699 | 86 |
| 4 | 0.0699 | 0.0717 | 0.0734 | 0.0752 | 0.0769 | 0.0787 | 0.0805 | 0.0822 | 0.0840 | 0.0857 | 0.0875 | 85 |
| 5 | 0.0875 | 0.0892 | 0.0910 | 0.0928 | 0.0945 | 0.0963 | 0.0981 | 0.0998 | 0.1016 | 0.1033 | 0.1051 | 84 |
| 6 | 0.1051 | 0.1069 | 0.1086 | 0.1104 | 0.1122 | 0.1139 | 0.1157 | 0.1175 | 0.1192 | 0.1210 | 0.1228 | 83 |
| 7 | 0.1228 | 0.1246 | 0.1263 | 0.1281 | 0.1299 | 0.1317 | 0.1334 | 0.1352 | 0.1370 | 0.1388 | 0.1405 | 82 |
| 8 | 0.1405 | 0.1423 | 0.1441 | 0.1459 | 0.1477 | 0.1495 | 0.1512 | 0.1530 | 0.1548 | 0.1566 | 0.1584 | 81 |
| 9 | 0.1584 | 0.1602 | 0.1620 | 0.1638 | 0.1655 | 0.1673 | 0.1691 | 0.1709 | 0.1727 | 0.1745 | 0.1763 | 80° |
| 10° | 0.1763 | 0.1781 | 0.1799 | 0.1817 | 0.1835 | 0.1853 | 0.1871 | 0.1890 | 0.1908 | 0.1926 | 0.1944 | 79 |
| 11 | 0.1944 | 0.1962 | 0.1980 | 0.1998 | 0.2016 | 0.2035 | 0.2053 | 0.2071 | 0.2089 | 0.2107 | 0.2126 | 78 |
| 12 | 0.2126 | 0.2144 | 0.2162 | 0.2180 | 0.2199 | 0.2217 | 0.2235 | 0.2254 | 0.2272 | 0.2290 | 0.2309 | 77 |
| 13 | 0.2309 | 0.2327 | 0.2345 | 0.2364 | 0.2382 | 0.2401 | 0.2419 | 0.2438 | 0.2456 | 0.2475 | 0.2493 | 76 |
| 14 | 0.2493 | 0.2512 | 0.2530 | 0.2549 | 0.2568 | 0.2586 | 0.2605 | 0.2623 | 0.2642 | 0.2661 | 0.2679 | 75 |
| 15 | 0.2679 | 0.2698 | 0.2717 | 0.2736 | 0.2754 | 0.2773 | 0.2792 | 0.2811 | 0.2830 | 0.2849 | 0.2867 | 74 |
| 16 | 0.2867 | 0.2886 | 0.2905 | 0.2924 | 0.2943 | 0.2962 | 0.2981 | 0.3000 | 0.3019 | 0.3038 | 0.3057 | 73 |
| 17 | 0.3057 | 0.3076 | 0.3096 | 0.3115 | 0.3134 | 0.3153 | 0.3172 | 0.3191 | 0.3211 | 0.3230 | 0.3249 | 72 |
| 18 | 0.3249 | 0.3269 | 0.3288 | 0.3307 | 0.3327 | 0.3346 | 0.3365 | 0.3385 | 0.3404 | 0.3424 | 0.3443 | 71 |
| 19 | 0.3443 | 0.3463 | 0.3482 | 0.3502 | 0.3522 | 0.3541 | 0.3561 | 0.3581 | 0.3600 | 0.3620 | 0.3640 | 70° |
| 20° | 0.3640 | 0.3659 | 0.3679 | 0.3699 | 0.3719 | 0.3739 | 0.3759 | 0.3779 | 0.3799 | 0.3819 | 0.3839 | 69 |
| 21 | 0.3839 | 0.3859 | 0.3879 | 0.3899 | 0.3919 | 0.3939 | 0.3959 | 0.3979 | 0.4000 | 0.4020 | 0.4040 | 68 |
| 22 | 0.4040 | 0.4061 | 0.4081 | 0.4101 | 0.4122 | 0.4142 | 0.4163 | 0.4183 | 0.4204 | 0.4224 | 0.4245 | 67 |
| 23 | 0.4245 | 0.4265 | 0.4286 | 0.4307 | 0.4327 | 0.4348 | 0.4369 | 0.4390 | 0.4411 | 0.4431 | 0.4452 | 66 |
| 24 | 0.4452 | 0.4473 | 0.4494 | 0.4515 | 0.4536 | 0.4557 | 0.4578 | 0.4599 | 0.4621 | 0.4642 | 0.4663 | 65 |
| 25 | 0.4663 | 0.4684 | 0.4706 | 0.4727 | 0.4748 | 0.4770 | 0.4791 | 0.4813 | 0.4834 | 0.4856 | 0.4877 | 64 |
| 26 | 0.4877 | 0.4899 | 0.4921 | 0.4942 | 0.4964 | 0.4986 | 0.5008 | 0.5029 | 0.5051 | 0.5073 | 0.5095 | 63 |
| 27 | 0.5095 | 0.5117 | 0.5139 | 0.5161 | 0.5184 | 0.5206 | 0.5228 | 0.5250 | 0.5272 | 0.5295 | 0.5317 | 62 |
| 28 | 0.5317 | 0.5340 | 0.5362 | 0.5384 | 0.5407 | 0.5430 | 0.5452 | 0.5475 | 0.5498 | 0.5520 | 0.5543 | 61 |
| 29 | 0.5543 | 0.5566 | 0.5589 | 0.5612 | 0.5635 | 0.5658 | 0.5681 | 0.5704 | 0.5727 | 0.5750 | 0.5774 | 60° |
| 30° | 0.5774 | 0.5797 | 0.5820 | 0.5844 | 0.5867 | 0.5890 | 0.5914 | 0.5938 | 0.5961 | 0.5985 | 0.6009 | 59 |
| 31 | 0.6009 | 0.6032 | 0.6056 | 0.6080 | 0.6104 | 0.6128 | 0.6152 | 0.6176 | 0.6200 | 0.6224 | 0.6249 | 58 |
| 32 | 0.6249 | 0.6273 | 0.6297 | 0.6322 | 0.6346 | 0.6371 | 0.6395 | 0.6420 | 0.6445 | 0.6469 | 0.6494 | 57 |
| 33 | 0.6494 | 0.6519 | 0.6544 | 0.6569 | 0.6594 | 0.6619 | 0.6644 | 0.6669 | 0.6694 | 0.6720 | 0.6745 | 56 |
| 34 | 0.6745 | 0.6771 | 0.6796 | 0.6822 | 0.6847 | 0.6873 | 0.6899 | 0.6924 | 0.6950 | 0.6976 | 0.7002 | 55 |
| 35 | 0.7002 | 0.7028 | 0.7054 | 0.7080 | 0.7107 | 0.7133 | 0.7159 | 0.7186 | 0.7212 | 0.7239 | 0.7265 | 54 |
| 36 | 0.7265 | 0.7292 | 0.7319 | 0.7346 | 0.7373 | 0.7400 | 0.7427 | 0.7454 | 0.7481 | 0.7508 | 0.7536 | 53 |
| 37 | 0.7536 | 0.7563 | 0.7590 | 0.7618 | 0.7646 | 0.7673 | 0.7701 | 0.7729 | 0.7757 | 0.7785 | 0.7813 | 52 |
| 38 | 0.7813 | 0.7841 | 0.7869 | 0.7898 | 0.7926 | 0.7954 | 0.7983 | 0.8012 | 0.8040 | 0.8069 | 0.8098 | 51 |
| 39 | 0.8098 | 0.8127 | 0.8156 | 0.8185 | 0.8214 | 0.8243 | 0.8273 | 0.8302 | 0.8332 | 0.8361 | 0.8391 | 50° |
| 40° | 0.8391 | 0.8421 | 0.8451 | 0.8481 | 0.8511 | 0.8541 | 0.8571 | 0.8601 | 0.8632 | 0.8662 | 0.8693 | 49 |
| 41 | 0.8693 | 0.8724 | 0.8754 | 0.8785 | 0.8816 | 0.8847 | 0.8878 | 0.8910 | 0.8941 | 0.8972 | 0.9004 | 48 |
| 42 | 0.9004 | 0.9036 | 0.9067 | 0.9099 | 0.9131 | 0.9163 | 0.9195 | 0.9228 | 0.9260 | 0.9293 | 0.9325 | 47 |
| 43 | 0.9325 | 0.9358 | 0.9391 | 0.9424 | 0.9457 | 0.9490 | 0.9523 | 0.9556 | 0.9590 | 0.9623 | 0.9657 | 46 |
| 44 | 0.9657 | 0.9691 | 0.9725 | 0.9759 | 0.9793 | 0.9827 | 0.9861 | 0.9896 | 0.9930 | 0.9965 | 1.0000 | 45 |
| | °1.0 | °0.9 | °0.8 | °0.7 | °0.6 | °0.5 | °0.4 | °0.3 | °0.2 | °0.1 | °0.0 | Deg. |

NOTE: For cotangents, use right-hand column of degrees and lower line of tenths.

| Deg. | °0.0 | °0.1 | °0.2 | °0.3 | °0.4 | °0.5 | °0.6 | °0.7 | °0.8 | °0.9 | °1.0 | |
|---|---|---|---|---|---|---|---|---|---|---|---|---|
| 45 | 1.0000 | 1.0035 | 1.0070 | 1.0105 | 1.0141 | 1.0176 | 1.0212 | 1.0247 | 1.0283 | 1.0319 | 1.0355 | 44 |
| 46 | 1.0355 | 1.0392 | 1.0428 | 1.0464 | 1.0501 | 1.0538 | 1.0575 | 1.0612 | 1.0649 | 1.0686 | 1.0724 | 43 |
| 47 | 1.0724 | 1.0761 | 1.0799 | 1.0837 | 1.0875 | 1.0913 | 1.0951 | 1.0990 | 1.1028 | 1.1067 | 1.1106 | 42 |
| 48 | 1.1106 | 1.1145 | 1.1184 | 1.1224 | 1.1263 | 1.1303 | 1.1343 | 1.1383 | 1.1423 | 1.1463 | 1.1504 | 41 |
| 49 | 1.1504 | 1.1544 | 1.1585 | 1.1626 | 1.1667 | 1.1708 | 1.1750 | 1.1792 | 1.1833 | 1.1875 | 1.1918 | 40° |
| 50° | 1.1918 | 1.1960 | 1.2002 | 1.2045 | 1.2088 | 1.2131 | 1.2174 | 1.2218 | 1.2261 | 1.2305 | 1.2349 | 39 |
| 51 | 1.2349 | 1.2393 | 1.2437 | 1.2482 | 1.2527 | 1.2572 | 1.2617 | 1.2662 | 1.2708 | 1.2753 | 1.2799 | 38 |
| 52 | 1.2799 | 1.2846 | 1.2892 | 1.2938 | 1.2985 | 1.3032 | 1.3079 | 1.3127 | 1.3175 | 1.3222 | 1.3270 | 37 |
| 53 | 1.3270 | 1.3319 | 1.3367 | 1.3416 | 1.3465 | 1.3514 | 1.3564 | 1.3613 | 1.3663 | 1.3713 | 1.3764 | 36 |
| 54 | 1.3764 | 1.3814 | 1.3865 | 1.3916 | 1.3968 | 1.4019 | 1.4071 | 1.4124 | 1.4176 | 1.4229 | 1.4281 | 35 |
| 55 | 1.4281 | 1.4335 | 1.4388 | 1.4442 | 1.4496 | 1.4550 | 1.4605 | 1.4659 | 1.4715 | 1.4770 | 1.4826 | 34 |
| 56 | 1.4826 | 1.4882 | 1.4938 | 1.4994 | 1.5051 | 1.5108 | 1.5166 | 1.5224 | 1.5282 | 1.5340 | 1.5399 | 33 |
| 57 | 1.5399 | 1.5458 | 1.5517 | 1.5577 | 1.5637 | 1.5697 | 1.5757 | 1.5818 | 1.5880 | 1.5941 | 1.6003 | 32 |
| 58 | 1.6003 | 1.6066 | 1.6128 | 1.6191 | 1.6255 | 1.6319 | 1.6383 | 1.6447 | 1.6512 | 1.6577 | 1.6643 | 31 |
| 59 | 1.6643 | 1.6709 | 1.6775 | 1.6842 | 1.6909 | 1.6977 | 1.7045 | 1.7113 | 1.7182 | 1.7251 | 1.7321 | 30° |
| 60° | 1.7321 | 1.7391 | 1.7461 | 1.7532 | 1.7603 | 1.7675 | 1.7747 | 1.7820 | 1.7893 | 1.7966 | 1.8040 | 29 |
| 61 | 1.8040 | 1.8115 | 1.8190 | 1.8265 | 1.8341 | 1.8418 | 1.8495 | 1.8572 | 1.8650 | 1.8728 | 1.8807 | 28 |
| 62 | 1.8807 | 1.8887 | 1.8967 | 1.9047 | 1.9128 | 1.9210 | 1.9292 | 1.9375 | 1.9458 | 1.9542 | 1.9626 | 27 |
| 63 | 1.9626 | 1.9711 | 1.9797 | 1.9883 | 1.9970 | 2.0057 | 2.0145 | 2.0233 | 2.0323 | 2.0413 | 2.0503 | 26 |
| 64 | 2.0503 | 2.0594 | 2.0686 | 2.0778 | 2.0872 | 2.0965 | 2.1060 | 2.1155 | 2.1251 | 2.1348 | 2.1445 | 25 |
| 65 | 2.1445 | 2.1543 | 2.1642 | 2.1742 | 2.1842 | 2.1943 | 2.2045 | 2.2148 | 2.2251 | 2.2355 | 2.2460 | 24 |
| 66 | 2.2460 | 2.2566 | 2.2673 | 2.2781 | 2.2889 | 2.2998 | 2.3109 | 2.3220 | 2.3332 | 2.3445 | 2.3559 | 23 |
| 67 | 2.3559 | 2.3673 | 2.3789 | 2.3906 | 2.4023 | 2.4142 | 2.4262 | 2.4383 | 2.4504 | 2.4627 | 2.4751 | 22 |
| 68 | 2.4751 | 2.4876 | 2.5002 | 2.5129 | 2.5257 | 2.5386 | 2.5517 | 2.5649 | 2.5782 | 2.5916 | 2.6051 | 21 |
| 69 | 2.6051 | 2.6187 | 2.6325 | 2.6464 | 2.6605 | 2.6746 | 2.6889 | 2.7034 | 2.7179 | 2.7326 | 2.7475 | 20° |
| 70° | 2.7475 | 2.7625 | 2.7776 | 2.7929 | 2.8083 | 2.8239 | 2.8397 | 2.8556 | 2,8716 | 2.8878 | 2.9042 | 19 |
| 71 | 2.9042 | 2.9208 | 2.9375 | 2.9544 | 2.9714 | 2.9887 | 3.0061 | 3.0237 | 3.0415 | 3.0595 | 3.0777 | 18 |
| 72 | 3.0777 | 3.0961 | 3.1146 | 3.1334 | 3.1524 | 3.1716 | 3.1910 | 3.2106 | 3.2305 | 3.2506 | 3.2709 | 17 |
| 73 | 3.2709 | 3.2914 | 3.3122 | 3.3332 | 3.3544 | 3.3759 | 3.3977 | 3.4197 | 3.4420 | 3.4646 | 3.4874 | 16 |
| 74 | 3.4874 | 3.5105 | 3.5339 | 3.5576 | 3.5816 | 3.6059 | 3.6305 | 3.6554 | 3.6806 | 3.7062 | 3.7321 | 15 |
| 75 | 3.7321 | 3.7583 | 3.7848 | 3.8118 | 3.8391 | 3.8667 | 3.8947 | 3.9232 | 3.9520 | 3.9812 | 4.0108 | 14 |
| 76 | 4.0108 | 4.0408 | 4.0713 | 4.1022 | 4.1335 | 4.1653 | 4.1976 | 4.2303 | 4.2635 | 4.2972 | 4.3315 | 13 |
| 77 | 4.3315 | 4.3662 | 4.4015 | 4.4374 | 4.4737 | 4.5107 | 4.5483 | 4.5864 | 4.6252 | 4.6646 | 4.7046 | 12 |
| 78 | 4.7046 | 4.7453 | 4.7867 | 4.8288 | 4.8716 | 4.9152 | 4.9594 | 5.0045 | 5.0504 | 5.0970 | 5.1446 | 11 |
| 79 | 5.1446 | 5.1929 | 5.2422 | 5.2924 | 5.3435 | 5.3955 | 5.4486 | 5.5026 | 5.5578 | 5.6140 | 5.6713 | 10° |
| 80° | 5.6713 | 5.7297 | 5.7894 | 5.8502 | 5.9124 | 5.9758 | 6.0405 | 6.1066 | 6.1742 | 6.2432 | 6.3138 | 9 |
| 81 | 6.3138 | 6.3859 | 6.4596 | 6.5350 | 6.6122 | 6.6912 | 6.7720 | 6.8548 | 6.9395 | 7.0264 | 7.1154 | 8 |
| 82 | 7.1154 | 7.2066 | 7.3002 | 7.3962 | 7.4947 | 7.5958 | 7.6996 | 7.8062 | 7.9158 | 8.0285 | 8.1443 | 7 |
| 83 | 8.1443 | 8.2636 | 8.3863 | 8.5126 | 8.6427 | 8.7769 | 8.9152 | 9.0579 | 9.2052 | 9.3572 | 9.5144 | 6 |
| 84 | 9.5144 | 9,677 | 9.845 | 10.02 | 10.20 | 10.39 | 10.58 | 10.78 | 10.99 | 11.20 | 11.43 | 5 |
| 85 | 11.43 | 11.66 | 11.91 | 12.16 | 12.43 | 12.71 | 13.00 | 13.30 | 13.62 | 13.95 | 14.30 | 4 |
| 86 | 14.30 | 14.67 | 15.06 | 15.46 | 15.89 | 16.35 | 16.83 | 17.34 | 17.89 | 18.46 | 19.08 | 3 |
| 87 | 19.08 | 19.74 | 20.45 | 21.20 | 22.02 | 22.90 | 23.86 | 24.90 | 26.03 | 27.27 | 28.64 | 2 |
| 88 | 28.64 | 30.14 | 31.82 | 33.69 | 35.80 | 38.19 | 40.92 | 44.07 | 47.74 | 52.08 | 57.29 | 1 |
| 89 | 57.29 | 63.66 | 71.62 | 81.85 | 95.49 | 114.6 | 143.2 | 191.0 | 286.5 | 573.0 | ∞ | 0° |
| | °1.0 | °0.9 | °0.8 | °0.7 | °0.6 | °0.5 | °0.4 | °0.3 | °0.2 | °0.1 | °0.0 | Deg. |

# Identifying the Leads of a Three-Phase, Wye-Connected, Dual-Voltage Motor

The terminal markings of a three-phase motor are standardized and used to connect the motor for operation on 240 or 480 V. *Figure 1* shows these terminal markings and their relationship to the other motor windings. If the motor is connected to a 240-V line, the motor windings are connected parallel to each other as shown in *Figure 2*. If the motor is to be operated on a 480-V line, the motor windings are connected in series is shown in *Figure 3*.

As long as these motor windings remain marked with proper numbers, connecting the motor for operation on 240- or 480-V power line is relatively simple. If these numbers are removed or damaged, however, the lead must be reidentified before the motor can be connected. The following procedure can be used to identify the proper relationship of the motor windings.

1. Using an ohmmeter, divide the motor windings into four separate circuits. One circuit will have continuity to three leads, and the other three circuits will have continuity between only two leads *(see Figure 1).*

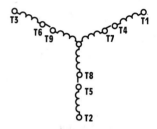

**Figure 1** Standard terminal markings for a three-phase motor

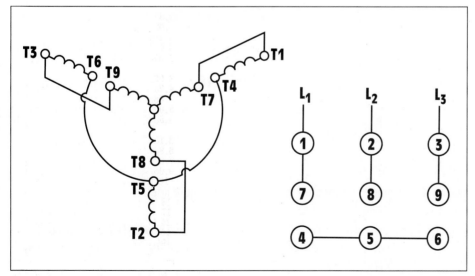

**Figure 2** Low-voltage connection

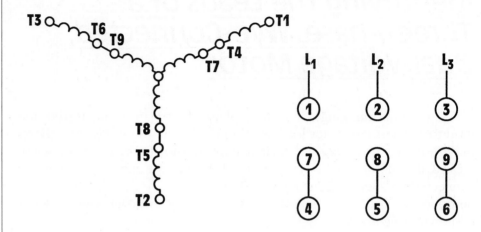

**Figure 3** High-voltage connection

C-4

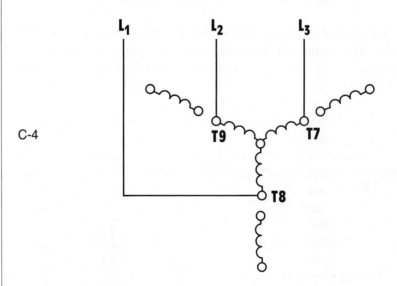

**Figure 4** T7, T8, and T9 connected to a three-phase, 240-V line

*Caution: the circuits that exhibit continuity between two leads must be identified as pairs, but do not let the ends of the leads touch anything.*

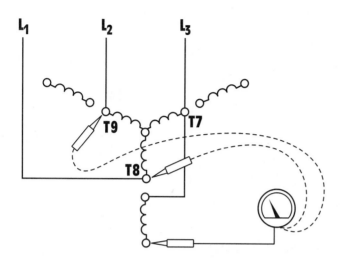

**Figure 5** Measure voltage from unconnected paired lead to T8 and T9.

2. Mark the three leads that have continuity with each other as T7, T8, and T9. Connect these three leads to a 240-V, three-phase power source (*Figure 4*). (Note: Since these windings are rated at 240 V each, the motor can be safely operated on one set of windings as long as it is not connected to a load.)

3. With the power turned off, connect one end of one of the paired leads to the paired leads to the terminal marked T7. Turn the power on, and using an AC voltmeter set for a range not less than 480 V, measure the voltage from the unconnected end of the paired lead to terminals T8 and T9 (*Figure 5*). If the measured voltages are unequal, the wrong paired lead is connected to terminal T7. Turn the power off, and connect another paired lead to T7. When the correct set of paired leads is connected to T7, the voltage readings to T8 and T9 are equal.

4. After finding the correct pair of leads, a decision must be made as to which lead should be labeled T4 and which should be labeled T1. Since an induction motor is basically a transformer, the phase windings act very similar to a multiwinding autotransformer. If terminal T1 is connected to terminal T7, it will operate similar to a transformer with its windings connected to form subtractive polarity. If an AC voltmeter is connected to T4, a voltage of about 140 V should be seen between T4 and T8 or T4 and T9 (*Figure 6*).

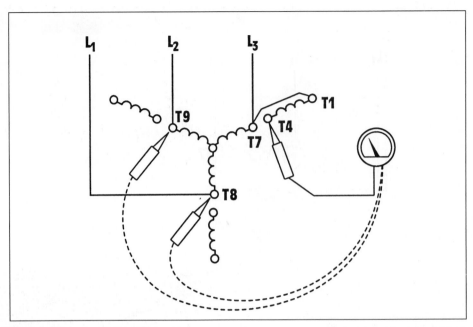

**Figure 6** T1 connected to T7

If terminal T4 is connected to T7, the winding will operate similar to a transformer with its windings connected for additive polarity. If an AC voltmeter is connected to T1, a voltage of about 360 V will be indicated when the other lead of the voltmeter is connected to T8 or T9 *(Figure 7)*.

Label leads T1 and T4 using the preceding procedure to determine which lead is correct. Then disconnect and separate T1 and T4.

5. To identify the other leads, follow the same basic procedure. Connect one end of the remaining pairs to T8. Measure the voltage between the unconnected lead and T7 and T9 to determine if it is the correct lead pair for terminal T8. When the correct lead pair is connected to T8, the voltage between the unconnected terminal and T7 or T9 will be equal. Then determine which is T5 or T2 by measuring for a high or low voltage. When T5 is connected to T8, about 360 V can be measured between T2 and T7 or T2 and T9.

6. The remaining pair can be identified as T3 or T6. When T6 is connected to T9, voltage of about 360 V can be measured between T3 and T7 or T3 and T8.

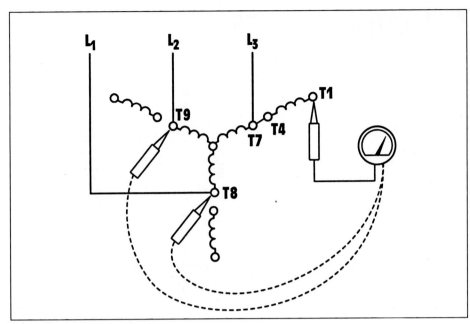

**Figure 7**  T4 connected to T7

# Alternating Current Formulas

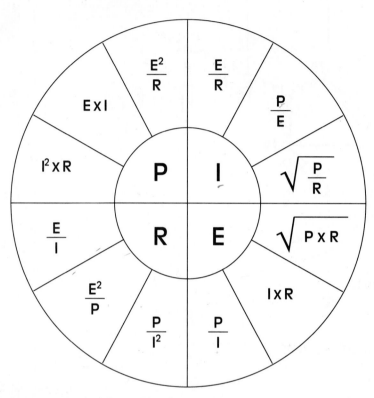

## Instantaneous and Maximum Values

The instantaneous value of voltage and current for a sine wave is equal to the peak, or maximum, value of the wave form times the sine of the angle. For example, a wave form has a peak value of 300 V. What is the voltage at an angle of 22°?

$$E_{(INST)} = E_{(MAX)} \times SIN \angle$$

$$E_{(INST)} = 300 \times 0.3746 \text{ (SIN of 22°)}$$

$$E_{(INST)} = 112.328 \text{ V}$$

$$E_{MAX} = \frac{E_{INST}}{SIN \angle}$$

$$SIN \angle = \frac{E_{INST}}{E_{MAX}}$$

## Changing Peak, RMS, and Average Values

| To change | To | Multiply by |
|---|---|---|
| Peak | RMS | 0.707 |
| Peak | Average | 0.637 |
| Peak | Peak-to-Peak | 2 |
| RMS | Peak | 1.414 |
| Average | Peak | 1.567 |
| RMS | Average | 0.9 |
| Average | RMS | 1.111 |

## Pure Resistive Circuit

$$E = I \times R \qquad I = \frac{E}{R} \qquad R = \frac{E}{I} \qquad P = E \times I$$

$$E = \frac{P}{I} \qquad I = \frac{P}{R} \qquad R = \frac{E^2}{P} \qquad P = I^2 \times R$$

$$E = \sqrt{P \times R} \qquad I = \sqrt{\frac{P}{R}} \qquad R = \frac{P}{I^2} \qquad P = \frac{E^2}{R}$$

## Series Resistive Circuits

$$R_T = R_1 + R_2 + R_3$$

$$I_T = I_1 = I_2 = I_3$$

$$E_T = E_1 + E_2 + E_3$$

$$P_T = P_1 + P_2 + P_3$$

## Parallel Resistive Circuits

$$R_T = \frac{1}{\frac{1}{R_1} + \frac{1}{R_2} + \frac{1}{R_3}}$$

$$R_T = \frac{R_1 \times R_2}{R_1 + R_2}$$

$$I_T = I_1 + I_2 + I_3$$

$$E_T = E_1 = E_2 = E_3$$

$$P_T = P_1 + P_2 + P_3$$

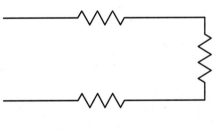

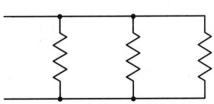

## Pure Inductive Circuits

In a pure inductive circuit, the current lags the voltage by 90°. Therefore, there is no true power, or watts, and the power factor is 0. VARs is the inductive equivalent of watts.

$$L = \frac{X_L}{2\pi F} \qquad\qquad X_L = 2\pi FL$$

$$E_L = I_L \times X_L \qquad I_L = \frac{E_L}{X_L} \qquad X_L = \frac{E_L}{I_L} \qquad VARS_L = E_L \times I_L$$

$$E_L = \sqrt{VARS_L \times X_L} \qquad I_L = \frac{VARS_L}{E_L} \qquad X_L = \frac{E_L^2}{VARS_L} \qquad VARS_L = I_L^2 \times X_L$$

$$E_L = \frac{VARS_L}{I_L} \qquad I_L = \sqrt{\frac{VARS_L}{X_L}} \qquad X_L = \frac{VARS_L}{I_L^2} \qquad VARS_L = \frac{E_L^2}{X_L}$$

## Series Inductive Circuits

$$E_{LT} = E_{L1} + E_{L2} + E_{L3}$$

$$X_{LT} = X_{L1} + X_{L2} + X_{L3}$$

$$I_{LT} = I_{L1} = I_{L2} = I_{L3}$$

$$VARS_{LT} = VARS_{L1} + VARS_{L2} + VARS_{L3}$$

$$L_T = L_1 + L_2 + L_3$$

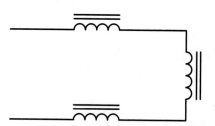

## Parallel Inductive Circuits

$$E_{LT} = E_{L1} = E_{L2} = E_{L3}$$

$$I_{LT} = I_{L1} + I_{L2} + I_{L3}$$

$$X_{LT} = \frac{X_{L1} \times X_{L2}}{X_{L1} + X_{L2}}$$

$$X_{LT} = \frac{1}{\dfrac{1}{X_{L1}} + \dfrac{1}{X_{L2}} + \dfrac{1}{X_{L3}}} \qquad L_T = \frac{1}{\dfrac{1}{L_1} + \dfrac{1}{L_2} + \dfrac{1}{L_3}} \qquad L_T = \frac{L_1 \times L_2}{L_1 + L_2}$$

$$VARS_{LT} = VARS_{L1} + VARS_{L2} + VARS_{L3}$$

## Pure Capacitive Circuits

In a pure capacitive circuit, the current leads the voltage by 90°. For this reason there is no true power, or watts, and no power factor. VARs is the equivalent of watts in a pure capacitive circuit.

The value of C in the formula for finding capacitance is in farads and must be changed into the capacitive units being used.

$$C = \frac{1}{2\pi F X_C} \qquad X_C = \frac{1}{2\pi F C}$$

$$E_C = I_C \times X_C \qquad I_C = \frac{E_C}{X_C} \qquad X_C = \frac{E_C}{I_C} \qquad VARS_C = E_C \times I_C$$

$$E_C = \sqrt{VARS_C \times X_C} \qquad I_C = \frac{VARS_C}{E_C} \qquad X_C = \frac{E_C^2}{VARS_C} \qquad VARS_C = I_C^2 \times X_C$$

$$E_C = \frac{VARS_C}{I_C} \qquad I_C = \sqrt{\frac{VARS_C}{X_C}} \qquad X_C = \frac{VARS_C}{I_C^2} \qquad VARS_C = \frac{E_C^2}{X_C}$$

## Series Capacitive Circuits

$$E_{CT} = E_{C1} + E_{C2} + E_{C3}$$

$$I_{CT} = I_{C1} = I_{C2} = I_{C3}$$

$$X_{CT} = X_{C1} + X_{C2} + X_{C3}$$

$$VARS_{CT} = VARS_{C1} + VARS_{C2} + VARS_{C3}$$

$$C_T = \frac{1}{\dfrac{1}{C_1} + \dfrac{1}{C_2} + \dfrac{1}{C_3}} \qquad C_T = \frac{C_1 \times C_2}{C_1 + C_2}$$

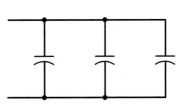

## Parallel Capacitive Circuits

$$E_{CT} = E_{C1} = E_{C2} = E_{C3}$$

$$X_{CT} = \frac{1}{\dfrac{1}{X_{C1}} + \dfrac{1}{X_{C2}} + \dfrac{1}{X_{C3}}} \qquad X_{CT} = \frac{X_{C1} \times X_{C2}}{X_{C1} + X_{C2}} \qquad I_{CT} = I_{C1} + I_{C2} + I_{C3}$$

$$C_T = C_1 + C_2 + C_3 \qquad VARS_{CT} = VARS_{C1} + VARS_{C2} + VARS_{C3}$$

# Resistive-Inductive Series Circuits

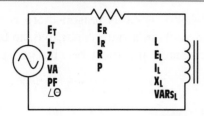

To find values for the resistor, use the formulas in the Pure Resistive section.

To find values for the inductor, use the formulas in the Pure Inductive section.

$$VA = E_T \times I_T$$

$$L = \frac{X_L}{2\pi F}$$

$$E_T = \sqrt{E_R{}^2 + E_L{}^2}$$

$$Z = \sqrt{R^2 + X_L{}^2}$$

$$VA = I_T{}^2 \times Z$$

$$I_T = I_R = I_L$$

$$E_T = I_T \times Z$$

$$Z = \frac{E_T}{I_T}$$

$$VA = \frac{E_T{}^2}{Z}$$

$$I_T = \frac{E_T}{Z}$$

$$E_T = \frac{VA}{I_T}$$

$$Z = \frac{VA}{I_T{}^2}$$

$$VA = \sqrt{P^2 + VARS_L{}^2}$$

$$I_T = \frac{VA}{E_T}$$

$$E_T = \frac{E_R}{PF}$$

$$Z = \frac{R}{PF}$$

$$Z = \frac{E_T{}^2}{VA}$$

$$VA = \frac{P}{PF}$$

$$PF = \frac{R}{Z}$$

$$P = E_R \times I_R$$

$$E_R = I_R \times R$$

$$I_R = I_T \times I_L$$

$$PF = \frac{P}{VA}$$

$$P = \sqrt{VA^2 - VARS_L{}^2}$$

$$E_R = \sqrt{P \times R}$$

$$I_R = \frac{E_R}{R}$$

$$PF = \frac{E_R}{E_T}$$

$$P = \frac{E_R{}^2}{R}$$

$$E_R = \frac{P}{I_R}$$

$$I_R = \frac{P}{E_R}$$

$$PF = COS \angle\theta$$

$$P = I_R{}^2 \times R$$

$$E_R = \sqrt{E_T{}^2 - E_L{}^2}$$

$$I_R = \sqrt{\frac{P}{R}}$$

$$P = VA \times PF$$

$$E_R = E_T \times PF$$

$$R = \sqrt{Z^2 - X_L{}^2}$$

$$E_L = I_L \times X_L$$

$$I_L = I_R = I_T$$

$$X_L = \sqrt{Z^2 - R^2}$$

$$R = \frac{E_R}{I_R}$$

$$E_L = \sqrt{E_T{}^2 - E_R{}^2}$$

$$I_L = \frac{E_L}{X_L}$$

$$X_L = \frac{E_L}{I_L}$$

$$R = \frac{E_R{}^2}{P}$$

$$E_L = \sqrt{VARS_L \times X_L}$$

$$I_L = \frac{VARS_L}{E_L}$$

$$X_L = \frac{E_L{}^2}{VARS_L}$$

$$R = \frac{P}{I_R{}^2}$$

$$E_L = \frac{VARS_L}{I_L}$$

$$I_L = \sqrt{\frac{VARS_L}{X_L}}$$

$$X_L = \frac{VARS_L}{I_L{}^2}$$

$$R = Z \times PF$$

$$X_L = 2\pi FL$$

$$VARS_L = \sqrt{VA^2 - P^2}$$

$$VARS_L = E_L \times I_L$$

$$VARS_L = \frac{E_L{}^2}{X_L}$$

$$VARS_L = I_L{}^2 \times X_L$$

## Resistive-Inductive Parallel Circuits

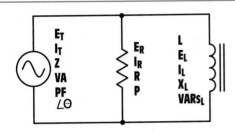

$E_T = E_R = E_L$

$E_T = I_T \times Z$

$L = \dfrac{X_L}{2\pi F}$

$E_T = \dfrac{VA}{I_T}$

$E_T = \sqrt{VA \times Z}$

$Z = \dfrac{1}{\sqrt{\left(\dfrac{1}{R}\right)^2 + \left(\dfrac{1}{X_L}\right)^2}}$

$Z = \dfrac{VA}{I_T^2}$

$I_T = \sqrt{I_R^2 + I_L^2}$

$I_T = \dfrac{VA}{E_T}$    $I_T = \sqrt{\dfrac{VA}{Z}}$

$Z = \dfrac{E_T}{I_T}$    $Z = \dfrac{E_T^2}{VA}$

$Z = R \times PF$

$I_T = \dfrac{E_T}{Z}$

$I_T = \dfrac{I_R}{PF}$

$VA = E_T \times I_T$

$PF = \dfrac{Z}{R}$

$E_L = I_L \times X_L$

$I_L = \sqrt{I_T^2 - I_R^2}$

$VA = I_T^2 \times Z$

$PF = \dfrac{P}{VA}$

$E_L = E_T = E_R$

$I_L = \dfrac{E_L}{X_L}$

$VA = \dfrac{E_T^2}{Z}$

$PF = \dfrac{I_R}{I_T}$

$E_L = \sqrt{VARS_L \times X_L}$

$I_L = \dfrac{VARS_L}{E_L}$

$VA = \sqrt{P^2 + VARS_L^2}$

$PF = COS \angle\theta$

$E_L = \dfrac{VARS_L}{I_L}$

$I_L = \sqrt{\dfrac{VARS_L}{X_L}}$

$VA = \dfrac{P}{PF}$    $VA = \dfrac{E_T^2}{Z}$

$VARS_L = \sqrt{VA^2 - P^2}$

$VARS_L = E_L \times I_L$

$VARS_L = \dfrac{E_L^2}{X_L}$

$VARS_L = I_L^2 \times X_L$

$E_R = I_R \times R$

$I_R = \sqrt{I_T^2 - I_L^2}$

$X_L = \dfrac{E_L}{I_L}$

$X_L = \dfrac{1}{\sqrt{\left(\dfrac{1}{Z}\right)^2 - \left(\dfrac{1}{R}\right)^2}}$

$E_R = \sqrt{P \times R}$

$I_R = \dfrac{E_R}{R}$

$X_L = \dfrac{E_L^2}{VARS_L}$

$X_L = 2\pi FL$

$E_R = \dfrac{P}{I_R}$

$I_R = \dfrac{P}{E_R}$

$X_L = \dfrac{VARS_L}{I_L^2}$

$R = \dfrac{E_R}{I_R}$

$E_R = E_T = E_L$

$I_R = \sqrt{\dfrac{P}{R}}$

$R = \dfrac{P}{I_R^2}$

$I_R = I_T \times PF$

## Resistive-Inductive Parallel Circuits (continued)

$$R = \sqrt{\dfrac{1}{\left(\dfrac{1}{Z}\right)^2 - \left(\dfrac{1}{X_L}\right)^2}}$$

$$R = \dfrac{E_R^2}{P}$$

$$R = \dfrac{Z}{PF}$$

$$P = \sqrt{VA^2 - VARS_L^2}$$

$$P = E_R \times I_R$$

$$P = \dfrac{E_R^2}{R}$$

$$P = I_R^2 \times R$$

$$P = VA \times PF$$

## Resistive-Capacitive Series Circuits

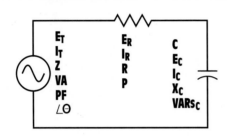

$$E_T = \sqrt{E_R^2 + E_C^2}$$

$$E_T = I_T \times Z$$

$$I_T = I_R = I_C$$

$$E_T = \dfrac{VA}{I_T}$$

$$I_T = \dfrac{E_T}{Z}$$

$$E_T = \dfrac{E_R}{PF}$$

$$I_T = \dfrac{VA}{E_T}$$

$$Z = \sqrt{R^2 + X_C^2}$$

$$Z = \dfrac{VA}{I_T^2}$$

$$Z = \dfrac{E_T^2}{VA}$$

$$Z = \dfrac{E_T}{I_T}$$

$$Z = \dfrac{R}{PF}$$

$$C = \dfrac{1}{2\pi F X_C}$$

$$VA = E_T \times I_T$$

$$PF = \dfrac{R}{Z}$$

$$P = E_R \times I_R$$

$$E_R = I_R \times R$$

$$VA = I_T^2 \times Z$$

$$PF = \dfrac{P}{VA}$$

$$P = \sqrt{VA^2 - VARS_C^2}$$

$$E_R = \sqrt{P \times R}$$

$$VA = \dfrac{E_T^2}{Z}$$

$$PF = \dfrac{E_R}{E_T}$$

$$P = \dfrac{E_R^2}{R}$$

$$E_R = \dfrac{P}{I_R}$$

$$VA = \sqrt{P^2 + VARS_C^2}$$

$$PF = \cos \angle\theta$$

$$P = I_R^2 \times R$$

$$E_R = \sqrt{E_T^2 - E_C^2}$$

$$VA = \dfrac{P}{PF}$$

$$R = \sqrt{Z^2 - X_C^2}$$

$$P = VA \times PF$$

$$E_R = E_T \times PF$$

$$I_R = I_T = I_C$$

$$R = \dfrac{E_R}{I_R}$$

$$E_C = I_C \times X_C$$

$$I_C = I_R = I_T$$

$$I_R = \dfrac{E_R}{R}$$

$$R = \dfrac{E_R^2}{P}$$

$$E_C = \sqrt{E_T^2 - E_R^2}$$

$$I_C = \dfrac{E_C}{X_C}$$

## Resistive-Capacitive Series Circuits (continued)

$$I_R = \frac{P}{E_R}$$

$$R = \frac{P}{I_R^2}$$

$$E_C = \sqrt{VARS_C \times X_C}$$

$$I_C = \frac{VARS_C}{E_C}$$

$$I_R = \sqrt{\frac{P}{R}}$$

$$R = Z \times PF$$

$$E_C = \frac{VARS_C}{I_C}$$

$$I_C = \sqrt{\frac{VARS_C}{X_C}}$$

$$X_C = \sqrt{Z^2 - R^2}$$

$$X_C = \frac{E_C^2}{VARS_C}$$

$$X_C = \frac{1}{2\pi FC}$$

$$VARS_C = I_C^2 \times X_C$$

$$X_C = \frac{E_C}{I_C}$$

$$X_C = \frac{VARS_C}{I_C^2}$$

$$VARS_C = E_C \times I_C$$

$$VARS_C = \frac{E_C^2}{X_C}$$

$$VARS_C = \sqrt{VA^2 - P^2}$$

## Resistive-Capacitive Parallel Circuits

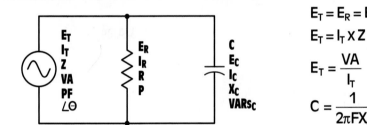

$$E_T = E_R = E_C$$

$$E_T = I_T \times Z$$

$$E_T = \frac{VA}{I_T}$$

$$C = \frac{1}{2\pi FX_C}$$

$$E_T = \sqrt{VA \times Z}$$

$$Z = \frac{1}{\sqrt{\left(\frac{1}{R}\right)^2 + \left(\frac{1}{X_C}\right)^2}}$$

$$Z = \frac{VA}{I_T^2}$$

$$Z = \frac{E_T}{I_T}$$

$$Z = \frac{E_T^2}{VA}$$

$$Z = R \times PF$$

$$I_T = \sqrt{I_R^2 + I_C^2}$$

$$VA = E_T \times I_T$$

$$PF = \frac{Z}{R}$$

$$P = E_R \times I_R$$

$$I_T = \frac{E_T}{Z}$$

$$VA = I_T^2 \times Z$$

$$PF = \frac{P}{VA}$$

$$P = \sqrt{VA^2 - VARS_C^2}$$

$$I_T = \frac{VA}{E_T}$$

$$VA = \frac{E_T^2}{Z}$$

$$PF = \frac{EI}{ER}$$

$$P = \frac{E_R^2}{R}$$

$$I_T = \frac{I_R}{PF}$$

$$VA = \sqrt{P^2 + VARS_C^2}$$

$$PF = \cos \angle\theta$$

$$P = I_R^2 \times R$$

## Resistive-Capacitive Parallel Circuits (continued)

$$I_T = \sqrt{\frac{VA}{Z}}$$

$$VA = \frac{P}{PF}$$

$$P = VA \times PF$$

$$E_R = I_R \times R$$

$$I_R = \sqrt{I_T^2 - I_C^2}$$

$$R = \frac{1}{\sqrt{\left(\frac{1}{Z}\right)^2 - \left(\frac{1}{X_C}\right)^2}}$$

$$E_C = I_C \times X_C$$

$$E_R = \sqrt{P \times R}$$

$$I_R = \frac{E_R}{R}$$

$$R = \frac{E_R}{I_R}$$

$$E_C = E_T = E_R$$

$$E_R = \frac{P}{I_R}$$

$$I_R = \frac{P}{E_R}$$

$$R = \frac{E_R^2}{P}$$

$$E_C = \sqrt{VARS_C \times X_C}$$

$$E_R = E_T = E_C$$

$$I_R = \sqrt{\frac{P}{R}}$$

$$R = \frac{P}{I_R^2}$$

$$E_C = \frac{VARS_C}{I_C}$$

$$E_R = E_T \times PF$$

$$R = \frac{Z}{PF}$$

$$VARS_C = I_C^2 \times X_C$$

$$I_C = \sqrt{I_T^2 - I_R^2}$$

$$X_C = \frac{E_C^2}{VARS_C}$$

$$VARS_C = \frac{E_C^2}{X_C}$$

$$I_C = \frac{E_C}{X_C}$$

$$X_C = \frac{1}{\sqrt{\left(\frac{1}{Z}\right)^2 - \left(\frac{1}{R}\right)^2}}$$

$$X_C = \frac{VARS_C}{I_C^2}$$

$$VARS_C = E_C \times I_C$$

$$I_C = \frac{VARS_C}{E_C}$$

$$X_C = \frac{E_C}{I_C}$$

$$X_C = \frac{1}{2\pi FC}$$

$$VARS_C = \sqrt{VA^2 - P^2}$$

$$I_C = \sqrt{\frac{VARS_C}{X_C}}$$

## Resistive-Inductive-Capacitive Series Circuits

$$E_T = \sqrt{E_R^2 + (E_L - E_C)^2}$$

$$E_T = \frac{VA}{I_T}$$

$$Z = \sqrt{R^2 + (X_L - X_C)^2}$$

$$Z = \frac{VA}{I_T^2}$$

$$E_T = I_T \times Z$$

$$E_T = \frac{E_R}{PF}$$

$$Z = \frac{E_T}{I_T}$$

$$Z = \frac{R}{PF}$$

## Resistive-Inductive-Capacitive Series Circuits (continued)

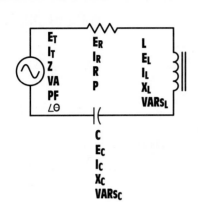

$$I_T = I_R = I_L = I_C$$

$$I_T = \frac{VA}{E_T}$$

$$I_T = \frac{E_T}{Z}$$

$$I_T = \sqrt{\frac{VA}{Z}}$$

$$VA = E_T \times I_T$$

$$VA = \frac{P}{PF}$$

$$VA = I_T^2 \times Z$$

$$VA = \frac{E_T^2}{Z}$$

$$VA = \sqrt{P^2 + (VARS_L - VARS_C)^2}$$

$$PF = \frac{R}{Z}$$

$$P = E_R \times I_R$$

$$P = VA \times PF$$

$$E_R = I_R \times R$$

$$PF = \frac{P}{VA}$$

$$P = \sqrt{VA^2 - (VARS_L - VARS_C)^2}$$

$$E_R = \frac{P}{I_R}$$

$$PF = \frac{E_R}{E_T}$$

$$P = \frac{E_R^2}{R}$$

$$E_R = \sqrt{P \times R}$$

$$E_R = \sqrt{E_T^2 - (E_L - E_C)^2}$$

$$PF = COS \angle\theta$$

$$P = I_R^2 \times R$$

$$E_R = E_T \times PF$$

$$I_R = I_T = I_C = I_L$$

$$R = \sqrt{Z^2 - (X_L - X_C)^2}$$

$$E_C = I_C \times X_C$$

$$I_R = \frac{E_R}{R}$$

$$R = \frac{E_R}{I_R}$$

$$R = Z \times PF$$

$$E_C = \sqrt{VARS_C \times X_C}$$

$$I_R = \frac{P}{E_R}$$

$$R = \frac{E_R^2}{P}$$

$$R = \frac{P}{I_R^2}$$

$$E_C = \frac{VARS_C}{I_C}$$

$$I_R = \sqrt{\frac{P}{R}}$$

$$X_C = \frac{1}{2\pi FC}$$

$$C = \frac{1}{2\pi F X_C}$$

$$I_C = I_R = I_T = I_L$$

$$X_C = \frac{E_C^2}{VARS_C}$$

$$X_C = \frac{VARS_C}{I_C^2}$$

$$VARS_C = E_C \times I_C$$

$$I_C = \frac{E_C}{X_C}$$

$$X_C = \frac{E_C}{I_C}$$

$$L = \frac{X_L}{2\pi F}$$

$$X_L = 2\pi FL$$

$$I_C = \frac{VARS_C}{E_C}$$

$$VARS_C = I_C^2 \times X_C$$

$$X_L = \frac{E_L}{I_L}$$

$$X_L = \frac{VARS_L}{I_L^2}$$

$$I_C = \sqrt{\frac{VARS_C}{X_C}}$$

$$VARS_C = \frac{E_C^2}{X_C}$$

$$X_L = \frac{E_L^2}{VARS_L}$$

## Resistive-Inductive-Capacitive Series Circuits (continued)

$$E_L = I_L \times X_L$$

$$I_L = I_R = I_T = I_C$$

$$I_L = \frac{VARS_L}{E_L}$$

$$VARS_L = E_L \times I_L$$

$$E_L = \sqrt{VARS_L \times X_L}$$

$$I_L = \frac{E_L}{X_L}$$

$$I_L = \sqrt{\frac{VARS_L}{X_L}}$$

$$VARS_L = \frac{E_L^2}{X_L}$$

$$E_L = \frac{VARS_L}{I_L}$$

$$VARS_L = I_L^2 \times X_L$$

## Resistive-Inductive-Capacitive Parallel Circuits

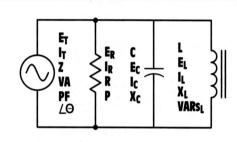

$$E_T = E_R = E_L = E_C$$

$$E_T = I_T \times Z$$

$$E_T = \frac{VA}{I_T}$$

$$E_T = \sqrt{VA \times Z}$$

$$Z = \frac{1}{\sqrt{\left(\frac{1}{R}\right)^2 + \left(\frac{1}{X_L} - \frac{1}{X_C}\right)^2}}$$

$$Z = \frac{VA}{I_T^2}$$

$$I_T = \sqrt{I_R^2 + (I_L - I_C)^2}$$

$$I_T = \frac{VA}{E_T}$$

$$I_T = \sqrt{\frac{VA}{Z}}$$

$$Z = \frac{E_T}{I_T}$$

$$Z = \frac{E_T^2}{VA}$$

$$Z = R \times PF$$

$$I_T = \frac{E_T}{Z}$$

$$I_T = \frac{I_R}{PF}$$

$$VA = E_T \times I_T$$

$$PF = \frac{Z}{R}$$

$$PF = COS \angle\theta$$

$$E_L = E_T = E_R = E_C$$

$$VA = I_T^2 \times Z$$

$$PF = \frac{P}{VA}$$

$$E_L = I_L \times X_L$$

$$E_L = \sqrt{VARS_L \times X_L}$$

$$VA = \frac{E_T^2}{Z}$$

$$PF = \frac{I_R}{I_T}$$

$$E_L = \frac{VARS_L}{I_L}$$

$$L = \frac{X_L}{2\pi F}$$

$$VA = \sqrt{(VARS_L^2 - VARS_C)^2}$$

$$X_L = \frac{E_L}{I_L}$$

$$X_L = 2\pi FL$$

$$VARS_L = \frac{E_L^2}{X_L}$$

$$VA = \frac{P}{PF}$$

$$I_L = \frac{VARS_L}{E_L}$$

$$X_L = \frac{E_L^2}{VARS_L}$$

$$VARS_L = I_L^2 \times X_L$$

$$I_L = \frac{E_L}{X_L}$$

$$I_L = \sqrt{\frac{VARS_L}{X_L}}$$

$$X_L = \frac{VARS_L}{I_L^2}$$

$$VARS_L = E_L \times I_L$$

## Resistive-Inductive-Capacitive Parallel Circuits (continued)

$$E_R = I_R \times R$$

$$I_R = \sqrt{I_T^2 - (I_L - I_C)^2}$$

$$I_R = I_T \times PF$$

$$E_R = \sqrt{P \times R}$$

$$I_R = \frac{E_R}{R}$$

$$R = \frac{E_R}{I_R}$$

$$R = \frac{1}{\sqrt{\left(\frac{1}{Z}\right)^2 + \left(\frac{1}{X_L} - \frac{1}{X_C}\right)^2}}$$

$$E_R = \frac{P}{I_R}$$

$$I_R = \frac{P}{E_R}$$

$$E_R = E_T = E_L = E_C$$

$$I_R = \sqrt{\frac{P}{R}}$$

$$R = \frac{P}{I_R^2} \qquad R = \frac{Z}{PF} \qquad R = \frac{E_R^2}{P}$$

$$P = \sqrt{VA^2 - (VARS_L - VARS_C)^2}$$

$$P = E_R \times I_R$$

$$P = I_R^2 \times R$$

$$E_C = \frac{VARS_C}{I_C}$$

$$P = VA \times PF$$

$$P = \frac{E_R^2}{R}$$

$$X_C = \frac{1}{2\pi FC}$$

$$E_C = I_C \times X_C$$

$$I_C = \frac{E_C}{X_C}$$

$$X_C = \frac{E_C}{I_C}$$

$$VARS_C = E_C \times I_C$$

$$E_C = E_T = E_R = E_L$$

$$I_C = \frac{VARS_C}{E_C}$$

$$X_C = \frac{E_C^2}{VARS_C}$$

$$VARS_C = I_C^2 \times X_C$$

$$E_C = \sqrt{VARS_C \times X_C}$$

$$I_C = \sqrt{\frac{VARS_C}{X_C}}$$

$$X_C = \frac{VARS_C}{I_C^2}$$

$$VARS_C = \frac{E_C^2}{X_C}$$

## Transformers

$$\frac{E_P}{E_S} = \frac{N_P}{N_S} \qquad \frac{E_P}{E_S} = \frac{I_S}{I_P} \qquad \frac{N_P}{N_S} = \frac{I_S}{I_P} \qquad Z_P = Z_S \left(\frac{N_P}{N_S}\right)^2 \qquad Z_S = Z_P \left(\frac{N_P}{N_S}\right)^2$$

$E_P$—Voltage of the primary
$E_S$—Voltage of the secondary
$I_P$—Current of the primary
$I_S$—Current of the secondary
$N_P$—Number of turns of the primary
$N_S$—Number of turns of the secondary
$Z_P$—Impedance of the primary
$Z_S$—Impedance of the secondary

## Three-Phase Connections

### Wye Connection

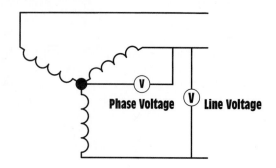

Phase Voltage    Line Voltage

In a **wye** connection, the line current and phase current are the same.

$$I_{(LINE)} = I_{(PHASE)}$$

In a **wye** connection, the line voltage is higher than the phase voltage by a factor of the square root of 3.

$$E_{(LINE)} = E_{(PHASE)} \times \sqrt{3}$$

$$E_{(PHASE)} = \frac{E_{(LINE)}}{\sqrt{3}}$$

### Delta Connection

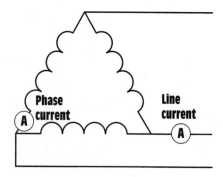

Phase current    Line current

In a **delta**-connected system, the line voltage and phase voltage are the same.

$$E_{(LINE)} = E_{(PHASE)}$$

In a **delta**-connected system the line current is higher than the phase current by a factor of the square root of 3.

$$I_{(LINE)} = I_{(PHASE)} \times \sqrt{3}$$

$$I_{(PHASE)} = \frac{I_{(LINE)}}{\sqrt{3}}$$

### Open Delta Connection

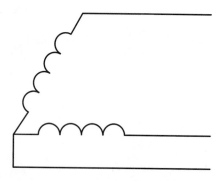

**Open delta** connections provide 87% of the sum of the power rating of the two transformers. Example: The two transformers show are rated at 60 kVA each. The total power rating for this connection would be

**60 + 60 = 120 kVA**

**120 kVA x 0.87 = 104.4 kVA**

All values of voltage and current are computed in the same manner as for a closed delta connection.

### Apparant and True Power

$$VA = \sqrt{3} \times E_{(LINE)} \times I_{(LINE)}$$

$$VA = 3 \times E_{(PHASE)} \times I_{(PHASE)}$$

$$P = \sqrt{3} \times E_{(LINE)} \times I_{(LINE)} \times PF$$

$$P = 3 \times E_{(PHASE)} \times I_{(PHASE)} \times PF$$

# Greek Alphabet

| Name of letter | Upper case | Lower case | Designates |
|---|---|---|---|
| Alpha | A | α | Angles |
| Beta | B | β | Angles; flux density |
| Gamma | Γ | γ | Conductivity |
| Delta | Δ | δ | Change of a quantity |
| Epsilon | E | ε | Base of natural logarithms |
| Zeta | Z | ζ | Impedance; coefficients; coordinates |
| Eta | H | η | Hysteresis coefficient; efficiency |
| Theta | Θ | θ | Phase angle |
| Iota | I | ι | |
| Kappa | K | κ | Dielectric constant; coefficient of coupling; susceptibility |
| Lambda | Λ | λ | Wavelength |
| Mu | M | μ | Permeability; the prefix micro-, amplification factor |
| Nu | N | ν | Change of a quantity |
| Xi | Ξ | ξ | |
| Omicron | O | o | |
| Pi | Π | π | 3.1416 |
| Rho | P | ρ | Resistivity |
| Sigma | Σ | σ | |
| Tau | T | τ | Time constant; time phase displacement |
| Upsilon | Y | υ | |
| Phi | Φ | φ | Angles; magentic flow |
| Chi | X | χ | |
| Psi | Ψ | ψ | Dielectric flux; phase difference |
| Omega | Ω | ω | Resistivity |

# Appendix F

## Metals

| Metal | Sym. | Spec. grav. | Melt Point °C | Melt Point °F | Elec. Cond. % copper | Lb/in.$^3$ |
|---|---|---|---|---|---|---|
| Aluminum | Al | 2.710 | 660 | 1220 | 64.9 | 0.0978 |
| Antimony | Sb | 6.620 | 630 | 1167 | 4.42 | 0.2390 |
| Arsenic | As | 5.730 | | | 4.90 | 0.2070 |
| Beryllium | Be | 1.830 | 1280 | 2336 | 9.32 | 0.0660 |
| Bismuth | Bi | 9.800 | 271 | 512 | 1.50 | 0.3540 |
| Brass (70–30) | | 8.510 | 900 | 1652 | 28.0 | 0.3070 |
| Bronze (5% Sn) | | 8.870 | 1000 | 1382 | 18.0 | 0.3200 |
| Cadmium | Cd | 8.650 | 321 | 610 | 22.7 | 0.3120 |
| Calcium | Ca | 1.550 | 850 | 1562 | 50.1 | 0.0560 |
| Cobalt | Co | 8.900 | 1495 | 2723 | 17.8 | 0.3210 |
| Copper | Cu | 8.900 | 1083 | 1981 | 100. | 0.3210 |
| Gold | Au | 19.30 | 1063 | 1945 | 71.2 | 0.6970 |
| Graphite | | 2.250 | 3500 | 6332 | 0.001 | 0.0812 |
| Indium | In | 7.300 | 156 | 311 | 20.6 | 0.2640 |
| Iridium | Ir | 22.40 | 2450 | 4442 | 32.5 | 0.8090 |
| Iron | Fe | 7.200 | 1400 to 1500 | 2552 to 2732 | 17.6 | 0.2600 |
| Malleable | | 7.200 | 1600to 1500 | 2912to 2732 | 10.0 | 0.2600 |
| Wrought | | 7.700 | 1600 | 2912 | 10.0 | 0.2780 |
| Lead | Pb | 11.40 | 327 | 621 | 8.35 | 0.4120 |
| Magnesium | Mg | 1.740 | 651 | 1204 | 38.7 | 0.0628 |
| Manganese | Mn | 7.200 | 1245 | 2273 | 0.90 | 0.2600 |
| Mercury | Hg | 13.65 | –38.9 | –37.7 | 1.80 | 0.4930 |
| Molybdenum | Mo | 10.20 | 2620 | 4748 | 36.1 | 0.3680 |
| Monel (63–37) | | 8.870 | 1300 | 2372 | 3.00 | 0.3200 |
| Nickel | Ni | 8.90 | 1452 | 2646 | 25.0 | 0.3210 |

| Metal | Sym. | Spec. grav. | Melt Point °C | °F | Elec. Cond. % copper | Lb/in.$^3$ |
|-------|------|-------------|---------------|-----|----------------------|------------|
| Phosphorus | P | 1.82 | 44.1 | 111.4 | $10^{-17}$ | 0.0657 |
| Platinum | Pt | 21.46 | 1773 | 3221 | 17.5 | 0.0657 |
| Potassium | K | 0.860 | 62.3 | 144.1 | 28.0 | 0.0310 |
| Selenium | Se | 4.81 | 220 | 428 | 14.4 | 0.1740 |
| Silicon | Si | 2.40 | 1420 | 2588 | $10^{-5}$ | 0.0866 |
| Silver | Ag | 10.50 | 960 | 1760 | 106 | 0.3790 |
| Steel (carbon) |  | 7.84 | 1330 to 1380 | 2436 to 2516 | 10.0 | 0.2830 |
| Stainless |  |  |  |  |  |  |
| (18–8) |  | 7.92 | 1500 | 2732 | 2.50 | 0.2860 |
| (13–Cr) |  | 7.78 | 1520 | 2768 | 3.50 | 0.2810 |
| (18–Cr) |  | 7.73 | 1500 | 2732 | 3.00 | 0.2790 |
| Tantalum | Ta | 16.6 | 2900 | 5414 | 13.9 | 0.5990 |
| Tellurium | Te | 6.20 | 450 | 846 | $10^{-5}$ | 0.2240 |
| Thorium | Th | 11.7 | 1845 | 3353 | 9.10 | 0.4220 |
| Tin | Sn | 7.30 | 232 | 449 | 15.0 | 0.2640 |
| Titanium | Ti | 4.50 | 1800 | 3272 | 2.10 | 0.1620 |
| Tungsten | W | 19.3 | 3410 | 6170 | 31.5 | 0.6970 |
| Uranium | U | 18.7 | 1130 | 2066 | 2.80 | 0.6750 |
| Vanadium | V | 5.96 | 1710 | 3110 | 6.63 | 0.2150 |
| Zinc | Zn | 7.14 | 419 | 786 | 29.1 | 0.2580 |
| Zirconium | Zr | 6.40 | 1700 | 3092 | 4.20 | 0.2310 |

# Appendix  *G*

# Scientific Notation

Scientific notation is used in almost all scientific calculations. It was first used for making calculations with a slide rule. A slide rule is a tool that can perform mathematical operations such as multiplying, dividing, finding square roots, finding logarithms of numbers, and finding sines, cosines, and tangents of angles. On a slide rule, only the actual digits in the number are used; decimal points are not used, and zeros are used only when they come between two other digits, as in 102. To a slide rule, the numbers 0.000012, 0.0012, 0.12, 1.2, 12, 120, 12,000, and 12,000,000 are all the same number: 12.

Once you realize that the slide rule recognizes only the basic digits of any number, you can imagine the problem of determining where to place a decimal point in an answer. As long as only simple calculations are done, there is no problem. For example when you multiply 12 by 20 it is obvious where the decimal point should be placed.

### 240.00

Now assume that the following numbers are to be multiplied:

### $0.000041 \times 380,000 \times 0.19 \times 720 \times 0.0032$

In this problem, it is not obvious where the decimal point should be placed. Scientific notation can be used to simplify the numbers so that an estimated answer can be obtained. Scientific notation is simply dividing or multiplying numbers by a power of 10. Any number can be multiplied by 10 by moving the decimal point one place to the right. Any number can be divided by 10 by moving the decimal point one place to the left. For example, the number 0.000041 can be changed to 4.1 by multiplying it by 10 five times. Therefore, if the number 4.1 is divided by 10 five times it will be the same as the original number, 0.000041. The number 0.000041 can be changed to 4.1 by multiplying it by 10 five times using scientific notation. The new number if 4.1 times 10 to the negative fifth.

### $4.1 \times 10^{-5}$

It is a common practice to use the letter $E$ to indicate the exponent in scientific notation. The above notation can also be written:

### 4.1 E–05

The number 380,000 can be reduced to 3.8 by dividing it by 10 five times. The number 3.8 must, therefore, be notated to indicate that the original number is actually 3.8 multiplied by 10 five times.

### 3.8 x 10⁵, or 3.8 E05

The other numbers in the problem can also be changed to simpler numbers using scientific notation:

> 0.19 becomes 1.9 E–01
> 720 becomes 7.2 E02
> 0.0032 becomes 3.2 E–03

Once the numbers have been simplified using scientific notation, we can estimate the answer by rounding off the numbers, multiplying them together, and adding the exponents: 4.1 is about 4; 3.8 is about 4; 1.9 is about 2; 7.2 is about 7; and 3.2 is about 3. The rounded-off numbers can be multiplied easily.

### 4 x 4 x 2 x 7 x 3 = 672

Then add the exponents:

### (E–05) + (E05) + (E–01) + (E02) + (E–03) = E–02

The estimated answer becomes 672 E–02. When the calculation is completed, the actual answer becomes

### 682.03 E–02, or 6.8203

## Using Scientific Notation with Calculators

In the early 1970s, scientific calculators, often referred to as *slide rule calculators*, became commonplace. Most of these calculators have the ability to display from 8 to 10 digits, depending on the manufacturer. Scientific calculations, however, often involve numbers that contain more then 8 or 10 digits. To overcome the limitation of an 8- or 10-digit display, slide rule calculators depend on scientific notation. When a number becomes too large for the calculator to display, scientific notation is used automatically. Imagine, for example, that it became necessary to display the distance (in kilometers) that light travels in one year (approximately 9,460,800,000,000 km). This number contains 13 digits. The calculator would display this number as:

| 9.4608 12 |
|---|

The number 12 shown to the right of 9.4608 is the scientific notation exponent. This number could be written 9.4608 E12, indicating that the decimal point should be moved to the right 12 places. A minus sign ahead of the scientific notation exponent indicates that the decimal point should be moved to the left. The number on the display shown below contains a negative scientific notation exponent.

$$7.5698 \; {-}06$$

This number could be rewritten as

### 0.0000075698, or 7.5698 E–06

## Entering Numbers in Scientific Notation

Slide rule calculators also have the ability to enter numbers in scientific notation. To do this, the exponent key must be used. There are two ways in which manufacturers mark the exponent key. Some are marked EXP and others are marked EE.

| EXP | | EE |

Assume that the number 549 E08 was to be entered. The following keystrokes would be used:

| 5 | 4 | 9 | EXP | 8 |

The calculator would display the following:

$$549 \; 08$$

If a number with a negative exponent is to be entered, the *change sign* (+/–) key should be used. Assume that the number 1.276 E–04 is to be entered. The following keystrokes would be used:

| 1 | . | 2 | 7 | 6 | EXP | 4 | +/– |

The display would show the following:

$$1.276 \; {-}04$$

## Setting the Display

Some calculators permit the answer to be displayed in any of three different ways. One of these ways is with *floating decimals* (FD). When the calculator is set for this mode of operation, the answers will be displayed with the decimal point appearing in the normal position. The only exception is if the number to be displayed is too large. In that case, the calculator will automatically display the number in scientific notation.

In the *scientific mode* (Sci), the calculator will display all entries and answers in scientific notation.

When set in the *engineering mode* (Eng), the calculator will display all entries and answers in scientific notation, but only in steps of three. When displayed in steps of three, the notation corresponds to standard engineering notation units such as kilo, mega, giga, milli, micro, and so on. For example, assume that a calculator is set in the scientific mode and displays the number

| 5.69836 05 |
| --- |

Now assume that the calculator is reset to the engineering mode. The number would now be displayed as

| 569.836 03 |
| --- |

The number could now be read as "569.836 kilo," because kilo means one thousand, or 1 E03.

# Answers to Practice Problems

### Unit 2  Ohm's Law

| Volts (E) | Amps (I) | Ohms (R) | Watts (P) |
|-----------|----------|----------|-----------|
| 153 | 0.056 | 2732.14 Ω | 8.568 |
| 305.5 | 0.65 | 470 Ω | 198.57 |
| 24 | 5.167 | 4.645 Ω | 124 |
| 3.59 | 0.00975 | 368.179 Ω | 0.035 |
| 76.472 | 0.0112 | 6.8 kΩ | 0.86 |
| 460 | 6.389 | 72 Ω | 2938.889 |
| 48 | 1.2 | 40 Ω | 57.6 |
| 123.2 | 154 | 0.8 Ω | 18,972.8 |
| 277 | 2.744 | 100.959 Ω | 760 |
| 14.535 | 0.0043 | 3380.205 Ω | 0.0625 |
| 54.083 | 0.000416 | 130 kΩ | 0.0225 |
| 96 | 0.0436 | 2.2 kΩ | 4.189 |

### Unit 5  Resistors

| 1st band | 2nd band | 3rd band | 4th band | Value | %Tol |
|----------|----------|----------|----------|-------|------|
| Red | Yellow | Brown | Silver | 240 Ω | 10 |
| Blue | Gray | Red | Gold | 6800 | 5 |
| Orange | Orange | Orange | Gold | 33 kΩ | 5 |

| 1st band | 2nd band | 3rd band | 4th band | Value | %Tol |
|---|---|---|---|---|---|
| Brown | Red | Black | Red | 12 Ω | 2 |
| Brown | Green | Silver | Silver | 0.15 Ω | 10 |
| Brown | Gray | Green | Silver | 1.8 MΩ | 10 |
| Brown | Black | Yellow | None | 100 kΩ | 20 |
| Brown | Black | Orange | Gold | 10 kΩ | 5 |
| Violet | Green | Black | Red | 75 Ω | 2 |
| Yellow | Violet | Red | None | 4.7 kΩ | 20 |
| Gray | Red | Green | Red | 8.2 MΩ | 2 |
| Green | Blue | Gold | Red | 5.6 Ω | 2 |

## Unit 6 Series Circuits

$E_T$ 120 V    $E_1$ 16.34    $E_2$ 13.68    $E_3$ 28.5    $E_4$ 38    $E_5$ 23.56
$I_T$ 0.038    $I_1$ 0.038    $I_2$ 0.038    $I_3$ 0.038    $I_4$ 0.038    $I_5$ 0.038
$R_T$ 3160 Ω    $R_1$ 430 Ω    $R_2$ 360 Ω    $R_3$ 750 Ω    $R_4$ 1000 Ω    $R_5$ 620 Ω
$P_T$ 4.56    $P_1$ 0.621    $P_2$ 0.52    $P_3$ 1.083    $P_4$ 1.444    $P_5$ 0.894

$E_T$ 50    $E_1$ 6    $E_2$ 16.5    $E_3$ 11    $E_4$ 9    $E_5$ 7.5
$I_T$ 0.005    $I_1$ 0.005    $I_2$ 0.005    $I_3$ 0.005    $I_4$ 0.005    $I_5$ 0.005
$R_T$ 10 kΩ    $R_1$ 1.2 kΩ    $R_2$ 3.3 kΩ    $R_3$ 2.2 kΩ    $R_4$ 1.8 kΩ    $R_5$ 1.5 kΩ
$P_T$ 0.25    $P_1$ 0.03    $P_2$ 0.0825    $P_3$ 0.055    $P_4$ 0.045    $P_5$ 0.0375

$E_T$ 340 V    $E_1$ 44 V    $E_2$ 94 V    $E_3$ 60 V    $E_4$ 40 V    $E_5$ 102 V
$I_T$ 0.002    $I_1$ 0.002    $I_2$ 0.002    $I_3$ 0.002    $I_4$ 0.002    $I_5$ 0.002
$R_T$ 170 kΩ    $R_1$ 22 kΩ    $R_2$ 47 kΩ    $R_3$ 30 kΩ    $R_4$ 20 kΩ    $R_5$ 51 kΩ
$P_T$ 0.68    $P_1$ 0.088    $P_2$ 0.188    $P_3$ 0.12    $P_4$ 0.08    $P_5$ 0.204

2.
$E_1$ = 18.46 V    $E_2$ = 40.62 V    $E_3$ = 33.23 V    $E_4$ = 27.69 V

## Unit 7 Parallel Circuits

**1.**

| | | | | |
|---|---|---|---|---|
| $E_T$ 120.03 | $E_1$ 120.03 | $E_2$ 120.03 | $E_3$ 120.03 | $E_4$ 120.03 |
| $I_T$ 0.942 | $I_1$ 0.176 | $I_2$ 0.146 | $I_3$ 0.255 | $I_4$ 0.364 |
| $R_T$ 127.412 Ω | $R_1$ 680 Ω | $R_2$ 820 Ω | $R_3$ 470 Ω | $R_4$ 330 Ω |
| $P_T$ 113.068 | $P_1$ 21.187 | $P_2$ 17.57 | $P_3$ 30.653 | $P_4$ 43.658 |

**2.**

| | | | | |
|---|---|---|---|---|
| $E_T$ 277.153 | $E_1$ 277.153 | $E_2$ 277.153 | $E_3$ 277.153 | $E_4$ 277.153 |
| $I_T$ 0.00639 | $I_1$ 0.00231 | $I_2$ 0.00139 | $I_3$ 0.00154 | $I_4$ 0.00115 |
| $R_T$ 43.373 kΩ | $R_1$ 120 kΩ | $R_2$ 200 kΩ | $R_3$ 180 kΩ | $R_4$ 240 kΩ |
| $P_T$ 1.771 | $R_1$ 0.640 | $P_2$ 0.384 | $P_3$ 0.427 | $P_4$ 0.32 |

**3.**

| | | | | |
|---|---|---|---|---|
| $E_T$ 48 | $E_1$ 48 | $E_2$ 48 | $E_3$ 48 | $E_4$ 48 |
| $I_T$ 13.4 | $I_1$ 3 | $I_2$ 4.8 | $I_3$ 3.2 | $I_4$ 2.4 |
| $R_T$ 3.582 Ω | $R_1$ 16 Ω | $R_2$ 10 Ω | $R_3$ 15 Ω | $R_4$ 20 Ω |
| $P_T$ 643.2 | $P_1$ 144 | $P_2$ 230.4 | $P_3$ 153.6 | $P_4$ 115.2 |

**4.**

| | | | | |
|---|---|---|---|---|
| $E_T$ 240.267 | $E_1$ 240.267 | $E_2$ 240.267 | $E_3$ 240.267 | $E_4$ 240.267 |
| $I_T$ 0.0143 | $I_1$ 0.00293 | $I_2$ 0.0032 | $I_3$ 0.00492 | $I_4$ 0.00387 |
| $R_T$ 16,802 kΩ | $R_1$ 82 kΩ | $R_2$ 75 kΩ | $R_3$ 56 kΩ | $R_4$ 62 kΩ |
| $P_T$ 3.436 | $P_1$ 0.704 | $P_2$ 0.78 | $P_3$ 1.031 | $P_4$ 0.931 |

**5.**

$R_T = 83.735$ Ω     $I_1 = 0.0116$ A     $I_2 = 0.00891$ A
$I_3 = 0.0140$ A     $I_4 = 0.0155$ A

**6.**

$R_T = 53,171.514$ Ω    $I_1 = 1.18$ mA     $I_2 = 0.886$ mA
$I_3 = 0.742$ mA     $I_4 = 3.19$ mA

## Unit 8 Combination Circuits

**1.**

| | | | | |
|---|---|---|---|---|
| $I_T$ 0.0241 | $E_1$ 36.183 | $E_2$ 12.103 | $E_3$ 26.6 | $E_4$ 38.651 |
| $R_T$ 3109.216 | $I_1$ 0.0241 | $I_2$ 0.0133 | $I_3$ 0.0133 | $I_4$ 0.0107 |

**2.**

| | | | | |
|---|---|---|---|---|
| $I_T$ 0.00946 | $E_1$ 208.108 | $E_2$ 85.135 | $E_3$ 56.757 | $E_4$ 141.892 |
| $R_T$ 37 kΩ | $I_1$ 0.00946 | $I_2$ 0.00473 | $I_3$ 0.00473 | $I_4$ 0.00473 |

3.

| | | | | |
|---|---|---|---|---|
| $I_T$ 0.0616 | $E_1$ 5.053 | $E_2$ 5.451 | $E_3$ 7.496 | $E_4$ 12.947 |
| $R_T$ 292.118 | $I_1$ 0.0616 | $I_2$ 0.0341 | $I_3$ 0.0341 | $I_4$ 0.0275 |

4.

| | | | | |
|---|---|---|---|---|
| $E_T$ 712.949 | $E_1$ 712 949 | $E_2$ 295.471 | $E_3$ 417.472 | $E_4$ 417.472 |
| $R_T$ 891.187 | $I_1$ 0.475 | $I_2$ 0.325 | $I_3$ 0.209 | $I_4$ 0.116 |

5.

| | | | | |
|---|---|---|---|---|
| $E_T$ 8033.068 | $E_1$ 8033.068 | $E_2$ 4272.908 | $E_3$ 3760.159 | $E_4$ 3760.159 |
| $R_T$ 12,358.56 | $I_1$ 0.365 | $I_2$ 0.285 | $I_3$ 0.171 | $I_4$ 0.114 |

6.

| | | | | |
|---|---|---|---|---|
| $E_T$ 58.551 | $E_1$ 58.551 | $E_2$ 19.708 | $E_3$ 38.843 | $E_4$ 38.843 |
| $R_T$ 48.792 | $I_1$ 0.781 | $I_2$ 0.419 | $I_3$ 0.176 | $I_4$ 0.243 |

7.

| | | | | |
|---|---|---|---|---|
| $E_T$ 250 | $E_1$ 55 | $E_2$ 50 | $E_3$ 18.8 | $E_4$ 11.2 |
| $I_T$ 0.25 | $I_1$ 0.25 | $I_2$ 0.1 | $I_3$ 0.04 | $I_4$ 0.04 |
| $R_T$ 1000 | $R_1$ 220 | $R_2$ 500 | $R_3$ 470 | $R_4$ 280 |
| $P_T$ 62.6 | $P_1$ 13.75 | $P_2$ 5 | $P_3$ 0.752 | $P_4$ 0.448 |

| | | | | |
|---|---|---|---|---|
| $E_5$ 40 | $E_6$ 30 | $E_7$ 52.5 | $E_8$ 67.5 | $E_9$ 75 |
| $I_5$ 0.1 | $I_6$ 0.06 | $I_7$ 0.15 | $I_8$ 0.15 | $I_9$ 0.25 |
| $R_5$ 400 | $R_6$ 500 | $R_7$ 350 | $R_8$ 450 | $R_9$ 300 |
| $P_5$ 4 | $P_6$ 1.8 | $P_7$ 7.875 | $P_8$ 10.125 | $P_9$ 18.75 |

8.

| | | | | |
|---|---|---|---|---|
| $E_T$ 24 | $E_1$ 3.36 | $E_1$ 4.8 | $E_3$ 0.48 | $E_4$ 1.248 |
| $I_T$ 0.024 | $I_1$ 0.024 | $I_2$ 0.0096 | $I_3$ 0.00384 | $I_4$ 0.00384 |
| $R_T$ 1000 | $R_1$ 140 | $R_2$ 500 | $R_3$ 125 | $R_4$ 325 |
| $P_T$ 0.576 | $P_1$ 0.0806 | $P_2$ 0.0461 | $P_3$ 0.00184 | $P_4$ 0.00479 |

| | | | | |
|---|---|---|---|---|
| $E_5$ 2.112 | $E_6$ 1.728 | $E_7$ 3.6 | $E_8$ 5.04 | $E_9$ 12 |
| $I_5$ 0.0096 | $I_6$ 0.00576 | $I_7$ 0.0144 | $I_8$ 0.0144 | $I_9$ 0.024 |
| $R_5$ 220 | $R_6$ 300 | $R_7$ 250 | $R_8$ 350 | $R_9$ 500 |
| $P_5$ 0.0203 | $P_6$ 0.00995 | $P_7$ 0.0518 | $P_8$ 0.0726 | $P_9$ 0.288 |

To solve the following Kirchhoff's Law problems, refer to the circuit shown in *Figure 8-59.*

9.

| | | | |
|---|---|---|---|
| $E_{S1} = 12V$ | $E_1 = 0.898$ | $E_2 = 20.9$ | $E_3 = 11.1$ |
| $E_{S2} = 32V$ | $I_1 = 0.00132$ | $I_2 = 0.0209$ | $I_3 = 0.0222$ |
| $R_1 = 680\Omega$ | $R_2 = 1000\Omega$ | $R_3 = 500\Omega$ | |

10.

| | | | |
|---|---|---|---|
| $E_{S1} = 3V$ | $E_1 = 1.822$ | $E_2 = 0.322$ | $E_3 = 1.18$ |
| $E_{S2} = 1.5V$ | $I_1 = 0.00911$ | $I_2 = 0.00268$ | $I_3 = 0.0118$ |
| $R_1 = 200\Omega$ | $R_2 = 120\Omega$ | $R_3 = 100\Omega$ | |

11.

| | | | |
|---|---|---|---|
| $E_{S1} = 6V$ | $E_1 = -22.64$ | $E_2 = 31.333$ | $E_3 = 28.8$ |
| $E_{S2} = 60V$ | $I_1 = -0.0141$ | $I_2 = 0.0261$ | $I_3 = 0.012$ |
| $R_1 = 1.6K\Omega$ | $R_2 = 1.2K\Omega$ | $R_3 = 2.4K\Omega$ | |

To answer the following questions, refer to the circuit shown in *Figure 8-54.*

12.

Find the Thevenin equivalent voltage and resistance across terminals A and B.

$E_S = 32V$ $\qquad$ $R_1 = 4\Omega$ $\qquad$ $R_2 = 6\Omega$ $\qquad$ $E_{TH} = 19.2$ $\qquad$ $R_{TH} = 2.4\ \Omega$

13.

$E_S = 18V$ $\qquad$ $R_1 = 2.5\Omega$ $\qquad$ $R_2 = 12\Omega$ $\qquad$ $E_{TH} = 14.896$ $\quad$ $R_{TH} = 2.07\ \Omega$

14.

Find the Norton equivalent current and resistance across terminals A and B.

$E_S = 10V$ $\qquad$ $R_1 = 3\Omega$ $\qquad$ $R_2 = 7\Omega$ $\qquad$ $I_N = 3.33$ $\qquad$ $R_N = 2.1\ \Omega$

15.

$E_S = 48V$ $\qquad$ $R_1 = 12\Omega$ $\qquad$ $R_2 = 64\Omega$ $\qquad$ $I_N = 4$ $\qquad$ $R_N = 10.1\ \Omega$

## Unit 9 Measuring Instruments

1. First find the voltage necessary to deflect the meter movement full scale.

$$E = 0.000100 \times 5000$$
$$E = 0.5\ V$$

The series resistor must have a full scale voltage value of 9.5 V with a current flow of 100 μA.

$$R = \frac{9.5}{0.000100}$$

$$R = 950\ k\Omega$$

2.

$$R_1 = \frac{14.5}{0.000100} = 145 \text{ k}\Omega$$

$$R_2 = \frac{45}{0.000100} = 450 \text{ k}\Omega$$

$$R_3 = \frac{90}{0.000100} = 900 \text{ k}\Omega$$

$$R_4 = \frac{150}{0.000100} = 1.5 \text{ M}\Omega$$

3. First find the amount of current that must flow through the shunt.

$$R_S = 2 - 0.000500$$
$$R_S = 1.9995 \text{ A}$$

Next find the amount of resistance necessary to produce a voltage drop of 50 mV when that amount of current flows through it.

$$R_S = \frac{0.050}{1.9995}$$
$$R_S = 0.025 \ \Omega$$

4. The first step is to find the resistance of the meter movement.

$$R_M = \frac{0.050}{0.000500}$$
$$R_M = 100 \ \Omega$$

The resistance of the total shunt will be found using the formula

$$R_S = \frac{I_M \times R_M}{I_T}$$

(NOTE: When the ammeter is set for a value of 0.5 A full scale, the total shunt is connected across the meter movement.)

$$R_S = \frac{0.000500 \times 100}{0.5}$$
$$R_S = 0.1 \ \Omega$$

The value of resistor $R_1$ will be computed next. Resistor $R_1$ is used to produce the 5-A scale. To find this value, use the formula

$$R_1 = \frac{I_M \times R_{SUM}}{I_T}$$

$$R_1 = \frac{0.000500 \times 100.1}{5}$$

$$R_1 = 0.01 \ \Omega$$

When the meter is connected to the 1-A range, the resistors $R_1$ and $R_2$ are connected in series with each other. The value of these two resistors can be found using the above formula.

$$R_1 + R_2 = \frac{0.000500 \times 100.1}{1}$$

$$R_1 + R_2 = 0.05 \ \Omega$$

Since the value of resistor $R_1$ is known, the value of $R_2$ can be found by subtracting the total of $R_1 + R_2$ from the value of $R_1$.

$$R_2 = 0.05 - 0.01 = 0.04 \ \Omega$$

The value of resistor $R_3$ can be found by subtracting the value of resistors $R_1$ and $R_2$ from the total value of the shunt.

$$R_3 = 0.1 - 0.04 - 0.01 = 0.05 \ \Omega$$

5.

$$R_S = \frac{2.5}{0.000010} = 250 \ k\Omega$$

## Unit 10 Using Wire Tables and Determining Conductor Sizes

1.

$$10.4 \times 450 = 4680$$

$$\frac{4680}{66,370} = .0705 \ \Omega$$

2.

$$1.817 \times 16,510 = 29,998.67$$

$$\frac{29,998.67}{500} = 60 \ (\text{Iron})$$

3.

    **430** (from Table 310-16) **X 0.71** (correction factor) **= 305.3 A**

4.

    **30 X 0.70 = 21 A** (310-16) (Note 8A)

5.

| | |
|---|---|
| Max voltage drop: | **480 X 0.05 = 24 V** |
| Length of conductor: | **300 X 2 X 0.866 = 519.6 ft** |
| Max resistance of conductors: | $\dfrac{24}{522}$ **= 0.046 Ω** |

$$\text{CM} = 10.4 \times \frac{519.6}{0.046} \qquad\qquad \text{CM} = 117{,}474.78$$

**Wire size = #2/0 AWG**

## Unit 14 Basic Trigonometry

| ∠X | ∠Y | Side A | Side B | Hyp. |
|------|------|---------|---------|---------|
| 40° | 50° | 5.142 | 6.128 | 8 |
| 57° | 33° | 72 | 46.759 | 85.85 |
| 37.1° | 52.9° | 38 | 50.249 | 63 |
| 52° | 38° | 17.919 | 14 | 22.94 |
| 48° | 42° | 173.255 | 156 | 233.138 |

## Unit 15 Alternating Current
### Sine Wave Values

| Peak volts | Inst. volts | Degrees |
|---|---|---|
| 347 | 208 | 36.8 |
| 780 | 536.9 | 43.5 |
| 80.36 | 24.3 | 17.6 |
| 224 | 5.65 | 1.44 |
| 48.7 | 43.99 | 64.6 |
| 339.4 | 240 | 45 |
| 87.2 | 23.7 | 15.77 |
| 156.9 | 155.48 | 82.3 |
| 110.4 | 62.7 | 34.6 |
| 1256 | 400 | 18.57 |
| 15,720 | 3268.37 | 12 |
| 126.86 | 72.4 | 34.8 |

### Peak, RMS, and Average Values

| Peak | RMS | Average |
|---|---|---|
| 12.7 | 8.979 | 8.09 |
| 76.073 | 53.8 | 48.42 |
| 257.3 | 182.262 | 164.2 |
| 1235 | 873.145 | 786.695 |
| 339.36 | 240 | 216 |
| 26.012 | 18.443 | 16.6 |
| 339.7 | 240.168 | 216.389 |
| 17.816 | 12.6 | 11.34 |
| 14.103 | 9.99 | 9 |
| 123.7 | 87.456 | 78.797 |
| 105.767 | 74.8 | 67.32 |
| 169.236 | 119.88 | 108 |

## Unit 16 Inductance in Alternating Current Circuits
### Inductive Circuits

| Inductance (H) | Frequency (Hz) | Induct. Rct. (Ω) |
|---|---|---|
| 1.2 | 60 | 452.4 |
| 0.085 | 400 | 213.628 |
| 0.75 | 1000 | 4712.389 |
| 0.65 | 600 | 2450.442 |
| 3.6 | 30 | 678.584 |
| 2.65 | 25 | 411.459 |
| 0.5 | 60 | 188.5 |
| 0.85 | 1200 | 6408.849 |
| 1.6 | 20 | 201.062 |
| 0.45 | 400 | 1130.973 |
| 4.8 | 80 | 2412.743 |
| 0.0065 | 1000 | 40.841 |

## Unit 17 Resistive-Inductive Series Circuits

1.

| | | |
|---|---|---|
| $E_T$ 480 | $E_R$ 247.136 | $E_L$ 411.509 |
| $I_T$ 20.595 | $I_R$ 20.595 | $I_L$ 20.595 |
| Z 23.307 | R 12 | $X_L$ 19.981 |
| VA 9885.442 | P 5089.686 | $VARs_L$ 8474.751 |
| PF 51.5% | $\angle\theta$ 59° | L 0.053 |

2.

| | | |
|---|---|---|
| $E_T$ 130 | $E_R$ 78 | $E_L$ 104 |
| $I_T$ 6.5 | $I_R$ 6.5 | $I_L$ 6.5 |
| Z 20 | R 12 | $X_L$ 16 |
| VA 845 | P 507 | $VARs_L$ 676 |
| PF 60% | $\angle\theta$ 53.13° | L 0.0424 |

3.

| | | |
|---|---|---|
| $E_T$ 120 | $E_R$ 96 | $E_L$ 72 |
| $I_T$ 1.2 | $I_R$ 1.2 | $I_L$ 1.2 |
| Z 100 | R 80 | $X_L$ 60 |
| VA 144 | P 115.2 | $VARs_L$ 86.4 |
| PF 80% | $\angle\theta$ 36.87° | L 0.15915 |

4.

| | | |
|---|---|---|
| $E_T$ 240 | $E_R$ 187.408 | $E_L$ 149.927 |
| $I_T$ 1.562 | $I_R$ 1.562 | $I_L$ 1.562 |
| Z 153.675 | R 120 | $X_L$ 96 |
| VA 374.817 | P 292.683 | $VARs_L$ 234.146 |
| PF 78% | $\angle\theta$ 38.66° | L 0.0382 |

## Unit 18 Resistive-Inductive Parallel Circuits

1.

| | | |
|---|---|---|
| $E_T$ 240 | $E_R$ 240 | $E_L$ 240 |
| $I_T$ 34.553 | $I_R$ 17.143 | $I_L$ 30 |
| Z 6.946 | R 14 | $X_L$ 8 |
| VA 8292.72 | P 4114.286 | $VARs_L$ 7200 |
| PF 49.6% | $\angle\theta$ 60.25° | L 0.02122 |

2.

| | | |
|---|---|---|
| $E_T$ 277 | $E_R$ 277 | $E_L$ 277 |
| $I_T$ 39 | $I_R$ 15 | $I_L$ 36 |
| Z 7.103 | R 18.467 | $X_L$ 7.694 |
| VA 10,803 | P 4155 | $VARs_L$ 9972 |
| PF 38.46% | $\angle\theta$ 67.38° | L 0.0204 |

3.

| | | |
|---|---|---|
| $E_T$ 72 | $E_R$ 72 | $E_L$ 72 |
| $I_T$ 2 | $I_R$ 1.6 | $I_L$ 1.2 |
| Z 36 | R 45 | $X_L$ 60 |
| VA 144 | P 115.2 | $VARs_L$ 86.4 |
| PF 80% | $\angle\theta$ 36.87° | L 0.15915 |

4.

| | | |
|---|---|---|
| $E_T$ 150 | $E_R$ 150 | $E_L$ 150 |
| $I_T$ 2.5 | $I_R$ 1.949 | $I_L$ 1.562 |
| Z 60.029 | R 76.956 | $X_L$ 96 |
| VA 374.817 | P 292.373 | $VARs_L$ 234.553 |
| PF 78% | $\angle\theta$ 38.74° | L 0.382 |

## Unit 19 Capacitors
## RC Time Constants

| Resistance | Capacitance | Time Constant | Total Time |
|:---:|:---:|:---:|:---:|
| 150 kΩ | 100 µF | 15 s | 75 s |
| 350 kΩ | 20 µF | 7 s | 35 s |
| 142,857,142.9 | 350 pF | 0.05 s | 0.25 s |
| 40 MΩ | 0.05 µF | 2 s | 10 s |
| 1.2 MΩ | 0.47 µF | 0.564 s | 2.82 s |
| 4166.67 Ω | 12 µF | 0.05 s | 0.25 s |
| 86 kΩ | 3.488 µF | 0.3 s | 1.5 s |
| 120 kΩ | 470 pF | 56.4 µs | 282 µs |
| 80 kΩ | 250 nF | 20 ms | 100 ms |
| 3.75 Ω | 8 µF | 30 µs | 150 µs |
| 100 kΩ | 1.5 µF | 150 ms | 750 ms |
| 33 kΩ | 4 µF | 132 ms | 660 ms |

## Unit 20 Capacitance in Alternating Current Circuits Capacitive Circuits

| Capacitance | $X_c$ | Frequency |
|:---:|:---:|:---:|
| 38 μF | 69.8 Ω | 60 Hz |
| 5.049 μF | 78.8 Ω | 400 Hz |
| 250 pF | 4.5 kΩ | 141.47 kHz |
| 234 μF | 0.068 Ω | 10 kHz |
| 13.263 μF | 240 Ω | 50 Hz |
| 10 μF | 36.8 Ω | 432.486 Hz |
| 560 nF | 0.142 Ω | 2 MHz |
| 176.84 nF | 15 kΩ | 60 Hz |
| 75 μF | 560 Ω | 3.789 Hz |
| 470 pF | 1.693 kHz | 200 kHz |
| 58.513 nF | 6.8 kΩ | 400 Hz |
| 34 μF | 450 Ω | 10.402 Hz |

## Unit 21 Resistive-Capacitive Series Circuits

1.

| | | |
|---|---|---|
| $E_T$ 480 | $E_R$ 288 | $E_C$ 384 |
| $I_T$ 24 | $I_R$ 24 | $I_C$ 24 |
| Z 20 Ω | R 12 Ω | $X_C$ 16 Ω |
| VA 11,520 | P 6912 | VARs$_C$ 9216 |
| PF 60% | $\angle\theta$ 53.13° | C 165.782 μF |

2.

| | | |
|---|---|---|
| $E_T$ 130 | $E_R$ 78 | $E_C$ 104 |
| $I_T$ 6.5 | $I_R$ 6.5 | $I_C$ 6.5 |
| Z 20 Ω | R 12 Ω | $X_C$ 16 Ω |
| VA 845 | P 507 | VARs$_C$ 676 |
| PF 60% | $\angle\theta$ 53.13° | C 165.782 μF |

3.

| | | |
|---|---|---|
| $E_T$ 318.584 | $E_R$ 254.867 | $E_C$ 191.15 |
| $I_T$ 0.452 | $I_R$ 0.452 | $I_C$ 0.452 |
| Z 704.832 Ω | R 563.865 Ω | $X_C$ 422.154 Ω |
| VA 144 | P 115.2 | $VARs_C$ 86.4 |
| PF 80% | $\angle\theta$ 36.87° | C 6.2833 μF |

4.

| | | |
|---|---|---|
| $E_T$ 185.874 | $E_R$ 126.394 | $E_C$ 136.285 |
| $I_T$ 1.614 | $I_R$ 1.614 | $I_C$ 1.614 |
| Z 115.163 Ω | R 78.311 Ω | $X_C$ 84.432 Ω |
| VA 300 | P 204 | $VARs_C$ 219.964 |
| PF 68% | $\angle\theta$ 47.156° | C 4.7125 μF |

## Unit 22 Resistive-Capacitive Parallel Circuits

1.

| | | |
|---|---|---|
| $E_T$ 120 | $E_R$ 120 | $E_C$ 120 |
| $I_T$ 10.463 | $I_R$ 8.571 | $I_C$ 6 |
| Z 11.469 Ω | R 14 Ω | $X_C$ 20 Ω |
| VA 1255.56 | P 1028.52 | $VARs_C$ 720 |
| PF 81.9% | $\angle\theta$ 35° | C 132.626 μF |

2.

| | | |
|---|---|---|
| $E_T$ 54 | $E_R$ 54 | $E_C$ 54 |
| $I_T$ 2.5 | $I_F$ 1.5 | $I_C$ 2 |
| Z 21.6 Ω | R 36 Ω | $X_C$ 27 Ω |
| VA 135 | P 81 | $VARs_C$ 108 |
| PF 60% | $\angle\theta$ 53.13 ° | C 14.736 μF |

3.

| | | |
|---|---|---|
| $E_T$ 307.202 | $E_R$ 307.202 | $E_C$ 307.202 |
| $I_T$ 140.99 | $I_R$ 65.6 | $I_C$ 124.8 |
| Z 2.17888 Ω | R 4.693 Ω | $X_C$ 2.461 Ω |
| VA 43.312 kVA | P 20.125 kW | $VARs_C$ 38.339 kVARs |
| PF 46.5% | $\angle\theta$ 62.27° | C 107.785 μF |

4.

| | | |
|---|---|---|
| $E_T$ 69.549 | $E_R$ 69.549 | $E_C$ 69.549 |
| $I_T$ 7.5 | $I_R$ 7 | $I_C$ 2.296 |
| Z 9.273 Ω | R 9.935 Ω | $S_C$ 30.291 Ω |
| VA 521.615 | P 486.75 | $VARs_C$ 187.5 |
| PF 94.9% | $\angle\theta$ 18.28° | C 5.254 μF |

# Unit 23 Resistive-Inductive-Capacitive Series Circuits

1.

| | | | |
|---|---|---|---|
| $E_T$ 120 | $E_R$ 72 | $E_L$ 200 | $E_C$ 104 |
| $I_T$ 2 | $I_R$ 2 | $I_L$ 2 | $I_C$ 2 |
| Z 60 Ω | R 36 Ω | $X_L$ 100 Ω | $X_C$ 52 Ω |
| VA 240 | P 144 | $VARs_L$ 400 | $VARs_C$ 208 |
| PF 60% | $\angle\theta$ 53.13 ° | L 0.265 | C 51 μF |

2.

| | | | |
|---|---|---|---|
| $E_T$ 35.678 | $E_R$ 24 | $E_L$ 21.6 | $E_C$ 48 |
| $I_T$ 0.6 | $I_R$ 0.6 | $I_L$ 0.6 | $I_C$ 0.6 |
| Z 59.464 Ω | R 40 Ω | $X_L$ 36 Ω | $X_C$ 80 Ω |
| VA 21. 407 | P 14.4 | $VARs_L$ 12.96 | $VARs_C$ 28.8 |
| PF 67.3% | $\angle\theta$ 47.7° | L 0.0143 | C 5 μF |

3.

| | | | |
|---|---|---|---|
| $E_T$ 24 | $E_R$ 14.993 | $E_L$ 74.963 | $E_C$ 56.222 |
| $I_T$ 1.249 | $I_R$ 1.249 | $I_L$ 1.249 | $I_C$ 1.249 |
| Z 19.209 | R 12 Ω | $X_L$ 60 Ω | $X_C$ 45 Ω |
| VA 29.985 | P 18.732 | $VARs_L$ 93.658 | $VARs_C$ 70.244 |
| PF 62.5% | $\angle\theta$ 51.34° | L 0.159 | C 64.6 μF |

4.

| | | | |
|---|---|---|---|
| $E_T$ 350 | $E_R$ 185 | $E_L$ 740 | $E_C$ 444 |
| $I_T$ 0.148 | $I_R$ 0.148 | $I_L$ 0.148 | $I_C$ 0.148 |
| Z 2358.495 Ω | R 1250 Ω | $X_L$ 5000 Ω | $X_C$ 3000 Ω |
| VA 51.8 | P 27.38 | $VARs_L$ 109.52 | $VARs_C$ 65.712 |
| PF 52.86% | $\angle\theta$ 58.1° | L 0.796 | C 0.053 μF |

# Unit 24 Resistive-Inductive-Capacitive Parallel Circuits

1.

| | | | |
|---|---|---|---|
| $E_T$ 120 | $E_R$ 120 | $E_L$ 120 | $E_C$ 120 |
| $I_T$ 3.387 | $I_R$ 3.333 | $I_L$ 3 | $I_C$ 2.4 |
| Z 35.43 Ω | R 36 Ω | $X_L$ 40 Ω | $X_C$ 50 Ω |
| VA 406.44 | P 400 | $VARs_L$ 360 | $VARs_C$ 288 |
| PF 98.4% | $\angle\theta$ 10.2 ° | L 0.106 | C 53.05 μF |

2.

| | | | |
|---|---|---|---|
| $E_T$ 240 | $E_R$ 240 | $E_L$ 240 | $E_C$ 240 |
| $I_T$ 22.267 | $I_R$ 16 | $I_L$ 8 | $I_C$ 24 |
| Z 10.607 Ω | R 15 Ω | $X_L$ 30 Ω | $X_C$ 10 Ω |
| VA 5430.58 | P 3840 | $VARs_L$ 1920 | $VARs_C$ 5760 |
| PF 70.7% | ∠θ 45° | L 0.0119 | C 39.79 μF |

3.

| | | | |
|---|---|---|---|
| $E_T$ 24 | $E_R$ 24 | $E_L$ 24 | $E_C$ 24 |
| $I_T$ 2.004 | $I_R$ 2 | $I_L$ 0.4 | $I_C$ 0.533 |
| Z 11.973 Ω | R 12 Ω | $X_L$ 60 Ω | $X_C$ 45 Ω |
| VA 48.106 | P 48 | $VARs_L$ 9.6 | $VARs_C$ 12.8 |
| PF 99.8% | ∠θ 3.81° | L 0.159 | C 58.94 μF |

4.

| | | | |
|---|---|---|---|
| $E_T$ 480 | $E_R$ 480 | $E_L$ 480 | $E_C$ 480 |
| $I_T$ 100 | $I_R$ 60 | $I_L$ 150 | $I_C$ 70 |
| Z 4.8 Ω | R 8 Ω | $X_L$ 3.2 Ω | $X_C$ 6.857 Ω |
| VA 48 kVA | P 28.8 kW | $VARs_L$ 72 kVARs | $VARs_C$ 33.6 kVARs |
| PF 60% | ∠θ 53.14° | L 0.509 mH | C 23.2 μF |

## Unit 25 Three-Phase Circuits

1.

| | |
|---|---|
| $E_{P(A)}$ 138.57 | $E_{P(L)}$ 240 |
| $I_{P(A)}$ 34.64 | $I_{P(L)}$ 20 |
| $E_{L(A)}$ 240 | $E_{L(L)}$ 240 |
| $I_{L(A)}$ 34.64 | $I_{L(L)}$ 34.64 |
| P 14,399.16 | $Z_{(PHASE)}$ 12 Ω |

2.

| | |
|---|---|
| $E_{P(A)}$ 4160 | $E_{P(L)}$ 2401.85 |
| $I_{P(a)}$ 23.11 | $I_{P(L)}$ 40.03 |
| $E_{L(A)}$ 4160 | $E_{L(L)}$ 4160 |
| $I_{L(A)}$ 40.03 | $I_{L(L)}$ 40.03 |
| P 288,420.95 | $Z_{(PHASE)}$ 60 Ω |

3.

| | | |
|---|---|---|
| $E_{P(A)}$ 323.33 | $E_{P(L1)}$ 323.33 | $E_{P(L2)}$ 560 |
| $I_{P(A)}$ 185.91 | $I_{P(L1)}$ 64.67 | $I_{P(L2)}$ 70 |
| $E_{L(A)}$ 560 | $E_{L(L1)}$ 560 | $E_{L(L2)}$ 560 |
| $I_{L(A)}$ 185.91 | $I_{L(L1)}$ 64.67 | $I_{L(L2)}$ 121.24 |
| P 180,317.83 | $Z_{(PHASE)}$ 5 Ω | $Z_{(PHASE)}$ 8 Ω |

4.

| | | | |
|---|---|---|---|
| $E_{P(A)}$ 277.14 | $E_{P(L1)}$ 277.14 | $E_{P(L2)}$ 480 | $E_{P(L3)}$ 277.14 |
| $I_{P(A)}$ 33.49 | $I_{P(L1)}$ 23.1 | $I_{P(L2)}$ 30 | $I_{P(L3)}$ 27.71 |
| $E_{L(A)}$ 480 | $E_{L(L1)}$ 480 | $E_{L(L2)}$ 480 | $E_{L(L3)}$ 480 |
| $I_{L(A)}$ 33.49 | $I_{L(L1)}$ 23.1 | $I_{L(L2)}$ 51.96 | $I_{L(L3)}$ 27.71 |
| VA 27,843.39 | $R_{(PHASE)}$ 12 Ω | $X_{L(PHASE)}$ 16 Ω | $X_{C(PHASE)}$ 10 Ω |
| | P 19, 204.42 | $VARs_L$ 43,197.47 | $VARs_C$ 23.037 |

# Unit 26 Single Phase Transformers

1.

| | |
|---|---|
| $E_P$ 120 | $E_S$ 24 |
| $I_P$ 1.6 | $I_S$ 8 |
| $N_P$ 300 | $N_S$ 60 |
| Ratio 5:1 | $Z = 3\ \Omega$ |

2.

| | |
|---|---|
| $E_P$ 240 | $E_S$ 320 |
| $I_P$ 0.853 | $I_S$ 0.643 |
| $N_P$ 210 | $N_S$ 280 |
| Ratio 1:1.333 | $Z = 500\ \Omega$ |

3.

| | |
|---|---|
| $E_P$ 64 | $E_S$ 160 |
| $I_P$ 33.333 | $I_S$ 13.333 |
| $N_P$ 32 | $N_S$ 80 |
| Ratio 1:2.5 | $Z = 12\ \Omega$ |

4.

| | |
|---|---|
| $E_P$ 48 | $E_S$ 240 |
| $I_P$ 3.333 | $I_S$ 0.667 |
| $N_P$ 220 | $N_S$ 1100 |
| Ratio 1:5 | $Z = 360\ \Omega$ |

5.

| | |
|---|---|
| $E_P$ 35.848 | $E_S$ 182 |
| $I_P$ 16.5 | $I_S$ 3.25 |
| $N_P$ 87 | $N_S$ 450 |
| Ratio 1:5.077 | $Z = 56\ \Omega$ |

6.

| | |
|---|---|
| $E_P$ 480 | $E_S$ 916.346 |
| $I_P$ 1.458 | $I_S$ 0.764 |
| $N_P$ 275 | $N_S$ 525 |
| Ratio 1:1.909 | $Z = 1.2\ k\Omega$ |

7.

| | | | |
|---|---|---|---|
| $E_P$ 208 | $E_{S1}$ 320 | $E_{S2}$ 120 | $E_{S3}$ 24 |
| $I_P$ 11.93 | $I_{S1}$ 0.0267 | $I_{S2}$ 20 | $I_{S3}$ 3 |
| $N_P$ 800 | $N_{S1}$ 1231 | $N_{S2}$ 462 | $N_{S3}$ 92 |
| | Ratio 1 1:1.54 | Ratio 2 1.73:1 | Ratio 3 1:8.67 |
| | $R_1$ 12 k$\Omega$ | $R_2$ 6 $\Omega$ | $R_3$ 8 $\Omega$ |

8.

| | | | |
|---|---|---|---|
| $E_P$ 277 | $E_{S1}$ 480 | $E_{S2}$ 208 | $E_{S3}$ 120 |
| $I_P$ 8.93 | $I_{S1}$ 2.4 | $I_{S2}$ 3.47 | $I_{S3}$ 5 |
| $N_P$ 350 | $N_{S1}$ 606 | $N_{S2}$ 263 | $N_{S3}$ 152 |
| | Ratio 1 1:1.73 | Ratio 2 1.33:1 | Ratio 3 2.31:1 |
| | $R_1$ 200 $\Omega$ | $R_2$ 60 $\Omega$ | $R_3$ 24 $\Omega$ |

## Unit 27 Three-Phase Transformers

1.

| | | |
|---|---|---|
| $E_P$ 2401.8 | $E_P$ 440 | $E_P$ 254.04 |
| $I_P$ 7.67 | $I_P$ 41.9 | $I_P$ 72.58 |
| $E_L$ 4160 | $E_L$ 440 | $E_L$ 440 |
| $I_L$ 7.67 | $I_L$ 72.58 | $I_L$ 72.58 |
| Ratio 5.46:1 | Z 3.5 $\Omega$ | |

2.

| | | |
|---|---|---|
| $E_P$ 4157.04 | $E_P$ 240 | $E_P$ 138.57 |
| $I_P$ 1.15 | $I_P$ 20 | $I_P$ 34.64 |
| $E_L$ 7200 | $E_L$ 240 | $E_L$ 240 |
| $I_L$ 1.15 | $I_L$ 34.64 | $E_L$ 34.64 |
| Ratio 17.32:1 | Z 4 $\Omega$ | |

3.

| | | |
|---|---|---|
| $E_P$ 13,800 | $E_P$ 277 | $E_P$ 480 |
| $I_P$ 6.68 | $I_P$ 332.54 | $I_P$ 192 |
| $E_L$ 13,800 | $E_L$ 480 | $E_L$ 480 |
| $I_L$ 11.57 | $I_L$ 332.54 | $I_L$ 332.54 |
| Ratio 49.76:1 | Z 2.5 $\Omega$ | |

4.

| | | |
|---|---|---|
| $E_P$ 23,000 | $E_P$ 120 | $E_P$ 208 |
| $I_P$ 0.626 | $I_P$ 120.08 | $I_P$ 69.33 |
| $E_L$ 23,000 | $E_L$ 208 | $E_L$ 208 |
| $I_L$ 1.08 | $I_L$ 120.08 | $I_L$ 120.08 |
| Ratio 191.66:1 | Z 3 $\Omega$ | |

# Glossary

## A

**AC (alternating current)** current that reverses its direction of flow periodically. Reversals generally occur at regular intervals.

**Across-the-line** a method of motor starting that connects the motor directly to the supply line on starting or running. Also known as full voltage starting.

**Air gap** the space between two magnetically related components.

**Alternator** a machine used to generate alternating current by rotating conductors through a magnetic field.

**Ambient temperature** the temperature surrounding a device.

**American Wire Gauge (AWG)** a measurement of the diameter of a wire. The gauge scale was formerly known as the Brown and Sharp scale. The scale has a fixed constant of 1.123 between gauge sizes.

**Ammeter** an instrument used to measure the flow of current.

**Amortisseur winding** a squirrel-cage winding on the rotor of a synchronous motor used for starting purposes only.

**Ampacity** the maximum current-carrying capacity of a wire or device.

**Ampere (A)** a unit of measure for the rate of current flow. One amp equals one coulomb per second.

**Ampere-turns** a basic unit for measurement of magnetism. The product of number of turns of wire times current flow.

**Amp-hour (A-hr)** a unit of measure for describing the current capacity of a battery or a cell.

**Amplifier** a device used to increase a signal.

**Amplitude** the highest value reached by a signal, voltage, or current.

**Analog voltmeter** a voltmeter that uses a meter movement to indicate the voltage value. Analog meters use a pointer and scale.

**Anode** the positive terminal of an electrical device.

**Apparent power** the value found by multiplying the applied voltage by the total current of an AC circuit. Apparent power is measured in volt-amps (VA) and should not be confused with true power, measured in watts.

**Applied voltage** the amount of voltage connected to a circuit or device.

**Armature** the rotating member of a motor or generator. The armature generally contains windings and a commutator.

**Armature reaction** the twisting or bending of the main magnetic field of a motor or generator. Armature reaction is proportional to armature current.

**ASA** American Standards Association

**Atom** the smallest part of an element that contains all of the properties of that element.

**Attenuator** a device that decreases the amount of signal voltage or current.

**Automatic** self-acting, operation by its own mechanical or electrical mechanism.

**Autotransformer** a transformer that uses only one winding for both primary and secondary.

## B

**Back-voltage** the induced voltage in the coil of an inductor or generator that opposes the applied voltage.

**Base** the semiconductor region between the collector and emitter of a transistor. The base controls the current flow through the collector-emitter circuit.

**Battery** a device used to convert chemical energy into electrical energy. A group of voltaic cells connected together in a series or parallel connection.

**Bias** a DC voltage applied to the base of a transistor to preset its operating point.

**Bimetallic strip** a strip made by bonding two unlike metals together. The metals expand at different temperatures when heated, causing a bending or warping action.

**Branch circuit** that portion of a wiring system that extends beyond the circuit protective device, such as a fuse or circuit breaker.

**Breakdown torque** the maximum amount of torque that can be developed by a motor at rated voltage and frequency before an abrupt change in speed occurs.

**Bridge circuit** a circuit that consists of four sections connected in series to form a closed loop.

**Bridge rectifier** a device constructed with four diodes that converts both positive and negative cycles of AC voltage into DC voltage. The bridge rectifier is one type of full-wave rectifier.

**Brush** a sliding contact, generally made of carbon, used to provide connection to rotating parts of machines.

**Bus way** an enclosed system used for power transmission that is voltage- and current-rated.

## C

**Capacitance** the electrical size of a capacitor.

**Capacitive reactance** the current-limiting property of a capacitor in an alternating current circuit.

**Capacitor** a device made with two conductive plates separated by an insulator or dielectric.

**Capacitor-start induction-run motor** a single-phase induction motor that uses a capacitor connected in series with the start winding to increase starting torque.

**Cathode** the negative terminal of an electrical device.

**Charging current** the current flowing from an electrical source to a capacitor.

**Center-tapped transformer** a transformer that has a wire connected to the electrical midpoint of its winding. Generally the secondary winding is tapped.

**Choke** an inductor designed to present an impedance to AC current or to be used as the current filter of a DC power supply.

**Circuit** an electrical path between two points.

**Circuit breaker** a device designed to open under an abnormal amount of current flow. The device is not damaged and may be used repeatedly. Rated by voltage, current, and horsepower.

**Clock timer** a time-delay device that uses an electric clock to measure the delay time.

**Coercive force** a material's ability to retain magnetism. Also called retentivity.

**Collapse of a magnetic field.** Occurs when a magnetic field suddenly changes from its maximum value to a zero value.

**Collector** a semiconductor region of a transistor that must be connected to the same polarity as the base.

**Commutating field** a field used in direct current machines to help overcome the problems of armature reaction. The commutating field connects in series with the armature and is also known as the interpole winding.

**Commutator** strips or bars of metal insulated from each other and arranged around one end of an armature. They provide connection between the armature windings and the brushes. The commutator is used to ensure proper direction of current flow through the armature windings.

**Comparator** a device or circuit that compares two like quantities such as voltage levels.

**Compensating winding** the winding embedded in the main field poles of a DC machine. The compensating winding is used to help overcome armature reaction.

**Compound DC machine** a generator or motor that uses both series and shunt fields windings. DC machines may be connected long shunt compound, short shunt compound, cumulative-compound, or differential-compound.

**Conduction level** the point at which an amount of voltage or current will cause a device to conduct.

**Conductor** a device or material that permits current to flow through it easily.

**Contact** a conducting part of a relay that acts as a switch to connect or disconnect a circuit or component.

**Continuity** a complete path for current flow.

**Conventional current flow** a theory that considers current to flow from the most positive source to the most negative source.

**Copper losses** power loss due to current flowing through wire. Copper loss is proportional to the resistance of the wire and the square of the current.

**Core** magnetic material used to form the center of a coil or transformer. The core may be made of a nonmagnetic conductor (air core), iron, or some other magnetic material.

**Core losses** the power loss in the core material caused by eddy current induction and hysteresis loss.

**Cosine** in trigonometry, it is the ratio of the adjacent side of the angle and the hypotenuse.

**Coulomb** a quantity of electrons equal to 6.25 x $10^{18}$.

**Counter-EMF (CEMF)** the voltage induced in the armature of a DC motor that opposes the applied voltage and limits armature current.

**Countertorque** the magnetic force developed in the armature of a generator that makes the shaft hard to turn. Countertorque is proportional to armature current and is a measure of the electrical energy produced by the generator.

**Current** the rate of flow of electrons.

**Current rating** the amount of current flow a device is designed to withstand.

**Current relay** a relay that is operated by a predetermined amount of current flow. Current relays are often used as one type of starting relay for air conditioning and refrigeration equipment.

**Cycle** one complete AC wave form.

**D**

**D'Arsonval meter** a meter movement using a permanent magnet and a coil of wire. The basic meter movement used in many analog type voltmeters, ammeters, and ohmmeters.

**DC (direct current)** current that does not reverse its direction of flow.

**Delta connection** a circuit formed by connecting three electrical devices in series to form a closed loop. It is used most often in three-phase connections.

**Diac** a bidirectional diode.

**Diamagnetic** a material that will not conduct magnetic lines of flux. Diamagnetic materials have a permeability rating less than that of air (1).

**Dielectric** an electric insulator.

**Dielectric breakdown** the point at which the insulating material separating two electrical charges permits current to flow between the two charges. Dielectric breakdown is often caused by excessive voltage. excessive heat, or both.

**Digital device** a device that has only two states of operation, on or off.

**Digital logic** circuit elements connected in such a manner as to solve problems using components that have only two states of operation.

**Digital voltmeter** a voltmeter that uses direct-reading numerical display as opposed to a meter movement.

**Diode** a two-element device that permits current to flow through it in only one direction.

**Disconnecting means (disconnect)** a device or group of devices used to disconnect a circuit or device from its source of supply.

**Domain** a group of atoms aligning themselves north and south to create a magnetic material.

**Dot notation** dots placed beside transformer windings on a schematic to indicate relative polarity between different windings.

**DVM** abbreviation for digital voltmeter.

**Dynamic braking** (1) Using a DC motor as a

generator to produce countertorque and thereby produce a braking action. (2) Applying direct current to the stator winding of an AC induction motor or cause a magnetic braking action.

# E

**Eddy current** circular induced current contrary to the main currents. Eddy currents are a source of heat and power loss in magnetically operated devices.

**Electrical interlock** when the contacts of one device or circuit prevent the operation of the some other device or circuit.

**Electric controller** a device or group of devices used to govern in some predetermined manner the operation of a circuit or piece of electrical apparatus.

**Electrodynamometer** a machine used to measure the torque developed by a motor or engine for the purpose of determining output horsepower.

**Electrolysis** the decomposition of a chemical compound or metals caused by an electric current.

**Electrolyte** a chemical compound capable of conducting electric current by being broken down into ions. Electrolytes can be acids or alkalis.

**Electromotive force (EMF)** electrical pressure, the force that pushes electrons through a wire.

**Electron** one of the three major parts of an atom. The electron carries a negative charge.

**Electronic control** a control circuit that uses solid state devices as control components.

**Electrostatic field** the field of force that surrounds a charged object. The term is often used to describe the force of a charged capacitor.

**Element** (1) One of the basic building blocks of nature. An atom is the smallest part of an element. (2) One part of a group of devices.

**Emitter** the semiconductor region of a transistor that must be connected to a polarity different from that of the base.

**Enclosure** mechanical, electrical, or environmental protection for components used in a system.

**Eutectic alloy** a metal with a low and sharp melting point used in thermal overload relay.

**Excitation current** the direct current used to produce electromagnetism in the fields of a DC motor or generator, or in the rotor of an alternator or synchronous motor.

# F

**Farad (F)** the basic unit of capacitance.

**Feeder** the circuit conductor between either the service equipment or the generator switchboard of an isolated plant and the branch circuit overcurrent protective device.

**Femto-** a metric prefix corresponding to $10^{-15}$.

**Ferromagnetic** a metal that will conduct magnetic lines of force easily, such as iron (ferrum). Ferromagnetic materials have a permeability much greater than that of air (1).

**Field loss relay (FLR)** a current relay connected in series with the shunt field of a direct current motor. The relay causes power to be disconnected from the armature in the event that field current should drop below a certain level.

**Filter** a device used to remove the ripple produced by a rectifier.

**Flashing the field** method used to produce residual magnetism in the pole pieces of a DC machine. It is done by applying full voltage to the field winding for a period of not less than 30 s.

**Flat-compounding** setting the strength of the series field in a DC generator so that the output voltage will be the same at full load as it is at no load.

**Flux** magnetic lines of force.

**Flux density** the number of magnetic lines contained in a certain area. The area measurement depends on the system of measurement.

**Frequency** the number of complete cycles of AC voltage that occur in 1 s.

**Full-load torque** the amount of torque necessary to produce the full horsepower of a motor at rated speed.

**Fuse** a device used to protect a circuit or electrical device from excessive current. Fuses operate by melting a metal link when current becomes excessive.

## G

**Gain** the increase in signal power produced by an amplifier.

**Galvanometer** a meter movement requiring microamperes to cause a full-scale deflection. Many galvanometers have a zero center, which permits them to measure both positive and negative values.

**Gate** (1) A device that has multiple inputs and a single output. There are five basic types of gates: *and, or, nand, nor,* and *inverter.* (2) One terminal of some electronic devices such as SCRs, Triacs, and field effect transistors (FETs).

**Gauss** a unit of measure in the CGS system. One gauss equals one maxwell per square centimeter.

**Generator** a device used to convert mechanical energy into electrical energy.

**Giga-** a metric prefix meaning one billion ($10^9$).

**Gilbert** a basic unit of magnetic in the CGS system.

## H

**Heat sink** a metallic device designed to increase the surface area of an electronic component for the purpose of removing heat at a faster rate.

**Henry (H)** the basic unit of inductance.

**Hermetic** completely enclosed. Airtight.

**Hertz (Hz)** the international unit of frequency.

**Holding contacts** contacts used for the purpose of maintaining current flow to the coil of a relay.

**Holding current** the amount of current needed to keep an SCR or Triac turned on.

**Horsepower** a measure of power for electrical and mechanical devices.

**Hydrometer** a device used to measure the spe-

cific gravity of a fluid, such as the electrolyte used in a battery.

**Hypotenuse** the longest side of a right triangle.

**Hysteresis loop** a graphic curve that shows the value of magnetizing force for a particular type of material.

**Hysteresis loss** power loss in a conductive material caused by molecular friction. Hysteresis loss is proportional to frequency.

## I

**Impedance** the total opposition to current flow in an electrical circuit.

**Incandescence** the ability to produce light as a result of heating.

**Induced current** current produced in a conductor by the cutting action of a magnetic field.

**Inductive reactance** the current-limiting property of an inductor in an alternating current circuit.

**Inductor** a coil.

**Input voltage** the amount of voltage connected to a device or circuit.

**Insulator** a material used to electrically isolate two conductive surfaces.

**Interlock** a device used to prevent some action from taking place in a piece of equipment or circuit until some other action has occurred.

**Interpole** small pole piece placed between the main field poles of a DC machine to reduce armature reaction.

**Ion** a charged atom.

**Isolation transformer** a transformer whose secondary winding is electrically isolated from its primary winding.

## J

**Joule (J)** a basic unit of electrical energy. A joule is the amount of power used when 1 A flows through 1 $\Omega$ for 1 s. A joule is equal to 1 W/s.

**Jumper** a short piece of conductor used to make connection between components or a break in a circuit.

**Junction diode** a diode that is made by joining together two pieces of semiconductor material.

# K

**Kick-back diode** a diode used to eliminate the voltage spike induced in a coil by the collapse of a magnetic field.

**Kilo-** a metric prefix meaning 1000 ($10^3$).

**Kinetic energy** the energy of a moving object, such as the energy of a flywheel in motion.

# L

**Lamination** one thickness of the sheet material used to construct the core material for transformers, inductors, and alternating current motors.

**LED (light-emitting diode)** a diode that will produce light when current flows through it.

**Leyden jar** a glass jar used to store electrical charges in the very early days of electrical experimentation. The Leyden jar was constructed by lining the inside and outside of the jar with metal foil. The Leyden jar was a basic capacitor.

**Limit switch** a mechanically operated switch that detects the position or movement of an object.

**Linear** when used in comparing electrical devices or quantities, signifies that one unit is equal to another.

**Load center** generally the service entrance. A point from which branch circuits originate.

**Locked rotor current** the amount of current produced when voltage is applied to a motor and the rotor is not turning.

**Locked rotor torque** the amount of torque produced by a motor at the time of starting.

**Lockout** a mechanical device used to prevent the operation of some other component.

**Long shunt compound** the connection of field windings in a DC machine where the shunt field is connected in parallel with both the armature and series field.

**Low-voltage protection** a magnetic relay circuit so connected that a drop in voltage causes the motor starter to disconnect the motor from the line.

# M

**Magnetic contactor** a contactor operated electro-mechanically.

**Magnetic field** the space in which a magnetic force exists.

**Magnetomotive force (mmf)** the magnetic force produced by current flowing through a conductor or coil.

**Maintaining contacts** used to maintain the coil circuit in a relay control circuit. The contact is connected in parallel with the start push button. Also known as holding or sealing contacts.

**Manual controller** a controller operated by hand at the location of the controller.

**Maxwell** a measure of magnetic flux in the CGS system.

**Mega-** a metric prefix meaning 1,000,000 ($10^6$).

**Mica** a mineral used as an electrical insulator.

**Micro-** a metric prefix meaning 1/1,000,000 ($10^{-6}$).

**Microprocessor** a small computer. The central processing unit is generally made from a single integrated circuit.

**Mil** a unit for measuring the diameter of a wire equal to 1/1000 of an inch.

**Mil-foot** a standard for measuring the resistivity of wire. A mil-foot is the resistance of a piece of wire one mil in diameter and one foot in length.

**Milli-** a metric prefix for 1/1000 ($10^{-3}$).

**Mode** a state or condition.

**Motor** a device used to convert electrical energy into mechanical energy.

**Motor controller** a device used to control the operation of a motor.

**Multispeed motor** a motor that can be operated at more than one speed.

# N

**Nano-** metric prefix meaning one-billionth ($10^{-9}$).

**Negative** one polarity of a voltage, current, or charge.

**NEMA** National Electrical Manufacturers Association.

**NEMA ratings** electrical control device ratings of voltage, current, horsepower, and interrupting capability given by NEMA.

**Neutron** one of the principal parts of an atom. The neutron has no charge and is part of the nucleus.

**Node** a joining point where electrical connections are made.

**Noninductive load** an electrical load that does not have induced voltages caused by a coil. Noninductive loads are generally considered to be resistive, buy they can be capacitive.

**Nonreversing** a device that can be operated in only one direction.

**Normally closed** the contact of a relay that is closed when the coil is deenergized.

**Normally open** the contact of a relay that is open when the coil is deenergized.

# O

**Off-delay timer** a timer that delays changing its contacts back to their normal position when the coil is deenergized.

**Ohm ($\Omega$)** the unit of measure for electrical resistance.

**Ohmmeter** a device used to measure resistance.

**On-delay timer** a timer that delays changing the position of its contacts when the coil is energized.

**Operational amplifier (OP AMP)** an integrated circuit used as an amplifier.

**Opto-isolator** a device used to connect different sections of a circuit by means of a lightbeam.

**Oscillator** a device used to change DC voltage into AC voltage.

**Oscilloscope** a voltmeter that displays a wave form of voltage in proportion to its amplitude with respect to time.

**Out-of-phase** the condition in which two components do not reach their positive or negative peaks at the same time.

**Overcompounded** the condition of a DC generator when the series field is too strong. It is characterized by the output voltage being greater at full load than it is at no load.

**Overexcited** a condition that occurs when the DC current supplying excitation current to the rotor of a synchronous motor is greater than necessary.

**Overload relay** a relay used to protect a motor from damage caused by overloads. The overload relay senses motor current and disconnects the motor from the line if the current is excessive for certain length of time.

# P

**Panelboard** a metallic or nonmetallic panel used to mount electrical controls, equipment, or devices.

**Parallel circuit** a circuit that contains more than one path for current flow.

**Paramagnetic** a material that has a permeability slightly greater than that of air (1).

**Peak-inverse/peak-reverse voltage** the rating of a semiconductor device that indicates the maximum amount of voltage in the reverse direction that can be applied to the device.

**Peak-to-peak voltage** the amplitude of AC voltage measured from its positive peak to its negative peak.

**Peak voltage** the amplitude of voltage measured form zero to its highest value.

**Permalloy** an alloy used in the construction of electromagnets; approximately 78% nickel and 21% iron.

**Permeability** a measurement of a material's ability to conduct magnetic lines of flux. The standard is air, which has a permeability of 1.

**Phase shift** a change in the phase relationship between two quantities of voltage or current.

**Photoconductive** a material that changes its resistance due to the amount of light.

**Pico-** a metric prefix for one-trillionth ($10^{-12}$).

**Piezoelectric** the production of electricity by applying pressure to a crystal.

**Pilot device** a control component designed to control small amounts of current. Used to control larger control components.

**Pneumatic timer** a device that uses the displacement of air in a bellows or diaphragm to produce a time delay.

**Polarity** the characteristics of a device that exhibits opposite quantities within itself: positive and negative.

**Potentiometer** a variable resistor with a sliding contact that is used as a voltage divider.

**Power factor** a comparison of the true power (watts) to the apparent power (volt-amps) in an AC circuit.

**Power rating** the rating of a resistor that indicates the number of watts that can be permitted without damage to the device.

**Pressure switch** a device that senses the presence or absence of pressure and causes a set of contacts to open or close.

**Primary cell** a voltaic cell that cannot be recharged.

**Prime mover** the device supplying the turning force necessary to turn the shaft of a generator or alternator. Can be a steam turbine, diesel engine, water wheel, and so on.

**Printed circuit** a board on which a predetermined pattern of connections has been printed.

**Proton** one of the three major parts of an atom. The proton has a positive charge.

**Push button** a pilot control device operated manually by being pushed or pressed.

# R

**Reactance** the opposition to current flow in an AC circuit offered by pure inductance or pure capacitance.

**Rectifier** a device or circuit used to change AC voltage into DC voltage.

**Regulator** a device that maintains a quantity at a predetermined level.

**Relay** a magnetically operated switch that may have one or more sets of contacts.

**Reluctance** resistance to magnetism.

**Remote control** controls the functions of some electrical device from a distant location.

**Residual magnetism** the amount of magnetism left in an object after the magnetizing force has been removed.

**Resistance** the opposition to current flow in an AC or DC circuit.

**Resistance-start inductance-run motor** one type of split-phase motor that uses the resistance of the start winding to produce a phase shift between the current in the start winding and the current in the run winding.

**Resistor** a device used to introduce some amount of resistance into an electrical circuit.

**Retentivity** a material's ability to retain magnetism after the magnetizing force has been removed. Also called coercive force.

**Rheostat** a variable resistor.

**RMS value** the value of AC voltage that will produce as much power when connected across a resistor as a like amount of DC voltage.

**Rotor** the rotating member of an alternating current machine.

# S

**Saturation** the maximum amount of magnetic flux a material can hold.

**Schematic** an electrical drawing showing components in their electrical sequence without regard for physical location.

**SCR (silicon-controller rectifier)** a semiconductor device that can be used to change AC voltage into DC voltage. The gate of the SCR must be triggered before the device will conduct current.

**Sealing contacts** contacts connected in parallel with the start button and used to provide a continued path for current flow to the coil of the contactor when the start button is released. Also called holding contacts and maintaining contacts.

**Secondary cell** a voltaic cell that can be recharged.

**Semiconductor** a material that contains four valence electrons and is used in the production of solid state devices.

**Sensing device** a pilot device that detects some quantity and converts it into an electrical signal.

**Series circuit** a circuit that contains only one path for current flow.

**Series field** a winding of large wire and few turns designed to be connected in series with the armature of a DC machine.

**Series machine** a direct current motor or generator that contains only a series field winding connected in series with the armature.

**Service** the conductors and equipment necessary to deliver energy from the electrical supply system to the premises served.

**Service factor** an allowable overload for a motor indicated by a multiplier that, when applied to a normal horsepower rating, indicates the permissible loading.

**Shaded-pole motor** an AC induction motor that develops a rotating magnetic field by shading part of the stator windings with a shading coil.

**Shading coil** a large copper wire or band connected around part of a magnetic pole piece to oppose a change of magnetic flux.

**Short circuit** an electrical circuit that contains no resistance to limit the flow of current.

**Shunt field** a coil wound with small wire and having many turns designed to be connected in parallel with the armature of a DC machine.

**Shunt machine** a DC motor or generator that contains only a shunt field connected in parallel with the armature.

**Sine-wave voltage** a voltage wave form whose value at any point is proportional to the trigonometric sine of the angle of the generator producing it.

**Slip** the difference in speed between the rotating magnetic field and the speed of the rotor in an induction motor.

**Sliprings** circular bands of metal placed on the rotating part of a machine. Carbon brushes riding in contact with the sliprings provide connection to the external circuit.

**Snap-action** the quick opening and closing action of a spring-loaded contact.

**Solenoid** a magnetic device used to convert electrical energy into linear motion.

**Solenoid valve** a valve operated by an electric solenoid.

**Solid state device** an electronic component constructed from semiconductor material.

**Specific gravity** the ratio of the volume and weight of a substance to an equal volume and weight of water. Water has a specific gravity of 1.

**Split-phase motor** a type of single-phase motor that uses resistance or capacitance to cause a shift in the phase of the current in the run winding and the current in the start winding. The three primary types of split-phase motors are resistance-start induction-run, capacitor-start induction-run, and capacitor-start capacitor-run motors.

**Squirrel-cage rotor** the rotor of an AC induction motor constructed by connecting metal bars together at each end.

**Star connection** another name for wye connection.

**Starter** a relay used to connect a motor to the power line.

**Stator** the stationary winding of an AC motor.

**Step-down transformer** a transformer that produces a lower voltage at its secondary than is applied to its primary.

**Step-up transformer** a transformer that produces a higher voltage at its secondary than is applied to its primary.

**Surge** a transient variation in the current or voltage at a point in the circuit. Surges are generally unwanted and temporary.

**Switch** a mechanical device used to connect or disconnect a component or circuit.

**Synchronous speed** the speed of the rotating magnetic field of an AC induction motor.

**Synchroscope** an instrument used to determine the phase angle difference between the voltages of two alternators.

**T**

**Temperature relay** a relay that functions at a predetermined temperature. Generally used to protect some other component from excessive temperature.

**Tera-** a metric prefix meaning one trillion ($10^{12}$).

**Terminal** a fitting attached to a device for the purpose of connecting wires to it.

**Tesla (T)** a unit of magnetic measure in the MKS system. 1 Tesla = 1 weber per square meter.

**Thermistor** a resistor that changes its resistance with a change of temperature.

**Thyristor** an electronic component that has only two states of operation, on and off.

**Time constant** the amount of time required for the current flow through an inductor or for the voltage applied to a capacitor to reach 63.2% of its total value.

**Toroid** a doughnut-shaped electromagnetic.

**Torque** the turning force developed by a motor.

**Transducer** a device that converts one type of energy into another type of energy. For example, a solar cell converts light into electricity.

**Transformer** an electrical device that changes one value of AC voltage into another value of AC voltage.

**Transistor** a solid state device made by combining three layers of semiconductor material together. A small amount of current flow through the base-emitter can control a larger amount of current flow through the collector-emitter.

**TRIAC** a bidirectional thyristor used to control AC voltage.

**Troubleshoot** to locate and eliminate problems in a circuit.

**U**

**Undercompounded** the condition of a direct current generator when the series field is too weak. The condition is characterized by the fact that the output voltage at full load will be less than output voltage at no load.

**Unity power factor** a power factor of 1 (100%). Unity power factor is accomplished when the applied voltage and circuit current are in phase with each other.

**V**

**Vacuum tube voltmeter (VTVM)** a voltmeter that uses the grid of a vacuum tube to produce a very high input impedance. VTVMs are used in electronic circuits to prevent loading the circuit that is being tested. Modern VTVMs replace the vacuum tubes with field effect transistors (FETs).

**Valence electrons** electrons located in the outer orbit of an atom.

**Variable resistor** a resistor whose resistance value can be varied between its minimum and maximum values.

**Varistor** a resistor that changes its resistance value with a change of voltage.

**Vectors** lines having a specific length and direction.

**Volt (V)** a measure of electromotive force. The potential necessary to cause one coulomb to produce one joule of work.

**Voltage** an electrical measurement of potential difference, electrical pressure, or electromotive force (EMF).

**Voltage drop** the amount of voltage required to cause an amount of current to flow through a certain resistance.

**Voltage rating** a rating that indicates the amount of voltage that can be safely connected to a device.

**Voltage regulator** a device or circuit that maintains a constant value of voltage.

**Voltaic cell** a device that converts chemical energy into electrical energy.

**Voltmeter** an instrument used to measure a level of voltage.

**Volt-ohm-milliammeter (VOM)** a test instrument designed to measure voltage, resistance, or milliamperes.

**VTVM** abbreviation for vacuum tube voltmeter.

## W

**Watt (W)** a measure of true power.

**Watt-hour (W-hr)** a unit of measure for describing the current capacity of a battery or a cell.

**Wave form** the shape of a wave as obtained by plotting a graph with respect to voltage and time.

**Weber (Wb)** a measure of magnetic lines of flux in the MKS system. 1 Wb = 100 million lines of flux.

**Windage loss** the losses encountered by the armature or rotor of a rotating machine caused by the friction of the surrounding air.

**Wiring diagram** an electrical diagram used to show components in their approximate physical location with connecting wires.

**Wound rotor motor** a three-phase motor containing a rotor with windings and sliprings. This rotor permits control of rotor current by connecting external resistance in series with the rotor winding.

**Wye connection** a connection of three components made in such a manner that one end of each component is connected. This connection is generally used to connect devices to a three-phase power system.

## Z

**Zener diode** a diode that has a constant voltage drop when operated in the reverse direction. Zener diodes are commonly used as voltage regulators in electronic circuits.

**Zener region** the region of a semiconductor device that is reached when current flows through it in the reverse direction.

**Zenith** a culmination or high point.

**Zone** an area, region, or division distinguished from adjacent parts or objects by some feature, characteristic, or barrier.

# Index